Automated Diagnostics and Analytics for Buildings

Automated Diagnostics and Analytics for Buildings

Editors

Barney L Capehart PhD, CEM
and
Michael R Brambley, PhD

Routledge
Taylor & Francis Group

LONDON AND NEW YORK

Published 2020 by River Publishers
River Publishers
Alsbjergvej 10, 9260 Gistrup, Denmark
www.riverpublishers.com

Distributed exclusively by Routledge
4 Park Square, Milton Park, Abingdon, Oxon OX14 4RN
605 Third Avenue, New York, NY 10017, USA

First issued in paperback 2023

Library of Congress Cataloging-in-Publication Data

Capehart, Barney L.
    Automated Diagnostics and Analytics for Buildings/ Barney Capehart.
        pages cm
    Includes bibliographical references and index.
    ISBN 0-88173-732-1 (alk. paper) -- ISBN 978-8-7702-2321-8 (electronic) -- ISBN 978-1-4987-0611-7
(Taylor & Francis distribution : alk. paper)
    Library of Congress Control Number: 2014946700

Automated Diagnostics and Analytics for Buildings by Barney Capehart and Michael R. Brambley

First published by Fairmont Press in 2015.

Routledge is an imprint of the Taylor & Francis Group, an informa business

Publisher's Note
The publisher has gone to great lengths to ensure the quality of this reprint but points out that some imperfections in the original copies may be apparent.

ISBN 13: 978-8-7702-2928-9 (pbk)
ISBN 13: 978-0-88173-732-5 (The Fairmont Press, Inc.)
ISBN 13: 978-1-4987-0611-7 (hbk)
ISBN 13: 978-8-7702-2321-8 (online)
ISBN 13: 978-1-0031-5190-6 (ebook master)

While every effort is made to provide dependable information, the publisher, authors, and editors cannot be held responsible for any errors or omissions.

The views expressed herein do not necessarily reflect those of the publisher.

Contents

SECTION II
Current Technology, Tools, Products, Services, and Applications

SECTION III
Methodology and Future Technology

SECTION IV
Current Technology, Tools, Products, Services, and Applications

SECTION V
AFDD for HVAC Systems and Equipment—
Methodology and Future Technology

SECTION VI
Conclusion and Author Bios

Foreword

This book will help you explore the new world of Automated Diagnostics and Analytics for Buildings and provide insight and connection into the industry thought leaders that are taking big data into a new reality. "Dynamic Data Fuels Deep Analytics" speaks to the importance of the next level of deep analytics of almost everything will have and how we as an industry will provide a new level of deeper analytics connecting inquiring minds to almost everything with low cost real time data. The journey will be driven by the first wave of online analytics that will point to the potential of looking further into building operation opportunities, but further analytics will be required to factually quantify these opportunities. We all know analytics begat analytics.

Over the recent past, the best use of an analytic software application for building systems has been fault detection and diagnostics (FDD). FDD techniques are typically equipment or device centric and characterized by pre-defined rules based on an engineering model of a piece of equipment. Despite the impressive progress with FDD, the industry is in its infancy of utilizing data analytic applications in buildings. If analytics for the HVAC system has provided outstanding outcomes, we need to take that template to other building systems.

Several of the chapter authors are regular contributors to our free online magazine so understanding their thoughts and coming to know them in the following chapters will bring this book alive and make it relevant for many years to come. Once you know the industry thought leaders assembled in this book you can start following them and their most recent evolving thoughts in our and other online resources their blogs and industry news feeds. The transition in the last few years has been amazingly rapid. In our magazine's 15 year history we have talked about the possible but it is only in the last few years and even more accelerated in the last few months that the possible has transitioned into the plausible and our new reality.

Bring Your Own Device BYOD Mobility coupled with the cloud has created an industry of large building automation folks trying to rapidly understand the big data transition. Cloud based Big Data Projects are truly morphing into a dynamic collection of people, things, and internet interactions; a collaborator, not just a project. A "collaboratory" is more than an elaborate collection of information and communications technologies; it is a new networked organizational form that also includes social processes; collaboration techniques; formal and informal communication; and agreement on norms, principles, values, and rules" (Cogburn, 2003, p. 86). You will see in most articles that Ownership of the Collaboratory is an important piece of the total success of Automated Diagnostics and Analytics for Buildings.

A clear component of every successful energy integration Diagnostics and Analytics project is a team of champions who asserted ownership of the project collaboratory. The importance of keeping our data free inside the collaboratory needs to be highlighted; a lesson we learned in the past but somehow need to keep relearning. The data not only needs to be free, it needs to be named and organized in a predicable agreed on format.

It is not just the naming of data but a consistent data model that allows us to free our data to a world of dynamic dimensions for our own purposes. No longer must data be predefined before use if an accurate self-discoverable model is present. This new way of viewing data allows us a new world in which data can be used in several different ways as a dynamic subset of many scenarios.

I am very pleased that Barney and Mike asked me to provide my thoughts in this foreword for their new book. They have done an amazing job of capturing and assembling the new evolving frontier of Automated Diagnostics and Analytics for Buildings now occurring as part of the Internet of Everything (IOE).

Ken Sinclair Publisher/Owner
www.AutomatedBuildings.com
Sinclair@automatedbuildings.com
March 2014 Vancouver Island, Canada

Foreword

Over the past four decades the United States has made dramatic improvements in energy efficiency, meeting 79 percent of the increase in energy demand according to analysis by ACEEE (Figure 1). A broad array of measures have contributed to these improvements including improved efficiencies of energy using technologies, appliance and equipment standards and improvements in building codes and construction practices and materials. In the past few years, some in the energy efficiency community have begun to acknowledge that the component-efficiency approach that has yielded many of the largest savings over the past decades is likely to produce diminishing returns in the future (see Lowenberger et al. 2012). Going forward, the largest opportunities will likely result from optimization of energy using systems. These systems can range from discrete subsystems to whole buildings or even campuses. It is not that experts in energy efficiency have been unaware of the potential for systems optimization, but rather than the technologies needed to implement practical and cost effective system optimization have not been available and affordable.

The past decade has seen dramatic increases in the capacity for energy using systems simulation, coupled with advances in communications and sensors technologies that have improved performance and dramatically decreased costs. These technology advances have enabled automation systems that go beyond the limited set-point controls that have dominated this space for over a century, allowing for practical implementation of what ACEEE has come to call intelligent efficiency. These "smart" sensor and control systems allow for continuous optimization of energy using system performance that adapts to the changing needs of a building or group of buildings. The technology implementations profiled in this book will enable energy efficiency to continue to meet a significant share the future energy needs of our country into the foreseeable future. A study by ACEEE suggested that about half of the future energy needs in the United States could be meet through energy efficiency, which coupled with renewable energy resources could result in the effective de-carbonization of our economy by the middle of this century (Laitner et al. 2012). ACEEE's analysis suggests that it is only through the application of a systems approach to energy efficiency, enabled by the practices discussed in this volume that these energy savings opportunities will be realized.

References

Laitner, John A. "Skip", Steven Nadel, R. Neal Elliott, Harvey Sachs, and A. Siddiq Khan. January 11, 2012. The Long-Term Energy Efficiency Potential: What the Evidence Suggests, ACEEE Research Report E121. Washington, DC: American Council for an Energy-Efficient Economy.

Lowenberger, Amanda, Joanna Mauer, Andrew deLaski, Marianne DiMascio, Jennifer Amann, Steven Nadel. March 8, 2012 The Efficiency Boom: Cashing In on the Savings from Appliance Standards, ACEEE Research Report A123. Washington, DC: American Council for an Energy-Efficient Economy.

R. Neal Elliott, Ph.D., P.E.
Associate Director for Research
ACEEE
March 2014

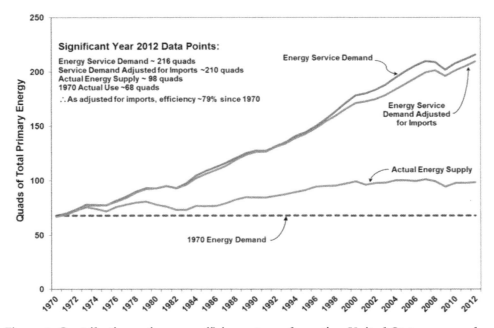

Figure 1. Contributions of energy efficiency toward meeting United States energy demand 1970 to 2012 (Source ACEEE 2014)

Foreword

This book presents state-of-the-art approaches for automated diagnostics and other analytics applied to buildings. It includes the results of a variety of efforts carried over the past few decades in response to the proliferation of low-cost computing, sensing, and network communication along with the increased awareness of the importance of maintaining building performance as part of improving energy efficiency and reducing environmental impacts. In addition to improved energy efficiency, automated fault detection and diagnostic (AFDD) systems have the potential to reduce equipment downtime, service costs, and utility costs. However, the commercialization of AFDD for building and HVAC&R applications is still in its infancy compared to other applications such as nuclear power plants, aircraft, chemical process plants, and automobiles. Relatively few commercial products exist and the ones that do exist are very specialized. This is undoubtedly because the customer benefits of AFDD for building systems are lower than for critical applications such as nuclear power plants or aircraft or for production facilities such as chemical process plants. In order to penetrate the marketplace, AFDD solutions need to have low implementation costs.

The need for low cost solutions has driven many of the developments that are described in this book. In particular, it is important to minimize the use of additional sensors and the time required to configure AFDD approaches. For adding AFDD to existing equipment and systems, many of the methods use data-driven approaches applied to measurements already available through an energy management system. Another approach for reducing the cost of realizing AFDD is to embed diagnostics within devices (VAV boxes, rooftop units, etc.) coming from the factory. This allows the use of manufacturer specific performance characteristics along with reductions in the cost of additional sensors through mass production.

Embedded diagnostics are just beginning to appear in the marketplace and may be poised for dramatic growth as costs come down and expectations for these capabilities grow. For commercial and residential buildings, embedded diagnostics will allow a service technician or owner to monitor and manage service using a handheld or local computer. As customer confidence and expectations for better economic performance grow, there could be a dramatic expansion of contracts with service providers who manage maintenance and repair using remote monitoring connected across a network to intelligent HVAC&R equipment. In addition to embedded diagnostics, a chip on board the equipment might store all of the documentation necessary for installation, commissioning, operation, maintenance, warranty, and repair. This could dramatically reduce the chances of losing this information and improve the prospects for proper installation and operation.

There is no doubt that future building systems and equipment will be able to self-diagnose problems and automatically provide information to service providers so that parts can be ordered and service scheduled. This will allow a small support staff to monitor and maintain a large number of different buildings from a remote, centralized location and thereby reduce the cost of maintaining building performance. Automated diagnostics and advanced analytics are coming to the building industry. It's just a matter of when. This book is part of a growing body of work that is enabling their development and deployment.

Jim Braun
Herrick Professor of Engineering
Ray W. Herrick Laboratories, Purdue University
March 2014

Introduction

The purpose of this book is to promote and document energy savings from the relatively new technology of Advanced Energy Information Systems called Automated Fault Detection and Diagnostics (AFDD) and Analytics for buildings and facilities. A number of studies have shown that commercial buildings in the United States (U.S.) waste as much as 15% to 30% of the energy they use (Katipamula and Brambley 2005). Analysis of HVAC and other building energy use data, along with whole-building utility data, sub-metered end-use data, and data from the building automation system (BAS) through the use of AFDD and building analytics can help identify opportunities to improve building operations and efficiency, and ultimately reduce energy and operating costs.

We will discuss the latest technologies available for fault detection, diagnostics, and building analytics, and operational experience with stand alone and web based systems for fault detection, diagnostics and analytics in currently operating buildings and facilities, and in varied applications, and to show how new opportunities have developed for energy and facility managers to quickly and effectively control and manage their operations more efficiently, with less energy use and cost, and experience improved energy system performance. You'll find information on what is actually available using this technology, what products and services are available at this time, and how they are being used at other buildings and facilities, and see what is involved for current and future installations of internet-based technologies. The material in this book on automated fault detection, diagnostics and analytics should greatly assist energy, facility and maintenance managers, as well as consultants and control systems development engineers. Chapters on methodology and future technological features should also assist those involved in research and development of these new technologies in AFDD and analytics for buildings.

The editors wish to thank each of the authors for graciously giving their time to write chapters which help provide so much valuable information for our readers.

Barney L Capehart
Michael R Brambley

SECTION I

The Case for Automated Tools for Facility Management and Operation

Chapter 1

Introduction

Barney L Capehart and Michael R Brambley

At the same time as energy efficiency and sustainability have gained traction among the public, government and the buildings sector, progress in information technology has continued to advance at a rapid pace. This provides a technological basis on which smart, energy-efficient building systems can be deployed. New opportunities for operating buildings and facilities more efficiently have arisen from the availability of cheap, high-speed, high-capacity personal computers, microprocessors and communication networking. As the internet expands, and with the addition of cloud storage, these innovations will combine with affordable smart sensors, smart meters, and modern software tools to analyze data, automatically interpret them, and provide useful information as never before possible. These capabilities provide the engines of growth for rapid expansion in use of both simple and sophisticated techniques for energy management, systems operation and maintenance, control, and automated fault detection and diagnostics. They will drive value by helping building and facility owners and operators collect and process large amounts of data to understand the condition and performance of their buildings and facilities as means to reduce their energy use and costs, and to improve their indoor environmental quality.

Energy information and automation systems are central to managing energy use and efficient operation in our buildings and facilities. The old adage is as true today as it ever has been – You can't manage what you don't measure! Metering, monitoring and data collection form the most basic parts of any energy information and control systems, which will likely become the infrastructures in which automated fault detection and diagnostics and building analytics are ultimately deployed.

Modern Information Technology provides high speed big data processing and data storage to convert the data to useful actionable information for operating buildings and facilities. It will also be used to display information using highly sophisticated graphical user interfaces and visually appealing dashboards for operational staff to achieve the goals of efficient and effective operation of the systems in buildings and facilities. Using data from smart devices such as building automation systems, smart meters, smart sensors, etc., is currently one of the hottest topics in the industry. The goal is to use this data to better understand the operation of building systems and identify ways to improve their operation.

Automated fault detection and diagnosis (AFDD) is one of the key technologies for understanding the condition and operation of building systems without placing undue burden on building operation staff. With the continually decreasing cost of computing technology, the general availability of both wired and wireless networking, the emergence of smart sensors and meters, and the proliferation of smart cellular phone and modem technology at rapidly decreasing cost, building control systems will soon transition from a field with a dearth of data to one awash in Big Data. Already, many buildings are installing more than the rudimentary sensors required for control to enable monitoring the energy consumption and the condition of systems. With most occupants carrying smart phones, the phones themselves could become sensors as well as interfaces through which occupants can provided input to building control systems and feedback on performance and see how the building is operating.

THE ENERGY AND SUSTAINABILITY CONNECTIONS

Over the last 40 years or so, since the first oil embargo in 1973, awareness of the fragility and limits of the non-renewable energy sources on which our national and world economies rely has continually increased. Furthermore, recognition of the negative impacts that energy consumption can have on the environment has also increased, first by the recognition that the conversion of fossil fuels such as coal by combustion to useful energy contributes significant quantities of pollutants (e.g., nitrogen oxides, sulfur dioxide, and particulates) to the atmosphere, which have local and regional impacts on air quality, but also that extraction of these fuels from natural deposits can lead to significant local environmental impacts on land use and local water resources. Although insightful scientists recognized the potential impacts on global climate of carbon dioxide emissions associated with the combustion of fossil fuels much earlier (e.g., Plass 1956, Revelle and Suess 1957, Keeling 1960, Manabe and Wetherald 1967), not until 1988 did the global warming impacts of these emissions via the greenhouse effect in the earth's atmosphere receive significant coverage by the popular (non-technical) press (Ungar 1992) or did U.S. and world political leaders recognize the importance of addressing this issue.

In 2010, buildings accounted for just over 41% of total U.S. primary energy consumption and had grown relatively steadily from about 36% in 1990 (U.S. DOE 2011). As such a significant fraction of total energy consumption, attention to the efficiency of buildings needs to be part of any attempt to reduce the impact of energy use on energy security, energy costs, global climate change and other environmental impacts.

Recognizing the importance of buildings on the environment, in 1993, the U.S. Green Building Council (USGBC), a non-profit private organization, was created to promote sustainability in the building and construction industries. The USGBC is best known for the green building certification program Leadership in Energy and Environmental Design (LEED), which was first released in 2000. LEED recognizes commercial and residential projects for environmental and health performance. The USGBC claims to currently have more than 54,000 active projects representing more than 10 billion square feet of space. Also, recognizing the importance of sustainable buildings, the American Society of Heating, Refrigerating and Air-Conditioning Engineers (ASHRAE) in 2006 identified sustainability as a key part of its mission stating in the first of 4 directions in the ASHRAE Strategic Plan that "ASHRAE will be a leader in advancing sustainable design, construction, and operations for new and existing built environments" (ASHRAE 2011). Notably, ASHRAE also has a Sustainability Roadmap (ASHRAE 2006) to guide its efforts in this area. The positions of these two well-recognized organizations in the buildings industry are indicators of a trend toward increasing recognition of the importance of sustainable buildings.

The purpose of this book is to document and promote awareness of the rapidly expanding field of building energy analytics, including the technology called

Automated Diagnostics, which although used in other industries is relatively new to the buildings sector. The authors discuss technologies available for fault detection and diagnostics, and operational experience with standalone and web based systems in actual facilities and in varied applications, and show how new opportunities have developed for energy and facility managers to quickly and effectively control and manage their operations more efficiently and at lower cost, while experiencing improved energy system performance. You'll find information on what is actually happening in the development of fault detection and diagnostics, what products and services are available at this time, how they are being used in facilities, and see what is involved for current and future installations. The material in this book on automated fault detection and diagnostics should greatly assist energy, facility and maintenance managers, as well as consultants, control systems development engineers and researchers.

Overall Benefits of Advanced Information and Control Technology

A number of studies have shown that commercial buildings in the United States (U.S.) could use as much as 15% to 30% less energy if they simply correct faults and improve operational practices (Katipamula and Brambley 2005; Brambley et al. 2005). Analysis of whole-building utility data, sub-metered end-use data, data from the building automation system (BAS) and wireless sensor networks can help identify opportunities to improve building performance and efficiency while reducing operating costs.

Many utilities in the U.S. and around the world are investing in advanced, smart metering devices that can record electricity consumption by time of use, typically at 15-minute intervals and in some cases at 5-minute intervals. Federal facilities in the U.S. are now required to have such meters in all larger Federal buildings. These meters are generally referred to as "smart" meters because in addition to recording consumption by time of use, they also provide communication between the utility and the meter. The data from these meters is commonly referred to as interval data. In some cases, the communication is in real time. In addition to the ability of the meter to communicate with the utility in real time, some meters also have the ability to communicate with the various devices in a home or a building. Although these new meters provide the capability for "smarter" building operations, the information that these meters provide is seldom used to identify opportunities to improve building performance. The high resolution data though have rich information that can be converted to actionable information either by fully automated processes or by systematic visual inspection of the data.

In addition to the interval metered data from "smart" meters, most large commercial buildings with floor areas of more than 100,000 square feet typically have BASs to manage operations and control building systems. The data from the BAS can be leveraged to identify operational problems, using processes that are discussed further in this book. However, nearly 90% of commercial buildings lack BASs. These buildings would benefit from analysis of whole-building interval data to identify inefficiencies and improve building operations, and ultimately reduce operating cost. If additional measured data, e.g., outdoor-air temperatures, are available, further analysis of the data is possible that reveals more opportunities for operational improvement (Taasevigen et al. 2011).

A major challenge for all commercial buildings and most manufacturing facilities lies in analyzing the data and identifying opportunities to improve operations and control strategies that result in energy savings. There are software tools available to help owners and operators analyze data to assist them in achieving their goals of energy savings across their buildings portfolios. While interval metered data are useful to identify overall building consumption trends, there are specific faults in heating, ventilating and air-conditioning systems that can also be identified from analysis of these data.

Automated Diagnostics

The concept of using software to comb through large amounts of data, real-time, near real-time, and historical, is relatively new to the building control and automation industry. Moreover, extracting meaningful, actionable information from the data requires specialized knowledge. Most engineering analysis tools produce additional data from input data. In the best cases, they provide the ability to generate plots of the data for use in interpreting its implications.

A new class of software tools for buildings known as automated diagnostics or automated fault detection and diagnostics (AFDD) both generates and interprets analytic results to produce easy-to-understand conclusions, which are often expressed in natural language (e.g., plain English). These tools go further than building energy analytics, which analyze data to produce numerical results. They interpret analytic results to provide conclusions and lend themselves to supporting ongoing monitoring-based re-commissioning. Although distinctions exist between automated fault detection and

automated diagnostics, for the purpose of this chapter, we'll simply use the terms automated diagnostics and automated fault detection and diagnostics synonymously to represent the class of tools with these sorts of capabilities.

There certainly have been numerous new technologies and products introduced in the past few years with claims to provide automated diagnostics. Benefits are usually expressed in terms of value to the business. Value is typically derived from energy cost avoidance; maintenance, repair, and operations (MRO) cost avoidance; and productivity savings. We know this about facilities management today:

- There are fewer dedicated facility managers than in the past. This responsibility may now reside under some other function of the business and be performed by someone with little or no domain knowledge.

- Facility managers are managing more with less. They are being asked to manage more buildings, more square footage, and more assets with fewer resources – financial and human.

- The skill of the personnel responsible for managing and operating facilities may be declining or becoming mismatched with the newer systems being installed. This is due in large measure to time constraints, overburdening of personnel, and a lack of commitment by businesses to develop their people.

- Operating and maintaining facilities is a "necessary evil" for most businesses. It is not generally one of their core competencies; it is not viewed as strategic to their operations; and, in the fight for capital funding, the winners are typically core business functions, not facilities or the equipment to run them.

Automated diagnostics can help address many of these issues. First, automated diagnostics do things that no human can do, that is, continuously look at thousands, if not millions, of data points, to identify anomalies (or differences between actual behavior and the behavior expected when properly operating) that indicate equipment and system performance issues.

Second, automated diagnostic solutions go beyond the capacity of a typical building automation system (BAS) and the operator's knowledge of it and the building systems. The ordinary functionality of a BAS does not include capabilities to extract conclusions on the operating state of systems and equipment from vast sums of data by processing diagnostic rules, models, statistical methods, genetic algorithms, and combinations of them that are common in automated diagnostic solutions. Many of the automated diagnostic processes go well beyond the diagnostics and troubleshooting that can be done by humans, simultaneously considering large numbers of variables, changes in their values over time, and models that capture relationships among them. As such, automated diagnostics provide a level of visibility into operations that far surpasses that of the typical BAS in the hands of even the most knowledgeable operators.

Third, automated diagnostic functions can be provided via software tools that exchange data with BASs and other data sources or as new software modules integrated into BASs.* When designed well, such software can provide information and guidance usable by operation and maintenance staff and managers, who do not need to possess expertise in the diagnostics used by the software, to make decisions; however, being a new technology, trust develops relatively slowly.

Fourth, another approach to deliver value to building owners is using a managed services delivery model for automated diagnostics. In this case, companies delivering monitoring and diagnostic services can use automated diagnostic tools to identify faults and opportunities to improve building operation and lower operating costs. The challenge of using a business' resources (financial and human) to manage something that is not core to the business of the organization occupying the building is largely eliminated. In this model, value is delivered as a service, treated as an expense, and can scale according to the changing needs of the business. This eliminates the need to make large capital investments in technology and its maintenance or to train staff, which frees up resources for functions that are core to the business.

Energy cost avoidance is achieved by identifying practices and situations that are consuming excessive energy. Examples include lights that remain on when the building is unoccupied, failed economizers, heating and cooling at the same time, and so on. We use the term energy cost avoidance because unless the condition is corrected, excessive energy consumption and its asso-

ciated costs continue and savings are lost, while when practices are implemented that decrease energy consumption, a portion of the energy cost is avoided.

Maintenance, repair, and operations (MRO) cost avoidance is a result of identifying and correcting issues before further damage or total failure. These types of failures can severely impact production and personnel and inevitably result in significantly higher cost. MRO savings are more a result of a process than a technology. It is an effective process that allows for timely corrective action in anticipation of its need. An effective MRO strategy allows you to determine if a specific issue needs immediate action or can be scheduled with the next preventive maintenance activity.

Productivity savings result from 1) implementing operational improvements that improve indoor environmental conditions in which personnel work, making them more productive, and 2) in the case where services are procured from an outside service provider (the managed services delivery model) with personnel that have a higher level of skill and deeper experience than in-house personnel, internal resources can be used more productively in core business functions. In the latter case, the technology may be the same, but the skill, focus, and experience of the implementers are vastly different.

The benefits of these tools are numerous, but facility managers are confronted with a wide range of products and features that can contribute to improved facility performance. From Alarms to Automated Fault Detection and Diagnosis (AFDD), Analysis Tools, and Automated Analytics, each has a place and offers specific capabilities and benefits. Systems Integrators and owners, however, are often faced with a comparison between "apples and oranges" as they try to evaluate the different tools. Parts of this book are specifically focused on providing information on existing building analytic and AFDD tools and services to help building and facility managers and operators sort through the range of commercial offerings to select the tools and services best suited to meet their needs.

Ensuring Analytic Tools and Services Meet Needs

Buildings analytics and automated diagnostics have the potential to dramatically reduce energy consumption in building systems while improving performance, extending equipment life, and ensuring that the occupants of buildings are comfortable and productive.

AFDD tools for building equipment entered the market in the mid-1990s but are not yet deployed on

---

*In the future AFDD capabilities may be integrated as embedded parts of BASs. Only the simplest of AFDD capabilities are currently starting to be integrated into some BASs.

a large scale across the buildings sector. Many factors have impeded progress, some associated with the design of AFDD tools and others tied to business factors. Primary among them may be the designs of the tools themselves and the slow pace of AFDD becoming embedded in building automation systems. Tools that provide only the detection or diagnose of faults but do not provide users information on the cost impacts of the faults don't provide the information most critical for building owners and managers to assess the importance of the faults and to prioritize repairing them. Building managers already have numerous demands for their attention, including the latest complaints from occupants, which ordinarily and deservedly in many cases receive priority. Therefore, AFDD tools need to provide the information most important for owners and managers to readily use their results. Furthermore, AFDD capabilities need to be built into building automation and control systems to make them conveniently available for system implementers to install and users to access in their routine use of control and monitoring.

If designed poorly, automated diagnostics also have the potential to overwhelm building maintenance personnel with a staggering number of repetitive, confusing, and ultimately useless messages. In short, they could become just as irrelevant as their forebears – alarms in BASs. To be useful, automated diagnostics need to focus on issues the customer really cares about, present information in an easily understandable format, identify the root causes of problems, and above all avoid alarming the customer with false alarms.

Tools need to be compatible with the personnel available to use them. As noted earlier, a business model that has gained traction over the last decade is delivering building energy analytics, performance monitoring, fault detection and diagnostics as a service to buildings (using the managed services delivery model). Rather than installing building analytics and AFDD in buildings, companies are collecting data, analyzing it at a central location, and reporting their findings to building managers. The analysis may be performed using a variety of tools, including analytics, experts visually sorting through analytic plots to identify operational issues, or automated tools, e.g., AFDD, that provide conclusions about faults in buildings automatically. Regardless of how the faults, opportunities and recommendations are developed at these service centers, the distinctive feature of this delivery model is that results are provided as a service, which will meet the needs of certain building owners, who choose to use outside experts rather than devoting their resources to developing their own ex-

perts in a field outside their core business.

In the last couple years, efforts have also emerged that are beginning to embed requirements for AFDD in heating, ventilating and air-conditioning equipment in codes, and a standard is under development to support verification of claims that AFDD capabilities are embedded in equipment. If these initial efforts are successful, such requirements could well significantly change the penetration of AFDD in the practice of building operation and management.

The emergence of cloud computing is also drastically changing the landscape of offerings in building analytics and AFDD. Applications are beginning to be offered from the cloud supported by cloud data sources. Use of the cloud could increase the availability of analytic and AFDD tools to commercial buildings, and decrease the cost for new vendors to enter the market. No longer do companies need to operate their own servers or even lease computers in server farms. With cloud services based on virtual machines, use of the computing and storage services can be expanded, contracted and distributed geographically as needed by business demands, only paying for the services actually used. This has enormous cost benefits to companies offering services that rely on access to networked computing with demand that fluctuates over time.

CONCLUSION

Energy information systems and building automation systems are central to managing energy use and efficient operation in our buildings and facilities. The old adage is as true today as it ever has been – You can't manage what you don't measure! The changing landscape of computing and communications, the drivers for increased energy efficiency, and the need to deliver faster, cheaper and better are increasing the need to use the new technologies of building analytics and automated fault detection and diagnostics to operate and maintain our buildings more efficiently and effectively, delivering services to meet the needs and expectations of occupants for comfort and productivity more cost-effectively and sustainably than ever possible before.

The book is organized into 6 sections, each with a collection of chapters relating to the section theme. Section I introduces the book (in this chapter) and establishes the case for automated tools for facility management and operation through chapters providing actual experience in case studies, perspectives on the value of analytics and automated diagnostics, and new ideas on energy smart buildings and intelligent efficiency. Section II comprises chapters on the spectrum of existing technology, tools, products, services and applications at the whole-building management level. This section will provide readers with an understanding of what's currently available and enable them to begin matching their needs with products and services. Section III focuses on methodologies for automated diagnostics and analytics primarily from a whole-building perspective. It includes chapters on monitoring and data collection, diagnostics and analytics for identifying opportunities for energy savings, commissioning, dealing with uncertainty, demand response, microgrids, and power monitoring. Section IV focuses on AFDD tools, technologies, products and services specifically for HVAC systems and equipment. Along with Section II, this section will help readers identify and begin sorting through current products and services for HVAC. It also provides information on the benefits of HVAC FDD. The collection of chapters in Section V addresses details of methodologies for HVAC AFDD and introduces some new technologies for the future. The chapters cover methods for detecting and diagnosing faults in HVAC systems, equipment and controls, introduce pattern recognition and virtual sensor based AFDD and self-correcting controls for HVAC, describe a versatile analytic software package, and describe new codes and standards that will influence the use of HVAC AFDD. Section VI provides the Editors' conclusions. Together the section provide a rather comprehensive look at the current state of building analytics and automated diagnostics and should provide a valuable resource for every stakeholder from the building owner or manager to the software developer or researcher.

Thanks to Paul Oswald from Environmental Systems, Inc., for some assistance in part of this introduction.

References

ASHRAE. 2006. *ASHRAE's Sustainability Roadmap*. American Society of Heating, Refrigerating and Air-Conditioning Engineers, Inc., Atlanta, GA.

ASHRAE. 2011. *ASHRAE Strategic Plan* (as amended June 26, 2011). American Society of Heating, Refrigerating and Air-Conditioning Engineers, Inc., Atlanta, GA.

Brambley M. R., P. Haves, S. C. McDonald, P. Torcellini, D.G. Hansen, D. Holmberg, and K. Roth. 2005. *Advanced Sensors and Controls for Building Applications: Market Assessment and Potential R&D Pathways*. PNNL-15149, Pacific Northwest National Laboratory, Richland, WA.

Katipamula, S. and M.R. Brambley. 2005. "Fault Detection, Diagnostics and Prognostics for Building Systems – A Review Part 2." *International Journal of Heating, Ventilating, Air-Conditioning and Refrigerating Research* 11(2), 169-187.

Keeling, C.D. 1960. "The Concentration and Isotopic Abundances of Carbon Dioxide in the Atmosphere." *Tellus* 12(2), 200-203.

Manabe, S. and R.T. Wetherald. 1967. "Thermal equilibrium of the atmosphere with a given distribution of relative humidity," *Journal of Atmospheric Sciences* 24(3): 241-259.

Plass, G.N. 1956. "The influence of the 15 μ carbon dioxide band on the atmospheric infra-red cooling rate." *Quarterly Journal of the Royal Meteorological Society* 82, 310-324.

Revelle, R., and H. Suess. 1957. "Carbon dioxide exchange between atmosphere and ocean and the question of an increase of atmospheric CO_2 during the past decades." *Tellus* 9, 18-27.

Ungar, S., 1992. "The rise and (relative) decline of global warming as a social problem." *The Sociological Quarterly* 33, 483–501.

Taasevigen, D.J., S. Katipamula and W. Koran. 2011. *Interval Data Analysis with the Energy Charting and Metrics Tool (ECAM)*. PNNL-20495, Pacific Northwest National Laboratory, Richland, WA.

U.S. DOE. 2011. *Buildings Energy Data Book*. U.S. Department of Energy, Washington, DC. Accessed at http://buildingsdatabook.eren.doe.gov/ on February 23, 2014.

Chapter 2

Energy-Smart Buildings
Demonstrating How Information Technology Can
Cut Energy Use and Costs of Real Estate Portfolios*

Accenture Consulting Staff and
Jessica Granderson, Lawrence Berkeley Laboratory

EXECUTIVE SUMMARY

Information Technology (IT) enables unprecedented efficiencies for businesses. Powerful analytics are helping firms better manage supply chains, improve resource allocation, detect fraud and optimize many core business functions. The real estate portfolio is no exception. Worldwide, buildings account for about 40 percent of total energy consumption and contribute a corresponding percentage to overall carbon emissions. Buildings used by businesses and public service organizations make up a large part of this footprint, whether they are office buildings, retail stores, hotels, schools or hospitals. In the US alone, businesses spend about US$100 billion on energy for their offices every year. In Asia, economic growth and a gradual shift toward service-based economies will expand the need for commercial buildings significantly over the coming years.

This provides scope for substantial cost savings. For the US, estimates predict that smarter buildings could save US$20-25 billion in annual energy costs. This opportunity is largely untapped today, as many building owners and operators are not yet aware of how data-driven optimizations can reduce energy consumption. Buildings may be equipped with hundreds of sensors and controls, but companies are leaving money on the table if they do not use this data more holistically to optimize their infrastructure. By applying analytics to make buildings smart (or energy-smart, to be more specific), companies can save billions and significantly reduce environmental impact.

This report, authored in collaboration between Microsoft, Accenture and the Lawrence Berkeley National Laboratory, examines how building owners, operators and occupants can achieve significant energy and cost savings through the use of smart building solutions. It is based on insights from a detailed case study of a smart building pilot program being conducted by Microsoft at its corporate headquarters campus.

What Was Done?

The pilot program by Microsoft's Real Estate & Facilities organization evaluated smart building applications from three vendors across 13 buildings within the company's main 118-building campus. In essence, these applications added an analytical layer on top of existing building management systems, without the need to replace existing infrastructure.

This new layer enables Microsoft to aggregate and analyze its building data to generate actionable insights that save energy and cut costs. In its initial stage, the program addresses energy consumption and cost in three specific ways:

- Fault detection and diagnosis to enable timely and targeted interventions in cases of faulty or under-performing building equipment.

- Alarm management to prioritize the many notifications generated by existing building systems and point engineers to the most impactful issues.

- Energy management through systematic tracking and optimization of building energy consumption and performance over time, while changing the behavior of building occupants with visual dashboards and benchmarks.

What Was Achieved?

Microsoft's experience thus far demonstrates that a smart building solution can be established with an upfront investment of less than 10 percent of annual en-

*This chapter is from Accenture Consulting, and is used here with permission from Mr. Darrell Smith of Microsoft Corporation. Darrell Smith is one of the co-authors of this report about their Microsoft smart buildings project.

ergy expenditure, with an expected payback period of less than two years.[1] By collecting and analyzing millions of data points (samples) per day, the company has been able to embark on multiple improvements that are reshaping the way its buildings are managed.

Microsoft's building engineers have become far more productive: instead of "walking around" to find issues, they are now "walking to" the problems that have the greatest impact on cost or comfort. By itself, the ability to continuously identify issues and optimize the performance of building equipment is expected to deliver annual savings of more than one million dollars. Furthermore, as building engineers can analyze data collected over time and occupants become more aware of individual energy use, Microsoft hopes to save several million dollars by optimizing base load (from building systems directly controlled by the building engineers) and by reducing plug load (from devices used by occupants) across its building portfolio.

How Can Others Replicate This?

Microsoft's pilot program demonstrates how corporate real estate organizations can collaborate successfully with IT, putting smart building technology to use in cutting costs and securing environmental benefits. Its experience has helped define a set of key design principles that can be used in any such rollout. These are outlined in more detail in this report, in summary they include:

Identify, collect and aggregate relevant data: This involves setting up automated aggregation of building, weather, utility and organizational data from building systems and other sources to feed into the smart building solution. Cloud computing, combined with on-site building management technology, can provide a powerful platform to gather, store, exchange, and process diverse datasets in a secure and scalable way.

Employ industry-leading analytics to identify savings: The core of the smart building solution is the analytics engine consisting of rules and algorithms that identify and prioritize interventions to maximize savings. Vendors differ in their approaches and capabilities and should thus be evaluated thoroughly.

Present results in a consumable and actionable form: The user experience needs to strike the right balance between ease of use and flexibility. Solutions need to improve an engineer's day-to-day productivity with better real-time information and access to data, while also providing a strong toolset for deeper analysis.

Centralize monitoring operations: A centralized operations center can effectively monitor building con-

ditions across a campus or multi-site portfolio and communicate directly with building engineers.

Engage the organization: Greater awareness of energy use and benchmarks, displayed via dashboards on the intranet or in hallways, can encourage employees and business leaders to save energy, reducing overall demand.

Avoid disruptive change: Existing building management systems do not need to be replaced. By deploying an analytics layer on top of these systems, prior investments can be significantly enhanced with minimal capital expenditure. Engineers can adopt new tools while still working directly with more familiar systems. Strong cross-organizational project management and a tailored change management approach are key to success.

INTRODUCTION:
THE CASE FOR SMART BUILDINGS

Building Efficiency Meets Information Technology

Buildings are the largest contributor to global carbon emissions, accounting for about 40 percent of the world's total carbon footprint.[2] In developed countries, commercial buildings alone represent close to 20 percent, about half of the total.[3] Commercial buildings are also costly. After salaries, buildings are one of the biggest operational expenses for organizations. Energy plays a significant part in this.[4] A more efficient building portfolio can improve the value of real estate assets, help the bottom line, cut emissions, and bolster the corporate image.

Executives have different choices to reduce the energy consumption of their building portfolio. Buildings can be designed more efficiently at the outset, but these opportunities are obviously limited. For existing infrastructure, the primary focus has usually been on retrofitting projects, which are often capital-intensive and disruptive to operations. But using software to ensure infrastructure runs more efficiently requires minimal capital investment and results in little or no disruption for occupants. From an economic standpoint, this should make it the preferred starting point for increased energy efficiency in a real estate portfolio.

Analytics software can help detect and address numerous sources of waste, such as:

- HVAC (heating, ventilation, and air conditioning) equipment that is simultaneously heating and cooling a given space due to a failed sensor or other fault.

- Technicians dealing with low priority or false alerts about building anomalies, while the notification system fails to highlight more impactful issues.

- Default configurations for all systems and pieces of equipment, meaning they run at suboptimal setpoints and are rarely updated after initial configuration.

- Lack of visibility and attention to energy waste on the part of occupants and building engineers.

- HVAC and lighting systems running at full capacity during periods when buildings are largely unoccupied.

In recent decades, most commercial buildings have been equipped with an increasing number of sensors, controls and other devices. Modern buildings have built-in control systems, referred to as building management systems (BMS) or building automation systems (BAS), allowing building engineers, facility managers and real estate management to control their infrastructure.

In this model, disparate building management systems and control panels are the access points to observe and manage building equipment, as illustrated in Figure 2-1 (left side). By introducing a smart building solution that provides an additional analytics layer (right side), a single data repository for all buildings is created and engineers are equipped with a powerful toolset to analyze data. In addition, this provides a foundation for

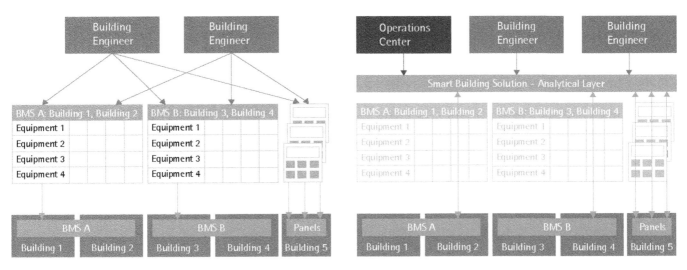

Figure 2-1: Building management: traditional approach (left) vs. smart building approach (right)

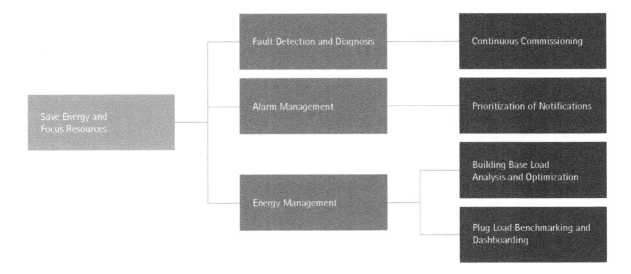

Figure 2-2: Basic smart building program components

tighter integration with a smart utility grid that manages energy supply and demand dynamically on a local or regional level.

How Analytics Can Cut Energy Waste

Fault detection and diagnosis: Through sophisticated fault detection and diagnosis rules, issues with building equipment across an entire real estate portfolio can be automatically identified and prioritized for building engineers. Conducting equipment maintenance on a continuous basis "so-called continuous commissioning" avoids waste and dramatically improves resource allocation. Engineers do not have to walk around looking for issues and money is spent where it is most needed. This also frees up engineers' time to address issues with smaller subsystems, which can add up to a large potential savings opportunity.

Alarm Management

By prioritizing and structuring the numerous notifications generated by building systems, a smart building solution focuses engineer's attention on the most critical events. They can concentrate on urgent and impactful interventions from the perspective of occupant comfort, energy consumption, cost and business impact.

Energy Management

Smart building solutions are able to centralize and correlate data from building systems, corporate data warehouses, and external sources, such as utilities and weather data feeds. Through analytics tools, building engineers can find anomalies and manage energy use holistically. Likewise, employees can be encouraged to save energy through information sharing in the form of dashboards, as well as energy benchmarks that create internal competition.

Microsoft's case study described in this report highlights the improvements made from each of these.

Successes in the Corporate Environment

Smart building programs are emerging as an effective solution for companies to save energy. Some early adopters have successfully transitioned to continuous commissioning, where building maintenance is conducted whenever the analytics engine detects a fault, rather than on recurring multiple-year cycles that rely on spot checks.

One example, featured in a study by the Lawrence Berkeley National Laboratory, comes from Sysco, a US$37bn food services company with over 140 facilities across North America. During a three-year energy effi-

ciency program launched in 2005-06,[5] the company used analytics as a critical component in cutting its portfolio energy use by 28 percent, a monthly saving of 18 million kWh. Importantly, the first 18 percent savings came from low or no cost fixes, with only the latter 10 percent requiring capital investment and upgrades.

Another study by the Berkeley Lab[6] benchmarked the impact of continuous, or monitoring-based, commissioning across 24 buildings, showing average energy savings of 10 percent, with as much as 25 percent in some cases.

In its work with corporate clients, Accentures smart buildings practice[7] observes that deployments usually pay back in 18-24 months, with energy savings in the 10-30 percent range. In markets where energy management is encouraged through policy incentives, the business case for deploying a smart building solution can be even better.

Inhibitors to Smart Building Adoption

Despite such benefits, corporate uptake of smart building implementations has remained relatively limited to date. This is due to a range of challenges that have inhibited adoption. These typically include:

Connectivity and Integration

The most immediate challenge lies in accessing the data from the existing building management systems. This is made more difficult by the disparity of systems, varying ages of the assets and different communication protocols. Also, as the smart building solution may be hosted externally, a secure connection may need to be established, which can complicate the data exchange and delay the rollout. Data volumes can be significant and can conflict with available capacities for extraction, transfer, storage and processing.

Depth and Breadth of Available Data

A second issue is collecting data that is sufficiently granular, both in quality and quantity. Firms need to ensure that all relevant equipment is networked and sends regular updates (e.g. in five-minute increments). Contextual information is also needed. For example, air conditioning usage needs to be mapped against weather conditions to distinguish savings simply due to favorable weather from real improvements. Internal data, such as the number of occupants, is similarly needed for meaningful analysis.

Usability

Usability challenges have been a common barrier to adoption, as many engineers have had limited exposure to advanced analytics tools. Some applications run the risk of overwhelming users with too many features, presented via a non-intuitive or unfamiliar user interface.

Organizational Support and Change Management

Implementing a smart building program can take several months and relies on many stakeholders. One particular challenge is the need for full commitment and close collaboration between all stakeholders â€" executives, building engineers, IT staff and external vendors. Engineers can be faced with conflicting workloads from both old and new responsibilities, which can inhibit uptake and delay payback periods.

Budget Challenges

Although the cost of implementing a smart building solution can be modest compared to the operating cost of the building, budgets are often tight and facilities teams may find it challenging to secure funds for such programs. Real estate leaders face the challenge of demonstrating both cost and sustainability benefits in their business case. Implementations for small portfolios are harder to justify as they lack economies of scale.

CASE STUDY—SMART BUILDINGS AT MICROSOFT

**Microsoft's Sustainability
Objectives in the Built Environment**

Data centers, employee travel and buildings are the top three contributors to Microsoft's overall carbon footprint, with buildings accounting for nearly 40 percent. As part of its efforts on environmental sustainability, the firm is addressing the first two by building energy-efficient data centers[8] and promoting the use of remote collaboration technology. To address the environmental impact of its building portfolio, Microsoft is investing in the deployment of new technologies to improve performance.[9] Beyond cutting its own emission footprint and reducing operational expense, Microsoft also sees its role in enabling the IT industry to develop smart building solutions. The underlying approach for the company is based on its fundamental belief that IT can help improve efficiency in all areas of energy and resource use.

As part of this, the firm's corporate headquarters in Redmond, Washington, is being used as a living lab to pilot several smart building solutions in parallel. Microsoft is working closely with vendors to test their solutions in a real-life setting, while applying a num-

ber of Microsoft's own products as part of the overall architecture. One key aim is to allow both vendors and Microsoft to improve their technologies for the broader market. Some highlights, outcomes and key knowledge from this pilot approach are shared here.

Introducing Smart Buildings at Microsoft

Microsoft's Real Estate and Facilities organization began its smart building program in 2009 with an initial analysis on how the company and its partners could use technology to significantly cut energy use in the built environment. In the first half of 2011, Microsoft rolled out smart building solutions from three different vendors at its corporate headquarters. The size of its main campus gives ample opportunity for simultaneously testing multiple solutions at scale. The campus consists of:

- 118 buildings with 14.9 million square feet (1.38 million m^2) of office space, approximately half of Microsoft's global real estate portfolio.

- 30,000 pieces of mechanical equipment to be maintained.

- 7 major building management systems used by engineers to manage equipment.

- Average daily consumption of 2 million kWh of energy, producing about 280,000 metric tons of carbon emissions annually.

In its initial pilot phase, Microsoft's smart building project focused on 13 buildings, representing 2.6 million square feet (about 240,000 m^2) of space, about the same floor area as the Empire State Building in New York. The age of the buildings varies from over twenty years old to almost new. For historic reasons, a variety of building management systems are in place, resulting in building engineers having to deal with a multitude of disparate systems.

Introducing a smart building solution that provides an analytical layer above existing building management systems creates a consolidated view of granular energy and operational data across Microsoft's building portfolio. This allows buildings to be managed holistically through a unified interface, instead of many disjointed systems (as illustrated in Figure 2-1). This approach does not replace existing building management systems, but aggregates and complements them with an analytical layer.

Program Components and Impacts

In its initial phase, Microsoft's smart building program focuses on the three main areas described in Figure 2-2:

- Fault detection and diagnosis
- Alarm management
- Energy management

At the time this report was published, the three solutions being piloted had introduced new capabilities and produced a range of promising results, but their evaluation was still in progress. Microsoft intends to publish more detailed data in the future. Nevertheless, the preliminary findings already provide useful insights for anyone considering the implementation of a smart building solution.

It's worth noting that Microsoft's corporate headquarters benefits from very low utility rates and carbon intensity thanks to the abundant hydropower supply from the nearby Columbia River. Savings in cost and carbon emissions achievable by businesses in other geographic regions may be several multiples higher than those on Microsoft's main campus.

Fault Detection and Diagnosis

One of the biggest single impacts the program has facilitated is the ability to identify building faults and inefficiencies in real-time by analyzing the data streams extracted from building systems. Most importantly, the software is able to quantify wasted energy from each identified fault in terms of dollars per year.

Previously, issues were typically found through periodic spot checks, an approach known as retro-commissioning. But with 30,000 pieces of equipment on the campus, this is a major effort, even if limited to only large HVAC systems. Historically, the firm's engineers spot-checked about one-fifth of the campus each year, or about 25 buildings. On average, a building would thus operate for five years before it got inspected and tuned again, despite degrading equipment and potential changes in use and occupancy. The saw tooth line in Figure 2-3 illustrates how building efficiency declines between retro-commissioning efforts.

Through this prior approach, Microsoft typically achieved energy savings of about 4 million kWh each year, cutting costs by about US$250,000. Now, the introduction of automated fault detection and diagnosis provides an entirely new tool for Microsoft's building engineers, enabling them to identify and prioritize faults as they occur.

The smart building solution provides engineers with a table similar to the simplified example in Figure 2-4, where equipment faults are prioritized and the cost of wasted energy is automatically estimated. Engineers can quickly decide which faults to address in which order and can predict whether the savings justify the expense for labor and spare parts. Fault detection also identifies issues that a conventional building management system would miss. One example encountered at Microsoft was an air handler's chilled water valve with a faulty control code issue. This meant that the valve was always 20 percent open, wasting several thousand dollars in energy. This issue was not easily visible before, but the analytics software was able to detect it immediately.

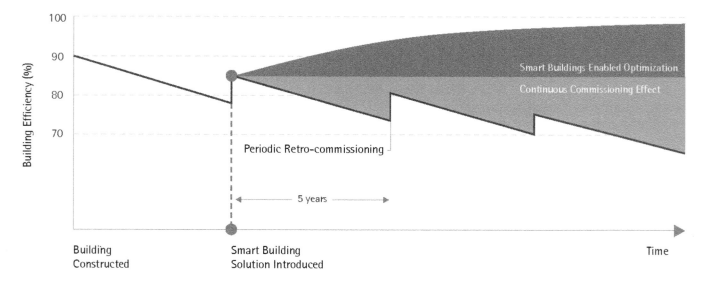

Figure 2-3: Continuous commissioning benefits (illustrative)

Building	Bldg. Cluster	Equipment	Fault and Diagnosis	Priority	Estimated Savings*
Bldg 58	Cluster E	AHU - 012	Leaking chilled water value	High	$11,291
Bldg 58	Cluster E	AHU - 003	Damper position fault	High	$4,782
Bldg 53	Cluster E	VAV - 022	Over cooling	High	$2,235
Bldg 58	Cluster E	CHI - 002	Changes to set points	Medium	$895
-	-	-	-	-	-
-	-	-	-	-	-
Bldg 54	Cluster E	VAV - 006	Air temperature sensor failure	Low	-

* Estimated savings potential, expressed an annual cost of wasted energy if not fixed.

Figure 2-4: Illustrative example of fault detection and diagnosis output (simplified)

As its smart building program evolves in coming years, Microsoft intends to quantify the exact benefits of this continuous commissioning approach. It is already evident that engineers save significant time on inspections and can adopt a highly focused approach to maintaining equipment. The software also allows them to detect smaller faults that could have been missed in traditional inspections.

Microsoft expects that interventions equivalent to a full 5-year retro-commissioning cycle for the entire campus can now be accomplished in just one year. Annual energy cost savings from continuous commissioning enabled by automated fault detection alone may thus exceed US$1 million.

Alarm Management

Microsoft's existing building management systems generate hundreds of alarms on a typical day, flooding engineer's email inboxes with automated notifications. Alarms range from major issues, such as a power outage, to insignificant messages, such as a notification that a self-test has started. Figure 2-5 shows sample statistics of Microsoft's building alarms over a 90-day period.

A key challenge is recognizing the importance of a given alarm, as well as correlations between messages from related events. Interpreting these requires deep knowledge of the building infrastructure and occupancy. Errors can lead to issues being missed and inadequate prioritization of interventions.

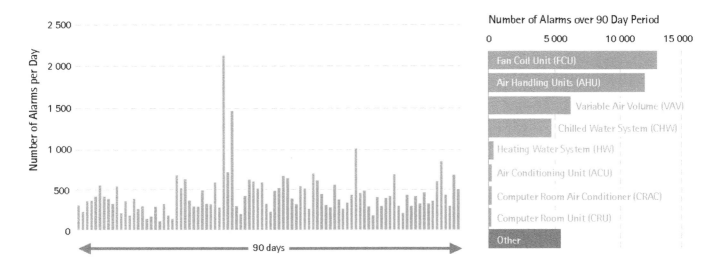

Figure 2-5: Alarms over 90-day period and distribution among equipment types

One example is when a casualty scenario occurs, a large-scale failure such as a recent substation fire in Redmond where several Microsoft buildings lost all or partial power. While recovering from such a scenario, alarm noise is excessive and prioritization is very difficult, with countless notification messages being generated by multiple systems trying to get back to a normal state. Without any automated grouping and prioritization of alarms, engineers may miss important alarms and inadvertently delay the response to an urgent issue.

Furthermore, analyzing thousands of alerts systematically to detect patterns over time is nearly impossible without a smart building solution. Here, analytics tools help engineers identify opportunities for efficiencies and cost savings.

Energy Management

The third strand in Microsoft's smart building rollout affects its ability to manage energy consumption more holistically. With the help of a smart building solution, engineers can optimize building base load, the power consumed by the major building systems, such as HVAC or lighting. Supported by analytics, they can tune setpoints and schedules, isolate wasteful equipment and address other opportunities by getting a much better understanding of energy use and trends across the building portfolio.

Microsoft anticipates that the smart building program will also help reduce the company's plug load, the electricity consumed by occupant's devices which accounts for about the same amount of energy as the building base load. To encourage better habits, Microsoft is planning to publicize energy consumption data internally, using dashboards that track and compare how much energy is consumed over time. For example, kilowatt-hours consumed per employee as a performance indicator can be benchmarked across organizational units and observed over time. Energy costs can be accurately broken down by organizational unit to define ownership and create incentives for managers to save energy. Grassroots efforts can more effectively educate and motivate employees when accurate energy consumption data is readily available. For example, Microsoft's internal Sustainability Champion Program[10] is expected to reduce plug load by 3-5 percent and will be leveraging the smart building solutions automated reporting features.

The foundation for this is the consolidated repos-

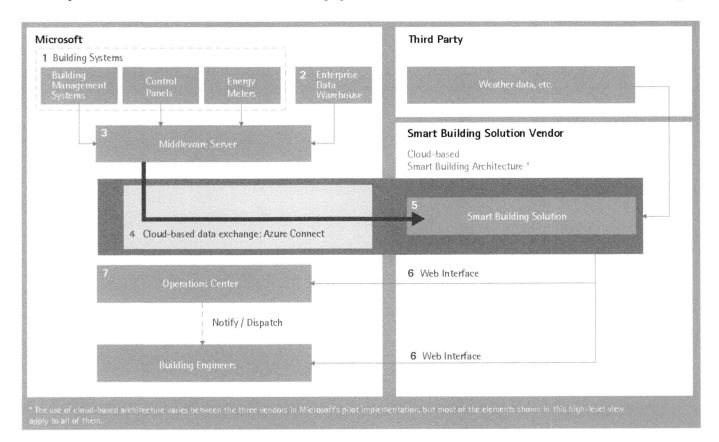

Figure 2-6: Microsoft's smart building architecture

itory of building and contextual data held by the smart building solution. Once Microsoft has collected several months of reliable data for energy management, measures to address both base load and plug load could contribute savings worth millions of dollars. This can be a significant factor in the overall business case for smart buildings.

Microsoft's Smart Building Architecture: A Technical Overview

The pilot program's architecture can be broken down as follows:

1. Equipment level data are collected and either sent directly from the control panel or from the BMS servers to a middleware server. For some buildings, this is done over an open protocol (BACnet). For most buildings, a protocol conversion was necessary, with additional scripting to extract the data.[11] Energy meters provide sub-metered electricity consumption data that complements the utility data (see point 5 below).

2. Microsoft's internal enterprise data warehouse provides a feed of contextual information, such as building type and headcount,[12] to the middleware server.

3. The middleware server acts as an aggregator for all on-site data. It also houses an Azure Connect endpoint to transmit the data to the relevant vendor application, hosted off-site.[13]

4. The Azure Connect service provides secure data transfer over the cloud. It is designed for applications that rely on a hybrid environment of both cloud-based and on-premise servers.

5. The vendor's application collects data from Microsoft and aggregates it with third party weather data and building-level electricity consumption data provided by the utility. Analytics are run by the vendor's software, applying the rules engine and algorithms against the collected data from multiple sources.

6. The output is shared with Microsoft's building engineers via an interactive graphical interface accessible via the web. Single sign-on through Active

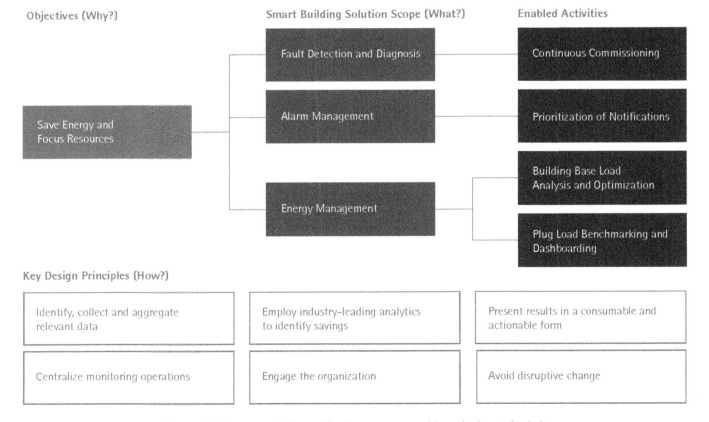

Figure 2-7: Smart building objectives, scope and key design principles

Directory Federation Services simplifies access to the externally hosted solutions. Future plans include mobile devices as endpoints.

7. In the newly established operations center, a team reviews the faults and alarms that have been identified and notifies engineers accordingly. The center is designed to support locations beyond Microsoft's main campus.

Key Design Principles of the
Smart Buildings Architecture

As outlined in section 2.4, introducing a smart building solution does not come without obstacles. In its approach, Microsoft has emphasized six design principles to overcome the most common barriers, while maximizing the program's impact. These principles underpin the smart building solution and its objectives.

Identify, Collect and Aggregate Relevant Data

The most important data in smart building architecture is collected either from existing building management systems or directly from installed control panels and energy meters. But to put this data into context, a range of additional private and public data is also needed. This includes building layouts, occupancy levels, and organizational information. For example, among the 45,000 employees[14] on the Microsoft main campus, more than 30,000 individual office moves can take place in one year. A daily data feed from the enterprise data warehouse can automatically keep track of building occupancy and other key parameters. Weather and utility information is gathered from third party providers. This collection of contextual information is necessary for normalizing energy consumption data, allowing for a better understanding of current performance and future potential. It also allows for demand forecasting and management.

Employ Industry-leading Analytics
to Identify Savings

A smart building solution's value depends directly on its analytics engine. Algorithms that detect faults, prioritize alarms and identify optimization opportunities amid vast amounts of data enable building engineers to unlock savings that are not addressable with traditional methods. Rules need to be customized by vendor resources or building engineers that are familiar with the software. Microsoft's pilot enabled it to experiment with three different analytics engines and determine what features were most important for its needs.

Present Results in a Consumable
and Actionable Form

Smart building solutions can produce very large volumes of complex data. To interpret the data, it is important to have an intuitive and customizable user interface that includes visual features, such as editable charts with drill-down functionality. Getting this right can make a major difference in enabling engineers to quickly identify irregularities. For example, by using Microsoft's Silverlight technology or the new HTML5 standard, rich browser-based user experiences can be created that work across multiple devices. This can provide engineers with user-friendly access to critical information wherever they are, on PCs and mobile devices.

Centralize Monitoring Operations

A key implementation challenge is the fact that building engineers often lack the time to familiarize themselves with analytics tools and make use of them in their daily routine. For large-scale deployments, one solution is to set up a central operations center that connects to engineers via their PCs and mobile devices. In this center, dedicated employees monitor the whole real estate portfolio, finding and prioritizing faults and dispatching building engineers accordingly. This is less disruptive to the engineer's role, and adds value by focusing their attention on high-priority issues. Keeping engineers engaged is essential to a smart building deployment, especially during an introductory phase, when the analytics engine is being fine-tuned. For smaller firms, a remote monitoring provider that communicates directly with on-site engineers can be an alternative.

Engage the Organization

One aim of a smart building program should be to influence occupant behavior, by providing employees and other stakeholders with information about their energy footprint. Visual benchmarks or graphical renderings showing consumption trends help make such metrics understandable for employees and management, and drive behavior change. Microsoft uses SharePoint to publish metrics internally, allowing tailored dashboards for specific user groups.

Avoid Disruptive Change

As an additional analytics layer rather than a replacement of an existing system, a smart building solution constitutes a low-risk IT project with little disruption to ongoing business. Given its cross-organizational nature, smart building programs need strong project management functions and executive buy-in. New tools come with a learning curve requiring training and expectation management. With appropriate change management efforts, the adoption of the new toolset can be accelerated and processes adjusted. Running an extensive pilot program helped Microsoft's organizations get familiar with the technological and operational aspects of smart buildings prior to full rollout.

FUTURE SMART BUILDING OPPORTUNITIES

A Cloud-based Approach

Cloud computing is set to transform information technology, by making third-party applications readily available as a service over the internet. For smart building solutions, this will deliver several benefits:

Accessibility

Aggregating and analyzing data from disparate sources is at the core of any smart building solution. The cloud is ideally suited to providing a universally available platform for managing building data mashed up with contextual information and made accessible for a variety of users and devices.

Scalability

With thousands of sensors and controls, modern buildings generate large volumes of data. Microsoft' main campus generates about half a billion data records per day from two million data points across 118 buildings. As the data volume and diversity of sources increases, a cloud-based architecture provides the scalability required to process massive volumes of data at an affordable cost. Abundant computing power allows complex modeling, such as correlating external temperature, cloud cover, and wind conditions with building access activity to refine heating, air conditioning and lighting patterns.

Ease of Deployment

Cloud technology can provide a secure and uncomplicated connection between off-site servers and on-site equipment. With Microsoft's Azure Connect the IT team was able to exchange data with the vendor solutions within minutes, avoiding the complexities associated with setting up a VPN.[15] Once connected, the cloud's elastic capacity enables vendors to serve new customers and increase the number of buildings managed without installing physical servers.

For two of the three vendors that are part of Microsoft's pilot, the cloud already plays an essential part in enabling a secure data exchange through Azure Connect. One of them hosts its solution on a public cloud, while the others are currently hosted on the vendor's premises and at Microsoft.

Automation and Real-time Analytics

We anticipate that the next generation of smart building solutions will allow organizations to automatically adjust building controls based on real-time data. For example, by monitoring the security badge access information for a building (as a proxy for the number of employees present), HVAC systems could be automatically adjusted to account for increased or decreased conditioning requirements. As an alternative, location and presence data from laptops or mobile phones could be used. Such solutions will rely on real-time analytics applied to incoming data streams, along with complex event processing, to execute automated adjustments to building controls. Microsoft StreamInsight, a component of the SQL Server software, can accomplish this task and is part of one pilot vendor's solution.

Future tools might even use machine learning to optimize algorithms over time, realizing even greater energy savings. For example, statistical analysis, simulation and predictive modeling can be applied to determine how many chillers need to be turned on, based on forecast occupancy levels and outside weather conditions. Researchers are also working on solutions that shape electricity demand curves of HVAC systems by using buildings as a form of energy storage and applying complex algorithms to optimize energy consumption through the course of the day.[16]

Integration with Utilities and City Infrastructure

With increasing adoption of smart building solutions, the built environment will achieve new efficiencies in energy use and improvements in occupant comfort. But this is only part of the story. Electricity grids are being upgraded with intelligent controls and two-way communication. As individual nodes of the smart grid, buildings will become active participants in managing energy demand and supply in a connected environment that includes power plants, transmission infrastructure and even electric vehicles.

For example, if a substantial share of Microsoft's employees were to use electric cars that are plugged in during the day, the campus could use the combined battery capacity to lower peak demand drawn from the utility at certain times during the day.[17] Likewise, demand response technology[18] can be used to shed loads in buildings when electricity consumption peaks, saving cost for utilities and building managers.

As buildings become increasingly networked, they play a crucial role in the development of energy-smart cities that unify the concepts of resource management and information technology on a municipal level.[19] A corporate campus like Microsoft's can serve as a testbed for many technologies that will shape the sustainable cities of the future.

CONCLUSION

Aggregated data and powerful analytics that add intelligence to existing building infrastructure have the potential to transform the way in which companies manage energy across their real estate portfolio. In particular, building engineers can be empowered to take a more targeted, data-driven approach to their work while automation improves their productivity. This delivers substantial cost savings, while helping firms achieve carbon reduction targets with relatively low capital investments.

Microsoft's smart buildings pilot program shows that while various adoption barriers remain, these can be overcome by following a set of key design principles. Most importantly, the underlying technologies are now more widely available and easier to implement. By sharing its experience with the public, Microsoft hopes to contribute to the evolution of the technology and encourage other companies to implement programs of their own.

The potential for information technology to improve building energy efficiency is huge. The Global eSustainability Initiative (GeSI)[20], a consortium of leading high-tech companies, estimates that smart building technology has the potential to reduce carbon emissions in the US by 130-190 million tons of CO_2 equivalent to the annual emissions of about 30 million passenger vehicles.[21] The related electricity cost savings amount to US$20-25 billion.[22] Quite simply, firms seeking to enhance their bottom line need look no further than the offices they are sitting in.

APPENDIX

About Microsoft

Founded in 1975, Microsoft is the worldwide leader in software, services, and solutions that help people and businesses realize their full potential. With 90,000 employees across its business divisions and global subsidiaries, the company generated revenues of US$ 69.9 billion for the fiscal year ended June 30, 2011. Its home page is www.microsoft.com.

Microsoft's Real Estate & Facilities organization is responsible for planning, delivery and operations of Microsoft's worldwide real estate portfolio, which comprises 33 million square feet (3.1 million m^2) and over 600 facilities across 110 countries.

About Accenture

Accenture is a global management consulting, technology services and outsourcing company, with approximately 236,000 people serving clients in more than 120 countries. Combining unparalleled experience, comprehensive capabilities across all industries and business functions, and extensive research on the world's most successful companies, Accenture collaborates with clients to help them become high-performance businesses and governments. The company generated net revenues of US$25.5 billion for the fiscal year ended Aug. 31, 2011. Its home page is www.accenture.com.

Accenture Smart Building Solutions (as part of Accenture Sustainability Services) enables commercial building owners to cost-effectively reduce energy usage and improve occupant comfort by managing and analyzing data to drive operational efficiency. The Accenture Real Estate Solutions practice helps organizations optimize their real estate portfolios and leverage their significant investment in space to support broader business goals and objectives.

About Lawrence Berkeley
National Laboratory

Berkeley Lab is a U.S. Department of Energy (DOE) national laboratory that conducts a wide variety of unclassified scientific research for DOE's Office of Science. Located in Berkeley, California, Berkeley Lab is managed by the University of California, and the director is Dr. Paul Alivisatos. Berkeley Lab has a total of 4,200 employees, including 1,685 scientists and 475 postdoctoral fellows. Its budget for fiscal year 2011 is US$853 million.

Berkeley Lab's Environmental Energy Technologies Division performs research and development leading to better energy technologies that reduce adverse energy-related environmental impacts. Researchers in the Division's Building Technologies Department work closely with industry to develop efficient technologies for buildings that increase energy efficiency, and improve the comfort, health and safety of building occupants.

Microsoft Project Sponsors and Contributors

Darrell Smith
    Senior Operations Manager, Real Estate & Facilities

Josh Henretig
    Director, Environmental Sustainability

Jay Pittenger
    Senior Director, Real Estate & Facilities

Rob Bernard
    Chief Environmental Strategist

Accenture Authors and
    Key Contributors

Andri Kofmehl
    Senior Manager, Strategy & Sustainability

Abigail Levine
    Senior Manager, Real Estate Solutions

Gregory Falco
    Consultant, Smart Building Solutions

Kreg Schmidt
    Senior Executive, Smart Building Solutions

Lawrence Berkeley National
    Laboratory Contributors

Jessica Granderson
    Research Scientist

Mary Ann Piette
    Staff Scientist and Deputy of the
    Building Technologies Department

References

1. Business case numbers for the deployment across all buildings of Microsoft's main campus, based on actual data from the pilot phase. The cost of deployment correlates with the number of buildings, so for smaller deployments, the percentage may stay in a similar range. For the payback period, the time required for installation, testing and tuning has been factored in, otherwise it could be less than one year.
2. Energy Efficiency in Buildings, World Business Council for Sustainable Development, August 2009.
3. ENERGY STAR and other climate protection partnerships, 2009 annual report, US Environmental Protection Agency, December 2010.
4. Commercial Buildings Energy Consumption Survey, Energy Information Administration, December 2006.
5. Building energy information systems: user case studies, Springerlink, Granderson, et al, May 2010.
6. Monitoring-Based Commissioning: Benchmarking Analysis of 24 UC/CSU/IOU Projects, Mills and Mathew, June 2009.
7. Accenture Smart Building Solutions (ASBS) is part of Accenture's Sustainability Services group.
8. A Holistic Approach to Energy Efficiency in Datacenters, Microsoft, 2010.
9. New workplace models constitute another strategy Microsoft employs to optimize its real estate footprint.
10. As of 2011, Microsoft's Sustainability Champions program included over 430 employees working in 73 different buildings across the main campus. Its objective is to reduce the company's environmental impact through employee engagement. Newsletters, events, training, green bag lunches and monthly building energy consumption reports by floor or wing are the key elements aimed at reducing employee-controlled electricity consumption.
11. For example, engineers collaborated with Siemens to convert data from the proprietary Apogee format.
12. Additional data elements include square footage, floors, types of rooms, location, and whether a building is owned or leased.
13. One of the three pilot software solutions is hosted on Microsoft's premises, whereas the other two are hosted off-site. The on-premise solution does not use Azure Connect as data is not transmitted outside Microsoft.
14. Including on-site vendor resources.
15. Virtual Private Network.
16. Research rock star: Efficiency star, Coloradobiz, 2011.
17. Microsoft's corporate headquarters already has a large fleet of company-operated hybrid cars to transport employees between buildings on campus. The company recently started installing charging stations for employee-owned electric vehicles.
18. Berkeley Lab Demand Response Research Center, www.drrc.lbl.gov.
19. The central role of cloud computing in making cities energy-smart, Microsoft, 2011.
20. GeSI counts more than 30 leading IT and communications companies among its members, including Microsoft. More information available at www.gesi.org.
21. SMART 2020: Enabling the low carbon economy in the information age, United States Report Addendum, GeSI, 2008.
22. Greenhouse Gas Equivalencies Calculator, US Environmental Protection Agency.

Chapter 3

Perspective on Building Analytics Adoption

Tom Shircliff and Rob Murchison
Intelligent Buildings

INTRODUCTION

Building analytics and primarily building energy analytics is poised to have a profound effect on facility and property management. Building analytics are an important part of an industry moving towards data driven buildings whereby not only operational methodology but even capital investment decision-making is determined by data. This is the manifestation of "big data" and "cloud" in the real estate industry, which is part of a trend in nearly all industries of leveraging softer, less expensive technology to lower operational costs. This is especially notable in real estate because of a general reluctance to mix information technology (IT) into traditional design, development and operations.

Paul Otellini, President and CEO at Intel, was asked by talk show host Charlie Rose "What is going to be obsolete next?" Otellini responded "Ignorance." This applies well to the age of so called data driven real estate where we no longer have to be ignorant of the massive amounts of data in our buildings that can be processed into information and knowledge.

The changing, digital nature of building controls systems is hastening the opportunities to leverage building analytics, even while there are still many organizational and cultural barriers to wider industry adoption. There are three pillars to building analytics as there are to nearly any real estate technology movement: Buildings, people and technology. Most people only focus on the technology itself when searching for or considering advanced real estate and facility solutions. However, the other two pillars can make or break a technology solution. If the building is not properly equipped to connect or enable the new technology then it will be limited or even completely ineffective.

That notwithstanding, the larger issue is how and where in an organization new technologies get purchased and what commitment the organization makes to change the way they do business to best leverage the new technology. Additionally, stakeholder value propo-

sitions must be considered throughout the organization. What sounds like a good idea to senior management may not be understood or implemented by front line facility managers or operations and maintenance vendors may not have aligned incentives or even contractual latitude to change management processes. In the following pages we will briefly describe:

1. History of IT in real estate
2. General technology issues
3. Building requirements
4. Analytics types
5. People dynamics that surround today's building analytics

HISTORY

Historically, the real estate industry has been slow to adopt current-day IT best practices into the facilities themselves. Technology in real estate has usually meant back office software such as property management, work order management and asset management. However, since the new millennium, IT has crept out of the back office and into building controls systems to the point that nearly every major controls systems uses an IT foundation for its very functionality. In other words, controls systems such as HVAC, lighting, elevators, advanced metering, security, parking and others use computer servers, category cabling, protocols, distributed digital controllers, end (or edge) devices and remote access to connect and operate.

This IT architecture is sometimes referred to as "digital" as a short hand description. Digital systems are computer controlled, programmable and produce many data points that have the potential to be acquired and analyzed which are the first two functional steps of building analytics which include: Acquiring, Analyzing and Acting.

TECHNOLOGY

With digital, IT-based systems you indeed have more accessible data points albeit with some limitations. While we can usually gather the data in so-called "digital" systems it is in many cases still an expensive, clumsy process. The reason is generally attributable to proprietary protocols, which are often referred to as "closed systems." These limitations lead to an open protocol movement as well as faster development and adoption of data normalizing middleware as a sort of stopgap measure.

At first open protocols happened within respective building system categories. For example the HVAC industry adopted both BACnet and LonWorks and the electrical metering industry segment leveraged Modbus. There were other open protocols in different industry segments as well as other segments borrowing existing protocols such as the three aforementioned. Even with industry segment openness, different segments (or different types of controls systems) still used different communication protocols. For example BACnet did not communicate directly with Modbus.

Middleware, is the preverbal black box that, when equipped with the necessary drivers can allow communication between these disparate protocols and also unlock proprietary protocols. This has been a key enabler for analytics and provides three primary benefits:

1. Interoperability between disparate systems in both the building and back office. An example of which would be a security card triggering a reaction in an HVAC system, which could in turn create a back office software billing event (see Figure 3-1).

2. Data normalization of proprietary systems for enterprise management of different systems, such as managing multiple HVAC controls by different manufacturers (see Figure 3-2).

3. Access to proprietary data points for building analytics which might include identification of energy consuming system faults or "energy faults." These energy faults are a myriad of system irregularities or problems that don't necessarily generate an alarm and may not even cause discomfort but do in fact cause energy and financial waste.

These types of faults are the same faults that generally get resolved during a re-commissioning or retro commissioning process. The limitation with retro commissioning is that its expensive and only occurs one time every 4-6 years at best. Analytics of this type offer the value proposition of continually commissioning to eliminate the faults and keep them from building up (see Figure 3-3).

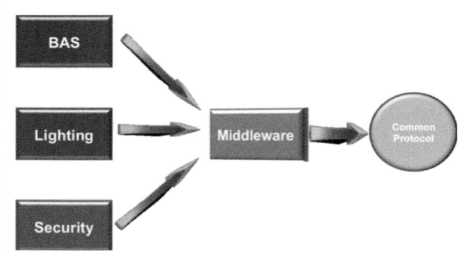

Figure 3-1.

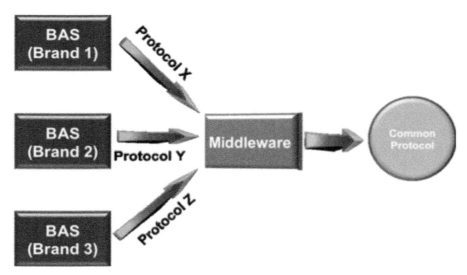

Figure 3-2.

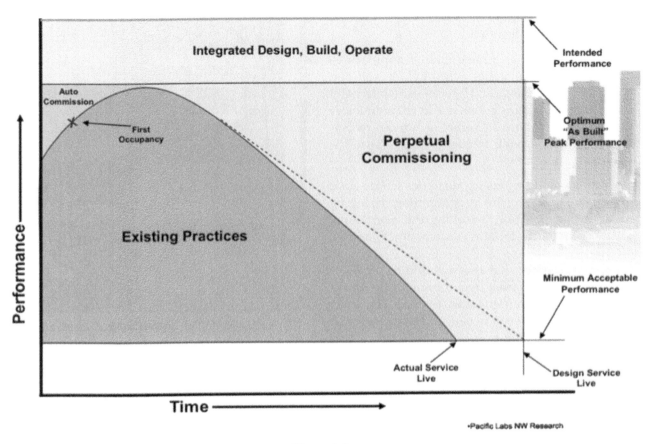

Figure 3-3.

To identify these faults through energy analytics you must have several key components which are: Access to the raw point data from HVAC and other key systems, normalization of the data so it can be consumed and an analytics platform with algorithms and rules that accurately identify faults.

Rules are a much-debated aspect of analytics and some believe rules are an ever-evolving "approach" to analytics that should be refined and shared and others believe they are intellectual property that will separate one provider from another. The answer is likely somewhere in between. There are surely those who have experience and expertise that will be first to certain ways (rules) to analyze building data, however, in the end it may well be hard to prevent workarounds to achieve the same results.

BUILDING ENERGY ANALYTICS TYPES

There are now hundreds of building analytics companies that call themselves "energy analytics" solution providers, which is contributing to the white noise and

confusion in the marketplace. While still very young this building analytics industry segment that focuses on energy has begun to separate itself into four category types as follows:

1. Dashboards
2. Platforms
3. Services
4. Operational

In Figure 3-4, the two axes show a range of service vs. software and a range of defined vs. customizable. There is no "best" quadrant or placement on the diagram. Rather, its purpose is to help match customer profile to service provider service type.

1. Dashboards provide a consolidated look at the building's energy usage and trends—and in some cases provide insights derived from data patterns.

2. Platforms are software platforms that offer very flexible and open-ended use of their technology to store information and build rules. Platforms offer, owned data licenses, less structure and more flex-

ibility but often require intensive integration and expertise to make them work.

3. Services are more of a handholding approach to building energy analytics whereby the solution provider gathers data and processes it internally apart from the building owner and communicates the recommendations in verbal or written recommendations at varying frequency such as weekly, monthly or quarterly.

4. Operational building energy analytics seek to have more system control for programming, changing routines and generally tweaking the operational functionality of the building automation system.

It is not uncommon that a savvy building manager might employ solutions from two or more of the above categories without any consequential overlap. However, there are very few property owners and managers that have gotten past selecting even one key building (energy) analytics company and most are still dizzied by the array of choices, technology lingo and hype. Therefore, it is important that building owners and managers self examine to determine their analytics profile—either internally or with outside consulting assistance. Then, a more accurate solution search can be engaged with a narrowed field of providers.

For example, if corporate policy exists that says all software used must be licensed solutions that function from company data centers then a platform solutions may be the only practical way to gather and analyze millions of data points. On the other hand, if an owner or manager acknowledges that there are no staff or resources to fully implement such an approach then a services solution may be the shortest path to value. Likewise, if there is less active property of facility management then operational solutions may provide needed system optimizations through analytics. Finally, if the three previous examples are more in depth than is palatable a more simple approach involving dashboards for energy trend data to drive awareness and behaviors may be all that is required.

BUILDING REQUIREMENTS

As previously mentioned, all of the opportunity available in building analytics is predicated on getting the data points to start with. Therefore, there are certain requirements in both new and existing buildings to get to the data. For new buildings, it is strongly recommend-

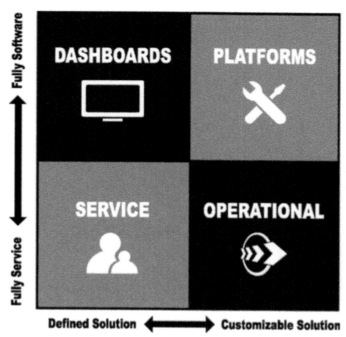

Figure 3-4.

ed that building standards and building specifications be developed that specifically define and require open protocols for all building controls so that a new property starts out ready for data consumption and analytics.

For the more predominant issue of existing buildings, there are several ways to get to the data.

First, a set of building standards and specifications like the previously mentioned new building standards should be developed for all building system replacements or scheduled refreshes. Then, as the systems get replaced in an aging building, that building will move towards more open and even converged systems as it reshapes its ability to cost effectively make data available.

Second, identifying existing digital systems that may be "open" or otherwise able to produce consumable data, especially with certain solution providers would be an immediate opportunity.

Finally, implementing a middleware solution that can function as a protocol "blanket" providing broad data normalization and the most flexibility for using building data is an approach with multiple benefits.

PEOPLE DYNAMICS

Even with maturing analytics technology, readily available middleware and more prevalent digital systems in buildings the organizational and overall "people" aspects of building analytics are and will remain the driving force for the pace of adoption. The key people aspects are:

1. Senior level support and linkage to high-level goals and priorities

2. Aligned incentives internally and externally

3. Individual stakeholder value proposition

Senior Level Support and Linkage to High-Level Goals and Priorities

Those of us who focus on operational efficiency including energy analytics and other strategies may find it hard to understand why anyone wouldn't want to save energy or money. While it is usually not the case that they don't want to save, it is frequently the case that it is a smaller financial or strategic issue in the scheme of things. However, even when energy and operational cost are not the priority it is probable that they can be meaningfully linked to support other organizational goals and priorities. For example, energy efficiency can be linked to sustainability, productivity, shareholder or taxpayer value and occupant experience.

A comprehensive analytics approach will not only fight for capital budget dollars but will also fight for attention in many organizations. We in the industry must understand the building owner and manager's organization and their priorities in order to most responsibly communicate the value of building analytics.

Additionally, since analytics can span several departmental issues that cover IT and facilities among others there must be a senior level sponsor for a proper analytics program. An analytics program will necessarily change the way you do business. As mentioned, technology and buildings do not fix energy faults by themselves and hence an organized, respectful but mandatory process must be instituted with formal measurement and management.

Since senior level support is important it can be more difficult for solution providers to know who the actual buyer is. Is it the director of IT, the VP of Facilities, the CFO or the President? Or even more perplexing is the question of whether the customer is organizationally aligned to purchase a next generation analytics solution at all because the solution spans multiple realms of responsibility. Said in short if everybody is in charge nobody is in charge.

Identifying the organizational decision making structure is a key first step for an analytics solution provider.

Aligned Incentive Internally and Externally

Aligning incentives is critical since just about all groups in a real estate or facilities organization has actual or de facto veto power in the supply chain of buildings analytics.

In new construction developments there may not be an incentive for the development manager to be concerned with future energy and operational costs and that could yield more proprietary systems that are harder to get data from and more expensive to maintain.

In existing buildings, IT departments may be incented to have secure systems without respect for costs or tangential issues. Operations and maintenance vendors may be incented only to maintain comfort. Sustainability managers may have narrowly defined targets that are not related to or don't leverage core energy opportunities.

It is understandable that with only a series of narrow departmental goal and worse yet misaligned goals between departments that the respective groups in an organization have to press on with their own agendas. There are many examples like this that are very inefficient and expensive from an enterprise perspective.

With the senior level support described in "PEOPLE DYNAMICS #1" you can align and manage the incentives for internal departments and external vendors to a mutually beneficial result.

Individual Stakeholder Value Proposition

Similar to the veto power of departments, individual stakeholders can scuttle even the best analytics efforts. The reality is that the results of good building energy analytics will include gains for many different stakeholder types including IT, sustainability, facility operations, branding, finance et al.

However, the fact is that there are many different stakeholder benefits and that value proposition should be communicated in a tailored way to each stakeholder type.

CONCLUSION

Real estate technology is materially different than it was less than a generation ago and that presents both risks and opportunities for building owners and managers. The way buildings are built and managed must change as a result in order to avoid significant negative impact to cost structure, performance degradation and an uninspiring occupant experience. Thus, the risk of not changing has exceeded the risk of changing.

However, with that change comes new opportunities to significantly lower cost structure, continually and perpetually optimize building performance, improve experience and enhance facility related careers. These new opportunities are built around building analytics. With a proper strategy that embraces the technology change as a chance to get, analyze and act on building data, owners and managers will enter the new age of cost efficiency and performance that also lowers risks through informed decision making.

Chapter 4

Delivering Successful Results from Automated Diagnostics Solutions

It's Not Just about the Technology

Paul Oswald, President
Environmental Systems, Inc.

ABSTRACT

As the title suggests, recent developments in automated diagnostics technology have provided opportunities for facility managers and owners to significantly reduce their operating costs. But, it's not just about the technology. There are many other factors that need to be considered when evaluating facility improvement initiatives.

Beyond specific platform considerations, owners looking to reduce costs and improve building performance and efficiency need to take a hard look at their current situation. Do you have the resources and expertise required to achieve the results for which you are looking? Many conclude that they do not and look to a managed services partner to help guide them to the way of continuous improvement.

INTRODUCTION

There certainly have been numerous new technologies and products introduced in the past three years with claims to provide automated diagnostics. In our experience however, if the only consideration given to a solution for automated diagnostics is the underlying technology, the outcome will likely miss its intended target and the project/solution will fail to achieve its desired results.

Generally speaking, analytics platforms and products fall into two categories: analytics based on a rules engine and analytics that use predictive modeling. While it is beyond the scope of this chapter to discuss details of the technologies themselves, for the most part, current platforms are rules-based.

In the following sections, we'll look at specific factors that can significantly influence the success (or not) of a facility improvement initiative. Included in these factors are skills, talent, time, and more. But, it's not (just) about the technology.

OVERVIEW

There are many contributing factors that need to be understood, discussed, and planned for in order to achieve a successful result. While technology is certainly one of them, for without robust underlying technology there is no solution, it is not the only consideration. Yet, time and again we see RFP's, papers, presentations, and proposals focused solely on technology. The mindset is too often that if you select the right platform, you'll achieve the desired results.

It is true that a rich and flexible technology platform is fundamental. At a minimum, it should be both flexible and robust in these areas:

- The type of data it supports
- How data is on-boarded
- The rules engine
- Reporting capabilities

Beyond the platform, other considerations that need to be addressed during the planning process include:
- Time
- Skill
- Infrastructure
- Quality of the data
- Budget
- Current state or condition
- Delivery method

TIME

How many current resources are available to work with the solution? This includes the time to review the output of the analytics, prioritize corrective actions, initiate corrective actions, monitor progress and ensure actions are completed, and measure the results. Depending on the size of the organization, this can be a full-time job for one or more people. Too often we've seen technology that's been implemented and never used due to a lack of focused resources.

SKILL

Who will be using the solution and do they have sufficient skills and expertise with buildings, mechanical equipment, electrical systems and rate structures, and best practices in facility operation and optimization? If the people responsible for the operation of the automated diagnostics solution are assigned to the role on a part time basis, it is highly unlikely any meaningful results will be achieved. Furthermore, if the people assigned lack the necessary background and experience, results will be significantly diluted. One other thing to keep in mind, having a basic set of skills is one thing, maintaining that skill over time is quite another. Is the organization prepared to invest in the on-going education and development of the resources that will be using the solution? This is key to ensure best practices are being applied and the capability of the technology does not quickly outpace the skills of the people using it.

INFRASTRUCTURE

What is the infrastructure like? By infrastructure we mean the data sources and pipeline or method of moving data from the data source to the automated diagnostics platform. There are many questions that come into play:

- Is there such a pipeline or method to move the data? If not, what are the options for getting the data from the source to the automated diagnostics platform?

- What update frequency or intervals are required to drive an effective result and can you meet those requirements?

- Are there security or network issues that need to be addressed?

- Is the pipeline robust or are their weaknesses that could lead to gaps in the data, which will ultimately limit the effectiveness of the solution?

- If the existing infrastructure is not sufficient to support the desired result, what will appropriate renovations cost and is it budgeted?

QUALITY OF THE DATA

How good are the data? This may seem like an obvious question, but it is often surprising how inconsistent system data can be. One of the most significant issues that need to be addressed in any automated diagnostics implementation is data tagging. On-boarding data with common tagging and consistent naming conventions are critical to achieve quality output and related results. Even when working with data from a single system or building, there can be conflicting tagging and naming practices for common assets such air handling units, VAV terminal units, and central plant equipment. This is typically due to the personal naming practices used by the engineer who programmed the system. The problem is compounded when dealing with multiple systems and multiple buildings within a campus or portfolio.

Not only is data tagging a significant issue to resolve, but gaps in data present problems. If the histories that provide source data are not consistent or are fragmented, the output and results will be compromised. There can be several reasons for fragmentation including:

- Histories are programmed with intervals that are too short, which cause memory overruns and premature termination.

- Improper file maintenance and/or archiving practices.

- Ineffective data management resulting in corrupt or missing data.

BUDGET

How much money is allocated for corrective action? This too may seem obvious, but all the money spent on automated diagnostics won't return a penny to the bottom line if the issues identified by the solution are not actually addressed. It is frequently assumed that by simply implementing an automated diagnostics solution will produce results. If an organization's maintenance strategy is "run to failure," it is impossible for any automated diagnostics implementation to achieve the desired results. Sufficient money and resources must be allocated to correct the issues identified by the solution.

How Important is Process?

Corrective action needs to be implemented in the context of a clearly defined process. The key to a successful implementation that is aligned with an organization's budget and resources is to build a process around the technology. The process manages the rate at which corrective action is taken to reflect that which the organization can effectively support. Identifying more issues than can be realistically addressed only serves to frustrate stakeholders and undermine support for the solution. The ultimate solution is built around a process that effectively triages, prioritizes, and ultimately corrects the issues that have the greatest impact on the organization's bottom line without placing an undue burden on its budget and resources.

What Gets the Priority?

There are a number of methods for prioritizing the issues identified by an automated diagnostics solution.

- Frequency—how many times did the fault occur?

- Duration—How long did the fault condition last?

- Cost impact—How much did the fault cost. Or, how much could the organization have saved if the fault was corrected? This is typically expressed in per hour units.

Frequency and duration are easily understood and easy to quantify. However, cost impact is another story. There is very little quantifiable data available to measure cost impact; there are just so many different variables that influence value. Consider for example economizer failure. To have an accurate cost impact value requires us to consider (at least) the following:

- Equipment age and condition
- Climate zone
- Hours of operation
- Energy cost
- Type of occupancy

For a single site implementation, this may not seem too difficult and would only need to be tailored for each asset to which this rule applies. However, when implementing this rule across hundreds or thousands of sites, the effort required to capture and input this data is a formidable task. As such, an approach that is effective in these situations is to apply some level of generalization. This provides a reasonable level of accuracy for the amount of time required to set up the data. Is it completely accurate for each piece of equipment? No. However, keep in mind that we are using this to prioritize corrective actions, not to predict absolute cost avoidance. Regardless of the method chosen, it should be something that is mutually accepted by the service provider and the customer. In this way, actions and results are transparent and understood by all stakeholders.

How Are Corrective Actions Implemented?

Once prioritized, how are corrective actions implemented? Typically, they'll fall into one of three buckets: immediate action, scheduled, or an aggregate. Immediate action items are those that are both urgent and important. Scheduled items are those that are important, but not necessarily urgent. And, other corrective actions can be aggregated together to form more strategic improvement projects. Regardless of the approach, it is imperative that the corrective action process keep pace with the issues that are identified.

Think of the process as a lens. In its initial deployment, we expect to find many issues. The rules give us a broad view of our portfolio, our building, and our assets. As issues are identified, prioritized, and corrected our focus becomes more concentrated. Simply put, we move from looking at the low hanging fruit to searching for change in the couch. If the process fails to correct the obvious, we'll never find the change in the couch and our efforts will be stymied. This is one of the primary purposes of the process; to "throttle" the amount of issues to a level that is consistent with that which the organization can handle from a staffing and/or budget availability standpoint.

In summary, our budget must provide for both the automated diagnostics solution and corrective action. It is rather straightforward to establish a budget for implementing the automated diagnostics solution. However, consideration must also be given to budgeting for continuous corrective action and the level of "throttling" that needs to take place to make the overall process sustainable. Identifying more than can be corrected is of little value and could ultimately be counter-productive to the effort.

CURRENT STATE OR CONDITION

What is the current condition of the equipment and is there an active program for maintenance and improvements? Once the automated diagnostic solution is implemented, what is it likely to find? A building or portfolio of buildings whose equipment is well maintained and has a regular replacement cycle in place will likely result in fewer "major" issues discovered by the automated diagnostics solution.

However, applying automated diagnostics to equipment that has not been properly maintained will likely result in significant findings that could include major repairs. This needs to be considered when planning an automated diagnostics solution because the findings could be overwhelming. This is not to say that it shouldn't be done or that equipment that has not been maintained properly is not a good candidate for an automated diagnostic solution. Quite to the contrary, these are excellent situations for automated diagnostics and the opportunity for meaningful results is significant. Just make sure there is a clear understanding that the problem wasn't created overnight and it won't be corrected overnight.

Issues that are identified will be corrected with the understanding that this is not a discrete project. Rather, it is a continuous process where issues are identified and corrected in a prioritized manner, within the limits of our staff and budget. When those tasks are completed another level of issues are identified and addressed. As each successive layer of issues is resolved, the process narrows its focus and moves to greater levels of detail, which continues to drive greater levels of efficiency and ultimately cost savings. Keep in mind though there is no "silver bullet" to achieve success. It's an on-going commitment to continuous improvement.

DELIVERY METHOD

We've addressed the issues that are critical to successfully implementing an automated diagnostics solution. Two of these, time and skill deserve further discussion. One key question that needs to be addressed is, "Does the business or organization have the capacity to successfully manage the automated diagnostics solution?" We oftentimes find that organizations have implemented the automated diagnostics solution and have been trained to use it, only to return in six months finding that no one has devoted the time needed to produce meaningful results and lay the groundwork for continuous improvement.

Beyond time, it takes a certain skill set to keep you on course. And, the learning needed to keep you abreast of ever changing technology requires an on-going commitment to continuous education and best practices. We've seen many organizations completely overlook the hidden costs of continuous learning. Moreover, in-house management of these efforts requires additional overhead and capital expense.

Is that to say that organizations should turn their backs on these types of initiatives? Certainly not. For many organizations, one solution to this is to have the process managed for them. Commonly referred to as managed services, this delivery model eliminates the challenges presented by a lack of manpower and experience. Furthermore, managed services eliminate the need for on-going internal training and skill development and can lead to cost savings and operational efficiencies. They allow organizations to focus on its core competencies.

Key benefits of the managed services delivery model include:

- Improved workforce productivity. The organization can focus on their core business, not transactional tasks.

- Enhanced access to technology. The organization benefits from the latest in technology and maintains a solution that is current.

- Continuous monitoring and improvement. The solution continues to produce results through on-going M&V to metrics and performance measurements.

- Improved strategic spending. The organization frees up working capital for core business tasks and functions.

Improving Workforce Productivity

Partnering with an experienced firm can help solve a problem many businesses face—the gap in skill and talent. Today's buildings and building systems require a vast range of knowledge and skills to operate facilities at peak performance, while staying abreast of the latest in codes, regulations, and best practices. Under a managed services delivery model, you have access to the expertise needed to efficiently run your facility, which allows your staff to concentrate on your core business.

Enhanced Access to Technology

Automated diagnostics are a proven way to reduce facility costs and increase efficiency. But, the time, expense, and risk involved in selecting and implementing a solution on your own can be daunting or even prohibitive. The managed service delivery model provides quicker, more cost effective, and less risky access to all the latest tools and technology available to reduce operating costs.

Ensuring Results

The managed services delivery model should include a transparent performance measurement and tracking plan to allow all of the stakeholders to see and understand how the process is working and where adjustments are needed. This allows stakeholders to clearly see inefficiencies and make informed, strategic decisions about spending and resource allocation.

Concentrating on Core Competencies

Businesses can be inundated with time-consuming and complex issues that fall outside of their true mission. Using the managed services delivery model enables service providers that excel in the areas of facility management to drive increased efficiency and cost savings, while allowing the organization to focus their time, resources, and capital on their core business. It is incumbent on business leadership to allocate scarce capital wisely—whether acquiring new, replacing old, or maintaining existing property, plant, and equipment. But, prioritizing facility budgets, including how much to allocate for things like repair and replacement, can be challenging. The managed services delivery model provides qualified facilities experts to proactively analyze and plan for short- and long-term needs, which result in significant cost savings and preservation of valuable assets.

One needs only to look at the growing trend to outsource IT and other back-office and transactional functions. The challenges of staying ahead of hardware and application advances, as well as turnover in the ranks of facilities professionals, makes outsourcing to organizations with the sufficient expertise a smarter strategy. It allows facilities, energy, and sustainability departments to concentrate on growing their businesses rather than keeping their buildings up and running.

SUMMARY

Automated diagnostics can drive significant savings for virtually any organization. This is true if, and only if, we keep in mind that the underlying technology is only one part of the formula for success. The process by which the technology is used and its method of delivery is critical to achieving the desired results.

Chapter 5

Intelligent Efficiency: Opportunities, Barriers, and Solutions

Ethan A. Rogers, R. Neal Elliott, Sameer Kwatra,
Dan Trombley, and Vasanth Nadadur

INTRODUCTION

A number of stakeholders, including people in both the public and private spheres, agree that intelligent efficiency will generate massive economic benefits—$55 billion in annual energy cost savings by our estimate—in the near future. Intelligent efficiency is the term the American Council for an Energy-Efficient Economy (ACEEE) uses for the deployment of inexpensive next-generation sensor, control, and communication technologies that collectively enhance our ability to gather, manage, interpret, communicate, and act upon large volumes of data to improve device, process, facility, or organization performance and achieve new levels of energy efficiency. Equipment and systems used in buildings, transportation, and manufacturing are becoming adaptive to environmental inputs, anticipatory in their performance, and networked to other devices and systems. The empowering hardware and software products that form the backbone of intelligent efficiency are information and communication technologies (ICT), and these are making possible analysis and levels of performance that could not be achieved as recently as ten years ago. Building management systems (BMSs) can now determine immediately when a boiler or chiller has begun to operate outside of normal parameters and dispatch a service technician to address the problem, and production management systems can slow down or turn off equipment in response to the production demands of the day, or even the hour (Fernandez et al. 2009). In the next two to three decades, these new capabilities will affect every sector of the economy and bring about efficiencies that we are only beginning to understand. How great might the benefits be when business offices and production departments of companies throughout a supply chain—from raw material suppliers to manufacturers to transportation companies to retail establishments—are all networked so that performance of and demands on one are communicated in real time to all the others in terms that enable each of them to adjust its performance accordingly? Or if the performance over the past ten years of the air-conditioning systems of several dozen similar buildings can be combined with the weather forecast for the next week to optimize the setting of a an individual building's air conditioning system for the next day's operations?

ACEEE first began to define intelligent efficiency in the report, A Defining Framework for Intelligent Efficiency (Elliott, Molina and Trombley 2012), offering examples and case studies and identifying steps policymakers could take to accelerate its adoption. Since the report's release, other organizations such as the Center for Climate and Energy Solutions (C2ES) and the Global e-Sustainability Initiative (GeSI) have expanded on this subject, examining the potential of intelligent efficiency to reduce government agencies' energy expenses and reduce emissions of greenhouse gases around the world. This report represents the next stage, to transition from definition and description to a focus on more strategic forms of analysis. What is the economic potential of intelligent efficiency, and how can consumers, policymakers, and the energy sector embrace it? What are the opportunities to accelerate its adoption? What are the challenges to making that happen?

Since the release of the first report, we have come to understand that the best near-term opportunities for the application of intelligent efficiency are in the commercial and industrial sectors. These sectors embrace automation more rapidly than do the public, transportation, and residential sectors due to the need for businesses and manufacturers in a competitive environment to control their operating costs. The companies that supply the commercial sector with automation—the building automation industry—are estimated to do $43 billion in sales by 2018 (ABI 2013). The growth for automation of the manufacturing sector is estimated to be even greater, reaching over $120 billion by 2020 (Cullinen 2013). With significant growth in automation of the commercial and industrial sectors anticipated over the next ten years, we can anticipate corresponding gains in energy savings. The "intelligent" or "smart" automation of future investments will not be mechanical or even just programmable controls, but a combination of sensors that provide bi-directional communication between devices and controls; remote access through the internet; networks within processes, buildings, and organizations; and new software programs that can manage large quantities of data that when combined together enable self-correcting and anticipatory capabilities that yield new capabilities and additional productivity, and energy savings.

With the ability of intelligent efficiency to generate the next-step change in energy savings, multiple economic benefits are possible as energy efficiency catalyzes increased economic activity: Direct benefits accrue from the avoidance of energy use due to greater efficiency; non-energy benefits stem from system optimization, including better services and, in industry, better quality control; and lower operating costs free up capital, making it available for additional investments in productivity and capacity. Because the implementation of automation systems based on intelligent efficiency are so cost-effective and pay back so rapidly, we anticipate that a great deal of economic activity will happen with little or no influence from the public sector (M2M. WorldNews 2012). However, there is an importunate opportunity to leverage intelligent efficiency for public policy goals.

The best near-term opportunities for polices that promote intelligent efficiency are at the state, utility, and local levels, where most energy policy and programmatic activity takes place. Each state has at least one agency with responsibility for regulating electric and natural gas utilities, and these agencies are shaped by state legislatures. Additionally, many municipalities have their own utilities, which are part of or in some fashion answerable to the local government. The states function as laboratories in which many different policies and programs are tested and refined, and from these experiments federal policy will likely be constructed.

The constantly evolving policies and programs at the state and municipal levels means that great opportunities for action exit there. To wit: A majority of states have a requirement for utilities to encourage energy efficiency, and as a result there are programs of all shapes and sizes designed to help customers use energy resources more effectively. Each of these programs has one or more performance goals for which the program administrator is held accountable. Many of these programs focus on medium and large buildings and manufacturing facilities, and depending upon the details of the program, intelligent efficiency could be a mechanism that enables them to not only meet their performance goals, but to do so at a lower cost per unit of energy savings than before.

Given the great potential for intelligent efficiency to affect widespread declines in energy use and to bring a range of economic benefits, in this report we continue our examination of challenges to its broader acceptance. We give special attention to structural barriers in the electric utility sector that affect commercial and industrial customers by making it difficult for utilities to invest in automation as part of their efficiency programs. These barriers include existing business and government accounting practices, utility regulations that focus on service territories rather than an organization's energy footprint, and efficiency program policies that favor hardware over software and products over services. These are also the barriers that energy sector policymakers and stakeholders—who determine what can and cannot be included in an efficiency program and specify how its performance is evaluated—have the greatest ability to change. We identify specific actions that can be taken by utilities, public utility commissions, government agencies, and energy consumers to reduce or eliminate these barriers. These policy and programmatic recommendations focus on creating an environment in which governments lead by example, design experimentation can take place in utility efficiency programs, and utilities support investments in intelligent efficiency.

These actions on the part of governments and utilities will be justified by the magnitude of the economic impact of intelligent efficiency on the commercial and industrial sectors—up to $90 billion a year in energy cost savings. In this report we attempt to quantify the electricity savings in these sectors that could result from the recommended policies and program activities that we present. We focus on electricity because there are many more efficiency programs for electricity than for other fuels. We examine the energy savings of a set of key individual intelligent efficiency measures, each of which has a high likelihood of significant market penetration and energy savings in the near term, and we then project the potential energy and energy cost savings over the next 20 years.

The economic analysis brings us full circle. It is the reason to care and the reason to act. There is an overarching economic need at the national level to embrace intelligent efficiency as it can produce a next-step change in the efficient use of energy in all economic sectors. There is also a targeted need within the utility sector to provide more effective energy efficiency programs. However, before each of these opportunities can be realized, it is necessary to overcome existing and potential barriers to greater market acceptance of intelligent efficiency and to have a plan for accomplishing this. To that end, the report concludes with a set of recommendations for policymakers, public utility commissions, utilities, suppliers of automation equipment for buildings and manufacturing, and energy efficiency program administrators to guide them in addressing these challenges and taking full advantage of the opportunities brought by the widespread implementation of intelligent efficiency.

INTELLIGENT EFFICIENCY DEFINED

Intelligent efficiency has resulted from the convergence of several new technologies and analytical capabilities that now enable another step change in energy efficiency. Intelligent efficiency is a concept, or a capability. Much like information technology is the capability to manage information with computers, software, and networks, intelligent efficiency is a new ability to save energy that arises from our ability to gather large volumes of data and to manage, interpret, communicate, and act upon it in ways that increase the energy efficiency of complex systems.

The hardware and software products that enable intelligent efficiency are information and communication technologies. This combination of enabling technologies—sensors, computers, data storage, networks, cloud computing—allows users access to real-time information, historical information, and analytical capabilities that, when combined, enable the users to determine the most efficient method to operate a device, system, process, facility, or even network of facilities. Many of these systems have the ability to study historical information and to use that information in combination with information about ambient conditions in order to evaluate multiple possible operating scenarios before selecting and implementing a final decision. It is be-

cause of this ability to learn and improve over time that these systems are often referred to as "intelligent" or "smart." For example, a chemical manufacturing plant with an intelligent process control system could use information from multiple previous operating scenarios—production volumes, process speeds, equipment set point, outdoor temperature, and humidity—to recommend operating conditions that would improve future performance.

There are many different ways in which intelligent efficiency can be leveraged to improve a product or service, opportunities that exist along a continuum with technology and human behavior at either end (see Table 5-1). Increased "intelligence" along this spectrum falls into three broad categories (Elliott, Molina, and Trombley 2012):

- People-centered efficiency provides consumers with greater access to information about their energy use as well as the tools to reduce energy use. In this type of intelligent efficiency, technology makes individuals' energy use visible, thus guiding them toward making major efficiency gains. An example is a dashboard display on a computer screen that provides facility or process managers timely and actionable information on energy use.

- Service-oriented efficiency, often referred to as "substitution" or "dematerialization," provides individuals with the option to substitute one material-based service for one that is not material-based. An example is the replacement of physical compact discs by digital music. In this type of intelligent efficiency, the end users choose the degree to which they will utilize information and communications technologies—which use less materials—to accomplish a goal. One of the greatest emerging manifestations of this is cloud computing. No longer must every company have its own servers and IT departments. Instead, they rent space in the "cloud" and subscribe to IT services.

- Technology-centered efficiency encompasses "smart" technologies that optimize energy systems in buildings, industries, and transportation

Table 5-1: Types of Intelligent Efficiency within the Continuum

Types of Intelligent Efficiency Technologies	Types of Intelligent Efficiency Solutions
People-Centered	Interfaces
Technology-Centered	Control Systems
Service-Oriented	Substitution

systems. Here, automated systems optimize energy use and anticipate energy needs, and human engagement is largely limited to the initial programming and commissioning of the system. A building's heating and cooling system that might have required routine adjustments to a thermostat by a person now responds to inputs from occupancy and temperature sensors, on-line weather forecasts, and stored information on the occupants' preferences.

The main focus of this report is technology-centered efficiency, with the other two types playing more minor roles. Intelligent efficiency encompasses a broad array of hardware, software, data storage, and analytical components with which a situation is analyzed automatically and the most efficient operating conditions determined. However, these systems seldom operate without some level of human interaction, and ideally these interfaces employ people-centered intelligent efficiency as well. And with much of the data storage and analytically capability remotely located, substitution, or service-oriented efficiency, is also part of the mix. Ultimately, intelligent efficiency enables workers to be more effective and managers to make more informed decisions. These are the compelling reasons for organizations to invest in this emergent technology.

Potential of Intelligent Efficiency to Save Energy

Intelligent efficiency saves energy in three fundamental ways:
1.   Improved management of businesses or production processes through a systems approach
2.   Elimination of the degradation of energy savings
3.   Substitution of [x technology] for [y technology]

A "SYSTEMS APPROACH" TO ENERGY SAVINGS

A key feature of intelligent efficiency is that it achieves energy savings at the system level and above rather than just at the device level. Understanding this distinction and what is meant by taking a "systems approach" is fundamental to comprehending the potential of intelligent efficiency to save energy in the commercial and industrial sectors.

A traditional engineering approach operates at the level of the device: It breaks processes down into their individual components and scrutinizes them for incremental improvements. For example, every energy-consuming piece of equipment in a manufacturing

setting, whether it be a lamp, a motor, or a steel melting oven, converts input energy (e.g., natural gas, electricity, gasoline) in to useful work (output) and does so at some level of efficiency that is less than 100%. A small electric motor might be 80% efficient at converting electricity to mechanical motion, and a pump might be 50% efficient at turning electricity into hydraulic energy. Boilers are often 75 to 85% efficient at turning the energy in natural gas into thermal energy in the form of steam. These are component or device efficiency levels, and energy is saved through the use of a more efficient motor, pump, or boiler.

Each of these devices is part of a larger system. The motor is connected to a pump, the pump to a piping system, and the piping system to a production line that is supplied steam produced by a boiler. In this traditional engineering approach, increasing the efficiency of the system means increasing the efficiency of each component part. However, even when every individual device is operating at its highest efficiency, the larger system is usually not. The component parts often operate at full capacity even when not needed, requiring water to be constantly recirculated and steam to be vented to the atmosphere. Their operation does not vary in response to environmental stimuli. And their operation is not informed by the most likely future conditions, input that might have been able to guide their most efficient use in the present.

A systems approach, in contrast, takes into account the behavior of the components of a system relative to one another, specifically, requiring the component parts to modulate their operation according to the needs and demands of other system components. Intelligent efficiency, as a systems approach, involves analyzing the behavior of a suite of individual technologies that are integrated together to function as a system.

In an intelligent efficiency approach, the system optimization means operating each component in concert with all other components and toward the goals of the entire system. Motors are slowed down or turned off when less or no water is needed. Boilers operate when production requires them, and the entire production line operates in response to customer demands. The system and its components respond to real-time demands and environmental conditions, rather than to an estimate of future demand and with no regard to past, current, or future environmental factors. A systems approach to efficiency can start with optimizing the pumping system and then moving to the production line, the entire factory, and the entire company. A systems approach means optimizing the entire supply chain from raw material to the end user, producing the product in response to real-time demand and ensuring that the elements of the production process are not only highly efficient individually, but that they operate only when needed and at the level necessary. Appliance manufacturers have made great strides in a systems approach. When an order is placed at a retail store, component parts from suppliers are schedule for delivery to the appliance assembly plant the next day. The appliance is not built until it is ordered and yet the customer still gets it shipped to their house in the same time as if it were in inventory at the store.

Smart Manufacturing and the Internet of Things

One of the single greatest manifestations of intelligent efficiency is the emergence of what is being called the "Internet of Things" or the "Industrial Internet" in which all of the components of a system have the ability to inform other parts of the system of their situation and react to the same information from other parts of the system. The more connected the components, the more powerful the network.

Embedded with sensors and communication capabilities, objects as small as shipping labels and as large as factories will communicate current data about various attributes that will enable other components and systems to react to situational changes. These "smart" devices and systems will make processes more efficient, give products new capabilities, and bring about new business models (Manyika and Roxburgh 2011).

A new generation of smart technology is already making its way into the production environment. Industrial motors use approximately half (EIA 2006) of all the electricity consumed in the United States. The performance of individual motors can be communicated and analyzed in real time. Unlike devices in the residential sector that can simply be plugged into an electrical outlet, industrial motors, the machines that turn electricity into mechanical motion that drives pumps, fans, and compressors must be connected to electrical drives, which condition the power for proper motor operation.

Many new drives can vary the speed of the motor and are embedded with the capability to report back to a plant's control system the energy use of a motor in real time. That control system can in turn communicate to the company energy management system. These built in meters eliminate the need to invest in meters for each major piece of equipment, a common recommendation of energy audits. This is important because a meter cannot produce an energy savings but must instead be purchased on the faith that it will produce data that management can use to identify projects that can save

energy. The ubiquity of device and system performance data coming from the plant floor will usher in a new era of continuous improvement.

The technology that facilitates this connectivity is most commonly referred to as "Machine-to-Machine' or M2M. Machines are collecting, sharing, and acting upon data without human intervention. The M2M industry is projected to maintain 23% annual growth rate over the next decade expanding from a $121 billion business in 2010 to almost a trillion dollars in 2020. (Cullinen 2013) and adding $10 to $15 trillion to the global economy over the next twenty years (Evans & Annuziata 2012).

The influence of ICT will be felt throughout the industrial sector. Systems and components of systems will be embedded with smart technology that will enable information exchange between the system control level and the facility operation level. The full integration of smart technology will connect facility operations to corporate enterprise management and a corporation's system will be linked with similar systems throughout supply chains. Not only will this linkage resolve the challenges of coordinating a facility's operational objectives with its corporation's corporate financial objectives, it will connect both with the corporation's energy and sustainability objectives. This emerging ability will wring new levels of efficiency out of the manufacturing process by networking devices, systems, and facilities has come be known as "Smart Manufacturing."

The technologies that make up Smart Manufacturing that are now being applied in limited fashion to specific processes are predicted to become the backbone for the industrial environment. In the future, modeling and simulation systems will be used in initial product development and design as well as the development of integrated facilities and processing operations. Process design will drive capital projects and investments as the application of intelligent efficiency moves from tactical to strategic. Ultimately, we can expect to see data analytics used to optimize the allocation and scheduling of a company's assets and production capabilities (Davis 2009).

ELIMINATING THE DEGRADATION OF ENERGY SAVINGS

One of the more vexing challenges in the energy efficiency sector is ensuring that the savings that result from the implementation of an efficiency measure persist over time. The ability of intelligent efficiency to prevent the degradation of energy savings is its second fundamental contribution.

Operators of complex production processes or managers of facilities that are heated, cooled, and ventilated have become accustomed to the decline in energy savings that typically occurs in the months and years following the implementation of energy efficiency measures. Some energy measures are more sustainable than others. Replace a low-efficiency industrial motor with a high-efficiency motor and it is very likely that for every hour of the new motor's use, less energy will be used than had the old motor not been replaced.

The confidence in savings over time, however, tends to decline with the complexity of a system. Since a system is interacting with numerous elements of its surroundings, changes to those surroundings can easily cause a system optimized for efficiency to lose that optimization. A building's occupants change, and the automation system does not serve the new occupants well and is disabled. Changes may be made to the building itself, causing the automation settings to be no longer optimal. Or, major problems may occur in the building, also causing the settings to no longer be optimizing efficiency. A common example confined to one device is a programmable thermostat. When first programmed, it is likely to reduce energy use because it enables the user to reduce the level of cooling or heating during hours when a building is not occupied. But this level of savings is likely to be lost once the use of the building changes hands because the new operator, rather than reprogram the thermostat, is likely to bypass the programming and set the temperature manually. Now, some or all of the savings have been lost.

This degradation of savings is a common issue at the whole-building level with building automation systems or BMS (Figure 5-1). When buildings are first put into service, they are often commissioned. All systems—heating, ventilating, and air-conditioning (HVAC); lighting; elevators; security systems; and others—are tested and adjusted and put into service. Boilers and chillers are tuned, louvers on ventilation systems are adjusted, cooling tower water fans are adjusted; all of the necessary steps are taken to ensure that the building operates properly for its new owners and occupants. If the building has a BMS that controls some or all of these systems, it is programmed to operate the building in a fashion that optimizes occupant comfort while also minimizing operating expenses. This could include turning lights on and off at certain times of the day and week, and turning up and down the heating or air conditioning at certain times of the day, week, and year. These systems provide building operators with routine information on the operating conditions of major building systems and can flag when something is not operating properly. A properly designed and operated system will—at least initially—save energy and lower operating costs (Capehart and Capehart 2008).

As the years go by, building tenants change, old walls come down, and new walls go in. The ductwork on a floor that supplied air to one large room now supplies air to several small offices. New tenants come with different businesses, different office equipment, and different hours of operation, all of which creates new expectations of the HVAC systems and the BMS that controls it. Without changes to the HVAC design and operating set points, the system may operate against the interests of the new tenants. Perhaps they suffer with it for a while, but ultimately they will likely begin to adjust thermostats and install work-arounds such as closing off air supply vents, overriding programs

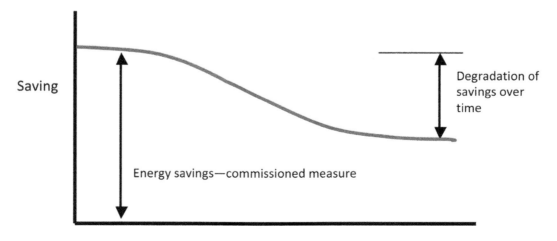

Figure 5-1: Degradation of Savings

on programmable thermostats, and opening windows. These changes induce a response from the system that may exacerbate the situation. A negative feedback loop can be established, and each iteration moves the system further away from satisfying the needs of all the building tenants in the most energy-efficient manner.

If the building management company is proactive, it may see less degradation of the energy savings provided by the BMS. This may require reprogramming the BMS with the build-out for each new tenant and having a properly staffed maintenance department that routinely checks all the building's mechanical systems and catches minor problems (for example, a plugged air filter on an air handler) before they become major problems (such as a broken air handler) that waste a great deal of energy and require expensive repairs. However, this is often not the case, and the building's performance will start to drift away from optimal, and the savings provided by the BMS, or any energy measure for that matter, will degrade (Figure 5-1) (Fernandez et al. 2009).

Intelligent efficiency improves upon existing efficiency technologies by reducing or eliminating the degradation of savings. It is even possible for existing energy efficiency technologies such as building automation to not only maintain initial levels of performance, but to actually improve on it over time—using an intelligent efficiency approach. Advanced BMSs learn from past performance and incorporate the information into future performance opens up new energy savings potential (Figure 5-2). That intelligent efficiency measures are adaptive and anticipatory is fundamental to how they are different from the conventional energy efficiency measures and the energy benefits of these attributes becomes more profound over time.

An illustrative example can be seen in the differences between the re-commissioning of buildings and its intelligent analog: continual optimization. In re-commissioning (or retro-commissioning), building systems are reviewed and adjusted through an intensive process that is intended to increase energy efficiency and overall performance. Essentially, re-commissioning is the recognition of the typical situation in which buildings drift out of optimal performance. Buildings are re-tuned to current operating conditions so that their systems will operate more effectively. This approach has proven very effective in improving energy efficiency (Mills 2009), and has garnered wide acceptance.

A limitation with this approach is that the building must be re-commissioned regularly. With each re-commissioning, improvements in technology and management practices are put in place and new levels of performance are achieved; however, the benefits are short-lived, as the operation and management of building systems move away from the most efficient settings after each re-commissioning such that the building ultimately spends more time out of optimization than in Figure 5-3.

An advanced BMS—an intelligent efficiency approach—continuously collects building information and combines it with other data streams such as utility real-time pricing and weather forecasts. These data feed into a computer modeling simulation that designs and implements a plan to control the building HVAC and lighting systems and perhaps even automated window shades. With time, this building simulation will be refined so that it gets better at predicting how the building will operate given any set of circumstances. This will result in the building staying at or near optimal operating conditions the majority of the time. In this report we

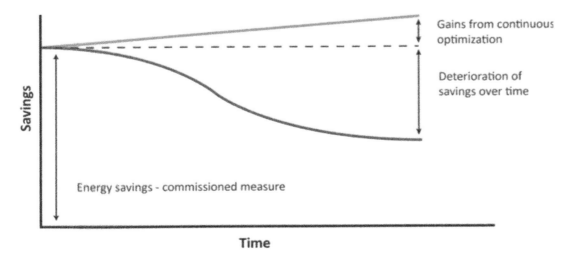

Figure 5-2: Eliminating the Degradation of Savings

refer to this as continual optimization as the system is continually seeking optimal operating conditions.

Contrasting re-commissioning with continual optimization in Figure 3, we see with re-commissioning the periodic optimization of the building systems followed by a drift out of optimization, resulting in increasing energy use and the need for repeated re-commissioning. With continual optimization, we achieve similar savings to re-commissioning a building, but the savings increase with time (and energy use decreases) as the intelligent systems learn and these continual refinements ensure that the building systems remain optimized.

The attributes seen in Figure 5-3 are reflected in many other intelligent efficiency measures considered in this report including intelligent compressed air, pump, and fan systems, and industrial process optimization. This ability to capture energy savings that would normally be lost to saving degradation is one of the key attributes of intelligent efficiency.

ENERGY SAVINGS FROM SUBSTITUTION

The third way that intelligent efficiency saves energy is by outright eliminating the need for energy consuming equipment or by replacing it with something that uses much less energy. The primary example is "cloud computing," which eliminates the need for every office and factory to have its own energy-consuming servers. For most businesses and other organizations, the traditional practice is to support desktop and laptop computers with a dedicated, local computer network. This requires considerable costs, both initial investment and operating costs. The servers use a lot of energy and require consistent air conditioning, which also requires considerable energy. A growing alternative to this is for organizations to provide many IT services through cloud computing, which involves large servers located in data centers that provide computing, storage, and software services connected to the user via the Internet. Two approaches to cloud computing are (1) a "private cloud," where IT resources are shared among different business units in the same organization, and (2) a "public cloud," where IT resources are delivered to multiple organizations through third parties with the ability to maximize the use of their equipment.

The cloud platform achieves these savings by enabling higher utilization of servers, more efficient matching of server capacity to server demand, and "multitenancy" to serve thousands of organizations with one set of shared equipment. When not all servers are needed, some can be turned off entirely, and loads can be rerouted to other servers; with significant energy savings. The public cloud platform provided by Salesforce.com, for example, enables greater efficiency over both on-premises network and private cloud options (Salesforce.com 2012). When a user switches to Saleforce.com's public cloud platform from a private network, the carbon savings are estimated to be 95% of per-use carbon emissions, and when users switch to the public cloud option from an on-premises cloud option, the carbon savings are on average 64% (Salesforce.com 2012).

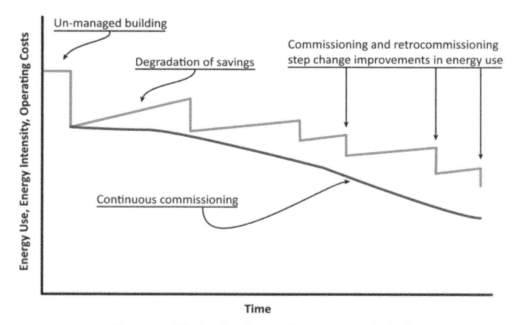

Figure 5-3: Eliminating the Need for Re-Commissioning

NON-ENERGY BENEFITS: SAFETY, QUALITY, JOB CREATION

The benefits of energy savings from intelligent efficiency include direct reductions in energy bills for consumers complemented by other "co-benefits" such as increased comfort, quality of life, productivity, and product quality. Smart technologies will also enable the more efficient use of raw materials, people's time, and capital assets. They make existing tasks easier and open up possibilities for businesses and manufacturers to provide new services. In the manufacturing sector, the non-energy benefits are likely to exceed energy benefits, as improved production performance is the primary driver of investments in manufacturing automation (Burgoon 2013).

In industry, where wasted energy can make a job unpleasant or even dangerous, less waste heat in a manufacturing process can mean a more comfortable and safe work environment. Intelligent process management systems that control plant utilities such as compressed air and steam not only save energy by operating them at lower pressures, but also make them safer. These systems also have the potential to reduce maintenance costs, as motors that run in an energy-efficient mode, more slowly, or less frequently, also tend to run cooler and break down less often.

On a national level, the benefits of intelligent efficiency go far beyond direct, end user savings of dollars and therms. The expanded deployment of intelligent efficiency will increase economic productivity and job creation, economic benefits that transcend the amount of money saved through lower energy usage and rest on how the money saved is ultimately spent. Money saved through energy efficiency moves consumer spending from the energy utility sector to other sectors of the economy that are much more labor intensive. For example, whereas $1 million spent on energy bills supports about ten jobs, if that money were spread throughout the economy, it could support more than 17 jobs (ACEEE 2013).* Moreover, because energy savings is often the result of purchasing energy efficiency services on an ongoing basis (as opposed to one-time purchases of efficient equipment), the trend of increased jobs tends to be sustained. Because of this, jobs induced through energy efficiency tend to dwarf any reductions in net jobs due

to initial investments. These reductions in consumption and demand also offer the prospect of reducing future energy prices for all consumers, as the need for expensive future upgrades to the energy infrastructure (Elliott, Gold, and Hayes 2011).

The Promise of Intelligent Efficiency

Intelligent efficiency offers the new and promising ability to capture savings at the systems and enterprise levels that have historically been difficult to secure. Focusing on component or device efficiency has left a great deal of efficiency uncaptured. Even the most efficient devices waste energy when not properly used or when operated irrespective of the need for them. With smart technologies, systems can adjust in real time to meet the needs of the moment, thus eliminating that waste. But even more importantly, intelligent efficiency enables savings to be captured at the process, facility, and enterprise levels. Smart ICT devices networked together share information about respective current conditions, and each unit has the ability to evaluate its options. In some instances, the machines make the decisions. In others, people are provided with options, including the potential implications.

Savings from many traditional energy measures tend to disintegrate because operations change or because equipment is not properly maintained. Even with re-commissioning, there can be a decline in savings over time as systems drift out of optimization. By contrast, smart technologies take greater volumes of information into consideration in determining optimal operating conditions and can recognize when systems are not operating in accordance with their specifications. More frequent adjustments can be made automatically that achieve additional energy savings that are significant in the aggregate and over time.

Intelligent efficiency also enables the remote location of analytical capabilities. While each office building owner may not be able to afford or justify an on-site technician to monitor and maintain his or her HVAC system, it is more likely that a property management company that oversees a dozen buildings, each with its HVAC system controlled by an advanced BMS networked to a central location could justify an off-site technician to monitor and optimize its fleet of buildings. The energy cost savings that the company realizes can justify the salary of that technician. Now, instead of the technician spending time searching for problems, the advanced BMS identifies and prioritizes them and the technician travels to each building as needed.

---

*This is a simple explanation of how energy efficiency creates jobs. For more detail, please see http://www.aceee.org/files/pdf/fact-sheet/ee-job-creation.pdf.

Case Studies

The types of benefits that organizations can expect from intelligent efficiency are best explained through examples. In this section, we highlight several recent examples of deployments of advanced BMSs and "smart manufacturing." As these examples demonstrate, even already-efficient operations can benefit from intelligent efficiency.

HARVESTING THE POTENTIAL OF BUILDING AUTOMATION SYSTEMS

Building automation systems have been around since the earliest computers. But what makes the advanced BMSs more powerful is that the new generation of sensors and controls are self-configuring and can self-diagnose without human intervention. With all of the system's components connected through wireless communications that allow two-way data transfers, a step change in new efficiencies can be realized. No longer is the focus on the device, but rather on the system and, even more so, on the building. Facility managers no longer need to walk around their buildings looking for problems; instead the system identifies a fault and either self corrects or directs the manager to the problem. A fault is a signal of something wrong. It could be a control in an improper setting, a device not responding to a signal, or device returning a signal outside of expected parameters.

With buildings constantly evaluating their performance against historical data and parallel simulations, the degradation of savings from any given efficiency measure decreases. Below, we highlight several case studies that have explored this phenomenon and documented the resulting savings. These lessons have broad applicability to many building types and demonstrate the scope of savings that is possible nationwide.

Microsoft Headquarters in Redmond, Washington*

Improving the efficiency of existing commercial buildings can be challenging. Most available options often fall between simple equipment replacement programs that are low cost but yield only a part of the total available savings, or more comprehensive retrofits that achieve deeper savings but are often capital-intensive. However, experience from a pilot conducted by Mi-

crosoft at its Redmond campus suggests that there is potential to save energy through the use of analytical software that is not just low cost but also yields deep energy savings. Moreover, this software solution is not disruptive of existing building operations and works well with existing infrastructure.

The 118 buildings at Microsoft headquarters include 14.9 million square feet of office space and deploy 30,000 pieces of mechanical equipment. Historically, major equipment inspections at each building were performed on a five-year cycle—engineers inspected about 25 buildings every year. These interventions achieved energy savings of about 4 million kilowatt-hours (kWh) each year. Collectively, these buildings had seven different kinds of BMSs that the engineers had to deal with to manage the equipment.

In 2011, Microsoft's facilities team started a pilot with an analytical software program, the Smart Building Solution, which is able to "talk to" these disparate BMSs. The software then acquires, aggregates, and analyzes the energy use data for different buildings to give a standardized output that is easier to act upon. Initially, 13 buildings were selected for the pilot. Later, encouraged by the success, Microsoft added more buildings, and soon the company plans to deploy the software across all the buildings on its campus (Warrick 2013) as well as with 2 million square feet of commercial property at other businesses in Seattle (Clancy 2013).

As shown in Figure 5-4, equipment and devices have sensors that record and send performance data to the BMSs. The analytical software communicates with the seven different building systems and integrates the data across buildings. These data are then combined with other information such as weather data, occupancy, and special occasions that alter energy use from the norm such as events or holidays in order to identify trends, patterns, and anomalies. The software collects 500 million data transactions every 24 hours. These data are then analyzed and transmitted in the form of various reports to the central operations. These reports focus on three main areas:

1. Continuous monitoring. By analyzing the data streams from the BMSs, faults such as leakages, overcooling, and sensor failure can be detected in real time. Analyzing larger spatial and temporal patterns often means that the software is able to identify more anomalies than can a conventional BMS. One example encountered at Microsoft was an air handler's chilled water valve with a faulty control code issue. The valve was always 20%

---

*This case study was taken in its entirety from a publication released by Microsoft and summarized by articles in Warrick 2013, Microsoft 2013, and Clancy 2013.

open, wasting several thousand dollars annually in energy used to chill the water. This issue had not been visible, but the analytics software detected it immediately (Microsoft 2013).

2. Prioritization. The software prioritizes the faults by estimating the cost of the inefficiency so that the engineers can focus their time and efforts on the more important tasks. Given the scale of operations, the central operations team may receive hundreds of notifications in a single day, not all of which are equally important. Analytical algorithms help the engineers to prioritize items based on multiple considerations, not just the financial cost. For instance, a fault that affects occupants in a highly critical operation—such as a hospital operating room— needs to take precedence over other faults, even if in purely economic terms it does not yield as great of savings. The software also enables better correlation of messages from related events, thus improving response time for larger or more critical problems.

3. Energy management. The third feature of Smart Building Solution is its ability to manage energy

consumption more holistically. With the analytical support provided by the software, the engineers can optimize major building systems like heating, cooling, ventilation, and lighting. They can fine tune set points and schedules, identify wasteful equipment, and act on other energy saving opportunities through a better understanding of energy use trends across the entire portfolio. The software also helps to reduce occupant-dependent plug loads such as computers, printers, and kitchen appliances, which are comparable to the base building energy use. The software benchmarks plug load data across and within buildings, which is displayed through dashboards for internal comparisons. Energy costs are broken down by organizational unit, and metrics such as kWh/employee define ownership and create incentives for a unit to outperform its peers.

Taking Building Energy Management to the Cloud

As the Microsoft case suggests, the full utilization of BMSs can bring significant energy savings. However, the information from a BMS is limited to equipment, appliances, and sensors. To truly unleash the potential of intelligent efficiency, this information needs to be

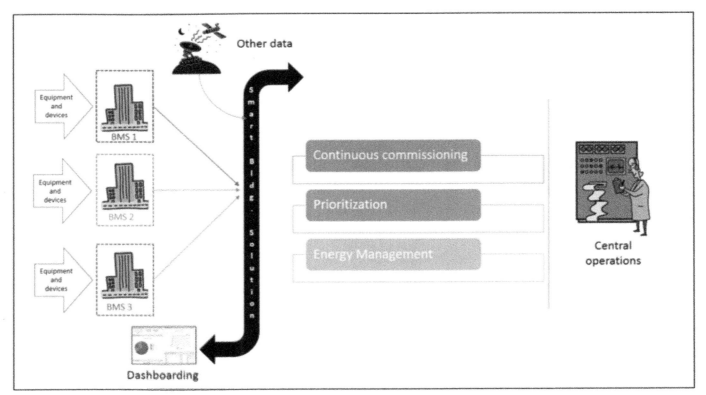

Figure 5-4. Intelligent Efficiency at Microsoft
Source: ACEEE graphic based on Accenture 2011

placed in context. The energy footprint of a building is determined to a large extent by extrinsic factors, including as building type, number of occupants, operating hours, nature of business, nature of ownership, climate zone, and many others. Knowledge of these variables helps building managers identify opportunities to reduce costs, one of which is energy. However, information about these extrinsic factors often exists in an array of different formats—utility bills, employee records, spreadsheets, and web sources.

Applications are now available that help in integrating different data sources in different formats and make them more amenable to analysis. Schneider Electric offers EcoStruxture Solutions (Schneider Electric 2013), and Johnson Controls offers their Panoptix system (JCI 2013), a set of applications for monitoring, diagnosing, and analyzing buildings' energy efficiency. Both of these cloud-based data storage and analytical programs pull data from building systems, meters, equipment, and utilities, and, through specific applications, provide building efficiency solutions. The cloud-based "apps" can be accessed by any computer, tablet, or smartphone and allow users to monitor and control different devices, systems, and sub-systems. Johnson Controls has made the application programming interfaces (APIs)—software specifications that help perform a specific function, for instance pulling data from different HVAC equipment—freely available on the web, thus creating a platform for third-party app developers. Johnson Controls' Panoptix also provides a marketplace where users can browse and purchase apps of interest. Figure 5-5 from their website summarizes how the cloud

bases efficiency solution works.

Some of the apps currently available in the market place include:

- Energy Performance Monitor, which helps conduct baseline energy assessments, set up savings targets, and monitor and measure energy savings from different energy efficiency projects, presenting them to building operators in a contextualized manner that simplifies their decision making process.

- Continuous Diagnostics Advisor, which constantly monitors building systems and automates the detection of problems in chillers, packaged HVAC units, air handlers, variable air volume boxes, terminal units, and boilers.

Other apps are available to calculate carbon emissions from a building, compare different buildings or different pieces of equipment, and display information about energy use through dashboards located at various facilities. Each of these apps provides end users a greater ability to manage their energy use and their businesses.

Going from Re-Commissioning to Continual Optimization

Below we present two case studies in which existing buildings with conventional automation systems were upgraded with advanced BMSs and realized significant energy savings. The first analysis was conducted by the National Resources Defense Council on three

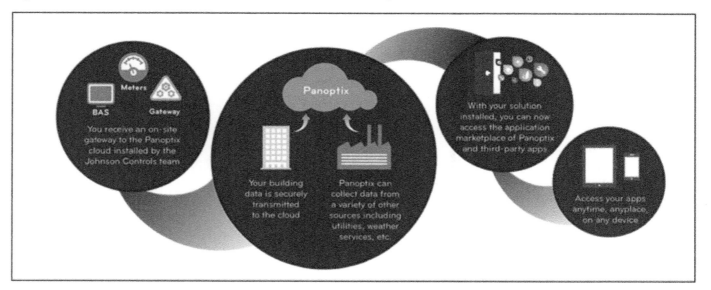

Figure 5-5. Panoptix System by Johnson Controls
Source: Panoptix website (https://whatspossible.johnsoncontrols.com/community/panoptix/apps)

of its own buildings in Washington, DC. The second is an analysis by an engineering firm using the Continuous Commissioning® protocol* for continually evaluating and adjusting the operating conditions of a building.

Incremental Gains from BMS Optimization of Three ENERGY STAR® Buildings

The National Resources Defense Council conducted an analysis of the potential of advanced building management systems to reduce energy use in already well-designed buildings. In its 2013 report *Real-Time Energy Management. A Case Study of Three Large Commercial Buildings in Washington, D.C.* (Henderson and Waltner 2013), it describes an analysis that demonstrated that an advanced BMS can achieve an incremental amount of savings beyond what is possible through the use of individual enabling technologies alone. NRDC worked with Tower companies, Inc. (Tower), to improve the energy management of three large, multi-tenant office buildings in Washington, D.C.

The three buildings chosen for this analysis were

already high-performing buildings having ENERGY STAR Portfolio Manager Scores of 71, 78 and 86;[†]therefore, it was reasonable to expect that there might not be much room for improvement. However, after a 12-month study period, electricity savings of 23%, 7%, and 17%, respectively were achieved by employing a continual optimization process to building management. The individual actions taken were not different than before, only they were done on an as-needed basis and in response to recommendations made by the advanced BMS. NRDC reports that the actions taken are highly replicable in other commercial office buildings. This study suggests that not only are significant savings possible in buildings without existing automation systems, they are also possible for buildings with conventional BMSs (Henderson and Waltner 2013).

The findings of the report indicate that even though the buildings already had building automation systems installed, the data were not being used to their fullest potential. The additional performance was possibly through a service provided by At Site, Inc., whose technical experts provided energy management services remotely. At Site managed the installation of meters to provide real-time energy use data and performed an assessment of each building. Their automated monitoring of the buildings revealed opportunities that were turned into actionable recommendations to the building engineers.

---

*Continuous Commissioning ® is trademarked by Texas A&M, Energy Systems Laboratory, Engineering Experiment Station: http://www-esl.tamu.edu/continuous-commissioning
†Portfolio Manager is an interactive energy management tool that can be used to track and assess energy and water consumption across a portfolio of buildings.

Table 5-2: Energy Savings for Three Large Commercial Buildings

	Square Feet	2012 Occupancy	KWH Used		Study Period Savings	
			2011	2012	%	$
1707	109,926	302	1,965,135	1,516,274	23%	$ 58,352
1828	332,928	928	5,590,937	5,227,183	7%	$ 47,288
1909	239,128	462	5,197,305	4,327,589	17%	$ 113,063
Total			**12,753,377**	**11,071,046**	**13.2%**	**$ 218,703**

Notes: The numbers in the first column are building names/numbers. Energy savings were determined using a whole-building, year-over-year method. Results are normalized for weather and occupancy. The 12-month study period was January to December 2012, and the 12-month baseline period was January to December 2011.[Tower's average total electricity rates for 2012 were about $0.13 per kWh (total cost with demand charges and all applicable fees and adjustments). In cities with different electricity rates, different savings and total return would be expected.]

From Retro-Commissioning to
Continuous Commissioning®

In 2009, SSRCx, LLC, an engineering firm in Nashville, Tennessee, demonstrated conclusively that more routine commissioning can result in another step change in energy savings. The company studied the energy use of a 320,000-square-foot commercial building that was built in 1999, retrofitted to U.S. Green Building Council Leadership in Energy and Environmental Design Existing Building (LEED EB) standards in 2008, and outfitted with a continuous commissioning BMS in 2009 (McCown 2009). The total unadjusted energy savings for January 2008 through August 2009 was 10% relative to the previous two years. Additional savings from the continual commissioning process during the cooling season were as high at 16% for the month of August 2009, making the combined savings for this month 28% over the 2006–07 baseline. The energy use index in January 2008 was 107.5 thousand British thermal units (kBtu) per square foot per year and was 95.2 in August 2009. This case study is an excellent example of how an advanced BMS can save considerably more energy than a conventional BMS.

SMART MANUFACTURING AND
ENERGY EFFICIENCY

All of the data coming in from the production floor has the ability to overwhelm management if not properly managed. With so much data, any number of questions can be answered. But which ones are most important to meeting business goals? Without experience determining which data to collect and what questions to answer, manufacturers are at risk of investing considerable time and money to answer the wrong question or to answer the right question incorrectly. In many cases, a manufacturer will begin to incorporate information feedback and controls technologies to improve the operation of the system, and will continue to add more advanced technology in order to further reap the benefits of intelligent efficiency. At some point though, they will reach a scenario in which they have more data than can be effectively managed.

Additional Savings by Closing the Control Loop

An example of how one company dealt with this can be seen in the Air Liquide facility in Bayport, Texas (Reid 2008). Air Liquide is a global corporation specializing in cryogenic liquids and industrial gases, and their Bayport plant is one of the largest industrial gas suppliers in the world, manufacturing oxygen, nitrogen, and hydrogen for use in other industries. Producing these gases requires a lot of steam heat, which is provided by seven large boilers (four of which are fired by the exhaust of gas turbines used to cogenerate electricity). The facilities' boilers operate under several performance variables, such as production volume, reliability, energy cost, and emissions. Prior to implementing closed loop control in which the generation of and response to performance data occur all within the control system, Air Liquide had been using Visual MESA software to track and optimize against these key indicators and provide open-loop feedback to operators to guide them in optimizing their boiler systems. This use of data tracking and analysis of operations is a best practice in the industry and is an example of one type of intelligent efficiency. Operators receive data and analysis results every 15 minutes.

Air Liquide then took optimization to the next level by closing the loop between the data feedback system and the boiler control system. Instead of relying on operators to adjust the system a few times a day, the new system analyzes process variables and adjusts the system immediately, allowing it to update the boiler control settings every 15 to 30 minutes.

The data feedback system in the open-loop optimization configuration that Air Liquide had been using takes ten to 12 months to install. Upgrading to closed-loop control can take another six to 12 months, but the energy savings alone are estimated to pay for the project in just one year. There are also additional sources of savings, such as increased system productivity and the benefits of freeing up operator time for other work. Experts working on the project estimate that as more closed-loop systems are installed and the savings are properly verified, more companies and manufacturing plants will choose to install the data feedback systems with closed-loop control at the outset, bypassing the open loop configuration entirely. This would allow the entire installation to take about 12 months, and the simple payback based on total cost savings could drop to less than a year.

Networking the Supply Chain, Reducing
Product Variability, and Saving Energy

General Mills Inc., one of the nation's largest food product companies, has committed to a 20 percent reduction in its energy usage rate by 2015 from a 2005 baseline. In order to reach this goal the company is investing in renewable resources such as using biomass as fuel to create steam at its manufacturing facilities, and embedding smart energy-saving technology into its manufacturing culture (General Mills 2013). This effort is part of their holistic margin management (HMM) initiative that was launched in 2005 to guide all of their decision making. General Mills had discovered that as its supply chain and product lines expanded, input costs increased and that the typical 2 to 3 percent annual increases in productivity they had come to expect was not sufficient to compensate (NYSE 2011). They needed to do a better job of incorporating farming, supply chain, manufacturing, and distribution. Working with the Smart Manufacturing Leadership Coalition (SMLC) and Rockwell Automation, they embarked on an effort to remove variability and by extension costs from their product lines. The result is simplified product development, collaborative manufacturing between General Mills and its suppliers, and networks that connect manufacturing control systems with their enterprise resource management system. Food quality tracking for the U.S. Food and Drug Administration (FDA) is included in this automation as well (Davis and Edgar 2013).

General Mills expected HMM driven changes to yield savings of $1 billion through fiscal year 2012 and $4 billion by 2020. (NYSE 2011) Though most of these gains are not energy costs, some of them are and General Mills reported an 11 percent reduction in its energy consumption in fiscal year 2011 from the 2005 baseline (General Mills 2013).

Economic Analysis

In order to quantify the potential economic benefits of intelligent efficiency if implemented nationwide, we calculated the estimated effects of a select group of "smart" energy efficiency measures that have the most promise for near- and medium-term implementation in the commercial and manufacturing sectors. For the purpose of our analysis, we estimate that half commercial and manufacturing sector will be able to benefit from intelligent efficiency. Our focus on these sectors is driven by their readiness for implementation of intelligent efficiency projects. They tend to be large energy consumers, have high levels of broadband communications interconnection, and have relatively wide implementation of sensors and controls, which represent important enabling infrastructure within which intelligent efficiency projects can successfully be integrated

Larger energy users are also more likely to have the financial and technical capability to implement intelligent systems. Some large manufacturing firms and commercial operations with energy consumption in the hundreds of thousands if not millions of dollars per year are already investing in these systems, and we anticipate that as the cost decreases and a greater variety of products and services become available, the market for intelligent efficiency will diversify and expand. Our focus in this analysis is on technologies that have the most promise for near and medium term adoption in these two sectors.

Utility energy efficiency programs are also poised to invest in intelligent efficiency. While not all of these large consumers of energy are served by utility programs, many of them are, and in regions where they are not, it is possible or likely that energy efficiency programs will soon be offered (e.g., Louisiana and Mississippi have recently decided to begin energy efficiency programs). Programs targeting larger customers have produced some of the lowest cost savings to date (Bradbury et al. 2013).

Our analysis focuses on electricity, driven by the predominance of electricity energy efficiency programs in the North America and by the large proportion of U.S. electricity consumption that occurs in the commercial and industrial sectors (Foster et al. 2012). In addition, the electric utility industry's current focus on the smart grid represents an opportunity to complement the benefits of intelligent efficiency, as it can provide real-time two-way communication between the consumer and the utility that is related to the volume and other quantifying characteristics of the electricity it accompanies. While to quantify the overall benefits of a national smart grid is beyond the scope of this report, we do consider the smart grid a component of intelligent efficiency and anticipate that it, much like the Internet, will be part of the supporting infrastructure.

ENERGY MEASURES CONSIDERED IN
THE ECONOMIC ANALYSIS

In order to estimate the potential energy savings available from intelligent efficiency, we looked at projections of energy savings resulting from a number of key enabling technologies that were selected for their ability to produce significant savings in a significant segment of the commercial and industrial sectors. Our threshold for what to treat as significant was shaped by (1) our estimation that savings would be in the billions of dollars, (2) our ability to discern whether or not a given energy measure would contribute to such a total, and (3) the availability of data. Our calculation of net savings took into account the initial cost of implementing the intelligent efficiency measures and ongoing costs of operating and maintaining the systems.

Most of the measures we identified target the building sector. The building automation market is already mature, with dozens of vendors providing a great variety of products and services covering a wide range of capabilities and price points. Most of the energy use in buildings is for environmental control and lighting; therefore, not surprisingly, most of the measures analyzed target those systems. We also wanted to capture the savings from office equipment and the many miscellaneous energy-consuming systems such as servers, elevators, and transformers, since intelligent efficiency is enabling savings in these areas that were not possible before.

In the industrial sector we focused on the ability of "smart manufacturing" to produce systemic change in the use of energy, both within production processes and throughout the plant.

DISTINGUISHING INTELLIGENT EFFICIENCY MEASURES

This report examines individual technologies that are integrated into a system that is greater than the sum of its parts. This raises the need for a heuristic to determine when an energy measure should be considered enabling technology—sensors, controls, and IT—and when it reaches the level of intelligent efficiency. Since the energy savings brought about by technologies that optimize an entire system (whether a commercial building or a manufacturing supply chain) are often in addition to savings already achieved by the enabling technologies, we have an attribution issue to resolve as well.

To address the attribution issue, we developed a methodology for determining the part of the overall energy savings that should be attributed to the intelligent efficiency measure— the difference between what is possible with enabling technology alone and what is possible with an intelligent efficiency approach. That methodology required a heuristic for determining

when an energy measure should be considered enabling technology and when it reached the level of intelligent efficiency. To aid in this determination and to help the reader categorize and compare technologies along this evolutionary scale, we devised a simple hierarchy. The levels connote complexity rather than additional energy savings, although energy savings generally increase as we move toward Level 4.

The challenge of determining the energy cost savings from intelligent efficiency pivots on what is considered the baseline. An easy-to-understand example is lighting. The baseline for controlling the amount of light in a room is a simple manual on/off switch. As there is no automation at this baseline level, we are calling it Level 0. A more complex example is an HVAC system, where the baseline could be a simple switch to choose heating, fan, or air-conditioning. The amount of heat provided by the old hot water or steam radiators was usually a knob that the operator manually adjusted— often with a little hope and a lot of fear of what might happen next.

The next level (1) is to have a reactive control such as a motion sensor that turns the lighting on and off automatically or a temperature switch in a thermostat that turns a system on and off when certain set points are reached. In terms of complexity, the next level (2) is programmability. Examples are a building's security lighting that turns on and off at different times of the day and week, or a programmable home thermostat. These technologies do not necessarily result in greater energy savings than Level 1, but the system is more complex.

Level 3 incorporates variable response. The lights are on a dimmer, the amount of light is determined by a sensor (motion or daylight), and a controller adjusts the amount of light produced. Again, this is not necessarily a guarantee of more savings, but reflects additional complexity. In our HVAC example, a Level 3 variable response might be the ability to ramp fan speeds up and down.

Level 4 is the full integration of all of these enabling technologies with an additional software component that analyzes past performance and adjusts system

Table 5-3: Five Levels of Energy Management

Level	Technology
Level 0	Manual On / Off
Level 1	Reactive On / Off
Level 2	Programmable On / Off
Level 3	Variable Response
Level 4	Intelligent Controls

outputs in anticipation of future performance. At this level, additional savings are possible because the advanced BMS is proactive and not just reactive. It has the ability to continually optimize and even improve performance over time. Much of the savings between Level 3 and 4 is achieved by reducing or even eliminating the degradation of savings that often happens following implementation of an energy measure.

The higher levels do not automatically translate into greater energy savings; for example, a reactive control for lighting in some applications will save as much or even more energy than a programmable system. Rather, the level speaks to the complexity of the system. In our selection of technologies to include in the economic analysis where the line between levels was blurred, or the incremental savings between levels was hard to discern, we grouped the levels together. An example of this is our treatment of lighting at Level 3 and 4: A system organized around intelligent efficiency would provide only as much light as needed in the locations needed and only at the times needed by workers to accomplish the required tasks. However, the difference in energy savings that would result from such a system compared to a system without predictive capabilities is hard to determine, so Levels 3 and 4 are combined in the analysis for lighting.

INTELLIGENT ENERGY MEASURES INCLUDED IN THE ECONOMIC ANALYSIS

As previously discussed, we aimed to quantify the marginal energy savings attributable to intelligent efficiency. Though these technologies improve upon the previously realized efficiency gains made possible through more efficient devices and automated building controls, we aimed to disentangle the energy savings that accrue specifically from the features that define a system as having an intelligent efficiency approach.

We analyzed over two dozen technologies for their ability to affect energy use in buildings in the commercial and manufacturing sectors. Each of the Level 4 energy measures (EM) considered has broad applicability, a likelihood of reaching more than 25% of its respective market by 2035, and the ability to produce savings that can be sustained for the life of the product. The analysis assumes a relatively modest increase in investments of 1% per year early in the twenty year period and finishing at 2%.

The commercial sector analysis included a dozen energy measurers that, for reasons explained below,

were grouped by the buildings systems whose energy use they are intended to impact. A different approach was taken in the analysis for the manufacturing sector. This sector is not as homogenous as the commercial sector; therefore, our ability to discern large-scale impacts of specific smart technologies is limited. Instead, the analysis applies broadly the potential of smart manufacturing overall to reduce the variable costs of production, one of which is energy, throughout the manufacturing sector.

Motivating our analysis is the notion that these projects are justified on the basis of potential energy cost savings. Conversations with people in the manufacturing automation sector, however, indicated that this is not often the case. Rather, primary motivations for investment in the industrial sector are to improve production efficiency, product quality, safety, and regulatory compliance, with energy savings often perceived to be an ancillary benefit. The energy cost savings however can justify the supporting investment, therefore it likely that there will be much greater investment in smart manufacturing over the next 20 years and our estimates are conservative.

Given the premise that intelligent efficiency investments can indeed pay for themselves through energy cost savings, the analysis assumed that investments in the commercial sector are made with an expectation of a 20% annual return, or a five-year payback and that investments in the manufacturing sector are made with an expectation of a 50% return or 2-year payback. The first year cost of any licensing and services fees are built into these cost estimates.

INTELLIGENT EFFICIENCY MEASURES FOR THE COMMERCIAL SECTOR

Buildings consume 23 percent (Navigant 2013a) of all electricity globally. The six major uses of energy in buildings are space heating, ventilating, and air-conditioning (HVAC); water heating; lighting; and plug load (all office equipment and other machinery "plugged" into the building). We placed intelligent efficiency measures into the following categories: smart building components, smart lighting, smart HVAC components, advanced BMSs, user interfaces, smart grids, office equipment and cloud computing, and miscellaneous. For each measure, an estimate was made of the average amount of energy saved. These estimates were based on one or more references. Each estimate number was applied to the energy use categories used by the Energy Information Agency (EIA) Commercial Building Energy Consumption Survey (CBECS). A second estimate of the percent of commercial buildings that could use the energy measure. A more detailed explanation of the analysis is contained later in the narrative and in the Appendix. These three variables are pointed out here because they are referred to in the energy measure subsections below.

Smart Building Components (Smart Windows) New developments in material science are giving us materials that are reactive to environmental conditions and can be designed to make a building more energy efficient. For example, it is now possible for windows to lighten and darken depending upon the intensity of sunlight. This reduces the air-conditioning load in the summer and the heating load in the winter. It also can improve the work environment by reducing glare. A recent study by the Lawrence Berkeley National Laboratory found that smart windows alone have the potential to reduce energy use for cooling by 19 to 26% and lighting by 48 to 67% (Lee 2007). For the purpose of this analysis, the efficiency gains of smart windows are used as a proxy for the average collective gains from all existing and future smart building components and that at least some of these energy measures will be of use to half of all commercial buildings and that it will reduce energy for heating and ventilating by 5%, air conditioning by 10% and lighting by 20%.

Lighting Automation

There continue to be incremental gains in the area of lighting. When a BMS has information related to current and future occupancy and current and future weather, it can not only bring lights on and off at optimal times and luminosities, it can do a comparative analysis of whether the impact on HVAC energy use that results from lightening smart windows and letting in sunlight will be less or more than darkening the windows and turning the lights up. A report by the Northwest Energy Efficiency Alliance (NEEA 2013a) and claims by OsramSylvania (OsramSylvania 2013) indicate potential savings of 40 to 75% beyond what is possible with standard occupancy-based lighting controls. Recognizing that many buildings already have some level of enabling technologies such as sensors and time of day programming, for the purpose of our analysis, we estimated that about half of this (35%) will be the average efficiency gain for a commercial or manufacturing building and applied it to the CBECS "Lighting" category. We also estimated that 75% of buildings will be able to utilize this technology.

Smart HVAC Components

Technologies are coming on the market now that enable each subset of a BMS to self-optimize. For example, motors can monitor their own performance and adjust their operation, as well as send information to the BMS that can be aggregated to get a global picture of the system's performance (Wang 2010). Although the savings from each component is small, in aggregate they can reduce the energy use of an HVAC system by 10 to 30% (Sinopoli 2010) and enable the holistic management of the system by advanced BMSs. We estimated that half of all buildings could get value from such measures and that on average they will see a 10% reduction in energy use by heating and ventilating systems and a 15% reduction in air conditioning and refrigeration systems.

Intelligent Building Management Systems (Advanced BMS)

The difference between a Level 3 BMS and a Level 4 BMS is the ability of the latter to perform continual optimization. The energy savings due to intelligent efficiency is the difference between a system that is occasionally optimized and one that is always optimized and continuously improving. Therefore, these savings will not be realized in the first year of deployment but over time as a system's self-correcting capability starts to pay dividends. Research done by the Pacific Northwest National Laboratory, the National Institute of Standards and Technology, the Natural Resource Defense Fund, and Energy Design Resources found savings ranging from 24 to 46% for enabling Level 3 BMSs and an additional 10 to 30% with intelligent Level 4 systems that have fault detection, historical analysis, and predictive capabilities (Wang et al. 2011, Sinopoli 2010, Henderson and Waltner 2013). In the analysis we estimated that advanced BMSs will be of value to three-fourths of all commercial buildings and that they will provide a marginal increase of 10% savings for cooling and ventilating and a 20% improvement in electric heating.

Smart Grids

Still in its infancy, the smart grid is an interactive electricity grid that will be able to communicate information between the utility and individual buildings, and possibly even systems within buildings and facilities, in real time. This information, likely a locational and time-of-day price signal, can be used by the BMS to determine how to run the building to minimize energy costs. Although the BMS will likely be programmed to prioritize energy costs over energy use, it is still likely to reduce overall energy use because much of the cost savings for buildings will result from reducing energy use when energy prices are highest, usually during the hottest time of the day. For the most part, these loads cannot be shifted to other times of the day, so on net, there will be savings. A study by Pacific Northwest National Laboratory indicated that electric utility customers could realize 10% energy savings through transactive controls (Katipamula et al. 2006).

In addition to utility build-out of transmission and distribution smart grids, large commercial and industrial customers are building their own smart grids within their facilities. To do so, the install smart meters which make visible their energy consumption at a system level. The cost of such meters has fallen 30% in the last two years (SmartGridNews.com 2013). Energy service companies such as Building IQ, Ecova, EnerNOC, Schneider Electric, and Siemens are expanding into this market, currently estimated to be $6.2 billion in sales (Smart-GridNews.com 2013), with services that can leverage these smart meters into energy management systems that provide clients with real-time energy management capabilities. In our analysis, we estimated that 75% of all commercial buildings could get value from connection to a smart grid and that it would reduce electricity use by HVAC systems by 10 percent.

Dashboards and other User Interfaces

The common method of conveying performance information today is through a computer screen. Displaying the raw data though is seldom useful to an operator or manager. The information needs context. Is a device running within its normal operating parameters? How does the performance of one device compare to another identical device? User interfaces conveying information to end users. A new generation of interfaces called "dashboards," named after automobile dashboards, attempt to convey pertinent information in a contextual manner that aids decision making.

Energy dashboards communicate to workers, technicians and managers energy information in a way that is instructive and actionable. We described the benefits of the Envision Charlotte (Envision Charlotte 2010, Downey 2012) piloting of this technology in our first intelligent efficiency report. The potential energy savings are anticipated to be as high as 20% for some participants who may be focused on energy use for the first time. For the purpose of our analysis, we estimate that half of this will be the average efficiency gain and that a fifth of all commercial buildings' could find value in it. Energy savings will come from HVAC systems, computers and other miscellaneous loads.

Office Equipment and Cloud Computing

Electricity used to power office computers, copiers, and servers can be saved through any of the three types of intelligent efficiency: technology-centered, people-centered, and service-oriented. In the technology-centered approach, intelligent controls turn office equipment on or off, or put in idle mode, according to historical trends and current conditions. ENERGY STAR estimates these savings to be approximately 5% of office equipment's energy demand (ENERGY STAR 2013).

In the people-centered approach, employees can optimize their use of equipment in ways that save energy. As described above with reference to the Envision Charlotte project, savings of up to 20% are possible, although we assume only half of that in our analysis.

The service-oriented approach likely produces the greatest savings. Most offices have a series of servers that handle their accounting, payroll, email, and other IT needs. These servers are seldom operating at peak capacity and yet use almost as much energy at partial load as at full load. When an organization switches to cloud-based computing, many of these servers—along with the energy they use and the air-conditioning they require—can be completely eliminated. A report for the General Services Administration estimated that cloud computing alone could reduce federal IT budgets by 25%, or $20 billion per year. (Kundra 2011)

In our first report on intelligent efficiency, we identified analyses conducted by Google, Holler, and the Carbon Disclosure Project that on average estimated that a company could reduce its IT energy demands by 50% (Google 2011, Holler 2010, Carbon Disclosure Project 2011) by switching to cloud-based computing. Salesforce.com estimates that carbon reductions greater than 80% are possible (Salesfore.com 2013). Though not always correlated 1 to 1, energy use and carbon emissions are directly related, and therefore it would not be unreasonable to expect that energy use might decrease by this amount.

For the purpose of our analysis, we estimated that 50% of all buildings will implement some mix of these energy measures and that the average benefit will be a reduction in the energy use that falls under the CBECS "Other" category by 50%.

Refrigeration Energy Management Systems

For many restaurants and food service facilities, walk-in refrigerators are the largest single energy-consuming system. They run 24 hours a day and every day of the year. Fortunately, there are control systems available that turn fans and chillers up or down in response to demand. The sophistication of these systems is increasing, and components will be able to be integrated into a holistic building automation system with the ability to respond proactively to environmental inputs (Baxter 2004). In the analysis, an estimate of 30% was used and applied to the amount refrigeration energy use and that it will be of value to 75% of all buildings with large refrigeration loads.

Smart Fume Hoods

Smart fume hoods for kitchens and laboratories can adjust the volume of air evacuated so as to reduce the energy used by the hood fan and the room's heating or cooling system. The food service sector uses hoods in kitchen areas, and research organizations use them to extract dangerous fumes safely from laboratories. While these devices represent a small portion of the commercial sector's energy consumption, they are often major energy consumers for the facilities that have them. Therefore, it is likely that the new technology that regulates the speed of hood fans in response to need will gain broad market acceptance. Since it is important that these systems always work when needed, higher levels of automation will be required. Technologies examined by the Lawrence Berkeley National Laboratory have the potential to reduce energy use by 10–30% (Desroches and Garbesi 2011). For our analysis, we estimated that these measures could be used by 10% of the buildings covered in the Commercial Buildings Energy Consumption Survey. Buildings that have such equipment could on average realize a 15% energy reduction in energy use by cooking and ventilating systems.

Miscellaneous Intelligent Efficiency Measures

In the search for intelligent efficiency measures to be considered in the economic analysis, several technologies were identified for which insufficient economic energy consumption data was available. In this section a few of those technologies are highlighted to bring attention to the breadth of impact intelligent efficiency will have on the economy. To reflect them in the analysis, we estimated collectively they might reduce by 2% the amount of energy used that falls under the CBECS "Other" category for all buildings.

Smart escalators can turn off or slow down when no one is onboard and speed up once an approaching person is detected. Smart elevators can coordinate with one another and can be programmed with the ability to optimize scheduling during peak demand based on historical performance data. Collectively, these are not major uses of energy within a building, but they are sig-

nificant and research has indicted a potential to reduce energy consumption at the system level by 20 to 46% (Otis 2011, Schindler 2013, KONE 2013).

The efficiency of vending machines continues to improve, and a recent study done for the U.S. Department of Energy indicates the next generation of vending machines could reduce energy use by 40 to 50% over standard equipment. An additional 20% is the potential gain on top of already realized efficiencies with the next level of technology. (McKenney et al. 2010).

Healthcare facilities are among the most energy-intensive commercial buildings (Singer and Tschudi 2009). In particular, the energy consumption of medical equipment such as magnetic resonance imaging (MRI) and computerized tomography (CT) equipment has grown considerably as more powerful technology provides better resolution and advanced diagnostics. Many types of medical equipment have a very high power draw and are often left in standby mode when not in use (McKenney 2010). The standby power draw of an MRI machine could be half of the 14,000 Watts it consumes when in use (McKenney et al. 2010). This suggests that the use of intelligent controls to run the equipment only when needed could save significant amounts of energy. With over 30 million pieces of medical equipment in the country (McKenney et al. 2010), the healthcare sector presents a new avenues of savings through deployment of intelligent efficiency.

ACCOUNTING FOR INTERACTIONS BETWEEN ENERGY MEASURES IN THE COMMERCIAL SECTOR

The challenge in a quantitative analysis of energy savings that results from multiple energy measures, each of which impacts the energy use of one or more systems in a building or manufacturing plant, is parsing out the savings attributable to given energy measures and individual building systems.

First, assuming no amplifying interactions between multiple measures operating simultaneously, the energy savings are not additive but factorial. For example, five energy measures targeting the energy use of an HVAC system are implemented (e.g., motors are replaced with more efficient motors, fans are replaced with more efficient fans, duct work is upgraded to produce less resistance, and boilers and chillers are tuned), and each measure has the potential to reduce energy use by 10%. The net benefit of these five measures is not a 50% reduction in energy use by the HVAC systems.

Table 5-4: Intelligent Energy Measures for Commercial Sector

Energy Measure	Range of Savings from Literature Search	Estimate Use in Economic Analysis
Smart Building Components	5–20%	10%
Smart Lighting	Up to 75%	35%
Smart HVAC Components	15%	10–15%
Advanced BMS	10–30%	10–20%
Smart Grid	10%	10%
User Interfaces	10—20%	10%
Office Equipment and Cloud Computing	2–50%	50%
Refrigeration Energy Management	30%	30%
Smart Fume Hoods	10–30%	15%
Miscellaneous Measures	20–50%	2%

Rather, the first measure will save 10% of the original energy use, the second measure 10% of the remaining energy use, and so on. The total savings will be 41%, since each measure subtracted 10% of a number that was growing smaller and smaller.

Total Energy Savings =
    [1 – (1-EM1%) x (1-EM2%) x
    (1-EM3%) x (1-EM4%) x (1-EM5%)]

In our analysis of HVAC systems, we used this methodology to determine the net savings from multiple energy measures. Five intelligent efficiency measures affect the energy use of an HVAC system: smart building components, smart HVAC components, advanced BMSs, user interface, and smart grid. The impacts of these energy measures on other building systems such as lighting was dealt with separately in the analysis.

In Table 5-5 we show the energy savings possible for three energy end uses, space heating, space cooling, and ventilation, for each of the five intelligent efficiency measures mentioned above. The three categories for buildings' energy use are taken from categories used by the U.S. Energy Information Administration's (EIA) in its Commercial Buildings Energy Consumption Survey.

We have adopted the EIA's categories in order to be able to use EIA data in our analysis when we determine energy savings on a national level.

Table 5-5 demonstrates that if the net savings were additive, we could potentially see energy savings exceeding 100% of energy used, which is of course not possible. Instead, by using the equation above, a more realistic number is determined for each of the three categories.

INTELLIGENT EFFICIENCY MEASURES FOR THE INDUSTRIAL SECTOR

The industrial sector is not as homogeneous as the commercial sector; therefore, it is much more difficult to identify a limited number of specific types of efficiency measures that will have broad applicability. A next-generation manufacturing process for plastic injection molding operations, for example, will not have applicability in metal casting or fabrication. With this limitation in mind, we identified generic smart technologies within manufacturing that, much like advanced BMSs in the commercial sector, have broad applicability across multiple manufacturing sectors, construction,

Table 5-5: Interaction of Energy Savings Measures and Determining Net Savings

Energy Measure / System	Space Heating	Space Cooling	Ventilation
Smart Building Components	5%	10%	5%
Smart HVAC Components	10%	15%	10%
Advanced BMS	20%	10%	10%
User Interface	10%	10%	10%
Smart Grid	0%	10%	10%
Cumulative Savings (Additive)	45%	55%	45%
Net Savings (Factorial)	38%	44%	38%

agriculture, and mining.

Using this criterion, we examined the five levels at which intelligent efficiency can be implemented in the manufacturing sector, within specific processes and within the organization overall.

- Device level: motor, pump, fan, compressor
- System level: pumping system, air handling system
- Production level: individual product line
- Facility level: manufacturing facility
- Enterprise level: corporation

As described earlier in this report, motors and the systems they drive can communicate with production-level control systems. The production control system provides plant management with up-to-date information on the state of the manufacturing process and progress against production targets. The production control system can interact with the facility's BMS and both of them communicate information to the engineering and maintenance departments to alert them to devices and systems that are not functioning properly. The production system communicates with the enterprise resource planning system, which provides corporate management with the information it uses to direct the company.

All of this communication between levels improves energy use throughout an organization. Prior to the advent of automation, information had to be recorded manually and communicated often in person. This was time consuming and inefficient. It is for this reason that manufacturing and automation go hand in hand and why it is a natural fit for intelligent efficiency to take root in manufacturing. The term "smart manufacturing" has come to represent the combination of capabilities that result from integrating ICT into the production process.

The type of manufacturing will dictate which specific intelligent efficiency measures are incorporated into production and how they are integrated within an organization; however, each segment of manufacturing will have some version of the categories described above into which common energy measures fit.

PLANT AUTOMATION AND CONTROL

There are different types of plant-wide automation and control. Some factories use distributed control systems* that are able to connect to analog equipment and

systems, while others purchase equipment that is already embedded with programmable logic controllers† that can be networked in a plant control system. Either process control methodology can benefit from intelligent efficiency. Research by the Smart Manufacturing Leadership Council has indicated that ICT-enabled smart process and production control technologies have the potential to improve operating efficiency by 10%, water usage by 40%, and energy usage by 25%. For the purpose of our analysis we estimated the average savings realized by to be 20% (SMLC 2013).

Determining the marginal energy efficiency savings that result from the smart manufacturing/intelligent efficiency overlay to the plant-wide control system is more challenging than for a BMS because it is harder to discern the break between Level 3 and Level 4 technologies. Most production devices and systems have embedded intelligence in the form of "firmware" that governs their operation and enable connectivity with other ICT-enabled systems. A typical factory has dozens if not hundreds of discrete software packages each with its own specific functionality. These programs may be commercial off-the shelf products, purchased from an equipment vendor or developed in-house. With this complexity it is difficult to determine at what point a manufacturing process or facility becomes "smart;" therefore, our solution in the economic analysis was to assume that all ICT-enabled devices are indeed intelligent efficiency measures and should be included.

In addition to initial capital costs, these automated systems also have recurring costs such as licensing or subscription fees, service contracts, preventive maintenance, and other fixed operating costs (Navigant 2013b). Based on conversations with vendors of manufacturing automation systems, in our analysis we assumed 20% of the original investment as the cost of recurring subscription and services contracts.

INDUSTRIAL BUILDING AUTOMATION

These systems are no different from those used in the commercial sector, therefore, we estimated the same level of potential savings as we used in the commercial sector. The one major distinction is that we assumed that these investments would have a simple payback of two years versus four years for the commercial sector,

---

*DCS is a control system that collects data from the field for use for current and future control decisions.

†PLCs are computers used in automation of production processes. The devices make control decisions based on information provided by one or more signal inputs and affect control via one or more outputs.

because investment hurdle rates in industry tend to be much higher than the commercial sector.

RESULTS: CUMULATIVE POTENTIAL ENERGY SAVINGS FROM INTELLIGENT EFFICIENCY

Having developed an estimate for the amount of savings likely for each of the selected intelligent efficiency measures, we then determined the ratio of energy that could be saved for several end uses. The savings of each of the intelligent efficiency measures for the commercial sector were put into a matrix with the EIA's Commercial Buildings Energy Consumption Survey to determine the amount of energy savings that might be possible by building type and by energy use. The output of the matrix, described in detail in the appendix, was a percent of potential energy savings that any investment in intelligent efficiency in the commercial sector can be expected to achieve.

Next, we developed a projection of energy savings using EIA 2013 Annual Energy Outlook forecast data and, based on prior ACEEE research (Nadel et al. 1994), selected 50% as the ratio of all commercial building space will adopt at least some level of intelligent efficiency by 2035. A sensitivity analysis was performed with an estimate that the error of the 50% target is in the range of +/- 50%, and these three scenarios are presented in Graph A as the low, mid, and high scenarios. The

three of them represent the range of potential energy cost savings that we estimate is possible in the commercial sector. The analysis also assumes a relatively modest increase in investments of 1% per year early in the twenty year period and finishing at 2%.

Investments were assumed to be made with an expectation of a 20% annual return, or a five-year payback. The first year cost of licensing and services contracts are built into these cost estimates.

In the analysis of the industrial sector, the scope of this report did not allow for the analysis of dozens of individual energy measures as was done in the commercial sector. This is an area ripe for additional research, as the number of emerging technologies and their potential to affect changes is great. To determine the potential energy savings likely in the industrial sector, four fundamental assumptions were made. The first assumption was that 80% of all energy use in the manufacturing sector is attributable to manufacturing processes, a number supported by EIA data and previous ACEEE research (MECS 2006, Elliott et al. 2000). The second assumption was that the average energy savings realized by manufacturing facilities adopting intelligent efficiency would be 20%. This value is based on past ACEEE analysis, reports by the Smart Manufacturing Leadership Council (SMLC 2011) and conversations with others investigating smart manufacturing. The third assumption was that the balance of energy use in manufacturing, 20%, is consumed by buildings

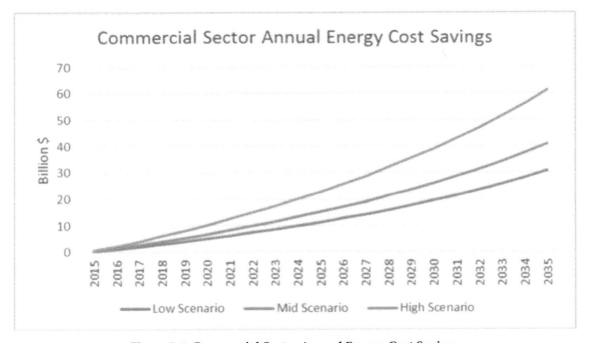

Figure 5-6: Commercial Sector Annual Energy Cost Savings

and that they have the same opportunities for energy savings as commercial buildings. The fourth assumption is tied to data reported by the EIA Manufacturing Energy Consumption Survey (MECS). Thirty-nine percent of energy used in the manufacturing sector was not attributed to specific end uses in responses to the End Uses Fuel Consumption 2010 survey. To compensate for this omission, we assumed that the breakdown of that 39% mirrored that of the 61% reported and increased the values of energy used in manufacturing processes and buildings accordingly.

The balance of the analytical method followed the analysis of the commercial sector. A target of 50% of all manufacturing electrical load is assumed be influenced by intelligent efficiency by 2035. Investments will increase initially at 1% per year and rise over the 20-year period to culminate at 3% per year. Investments will produce a 50% return (two-year payback), and licensing and service contracts will equal 20% of annual investments.

A similar sensitivity analysis was done estimating that the error of our original estimate is +/- 50%. Built into all of this analysis is that these projects are justified on potential energy cost savings. Conversations with people in the manufacturing automation sector indicated that this is seldom the case. Primary motivations for investment in the industrial sector are to improve production efficiency, product quality, safety, and regulatory compliance. Energy savings is often perceived to be an ancillary benefit. This detail though does not deter us from our findings as the energy cost savings

still can justify the supporting investment. If anything, it supports the perspective that our estimate is conservative and that much greater investment is likely in smart manufacturing over the next twenty years.

Our analysis indicates that the potential energy cost savings from intelligent efficiency measures for these two sectors could exceed $55 billion annually by 2035.

Even though energy consumption in both sectors is similar, a greater amount of savings is forecast for the commercial sector. This is because the commercial sector accepts a lower rate of return on its investments and because most of the energy in this sector is consumed in heating, cooling, ventilation, and lighting—areas that are fairly easy to automate. The same is not true for the manufacturing sector, which is much more complex and difficult to automate. Nevertheless, the potential for dramatic energy and cost savings in both sectors is very large and worthy of further attention and analysis. Policymakers and organizations with an interest in the energy sector need to take notice of intelligent efficiency approaches—cost-effective and broadly applicable to companies and industries nationwide—and develop ways to advance and encourage them.

THE CHALLENGES OF BIG DATA AND DATA ANALYTICS

The revolution we have seen in recent years in sensors, data collection, data storage, and computa-

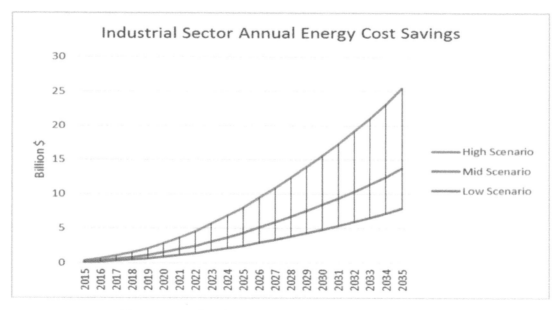

Figure 5-7: Industrial Sector Annual Energy Cost Savings

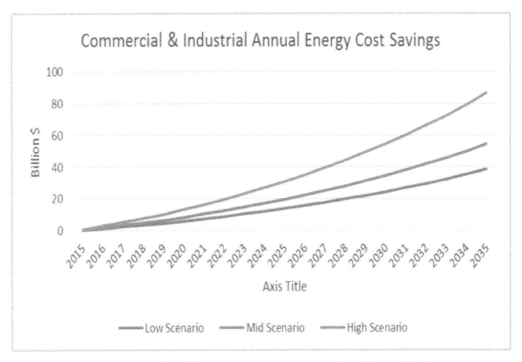

Figure 5-8: Combined Energy Cost Savings from Intelligent Efficiency

tional capabilities has transformed the way we think about data and has ushered in an era of "big data." Today, we can afford to collect massive amounts of data in real time, and with the dramatic decrease in data storage costs, we can afford to keep those data indefinitely (Economist 2012). These technology innovations have transformed many markets from having the limiting factor being the cost of collecting data to being the need for understanding of data analytics and techniques to discern meaningful information from a flood of data.

Big data and data analytics are fundamental to intelligent efficiency as they enable all of the computational and optimizing abilities described thus far in this report. But before the BMSs or production control systems can provide management better information, or improve the effectiveness and efficiency of devices and systems, they must be programmed to gather the most pertinent data and to use that massive amount of data to produce useful information. Distinguishing the information the end user needs, the "signal," from other non-relevant information "noise" requires considerable computation capabilities and this is where data analytics come in. They enable the parsing of key information such as a fault of a minor component of an HVAC system, or trends such as compressed air usage spikes on production line #1 in on mornings when product "A" is being made on production line #2.

Once pertinent data are collected, operators of buildings and processes face a new problem in how to manage those data and how to turn them into actionable information. The sheer volume of data taxes the IT systems of many small and medium size businesses. Larger operations may have the ability to store the data but are, again, challenged to turn it into useful information. Companies need help determining what data to collect, which data to discard, and what questions to ask of the data, and how best to get actionable answers.

The challenges of dealing with big data are not unique to energy efficiency, however the interest in harnessing the ability of big data and intelligent efficiency to produce economic growth, energy savings, and associated environmental benefits has produced efforts to quickly overcoming these barriers at the highest levels of government.

Big Data Management Demonstrations and Education

The federal government has the largest ICT infrastructure in the world and therefore has the greatest potential to benefit from implementing best practices. Government can demonstrate what big data and data analytics can do with projects like traffic control, fleet management, and building energy management to demonstrate what can work for large complex organizations. A product of these demonstration projects will be information on what data to collect and how to mine it for actionable information.

An example of this is the Open Data Format initiative that will make available energy consumption and intensity data from all government buildings (Whitehouse.gov 2011). From these data, not only have building energy consumption trends be identified, but new data mining efficiency products and services have been created (Whitehouse.gov 2013). These new services are now or will shortly be available to help organizations that are finding it challenging to benefit from big data.

Utilities also have a role to play in demonstrating the power of big data to help customers reduce energy usage. They have experience processing large volumes of energy information for their own use. New smart grid systems will only increase their ability to identify trends in customer consumption patterns. Some utilities and the commissions that regulate them are already considering how this might evolve. The California Public Utility Commission (CPUC) held an information exchange event in April of 2013 to educate commissioners and staff on how energy consumption data can be utilized to evaluate the effectiveness of policies and programs and the availability of such data (CPUC 2013). Utilities can also help customers understand their energy data and how different responses, in the form of changes to consumption patterns, can impact not only their use of energy, but also their energy expenditures. Since it is unlikely utilities will have all the answers at first, such customer engagements can evolve as smart grids are rolled out and the ability to decipher meaningful information from the exchange of data across them.

Several utilities and grid operators have come together to form the Open Automated Demand Response Alliance and a new protocol for communicating demand response information. Demand response is the curtailment of load at the customer level at the request of its supplying utility. OpenADR is an open and standardized way for electricity providers and electricity system operators to communicate demand response signals with each other and with their customers using a common language over an existing network such as the Internet (OpenADR Alliance 2013). Before a request for demand reduction is sent, the need must first be identified. This requires analyzing large volumes of data and determining the optimum response. This collective effort and common protocol could be a foundation for utilities and their customers to analyze, manage, and communicate energy data.

A COMMON INTERCONNECTION TO COMMUNICATE ENERGY DATA

Hundreds of manufacturers are involved in building and manufacturing automation and many of them have their own software programs. These programs are often not consistent in how they communicate energy data (ODVA 2011). Process control and automation systems are often installed in piecemeal fashion (Burgoon 2013) and since most facilities do not have the option to choose only one vendor for all of their automation needs, they are left with the choice of either having systems that cannot communicate with each other or that do so through some type of translation process.

There are already several industry led efforts to develop interconnection standards for industrial equipment and systems, three of which are focused on the problem of communicating energy data.

Cisco Systems, Rockwell Automation, and Schneider Electric are working with ODVA, a global association of leading automation companies, to develop an international energy communication protocol, CIP Energy, based on the Common Industrial Protocol (CIP™) architecture, that is designed to transform the way manufacturers monitor and control energy usage by providing a common-command interface and network-visible data structure (Lydon 2011, Rockwell 2011). It is an extension the popular protocol at the heart of EtherNet/IP™ (ODVA 2011). The specification for CIP Energy includes attributes and services that help system designers reduce the cost and time to implement energy-improvement projects. CIP Energy makes operational energy consumption data available at the network level, enabling manufacturers to optimize energy usage during production and diagnose potential problems at the process or even machine level (Rockwell Automation 2011).

Another private sector effort to address the emerging issues around working with big data in general and energy and environmental data in particular is the Information Technology Industry Council, whose membership is composed of the world's leading software and electronics companies (ITIC 2013a). It was formed to lead the advocacy and policy discussions on a national level around the transformative potential of ITCs. The council has engaged the White House Council for Environmental Quality on multiple occasions as well as several energy legislation–focused senators and representatives.

The Information Technology Industry Council is working with National Institute of Standards and Technology, the White House Subcommittee on Standards, and other agencies to promote voluntary standards for ICT (ITIC 2013b). It is also working internationally with industry-led standards groups in Brussels, Beijing, and elsewhere. It has sponsored the InterNational Committee for Information Technology Standards, a forum of 1700 members interested in promoting voluntary ICT standards (ITIC 2013b).

The Energy Information Standards Alliance (EIS Alliance) is a trade association for companies that provide energy management and smart grid products and services and whose members work collaboratively to educate policymakers, utilities and other stakeholders on how energy management systems can help them reduce energy usage (EIS 2013). They have been involved in developing a common framework for customer equipment to use, generate, and communicate energy data. They have collaborated with the White House its initiative to implement simple consumer energy management system products under the "Green Button" label (EIS 2012).

The ODVA, ITIC, and EIS efforts are great examples of the private sector coming together to facilitate intelligent efficiency through development of common protocols. A second piece of the puzzle is development of a common protocol for determining energy savings.

COMMON PROTOCOLS FOR DETERMINING ENERGY SAVINGS

As more and more companies focus on energy, they are confronting challenges in determining and reporting on energy savings. The fundamental challenge determining energy savings is the need to measure a counterfactual. This can be challenging because in order to determine how much energy has been saved, an operator of a business or manufacturing plant must first determine how much would typically have been used—how much has been used for a given process in the past. This "baselining" often involves a regression analysis of multiple production and environmental factors for the purpose of determining what drives energy consumption. Then, once an energy measure is implemented, these mathematical models can be used to determine net savings.

The challenge is two-fold. First, this process is time-consuming and therefore expensive. The second is that energy use is dependent upon production levels. As product volumes and mixes change, so does the energy consumption profile. The resulting energy savings of any given energy measure can be difficult to determine

when no two days of production are the same. However, it is not necessary to determine the precise amount of energy saved at any given time by a specific energy measure, only the average amount over time, because energy use is observed in the aggregate. Accuracy in the determination of energy savings improves with time as more data is collected, collated and analyzed. What is needed is a common method that accommodates fluctuation in energy usage, perhaps by leveraging the increased volumes of data, to produce more accurate and reliable energy savings values.

New energy management protocols such as International Performance Measurement & Verification Protocol (IPMVP) (see the box on the next page) are giving companies the ability to measure energy savings with increased accuracy. This protocol, first published in 1996, encourages good measurement and verification (M&V) design and ongoing monitoring of performance. These protocols can be and in some case already are being incorporated into advanced BMS and smart manufacturing control systems. Intelligent efficiency platforms, such as the Panoptix Energy Performance Manager app highlighted in a case study above, can provide the analytics to measure and verify the savings of operations (IPMPV 2002).

Energy Management Standards

Some industrial companies are going the extra step of implementing a system to manage their energy use and guide their energy efforts that is compliant with international standards. One such system, Superior Energy Performance, is designed to provide companies with a transparent, globally accepted method for verifying energy performance improvements and management practices. A key component of Superior Energy Performance is that companies also implement the globally energy management standard, ISO 50001, which has additional requirements to achieve and document energy performance improvements. The program includes methodology to verify energy performance improvements, certified professionals to assist in implementation, system assessments and verify conformance (SEP 2013). U.S. Department of Energy (DOE) developed Superior Energy Performance in collaboration with the U.S. Council for Energy-Efficient Manufacturing (U.S. CEEM).

To date, few companies in the United States have implemented an ISO 50001 Certified energy management system or the Superior Energy Performance system as the up-front investment and long-term commitments are significant. Still, the value of them, and the

IPMVP is that organizations that abide by them produce energy savings values that other organizations can trust and by extension, might be willing to pay for. It is this second point that directs our analysis to focus on how energy efficiency programs might benefit from intelligent efficiency.

IPMVP Protocol

When IPMVP was first published it contained methodologies compiled by a technical committee from the United States, Mexico, and Canada. IPMVP's direction on M&V practice encourages the good design of energy management projects, providing participants in an energy project a common set of terms with which to discuss key M&V project-related issues and establishes methods that can be used in energy performance contracts. The protocol defines broad techniques for determining savings from and the facility overall as well as an individual technology. It is flexible enough to be relevant to a variety of facility types, including residential, commercial, institutional, and industrial buildings, and industrial processes. It provides an outline for applying procedures to similar projects throughout all geographical regions and that are internationally accepted, impartial, and reliable.

The IPMVP:

- Presents procedures, with varying levels of accuracy and cost, for measuring and/or verifying baseline and project installation conditions, and long-term energy savings.

- Provides a comprehensive approach to ensuring that buildings' indoor environmental quality issues are addressed in all phases of project design, implementation, and maintenance.

- Creates a living document that includes a set of methodologies and procedures that enable the document to evolve over time.

The protocol is intended for facility energy managers, project developers and implementers, energy service companies, water service companies, non-governmental organizations, finance firms, development banks, consultants, utility executives, environmental managers, researchers, and policymakers.

Example: Federal Energy Management Program (FEMP)

The U.S. Department of Energy's Federal Energy Management Program (FEMP) was established, in part, to reduce energy costs to the U.S. government from federal facilities. FEMP assists managers at federal facilities by identifying and procuring energy-saving projects. The FEMP M&V Guidelines follow the IPMVP and provides guidance and methods for measuring and verifying the energy and cost savings associated with federal agency performance contracts. It is intended for federal energy managers, federal procurement officers, and contractors implementing performance contracts at federal facilities. The FEMP M&V Guidelines have two primary uses:

- They serve as a reference document for specifying M&V methods and procedures in delivery orders, requests for proposals, and performance contracts.

- They are a resource for those developing project-specific M&V plans for federal performance contracting projects.

Many states in the United States have incorporated the IPMVP into their energy efficiency programs and their determination of energy savings from energy performance contracting. The New York State EnVest program, for example, is structured to be consistent with the IPMVP, and the New York State Energy Research and Development Authority strongly recommends the use of IPMVP for institutional projects. Other states that have incorporated the IPMVP in state energy performance contracting and other energy efficiency programs are California, Colorado, Oregon, Texas, and Wisconsin.

The Opportunity for Intelligent Efficiency in Energy Efficiency Programs

The typical base-load generation produces electricity at a cost of $0.073 to $0.135 per kWh while energy efficiency can achieve savings at an average cost of $0.03 per kWh saved (Friedrick et al. 2009). These economics have encouraged states utility commissions to view energy efficiency as the least-cost energy resource and consequently there has been an increase in the number of programs and respective targets for energy savings. (Chittum, Elliott, and Kaufman 2009). These energy efficiency programs provide technical assistance and financial incentives to encourage the purchase of energy savings equipment. Incentives in 2010 totaled more than $1 billion in the industrial sector alone (Chittum and Nowack 2012).

Utility-sector energy efficiency programs are intended to mitigate the need for utilities to invest in conventional generation and transmission by instead using funds to assist their customers in reducing their energy consumption. Energy efficiency programs within the utility sector are created by state governments and public utility commissions, the utilities that serve electricity and natural gas customers, and the administrators of those programs. Each stakeholder has its own goals and priorities and tries to incorporate them during the process of developing a program.

UTILITY SECTOR EFFICIENCY PROGRAMS

Many commercial sector programs targeting building efficiency focus on the building envelope, lighting, and the components of HVAC systems. When any of these components are upgraded, a certain reduction in existing energy use can be predicted with a fair level of certainty. However, upgrading the devices is only part of the solution. Lighting and HVAC systems do not run continuously but are turned on and off, and up and down throughout the work week. A light that is off uses less energy than a light that is on, so building operators and efficiency program administrators are always looking for methods to turn lighting and other equipment off. This can be accomplished through automation or worker training.

The industrial sector has been challenging for energy efficiency programs to penetrate due in large part to the heterogeneity of the sector (Chittum, Elliott, and Kaufman 2009). Where efficiency programs may be able to commoditize their offerings to residential and commercial sectors, the industrial sector often requires a more custom approach. In custom programs, the building operator or factory manager establishes a baseline of performance. Energy measures are evaluated and future energy use, based on past building use, or factory production, is forecasted. Using forecasted savings numbers, the program administrator commits to a specific financial incentive which the customer then uses to help finance the implementation of the identified energy measures. Though these projects can save a great deal of energy, they often require a considerable up-front investment in time.

Simplicity of administration is major reason that the majority of energy efficiency programs focus on increasing the efficiency of individual devices rather than larger systems. Installing a motor that is 3% more efficient results in 3% energy savings. The cost of such assets and the savings they provide are easy to measure and verify.

M&V is an important part of any efficiency program. Programs must ensure that the savings—energy efficiency resources as they are often called—they are claiming will be available some months or years in the future when needed. Evaluating energy savings attempts to measure something that does not exist—the energy that is not being used—and so program evaluators must rely on assumptions and estimates to make their determinations.

With simple projects such as replacing lighting or motors with more efficient units, the amount of energy saved can be established at the outset of the project. With more involved projects, engineering analysis, surveys, and analysis of energy bills may be required (Chittum 2012). One of the challenges with this approach is that a thorough engineering analysis to determine the exact energy savings realized over time by a large industrial project could be so expensive as to make the energy savings cost-prohibitive.

CHALLENGES FACING EFFICIENCY PROGRAMS

Program administrators of older energy efficiency programs may soon find it challenging to meet targets that continually rise while keeping down the programmatic cost per unit of energy saved. The first projects that a utility implements are often the lowest cost and the easiest to evaluate. As utilities get further into their energy savings journey, the projects become more complex, as does the determination of savings. For the program administrator with responsibility for effective use of ratepayer or taxpayer funds, this is an important issue and one about which many of their stakeholders have opinions.

Some program managers are realizing that greater savings are possible and at lower costs with investment in projects that focus on system optimization. For example, the Northeast Energy Efficiency Alliance (NEEA) focuses much of its engagement in the manufacturing sector on assistance with the implementation of energy management practices (NEEA 2013b). The companies engaged adopt a systematic approach to energy management, tracking energy use and implementing best practices on a continual basis. These engagements can be very effective, however determining the energy savings that results from them can be challenging.

To determine energy savings with accuracy, it is first necessary to establish a baseline of energy use at different times of the year prior to the installation of an energy measure. Energy use data from after the efficiency measure(s) are implemented can then be compared to that baseline and net savings determined. Beyond the challenge of determining the baseline, all of this can be time consuming and therefore costly. So even though energy efficiency has consistently proven to be cheaper than other types of resources (Friedrich et al. 2009), program administrators for efficiency programs can still be challenged to secure sufficient savings to meet their targets in a cost effective manner and with a high degree of confidence.

Two additional M&V challenges are those of attribution and energy intensity. If several energy measures are implemented at the same time and there isn't before and after system specific energy consumption data, it is very difficult to attribute energy savings to each measure using conventional analysis techniques. Energy intensity refers to the amount of energy needed to perform a specific task. If a company installs a new product line that employs cutting-edge technology, energy savings in terms of energy per unit of production (energy intensity) is likely to decrease as a result of more efficient components but also from a more efficient process. However, the new production line might produce twice as much product as the previous line and therefore use more energy overall. With all of these M&V issues it is clear that a more effective method of measuring and verifying energy savings is critically needed.

In an effort to overcome these challenges, some programs are considering projects involving automation and controls (Monsalves-Salazar 2013, Goldman 2013). These projects are a promising opportunity for efficiency programs not only because they provide programs a new set of energy saving assets to incent, but also because they may yield savings at a lower cost and with a higher level of confidence.

INCLUDING AUTOMATION AND CONTROLS
IN EFFICIENCY PROGRAMS

Leading efficiency programs are seeking new programmatic methods to gain greater volumes of energy savings from each customer and an emerging trend is to create programs that capture savings from multiple systems in one project such as whole-building retrofits and building automation. As described in the case studies earlier in this report, conventional BMSs have a proven ability to reduce energy consumption by 10 to 30 percent and advanced BMS even more so. What is promising about including smart automation and controls in efficiency programs is that if done right, it will not only provide additional savings, but also provide an improved measurement capability.

The computational power of data analytics enables

the establishment of a baseline much more easily and inexpensively than before, even with historical data that is not of the quality of current BMS generated data. Advanced BMS and manufacturing process control systems now have the ability to measure current performance, compare with past performance, and then forecast future performance. This solves the issues of attribution and energy intensity. The intelligent efficiency measures can track energy consumption at the device level, match that with facility use or production values, and provide both facility operators and efficiency program administrators energy performance data and forecasts that they can use to forecast future energy resource needs. And since this is an automated processes, the exchange of information can happen at or near real time and at a lower cost than conventional data collection and reporting.

CHALLENGES OF INCORPORATING AUTOMATION
AND CONTROLS INTO EFFICIENCY PROGRAMS

Though it may also be easy to understand how automation can save energy, it is more difficult to determine how it can be incorporated into an energy efficiency program. And though the net energy savings is not in dispute, this type of measure creates challenges for the conventional energy efficiency program.

Utility sector energy efficiency programs operate on a cycle determined by the utility and public utility commission. They may be as short as a year or more than five, but most programs tend to be two or three years in duration. At the end of a cycle they are subject to review and may or may not be renewed. This is problematic for commercial and industrial customers making long-term investment plans. Furthermore, many programs do not allow for applications throughout the year but instead have specific application windows. These cycles are not likely to be in sync with a company's capital investment cycle or compatible with the multi-year implementation period of larger capital projects. This can cause the financial incentive and the measurement of savings to be split between the year(s) of installation and the first year of operation (Chittum 2012). If a program totals the savings of a project at the end of its first year, it may capture only a fraction of possible savings. Due to performance pressures by program evaluators, program managers attempt to book savings as early as possible, which means that savings from larger long-term projects are not properly counted (Chittum 2012).

As described earlier regarding the benefits of continual optimization, with some intelligent efficiency technologies, it is not appropriate to calculate savings prior to installation since the ability of automation to wring savings from a system is influenced by many variables. Instead, programs wanting to influence the full use of the automation will seek a method to determine actual savings and let those values drive the amount of financial assistance. Programs may also want to provide assistance over a period of one to three years as a method to encourage customers to get full value of the investment.

Conventional energy efficiency programs have focused on component energy efficiency, which poses several challenges to advancing intelligent efficiency measures. The first is a connectional challenge—in a complex project involving multiple components and controls, the energy savings happens at the device level even though it is influenced at the control level. What, if any, portion of the energy savings realized can be attributed to the controls? When performing M&V, how should energy savings be attributed? For example, a building retrofit project might include new lighting, new HVAC components, replacement windows, and a new BMS. It is easy enough to verify that equipment has been installed and to measure the amount of energy saved, but determining the portion of energy saved attributable to the BMS in this situation will be difficult if not impossible.

And it would not be necessary if for not the fact that many programs are organized around encouraging the purchase of efficient equipment such as high efficiency lighting or motors. The reason for this is straight forward. Determining prior to implementation the maximum energy savings possible with these assets is straight forward. With a little more effort, reasonably accurate estimates of savings under common usage can be determined and applied broadly. For example, a program can estimate with confidence that for every hundred lighting projects, the average reduction from prior usage will be 20% and with this knowledge set an incentive at an amount related to the value of energy saved.

The administration of such a program is simplified by the ability to determine future savings for a given energy measure before implementation and then only need to verify implementation to satisfy M&V requirements. However, in some intelligent efficiency measures do not require the purchase of physical assets but instead involve the use of on-site software and/or off-site, online computational capabilities. Some utility programs have encountered challenges providing incentives to these types of projects because of what some in the intelligent efficiency community have come to characterize as the "asset tag" problem. What exactly is it that they pay for when they provide a financial incentive?

For example, a new software program that has improved diagnostic abilities due to its ability to process larger volumes of data could produce energy savings by providing building operators with improved building performance metrics that alert them to opportunities to adjust set points for more efficient HVAC operation. Such savings would not be possible without the software or the building operator training.

Another example of "asset tag" challenge is the opportunity for "virtualization" of data centers—the rooms filled with routers, servers and switches that have become ubiquitous throughout all sectors of our economy. Millions of these data centers exist across the country contributing to the growth in miscellaneous energy use that was documented in a recent ACEEE report (Kwatra, Amann, and Sachs 2013). In virtualization, a majority of the functions performed by local servers are migrated into the "cloud" resulting in a substantial reduction in net energy use since the energy intensity of the data centers that enable cloud computing are lower because of scale and the ability to manage loading across multiple data centers ensuring optimal use of resources, which is also results in lowest energy consumption (EMC2 2010).

In virtualization projects such as just described, no new equipment is acquired, and in fact existing equipment is often retired in the process. In its place an ongoing service is subscribed to and while onsite energy use is clearly reduced due to the retirement of equipment, some of the remaining energy use now occurs off site at a facility that may or may not be in the program service territory. In fact, it may occur anywhere in the world.

This inability to associate net savings with an asset requires a paradigm shift in how we think about an energy efficiency measure. Instead of paying for the assets, programs may start paying for actual savings. And in a very interesting and serendipitous development, intelligent efficiency may provide the very ability to do so.

Performance Based Efficiency Program

Intelligent efficiency provides an opportunity to move from energy efficiency programs that are device-based to programs that systems- and performance-based. Older programs that may be reaching the limits of what can be achieved with fixed rebates for purchasing specific items may find the concept of paying for performance of interest, especially if they are looking for new program ideas that will appeal to their larger industrial and commercial customers.

With the ability to determine current and future savings, a building operator or factory manager and the efficiency program administrator can begin a conversation on paying for performance. Once in place, the advanced BMS or the smart manufacturing system is able to compare current operating conditions with a previous baseline under similar operating conditions and determine the net energy savings. It may also have the ability to forecast future energy demands. Performance information is reported to the program administrator and the incentive paid is based on energy saved. Programs may provide a bulk of the incentive upfront based on forecasted energy savings and later, as actual performance is reported, the balance is released. That balance may increase or decrease depending whether more or less energy has been saved than forecasted and it may be released over a period of one or more years.

So long as the protocols for determining energy savings are agreed to by both parties at the beginning of the project, this arrangement has promise. As previously highlighted, the IPMVP has gained broad acceptance and is currently in use in many states. The ability of a smart technology to follow this protocol and report performance data on a timely basis lowers the cost of measurement and verification and ultimately the utility's cost of running an energy efficiency program and acquiring these energy efficiency resources.

Performance Contracting: a Model for Pay-for-Performance

An example of paying for performance is performance contracting. Energy service companies (ESCos) have been helping public sector facilities reduce energy consumption through performance contracts. In these arrangements, the ESCo makes the capital investment in upgrading the energy consuming equipment of a facility: lighting, heating, air-conditioning, hot water systems, etc. As a result of these investments, the facility's energy costs go down thereby freeing up cash flow for the facility repay the ESCo. The energy cost savings are essentially split between the facility and the ESCo so the more energy saved, the more the ESCo can potentially earn*.

The determination of savings of course requires the establishment of a baseline. Baselines and verification requirements are determined in the design phase of a project, and included in the contracts. To ensure that

savings continue after equipment installation, many performance contracts include service agreements. Inclusion of intelligent efficiency in performance contracts is increasing. The Department of Energy is investigating the use of performance contracts to fund upgrades in IT and data centers (C2ES). Johnson Controls includes BMSs in all of its performance contracts because of the additional energy savings they provide and because they simplify performance measurements (Nesler 2013.

Though these agreements do not usually include per unit energy savings payments, the methodology used to determine baselines, measurement, and verification is similar to that used by utilities in their custom programs. An important feature of these agreements is that they are focused on energy savings rather than specific assets. In fact, because they are paid for their performance, ESCos are motivated to achieve as much energy cost reduction for as little capital investment as possible. With such potential for a new, more effective method of securing energy savings, the idea of energy efficiency programs paying for performance is an area worth of more research.

Summary

Intelligent efficiency is making possible new levels of energy consumption analysis and energy management. This will have broad implications for building operations and manufacturing production management and control. Building operators now have the ability to learn immediately when systems start to operate outside of normal parameters, thereby enabling them to dispatch service technicians to address small problems before they become big problems, or at the very least, use energy unnecessarily. Manufacturers have the ability to network entire production lines, even supply chains, so that they can eke out marginal savings at every point in the system.

Over the next two to three decades we will see these new capabilities available to every sector of the economy. With the ability of intelligent efficiency to generate the next-step change in energy savings, multiple additional economic benefits are possible. Non-energy benefits stem from system optimization, including better services and, in industry, better quality control. Lower operating costs free up capital making it available for additional investments in productivity and capacity. Environmental benefits related to energy savings will be realized at the point of use and across the nation as the need for new generation decreases.

Many of these smart technologies are already cost-effective and therefore we can anticipate that a

---

*There are many types of performance contracts, each with different features and benefits. The example used here was chosen for its simplicity and relevance to the pay for performance concept.

great deal of economic activity will happen with little or no influence from the public sector; however, there is an importunate opportunity to leverage intelligent efficiency for public policy goals. With its potential to bring about new levels in energy savings nationwide, intelligent efficiency measures appear very likely to become part of state-level efforts to reduce energy consumption in the commercial and industrial sectors.

This previously unavailable method to save energy is attributable to intelligent efficiency systems' having the ability to determine the baseline energy consumption for multiple operating conditions, monitor energy consumption and production inputs and outputs, identify correlations that can be used to determine current energy savings, and forecast future energy use. Intelligent efficiency systems can also confirm these correlations by regularly comparing current performance with past predictions, adding even greater levels of confidence in reported savings numbers. Automated control systems can be programmed to follow energy savings determination protocols that are broadly accepted. This combination of analytical capabilities presents us with an opportunity to determine energy savings on a real-time basis. That capability in turn opens up the opportunity for energy efficiency programs to pay for performance rather than for implementation.

Adding the financial resources that are currently funding conventional utility investments and device-level energy efficiency investments into the total investment mix targeting intelligent efficiency means an accelerated adoption profile of intelligent efficiency measures. By our estimate, it could reach $55 billion by 2035. This is an opportunity that federal and state policymakers, utility regulators, energy efficiency program administrators and evaluators, and vendors of ICT products and services should embrace. With that goal in mind, we offer the following recommendations.

RECOMMENDATIONS

In this report, we have recommended several actions that different stakeholders can pursue to facilitate the implementation of intelligent efficiency approaches across the commercial and manufacturing sectors. These recommendations are not exhaustive but are intended as jumping off points to more in-depth discussions, research, and analysis. The potential economic impacts are clear and the barriers manageable. Here we outline key actions by key stakeholders that will significantly increase the likelihood of widespread adoption of intelligent efficiency throughout the U.S. economy.

Role for Government

With the potential to produce a step change in energy efficiency and the associated cost savings throughout the economy, intelligent efficiency is an ideal strategy for government policies and programs to encourage. The federal agencies that consume a great deal of energy can lead by example through incorporating smart BMSs and other intelligent efficiency measures into their buildings. This can be done through direct investments and energy service performance contracts. Specifications for performance contacts can include requirements for advanced building automation with the ability to self-correct and continuously optimize.

Government also has a role to play in catalyzing innovation by funding research, development, and demonstration projects. Current examples include the funding of pilot projects that include the software, firmware, network, and data analytic components of smart manufacturing at Department of Defense facilities, funding of smart grid research projects by the Department of Energy at its national laboratories, development of communication standards by National Institute of Standards and Technology, and demonstration of performance contracting by the General Services Administration (Ye and Seidel 2012, ITIC 2013b). As the technology continues to evolve, so too can the projects these agencies use to demonstrate and realize the benefits of intelligent efficiency.

Role for Utilities and Energy Efficiency Program Administrators

Program administrators for utilities sector energy efficiency programs should seek out opportunities to include intelligent efficiency measures as qualifying projects in their existing programs. They can pilot programs that target smart technologies and experiment with these technologies' ability to provide timely performance data.

Programs can also experiment with paying for performance, refining the approach as they learn what does and does not work, then gradually expanding to other appropriate larger customer pools. They would do well to participate in collaborative efforts to establish common energy management practices and energy savings determination protocols. Existing efforts to develop common protocols for demand response such as the Open ADR Alliance can be leveraged and expanded to communicate energy data between utilities and their customers.

As utilities install smart grids, they can work with their commercial and manufacturing customers to integrate the ability of a smart grid to communicate the value of energy given the time of day and the customer's location to customer's advanced BMS or smart manufacturing systems. Each customer can then respond with changes in energy usage that reflects internal priorities, one of which may be reducing energy expenses.

Role for Public Utility Commissions

Public utility commissions should allow utilities' energy efficiency programs to do pilots in order to learn what works and what doesn't, as well as discover solutions to M&V challenges. There is no substitute for the learning that occurs through doing, and pilot projects enable this learning with a low level of risk. Programs may run into unanticipated barriers and will have the opportunity to work through them on a small scale. Once the concepts are proven and ICT performance standards are developed, public utility commissions can then authorize broader program acceptance of smart technologies and systems.

Suppliers of ICT Products

The many companies engaged in developing and selling intelligent efficiency products and services can seek opportunities to collaborate on non-competitive research and development as well as to education and create awareness of the benefits of ICT. Activities such as those by the Information Technology Industry Council (ITIC) to bring awareness to IT issues within policy circles; the Smart Manufacturing Leadership Coalition (SMLC) to form collaborative research, development, and implementation teams to develop common software platforms, standards, and approaches (SMLC 2013a); and the Energy Information Standards (EIS) Alliance to develop common communication framework for equipment to generate, communicate, and use energy data (EIS 2013) are all examples of what is helpful and necessary to move the adoption of intelligent technologies forward. Private sector leadership in this area is necessary as there is insufficient technical knowledge or capacity elsewhere. Only the companies engaged in this sector have the detailed understanding of the many unique software products that are needed to enable the level of interoperability that will facilitate greater market penetration of intelligent efficiency technologies.

Conclusion

Broad action on these recommendation will help to diminish—and eventually eliminate—the barriers

standing in the way of the U.S. economy's reaping intelligent efficiency's benefits. These actions, taken simultaneously by a diverse group of stakeholders, will advance the energy efficiency options of the commercial and industrial sectors to a level not seen before, helping those sectors to reduce their energy consumption and costs, improve product quality and employee satisfaction, and strengthen their resilience in the global economy.

Going forward, more research in the area of intelligent efficiency and utility section energy efficiency programs is warranted. Such research could lead to demonstrations of building or plant automation systems that provide real-time energy performance data, and eventually to utility efficiency programs that pay for energy saved rather than equipment installed.

The potential for intelligent efficiency technologies such as machine-to-machine and smart grid to bring about new efficiencies in manufacturing is only beginning to be understood. Additional research is required to gain a better understanding of this opportunity and its ramifications. Will this new level of automation, as we have seen in previous industrial revolutions, grow the size of the manufacturing sector? Will it bring about more and more satisfying jobs than it replaces? What will be required of workers if they are to successfully utilize smart technologies? These are but a few questions that would be useful to answer early in the journey to embracing intelligent efficiency.

More research is needed to understand the specifics of not just how data analytics can mine big data to facilitate efficiency gains within organizations, but also how external data can be harvested for the benefit of the supply chain. It is likely that this new level of connectivity will soon integrate customers into product and service design processes. It would be beneficial to understand the broad implications for energy consumption of such a streamlined process as it will likely have significant economic implications.

References

ABI. 2013. "Commercial Building Automation Market to Top $43 billion by 2018, Says ABI Research." Press Release. April 30.
http://finance.yahoo.com/news/commercial-building-automation-market-top-170600126.html.

[ACEEE] American Council for an Energy-Efficient Economy. 2013. "How Does Energy Efficiency Create Jobs?" http://www.aceee.org/files/pdf/fact-sheet/ee-job-creation.pdf

Baxter, Van D. 2004. Evaluation of Abbotly Technologies Compressor Optimization Control Product "ESM System 4000" as applied to Two Refrigeration Compressor Rack Systems at the ASDA/Wal-Mart Super Center in Sheffield, UK. Oak Ridge, TN: UT-Battelle, LLC/Oak Ridge National Laboratory. http://www.smartcool.net/documents/testingresults/ORNL_Refrigeration.pdf

Bradbury, James, Nate Aden, Achyut Shrestha, and Anna Chittum. 2013. "One Goal, Many Paths: Comparative Assessment of Industrial Energy Efficiency Programs." In Proceedings of the 2013 ACEEE Summer Study on Energy Efficiency in Industry. Washington, DC: American Council for an Energy-Efficient Economy. http://www.aceee.org/files/proceedings/2013/data/index.htm.

Burgoon, Mary (Rockwell Automation). 2013. Personal communication. August 9.

Capehart, Lynne, C. and Barney L. Capehart. 2008. Facility Energy Efficiency and Controls: Automobile Technology Applications. Encyclopedia of Energy Engineering and Technology, 671-679. Taylor & Francis.

Carbon Disclosure Project. 2011. Cloud Computing—The IT Solution for the 21st Century. Prepared by Verdantix. London, United Kingdom: Carbon Disclosure Project.

———. 2012. Meaningful Impact: Challenges and Opportunities in Industrial Energy Efficiency Program Evaluation. Research Report IE122. Washington, DC: American Council for an Energy-Efficient Economy.

Chittum, Anna, R. Neal Elliott, and Nate Kaufman. 2009. Trends in Industrial Energy Efficiency Programs: Today's Leaders and Directors for the Future. Research Report IE091. Washington, DC: American Council for an Energy-Efficient Economy.

Chittum, Anna and Seth Nowak. 2012. Money Well Spent: 2010 Industrial Energy Efficiency Program Spending. Research Report IE121. Washington, DC: American Council for an Energy-Efficient Economy.

Clancy, Heather. 2013. "Seattle, Microsoft team up to bring energy efficiency downtown." July 10. http://www.greenbiz.com/print/53109.

[CPUC] California Public Utilities Commission. 2013. "Thought Leaders Speaker Series—Utilizing Energy Consumption Data to Evaluate the Effectiveness of Policies and Programs." April 18. San Francisco, CA. http://www.californiaadmin.com/agenda.php?confid=CPUC_SS041813&dir=cpuc.

Cullinen, Matt, 2013. Machine to Machine Technologies: Unlocking the Potential of a $1 Trillion Industry. The Carbon War Room.

Davis, Jim, Tom Edgar, Yiannis Dimitratos, Jerry Gipson, Ignacio Grossmann, Peggy Hewitt, Ric Jackson, Kevin Seavey, Jim Porter, Rex Reklaitis and Bruce Strupp. 2009. "Smart Process Manufacturing. An Operations and Technology Roadmap." UCLA: Los Angeles, CA.

Davis, Jim, and Tom Edgar. 2013. "Smart Manufacturing as a Real-Time Networked Information Enterprise." Smart Manufacturing Coalition presentation. UCLA: Los Angeles, CA. http://egon.cheme.cmu.edu/ewocp/docs/DavisEdgarEWOWebinar12213v4.pdf.

Desroches, L.B. and K. Garbesi. 2011. Max Tech and Beyond Maximizing Appliance and Equipment Efficiency by Design. Berkeley, CA: Lawrence Berkeley National Laboratory.

Downey, J. 2012. "Duke Energy Wins Award for its Part in Envision: Charlotte." Charlotte Business Journal, January 24. http://www.bizjournals.com/charlotte/blog/power_city/2012/01/duke-energy-wins-award-for-its-part-in.html.

[EC ISM] European Commission Information Society and Media. 2009. ICT and Energy Efficiency. The Case for Manufacturing. Brussels, Belgium.

Economist. 2012. "Rise of the Machines. Moving from Hype to Reality in the Burgeoning Market for Machine-To Machine Communications." Economist Intelligence Unit.

[EIA] Energy Information Administration. 2003a. "Table A6. CBECS Building Size, Floorspace for All Buildings (Including Malls)." Washington, DC: U.S. Department of Energy.

———. 2003b. "Table 2.11 Commercial Buildings Electricity Consumption by End Use, 2003." Washington, DC: U.S. Department of Energy. http://www.eia.gov/totalenergy/data/annual/showtext.cfm?t=ptb0211.

———. 2006. "Table 2.2 Manufacturing Energy Consumption for All Purposes, 2006." Washington, DC: U.S. Department of Energy. http://www.eia.gov/totalenergy/data/annual/showtext.cfm?t=ptb0202.

———. 2010. Table 5.1 "End Uses of Fuel Consumption, 2010." Washington, DC: U.S. Department of Energy. http://www.eia.gov/consumption/manufacturing/data/2010/#r5.

———. 2013a. "Annual Energy Outlook 2013—Energy Consumption/Industrial Sector Key Indicators and Consumption/Reference Case." Washington, DC: U.S. Department of Energy. http://www.eia.gov/oiaf/aeo/tablebrowser/#release=AEO2013&subject=2-AEO2013&table=6-AEO2013&region=0-0&cases=ref2013-d102312a

———. 2013b. "Annual Energy Outlook 2013—Table E3A: CBECS Energy Consumption (Tbtu) AEO2013ER — Commercial Sector Key Indicators and Consumption Reference Case." Washington, DC: U.S. Department of Energy.

———. 2010 "Table 5.2 End Uses of Fuel Consumption, 2010." Washington, DC: U.S. Department of Energy.

EIS. 2012. "EIS Alliance Efforts Lead to "Green Button" for Customers." Morgan Hill, CA: Energy Information Standards Alliance. http://www.eisalliance.org/green-button.

———. 2013. "Our Mission." Morgan Hill, CA: Energy Information Standards Alliance. http://www.eisalliance.org/about-the-eis-alliance.

Elliott, R.N., J. Amann, A. Shipley, N. Martin, E. Worrell, M. Ruth, and L. Price. 2000. Emerging Energy-Efficient Industrial Technologies. Washington, DC: American Council for an Energy-Efficient Economy. http://aceee.org/research-report/ie003.

Elliott, R. Neal, Rachael Gold, and Sara Hayes. 2011. Avoiding the Train Wreck: Replacing Old Coal Plants with Energy Efficiency. Washington, DC: American Council for an Energy-Efficient Economy.

Elliott, R. Neal, Maggie Molina, and Dan Trombley. 2012. A Defining Framework for Intelligent Efficiency. Research Report E125. Washington, DC: American Council for an Energy-Efficient Economy.

EMC2. 2010. "Ms. Winkler Goes to Washington." http://www.emc.com/leadership/features/winkler-goes-to-washington.htm. From ON Magazine, (1) 2010.

ENERGYSTAR. 2013. "Energy Star. Power Management." Washington, DC: U.S. Environmental Protection Agency. http://www.energystar.gov/index.cfm?c=power_mgt.pr_power_mgt_low_carbon_join.

Envision Charlotte. 2010. "Energy Program." Accessed February 16, 2103. http://www.envisioncharlotte.com/energy-program/.

Evans, Peter C. and Marco Annuziata. 2012. "Industrial Internet: Pushing the Boundaries of Minds and Machines." Fairfield, CT: General Electric Company.

Fernandez, N. H. Cho, M.R. Brambley, J. Goddard, S. Katipamula and L. Dinh. 2009 Self-Correcting HVAC Controls Project Final Report. PNNL-19074. Richland, WA: PNNL.

Foster, Ben, Anna Chittum, Sara Hayes, Max Neubauer, Seth Nowak, Shruti Vaidyanathan, Kate Farley, Kaye Schultz, and Terry Sullivan. 2012. The 2012 State Energy Efficiency Scorecard. Washington, DC: American Council for an Energy-Efficient Economy.

Friedrich, K., M. Eldridge, D. York, P. Witte, and M. Kushler. 2009. Saving Energy Cost-Effectively: A National Review of the Cost of Energy Saved Through Utility-Sector Energy Efficiency Programs. Washington, DC: American Council for an Energy-Efficient Economy.

General Mills. 2013. "Global Responsibility 2013." http://www.generalmills.com/~/media/Files/CSR/2013_global_respon_report.ashx.

Goldman, Ethan (Vermont Energy Investment Corporation). 2013. Personal communication June 11.

Google. 2011. Google's Green Computing: Efficiency at Scale. http://static.googleusercontent.com/external_content/untrusted_dlcp/www.google.com/en/us/green/pdfs/google-green-computing.pdf

Henderson, Phil and Megan Waltner. 2013. Real-Time Energy Management. A Case Study of Three Large Commercial Buildings in Washington, D.C. National Resources Defense Council.

Holler, Anne. 2010. "The Green Cloud: How Cloud Computing Can Reduce Datacenter Power Consumption." Presentation at SustainIT10, Feb. 22, 2010, San Jose, CA. http://www.usenix.org/event/sustainit10/tech/slides/holler.pdf.

[IPMPV] International Performance Measurement & Verification. 2002. Protocol Concepts and Options for Determining Energy and Water Savings. Volume 1. http://www.nrel.gov/docs/fy02osti/31505.pdf.

[ITIC] Information Technology Industrial Council. 2013a. "ITI Background." http://www.itic.org/about/.

———. 2013b. "Standards." http://www.itic.org/public-policy/standards.

[JCI] Johnson Controls. 2013. "Panoptix by Johnson Controls." Accessed 9/25/2013. https://whatspossible.johnsoncontrols.com/community/panoptix.

Katipamula, S., D.P. Chassin, D.D. Hatley, R.G. Pratt and D.J. Hammerstrom. 2006. PNNL Transactive Controls: Market-Based GridWise™ Controls for Building Systems. PNNL-15921.

KONE. 2013. "EcoMod™: A Full-Replacement Solution for Existing Buildings." http://cdn.kone.com/www.kone.us/Images/kone-case-study-morgan-post-office.pdf?v=2

Kundra, Vivek. 2011. Federal Cloud Computing Strategy. Washington, DC: U.S. Office of Management and Budget. https://cio.gov/wp-content/uploads/downloads/2012/09/Federal-Cloud-Computing-Strategy.pdf.

Kwatra, Sameer, Jennifer Amann, and Harvey Sachs. 2013. Miscellaneous Energy Loads in Buildings. Report Number A133. Washington, DC: American Council for an Energy-Efficient Economy.

Laitner, S., S. Nadel, N. Elliott, H. Sachs, and S. Khan. 2012. Long Term Energy Efficiency Potential. Washington, DC: American Council for an Energy-Efficient Economy.

Lee, Eleanor. 2006. Advancement of Electrochomic Windows. LBNL PIER Final Project Report. CEC-500-2006-052. Lawrence Berkeley National Laboratory. http://www.lbl.gov/Science-Articles/Archive/sabl/2007/Jan/Advance-EC-Windows.pdf

Lydon, Bill. 2011. "ODVA Industrial Networks Energy Initiative." March 20. Automation.com. http://www.automation.com/automation-news/article/odva-industrial-networks-energy-initiative. Automation.Com

M2M.WorldNews. 2012. "Machine to Machine Connections to Hit 18 Billion in 2012 Generating $1.2 Trillion in Revenue." Machina Research. http://m2mworldnews.com/2012/11/29/28546-machine-to-machine-connections-to-hit-18-billion-in-2022-generating-usd1-2-trillion-revenue/.

Manyika, James and Charles Roxburgh. 2011. The Great Transformer: How the Internet Is Changing the Globe and its Citizens. McKinsey Global Institute.

McCown, Paul. 2009. "Continuous Commissioning® of a LEED-EB Gold Certified Office Building." In Proceedings of the Ninth International Conference for Enhanced Building Operations. Austin, TX.

McKenney, K., M. Guernsey, Ratcharit Panoum and Jeff Rosenfeld.

(2010). Commercial Miscellaneous Electric Loads: Energy Consumption Characterization and Savings Potential in 2008 by Building Type. Prepared for the U.S. Department of Energy. Washington, DC: TIAX.

Microsoft. 2013. "88 Acres. How Microsoft Quietly Built the City of the Future." http://www.microsoft.com/en-us/news/stories/88acres/88-acres-how-microsoft-quietly-built-the-city-of-the-future-chapter-1.aspx

Mills, Evan. 2009. A Golden Opportunity for Reducing Energy Costs and Greenhouse Gas Emissions. LBNL-3645E. Berkeley, CA: Lawrence Berkeley National Laboratory.

Monsalves-Salazar, Cristian (NStar). 2013. Personal communication. August 28.

Nadel, Steven, Miriam Pye, and Jennifer Jordan. 1994. Achieving High Participation Rates: Lessons Taught by Successful DSM Programs. Research Report U942. Washington, DC: American Council for an Energy-Efficient Economy.

Navigant. 2012. "Global Revenues for Commercial Building Automation Systems Will Reach $146 Billion by 2021." Feb. 7. http://www.navigantresearch.com/newsroom/global-revenues-for-commercial-building-automation-systems-will-reach-146-billion-by-2021.

———. 2013a. Commercial Building Automation Systems. http://www.navigantresearch.com/research/commercial-building-automation-systems.

———. 2013b. "Building Energy Management Systems: IT-Based Monitoring and Control Systems for Smart Buildings: Global Market Analysis and Forecasts." http://www.navigantresearch.com/research/building-energy-management-systems.

NEEA. 2013a. "Luminaire Level Lighting Controls." http://neea.org/initiatives/emerging-technology/luminaire-level-lighting-controls.

———. 2013b. "Industrial Initiatives." http://neea.org/initiatives/industrial.

Nesler, Clay (Johnson Controls, Inc.). 2013. Personal communication. September 9.

NYSE Magazine. 2011. Inside Smart Manufacturing. Second Quarter. http://www.rockwellautomation.com/resources/downloads/rockwellautomation/pdf/about-us/company-overview/ManufacturingIntelligence.pdf

ODVA. 2011. "Leading Industrial Suppliers, ODVA Unite to Outline Best Practices for Managing Energy Data." Ann Arbor, MI. Feb. 7. http://www.odva.org/Home/tabid/53/ctl/Details/mid/372/ItemID/73/lng/en-US/language/en-US/Default.aspx

OpenADR Alliance. 2013. "Overview." http://www.openadr.org/about-us.

OsramSylvania. 2013. "Osram Sylvania and Encelium Showcase Industrial Leading Light Management Systems at LIGHTFAIR International." http://www.sylvania.com/en-us/newsroom/press-releases/Pages/industry-leading-light-management-systems.aspx.

Otis. 2011. "Otis Elevator Company Introduces Energy-Efficient Escalator amid Significant Environmental Gains in its 'The Way to Green' Program." Oct 5. Press Release. Toronto. 2011/PRNewswire/http://www.prnewswire.com/news-releases/otis-elevator-company-introduces-energy-efficient-escalator-amid-significant-environmental-gains-in-its-the-way-to-green-program-131138313.html.

Reid, M., C. Harper, and C. Hayes. 2008. "Finding Benefits by Modeling and Optimizing Steam and Power Systems." Air Liquide Large Industries U.S. LP. http://svmesa.com/pdfs/air-liquide-finding-benefits-by-modeling-and-optimizing-steam-and-power-systems-visualmesa.pdf.

Rockwell Automation. 2011. "CIP Energy—Fact Sheet." http://www.sustainableplant.com/assets/CIP-Energy-Fact-Sheet.pdf. Milwaukee, WI: Rockwell Automation, Inc.

Salesforce.com. 2012. Sustainable Company Sustainable World: Salesforce.com Sustainability Report FY2012. Secure.sfdcstatic.com/assets/pdf/misc/SustainabilityReport.pdf.

Schindler. 2013. Schindler Escalator Energy Efficiency Manager. Reduced Consumption. Increased Savings. http://www.schindler.com/content/us/internet/en/modernization/escalator-upgrades/_jcr_content/rightPar/downloadlist/downloadList/161_1336059897618.download.asset.161_1336059897618/EscalatorEnergyEffiiencyMgr_061810%20%20EQN-1006.pdf,

Schneider Electric. 2013. "EcoStruxure." http://www.schneider-electric.com/solutions/ww/en/edi/4871808-ecostruxure.

[SEP] Superior Energy Performance. 2013. "Achieving Superior Energy Performance, Overview." http://www.superiorenergyperformance.net/.

Singer, B. C. and W. F. Tschudi. 2009. High Performance Healthcare Buildings: A Roadmap to Improved Energy Efficiency. Berkeley, CA: Lawrence Berkeley National Laboratory.

Sinopoli, Jim. 2010. "FDD Going Mainstream? Whose Fault Is It?" Building.Com. April http://www.automatedbuildings.com/news/apr10/articles/sinopoli/100329091909sinopoli.htm.

SmartGrid News.com. 2013. "The Biggest Overlooked Metering Market: It's behind the fence." Groom Energy Research. August 1. http://www.smartgridnews.com/artman/publish/Technologies_Metering/The-biggest-overlooked-metering-market-of-all-It-s-behind-the-fence-5919.html?utm_medium=email&utm_source=Act-On+Software&utm_content=email&utm_campaign=Consumers+really+don%27t+want+dynamic+pricing%3F+An+expert+answers&utm_term=T-The+biggest+overlooked+metering+market+of+all%3F+%28It%27s+behind+the+fence%29&cm_mmc=Act-On+Software-_-email-_-Consumers+really+don%27t+want+dynamic+pricing%3F+An+expert+answers-_-T-The+biggest+overlooked+metering+market+of+all%3F+%28It%27s+behind+the+fence%29#.UfqdRtLVDL9

[SMLC] Smart Manufacturing Leadership Coalition. 2011. Implementing 21st Century Smart Manufacturing. June 24. Los Angeles, CA: University of California Los Angles. https://smart-process-manufacturing.ucla.edu/about/news/Smart%20Manufacturing%206_24_11.pdf.

———. 2013a. "About SMLC." https://smartmanufacturingcoalition.org/about. Los Angeles, CA: University of California Los Angles.

———. 2013b. "Economic Benefit." https://smartmanufacturingcoalition.org/economic-benefit. Los Angeles, CA: University of California Los Angles.

Wang, Shengwei. 2010. Intelligent Buildings and Building Automation. Spon Press.

Ye, Jason and Stephen Seidel. 2012. Leading by Example: Using Information and Communications Technologies to Achieve Federal Sustainability Goals. Washington, DC: Center for Climate and Energy Solutions.

Wang, W., Y. Huang, S. Katipamula, M.R. Brambley. 2011. Advanced Control Strategies for Packaged Air-Conditioning Units with Gas Heat. Richland, WA: Pacific Northwest National Laboratory. PNNL 20955

Warrick, Jennifer. 2013. "88 Acres: How Microsoft Quietly Built the City of the Future." Realcomm, Advisory Topic: 13 (16). April 18.

Whitehouse.gov. 2011. "Modeling a Green Energy Challenge after a Blue Button." Posted by Aneesh Chopra on Sept. 15. http://www.whitehouse.gov/blog/2011/09/15/modeling-green-energy-challenge-after-blue-button.

———. 2013. "Green Button: Enabling Energy Innovation." Posted by Monisha Shah and Nick Sinai on May 2. http://www.whitehouse.gov/blog/2013/05/02/green-button-enabling-energy-innovation.

SECTION II

Current Technology, Tools, Products, Services, and Applications

This topic of Automated Diagnostics and Analytics for Buildings is one of the most exciting and dynamic areas of technological advancement in our field of energy and facility management today. With the widespread availability of high speed, high capacity microprocessors and microcomputers with high speed communications ability, big data from building sensors and smart meters, and sophisticated energy analytics software, the technology to support deployment of automated diagnostics is now available, and the opportunity to apply automated fault detection and diagnostics AFDD) to every system and piece of equipment in a facility, as well as for whole buildings, is imminent. The purpose of this section of the book is to share information with a broad audience on the state of automated fault detection and diagnostics and analytics for buildings applications, the benefits of these applications, examples of field deployments, automated diagnostic tools presently available, guidance on how to use automated diagnostics and analytics, and related issues.

Most of the chapters in this section of the book are from authors whose companies provide commercially available software tools for AFDD and Analytics for buildings. They have been gracious enough to share this information with us, and to tell us what their software or services will do to help us operate our buildings more energy efficiently and more cost effectively. In most cases they provide us with case studies and detailed examples of how existing users of their software or services have reduced the energy use of their buildings, and how they have located opportunities to correct problems and to improve the operation of their buildings with their sophisticated tools.

The editors hope that this information is helpful to energy managers, facility managers, building owners and operators, in assisting them in deciding on implementing this new technology to help them operate their buildings.

Chapter 6

Automated Diagnostics
This time for sure!

Steve Tom, Automated Logic, Corp

ABSTRACT

Automated diagnostics have the potential to dramatically reduce energy consumption in building HVAC systems while improving performance, extending equipment life, and ensuring that the occupants of these buildings are comfortable and productive. Automated diagnostics also have the potential to overwhelm building maintenance staffs with a staggering number of repetitive, confusing, and ultimately useless messages. In short, they could become just as irrelevant as their forebears—alarms. To be useful, automated diagnostics need to focus on issues the customer really cares about, present information in an easily understandable format, identify the root causes of problems, and above all avoid alarming the customer with false alarms.

INTRODUCTION

I have spent more time in the HVAC controls business than I care to admit, and during all those years one basic rule has remained unchanged: The first step in any HVAC energy program should be to get the system to run as designed. I say "should" because this step is often overlooked. After all, it's not very exciting. Who wants to clean a coil when you could be installing a new high efficiency cooling unit? HVAC managers don't get kudos when they spend a little money on routine maintenance to keep equipment from wearing out. They gain power and prestige when they have a huge budget to replace worn out equipment. And maintenance technicians don't get noticed when they replace a fan belt before it breaks. They become heroes by working overtime to replace it after it breaks on a hot August afternoon.

I had a friend who went into the energy service contracting business. He contracted with building owners to reduce their energy costs for a percentage of the savings. When he first started, he achieved huge savings by fixing stuck dampers, replacing leaking valves, and curing simple problems that were wasting energy. His customers were not happy. "You didn't do anything we couldn't have done!" they complained. This was during the dawn of the DDC age, so he began installing a rudimentary DDC system as well as fixing all the problems that were wasting energy. When he did that, his customers regarded him as a genius. He had installed a magic box with blinking lights, and they attributed all the energy savings to that magic box. They were happy because that was something they couldn't have done by themselves.

How much energy can you save by getting equipment to run the way it was designed to operate? That depends on how badly it's deteriorated. NIST estimates that effective fault detection and diagnostics of HVAC equipment could save 10% to 40%[1]. I have seen equipment with even greater potential for savings. Look closely at case studies on energy saving technologies and sometimes you can find unintended testimony to the effectiveness of maintenance and repair. One such study published during the early days of DDC claimed the installation of a "magic box" had cut energy consumption by 80%! If you read the report closely, however, you discovered the first thing they did was to clean coils, calibrate sensors, adjust fan speeds, and otherwise overcome years of neglect. This produced a 40% savings before they even installed their DDC controller. Installing the DDC unit produced an additional 40% savings, but if you looked at the control algorithms the only "new" concept they introduced was to turn the HVAC off on nights and weekends, when the building was unoccupied. The previous control system had a time clock that was supposed to do that but no one had ever installed the pins to make it work.

So, automated diagnostics have a tremendous potential to save energy. And don't overlook the fact that by ensuring equipment runs the way it was designed you're doing much more than just saving energy. The health and productivity of the people inside a building depend upon the HVAC system working as designed. An outside air damper that is stuck closed will save energy during extremely hot or cold weather, but it won't bring in the fresh air that's required for ventilation. Similarly, a DX coil that tripped off due to low refrigerant is saving energy, but it's making the building occupants sweat. Many studies have shown that the cost of lost productivity when workers are uncomfortable far outweighs the cost of all the energy used by the building.[2]

In short, automated diagnostics have the potential to dramatically improve the way buildings operate. Tools that are available today can automatically detect problems that have remained hidden for years, and bring these problems to the attention of the people who can fix them. The results will be significant energy savings, lower operating costs, a healthier and more productive environment for people inside the building, and sharp reductions in greenhouse gas emissions. As someone who has long advocated for better maintenance programs I should be ecstatic. Instead, I'm cautiously optimistic. Why? Because for years we've had a tool which could have accomplished at least some of this but in many cases we've failed to use it. That tool is called "Alarms."

Alarms are the "Rodney Dangerfield" of building automation. They get no respect. At best, they get ignored. Often they get disabled. The fault does not lie with the alarms themselves. Alarms are not nearly as sophisticated or as capable as automated diagnostics, but they can serve to call our attention to operational problems that are costing money and degrading performance. The fault lies with the way we as an industry have implemented alarms. Instead of designing alarms to help the customer focus on the most important problems, we have alarmed the things that were easy to alarm and ignored problems that were hard to detect. The result is that we have overwhelmed the user with meaningless alarms, often to the accompaniment of flashing text and annoying beeps.

What did you do the last time you heard a car alarm go off in a parking lot? Did you call 911? Did you rush toward the sound to see if you could chase away the thief? Or did you ignore it and go about your business, secure in the knowledge that it was probably a false alarm? If you ignored it, you reacted the same way many building operators react when they receive an alarm that says "It's too hot in room 243." Especially if it has gotten too hot in room 243 every day since they turned it from a private office to a copy/break room. That little VAV box just can't keep up with two copy machines, three mi-

crowaves, and an industrial coffee maker. Now imagine how excited that operator will be about room 243 if the chiller plant is down and every room in the building is sending him alarms because it's too hot.

Automated diagnostics are much more sophisticated than alarms, and they have the potential to be much more useful. However, like any computer tool they are subject to the "garbage in = garbage out" syndrome. If we want automated diagnostics to solve the problems that alarms never cured, we have to look at how alarms were misused and learn from those mistakes. I wish I could supply carefully researched, well documented treatise on "lessons learned" from alarming. Unfortunately, I am not aware that anyone has studied the failure of building automation alarms. In the absence of scientific data, I will present my own opinionated guidelines for intelligent diagnostics of building problems. These guidelines could apply to diagnostics algorithms that are scanning data uploaded from a building automation system (BAS) or to diagnostic algorithms that are running in the BAS controllers. The first method falls under the general heading of "Analytics" (a marvelously flexible term that means different things to different people) while the second method is more of a "hyper-intelligent alarm package." For simplicity I will describe the guidelines as though they were being executed in BAS controllers, but the principles are the same for Analytics or Alarms.

DON'T GENERATE FALSE ALARMS

The main purpose of automated diagnostics is to detect problems the building owner is not aware of which waste energy and cause substandard performance. These problems will not go away overnight. Tune your detection criteria to look for repeated or long-standing performance problems. Don't cry "wolf" at the first indication that there may be a problem. The building owner and maintenance crews will quickly lose confidence in the system if it sends them on a wild goose chase, trying to fix problems that don't exist. For example, when a DX coil turns on the discharge air temperature should drop. If the unit is just starting up on a hot day, however, a slug of hot air from the return duct may overwhelm the DX coil and cause the discharge air temperature to rise. Notifying the user of this "failure" will do nothing except erode confidence in the system. DX coils typically cycle on and off several times per day or per week, so if your diagnostic routine is looking for repeated failures a coil that is well and truly dead will eventually fail to cool the discharge air enough times to warrant reporting. Ideally, the automated diagnostic system would provide the user with a simple way to adjust the sensitivity so any diagnostic routine that is too sensitive can be turned down without being turned off.

The fact that a slug of hot air can "fool" diagnostics when equipment is first started is an example of why it is usually not a good idea to look for problems on start-up. Unless you're specifically looking for start-up problems, wait until conditions have stabilized before enabling diagnostic routines. "Stable" conditions also mean there haven't been any recent changes to a setpoint, operating mode, or other control parameters. An intelligent fault detection routine will include a set of "enabling conditions" that prevent its application during transient operations. Again, the goal is to identify equipment that is truly broken, not just in the process of changing from one mode to another. If the equipment is broken, it will still be broken after the operation stabilizes and the fault detection will catch it then.

A "filter" that can help eliminate false alarms is to look for repeated or prolonged fault conditions. Even carefully thought-out detection logic with intelligent enabling conditions can be fooled once in a while by unexpected circumstances. On the other hand, a broken piece of equipment will fail every time it tries to run. A diagnostic that only calls the operator's attention to a problem if it's occurred, say, 5 times in the last 7 days is less likely to generate a false alarm than one which calls out every time it senses a problem.

FOCUS ON PROBLEMS THE CUSTOMER CARES ABOUT

One of the easiest problems in the world to detect is a room that's too hot or too cold. It's been my experience that this is also one of the most commonly disabled alarms in a system. It's not that the facility staff doesn't care that rooms are uncomfortable, it's that alarms are often so poorly configured that they get hundreds of hot and cold zone alarms when the system is in fact running as designed. When they do have a major problem, such as a failed chiller plant, the hundreds of "hot zone alarms" that flood their screen do nothing except annoy and distract the technicians who are trying to diagnose the real problem. There may be a few high priority rooms that need to be alarmed, such as an Operating Room in a hospital or an Executive Boardroom in a business office, but for most zones the alarm the maintenance crew depends upon most is the voice of an unhappy occupant

on the telephone.

Ideally, you should talk to the people who will actually be responding to the alarms to find out what they're most interested in. While this may sometimes be possible, especially in a retrofit application, in many situations the people who are configuring a fault detection/alarm program have virtually no contact with the people who will be using it. If that is the situation, the best the designer can do is to focus on common problems which will remain undetected without this system and which can cause serious energy waste or occupant discomfort. Examples of this type of alarm include:

Stuck Dampers

Outside air dampers are particularly prone to getting stuck because they're exposed to extreme weather conditions which can lead to rust, but this fault can afflict any type of damper. Sometimes the dampers bind and refuse to move. Other times the actuator linkage comes loose. The net result is the same—the control logic commands the damper to move to a certain position but the damper doesn't move. Often the system graphics are keyed to the commanded position, so if you simply look at the graphic it looks like the damper is in the correct position, but it isn't. If there's a flow sensor downstream of the damper you can use fault detection logic that compares the flow with the commanded position. When the damper opens, the flow should increase and vice versa. If there's no flow sensor, it's often possible to detect damper problems by examining conditions when the damper is commanded to 100% or 0%. For example, an outside air economizer often commands the outside air damper to be 100% closed during morning warm-up. Under those conditions, the mixed air temperature should be equal to the return air temperature. If there's a significant difference, the outside air damper may be stuck open. Similarly, when the control logic calls for 100% outside air the mixed air temperature should be equal to the outside air temperature. If there's a significant difference, the outside air damper may not be opening fully.

Leaking Coils

Like dampers, coils can sometimes get stuck and not move to their commanded position. Perhaps more commonly, valve seats can erode with time and allow water or steam to pass even when the actuator has closed them. It's difficult to tell for certain that a valve isn't opening fully, but a coil which is chronically unable to maintain setpoint may be an indication of a slipped linkage or similar problem. A leaking valve, on the other hand, can best be detected when it's commanded shut. A significant temperature rise across a closed heating coil (or a drop across a closed cooling coil) typically indicates a leaking valve. Either the actuator isn't closing it completely, or the seat has eroded. This is the type of problem that can go unnoticed for years and waste a significant amount of energy, as the other coil will open further to compensate for the leaking valve. For example, a leaking hot water valve in an Air Handling Unit will not only waste energy by heating the air unnecessarily, it will also drive the unit into a cooling mode and force the cooling valve to open. Neither the system graphic nor the control logic will indicate the system is heating and cooling simultaneously, but that is in fact what is happening.

DX Stages Not Working

Multi-stage DX units turn cooling stages on and off as required to cool a building. There are many problems that can cause a DX stage not to run when it's commanded on. Low refrigerant, loose wiring, high head pressure, and a burned out compressor are typical causes for DX stage failures. When a stage fails in the spring or the fall, the controls simply turn on the next stage to provide the needed cooling and nobody notices. Or at least, they don't notice until the weather turns really hot. Then the remaining stages can't keep the building cool, and fixing the bad stage becomes an emergency. A good fault detection algorithm can spot the problem before it becomes an emergency. Conceptually this is an easy problem to look for, if the equipment interface lets you know when stages turn on or off. When a stage turns on, the discharge air temperature should drop say, ½ degree within 5 minutes. This test can be affected by factors such as sudden changes in the incoming air temperature, so it's a good idea to use a filter which only trips on repeated failures.

Rogue Schedules

A "rogue" schedule is an operating schedule which is significantly different than the actual occupancy schedule. The cause is more often human error than equipment failure, but the end result is an unnecessary waste of energy. Sometimes rogue schedules result from operational changes. A manufacturing area which once ran two shifts is now only running one shift, but no one thought to revise the HVAC schedule to remove the second shift. Or perhaps a special function was planned for one particular Saturday, and the person who entered the schedule accidentally programmed the system to run every Saturday. Sometimes equipment is overridden "on" for maintenance reasons and the person who set

the override forgot to remove it when the maintenance was complete. The equipment runs when no one is in the building, and since no one is there no one notices the equipment is running.

Detecting rogue zones can easily be accomplished by using a run time meter or an operating log to record when the equipment is running, and alerting the operator whenever the equipment runs more than "X" hours per day. The threshold "X" can be lower for weekends than for weekdays if the building is not normally occupied over the weekend. While this is a very simple test, the savings can be substantial. In particular, buildings with older building automation systems (BAS) or with no BAS often run 24/7 even though they are only occupied 8/5. Retrofit installations of new BAS often pay for themselves based on the savings obtained through scheduling alone, but if there is no automated system to detect rogue schedules the savings will soon dwindle. Entropy is a basic fact of life, even in something as simple as an operating schedule.

Sensor Calibration Errors

Descriptions of automated fault detection often list detection of sensor calibration errors as a way to save energy. While it is true that sensor errors can waste energy, detecting these errors can be very difficult. Detecting sensor errors generally requires multiple sensors which, under certain operating conditions, have a known relationship. For example, in an economizer the mixed air temperature should be between the outdoor and the return air temperatures. If it is not, one of the sensors is out of calibration. The question is, which one? That may not be an important question, since regardless of which sensor is in error the solution is to go to the air handler and use a calibrated instrument to get an accurate temperature reading of the airflow in question and recalibrate the installed sensor. Once you're at the air handler with a calibrated thermometer, you might as well recalibrate all the sensors, not just the one you suspect is in error. Other opportunities to detect sensor calibration errors arise when you have multiple pieces of equipment which should have similar readings. For example, if the VAV boxes in a particular system have discharge air temperature sensors, the discharge temperatures of all sensors fed by the same air handler unit should be similar. If one or two are significantly different (and no reheating is taking place) the accuracy of those sensors is suspect. Unfortunately, most systems also include flow sensors, humidity sensors, CO_2 sensors, and other critical sensors which cannot be compared to any other sensor in the system. Thus, while automated fault detection can detect some sensor errors, it does not remove the need for periodic recommissioning of equipment.

Simultaneous Heating and Cooling

This is another type of fault that is often included in descriptions of automatic fault detection, but in this case the goal is to find errors in the control algorithms rather than malfunctioning equipment. It is easy to monitor the status of a heating coil and a cooling coil in a single piece of equipment and notify the operator if they are both operating simultaneously. It is also easy to monitor the status of heating and cooling elements in related equipment to see if a system is simultaneously heating and cooling, as occurs when a VAV air handling unit is cooling air which is being supplied to VAV boxes with active reheat coils. It is a little more difficult to detect if unnecessary heating and cooling is taking place, as some systems are designed to simultaneously heat and cool. Dehumidification, for example, requires air to be cooled below its dew point to remove moisture, and this air may then need to be reheated to maintain an acceptable space temperature. The aforementioned VAV systems are another example where simultaneous heating and cooling may be intentional, especially so since many VAV air handling units either do not have heating coils or do not use them to operate in a heating mode. If some zones need heat, most commonly the exterior zones in a building, this system will intentionally provide simultaneous heating and cooling. Dedicated Outdoor Air Systems (DOAS) are another example of a system that is typically designed to provide cooling only, and which will intentionally perform simultaneous heating and cooling if any zones need heat.

If a system is designed to simultaneously heat and cool, an automated fault detection which alarms on this condition will do nothing except annoy the operator. The key is to detect conditions where the system is unnecessarily performing simultaneous heating and cooling. For example, if no VAV zones are above their cooling setpoint and some perimeter zones are actively heating, there is no reason for the air handler unit to provide mechanical cooling or to provide more than the minimum amount of outdoor air. A useful fault detection system would therefore need to look at all of the zones served by the air handler unit, determine if simultaneous heating and cooling was necessary, and alarm if it was occurring but not needed. If the control system is not difficult to program, it would also be advisable to apply this same logic to the control algorithms, using reset schedules and interlocks as required to prevent this situation from occurring in the first place.

These are just a few examples of how automated diagnostics can be applied to common problems. Obviously there are many more faults that can be detected. Hydronic systems, chiller plants, heat pumps—all have different conditions that can indicate alarms. The important thing is to focus on problems that are most common, that have a significant impact on energy use or system performance, and that are not likely to be discovered without automated diagnostics. Most customers do not have an unlimited budget to spend on programming diagnostic systems and they do not have an unlimited maintenance staff. Don't waste their money writing diagnostic routines to detect once-in-a-lifetime events, and don't overwhelm their maintenance staff by reporting problems with little or no impact.

USE HIERARCHICAL FAULT SUPPRESSION

Within a building or a campus it is very common to have multiple pieces of equipment working together as a system. If one piece of equipment isn't working right, it can affect the rest of the system. For example, a VAV box uses cold air from an air handler to cool the room beneath it. If a fan belt breaks in the air handler the cold air stops flowing to the VAV box and the room is going to get hot. There's no point in having automated diagnostics to tell the operator that the room is hot because the problem is not in the room. Every room supplied by that air handler is going to get hot, and if the system annoys the operator with complaints from every room that's too hot he's liable to miss the important diagnostic message, the one from the air handler that says the fan belt broke. Similarly, the air handler can't supply cold air unless the chiller plant is supplying it with chilled water. If the chiller plant goes down, every room supplied by that chiller is going to get hot. An unintelligent diagnostic system could overwhelm the operator with thousands of complaints about hot rooms. The one useful needle in that haystack, the diagnostic message that describes the problem in the chiller plant, is almost sure to get overlooked amid the torrent of useless messages.

One way to keep useless diagnostic reports from flooding the system is to implement hierarchical fault suppression. In the system just described, the chiller is at the top of the "hierarchy" of cooling equipment. The chiller plant diagnostics determine whether or not the chiller and its associated pumps are supplying cold water to the system. If they aren't, the chiller sends a "cold water failure" message to the equipment below it in the hierarchy. Not every problem in the chiller plant requires a "cold water failure" alert. The automated diagnostics might detect a vibration alarm in the chiller unit, a warning that it's time to change the belts in the cooling tower, or the fact that the primary pump failed to start and the system is running on the back-up pump. These indicate potentially serious problems in the chiller plant, but they don't prevent the chiller from supplying cold water so they wouldn't send a failure message to other equipment.

At the air handling unit, cooling diagnostics that depend upon chilled water are disabled if it receives a "cold water failure" message. Diagnostic routines that check heating operation, fan operation, and other non-cooling functions may still generate alarms, but the air handling unit will not generate cooling alarms if it is not receiving chilled water. It will, however, send a "cooling failure" message to the VAV boxes. It will also send this failure message if the fan is not running properly or if other problems prevent it from meeting its discharge air temperature and supply static pressure setpoints, even if chilled water is available. At the VAV boxes, cooling diagnostics will not generate alarms if a "cooling failure" message is received from the air handler.

Similar messages can be used in hot water systems, supply air systems, and other systems where the performance of one piece of equipment depends upon the performance of another. The goal is to only generate alarms from the root cause of a problem—not from the hundreds of symptoms that develop as a result of this problem.

PRESENT INFORMATION, NOT DATA

Once the automated diagnostics system has detected problems, it's important to present it to the user as useful information, not just random data. This is one area where conventional alarm packages often fall woefully short. A chronological list of alarms from all the buildings in the system provides very little guidance about what the operator ought to do first. At the very least, the operator ought to be able to sort the alarms by criticality and location. Alarm totals can be a useful prioritization tool, too. For example, a summary view that shows that during the past 30 days there were 7 critical alarms in AHU-1 and 3 critical alarms in AHU-2 is more helpful than expecting the user to scroll through a list of all the alarms in the system counting instances. Similarly, when the user looks at the alarms in AHU-1 show him that there were 3 times that it failed to maintain cooling setpoint, 2 times that it failed to maintain static pressure,

1 leaking heating valve and 1 leaking OA damper. Allow the operator to dive into a detail view if he wants to know when the 3 cooling alarms occurred, but at the summary level he just needs to know that there were 3 occurrences.

An estimate of the severity of the fault can also be very useful for prioritizing maintenance, but users need to be aware that this is only an estimate. Most users would love to have their automated diagnostics system tell them that their leaking OA damper was costing them $300 per month and be confident that if they replaced the damper they'd see their energy bill drop by that amount. The truth is that how much a leaking damper costs depends upon the weather conditions, the cooling load on the unit, and the cost of electricity. All of these conditions change throughout the day. Under some conditions, the leaking damper costs nothing. At other times it can be quite expensive. Ideally the diagnostics system would precisely measure the leakage; profile the unit's operating hours, cooling load, and other conditions over the past year, and use a sophisticated energy model to predict how much that leak costs per year. From a practical standpoint, a rough estimate based on typical operating conditions and average monthly run hours will provide the user with the information he or she needs to compare this fault to other faults in the system and fix the worst faults first. Just make certain the user understands this is only a rough estimate. Like the EPA mileage estimates, "your numbers may vary."

As mentioned previously, don't forget to estimate the human cost as well as the energy cost. An outside air damper that won't open may actually save energy most of the year, but it can cause serious productivity losses, and could lead to high absenteeism and "sick building syndrome" lawsuits. Similarly, when temperatures stray outside the average worker's comfort level, typically around 72°F - 75°F, productivity can fall by over 1% per degree F.[3] While that may not seem like a significant percentage, it's not unusual for the salaries of the people inside a building to be 100 times the total energy cost for the building. That means a 1% decrease in productivity will cost more money than could be saved by turning off all the lights, air conditioning, and other equipment. It's often difficult to get energy managers to consider these costs because they're hard to measure and they come out of someone else's budget, but a diagnostic report that estimates the energy cost of a stuck damper while ignoring the human cost is being "penny wise and pound foolish."

To the maximum extent possible, diagnostic systems should try to identify the root cause of problems.

An alarm that simply says "the room is too hot" doesn't provide much information. The first reaction of many people is to lower the setpoint, but chances are the room is too hot because it cannot maintain the current setpoint. Lowering the setpoint will do no good whatsoever. A more specific message like "the room is hot because the AHU is not in a cooling mode" lets the operator know the root cause of the problem is likely to be found in the AHU, not in the room. Why isn't the AHU in a cooling mode? Often there is more than one possible reason for a fault. For example, if an AHU is not in a heating or a cooling mode but the discharge temperature is significantly higher than the mixed air temperature it could be because the closed heating valve is leaking hot water into the coil, because the discharge air sensor is out of calibration, or because the mixed air sensor is out of calibration. If the sensors have recently passed the sensor calibration automated diagnostic test, then the most likely culprit is the heating valve. Conversely, if the boilers and hot water pumps are not running, a leaking valve will have no effect upon the temperature and the problem is sensor calibration. If the sensors have recently passed the test and the hot water system is turned off, the root cause is something unusual which required deeper investigation. (Is the fan motor on fire?)

The user interface should also allow the operators to place an alarm on "hold," so that it's not forgotten but is not aggressively demanding the operator's attention. Most maintenance organizations do not have the staffing to assign someone to correct every fault as it occurs. Instead, they use a combination of fault diagnostics and maintenance schedules to determine when to service a unit. A sensor that is slightly out of calibration may be deferred until the next scheduled maintenance, but a broken fan belt would require an immediate response. Once the technician has replaced the fan belt and resolved the emergency condition, he should be able to view the list of faults that are on hold and fix as many as possible while he is at that unit.

ADVANCED DIAGNOSTICS

An advanced diagnostic system should be able to look beyond the current conditions in a single piece of equipment. Sensor calibration is one area where diagnostics can be improved by taking a broader view. A single air handling unit probably only has one outdoor air sensor, but if the system includes multiple air handling units their readings should all be close. Similarly, if several air handlers serve a large common area, the return air temperatures should be similar. Comparing the operating efficiencies of similar pieces of equipment can also identify performance problems that may be caused by equipment faults, as can comparing the operating efficiency of a single piece of equipment over two similar time periods.

Advanced diagnostic routines could also take direct control of equipment and run through specific sequences designed to detect faults, rather than just waiting for the right conditions to occur naturally. For example, an air handler unit could close the outdoor air damper completely, turn off the heating and cooling coils, and check to insure the return air temperature, mixed air temperature, and discharge air temperature indicated the same temperature. It could then open the outdoor air damper, close the return air damper, and compare the outside air, mixed air, and discharge air temperature readings. It could open the heating valve and check to see if the discharge air temperature rose. It could similarly test the cooling coil, the fan speed control, and other components to insure they were working properly. Obviously it would be best to run this test when the building was unoccupied, say during a morning warm-up period. Tests involving the outdoor air damper should not be run during extraordinarily hot or cold weather, and other safety precautions would need to be observed. Within these limitations, however, an automatic self-test cycle could be extremely useful as an initial check prior to commissioning a piece of equipment, and it could also be run periodically to check for faults. Similarly, if the normal fault detection logic spotted a fault in an operating piece of equipment, an advanced diagnostics system could run a series of tests to more precisely identify the root cause of the fault.

In summary, automated diagnostics have the potential to dramatically improve the way buildings operate. They also have the potential to fail for the same reasons that alarms have failed. If we implement automated diagnostics intelligently, focusing on providing useful information about problems the customer cares about and not overwhelming the customer with meaningless data, trivial warnings, and false alarms we will have taken a giant step in the right direction.

References
1. A rule-based fault detection method for air handling units, Schein et al, National Institute for Standards and Technology, April 2006, http://fire.nist.gov/bfrlpubs/build07/PDF/b07023.pdf
2. Managing Energy and Comfort, Tom, Steve, ASHRAE Journal, June 2008
3. Olli Seppanen et al, Control of Temperature for Health and Productivity in Offices, Helsinki University of Technology, 2004

Chapter 7

Enabling Efficient Building Energy Management through Automated Fault Detection and Diagnostics

Kelsey Haas, Ezenics

INTRODUCTION

As building systems continue to become more intelligent, an increasing amount and variety of data can be extracted from them in order to enhance performance, increase efficiency, and decrease operating costs. Automated Fault Detection and Diagnostics (AFDD), is becoming increasingly popular within the intelligent buildings industry because providers of the technology are able to extract and normalize data, implement algorithms, and prioritize actionable results for their clients, which can generate significant ongoing savings throughout the life of the buildings that they manage. This chapter will review how today's AFDD providers harvest data from buildings, translate that data, manage assets, implement algorithms, and produce actionable, prioritized results for their clients. It will also explore how the integration of people, equipment, and processes can enhance these results and further increase savings.

HARVESTING DATA AND THE IMPORTANCE OF RELIABLE DATA STORAGE

Today's intelligent buildings contain more data-producing systems than ever before, and although connecting to and obtaining data from these systems was once an arduous task for engineers and facility managers, there are now many options (both traditional and non-traditional) available which can greatly automate and speed up the process.

The world of building automation and automated diagnostics is complicated, but without one process the application of complex analytics would not be possible. Reliable, continuous data storage is the foundation of all building systems analytics. Harvesting this data from a facility involves several key functions: establishing a reliable method of connecting to the various building systems that does not cause excess traffic or downtime on the client side, creating a continuous connection to the

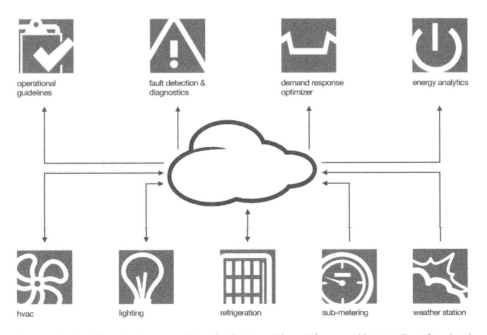

Figure 7-1: Facility Systems and Analysis Data Flow Diagram (© 2013 Ezenics, Inc.)

building systems data, and delivering a scalable, reliable data storage platform.

There are many types of data which can be harvested from a facility. Common sources of data include: HVAC, Refrigeration, Lighting, Power Meter, Utility, Financial, Transaction, Occupancy, and Weather. Traditionally, many AFDD providers have focused solely on HVAC-R and lighting system data available through Energy Management Systems (EMS), while power meter and utility data were utilized by energy managers, Demand Response (DR) Providers, and Energy Service Companies (ESCOs). However, with the variety of data types available to extract from a facility, it is necessary to aggregate these services into one platform in order to maximize the efficiency of the building and its inhabitants' processes. Implementing an AFDD system that utilizes a multitude of data and meta-data types to assess the building system allows facility managers to develop and employ processes to ensure the most efficient reaction to key performance indicators.

Connectivity: High-Level Overview

In a world where buzz words such as "big data" and "smart buildings" have become commonplace, we take for granted how complex the continuous transfer of a significant amount of data can become. The foremost questions that need to be addressed include: "what systems are installed in the building," and "which are connected to the Internet?" If a building system is not connected to the Internet or at least to a phone line, connectivity will not be possible without the installation of additional hardware. Conversely, if the devices are connected to the Internet, then establishing connectivity to them will be possible without having to install additional hardware or visiting the site. There are multiple ways of connecting to this data, whether it's through communicating with the device or controller directly, communicating with the EMS or building automation system (BAS), or communicating with another platform that is already extracting data from the facility.

The granularity of the data, e.g., the frequency at which the data are retrieved and stored, is an important consideration in building connectivity and analytics implementation. For building systems analytics, the higher the granularity of the data, the more descriptive and accurate the analytics will be. One minute storage frequency is considered "real-time" and is the best option for achieving the most accurate results. High data granularity is important when analyzing HVAC devices because it allows engineers to pinpoint precisely when devices turn on and off or commands are given to a vari-

able frequency drive (VFD). Energy data are typically stored and obtained at 15-minute intervals; however, at higher granularity, can enable peak load management, economic load control, or synchronized reserves events.

Connectivity Summary

There are many considerations and potential obstacles that need to be taken into account when connecting to and communicating with a building system, device, or platform, including:

Obtaining high granularity data from devices provides the ability to create more sensitive, advanced analytics and manage peak demand and energy usage more effectively.

the connection method and integration structure, building system types and availability, communication protocols, and granularity of data. With these considerations, there are many options and combinations applicable; however, there is one key success factor for any company that wishes to obtain data from a building: flexibility. AFDD and Energy Management vendors must be flexible enough to adapt to client security needs and existing hardware/software within their facilities, ensuring a cost-effective connectivity solution.

DATA NORMALIZATION

Once data from the facility has been extracted, there is a key step before applying analytics: data normalization. Data normalization is of the utmost importance because of the often high variability in data naming conventions. For instance, one system might label a data point as "Supply Air Temperature," while another system labels it as "Discharge Air Temperature." Each building system or device may label data differently. Therefore, the system that requests this data must be capable of translating it into a normalized form that makes sense to the client and facility management users.

One data normalization method is to first create a database of building system data that has already been identified and then apply standardized labels to new data that matches the previously encountered data. A variation on this method in-

Data translation provides normalization of incoming building systems data in a way that makes sense to users and allows for expedited, scalable setup of analytics.

cludes having an engineer review the data, work with the client to ensure accuracy and investigate questionable data, and then apply a standardized set of system labels.

Data normalization is not only important to the end user but it is also essential to expediting the setup process for building analytics. Regardless of the method used to normalize the data, a standardized system must be employed in order to ensure quick, scalable setup of facility data within a platform so that analytics can be applied.

ASSET MANAGEMENT

Another essential requirement for setting up AFDD analytics is asset management. Asset management allows the client as well as the provider of analytics to be able to keep track of valuable facility and equipment information, including: equipment type, size, number of compressors or pumps, brand, cooling and heating setpoints, schedules, zone, model number, serial number, minimum outside air damper position, refrigerant, etc. With this information, clients and providers alike can perform advanced queries in order to review specific pieces of equipment, establish patterns amongst facilities in the same region, review retrofits, and much more.

Asset management also provides the opportunity to gain vital information for more exact impact calculations and enables automatic exception and fault solving by utilizing pre-determined business rules. Furthermore, "what-if" scenarios can be run using this information to determine the energy impact of test cases before implementing strategies. With asset management tools, many tasks can be accomplished with building systems from providing a way for the facility manager or building owner to keep track of the current equipment within the building to allowing more exact savings figures to be provided from facility analytics.

OPERATIONAL GUIDELINES™ AND THE LOW-HANGING FRUIT

Once reliable, continuous data connectivity and storage have been configured and translation has been completed, algorithms can be implemented to detect anomalies in building operational conditions. These algorithms identify issues that are best described as

> *Operational Guidelines™ are a standard set of rules, as defined by the client, that specify how and when building systems should operate. When building systems operate outside of those rules, exceptions are reported, enabling significant realization of results through quick, easy fixes.*

"low-hanging fruit," i.e., quick, simple fixes that can result in significant realized savings.

Building use and conditions are constantly evolving, often requiring quick, temporary modifications to control settings in order to ensure occupant satisfaction, accommodate changing loads, and mitigate maintenance issues. If these changes are forgotten or left unchecked, they can potentially mask other issues, decrease energy efficiency, and persist long after the need for the temporary change has expired.

The Operational Guidelines (OG™) module performs continuous, automated commissioning of building systems and reports exceptions when it identifies that controls' programming does not comply with the approved optimization settings. This ensures that actual setpoints; schedules; Demand Management, ventilation, and humidity strategies; refrigeration setpoints such as temperature, defrost count, discharge pressure, door heater dew point; etc.; are implemented as designed, and that the devices are operating as intended. Being able to detect and correct setpoints that have been changed and identify devices that are operating outside of guidelines is critical to increasing efficiency and realizing savings.

Examples
Example 1

Henry, a facility manager, sets the unoccupied cooling setpoint for every zone on floor 3 of his building to an operational standard of 80°F in order to ensure minimal cooling and higher energy efficiency during the evening hours when there are no occupants in the facility.

One particularly hot and humid summer evening, a business manager, Barbara, needs to work late in her third floor office to meet a client deadline. Barbara is in a time crunch to get her work done, so she drinks a few cups of hot coffee and begins to feel uncomfortably warm in her office. She recalls that she can override the thermostat in her office in order to cool it down and make her office more comfortable, so she applies an override and alters the cooling setpoint to 72°F. Within about 10 minutes, her office has cooled down and become more comfortable, enabling her productivity to increase and allowing her to get her work done in time to meet her deadline. When she is finished, Barbara leaves the office at around 1:30am. Unfortunately, she forgot to reset the cooling setpoint in her office before she left. Over the next month, no one needs to be on the third floor office past the normal occupied time period, so there is no one there to notice that the unoccupied cooling setpoint in Barbara's office is still adjusting to

72°F every evening and wasting a significant amount of energy (and money).

With the OG™ module, the facility manager would have been alerted to the difference between the actual cooling setpoint and the standard operational guideline™ immediately after the change had been made. Henry could have utilized the OG™ module to assess what the operational standard setpoint was (80°F), what the actual setpoint was changed to (72°F), the impact that this change had on the facility's energy consumption, and how much money it cost the building owner. Furthermore, Henry would have been able to change the unoccupied cooling setpoint back to what it should be once Barbara was out of the office (or the OG™ could have been set up to correct the issue automatically according to key business rules) and instantly realize energy and monetary savings.

Example 2

During every holiday season at a large, centrally managed retail chain with 1,000+ buildings, the store hours are adjusted to allow them to stay open longer. Because the stores stay open longer, their unoccupied and occupied schedules, climate, and lighting are adjusted to accommodate the customers. After the holiday season is over, the store schedules must be set back to their normal operational standards. However, with more than 1,000 stores, it would be very difficult to manually keep track of what the normal and holiday schedules are and whether each store has its schedule set to the holiday schedule or the normal schedule. If a store's schedules are not properly set back to normal hours and setpoints, its lighting, HVAC, and refrigeration systems will continue to operate several hours longer than they should each evening, resulting in significant amounts of energy being wasted and higher utility bills.

By employing the OG™ module to automatically commission their 1000+ buildings, the large retail chain achieves several objectives:

1. Effective asset management, defining the operational standard schedules and special holiday schedules running at each store and on each device.

2. Automated identification of which stores or devices are running holiday schedules that deviate from the operational standard holiday schedule, as well as the energy consumption and monetary or comfort impact of these exceptions.

3. Automated identification of which stores or devices are still running the holiday schedules after the reversion to the normal operational schedules was supposed to take place and the energy consumption and monetary impact of these exceptions.

4. Manual or automated control of the stores and devices that are operating outside of the guidelines and realization of significant savings through correcting the exceptions.

Operational Guidelines™ Summary

Whether implemented for a single building or multi-national client with thousands of locations in their portfolio, the OG™ module provides scalable asset management capabilities and automated, continuous commissioning of building systems and devices. Because operational conditions within a facility constantly change, temporary overrides are made regularly to specific zones or equipment, therefore, the exceptions generated by the OG™ module can save valuable time, energy, money, and ensure occupancy comfort within facilities. Exceptions to the operational standards identified by the OG™ represent the "low-hanging" fruit of automated building system analytics, meaning that they can easily be identified and addressed within minutes of detection, providing significant realized savings.

> *The Operational Guidelines™ module provides asset management, identifies exceptions to operational standards, and enacts supervisory control to fix these issues and realize savings quickly and efficiently.*

AUTOMATED FAULT DETECTION AND DIAGNOSTICS

Automated Fault Detection and Diagnostics (AFDD) is the process of applying advanced analytics to building systems data and identifying the "invisible issues" occurring on a device or between devices within a location. The types of issues that AFDD identifies require complex algorithms to detect; they are not simple guideline exceptions or alarms being reported from a device but are issues that can only be found by analyzing the behavior and interactions of one or many data points over time. The prioritized results that AFDD analytics provide represent significant savings for facility managers in regards to reduction of maintenance costs, reduction of energy consumption and its associated costs, increased productivity, and extended equipment life.

Opportunity

Without AFDD insight into building data, equipment problems are often only pursued after a customer or employee makes a hot/cold call to the service personnel, signifying that the associated piece of equipment is not performing properly. Reactive maintenance is the practice of running equipment until it fails, and many executives believe it is the lowest cost maintenance strategy because there is no cost until failure occurs at the equipment level. However, reactive maintenance can actually be very destructive to ROI; increase maintenance and energy costs; and cause unnecessary downtime, occupancy comfort issues, and shortened equipment life.

Many of the problems that eventually contribute to equipment failure start as small, manageable issues that can often be solved by making quick, easy adjustments to machines or control strategies. Although the types of issues detected by AFDD do not immediately affect comfort or productivity (no one complains when they are too comfortable), they do immediately increase a facility's energy, maintenance, and equipment costs. Preventative maintenance strategies utilizing the automated, advanced analytics capabilities of AFDD software provide facility managers with the means to proactively detect and diagnose efficiency losses and equipment failure, allowing them to achieve significant energy and maintenance savings, minimal unnecessary downtime, and improved occupancy comfort.

Typical BAS systems provide some data about equipment performance and problems, but the primary "invisible issues" that cause those problems often go undetected or repaired if no AFDD solution is in place, which ultimately results in costly repairs, complaints from tenants, and eventually shorter life spans of key assets. Implementation of AFDD technology can extend equipment life 10 to 25 years, depending on the equipment type. For example, a small 6% refrigerant undercharge can cause compressors to work harder, effectively reducing equipment life by 10% and necessitating capital expenditure much earlier than expected.

Continuous automated commissioning through AFDD can reduce maintenance costs by an average of 25% through:

- Early, preventative detection and diagnosis of issues prioritized by their severity and financial impact

- Providing technicians with advance knowledge of the issues so that they will already be equipped with the correct tools and parts before going out to the site, allowing them to solve the issues quickly and efficiently

- Ensuring that the parts are ordered and maintenance is scheduled proactively rather than reactively on the hottest or coldest days of year when technicians are already busy and scheduled elsewhere

- Allowing maintenance to only be performed when needed and smoothing service technicians' workload

- Minimizing downtime

- Preventing emergencies and the spreading of issues

AFDD implementation can yield energy cost savings as well. Continuous analysis of equipment can ensure that energy costs are optimized through the reduction of peak loads and energy consumption through decreased run-times. In addition, it has been shown that undetected equipment issues account for a 15-30% increase in HVAC-R energy usage, and eliminating these issues can essentially prevent the increase and generate savings.

> *Implementing AFDD advanced analytics can enable recapturing significant savings through increased productivity, extended equipment life, and reduced maintenance and energy costs.*

Testing Faults in the Laboratory Setting

There are many ways to test AFDD algorithms, but there is no substitute for being able to verify results and fine-tune algorithms based on real-world conditions using the actual, physical equipment in a controlled laboratory setting. Only in a laboratory can environmental conditions be accurately manipulated, allowing engineers to replicate the exact conditions required to initiate faults and make adjustments to algorithms based on real results.

With funding from the California Energy Commission (CEC) in conjunction with academic partners at the University of Nebraska, Ezenics, Inc. has built a large, dual-climatic chamber laboratory designed specifically for AFDD validation that houses over 250 sensors. Being dual-climactic, the lab is able to simulate both external and internal conditions. Engineers continuously run experiments to define unknown variables, correlate data, develop new algorithms, and improve the accuracy and optimize the sensitivity of existing algorithms.

Virtual Data Points

Virtual data points are calculations which use existing sensor data, asset information, and runtime data to produce virtual values for sensors or sub-meters that do not exist. Virtual data points are extremely helpful because implementing them is a low-cost effort that allows for deployment of higher level analytics without having to add additional sensors or hardware to a facility. For example, a virtual mixed air temperature data point can be calculated from the combination of real data from existing supply, outside, and return air temperature sensors. With the virtual mixed air temperature sensor, engineers can gain additional valuable insight into compressor issues, outside air damper issues, etc. Another example of a constructive virtual data point is virtual sub-meter data. For most equipment, energy usage can be calculated from the combination of runtimes, machine operational data such as tonnage or wattage, and other variables. Virtual sub-meter data allows facility managers to view energy data continuously for all equipment and realize where the most energy is being wasted without installing costly sub-meters in the facility.

AFDD Applications & Examples

AFDD can be applied to systems such as HVAC, lighting, refrigeration, electrical, energy, and renewable energy sources and has demonstrated significant opportunity for application across a wide range of sectors, including industry, hospitals, retail, commerce, offices, campuses, multi-tenant spaces, etc.
The most basic AFDD algorithms detect data quality and sensor faults. For instance, if a generic roof top unit (RTU) relies on an outdoor air sensor in order to determine when it needs to economize but the sensor gets stuck, the unit will not properly economize, or it could even open its outside air damper to full open position and start bringing in outside air when the outside air temperature is very hot, causing it to expend excess energy.

Sensor placement can also greatly affect the performance of equipment. For example, a retail store may employ a control strategy to dehumidify and cool the sales floor in order to ensure occupancy comfort and product integrity. The dehumidification strategy enacts cooling on specific sales floor rooftop units once the store dew point sensor reaches a specific level in conjunction with the outside air temperature. However, if the store dew point sensor is placed near the main entrance to the store, it will react to the additional heat and humidity, and the dehumidification strategy will continually tell various units to run cooling, causing occupancy comfort and wasted energy issues due to the unnecessary overcooling in these zones. Furthermore, if the dew point sensor is placed underneath a supply air diffuser, it may report a lower dew point reading than actual because of the additional cooling that is running over it.

In addition to detecting sensor failures and other issues that would prevent efficient facility and equipment operation, AFDD also diagnoses inefficiencies that will eventually contribute to equipment failure. For example, Energy Recovery Ventilation Wheel (ERV Wheel) Inefficiency is a fault that indicates that an ERV wheel is not creating a sufficient temperature difference in the incoming airstream.

When the temperature difference between the return air and outside air is sufficient, the efficiency of the ERV wheel is calculated. If the efficiency of the wheel is too low, 10-20% as shown in the red regions of Figure 7-2, AFDD detects and reports an ERV wheel Inefficiency fault. The green region shown in Figure 7-2 illustrates acceptable ERV wheel efficiency between 40-50% when no fault is detected. The unit must be running the fan in order to detect faults.

An ERV wheel inefficiency diagnosis affects the equipment in the following ways:

- Energy—If the ERV wheel is operating inefficiently, the unit may be required to provide more mechanical cooling or heating than would have been otherwise necessary.

- Capacity—If the incoming air is not preconditioned, then the overall ability of the unit to condition its zone is reduced, but it should still be able to maintain the zone(s) load.

- Maintenance—Additional cooling or heating that is caused by the inefficient energy recovery wheel will add unnecessary runtime hours to that equipment, reducing its functional lifespan.

It is important to decipher whether or not this fault is an isolated event or a recurring issue that necessitates fixing. If this fault is detected under 25 times or its duration is less than 7 days, then it is likely an isolated event or maintenance is being performed. However, if this fault occurs more than 25 times and the duration is extended, then it is likely a reoccurring issue that must be fixed.

If the fault is a reoccurring issue, then it may necessitate fixing in order to avoid equipment failure or increased equipment wear over time. Potential solutions for fixing an ERV Wheel Inefficiency fault include:

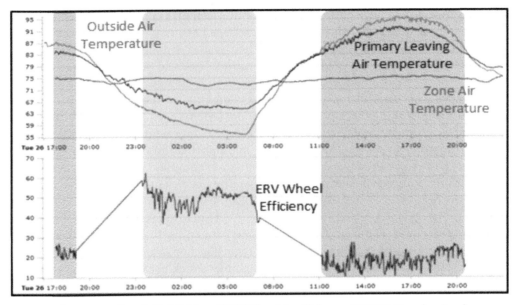

Figure 7-2: Graph Exhibiting ERV Wheel Inefficiency (© 2013 Ezenics, Inc.)

- If the primary leaving air temperature, outside air temperature, and zone air temperature sensors are not placed in ideal locations, then the sensor placement may be what is decreasing the efficiency of the ERV wheel and other systems. In this case, the sensors need to be relocated.

- If the ERV wheel's drive motor is stuck, damaged, blowing fuses, or the drive belt is loose or broken, then the drive motor needs to be repaired or replaced.

- If the ERV wheel drive belt is slipping or broken, then the drive belt will need to be replaced or adjusted.

AFDD in Action: The CEC PIER Project Case Study

As part of the Plug-n-Play Diagnostics & Optimization for Smart Buildings project sponsored by the CEC Public Interest Energy Research (PIER) program, Ezenics, Inc. deployed low-cost enterprise facility optimization analytics and operational guidelines™ solutions to 33 million square feet of commercial facilities across the State of California.

The PIER program awarded funding to a well-rounded team of companies and institutions lead by Ezenics, Inc., and Pacific Gas & Electric Company (PG&E), Southern California Edison, and Southern California Gas Company provided utility support for the project. Industry support was provided from Lennox, Automated Logic Controls, Amazon.com, Microsoft, Accenture, and Target Corporation. The University of Nebraska was an academic partner in the project.

Objective

The objective of the project was to develop, deploy, evaluate, and demonstrate advanced, low-cost, non-invasive, and plug-and-play enterprise diagnostics and optimization technologies. These technologies could be applied to energy, HVAC, refrigeration, and lighting systems to reduce building operating costs and carbon emissions.

Key components of the project include:

1. Deploy the solution across a very large number of existing facilities in California that contain multiple systems.

2. Prove that the technology can scale effectively and be quickly deployed at low cost.

3. Ensure accurate results.

4. Create significant value through the ability to create actionable information.

Opportunity

Automated, continuous commissioning with advanced analytics through AFDD technology is supported by a vast amount of research and is proven to be able to successfully identify operational issues in facilities across a wide range of industries. AFDD has gained momentum with support from utilities and businesses and is now matured to the point that it has been incorporated into the Title 24 building code in California.

Research supported by ASHRAE, the Department of Energy (DOE), and the CEC provides solid evidence that numerous problems in commercial buildings can be solved more efficiently and inexpensively through pro-active, AFDD-enabled maintenance strategies by ensuring that existing equipment and controls are working as intended than through reactive maintenance strategies, which ultimately result in large capital expenditures. Traditional, on-site commissioning generally consists of one-time fixes or full equipment replacements and can be very costly for companies to pursue. Traditional commissioning also must be performed at least once per year, and additional work is required to ensure that the issues have actually been resolved, which impacts the return on investment.

The current opportunity and value of AFDD for managing and maintaining building systems is clear, and there is plenty of room to grow. The article, "Detecting and Diagnosing the Faults of FDD," written by Lee Hamilton from Esource provides an excellent summary of AFDD solutions and articulates that opportunities for improvement exist because many AFDD solutions have either been too broad, looking at only the energy data for a wide variety of systems, or too narrow, focusing only on a single system or machine. Optimal AFDD solutions must be able to analyze and compare a wide variety of data over a wide variety of systems. HVAC and Refrigeration systems, for example, are often separate from one another, but these systems have an enormous effect on one another within a facility, and an ideal AFDD system would be able to utilize a broad spectrum of data from both and examine interactions between them in order to properly diagnose the issues. Some AFDD solutions are also embedded into hardware, which requires clients to purchase additional physical devices and sensors to obtain data or providers must physically visit each site, thus making the technology too difficult and costly to scale for the benefits received.

The CEC PIER-sponsored Plug-n-Play Diagnostics & Optimization for Smart Buildings project aims to prove that low-cost, rapidly deployable, multi-system AFDD technology, can more effectively scale the benefits of commissioning and fault detection and diagnostics. Scalable, quick deployment of AFDD analytics will lead to faster adoption of the technology, enabling substantial reduction in carbon emissions and energy consumption, resulting in significant realized financial savings.

Key Features of the Ezenics AFDD and OG™ Solution

Several key features of the Ezenics AFDD and OG™ solution were identified in order to make the technology scalable with minimized up-front and ongoing costs. First, with the Ezenics solution being hosted in the cloud

and accessible via web browser, clients are not required to install any new software. Because of this, no additional hardware needs to be installed on-site; there is no need for site visits to set up the solution; and results can be securely obtained from existing sensors, equipment, and systems over Ethernet through HTTPS-secured connections. Many fixes for AFDD faults and OG™ exceptions can be fixed remotely as well. The system provides valuable information directly to on-site technicians enabling them to complete maintenance and resolve issues quickly. The Ezenics solution decreases capital expenses and is flexible enough to optimize existing infrastructure.

The Ezenics solution is built to be "interoperable," meaning that there is a common user experience regardless of underlying equipment or systems, which enables quick user adoption across a diverse portfolio of locations. Additional internal and external data sources such as alarms, work orders, weather data, energy rates, and DR event information can be integrated into Ezenics' multi-system analytics. Results from the solution, including the diagnosis, normalized severity ranking, duration, quantity of occurrences, and financial impact are pre-filtered and prioritized based on clients' business rules. The business rules are driven by setting thresholds for comfort and energy usage, ensuring that the system will trigger only relevant, actionable results, avoiding an overload of information. The solution triages information by facility faults and exceptions and can automatically generate work orders based on priorities instead of overwhelming facility management staff, ensuring rapid adoption and maximized savings.

Project Key Components
Data Exchange Carrier

The Data Exchange Carrier was developed on a scalable, cloud-based infrastructure and can establish connectivity with an unlimited number of building automation systems and equipment in order to obtain, aggregate, store, share, and process data at near-zero cost. This is commonly referred to as the ability to deal with "Big Data."

Virtual Sensors

One of the project objectives is to produce valuable analytics from data obtained through the existing systems in commercial buildings. Due to this need, five low-cost, high-accuracy virtual data points or "virtual sensors" were deployed. These virtual data points were utilized to expand the onboard measurements and enable both existing and new diagnostics and optimization technologies.

Multi-system Interaction

The diagnostics and optimization technologies deployed in this project address the interactions among different systems of the same type and different types of systems such as HVAC, lighting, refrigeration, energy, weather, and rate structure data.

Large Demonstration

The Ezenics enterprise-level, plug-n-play diagnostics and optimization solution integrates existing and new technologies in order to enable smart buildings and can be deployed in a non-invasive, low-cost manner. For this project, the solution was deployed, tuned, and evaluated in 252 retail locations throughout California.

Figure 7-3: CEC Fault Verification

Results

The Ezenics solution was rapidly deployed and evaluated in 252 geographically diverse retail locations in California, covering over 33,000,000 square feet in multiple utility zones. Through various methods, 16,480 pieces of equipment and systems (an average of 65 per facility) have been connected without visiting the site or requiring additional hardware. Of the 16,480 pieces of equipment, 5,845 were categorized as HVAC machines, 4,742 were refrigeration components, and the remaining one-third of the devices consisted of lighting, sub-meters, utility data, and other miscellaneous systems data.

The Ezenics data exchange carrier communicates with and pulls data points at near real-time frequencies from all units (400,000 data points every minute) and stores it forever. Over the course of a month, 19 trillion data points are collected from the 252 California sites, which equates to roughly 1.8 GB per day.

Faults from the Ezenics AFDD are thoroughly validated and confirmed by certified contractors on-site. These certified contractors implement fixes to confirm that the faults identified in the Ezenics platform were solved. In addition to verifying faults at the site through a certified contractor, Ezenics built a large dual-climatic chamber laboratory specifically for AFDD validation that is utilized to test, tune, and confirm faults and severity levels. The results of the CEC project validate that the Ezenics solution detects faults and generates appropriate solutions with nearly 100% accuracy.

Validated savings for the 252 locations in California over the last twelve months as of when this chapter was written equated to a reduction of over 35.5 million kilowatt hours in energy usage, approximately $3.2 million in savings. A breakdown of the faults revealed that approximately $1.6 million of the $3.2 million

Figure 7-4: Ezenics' Dual-Climatic Chamber at the University of Nebraska

saved (approximately half the total savings), was saved from approximately 12,000 HVAC mechanical issues. The remaining savings was generated by 9,000 HVAC control faults ($912,000), 300 lighting faults ($574,000), and 3,000 refrigeration faults that prevented food quality issues (de-merchandizing events) in addition to energy savings of $82,000. AFDD results have also been found valuable through cross-platform coordination with demand management strategies. AFDD enables dynamic control strategies, which enhance the success of DR and peak load reduction programs while ensuring that such strategies do not impact comfort to an unacceptable degree. The project has already inspired implementations in office buildings, data centers, hospitals, and additional retail locations across the country that accounted for over $18.5 million in energy efficiency savings within the first half of 2013.

Realizing Savings with AFDD

After connectivity has been established, data are harvested, and the AFDD solution has identified and prioritized what must be done in order to achieve maximum savings and occupancy comfort, facility management still needs to take action in order to realize those savings. A variety of issues can automatically be resolved according to defined business rules; however, for the majority of issues that require a human presence, efficient processes and workflow management are essential. The faster and more efficiently that fixes are performed, the more savings that will be realized with the same amount of resources.

Ezenics Mobile: The Power of Enterprise-Level Analytics, Work Order Management, and Building Controls in the Palm of Your Hand

Utilizing the cloud-based, agnostically integrable Ezenics Mobile application with commonly available smart-phone technology, facility managers are now able to dramatically streamline service processes and leverage their AFDD analytics, asset management, workflow management, supervisory control, and other key enterprise-level systems to directly support the roles and tasks of technicians in the field.

 Mobile

© 2013 Ezenics, Inc.

AFDD mobile applications can effectively slash the time that it takes to locate, examine, and service a machine; respond to calls; and complete work orders to a fraction of what it would have otherwise taken using more traditional methods. All within a few thumb taps, technicians can scan a QR code or asset tag on a physical device and instantly know everything that there is to know about it, respond to and close out work orders, view and modify the points on a device using their phone as a remote control, open a report of the issues being diagnosed by the AFDD and OG™ along with a list of engineer-prescribed solutions, update or correct asset information, and much more. For technicians, this means increased productivity, easier access to critical information, unprecedented support from enterprise-level systems, and a seamless, all-in-one interface from which they can accomplish a large variety of tasks. For facility and service managers, the increased productivity of their technicians means that more work orders are being closed in the same amount of time, leading to significantly increased savings and client satisfaction.

Mobile Tool Utilization Example

For this example, "Sue" is an office worker on a Smart Campus, and "Matt" is a brand-new service technician. Sue arrives very early to prepare a meeting room and realizes that it's uncomfortably cold, so she asks the secretary if anything can be done to warm up the space. The secretary makes a quick call to the facility management call center, and they raise the temperature setpoint from 74°F to 78°F. After half an hour, Sue is still cold and her meeting is soon to take place, so she again alerts the secretary. In response to the heightened priority of resolving the issue before the meeting, the facility management firm quickly dispatches Matt to the building, smart phone in hand. Arriving at the location, Matt uses the GPS-enabled location detection feature and is presented with a street view of the building along with summary information and additional downloadable content such as floor plans and other available engineering documents.

On the first screen, Matt sees a red notification for an open work order assigned to him for the building via the integrated work order management system and taps "Maintenance" to bring up the details. Right away, he can identify which machine the work order is associated with (VAV-01), the reporter, the call-back number, a description of the situation, and an image of where the machine is located within the building.

Matt also notices a red issue notification for the VAV and taps "Issues" to open a list of faults detected by the integrated AFDD system. With the AFDD diagnosing the issues automatically, Matt has no need to manually diagnose the problem himself. On the Issues page, he sees that there are a few issues listed on the VAV, including a "Setpoint Not Met—ZAT—Overcool-

> *Even though he is a newly-hired technician who has never even been to the building, he already has an understanding of the situation, including the location of the machine responsible for making Sue's meeting room too cold.*

Figure 7-5: Ezenics Mobile Location Info Page ©2013 Ezenics, Inc.

Figure 7-6: Ezenics Mobile Work Order Details Page ©2013 Ezenics, Inc.

ing" fault and a "Supply Air Damper Unresponsive—On/Open" fault.

Matt opens the details pages for both faults, which provide him with full descriptions of the faults, how many times they have occurred, when they were first detected, their financial impact, severity ranks, and even a list of engineer-prescribed potential solutions. He finds that the "Setpoint Not Met—ZAT—Overcooling" fault means that even though the room is too cold, the machine is still cooling the room even further, wasting energy and causing a comfort problem. The "Supply Air Damper Unresponsive—On/Open" fault indicates that the unresponsive damper is to blame for the problem and needs to be repaired.

After identifying the problems, Matt taps on "Information" to learn the VAV's brand, model, serial number, and other information in the asset management system. With this information, he can determine which parts, tools, and equipment to take with him into the building.

After the repair is complete, Matt simply scans the QR code on the side of the machine to bring up the diagnostics and BMS access in order to verify that it is

now working properly. He notices that there is a cooling setpoint exception caused by the previous modification made by the call center, which if left unchecked would result in Sue's meeting room drifting to 80°F, invariably resulting in another service call. In order to correct the setpoint and resolve the exception, Matt simply taps "Remote" and is not only able to view real time data from the VAV but is also able to set the cooling setpoint back to its correct value via the secure BMS integration.

Without prior experience diagnosing this VAV and issues manually, Matt already knows what the problems are, how severe they are, potential solutions for how to fix them, and what he needs to take with him to repair the VAV before he even leaves his truck.

The only thing is left to do is to complete the work order by going back to "Maintenance" and changing its status to "Complete" through the integrated work order management system.

Rather than making multiple trips in and out of the building, spending time manually diagnosing the issue,

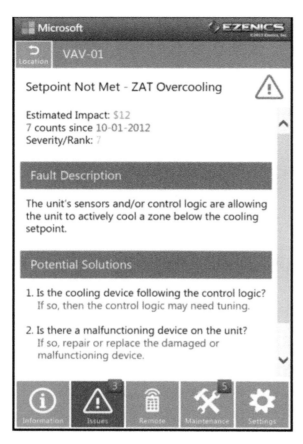

Figure 7-7: Ezenics Mobile Fault Details Page ©2013 Ezenics, Inc.

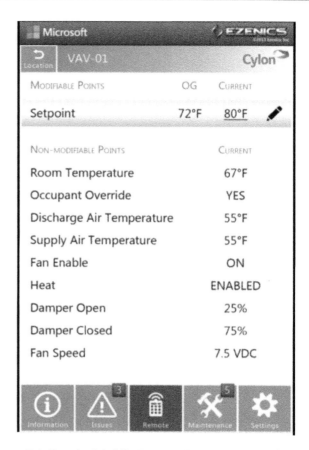

Figure 7-8: Ezenics Mobile Remote Page ©2013 Ezenics, Inc.

and juggling both his work order system and the BMS, Matt has managed to complete the job using a single, integrated software tool on his phone in just one trip. The agnostical-ly integrable mobile solution provided him with an unprecedented level of support and control, and for Matt, that's reduced frustration and time spent resolving the issue. For his supervisor at the facilities management company, it's time saved and another job squeezed in that day. For Sue, the conference room will be comfortable in time for her meeting.

> *Using the Remote control feature, Matt is able to effect change on the VAV using his phone and fixes the exception without having to call in or find another terminal to access the Building Management System.*

THE NEW FRONTIER OF AFDD: UTILIZING AFDD TO ENABLE DEMAND MANAGEMENT

The same concept of integrating various groups, platforms, and devices that provides the basis for the mobile application can be applied to maximize process-es, time, and monetary impact for other building systems applications as well. AFDD can also be utilized to make more effective decisions and efficient program-ming in energy management.

Traditionally, energy management and AFDD have been separate realms in the building systems world. Customarily, energy management companies utilize an energy management platform, which leverages power meter data to output energy usage, carbon, and cost metrics. Facility management staff, on the other hand, employ AFDD, which obtains data from a BAS; apply algorithms to data; and then output meaning-ful diagnoses for comfort, maintenance, and energy issues for the systems they monitor. Separately, these systems can provide signif-icant impact for the teams managing them; however, if these systems were integrat-ed, AFDD could also drive and maximize energy goals.

> *Energy management and AFDD have traditional-ly been two very separate realms in the building sys-tems world. Separately, these two systems can pro-vide significant impact for the teams managing them; however, if these systems were integrated, AFDD could also drive and maxi-mize energy goals.*

The Importance of Managing Demand

Table 7-1: Sample Demand Charge from 60 commercial buildings during February-May 2011

Utility	Demand % of Bill
Dominion VA&NC Power	53.5%-54.8%
Memphis Light, Gas & Water	24.1%-26.7%
Southern California Edison	16.8%-31.3%
Tampa Electric Company	21.8%-25.1%
Consolidated Edison of New York	19.9%-21.3%
Exelon Energy	12.4%-31.9%
Mid American	13.8%-21.4%

Electric demand is becoming increasingly imperative to manage for commercial energy consumers. Employing demand response, utilities incentivize or provide rebates to those curtailing load during peak times on the grid. The lesser-known consequence of the increasing importance of demand is that over the last 5 years, electricity consumption charges have decreased but demand charges have significantly risen in both cost and percentage of the monthly utility bill. The increase in demand charges on the monthly utility bill is not always obvious to consumers because taxes and other line item charges that were once based on kWh are now based on the monthly peak kW instead.

Unaware of the potential financial impact that demand can have, energy and facility managers often instead emphasize managing consumption. However, due to the importance of reducing demand on the grid during critical times and reducing peak demand to lessen the monthly utility bill, managing demand in a facility or portfolio can represent a significant opportunity to save energy costs.

Obstacles to Demand Management Adoption

Demand management strategies are often fixed, formulated based on operational assumptions, not the current actual operations of the facility. The actual operational conditions of a facility can make static demand management strategies ineffective, ultimately resulting in undesirable comfort issues and minimized energy reduction (sometimes, even an increase in demand). This fact has led to deploying conservative demand management strategies or none

Changing Operational Conditions

- Setpoints
- Schedules
- Loads
- Space conditions
- Equipment Issues
- Controls Issues

at all, leaving a significant opportunity to realize savings or earn incentives.

Static demand management strategies can be affected by a myriad of issues that can contribute to the failure of their deployment, including: standard cooling setpoints or schedules having been changed, offline controllers, "hands-off auto" (HOA) lighting switches being set to manual, preexisting comfort issues in a space, or equipment failure that causes increased zone temperature in the space and an adjacent zone (this would cause the unit in an adjacent zone to "work harder" than it normally would). These situational conditions are only a few of the issues that can potentially plague a static demand management event, preventing optimized performance. With numerous variables that must be accounted for in order to achieve building energy efficiency, it is difficult to account for all of them in a static demand strategy and still provide an optimized solution.

> *The potential for failure of most static demand management strategies due to unaccounted variables and other issues causes many facility managers to decide against deploying any strategies at all despite the fact that demand charges make up a significant portion of a facility's bill and participating in demand response events can yield incentives and rebates.*

> *High net savings can be achieved from an AFDD-driven, dynamic demand management solution because money does not need to be spent fixing faults in order for a strategy to work and achieve quantified results. By creating a strong, integrated interdependence between Demand Management and AFDD, the industry now has solutions which maximize the balance between energy and comfort, ultimately assisting in avoiding unnecessary utility charges.*

AFDD-Enabled Dynamic Demand Management

AFDD provides significant insight into building operations and greatly enhances the ability to quickly and reliably find and prioritize problems across a portfolio that contribute to increased electric demand. Often, identified faults do not have to be eliminated in order to achieve demand management results. Instead, AFDD can drive dynamic demand management strategies that take into account analytic outputs to adjust strategies accordingly, optimizing load shed and, therefore, monetary results.

Optimized Demand Response Deployment

The opportunity to get involved in DR programming is well known in the industry with the allure of monetary incentives from utilities and the ease of solutions provided by vendors. Even though DR programs are being adopted all over the United States as well as internationally, there are still many commercial buildings that are not participating or are not maximizing the opportunities associated with DR programming. Enterprise-level AFDD integrated with DR provides the ability to manage many events, vendors, and locations concurrently. The current state of equipment and

> *AFDD analytics promote confidence and enable energy managers to justify pursuing more aggressive curtailment strategies by assessing the current state of equipment and environmental conditions within a building in order to enable dynamic load curtailment strategies that provide maximized bidding potential and event performance for a building or portfolio of buildings.*

the environment within a building is known, so load shed strategies can be adjusted accordingly, providing dynamic load curtailment strategies that maximize event performance while ensuring that there are no adverse comfort issues. Thus, the integration of DR with AFDD promotes confidence and enables energy managers to justify pursuing a more aggressive solution which maximizes results.

The increase in charges based on the peak kW during the month combined with the ability to deploy a dynamic, AFDD driven strategy, has created a significant opportunity for savings on commercial buildings' monthly electric utility bills.

Peak Load Management in Addition to Demand Response

Often, commercial buildings that are successful in DR programming do not implement any Peak Load

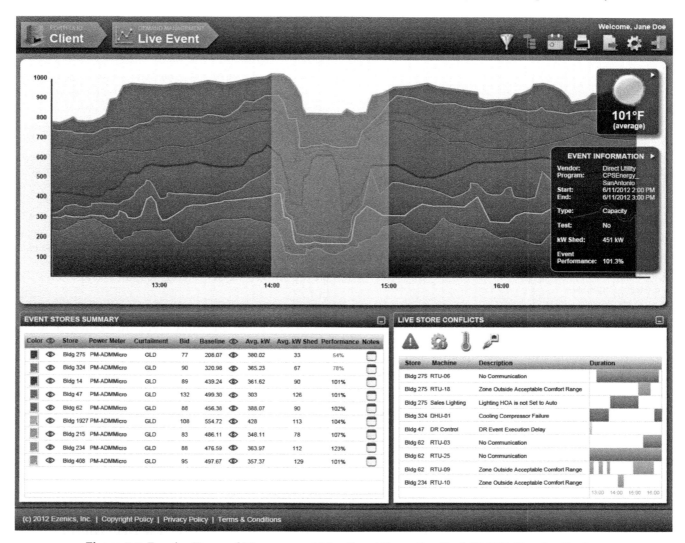

Figure 7-9: Ezenics Demand Management Live Event Reporting Tool (© 2013, Ezenics, Inc.)

Management strategies due to the varying conditions in environment, potential adverse impact on comfort in the space, and the difficulty of predicting peak demand times. Successful DR deployments prove that DR strategies are able to shed a determined amount of load while not affecting comfort, and these existing strategies can also provide the same benefits in limiting demand if peak demand times can be reliably predicted. These are variables that are already determined with DR, but are less pronounced in Peak Load Management activities.

Determining the Opportunity for Savings

Demand Response often occurs during the hottest days of the year when demand on the grid is the highest; however, many times, the greatest opportunities for limiting demand occur during months in which the environmental conditions create a few sharp peaks.

Many facilities have energy signatures that would require running a Peak Load Management strategy for only a small percentage of the time throughout a month in order to achieve significant savings by reducing peak demand. There are months in which the opportunity for limiting demand is low because the energy signature displays a smoothed load curve, which necessitates elongated Peak Load Management events in order to shed the amount of load necessary to affect the monthly peak. Elongated event periods in which HVAC and refrigeration loads are used to curtail can also cause significant comfort issues, nullifying monetary savings. Therefore, months with smooth peak curves, such as the curve displayed in Figure 7-10, are not optimal for implementing Peak Load Management programs.

However, there are often more months throughout the year in which the energy signature provides sharp peaks for only a few days throughout the monthly billing cycle. These sharp peaks provide the highest opportunity to curtail load because they require short periods of load shed that will not adversely affect comfort.

The Limitations of Traditional Demand Strategies Not Driven by AFDD

Traditional Peak Load Management can involve setting a kW threshold in a facility's control programming that will enact a scheduled set of actions to curtail load on specific pieces of equipment once the facility reaches the pre-determined threshold. It is commonplace to find facility managers not utilizing this feature because the control strategy often does not take into account comfort thresholds and the actual operational conditions of the facility. Another limitation of such an approach is that in months when the peak does not exceed the set limit, no demand reduction initiatives are initiated, and potential savings are unexploited.

AFDD-Driven Peak Load Management

Dynamic, AFDD-driven Peak Load Management offers significant improvement over traditional Peak Load Management because it adjusts based on the operational conditions of a facility. It utilizes equipment and systems data that are collected every minute in order to drive real-time analytics and provide equipment and operational conditions to prediction models that determine when to run Peak Load Management strategies.

> *AFDD greatly enhances the ability to quickly and reliably identify problems that contribute to the peak demand and provide prioritized, actionable solutions across a portfolio.*

Without the support of AFDD, if a strategy is enabled prematurely, then comfort thresholds may be breached before the peak occurs, which means that the peak will not be avoided, and potential monetary savings will not be realized. Even if the strategy is enabled at the correct time and the peak is avoided, the space may still be uncomfortable, and de-merchandizing could potentially occur. If a Peak Load Management strategy is enabled belatedly, the peak for the month may be missed entirely. In order to avoid poorly timed strategy deployment, AFDD-driven Peak Load Management employs prediction models that take AFDD results as inputs, allowing demand reduction to be dynamic according to the real-time operations of the facility. Thus, the Peak Load Management strategy is constantly tuned according to operational conditions, facilities demand less energy as a result, occupancy comfort is ensured, and peaks are avoided without having to fix identified faults. All data, AFDD analytics, Peak Load Management, and Economic Load Control are software-based and do not require additional hardware to be installed.

A significant benefit of applying AFDD analytics to equipment and controls data is that it can bring issues that are contributing to the demand peak for the month to light. Solving these issues can often be done with one-time fixes or through automated supervisory control. For example, poor equipment staging occurs frequently because equipment is often controlled with separate thermostats or even control systems that do not include efficient logic to avoid equipment running simultaneously; however, it can be avoided through alterations to a control strategy without causing any adverse impact on comfort. A common example of a staging issue is a non-optimal start sequence. Staggering runtimes

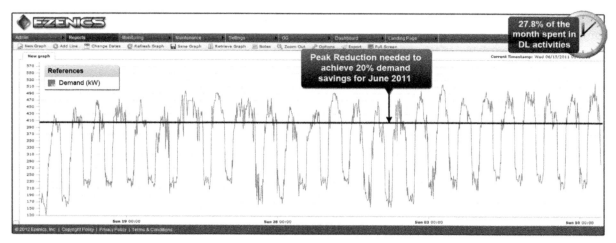

Figure 7-10: Typical Retail Facility Example for Reducing Peak Demand During Months of Consistent Usage (© 2013, Ezenics, Inc.)

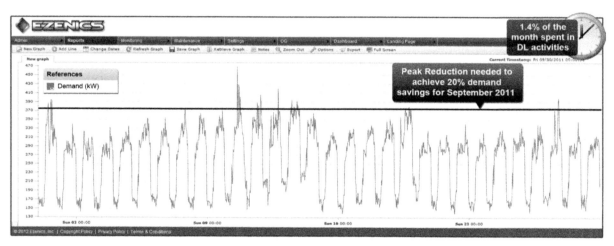

Figure 7-11: Typical Retail Facility Example for Reducing Peak Demand During Months of Sharp Energy Peaks (© 2013, Ezenics, Inc.)

on units will often result in one unit being able to meet the demand goal before others even need to start. Additional strategies can be put in place for refrigeration by optimizing defrost cycles with no effect on meeting the needed setpoints.

Economic Load Control

Peak load management can provide an effective means of saving significant money in facilities with a monthly electric utility bill that has a high demand charge. For energy managers who wish to go a step further to exploit real-time or wholesale energy markets, taking peak load management technology and adapting it to utilize real-time pricing signals can also provide significant benefit. In deregulated energy markets, energy managers can take advantage of employing an Energy Service Company (ESCO) in order to buy (and sell) electricity at

wholesale prices (also called Real Time Pricing), essentially getting the best value if they run their facility based on the constantly changing energy prices.

PJM Interconnection, a regional transmission organization (RTO), provides customers with the ability to opt into a wholesale energy pricing market which uses a model referred to as Location Marginal Pricing (LMP). The LMP model, a type of Real-time Pricing (RTP) model, bases pricing on available generation, transmission, and demand on each node or location. Each node is assessed a different cost for electricity each hour or five minute period. Therefore, facilities in nodes which have high transmission congestion or a lack of generation compared to the demand, will pay more in electricity than facilities those in areas where there is adequate generation and transmission to meet the needs of its customers.

PJM's LMP markets, as well as other wholesale

pricing markets, provide pricing to customers as day-ahead or hour-ahead real-time pricing. Day-ahead RTP pricing provides one hour interval electric pricing bids for the same hour of the following day. The real-time RTP pricing structure bids electricity prices for five or fifteen minute intervals (market dependent) one hour in advance. The process of responding to market pricing signals through load shed is referred to as Economic or Continuous Load control. Because of the short time available to optimize shifting or shedding load in order to take full advantage of wholesale energy prices, this process must be automated.

Traditionally, economic load control technologies have mirrored static Demand Response strategies with the exception of integrating real-time pricing signals in order to dictate when and how much load should be shed. However, because static DR strategies can often be ineffective, causing potential comfort issues or increased load, applying technology which incorporates AFDD inputs into energy management strategies (dynamic energy management) can be advantage for use in Economic Load Control programs as well. Dynamic, AFDD driven Economic Load Control strategies ensure that efficient load reduction occurs as a result of pricing, comfort is maintained during short or long time periods of load shed, and monetary results are maximized in a real-time pricing structure.

> *Continuous Load Management or Economic Load Control can be maximized through integrating real-time pricing feeds with predictive building modeling and AFDD analytics. These variables allow for an efficient, continuous means of control that adjusts to meet monetary and comfort goals.*

The Ezenics, Inc. platform performs Continuous/Economic Load Control by collecting equipment and systems data every minute, performing real-time AFDD analytics to provide equipment and operational conditions to predictive facility energy models, and then combines the models with real-time pricing signals in order to determine what actions should be taken based on these variables. Utilizing AFDD analytics and predictive models in conjunction with real-time pricing feeds, creates an automated, continuously optimized, load management platform.

Integration with Renewables

In addition to Demand Response, Peak Load Management, and Continuous/Economic Load Management, another sector driving AFDD-driven Energy Management technology is renewable energy. Renewable energy sources are clean, continuous sources of power generation that include wind, solar, tidal, biomass, geothermal, hydropower, etc. These resources are often intermittent, meaning that they are dependent upon sporadic variables and often produce power in surges rather than an uninterrupted, reliable stream. Because of the variability in the amount of power generated from renewable energy sources at any given point in time and the difficulty in predicting the production from these devices, it is challenging to employ them as means of powering an entire building, and many times they are only utilized to power non-critical or secondary systems.

Renewable energy sources can be tied into a facility's main power distribution system to provide it with clean energy resources and limit the need to use additional energy and its associated costs. However, the intermittency of the power produced by renewable sources makes this concept in its fullest form, difficult to achieve. To exploit renewable energy sources to their fullest, models can be created which take weather conditions such as wind speed, wind direction, average lumens, temperature, humidity, etc., into account in order to predict the renewable energy system's future generation capability. These predictive energy generation models can then be utilized in conjunction with predictive facility energy demand models to optimize a building's energy usage at any given point in time.

For example, during a 15-minute time period, clouds may be covering a facility's solar panel array, triggering a facility with AFDD-driven, dynamic demand management strategies to respond by instigating load shed across systems in order to maintain the same energy demand from the utility even though solar generation is limited (this is the same methodology used in spinning reserves DR programs that require short 5- to 15-minute periods of load shed from a facility). When the clouds move away and solar power is restored to its maximum production, the facility's HVAC controls automatically return back to normal operation because of the increased generation.

In addition, with AFDD technology, algorithms can also be applied to renewable energy sources to determine when a solar panel is not producing the expected generation and pinpoint if a diode is failing and which one it is, or identify whether a mechanical or electrical part

> *AFDD can be deployed on renewable energy sources in addition to traditional building systems and not only provide underlying performance issues and fixes but also a means with which to automatically control the facility based on renewable sources and operational conditions.*

in a wind turbine is causing a minor efficiency loss in the overall generating capacity. Armed with AFDD fault information and the predictive energy models, facilities can be flexible and bend their systems toward the sum of their parts, maintaining consistent, efficient energy usage and minimizing costs.

Proving the Potential of Peak Load Management

Ezenics has so far deployed its groundbreaking peak load management technology over approximately 4.2 million total square feet of building space with an average of 140,000+ ft^2 per Peak Load Management implementation site. These sites are located across the United States and Canada and have realized savings of $210,000, or a 10-20% reduction in energy costs per site up to the first half of 2013.

Summary

As utilities provide more incentives for demand management and energy costs continue to rise, managing demand will become more and more financially advantageous to facility and energy managers. However, the current static demand management model must be replaced by dynamic, AFDD-driven strategies and models that constantly adapt to changing operational conditions that would otherwise potentially derail outdated, static strategies. AFDD-driven demand management solutions provide insight into the real-time operational conditions of a building, allowing for the creation and deployment of dynamic curtailment strategies that do not adversely impact comfort. These can be easily scaled across a portfolio and managed centrally to provide significant monetary benefit to clients.

CHAPTER SUMMARY/CONCLUSIONS

This chapter has reviewed the concepts of connectivity, data harvesting and translation, algorithm application, Operational Guidelines™, AFDD, realizing savings from advanced analytics, asset management, integrated systems and processes, and using AFDD to drive Demand Management strategies.

Exceptions are analytics which pinpoint differences between the operational standards and actual operating conditions, providing a quick, efficient means of fixing these issues and realizing significant impact almost immediately. AFDD realizes savings through identification of equipment, control, or system inefficiencies and failures, providing benefit through increasing main-

tenance productivity, extending equipment life, and reducing maintenance and energy costs. An agnostically integrable mobile platform allows facility managers to optimize service processes and leverage AFDD analytics, asset management, workflow management, supervisory control, and other key enterprise-level systems to directly support the roles and tasks of technicians in the field. AFDD and asset management can also be utilized as variables in energy management programming, enabling dynamic curtailment strategies which react to real-time operational conditions and maximize energy and monetary results. As intelligent buildings continue to develop and evolve, these technologies will continue to advance as well, ensuring that buildings are constantly tuned to optimize energy, people, processes, and financial results.

ABOUT EZENICS

Ezenics was founded in Europe in 2004 and now has offices the United States, Dubai, and Argentina. Ezenics offers an intelligent, enterprise facility optimization platform that continuously analyzes sensor, machine, and energy data, creating prioritized, actionable information and intelligent control.

No hardware, software, or on-site visits are needed to reliably collect and analyze billions of data points from a large range of existing building equipment and systems. Scalable analytics results automate the optimization of facilities across global client portfolios.

Commercial buildings of all sizes and across all verticals utilize Ezenics' solutions to:

- Optimize energy consumption
- Reduce demand and time of use charges
- Decrease maintenance costs
- Reduce carbon emissions
- Extend equipment life and uptime
- Track and improve occupancy comfort

Ezenics provides multiple facility optimization solutions that focus on energy, HVAC, lighting, and refrigeration continuous optimization:

- Continuous Automated Commissioning
- Dynamic Energy & Demand Optimization
- Automated Fault Detection, Diagnostics, and Impact (AFDDI™)
- Operational Guidelines™ (including asset management)
- Maintenance and Alarm Management

Chapter 8

A Practical, Proven Approach to Energy Intelligence, Measurement, and Management

Gregg Dixon, EnerNOC

INTRODUCTION

Technology has been incredibly effective at increasing productivity and reducing operating costs for enterprises. Companies use advanced software systems across every business function to track and analyze data, forecast trends, maintain a competitive edge, and ultimately, make each of us more productive.

Consider these examples:

- Since 1980, costs per mile flown per U.S. airline passenger have dropped by 40%, according to Airlines for America.

- The once-stodgy U.S. manufacturing industry is in the midst of a "new industrial revolution" thanks to technology. With new technologies, Boston Consulting Group estimates that 30% of Chinese exports to the U.S. could be economically produced domestically by 2020.

- 64% of Salesforce customers say they've reduced sales, service, marketing, or other operational costs with the integration of its CRM software while increasing customer retention by 63% and enhancing cross-sell and up-sell opportunities by 59%.

- McKinsey states that with the continued integration of social technologies to the workplace for what it calls 'interaction workers,' productivity could increase 20-25%.

So why hasn't technology played a more significant role in helping U.S. enterprises effectively measure and manage the billions of dollars they spend annually on energy consumption? It's not like we don't have room to improve: despite the fact that the U.S. is the world's technological epicenter, we're not even among the top 25 most energy efficient countries.

My hypothesis: this stems from an inability to make sense of the largely fragmented, muddled landscape of energy management vendor offerings that leaves decision makers asking themselves; "Do I need an energy services company? A utility bill management solution? A demand response provider? Something else?" These questions, coupled with our constantly shrinking attention spans and healthy fear of the unknown (from not realizing true energy spend or having limited understanding of the benefits of energy intelligence), create a market ripe for transformation.

But I'd take this further, and posit that an even more important contribution to this market transformation is the crumbling notion of energy usage as a fixed cost, like rent. Thanks to the growing ubiquity of real-time energy data and a deeper understanding of minute-by-minute energy use, this 'fixed' cost is turning into a variable, controllable cost.

Until now, real-time energy and building systems data has been difficult and expensive to acquire (especially at scale). However, this dynamic is quickly changing as the energy management industry shifts to a more data-driven model thanks to an emerging technology category known as Energy Intelligence Software (EIS).

EIS is a hybrid of traditional energy expertise and technological innovation, enabling organizations to connect energy usage—how it's bought, when it's used, and how much is used—with important business metrics, all through the lens of big data and in real-time demand. In my view, there are four critical components of any viable EIS solution:

Real-time Data

It's plain and simple: you can't manage what you don't measure. Uploading data into spreadsheets and doing manual analysis after money has already been spent isn't intelligent energy management. A true EIS platform can measure billions of energy data points monthly, delivering surprising insights around potential energy savings, operational efficiency, relative performance, and overall improved visibility into how organizations use energy, all in real-time—a powerful combination that delivers a competitive business edge.

Actionable Insights

There's a reason few organizations use the term "carbon tracking" or "carbon accounting" anymore. Tracking alone does not deliver value or catalyze change; it's passive. EIS is about continuous improvement and driving productivity. High quality data coupled with intelligent analytics help energy managers ensure they're meeting or exceeding goals at any point in time—and alert them to opportunities around the bend.

Efficiency Matters, Too

Energy managers often wear several hats, many of which don't involve managing energy usage. Energy intelligence software brings together the desperate information needed to help them be more productive with their limited time.

With the billions of energy data points an EIS platform collects across many facility types, EnerNOC has developed algorithms that know how to identify anomalous energy usage—from simple things like when lights are left on over the weekend to more complex issues like when a battery charging bank is being used during the most costly hours of a time-of-use utility tariff. These small inefficiencies can cost millions of dollars a year to enterprise class customers—and hurt productivity, to boot.

And we're just scratching the surface. We know there's a tremendous opportunity to do more with big data, including the ability of predictive analytics that will give enterprises a crystal ball of sorts to more accurately forecast energy use and identify deeper savings opportunities.

Quantifiable Savings

There are many reasons to care about energy efficiency, but increased profitability is front and center. If you can't even measure the impact of your investment, getting organizational alignment around the value of energy management will be next to impossible.

Time and again, we see customers like Jackson Family Wines, California State University, and Glenborough Properties access their real-time data and use it toward transformative results. Their appetite for exploring additional opportunities to control energy usage and costs soars, and we partner in further optimizing their

portfolios, finding even more value. A key component of this approach is the use of automated diagnostics.

AUTOMATED DIAGNOSTICS AS A COMPONENT OF ENERGY INTELLIGENCE SOFTWARE

For more than a decade, EnerNOC has worked with customers around the globe to harness cutting-edge technology to manage their energy costs and impact the bottom line. Our suite of software currently supports more than 14,000 sites and collects data from more than 30,000 devices. More importantly, we've helped our customers achieve nearly $800M in savings and utility payments, and we've done it by selecting the appropriate technology that fits each customer's needs.

We have seen that scalable, repeatable, diagnostic solutions are critical to long-term success. However, it is also important to meet each organization where they are in the energy intelligence spectrum. Does an organization already have a centralized utility bill management and reporting system that is accessible to decision makers? Do they know how many meters are connected to their building? Have they evaluated their tariffs? Have they already captured the low-hanging fruit across their portfolio and need to address specific "problem children"? The answers to these questions will often determine the shortest path to quick wins.

What follows are two case studies that clearly demonstrate the importance of flexible and scalable diagnostics solutions. Both organizations have extracted significant energy and operational savings, one starting with utility and sub-meter data to identify and prioritize opportunities throughout a portfolio, the other peering deep inside its buildings to mine the mountains of data streaming from mechanical and lighting systems. In both scenarios, the customers were already actively engaged in energy management but lacked the necessary tools to further accelerate the identification, implementation, and verification of operational savings opportunities. Through EnerNOC's proven EfficiencySMART software, part of the company's EIS suite, each organization continues to be a leader in their space.

CASE STUDY: THE POWER OF EFFICIENCY-SMART INSIGHT'S METER-LEVEL ANALYTICS AT BEACON CAPITAL

Beacon Capital Partners, LLC is a private real estate investment firm that develops, owns, and operates commercial real estate assets. Beacon's investments total over $30 billion. The company's investment strategy focuses on office buildings located in a select number of target markets that are primarily urban, knowledge-based economies. Beacon has made a strong commitment to sustainable operations—over 22 million square feet of the Beacon Portfolio is LEED Certified and Beacon has a policy to achieve the ENERGY STAR label at 100% of its properties, meaning they perform in the top 25% of their peer group.

The Opportunity

Energy is a critical concern for Beacon as part of the company's sustainability initiatives and continuing drive to reduce costs. In order to effectively manage energy across the portfolio, Al Scaramelli, Senior Vice President, wanted to analyze real-time energy usage to find opportunities to reduce energy cost through low or no-cost changes.

The Solution

Beacon Capital harnessed EnerNOC's DemandSMART solution to optimize their facilities for demand response in the U.S., creating a valuable revenue stream in exchange for agreeing to reduce energy consumption during times of peak demand. But demand response was just the tip of the iceberg. The team at Beacon became power users of EnerNOC's powerful online energy profiling tool, which gave Scaramelli access to real time energy data at all Beacon properties. Harnessing this real-time data, Scaramelli identified energy-related anomalies and worked with his team to take corrective actions that lead to significant energy savings. Encouraged by the results, Scaramelli was eager to realize these savings find other savings opportunities throughout the portfolio, but he needed additional tools to scale the results and a team of people trained to identify the highest impact energy savings opportunities. Scaramelli selected EnerNOC's EfficiencySMART Insight to parlay energy savings ideas across a large portfolio of sites.

EfficiencySMART leverages powerful real-time data analytics, fault detection software, and ro-bust reporting capabilities to systematically identify potential energy savings across a portfolio of properties. To compliment these results, Beacon turned to EnerNOC's team of energy analysts to periodically review the information and highlight the highest impact operational Energy Efficiency Measures (EEMs). The EnerNOC team then works collaboratively with Beacon and the property managers and engineers to provide workable efficiency recommendations that are tracked through the implementation phase. This workflow has been very effective at ensuring results are achieved.

The Results

Using EnerNOC's real-time electricity data, Scaramelli and the EnerNOC team have identified significant energy savings. *"The results have been amazing. We've been able to save several million dollars a year, and it costs us almost nothing."* (Figure 8-1)

Three early examples of EEMs include:

1) Smart Morning Startup

While most Building Management Systems (BMS) are programmed for timed morning start-ups, Scaramelli found that they were not necessarily focused on optimal energy efficiency. "The 'optimum start' feature of BMS or EMS was actually, in many cases, non-optimal," said Scaramelli. For one of Beacon's buildings in particular, the HVAC system was coming on throughout the

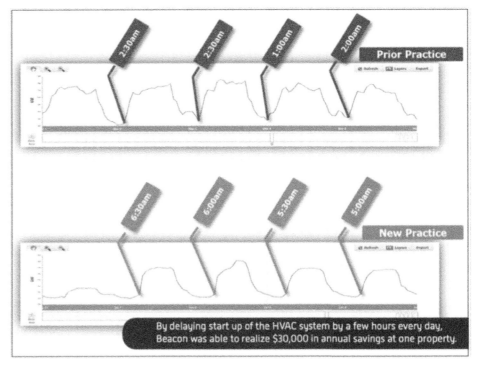

Figure 8-1

night, between 1 and 2am, and hitting the desired space temperature within 60 minutes. The tenants, however, were not arriving at the building until 7am. Scaramelli asked the building engineer to make an adjustment, and now the system kicks on much later, between 5 and 6am. Scaramelli continued to make this change across 80-90% of Beacon's buildings, and the savings have been significant. (Figure 8-2)

2) Energy Coasting

Like many class-A office buildings, Beacon's properties are contractually obligated to maintain room conditions within a certain temperature band to tenants during operating hours. Scaramelli experimented with shutting down cooling sources (such as chillers, cooling tower fans, and compressors) while continuing to provide ventilation through supply fan operation. What he found was that the zone temperature stayed well within the contracted band for a significant period of time, and the energy savings achieved by shutting down the cooling equipment earlier were extremely advantageous. For one building, Scaramelli saw 1050 kW savings for one hour, which amounted to $30,000 in annual savings for that property.

3) Eliminate Nighttime Base Load

Another Beacon building in Washington, DC had federal employees working in it from 7AM to 6PM. When looking at his real-time energy data, however, Scaramelli noticed a significant load plateau each day that lasted until 10PM, in spite of the fact that the building was empty most days by 7PM. After 10PM, the load dropped by 400 kW. There were cooling operations that went on during the night and emergency lighting added to nighttime load, however the kW did not seem to add up. With the electricity data in front of him, Scaramelli knew that something was not right. He asked the building engineer to check and he found that a chiller was actually set to run until 10pm every night.

As it turns out, a tenant had asked for some overtime HVAC a few months earlier, and the chiller's turn-off was set to 10pm, but then never set back to 6pm. Once corrected, Scaramelli could clearly see the next day's base load dropping at 6pm like it should have been all along. Within 30 days, over $200,000 of annual savings were identified as part of Scaramelli's key energy initiatives.

The Future

The Beacon team is resolving issues and achieving goals with a consistent process, based around the automated diagnostics of the EfficiencySMART Insight software and supported by regular meetings with the EnerNOC team. "Before this came along," said Scara-

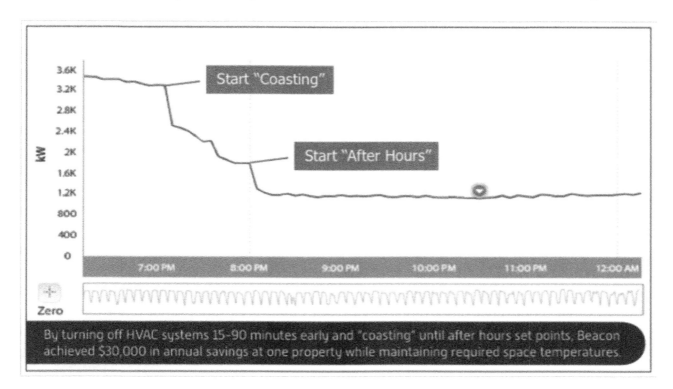

Figure 8-2

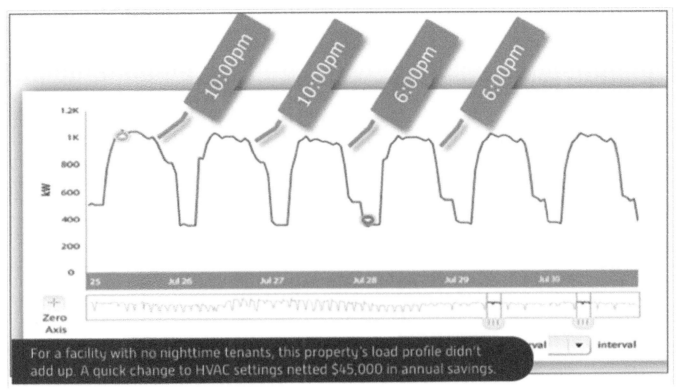

Figure 8-3

melli, "we were just blind. We had very limited data about what was going on inside our buildings." EfficiencySMART Insight is now in place throughout nearly all of Beacon's properties across the United States and continues to drive operational best practices and cost savings.

The on-going nature of the EfficiencySMART's performance monitoring ensures that savings persist over time.

CASE STUDY: MINING OPERATIONAL SAVINGS FROM BUILDING MANAGEMENT SYSTEMS WITH EFFICIENCYSMART AT WESTERN CONNECTICUT STATE UNIVERSITY

Western Connecticut State University (WCSU) is a leading higher educational institution in Connecticut, providing 6,000 undergraduate and graduate students with wide-ranging educational programs on its two campuses in Danbury. The school's leaders are aware of the need to reduce energy as a means to lower its overall environmental impact and work hard to incorporate sustainable business practices into everyday campus life.

In 2003, WCSU became the first state facility in Connecticut to reduce energy usage via demand re-sponse (DR). At the time, WCSU reduced an impressive 2,200 kilowatts (kW) of electricity per DR event, contributing to grid reliability in their region and earning approximately $100,000 each year in DR payments for their efforts.

In 2007, the university signed up for EfficiencySMART Insight to optimize its building operations. EfficiencySMART Insight delivers powerful visibility and analytics on campus energy use and enables facilities personnel to prioritize efforts to conserve. Using EfficiencySMART, *WCSU saved over $715,000 in the first three years of service* via new energy efficiency measures—and these savings are expected to expand as the software continues to identify new opportunities and prevent building drift.

Reducing Energy Use While Identifying New Energy Measures

WCSU invested the $100,000 in annual DR payments it receives from EnerNOC in automation systems and other programs designed to promote energy efficiency—including implementation of EfficiencySMART. "The transition from DR to optimizing our use of energy seems extremely natural to us," says Marcone. "EnerNOC gave us an opportunity to reinvest and take the next logical step—optimizing energy use throughout our building operations."

As part of the software setup and implementation, EnerNOC installed meters and monitoring equipment that collect and relay energy usage information from various building-level meters. During the next phase, EnerNOC worked with WCSU to integrate the EfficiencySMART Insight software with the university's existing Building Management System (BMS). Once deployed, EfficiencySMART software continuously monitors, compares, and analyzes mechanical system settings, operating schedules, lighting, and other operational components available through the BMS.

"EfficiencySMART has become a really valuable tool for us," says Marcone. "It's transparent and continues to help us improve energy efficiency on a day-to-day basis." EnerNOC's energy efficiency analysts identify and highlight operational anomalies that Marcone and his team can evaluate and take immediate action to fix— from simple issues (e.g., broken thermostats) to more complex equipment malfunctions (e.g., defective steam traps). "We have thousands of monitored points in our buildings," Marcone says. "We can't possibly monitor and verify them all. With EfficiencySMART, we know what's happening—and how we can improve energy efficiency."

Results That Are Protected

EfficiencySMART analysts identified 100 distinct, preventative measures with the potential to decrease energy costs by more than $385,000 each year. For example, the software uncovered scheduling issues relating to seasonal change, holidays, and changes in facility operation. Marcone and his staff evaluated these measures and weighed the potential savings against their effect on the campus.

Marcone sees the school's participation in DR and its energy efficiency efforts as part of the same challenge. "If large energy users like WCSU can continue to improve our energy efficiency, we won't find ourselves in emergency situations where we need to implement DR," he says. "Until then, we're doing what we can do to reduce energy use on our campus, while staying ready to curtail energy if and when a DR event happens."

WCSU's efforts already garnered the prestigious Energy Project Award from the New England chapter of the Association of Energy Engineers (AEE).

The Benefits

The ongoing nature of EnerNOC's automated diagnostic solution, EfficiencySMART Insight, is one of the key benefits it brings to WCSU. The system provides a clear, comprehensive view of campus operations, summarizes the energy profile of WCSU's facilities, and provides an at-a-glance review of opportunities to improve comfort, savings and preventative maintenance items. It highlights ongoing low-cost/no-cost energy efficiency opportunities that are identified by EnerNOC analysts who review operational data. These opportunities indicate annual avoided cost calculations, enabling WCSU to make informed decisions about which measures to implement.

"The system is extremely easy to read and interpret," says Marcone. "We see anomalies, differentials, energy consumption curves, and other information that lets us identify buildings and areas that aren't working efficiently—so we can reduce our energy use even more."

Other benefits of EnerNOC and EfficiencySMART include:

Fast Implementation

The EnerNOC team integrated EfficiencySMART Commissioning with WCSU's existing infrastructure quickly and easily, without disrupting operations.

Unparalleled Visibility

Before EfficiencySMART, WCSU had a steady flow of extensive building-level metering data, but no way to review or analyze it. EfficiencySMART now brings together meter and BMS data to identify energy efficiency opportunities automatically—without requiring more work from the university's busy engineers. It zeros in on areas that need attention and analyzes their impact and potential cost savings.

Informed Decision-Making

With EfficiencySMART data in hand, Marcone and his team can make informed decisions about which measures to implement. While replacing a malfunctioning thermostat is an easy decision, others are more complex and must balance any savings or increased efficiency against the impact on the infrastructure—as well as on the people who work and learn within it every day. In every case, WCSU can make decisions backed by data.

Better Use of Staff Time

EfficiencySMART helps WCSU monitor its buildings more vigilantly and automatically, freeing the staff to focus on more strategic assignments. It complements the capabilities of WCSU's facilities staff by providing critical data in a form that's easy to interpret.

Ongoing, Responsive Support

"From our initial DR project to implementing EfficiencySMART, we've continued to get a high level of service and support from EnerNOC," says Marcone. "They're our solid partner for all aspects of energy management."

The Future

WCSU's ultimate goal is to become carbon-neutral. This innovative university has signed the American College & University Presidents Climate Commitment—a national effort to neutralize greenhouse gas emissions and address climate change. As a first step, WCSU is already using EfficiencySMART data to help determine and map its current carbon emissions and footprint. As it continues to identify new energy efficiency measures, the school is looking for opportunities to reduce emissions, buy clean power, and reduce its overall environmental impact.

CONCLUSION

Proactive energy management requires operational visibility, actionable results, and quantified impacts. EIS is a powerful new tool that allows organizations to increase productivity and reduce costs through the power of real-time data and automated diagnostics. While the type of approach will vary from organization to organization, the phased method of increasing granularity has proven time and time again to be most effective.

Ultimately, energy management is linked to organizational business objectives. Whether you're a healthcare organization seeking to invest more in patient care, a commercial real estate firm looking to increase asset value or a public university seeking to lead by example and attract top talent, energy costs are variable and can be controlled. With EIS, the energy management bucket not only stops leaking, it fills with dollars.

Chapter 9

Di-BOSS: Research, Development & Deployment of the World's First Digital Building Operating System

Roger Anderson, Albert Boulanger, Vaibhav Bhandari, Jessica Forde,
Ashish Gagneja, Arthur Kressner, Ashwath Rajan, Vivek Rathod,
Doug Riecken, David Solomon, and Leon Wu, Columbia University

John Gilbert and Eugene Boniberger, Rudin Management

Mattia Cavanna, Willem Neiuwkerk, Bruce Sher, Nate Maloney, Selex ES

INTRODUCTION

Commercial and residential buildings are designed for tenant comfort and safety, with energy-efficient equipment managed by Building Management Systems (BMS). BMSs can integrate a number of components to assist building operators with maintenance and operation. BMS can be used to retrieve building energy-related data, such as data reading from electric sub-meters and space temperature sensors. Such systems can be operated efficiently to reduce costs of energy consumption while maintaining quality of comfort for tenants. For example, in the case of a commercial building, air condition systems can be regulated during business hours by lowering the speed of electric fans using variable frequency drives (VFDs) to reduce energy costs. However, BMS do not guarantee tenant comfort and reliable building operation because they do not integrate with lighting systems, tenant owned supplemental air conditioning and heating systems, elevator management systems, power systems, fire systems, security systems and the like. Accordingly, there is a need for improved techniques for optimizing the system-of-systems in larger buildings in order to simultaneously optimize comfort, safety, energy efficiency and equipment reliability in building operations and management.

Forty percent of all the energy usage in what we call "vertical cities," such as the great mega-cities New York, London, Shanghai, Hong Kong, Singapore, and Tokyo, to name just a few, is consumed in producing the comfort and safety of tenants in high-rise office buildings. In turn, forty percent of energy costs in these buildings go to maintaining and operating the HVAC air condi-

tioning and heating systems. High-Rise office buildings are by lease, obligated to provide comfortable spaces to the tenant companies that they house. The operation of these skyscrapers is of great importance, and is a fundamental baseline for the success of tenant companies and ultimately of the economies they drive. The buildings also occupy a very important business space, affecting the decisions of local government and power-grid companies, as well as providing necessary services of housing, security, power, and infrastructure to their tenants. As the world moves into the big-data age, the opportunity to optimize and build more secure and efficient energy systems is critically important.

Not only in large urban buildings, but also in university campuses, military bases, hospitals, industrial and manufacturing facilities, the capacity to reduce energy consumption by 2020 in these sectors of the global economy is more than 30% overall. But this improvement has to remain secondary to these operations' main functions: improved productivity while sustaining comfortable and safe operating environments. Such a system-of-systems approach requires that each building be treated as if it were like an organism, with data sensors providing the innervation throughout all critical components, and a central "Brain" needed to provide the identification of problems, evaluation of possible solutions, and prioritization of actions—all in real time. The brain must oversee all the subsystems of large buildings, like the BMS, HVAC, elevator control, fire, security and all their SCADA sensor networks that provide the information "innervating" the building. In addition, that brain should provide each building with an operating system (OS), like that in computers, tablets, and smart phones so that new applications (apps) can be easily drag-and-dropped into the central brain as new technologies spring up to serve subsystem needs.

THE ORIGIN OF DI-BOSS

The Digital Building Operating System (Di-BOSS) was conceived and designed to lower that percentage of energy consumed by using computer-aided Machine Learning (ML) techniques to form the brain and forward decisions to the building nervous system to take actions that optimize energy efficiency while maintaining tenant comfort and safety. It was born from a unique partnership among Rudin Management Company, which has state-of-the-art expertise in balancing tenant comfort with enhanced energy efficiency and Columbia University's Center for Computational Learning Systems (CCLS), which has state-of-the-art expertise in providing intelligence for control center operations, and Selex ES, a global security and systems integration technology company owned by Finmeccanica, the second largest conglomerate in Italy. Each party brought unique skills to the "3-Legged Stool" that now supports Di-BOSS.

The Di-BOSS collaboration began with the American Recovery and Reinvestment Act of 2009, which was intended to act as a stimulus to the recessionary national economy via DOE funding of improvements in infrastructure, including the Smart Electric Grid. Consolidated Edison of New York (Con Ed) had been collaborating with Columbia's CCLS on Smart Grid research since 2004, and together they jointly approached Rudin Management to join a partnership to propose a DOE demonstration project for the Smart Grid in New York City.

Con Ed and Columbia showed Rudin Management a 40-second warning of voltage and frequency sags that the Bulk Power Control Center recorded before the great northeastern blackout of 2003. At the time, the severity of the coming blackout was not fully realized. Columbia subsequently found that measurements of transmission voltage and frequency were being tracked at two different consoles on opposite sides of the bulk power control center. Not only should a Smart Grid be able to integrate systems within a control center, but also warnings could be sent out to customers of impending danger. Rudin was asked if a short 40-second warning before the 2003 blackout could have been useful. Absolutely yes was the immediate response: Elevators could be ordered to safe exit positions if smart computers could communicate directly between the Con Ed control center and Rudin's elevator management system (Figure 9-1). The DOE Smart Grid demonstration project was then created to develop such capabilities.

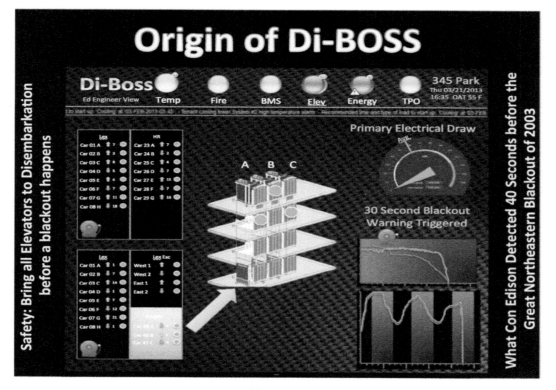

Figure 9-1

Con Ed, Rudin, and CCLS, along with several other partners, then won the largest award of all the DOE Demonstration Projects. Con Ed, in turn, decided to spend the federal funds elsewhere, and the funding of the Rudin/CCLS partnership was "Terminated for Convenience."

So the Rudin/CCLS team decided to work on their own, and beginning in 2009, they began to draw out the requirements for the development of the data-intensive machine learning tools necessary for ingesting big data sets, defining the innervation of buildings, identifying clusters of correlation, and marrying them to future forecasting so that the new brain (the nascent Di-BOSS) could deliver predictions of the future and recommend optimal actions to operators and managers to maximize energy savings while minimizing tenant disruptions in comfort and safety. The Total Property Optimizer prototype was built and installed in 2 Rudin buildings to see if its brain had any value. See Figure 9-2.

For over 100 years, Rudin has been the leading private owner/ manager of office space in Manhattan. They own and operate more than 10 million square feet in their 16-building commercial portfolio, and an additional 5 million sq. ft. of residential buildings. Several of their buildings are critical to the economic vitality of New York City, including perhaps the single most important building in the city, the original AT&T Long-

lines hub for all trans Atlantic telephone cables. That building is now central to almost all cell phone, HDTV, and internet-cloud traffic between NYC and the rest of the world. Another of their buildings houses the world headquarters of Thompson Reuters news service in Times Square, and their flagship on Park Avenue houses the worlds largest accounting firm and the National Football League. For more information visit www.rudin.com. See Figure 9-3.

First and foremost, Rudin Management has industry leading operational expertise in commercial real estate operations, and particularly, their skills are in energy efficiency. Since 2004, Rudin Management has improved the energy consumption of their entire portfolio by more than 15% for electricity, and by more than 40% for steam and natural gas (in Btu equivalents). They have won numerous national awards for both energy savings and environmental sustainability while lowering their overall energy costs from ~$60 million to less than $40 million per year. Year-over-year energy cost savings at Rudin have averaged 7% annually since 2008, with the Di-BOSS pilot in 2 million square feet saving $1 million of that since system testing began in the summer of 2012.

Second, the CCLS team encompasses Research, Development, Deployment and Engineering (RDD&E) of next generation real-time Machine Learning systems to control smart electric grids, manufacturing operations,

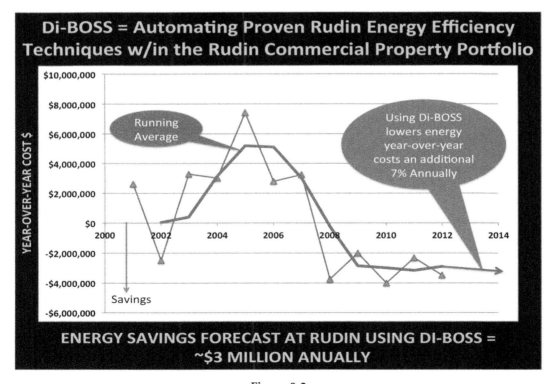

Figure 9-2

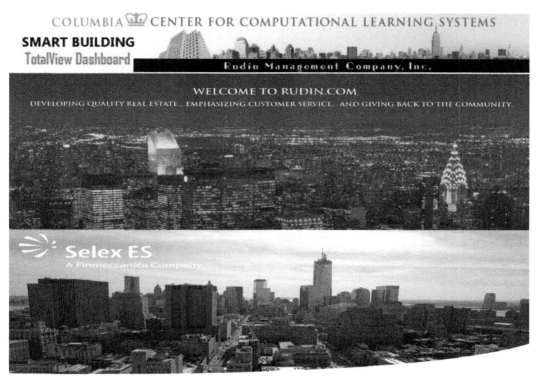

Figure 9-3

Smart Cities and the recharging of fleets of Electric Delivery Vehicles. The CCLS team specializes in adaptive stochastic control systems, in real options and portfolio management of the business being controlled, and in alternative energy options including solar, wind and large industrial battery storage for greening the business. The CCLS team is also affiliated with Columbia's Institute for Data Sciences and Engineering and the Lamont-Doherty Earth Observatory. For more information see http://ccls.engineering.columbia.edu.

However, neither Rudin nor Columbia market, install, or support commercial innovations. So they formed a 50/50 partnership and sought an additional partner to sell, deploy and support what would become Di-BOSS globally. After talking with the American suppliers of BMS and system-of-systems software, it became obvious that no out-of-the-box solution was required. In Finmeccanica, they found a subsidiary, Selex ES, which had the requisite electronics, systems integration, and security skills and global presence needed to market Di-BOSS. Selex ES was chosen to be the sales, marketing, and support partner. For more information see http://www.finmeccanica.com/business-mercati-markets/planet-inspired-solutions.

In October, 2012, Rudin and Columbia combined their intellectual property and jointly licensed a total of 17 background patents and all future intelligent building patents developed by the team, including TPO (Total Property Optimization), to Selex ES.

Selex ES immediately assigned a joint team from their American subsidiary Elsag NA operations (they do the license plate recognition software for the NYPD and other cities in the US), and from their Genova headquarters to integrate the TPO prototype into an industry-hardened Di-BOSS product. Selex ES had already globally marketed a Systems Integration Facility (SIF) that would become the middleware that gathers all the data from the disparate sources, historically archives it, and feeds the TPO and other alarming and recommending components of Di-BOSS in real time. Rudin then piloted the Di-BOSS™ system in two of its largest New York City properties. Rudin employees provided critical user feedback that influenced the system's user interface design, report formats, and analytical capabilities. Di-BOSS™ is now being rolled out in all 14 remaining Rudin commercial properties. Amazingly only 8 months later, Di-BOSS was announced as a commercial product at RealComm's Intelligent Buildings convention (IBcon) on June 10-12 in Orlando, FL. See http://diboss.selex-es.com.

THE DI-BOSS BUILDING OPERATING SYSTEM SOLUTION

"The Di-BOSS™ system's practical ease of use and ability to connect all of the building's systems are critical features that appealed greatly to Rudin's building managers and engineers. See Figure 9-4.

The feedback loops programmed in Di-BOSS™ also enable the system to predict adverse conditions such as power grid failures and allow building managers to act in advance to clear elevators and put other security measures into place to minimize the impact on occupants. In addition, because Di-BOSS™ tracks occupancy, building managers can provide headcounts by floor to emergency personnel if needed.

During the initial period of prototyping, the Rudin team would pour over each new ML and visualization capability, critiquing and correcting until the TPO was accurately forecasting electricity and steam consumption for each coming day of operations. From pre-heating in the winter, to startup each morning, to heating-up and cooling-down and ramp-down at the end of the day, recommendations were manually made by the TPO and acted on by the Rudin engineers, and energy consumption improvements tracked. See Figure 9-5

During the pilot project at the two Rudin buildings, savings were realized across the board from two buildings that had already increased their energy effi-

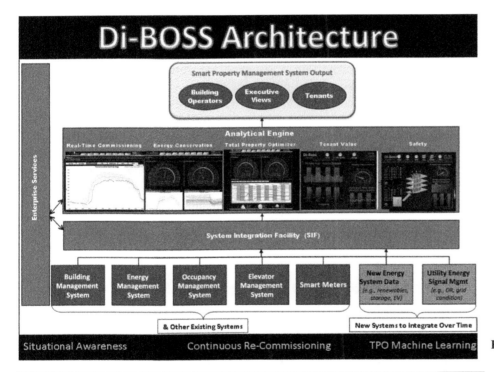

Figure 9-4

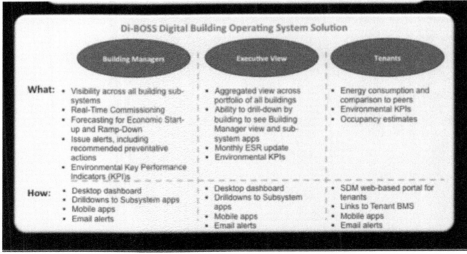

Figure 9-5

ciency by 30% since 2005. Return on Investment (ROI) was realized from better building and tenant practices, additional efficiencies in OPEX (operation expenses) and CAPEX(capital expenses), and replacement of reactive with anticipatory maintenance.

DI-BOSS EXECUTIVE COCKPIT

See Figure 9-6.

DI-BOSS BUILDING OPERATOR COCKPIT

See Figure 9-7.

DI-BOSS TENANT COCKPIT

The system also features the ability to analyze occupancy and energy consumption trends by tenant. Through secure online websites, tenants can check real-time occupancy and energy consumption data for their floors and can see their performance versus other tenants (Figure 9-8). Since tenants control roughly 60% of a building's energy consumption, Di-BOSS™ gives building managers the data to plan improvements that

result in real savings for tenants and for the total building. To realize optimal energy efficiency, it's clearly essential that owners, operators, tenants, and local utilities collaborate if all are to achieve optimal operation of buildings, campuses, bases, and facilities into the next century.

TPO, THE BRAIN OF DI-BOSS

CCLS developed an integrated intelligent building portfolio management and operation support system called the Total Property Optimization (TPO) system to provide the brain for Di-BOSS. TPO is a machine learning technology that analytically monitors and learns critical building environmental behaviors and makes recommendations to optimize performance by building operators to maintain correct environment levels while providing a baseline for continuous improvement in economic efficiency that does not compromise tenant comfort or safety. This novel approach combines human-in-the-loop expert control techniques with machine learning algorithms to better adapt to changing weather, building and tenant behavior over time. Continuous re-commissioning in real-time by Di-BOSS is the result.

TPO efficiently learns via analysis of massive data collected from a large range of building data sources

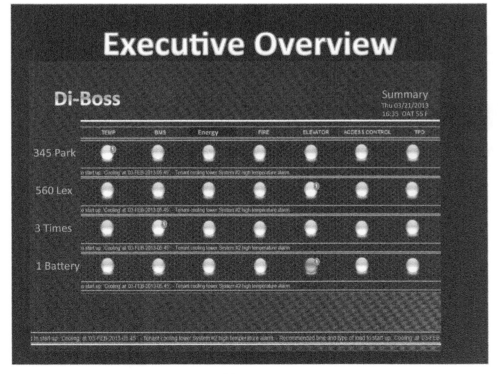

Figure 9-6

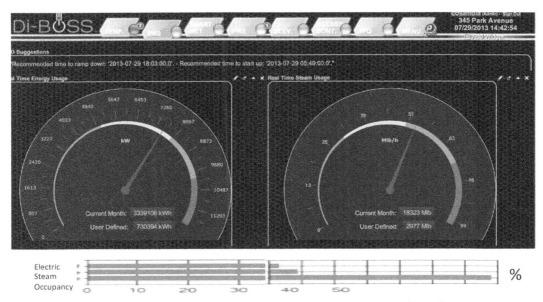

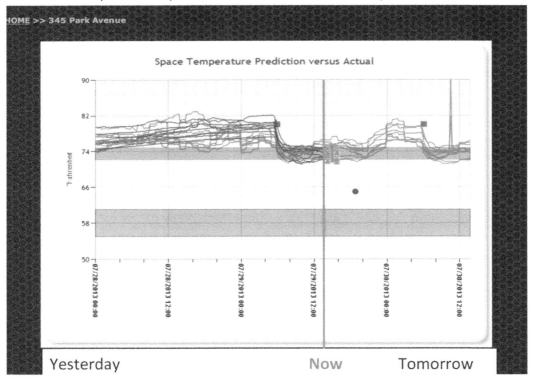

Figure 9-7a

and sensors distributed throughout each building in a company's Di-BOSS portfolio. TPO works as the integration component in the Di-BOSS architecture: 1) it receives BMS building SCADA along with occupancy data from the access control system, and weather history and forecasts; 2) it uses these archived historical datasets of covariates (predictors) to train Support Vector Machine learning tools to forecast day-ahead interior space temperatures, electrical consumption for each building as a whole and large tenants with smart meters, steam consumption, and occupancy. Each hour, the TPO recalculates these day-ahead forecasts; and 3) it transmits the forecasts and recommendations for preheating in the winter and daily startup and ramp-down times across

TPO communications architecture to the Di-BOSS System Integration Facility (SIF) which then feeds graphical User Interfaces (UI) for the executive, building operator, and tenant cockpits. The building operator TPO UI is shown in Figure 9-8.

TPO 24-hour-ahead Forecasting

A Support Vector Machine (SVM) model was selected for day-ahead forecasting of space temperature, electric and steam loads, and occupancy, after back-test experimentation with actual building and weather data. SVM is a classification/regression algorithm, which uses the concept of maximizing a dividing hyper-plane as the methodology to learn the functional mapping between the input covariates and output forecasts. Since there is a cyclical component in the building load profiles, covariates such as previous day space temperature, electrical, steam and occupancy loads, previous week load, previous day average, previous week average, time-of-the-day and day-of-the-week are incorporated in the learning model. Additionally to account for the weather variant HVAC load, an index called Humidex (composite of temperature and wet bulb dew point) is included as a covariate.

Each of the four forecasting components for electric (building and tenant), steam, space temperature and occupancy, is designed to perform two functions: prediction based on learning from past observations and optimization of learning/model parameter variables, with the optimization function performed daily and the prediction function performed hourly. That is, the parameter-optimized SVM prediction function is used hourly by TPO to calculate 24 different day-ahead SVM model forecasts for space temperatures of key floors, electrical and steam consumption for the building, electrical usage for tenants with smart meters, and building occupancy corresponding to each hour of each new day. See Figure 9-9.

For the daily parameter optimization, a grid search is used to find the optimal values of the parameters in the SVM model. In view of varying time series input data, a customized cross-validation algorithm was also implemented. TPO partitions the training data into two sets: 1) all available training data but one week is used to train the SVM model and 2) the left-out week's data that was not used for training is used to validate the model. The week that is left out is rotated among all available weeks in the training data and the procedure repeated. Minimizing the day-to-day variance of the error metric Mean Squared Error (MSE) is used as the performance score for parameter optimization. The parameter set corresponding to the minimum average MSE over the complete training data set is selected as the optimal one for all forecasting model runs until the results from the next optimization run are available. Retraining occurs in the early morning hours when the system load is expected to be low.

For the hourly forecasts, Figure 9-10 represents the basic internal processes performed for each respective component addressing one of the four forecasting functions for electric (building and tenant), steam, space temperature and occupancy. For example, the component for electric load prediction involves:

1. a process to "clean" the input data from the building systems and sensors; this data is processed to obtain covariates (variables exhibiting correlated variation),

2. the covariates are then applied by a task specific TPO SVM performing analytical machine learning,

3. the SVM learning generates a decision model update to be applied internally by the prediction/

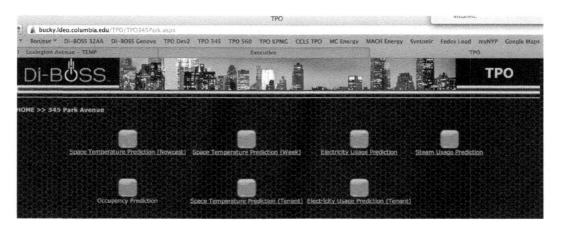

Figure 9-8

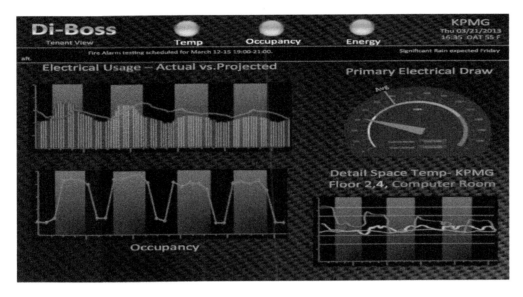

Figure 9-9

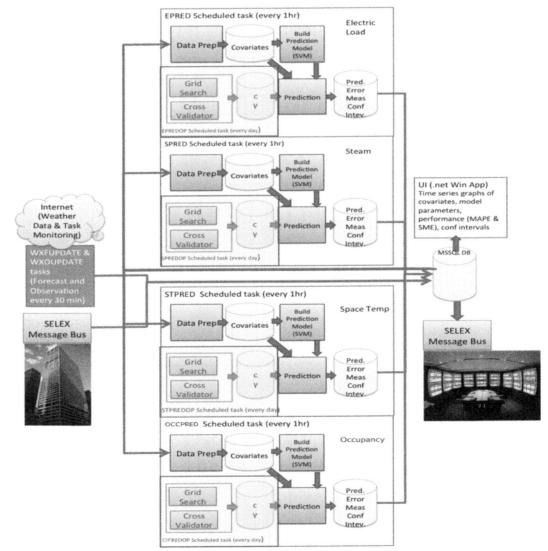

Figure 9-10

recommendation process; the prediction(s) are represented as data stored and forwarded to the Di-BOSS SIF which in turn posts the TPO output as updates to Di-BOSS subsystems for storage and UI for operations.

REAL-TIME PERFORMANCE METRICS

The scoring component for TPO forecasts was written to address the need to have an objective way to measure model performance and to also have a way to help decide whether any newly added covariate improves or degrades model performance. The scoring component uses the actual observations and compares them to predicted values and computes an error scores for each day. The magnitude of this daily number summarizes how the model performed on that day. This component runs in the early morning hours and computes the error score for the past day.

Mean Absolute Percentage Error (MAPE) is used to score forecast accuracy versus actual performance for Space Temperatures, Electric Consumption, and Occupancy. MAPE is used for periodically cyclical forecasts such as for sinusoidal variations over time. As you can

see from the real-time performance metrics automatically computed each day by TPO (Figure 9-11), temperatures are being forecast to less than 4% error, Electric to less than 2% error, and Occupancy to less than 15% error. Occupancy forecasting required conversion of in-only turnstiles to in-out counting turnstiles that took time to install, and it's forecasting was just begun in August of 2013. The error will come down as more time is accumulated in the database.

Root Mean Square Error (RMSE) is used for steam forecast scoring because the variation from day to day is more 'boxcar'-like than sinusoidal. That is, either on or off with the amplitude largely controlled by how many HVAC chillers are on at any given time. RMSE is measured in units of what is being compared to actual; and in this case steam units are Mlb/hr.

PREHEAT, STARTUP, AND RAMP-DOWN RECOMMENDATIONS

In June 2012, TPO began forecasting space temperatures for tenants at Rudin's 345 Park Avenue flagship (1.8 million sq ft), and in November, also at 560 Lexington Avenue (300,000 sq ft). By using the predic-

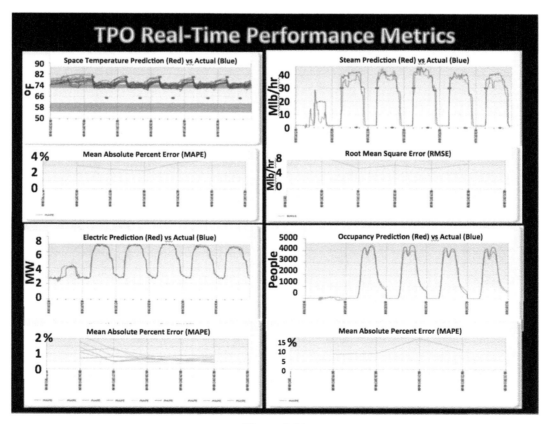

Figure 9-11

tions for space temperature, TPO is able to learn the past thermodynamic response of the normal operations of the building, floor-by-floor. This in turn is used by the preheat, startup and ramp-down models that make daily time recommendations. Valid preheat, startup and ramp down recommendations must be made by the end of business the day before, as a just-in-time recommendation would not serve the needs of building operators and managers. With this constraint on the problem, the TPO implementation optimization strategy assigns optimal preheat, startup and ramp down times to all past days (thus gaining access to the full variable set), and then learns the functional mappings between these optimal times.

Preheating recommendation

New York City's steam system supplies 27 billion pounds a year to heat, cool, and power Manhattan buildings. Many commercial buildings use steam to meet their space temperature requirements. Contractually, landlords must maintain a space temperature within a specific range during the workday. As a result, peak demand for steam in New York occurs during the colder months of the year. The provider of these steam services, Con Edison, charges an additional on-peak fee for steam demanded between the hours of 6 and 11 in the morning from December to March, as the workday begins. This on-peak-fee is equal to $1,629 times the maximum rate of steam, measured in million pounds per hour (Mlb\hr), demanded during on-peak hours within a billing cycle between December and March. This rate is the most expensive within Con Edison's billing structure, by far.

To minimize this charge during building start-up building managers attempt to heat the building using steam before 6 am. By storing energy generated by steam before 6 am, they "preheat" the building using an effective Hydro-Battery. That is they pump heated water into the riser circulation system of the building before 6am and return the hot water to the HVAC system after 6am at little additional cost. Specifically, if the maximum rate of steam demanded before 6 am is greater than the maximum rate demanded between 6 and 11 am, the building has been "preheated," and the cost of that off-peak-fee steam is 100 times cheaper.

An example of preheat during building start-up is shown in Figure 9-12. These graphs come from building data supplied by 345 Park Avenue. Typically, building managers at 345 Park preheat about every other day during the on-peak winter heating months. The two dates in each plot have similar weather based

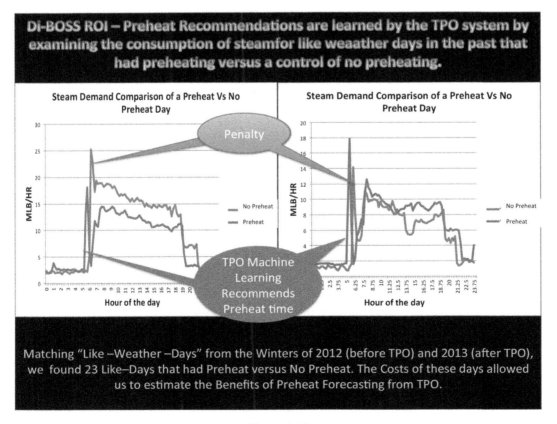

Figure 9-12

on the their heat indexes. In each plot, a spike occurs each day before 7 am. To store heat, building managers turn on water pumps to fill the vertical riser pipes with hot water. This sudden increase in demand for heat results in a steam demand spike. To release the stored heat, building managers turn on the HVAC fans that then circulate air heated by the hot water throughout the building. On preheat days, steam demand spikes occur before 6 am. One can observe that preheat does not always result in a greater daily steam consumption or spike in steam usage.

Given these temperature and peak demand penalties, building managers do not yet know the optimal method to reduce steam demand and steam costs during these on-peak months. The TPO preheat objective, therefore, is to compute the optimal means of heating the building that reduces steam demand and minimizes cost and recommend that the day before to building operators. TPO uses a Support Vector Machine model using covariates data going back as far as January 2012, which includes weather, predicted weather, internal temperature recordings, and water pump and fan indicators from the BMS.

A major conclusion from the analysis of past day preheating versus cost of operations is that the addition of the preheat cost of steam and electricity used for pumping the hot water up into the building risers is insignificant if penalties can be avoided between 6 and 11 am. The TPO Preheat calculator always recommends preheating during the winter season. Cost savings for that change in the winter of 2012 are given below, and preheating will occur every day during subsequent winters. This is a major operational policy change that TPO has recommended to Rudin Management.

Startup and Ramp-Down Recommendations

The startup and ramp-down recommendation generator employs tomorrow's weather forecast to select those learned days from the past that most closely fit tomorrow's forecast by day of the week. Every hour into the future, the actual weather is matched to the forecast weather to compute a corrected startup and ramp-down time 24 hours into the future from that new time.

High-rise office buildings are by contract, obligated to provide services to the tenant companies that they house. The operation of these skyscrapers is of great importance, and is a fundamental requirement for the success of tenant companies and ultimately of the economies they occupy. The buildings occupy a very important business space, affecting the decisions of local government and power-grid companies, as well as providing necessary services of housing, security, power, and infrastructure to their tenants. Though the buildings provide a wide array of services to these tenants, one of the most pertinent questions is
determining the operational hours of the building, from startup (building turn-on) to ramp down (building ramp down from full operations to turn-off) times.

The startup and ramp down recommendation generators of the TPO system both use advanced analytics to correlate 24-hour predictions and trends with recorded or calculated startup and ramp down times. Whereas deriving optimal startup and ramp down times requires access to a wide array of data, including values of HVAC control levers and current space temperatures, meaningful startup and ramp down recommendations must be made up to 24 hours ahead of time. The inability to meaningfully predict 24-hours expectations for all data in the building dataset makes this a fundamentally difficult question. This is where these TPO modules step in, allowing the mapping between an input space of 24-hour predictions and output "times" to be discovered. Each run of the startup and ramp down recommendation generators accepts as input temperature and electricity consumption data from the past 14 days, along with the predictions of these for tomorrow, and outputs the expected date and time of the next predicted startup and ramp down events, updated for 24 hours ahead.

CONVERGENCE
TO THE OPERATIONAL OPTIMAL

Because TPO is a human-in-the-loop system, and because it uses both 24-hour predictions and just-in-time optimization techniques, it produces "convergence to the optimal." This invention is unique to TPO, and manifests itself in the startup and ramp down recommendation generator. The idea of convergence to the optimal can be conceptualized as follows: Every hour, a new layer of 24-hour forward predictions is computed that provides TPO with accurate predictions of the behavior of the building system given the past actions of the building operators and exogenous variables like outside air temperature, the time of year, and whether the tenants are on holiday. However, this system weights more recent activity as more valuable than past variables, giving a temporal preference to more recent data.

Recall that the startup and ramp down recommendation systems discover the functional mapping between these hourly 24-hour predictions and provides a comparison and correction based on actual recorded versus calculated times from the recent past. The convergence to the optimal then happens as we calculate optimal startup and ramp down times, and provide these as the training labels for the recommendation engine. The recommendation engine outputs an updated recommendation for the next day's operation, hourly, and building operators act upon these optimal recommendations as morning and evening approach.

Over each following day there is a shift in the way the building startup and ramp down times are forecast. The degree of this shift depends on how sub-optimal the past startup and ramp down have been. TPO then computes from a shift in the temperature and energetic use of each day, optimal building operations for the next day. As each shift takes place, the 24-hour predictions begin to learn the new system, and adapt appropriately, predicting with each passing day, time-series data that represents the operation under the optimal conditions.

We escape any notion of circular logic by noting that the original optimization calculations used to identify past optimal startup and ramp down times begins to operate off of each newly past dataset. Since the optimization is based on a model that finds the thermodynamic response of each floor of the building to various similar startup and ramp down times from the past, and computes the cost-optimal solution for the future, it will either agree with the current history-based strategy, or discover a new optimality strategy. Thus, the layered nature of the TPO recommender system, and the nature of the TPO optimization will drive the system to always converge to the optimal building operation strategy. This is the phenomenon that we have termed "convergence to the optimal"

The startup and ramp down generator relies on new forms of input data:

1. 24-hour predictions: A separate energy forecasting module of our TPO machine learning suite uses a variety of covariates to predict 24-hour forecasts for space temperature, steam use, and electricity use, amongst others. By using this as our startup and ramp down covariate sets, we gain an understanding of how these 24-hour ahead predictions correlate and map to actual performance. Indeed, in using the TPO's 24-hour predicted values for energy as covariates, we are able to include an abstract covariate set into our predictions; intrinsically, we include covariates like occupancy, outside weather and holidays by use of the 24-hour predictions, since they use those variables in their forecasts.

2. A growing set of covariate generation methods: The nature of the data set as time series data allows for robust covariate generation. Current generation techniques look at trajectories (relative change over varying timescales), volatility, total change in values, percentage of maximum, and more in generating the parameter set to identify the classification function tying the input variables to the output time. Beyond also using such time series derived data, single variable covariates like time of day and the raw values of the past prediction accuracy are used.

3. A kernelized support vector machine (SVM) classification: To map the 24-hour prediction data to the provided startup times, we frame the learning task as a classification problem. Given times for startup and ramp down, we label all of the prediction data corresponding to the times before the startup and ramp down times as class = -1, and all of the values after as class = 1. We then arrive at our recommended times through a process of decision boundary discovery, where we use interpolation to find, at minute granularity, the recommended startup and ramp down times.

We employ SVM classification to make our estimate of the optimal decision boundary. SVM classification uses the concept of maximizing a dividing hyper-plane as the methodology to learn the functional mapping between the input and output spaces. Through use of a radial basis function (RBF) kernel, we are able to employ the 'kernel trick' to account for and discover the nonlinear relationship between our input variables and our output predictions. The current implementation runs a grid-search to find optimal parameters on every run, and uses such staples as covariate scaling and k-fold cross validation in the parameter optimization.

The choice to use SVM classification is innovative. Not only is the SVM well studied and validated in the literature, but its use of the "support vector" makes it an ideal candidate for dealing with time series data. Although this idea holds less rigorously in a kerneled infinite dimensional space like the one mapped to by the RBF kernel, we note that the decision boundary requires only a subset of data to be expressed. That is why the data points that define the decision boundary are called the "support vectors."

Startup Recommendation

Choosing an intelligent startup time for the building is very important: start too late, and operators will have to spike energy use and will struggle to meet contractual obligations of temperature settings within certain hours; start too soon, and needless energy and money will be wasted in powering a building needlessly. See Figure 9-13.

Ramp-Down Recommendation

As a part of building lease terms, building management operators are obligated to maintain ambient temperatures (space temperatures) within specific ranges while there are tenants actively using the space. The current operational paradigm does not capitalize on the savings that could be had by ramp-down of the HVAC in the building when tenants have left. See Figure 9-14.

The Ramp-down recommendation generator employs tomorrow's weather forecast, as does the startup recommender, to select those learned days from the past that most closely fit tomorrow's forecast by day of the week. Every hour into the future, the actual weather is matched to the forecast weather to compute a corrected ramp-down time 24 hours into the future from that new time. However, a dominant covariate is added to electric consumption and space temperature forecasts used for startup. Turnstile clicks from the Access Control system are used by TPO to forecast day-ahead occupancy ev-

ery hour. That, in turn, sees the exact time that tenants are beginning significant exiting of the building each afternoon sooner than either space temperatures begin to fall because each person is an equivalent heat source to a 100-watt light bulb. That loss of heating when a person leaves the building takes time to show up in overall floor space temperature sensors, and the same lag occurs from turning off their electrical equipment, which is often on a time-since-last-use switch.

Occupancy covariates allow the TPO SVM to forecast when tenants are going home each evening allowing ramp-down of HVAC systems earlier than previously detected resulting is significant savings (Figure above).

NOW-CAST

When Rudin requested short-term forecasting as well as 24-hr-ahead forecasting so that they could steer the building during rapidly changing conditions, CCLS created a 2-hour-ahead forecasting system called "Now-Casting" for the TPO. The Now-Cast space temperature trajectory suite of machine learning system sits atop the hourly 24-hour ahead forecasts, and gives insights into and makes predictions about the effects of

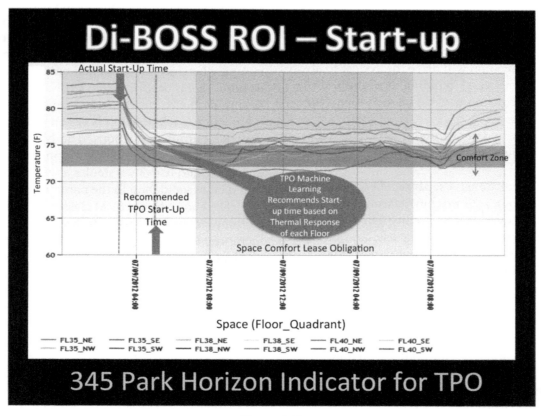

Figure 9-13

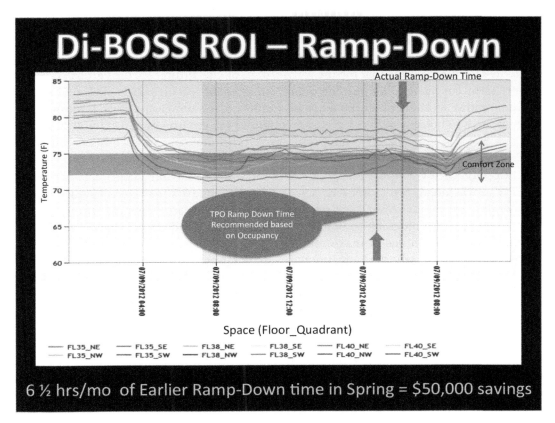

Figure 9-14

the current setting of the buildings temperature values and the values of the building operator's control levers on ambient space temperature. Utilizing both historical and predicted data, it uses a blend of relevant covariates to aid the building operators in ensuring their decisions will not deviate from tenant lease requirements. This, in short, takes any guesswork out of the building's operation. Each run of the Now-Cast suite provides temperature predictions from now to 2 hours ahead, resulting in 8 predictions (at 15 minute resolution) per floor.

The Now-Cast module is also a human-in-the-loop system, which uses SVM to provide building operators with the ability to steer the building to the most efficient energy comfort level floor-by-floor. HVAC setpoints and supply air temperatures are matched against this 2-hour look-ahead forecast to steer the floor space temperatures using the TPO Horizon Indicator. Though there are automated control systems at work in the BMS, ultimate control is left in the hands of highly skilled building operators at Rudin, who utilize specific control levers, most often in temperature setpoint values, to maintain the individual floor space temperatures. For example, if the Now-Cast predicts little change from the green band of best comfort, no change in supply air temperature setpoint is needed. If however, the Now-Cast suggests

movement upward to hotter conditions on the floor, setpoints should be lowered. Alternatively, if the Now-Cast suggests movement lower to colder conditions, setpoints should be raised.

The Now-Cast predicts a 2-hour-ahead space temperature trajectory using more time sensitive covariates than the 24-hour-ahead forecasting model:

1. Realtime space temperature values:

    In addition to the real-time BMS data feed which provides a view of the current temperature of the air in critical quadrants of each floor of the building, basic thermodynamic modeling of each floor's historical response to previous HVAC setpoint changes allows the Now-Cast to identify correlative relationships between the various air and water temperature HVAC settings and the ambient space temperatures.

2. Supply air and return air temperatures measured by the BMS and setpoints of specific fans feeding central air and heat to the appropriate floors:

    The real-time BMS data also provides a view of the current setpoint values for a variety of control systems. These are often in the form of thermostat setpoint values.

3. The forecast temperatures from the TPO

Adding the 24-hour forecast values for each floor, TPO is able to provide an abstract covariate set into the Now-Cast predictions that intrinsically includes short term covariates like current occupancy, outside weather and tenant workload (such as varying responses to holidays).

The Now-Cast space temperature trajectory suite uses kernelized Support Vector Regression (SVR) to supplement the TPO's SVM longer-term forecasts. Now-Cast covariate generation techniques look at trajectories (relative change over varying timescales), volatility, total change in values, and percentage of maximum to generate the parameter set to identify the SVR function tying all of the input variables and control levers to the output space temperatures from now to two-hours into the future.

Through use of a Radial Basis Function (RBF) kernel, the Now-Cast is able to employ the 'kernel trick' to account for and discover the nonlinear relationship between input variables and output predictions. The current implementation utilizes a daily grid-search to find optimal parameters, and uses such staples as k-fold cross validation in this parameter optimization.

CONCLUSION

Efficiency and safety will drive the design of buildings of the future – true vertical cities endowed with distributed intelligence, which will ensure, at the same time, cost savings, increased comfort and compliance with high environmental standards.

Technologies, like Di-BOSS, must be implemented not only in new building construction but in retrofitting and re-commissioning existing structures in order to meet the challenge of urban sustainability. Integrated technologies and skills are essential to improving the lives of people and the livability of large metropolitan areas.

Di-BOSS is big step forward in a totally integrated solution connecting all the "nerves" of a building to a central "brain" which can learn and forecast how the building engineering team should tailor the systems of the building to maximize efficiency while protecting tenant safety and comfort.

Di-BOSS, which is more traditional, albeit original in the way it is applied, consists in pooling all systems and sensors found in large scale buildings that host thousands of people every day. In this way, it was possible to ensure that all complex systems managing specific functions speak the same language. Distribution of

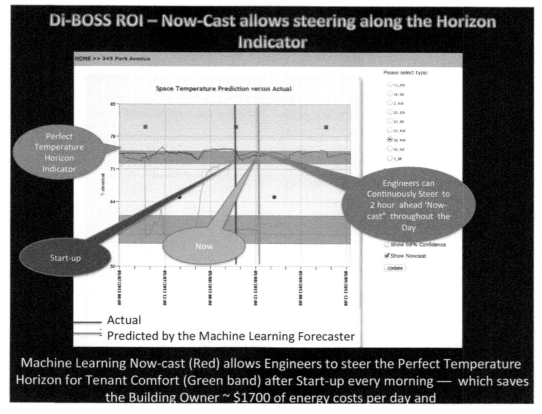

Figure 9-15

electricity and heat, management of the elevators, access control, fire control and the ICT backbone are all treated as ganglia of a single "central nervous system" within a smart living organism, controlled by a single "brain." These connections gave rise to matrices and algorithms used for energy monitoring.

In terms of integration, Di-BOSS effectively handles massive amounts of data related to the building's systems infrastructure. This information, which is gathered from heterogeneous and distributed systems, is processed and dispatched in real time to those who are responsible for the buildings' operations via Di-BOSS' "cockpit style" user interface.

However, Di-BOSS is more than a sophisticated analytical tool-as it has the ability "learn" from the behavior of the building, seen as a living organism, and then readjusts its functionalities according to needs and behaviors, with three major areas in mind: safety, comfort, and energy saving. For all practical purposes, this is an intelligent system that learns from experience, and is able to identify problems and propose solutions and recommendations.

A notable example is given by processes that allow the 'brain' to identify the optimal time for turning on air conditioning systems, varying it in accordance with behaviors recorded in the past and the number of users present in the building at a given time, depending on weather forecasts and the state of equipment on a specific day of the year. This allows for a reduction of energy consumption and greenhouse gas emissions—that's sustainability (see Figure 9-16). Additionally, Di-BOSS offers a module which allows the sharing of data directly with tenants so they can monitor individual consumption and participate at the tenant level with their own reduction planning.

Looking into the future, the next developments will involve the possibility of exchanging information between buildings– and comparable infrastructure–and the local power grid, a request already made by New York's grid operator, which is aiming to achieve collaborative demand-response management of power, improved responsiveness in the event of power failure, greater integration between renewable technologies and energy storage, and promotion of electric mobility.

The goal will be to turn large scale buildings into cells of a larger, more complex and intelligent living organism, that provide safe and comfortable environments for the occupants while reducing operating costs and meeting the demand of more sustainable energy consumption.

Chapter 10

Analytics, Alarms, Analysis, Fault Detection and Diagnostics
Making Sense of the Data-oriented Tools Available to Facility Managers

John Petze, Partner, SkyFoundry, LLC

Using data from smart devices such as building automation systems, smart meters, smart sensors, etc., is one of the hottest topics in the industry. The goal being to use this data to better understand the operation of building systems and identify ways to improve that operation.

Facility managers are confronted with a wide range of products and features that can contribute to improved facility performance. From Alarms to Fault Detection & Diagnosis (FDD), Analysis Tools, to Automated Analytics, each has their place an offers specific capabilities and benefits. Systems Integrators and owners, however, are often faced with a comparison between "apples and oranges" as they try to evaluate the different tools.

Comparing the technologies with set of criteria can help facility managers better understand the roles, capabilities and benefits of these tools so that they can assess the best fit for their needs. This paper will look at the range of tools from the perspective of:

- Time and Location of Implementation (when they are defined and where they run)
- Data Scope—the range of data items being analyzed
- Time Range of analysis
- Expressiveness of the tool—the tools available to describe the issue to be detected

ALARMS

Alarms are one of the fundamental tools that have been available in BAS systems since the early 1980's and remain an important tool. Often, when first introduced to advanced analytic tools people look to make comparisons with alarms. After all, doesn't an alarm programmed in a BAS tell me something is wrong? At a very basic level there is a similarity but if we look a little bit deeper we see that there are fundamental differences between alarms and more advanced analytic tools.

First of all, alarms require that you understood what you wanted to look for at the time you programmed the system. In other words you knew exactly what you wanted to look for and took the time to program that specific alarm definition into the system. This is fine for simple issues like a temperature going outside of a limit. There are many inter-relationships between equipment systems that may not be known at the time the control system was installed and commissioned, however.

One of the great benefits of analytics is that it enables you to find patterns and issues you weren't aware of at the outset of a project—providing results that show how your building systems are really operating vs. how you thought they were operating.

Given that alarms require that you could define the specific condition ahead of time, time of implementation is typically during the initial programming of the control system. This requirement fits a wide range of conditions that we want to identify in our control systems, but is also a limiting factor.

Data Scope

Alarms usually evaluate a sensor value vs. a limit. They may also include a time delay—i.e., the condition must be true for 5 minutes before an alarm is generated. Alarms are most often associated with a specific point. For example, one of the most common approaches is to set alarm limits for each individual point when it is configured. The data scope of alarms is also typically limited to the data in the local controller or other devices within the control system. Alarms do not typically evaluate enterprise data or data from other external sources.

Time Range

Alarms are typically evaluated "now." By this we mean the real time condition of the sensor vs. the alarm limit. This is a key point—very different techniques are needed to look back over months or years of data to identify conditions patterns and correlations.

Describing What Matters

The next difference to consider is the flexibility of expressing what you want to find. Alarms don't typically allow for sophisticated logic that interrelates multiple data items, conditions, data sources, etc. For example, an alarm definition might be: "Is the value of the Room Temp sensor above 76 degrees F right now?" An analytic evaluation on the other hand might be: "how me all the times when any room temperature was above 76 degrees in the last year for more than 5 minutes at a time during occupied hours, and totalize the number of hours by site."

Processing Location

Adding new alarms typically means modifying control logic or parameters in controllers. This means you need to have access rights to modify the controller logic to change or create alarms. This can be very limiting if you just want to "find things" in the data or are trying to analyze data from a system installed and managed by others.

The need to "reach into the controller" makes alarms "expensive" when trying to use them as an analysis tool. For example, could we justify reprogramming controllers in remote 500 sites because we have an idea of a data relationship we want to look for? Most likely this would be cost prohibitive. We might also ask whether the "analyst" should even have access rights to the configuration tools in an automation system. The important point is that there is significant "friction" involved in using basic alarm techniques for anything beyond limit-based relationships of individual points.

FDD—FAULT DETECTION & DIAGNOSIS

FDD techniques are typically equipment centric and characterized by pre-defined rules that are based on an engineering model of a piece of equipment—for example, FDD rules for a type of Packaged AHU.

Time of Implementation

There are two "implementation time" components to consider with FDD. Generally FDD requires that an engineering model of the equipment be developed before hand. In this respect they require significant pre-knowledge of the system. As such, FDD rules are often not flexible for use on custom, built-up central systems, etc. The fact that no two buildings are alike can further limit where FDD techniques can be applied.

Because of the dependence on predefined equipment models FDD is typically not a good fit for ad hoc analysis—e.g., "I have this idea about a behavior I want to detect." In addition, FDD rules can often be developed only by the software/service provider. The rules are "part of the product" versus being programmable on a project specific basis.

Processing Location

FDD solutions are typically applied as a separate software application that pulls data from the BAS system. The software may be installed locally or hosted in the cloud. Some FFD solutions can be programmed into BAS controllers. In this case they require "touching" the control system. As previously discussed, this can be a limiting factor in their application.

Data Scope

FDD rules are typically focused on the predefined points associated with a known piece of equipment. They may include data such as weather, but do not typically encompass external data, like age of building, historical energy consumption, type of facility, square footage, type of equipment, etc. or provide the ability to rollup and correlate data from hundreds of pieces of equipment.

Time Range

FDD rules typically look at real time conditions, but some have the ability to look at data from a sliding window of time—such as the last hour or day of operational data. Ad hoc analysis of random time periods (i.e., last August vs. this June) may not be available.

ANALYSIS TOOLS

Most often discussed in relation to energy meter data, analysis tools provide an experienced user with the ability to look at data with a range of charting tools and "slice and dice" that data with a range of tools to identify peaks, and other anomalies. Analysis tools typically include the ability to perform normalization against weather, building size, and other factors, etc.

The most significant characteristic of analysis tools

is that they require a knowledgeable user to be sitting in front of a screen to interpret that charts and graphs to identify the important issues—in other words, "wetware" is a key part of the issue identification process.

Data Scope

Most commercially available analysis tools focus on a specific type of data and application, for example, energy meter data. They integrate weather data (degree days as a minimum), occupancy schedules and building size, but do not integrate the full set of equipment data such as temperatures, pressures, speeds or rate of operation, equipment status, etc.

Time Range

From a time perspective, analysis tools provide the ability to analyze across a wide time range. As for "real time" data they can typically handle data "up to the last reading"—often a 15-minute sample, but are much less likely to be connected to a data feed that updates values every minute or second. They also support batch loads of historical data from meters, utility sources etc.

Processing Location

Analysis tools can be applied on top of existing systems as long as the data is available in some open format. They do not need to be part of the initial installation and typically do not require any changes to BAS programming. Analysis software can be hosted in the cloud or installed on-premise.

ANALYTICS

In many ways, analytics can be thought of as a superset of the other categories we have described. For example, analytics can be applied to "real-time" alarming situations and offer the ability to define more sophisticated alarm conditions to create "enhanced alarming."

FDD rules that diagnose equipment performance issues are a type of analytics as well. While most FDD solutions employ pre-written rules based on known models of equipment, programmable analytic tools enable experienced engineers to implement rules based on their knowledge—they are not limited to rules defined by the software provider.

In comparison to the other technologies analytics have the following characteristics:

Automated Processing

An analytics engine continuously processes data to look for the issues that an experienced engineer would normally look for manually. This ability to automatically process rules to identify important patterns and correlations is the hallmark of modern analytics solutions.

Time of Implementation

Analytic solutions can be implemented anytime, during initial installation or years after. They do not require reaching back into the control system to make programming changes for analysis. They do of course require that data be accessible (we will talk more about data availability in a moment.)

Expressiveness—Flexibility to Define Rules for Conditions to be Detected

While a typical alarm might evaluate a single item against a limit at a single point in time—analytic rules crunch through large volumes of time-series historical data to find patterns that are difficult or impossible to see when looking only at real-time data.

For example, while an alarm might tell us our building is above a specific kW limit right now, analytics tells us things like how many hours in the last 6 months did we exceed the electrical demand target? And how long were each of those periods of time, what time of the day did they occur and how were those events related to the operation of specific equipment systems, the weather or building usage patterns.

Analytic rule languages should enable sophisticated data transformations beyond limit checks. Examples include: Rollups across time periods, calculation of max, min, average, interpolation across missing data entries, linear regression, correlation of data sets to find patterns such as intersections (or lack thereof), etc.

One of the key characteristics of analytics is that they expose things you were not necessarily looking for, or even knew to look for. Analytic data presentations expose data relationships and correlations even without writing rules. And systems that offer user programmability enable new rules to be implemented as findings illuminate actual operating characteristics, and new priorities emerge due to changing energy costs, operating requirements or building usage patterns. In fact, the successful application of analytics is a journey with one discovery providing insight for additional analytic rules. More on this later.

A Wide Data Scope

Analytics enable multiple data sets from different sources, in different formats and with different time sampling frequencies. They are not limited to data within a controller or a control system. Examples might include:

- Energy, weather, and control system data
- Size of facility (sq ft/sq m)
- Age of building
- Type of building and systems (packaged HVAC vs. built up central systems)
- Equipment brand
- Service company

In many cases the analytic process starts with data available without establishing live connections to control systems, meters of other devices. More on this topic shortly.

Processing Location

Analytic tools can be applied on top of existing systems as long as the data is available in some open format. They do not require changes to the control system and do not need to be part of the initial installation. This is a key benefit, as they allow analysts to apply rules to the data without disturbing underlying systems.

Analytic software can be hosted in the cloud or installed on-premise. Each approach offers trade offs. For example, hosted solutions allow the software to be managed centrally, but require that the customer accept an external connection to their network for continuous access to data.

HIGHLIGHT THE DISTINCTIONS

An Alarm

Detect zone temperatures above 76 deg F when occupied.

An Analytic Rule

Look at signature of data associated with all sensors to indentify "broken" sensors or sensors out of calibration (Figure 10-1).

Alarm

Detect kW above a specified limit in real time

Analytic Rule

Identify periods of time demand is above a specified kW limit, calculate cost impact, make reports available showing, duration and even cost across any selected time frame, and provide continuous real time processing of the rule as new data are received (Figure 10-2).

An Analysis

Generate a graph of energy consumption across a specific time (Figure 10-3).

Analytics

Automatically correlate equipment operating status with energy consumption across a specific period of time (Figure 10-4).

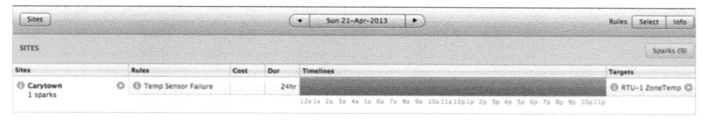

Figure 10-1.

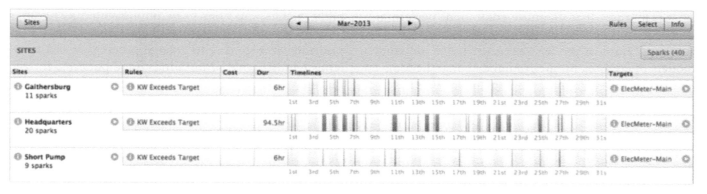

Figure 10-2.

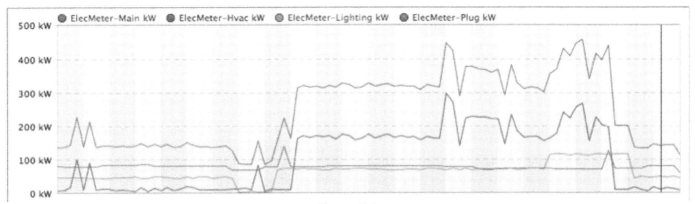

Figure 10-3.

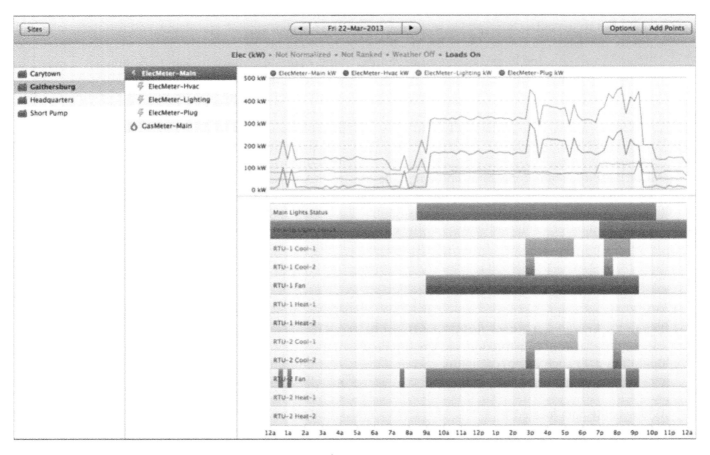

Figure 10-4.

CAN ANALYTICS EFFECT CONTROL ACTIONS?

Once analytics detects a pattern of interest can the system act on it? The answer is yes, analytics can be used to issue commands to control systems, but its important not to overestimate the applicability of this feature. Many issues found using analytics will not lend themselves to being corrected with a simple "command" to the control system. Two quick examples are illustrative:

1. Errors in control strategies. If an analytic rule detects conditions resulting from defects in a control sequence, the logic will need to be modified. An example might be simultaneous heating and cooling. While you could command the heating or cooling off when detected, the correction of the problem will require actual changes to the control sequence.

2. Physical equipment issues. If a damper linkage is broken or stuck, or a valve won't fully close, or a sensor is out of calibration or not reading correctly, there is no command to "fix' these issues.

This doesn't mean there is no use case for automated response to analytic results. Two examples of an automated response to analytic findings are:

1. Automatic generation of more intelligent work orders by the integration of analytics and CMMS tools.

2. Automated Demand Response. Demand response actions can be executed in response to energy use patterns detected (or predicted) using analytics. This provides more intelligence than more simple limit based demand response control.

DATA ACCESS—A KEY REQUIREMENT FOR ALL DATA-ORIENTED TOOLS

All of these tools are dependent on being able to access relevant data. Because alarms are processed locally in the control system data access is not an issue—the data "is there." In order to take advantage of the others tools, however, we need to assess the process to gain access to the data. A good way to start an assessment is by answering the following:

1. What data do you have?
    Examples: Energy meter data, facility data (size, location, type, year of construction, etc.), equipment operation data such as on/off status, sensor data, etc.

2. Where is the data located?
    Examples: BAS system, SQL database, utility company website, Excel spreadsheets, etc.

3. What method will be used to access it?
    Examples: Live collection of data via Bacnet or oBix, Haystack, Modbus, etc., data download from utility website via xml (perhaps Green Button data), CSV file import, SQL queries processed on a daily, weekly, hourly or minutely basis. The answers to these questions will vary dramatically based on characteristics of the specific project and customers needs.

ANALYTICS AS AN EXPLORATORY PROCESS

Analytics needs to be viewed as an exploratory process. Analytics show you how your buildings really operate, identifying where opportunities for savings exist, which assumptions are correct and which are not. Results from one rule or algorithm often identify behaviors that provide insight into other rules that should be implemented.

Because of this the analytic process is best applied incrementally, driving value to the owner with each step. We have a tendency to think of any energy conservation measure as a big capital expense project. Unlike energy conservation measures that involve the installation of major capital equipment, you can start small with analytics and generate returns in a very short period of time—you don't have to "do it all" at once to get value from analytics. The results from initial analytics generate the savings to go deeper into your operational data. This is a great advantage to building owners, in that it possible to start with a limited amount of data, and a low risk, low cost initial evaluation, to produce short term financial results.

An Example of a Shallow and Wide Approach. Basic interval energy data, combined with building occupancy, and weather data is a great place to start as shown below. Lets say the only data I can easily access is my interval meter data (kW demand), which is provided once per day by my utility company, and a list of occupancy schedule times in an Excel™ spreadsheet. We can gain any valuable insight from that limited amount of data. With just that limited amount of data SkySpark can identify:

*   Buildings starting early
*   Buildings running late
*   Buildings that operate continuously
*   Demand peaks that occur outside of occupied times
*   Peak Load, Annual & Monthly and Short Load Durations

Here's how. Using conventional energy analysis tools I can easily identify my kW demand pattern for each day. By looking at it manually I can determine whether the pattern follows occupancy times (Figure 10-5).

If I have a large number of buildings this type of manual review would be prohibitive. With analytics I can automate the process and have the software find these patterns for me. A rule can look for a percentage change in kW demand at the transition to and from occupancy. The rule generates "notifications" whenever the demand does not change by the expected amount within a defined period of time around the occupancy transition. In this way the system automatically identifies when buildings run late, and start early. A weekly (or monthly) view shows how many times the issue has occurred and the total cost. There's no need to hunt through the data manually (Figure 10-6).

DO THEY HAVE TO BE LIVE DATA?

Another common misconception is that you can't derive value from data unless they're live and continuously updating in real time. This simply isn't true. Live data are great, but by no means essential to get started with analytics. You can get tremendous value from running analytics on a snapshot of historic data. And, one of the benefits of using snapshot data is that you can avoid the costs and delays associated with gaining IT approval for network access to live systems. A great example is an initial portfolio assessment, but you can also do deep equipment analytics with historic data.

It's worth digging into the topic of "real time" data a bit more. We have found that there is a significant variation in what people mean when they use the term real time data. A few definitions can help clarify the topic and enable a clear understanding of how analytic tools can be used with "real time" and historic data.

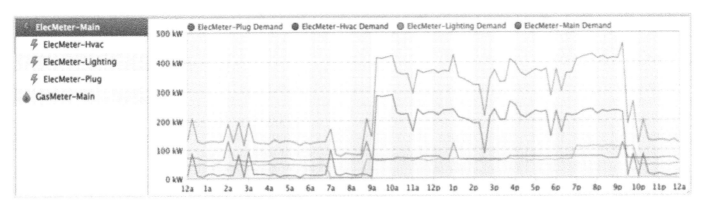

Figure 10-5

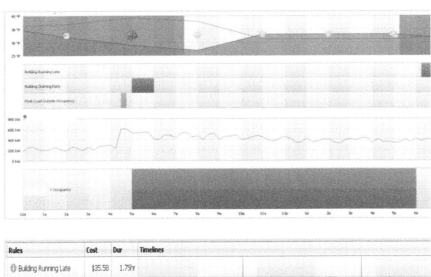

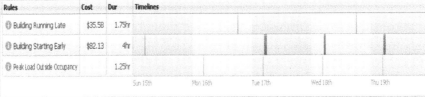

Figure 10-6

A Definition

Real time means fast enough for the application. So if we are controlling a piece of industrial manufacturing equipment that might mean we need a control loop response time of 10 milliseconds to meet the real time needs of the process. Many processes require even faster real time requirements—perhaps on the order of microseconds.

For a VAV box controlling temperature of a room, the control response time of the temperature control loop might be 30 seconds, a minute or longer. On that same VAV box, though, the control loop responsible for managing the airflow volume to maintain an airflow setpoint might have a response time requirement of 1 seconds.

At the other end of the spectrum, in energy metering applications real time often means 15-minute interval data. When we are looking at applying analytics to data there are a few facets of "time" to consider.

Resolution of Data—The Sampling Rate

One aspect of "real time" we need to consider is the resolution of the sensor data. By this we mean how frequently a system senses (and records) data values. For example, many BAS controllers can store data values on the second. In this case the data-sampling rate would be once per second.

The sampling rate of sensor data has a major impact on the volume of data created. For example a history record of zone temperatures once per minute will generate 1,440 values per day. That same sensor recorded on a sampling rate of 1 second will generate 86,400 sensor values. Sixty times as much data!

In most systems it's common that the control response capability and sampling rate are equal. For example, a controller capable of providing 1 second control response will often allow sensor values to be recorded once per second. One major limiting factor, however, is that most controllers will typically run out of storage space if you choose to record sensor values every second. So in reality they are often set to record values once per minute or longer, or need to upload the history data to a repository on a frequent basis.

Freshness of Data

The next concept to consider when discussing real time applications is the "freshness" of data or frequency of the updates to the software application consuming the data. A common example of this would be the update frequency of sensor data on an operator screen such as a graphic representation of an air handler. The

capability of a system to deliver updates to the user is affected by a variety of factors including:

- Bandwidth and speed of the communication network

- Computing resources of the controller that can be dedicated to communicating data to the UI application

- Efficiency of the software that presents that data to the user

- Computing resources of the computer that is displaying the data

Different systems vary considerably in this area and it's not uncommon for screen updates of 10 seconds to be considered "real time" updates. Going back to our definition of real time, the "application" here can be described as the ability (or need) of the human operator to read and respond to changing values he sees on the screen. Based on this definition a ten 10 second update frequency may be considered adequate.

The data source—the controller—may be operating its local control loops with a response time of 1 second or 100 milliseconds, but the fastest that new data will appear on the screen might be once every 10 seconds or longer. In some building automation systems it's not uncommon to see screen update times be much longer. That might be perfectly acceptable for the "application" of the human user (although most people find that a bit frustrating).

Application of These Principles with Analytics

When planning the application of an analytics solution all of these factors need to be considered. For example, knowing the sampling rate of the data will enable us to understand the richness that analytic rules will be able to work with—think about the different level of insight provided by 15 minute interval meter data versus monthly energy data. The sampling rate will also tell us about the volume of data we will need to store and manage. In many applications there will be different data sources, each with it's own sampling rate. The analytics application must be designed to handle the challenge of aligning and analyzing data with different sampling rates.

Next there is the question of how frequently data can be pulled from the source. Here again different data sources may have different update frequencies. Data Freshness is related to the application. The frequency with which you pull new data into the analytics database varies based on the needs of your application and limitations of your communication infrastructure. For example, you might have an application where you can only upload data from a site once per night at midnight (we've seen these restrictions placed on systems by IT departments). At that time, however, you might be pulling in data that has 1 second resolution. So in this case, the data freshness would be once every 24 hours, but the data sampling rate would be once per second. The sampling rate and the freshness are "decoupled." Then there is the question of the frequency of the rule processing. Ideally, rule processing is decoupled from the sampling rate and freshness of the data.

Next let's consider the user of the analytics application. Using the same example, we might have a situation where building operators want to see their analytic results (sparks) every morning. They could subscribe to a daily digest notification, which will provide a daily summary email. When they view their sparks they will be seeing all of the issues detected in the last 24 hours of data. Their "response time" would be on the order of a day (about the same time frame as the data freshness).

Too slow you say? It depends on the application and needs of the user—there is no one size fits all answer. For example, we have seen some examples where the data sampling rate is once per minute, the data freshness is once per 15 minutes, and yet the update provided to the user (the building operator) is once per week or once per month. That's right, building operators are informed of sparks once per week! Why might an operator find this level of "freshness" in sparks acceptable of even desirable?

It could be because the types of issues they are looking for can't really be addressed more quickly. Perhaps they need to be planned into their facility maintenance schedule, which is set on a weekly basis. Or it could be because they have so many issues requiring attention that they simply can't do anything with issues presented on a more frequent basis.

It could also be because the issues they are tracking take time to form—the patterns appear over a period of time. Some patterns form in a minute, some take hours, some take days and some might take a season or a year to form. This is a key point where people often confuse analytics with alarms. For example, the pattern that represents a defective sensor might require 12 or 24 hours of operation to detect. It can't be detected at any specific second in time—it is detected by interpreting a pattern in data that appears after some period of time has elapsed.

All of this leads us to appreciate that there is yet another "decoupled" response loop involved in the use of analytics—that of the corrective action response time. The limiting factor here is typically the human systems that will respond to analytic results. An issue detected by analytics might not be able to be addressed until the next planned service call to the facility.

A FAST MOVING FIELD

None of the examples presented is meant to be absolute, rather they are offered to help systems integrators and facility managers gain an understanding of these tools, their requirements and potential benefits. With the rapid advances in data-oriented facility management tools, there is some overlap between them, and the lines blur as vendors advance their technology.

As we stated in the introduction, each of these tools plays an important role in achieving an efficient building. Analysis tools help us gain insight into operating characteristics, which then support automated analytics to provide continuous detection of important issues relating to equipment operation and energy use.

Chapter 11

Using Custom Programs to Enhance Building Diagnostics and Energy Savings

Paul Allen, Walt Disney World
Rich Remke, Carrier Corporation
David Green, Green Management Services Inc.

ABSTRACT

Energy management system (EMS) control building heating, ventilation and air conditioning (HVAC) and is programmed to provide comfortable occupant temperatures/humidity levels and minimize energy savings. Keeping an EMS at optimal conditions is a difficult task. Changes in the building occupancy, equipment can result in changes in temperature/humidity and equipment schedules. As equipment ages, failures of sensors, actuators, dampers and valves can occur. Timely response to changes can keep an EMS operating at optimal conditions. By using the data from reports and functions already available in their EMS today, Users can better understand, diagnose and manage their building environmental conditions and energy usage better.

This chapter describes the several custom programs that aid in energy management system operation and diagnostics that result in lower energy usage and more comfortable building spaces. The custom programs showcased are: (1) the Facility Time Schedule Program and (2) the Building Tune-Up System, (3) the Carrier Alarm Notification program and (4) the Energy Management System Reports program.

THE FACILITY TIME SCHEDULE PROGRAM

One of the most fundamental features of an EMS is the ability to schedule the start/stop times of the equipment being controlled and provide setpoints for temperature/humidity and flows. The purpose of the Facility Time Schedule (FTS) program is to provide the Energy Manager a method to manage the time and setpoint schedules for a large campus facility in a master schedule database.

The FTS program automatically resets the EMS time schedules and setpoints to their optimal valves on a daily basis. Without this automatic reset feature that is created when the master schedules are downloaded each night, the time schedules and setpoints would eventually get changed from their optimal settings.

The FTS program is a custom client/server program that interfaces with the Carrier ComfortView EMS. The purpose of the FTS program is to provide the Energy Manager a method to manage the time and setpoint schedules for a large campus facility in a master schedule database. Time schedules can be set up as "relative schedules" that incorporate the facility opening/closing times, dusk/dawn times and by day of week. Each day, the FTS program determines the appropriate open/close/dusk/dawn times and calculates actual time schedules that are broadcast to the Carrier EMS controllers on the Carrier ComfortView Network (CCN).

The FTS program can also handle special events that occur after the normal open/close time schedules by sending additional time schedules that effectively increase the HVAC/Lighting equipment run-time to accommodate the special event. The FTS Program provides some additional operational features:

- Users can make local EMS panel adjustments to both time and setpoint schedules to respond to building conditions without worrying that the changes would be permanent. All schedule changes will revert to the master schedule at the programmed download time the next day.

- Setpoint schedules can be grouped into common areas or types and can be programmed with a bias offset. This offset can be used during loadshed conditions to change the setpoint low and high values to a user adjustable level, reducing energy consumption. Schedule groups can also be used to pre-cool or pre-heat an area during special functions. The setpoint bias will be removed from the setpoints after the next automatic download.

- These master schedules are sent automatically to each Carrier EMS controller on a daily basis. For each schedule, the system administrator can choose whether and when to send either or both time and setpoint schedules.

- If the facility open/close times are changed during a given day, the new time schedules are re-calculated and downloaded again to the EMS controllers to reflect these changes.

THE BUILDING TUNE-UP SYSTEM

The Building Tune-Up System (BTUS) is a web-based program that provides information on each air conditioning system and shows the time and setpoint schedules that are in effect. The purpose of the BTUS is to give Building Owners/Occupants a view into their buildings HVAC and lighting systems without accessing the EMS directly. The Building Owners provide the information on how their HVAC systems are controlled by establishing the time and setpoint schedules. This data is input in the BTUS database and provides a permanent record of the equipment time schedules and setpoints.

Because it is web-based, it also allows a broader audience to access it via its web browser interface.

The BTUS shows the following information for each HVAC system:

- Description of area serviced. A color-coded floor plan can be displayed if available.

- Time and setpoint schedules. Includes both desired schedules and a link to look at the most recent schedules broadcast by the FTS program.

- Shows equipment in need of repair.

- HVAC temperature, humidity and status trends can be graphically displayed if available

The main control on the BTUS is a drop-down menu that allows the user to select the area desired. The user is then presented with a list of buildings from which to pick. Once the user selects an individual building, the detailed data for each HVAC system is displayed. Figure 11-2 shows a screen shot that shows a list of the HVAC systems in the building PAVILION1. Links are available to show the detailed information on each HVAC system.

The BTUS displays the HVAC system time sched-

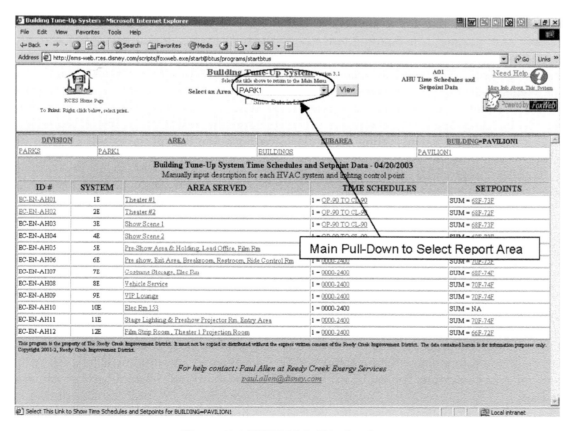

Figure 11-1. BTUS Main Display Screen

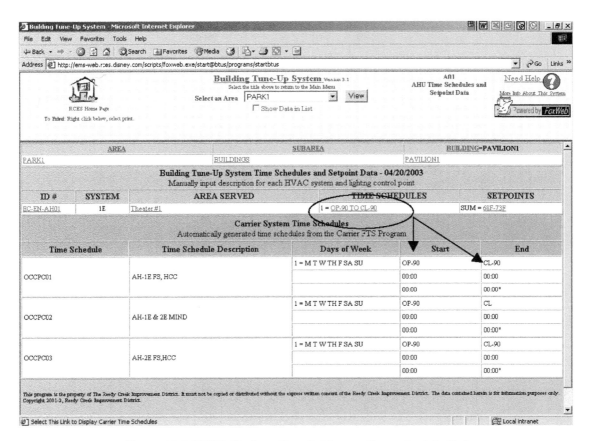

Figure 11-2. BTUS display after clicking on time schedule link

ules and setpoint schedules for each HVAC system. Clicking on the link for these schedules displays the latest actual schedules that were downloaded the previous night to the EMS controllers by the FTS program. Figures 11-3 and 11-4 show screen shots for these displays.

Another useful feature of the BTUS is to display a graphical floor plan that shows what each HVAC system covers. The floor plan is color coded to show the coverage areas for each HVAC system.

If an EMS trend report is available for a particular HVAC system, a link under the ID# will be highlighted. This makes an easy-to-use method to display trend data graphically.

If there is a link associated with the ID#, then this indicates that an EMS Trend reports is available for this AHU. After the user clicks on the ID# link, the EMS trend report is displayed for the previous day. Figure 11-5 shows a sample EMS trend report screen.

The EMS trend report is both informative and easy to use. This report uses extensive use of embedded links to sub-reports and graphs with a few clicks of the mouse. The EMS trend report shows the data in tabular format with embedded links on numbers and data labels that

the user can click on to either re-sort the data, produce detailed sub-reports or graphs showing the data various ways. This makes the use of the EMS trend reports intuitively easy for the user to navigate.

As an example, if the User clicks on the "AH-1E THEATER #1 TEMP" link in the report shown in Figure 11-5, a graph is displayed (Figure 11-6) showing the 24-hour maximum, minimum and average temperatures for this point. Similarly, if the user clicks on the "74.22" number in the report in Figure 11-5, a graph is displayed (Figure 11-7) showing the past 30-days average temperature for this point.

THE CARRIER ALARM NOTIFICATION PROGRAM

An EMS can be programmed to generate alarms to automatically notify maintenance personnel of equipment failures when they occur for immediate corrective response. Even with the best EMS, control equipment failures can and do occur over time. Using the EMS alarming function is an excellent means to let the EMS self-diagnose problems when they occur.

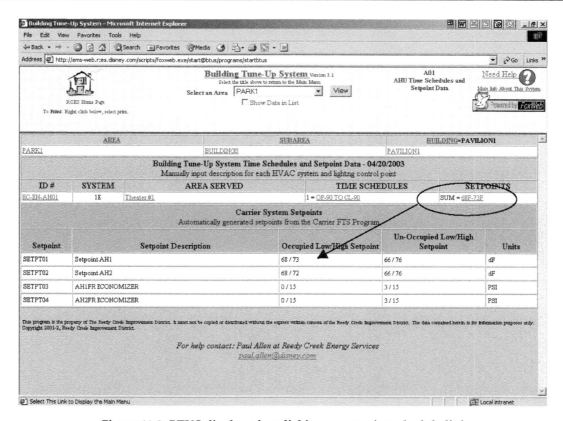

Figure 11-3. BTUS display after clicking on setpoint schedule link

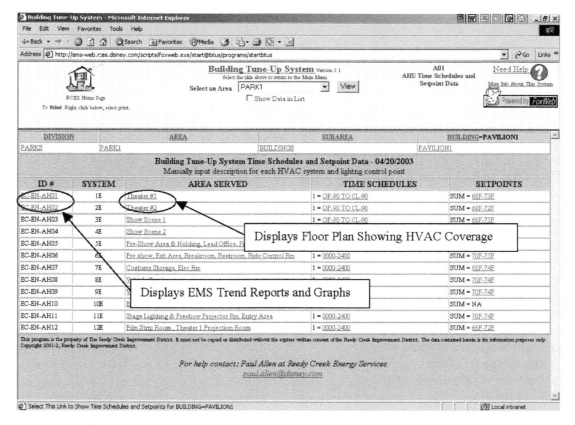

Figure 11-4. BTUS display showing EMS trend reports and floor plan links

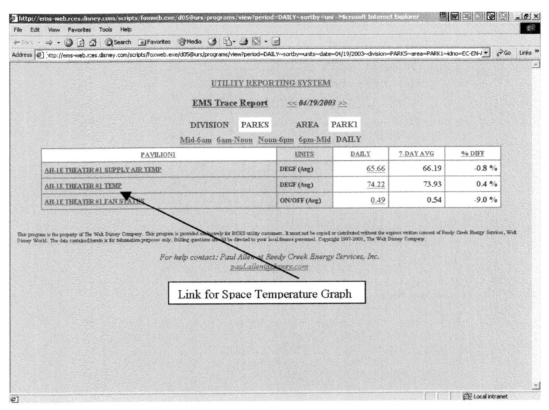

Figure 11-5. BTUS Trend Report for AH-1E in Pavilion1

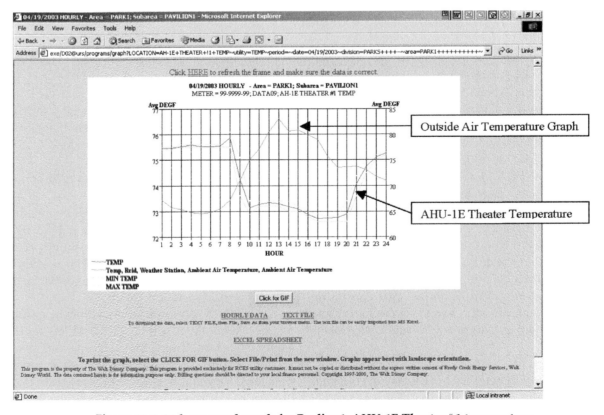

Figure 11-6. 24-hour trend graph for Pavlion1, AHU-1E Theater #1 temperature

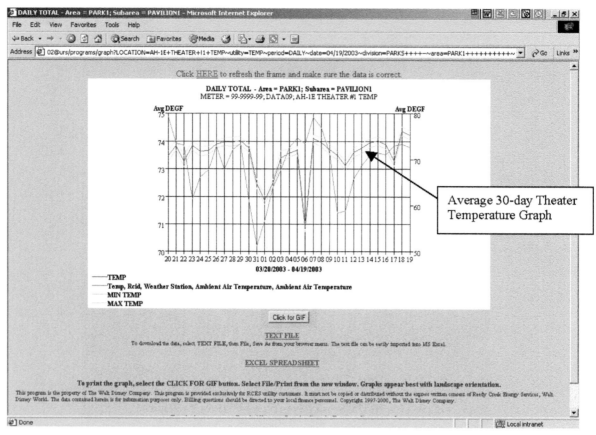

Figure 11-7. 30-day trend graph for Pavlion1, AHU-1E Theater #1 average temperature

The Carrier Alarm Notification (CAN) program enhances the EMS alarm capability by better managing the alarm data. When an EMS alarm occurs, the CAN program identifies the alarm and automatically sends the alarm via email/pages to the appropriate maintenance department. The alarms need to be prioritized to some extent, since the number of alarms could be overwhelming. The high priority alarms should be sent immediately by email/page to the appropriate maintenance personnel. An example of a high priority alarm would be a high temperature alarm for the chilled water system. This alarm would indicate something wrong with a chiller, chilled water pump or cooling tower that would require immediate attention.

There is another group of alarms that can notify the Maintenance Department when the HVAC/Lighting systems are not operating properly. Typical EMS Alarms are shown below:

• Chilled water valve closed, but display of cold supply-air temperature indicates that the valve is not closed.

• Fan that is commanded off, but the fan status shows that it remained on.

• Forcing of EMS controls to manual, rather than automatic control, which frequently suggests an operational or maintenance deficiency.

• EMS communication failures to EMS controllers.

Low priority alarms would not be emailed/paged immediately but would instead be logged into a database and included in a weekly summary report of all EMS alarms requiring attention. The weekly report is sent to the responsible maintenance department for follow-up corrective action.

The CAN program is a custom program that interfaces with the Carrier ComfortView energy management system. The CAN Program uses two separate programs in its operation. The first program runs on a dedicated server and is executes a program every 5 minutes to query the Carrier ComfortView SQL database Alarms Table. This table contains all of the alarms generated by the Carrier ComfortView program. This data is then compared to alarm definition table that contains the information about the alarms being monitored. If there is a match between the active alarms read from the Comfortview SQL Table and the Alarm Definition Table, the information is either emailed/paged out to the email/page distribution list assigned to the alarm or just logged to the CAN alarms history database.

The second program is a web-based system to display data in alarms history table along with data from other Carrier database tables containing manual overrides (Forces) and EMS Communication errors. This allows the User to quickly see all the EMS issues needing further attention. Clicking on the ALARMS button will display all of the alarms received within the dates and area selected. A sample report is shown in Figure 11-8.

In Figure 11-9, AH-1 SUPPLY FAN STATUS description means that AH-1 had a problem with the fan status. Either AH-1 is staying ON when commanded OFF by EMS (most likely), or AH-1 is OFF when commanded ON by EMS (least likely). This alarm occurred 6 times in the dates selected. To look at the details of this alarm, clicking on the "6" in the right most column to

generates the report shown in Figure 11-9.

The report shows that every night the same alarm occurs and shows "AH-1 SUPPLY FAN STATUS 1." The "1" indicates that the fan is staying ON when it should have been OFF. The most likely cause for this is that AH-1 is in MANUAL Control at the motor control center and needs to be put back into AUTO.

Figure 11-10 shows "AH-3 CHECK FOR DEFECTIVE CHW VALVE" which means that AH-3's chilled water valve was showing closed, but the supply air temperature was cold indicating the valve was probably not closing. This alarm occurred 11 times during the dates selected. This chilled water valve should be further tested by Maintenance Technicians and repaired or replaced to make sure it is operating properly.

Clicking on the FORCE button will display all of the EMS points that were overridden (or "FORCED") within the dates and area selected. The report shows who did the force, when the force was done, and to what point on the EMS was forced.

Clicking on the COMM button will display all of the EMS Controllers that were off-line and in EMS Communication failure within the dates and area selected.

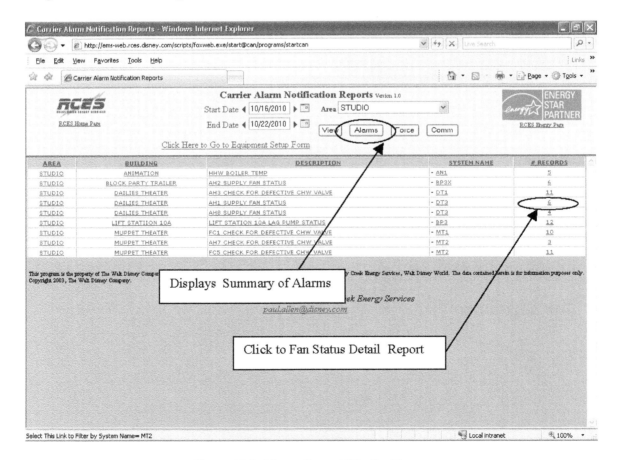

Figure 11-8. Alarms Report Display Page

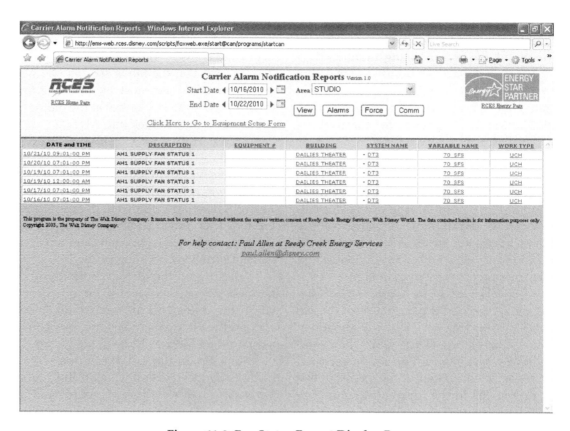

Figure 11-9. Fan Status Report Display Page

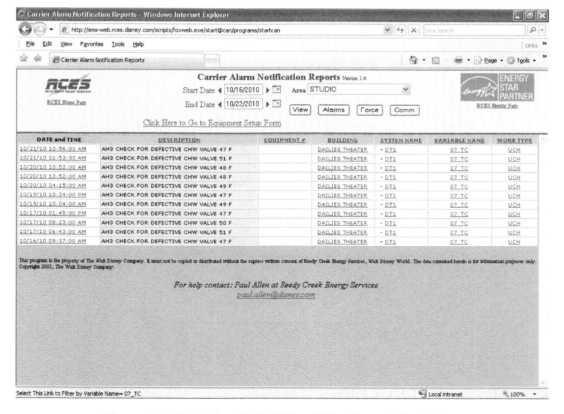

Figure 11-10. Defective Chilled Water Valve Report Display Page

The EMS communication to these controllers need to be investigated and brought back on-line. A common reason for controllers being off-line is lightning damage to EMS equipment. This report provides the Maintenance Technicians a list of EMS Controllers to look at and get the EMS Communications to the server working again.

EMS REPORTS

The EMS Reports Program was created to display the system activity report from the Carrier Comfort-WORKS Energy Management System. The EMS Reports program displays the actions each EMS User does when logged into the EMS to help in diagnosing problems that might have been caused by the EMS Users themselves.

Figure 11-11 shows the layout for the EMS Reports main screen.

EMS Reports program is intuitively easy to understand and use. The user simply selects the area they are interested in seeing then selects the VIEW button to display the report. The report defaults to the last area selected by the user and to yesterday's report date so

the user does not have to keep reselecting the areas they generally use. The program stores this information in a table by the user's IP address and looks up the last area selected every time the user starts up the program.

The report is displayed in chronicle order by the time and date of the events. The report shows the user's name that created the event, the type of action done, the controller and device that was affected and value if an override force was done. Typical events that are shown in the report are time and setpoint schedule changes, force and auto points, programming configuration changes and alarms. All system activity is shown associated with a user name, so that a history of events can be tracked to pinpoint system problem if they were to occur.

To further enhance the system easy-of-use, the user can either sort the report or filter the report data. To display only the events associated with a particular user, clicking on the user name would filter and re-display the report with only that user name. Figure 11-12 shows the data filtered on one user name.

The filter is released by clicking on the USER column title. The program is designed to filter on ANY data

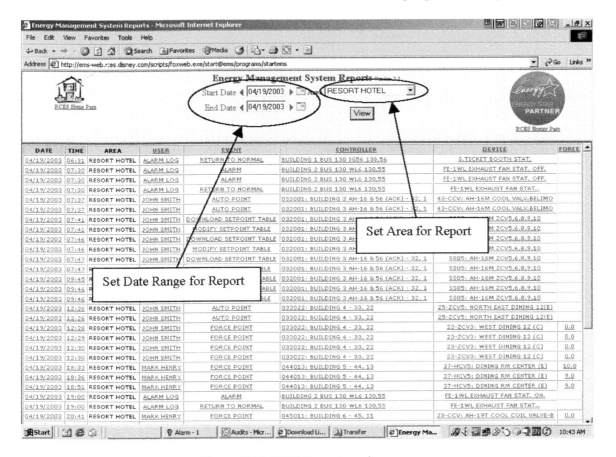

Figure 11-11. EMS Reports main screen

that are shown in the report by clicking on the link.

The report can also be sorted by clicking on a column title. This sort also works if the data has been previously filtered. Figure 11-13 shows the report that is sorted on the EVENT that was also filtered on one user.

The EMS Report program has provided the Engineering Services Departments with a simple tool to track the activities on their EMS. This has proven to be invaluable in troubleshooting EMS issues by continuously tracking all events and alarms.

CONCLUSION

Enhancing the existing reports and functions of an EMS can help diagnose problems and result in lower building energy usage and optimal EMS operation.

The custom programs described in this chapter keep the energy management system settings at their optimal state. The FTS program automatically resets the energy management system time schedules and setpoints to their optimal valves. The BTUS program allows all users to view the time schedules and setpoints from their own PCs using web browser software. The CAN program notifies the Maintenance Department when equipment gets out of normal operation so that timely repairs can be performed. The EMS Reports program identifies User changes to EMS programming.

Using existing data generated by the EMS, custom programs can reformat this data into information that lets Users diagnose their HVAC system operation through alarms and provides a method to prevent EMS degradation by auto-resetting time and setpoint schedules on a daily basis. Keeping HVAC at established company standards and EMS equipment working properly will result in lower energy consumption and costs.

References
(1)   Continuous Commissioning SM in Energy Conservation Programs, W. Dan Turner, Ph.D., P.E., Energy Systems Lab, Texas A&M University, 409-862-8480, e-mail: dturner@esl.tamu.edu

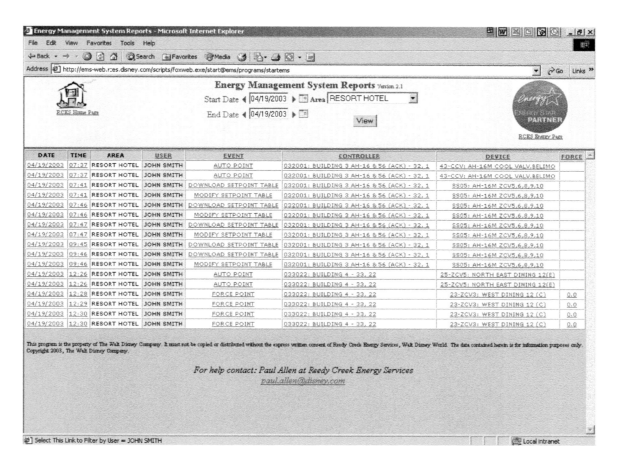

Figure 11-12. Report filtered on one user

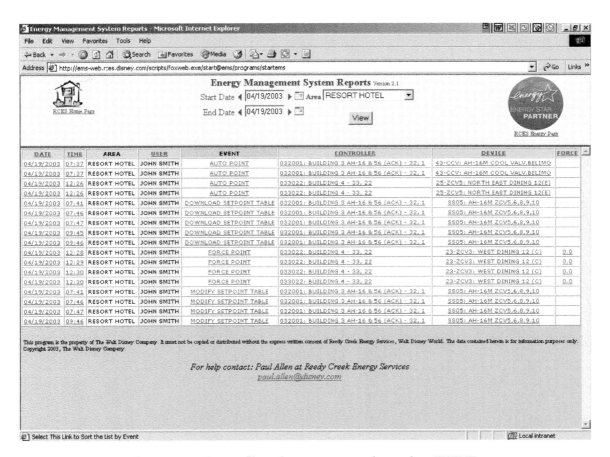

Figure 11-13. Report filtered on one user and sorted on EVENT

Chapter 12

88 Acres

How Microsoft Quietly Built the City of the Future*

Jennifer Warnick, Microsoft Corporation

A small, covert team of engineers at Microsoft cast aside suggestions that the company spend US$60 million to turn its 500-acre headquarters into a smart campus to achieve energy savings and other efficiency gains. Instead, applying an "Internet of Things meets Big Data" approach, the team invented a data-driven software solution that is slashing the cost of operating the campus' 125 buildings. The software, which is saving Microsoft millions of dollars, has been so successful that the company and its partners are now helping building managers across the world deploy the same solution. And with commercial buildings consuming an estimated 40 percent of the world's total energy, the potential is huge.

THE VISIONARY

"Give me a little data and I'll tell you a little.
Give me a lot of data and I'll save the world."

Darrell Smith
Director of Facilities and Energy—Microsoft

"This is my office," says the sticker on Darrell Smith's laptop, and it is. With his "office" tucked under his arm, Microsoft's director of facilities and energy is constantly shuttling between meetings all over the company's 500-acre, wooded campus in Redmond, Washington. But Smith always returns to one unique place. The Redmond Operations Center (often called "the ROC") is located in a drab, nondescript office park. Inside is something unique—a new state-of-the-art "brain" that is transforming Microsoft's 125-building, 41,664-employee headquarters into one of the smartest corporate campuses in the world.

Smith and his team have been working for more than three years to unify an incongruent network of sensors from different eras (think several decades of different sensor technology and dozens of manufacturers). The software that he and his team built strings together thousands of building sensors that track things like heaters, air conditioners, fans, and lights—harvesting billions of data points per week. That data has given the team deep insights, enabled better diagnostics, and has allowed for far more intelligent decision making. A test run of the program in 13 Microsoft buildings has provided staggering results—not only has Microsoft saved energy and millions in maintenance and utility costs, but the company now is hyper-aware of the way its buildings perform.

It's no small thing—whether a damper is stuck in Building 75 or a valve is leaky in Studio H—that engineers can now detect (and often fix with a few clicks) even the tiniest issues from their high-tech dashboard at their desks in the ROC rather than having to jump into a truck to go find and fix the problem in person. If the facility management world were Saturday morning cartoons, Smith and his team have effectively flipped the channel from "The Flintstones" to "The Jetsons." Instead of using stone-age rocks and hammers to keep out the cold, Smith's team invented a solution that relies on data to find and fix problems instantly and remotely. "Give me a little data and I'll tell you a little," he says. "Give me a lot of data and I'll save the world."

Smith joined Microsoft in December of 2008. His previous work managing data centers for Cisco had given him big ideas about how buildings could be smarter and more efficient, but until he came to Microsoft he lacked the technical resources to bring them to life. What he found at Microsoft was support for these ideas on all sides—from his boss to a handful of savvy facilities engineers. They all knew buildings could be smarter, and together they were going to find a way to make it so.

Smith has a finger-tapping restlessness that prevents him from sitting through an entire movie. His

---

*This chapter was excerpted and adapted from Jennifer Warnick's feature story "88 Acres: How Microsoft Quietly Built the City of the Future," which was originally published at www.microsoft.com/stories." Used here with permission from Mr. Darrell Smith, Microsoft Corporation

intensity comes paired with the enthusiastic, genial demeanor of a favorite bartender or a softball buddy (and indeed, he does play first base for a company softball team, the Microsoft Misfits). Ever punctual and an early riser, Smith lives near Microsoft headquarters and has taken to spending a few quiet hours at his desk on Sundays. "I call it my den because I live a mile away. I come here, I make coffee, I have the building to myself," Smith says.

His family and the people who know him best understand. Smart buildings are his passion, and everything in his life has been moving toward finding ways for companies the world over to get smarter about managing their buildings (which will help them save money and reduce their energy use). "Smart buildings will become smart cities," Smith says. "And smart cities will change everything.

88 ACRES IN A ONE-STOPLIGHT TOWN

Today Microsoft may have one of the smartest corporate campuses in the world, but in 1986, its headquarters was still a grass- and forest-covered 88-acre plot of land in Redmond, a sleepy, one-stoplight suburb of Seattle. Back then, there wasn't even a store in town to buy underwear, and from city hall you'd have to walk to the grocery store for lunch because the nearest fast food restaurant was too far away, says Judd Black, who has worked for the Redmond planning department for 26 years. "Redmond and Microsoft, we've grown up together, and we've learned from each other quite a bit," Black says. "We've both worked to create a great place for people to work and live."

The 88 acres of land Microsoft chose for its headquarters (the code name for the project was just that: 88 Acres) was originally supposed to be a shopping center, but that plan was bagged during hard economic times. Microsoft snapped up the land and quickly constructed its first office complex—four star-shaped buildings surrounding Lake Bill (a large pond affectionately named by employees for founder and then-CEO Bill Gates). Business was booming, and construction on campus followed—so quickly that Microsoft initially didn't have deeply defined construction standards. To meet demand, the company had to work with a variety of contractors and construction schedules, so consequently Microsoft's 125 buildings were constructed in a variety of styles and configurations. By the time Microsoft instituted comprehensive building standards in the 1990s, a large portion of the campus was already built.

Microsoft's campus would become the size of a small city—a city within a growing city. Redmond was expanding as well, and soon a quarter of its residents were Microsoft employees. "Redmond and Microsoft, we've grown up together, and we've learned from each other quite a bit" says Judd Black, City of Redmond. Today the campus spans 500 acres. There's a soccer field and cricket pitch, miles of wooded walking paths—and 14.9 million square feet of office space and labs that now function as one interconnected system.

It wasn't always that way. Until recently, Microsoft was using disparate building management systems to manage 30,000 unconnected, sensor-enabled pieces of equipment. Imagine a symphony orchestra, but with every musician playing from different sheet music. Then, imagine trying to conduct that symphony—to make sure the music was on tempo, in key, and starting and stopping as it should. Microsoft's buildings were experiencing data dissonance that would make the works of Igor Stravinsky sound like a barbershop quartet.

Smith's team was on a journey to find harmony. When Smith, Jay Pittenger (Smith's boss), and others started exploring ways to manage buildings smartly, they realized it would cost upward of $60 million to "rip and replace" enough equipment to get those 30,000 sensors to whistle the same tune. This would not only involve costly construction and equipment replacement, but it also would mean displacing employees and losing work while teams temporarily shut down labs. Smith and team knew there had to be a less pricey, less disruptive way to achieve data harmony, but after a whole lot of looking, they couldn't find one. So they invented one.

Smith's team enlisted the help of three vendors in the field of commercial building data systems and created a pilot program in 13 of the buildings on Microsoft's Redmond campus. The team developed an "analytical blanket" to lie on top of the diverse systems used to manage the buildings. The blanket of software finally enabled equipment and buildings to talk to each other, and to provide a wealth of data to building managers. "It hasn't been a bowl of cherries—my hair wasn't as gray before we started," Smith says. "The challenge with building systems is that they can create a lot of chatter from multiple systems, but there's value there if you connect and capture it. It's all about the data. If you can't get data out of the buildings, you're done."

The new tool did get data out of the buildings—great tidal waves of data that came cascading into the ROC, telling engineers about everything from wasteful lighting schedules to hugely inefficient (but up until then, silent and undetectable) battles being waged between air conditioners and heaters to keep temperatures pleasant. In one building garage, exhaust fans had been mistakenly left on for a year (to the tune of $66,000 of wasted energy). Within moments of coming online, the smart buildings solution sniffed out this fault and the problem was corrected.

In another building, the software informed engineers about a pressurization issue in a chilled water system. The problem took less than five minutes to fix, resulting in $12,000 of savings each year. Those fixes were just the beginning. Suddenly, the symphony of sensors was not only following the conductor, its musicians were all playing the same song. As buildings came online and data poured in, it created what engineers called a "target-rich environment" for problem solving. Smith and the team soon expanded the pilot to a handful of additional buildings, and by summer's end they plan to have the whole Redmond campus online.

The team now collects 500 million data transactions every 24 hours, and the smart buildings software presents engineers with prioritized lists of misbehaving equipment. Algorithms can balance out the cost of a fix in terms of money and energy being wasted with other factors such as how much impact fixing it will have on employees who work in that building. Because of that kind of analysis, a lower-cost problem in a research lab with critical operations may rank higher priority-wise than a higher-cost fix that directly affects few. Almost half of the issues the system identifies can be corrected in under a minute, Smith says.

The change has created groundbreaking opportunities for Smith and his team. "Our conversations have changed," Smith says. "Before, the calls we got were about buildings being too hot or too cold, or about work orders. Now we're talking about data points and building faults and energy usage. We're seeing efficiencies that we never even contemplated when we started this journey."

THE LIVING, BREATHING BUILDING

Tearle Whitson and Jonathan Grove stroll across the rooftop of Microsoft's Building 88. It's one of the first sunny days of spring, and singing birds accompany the picture-postcard view of Washington state's Cascade Mountains. "This is one of the perks of the job," says Whitson, taking a moment to survey the scene before climbing into the whirring, dark interior of one of the roof's large, white air handlers. As nice as they can be, now rooftop field trips like this are few and far between.

Whitson and Grove have experienced a seismic shift in their workday since helping to develop Microsoft's smart buildings tool. Two years ago the two spent a lot of time climbing over rooftops, inspecting pump rooms, and peering above ceiling tiles at variable air volume (VAV) boxes. "I used to spend 70 percent of my time gathering and compiling data and only about 30 percent of my time doing engineering," Grove says. "Our smart buildings work serves up data for me in easily consumable formats, so now I get to spend 95 percent of my time doing engineering, which is great."

Before Microsoft's buildings leapt up the IQ curve, the duo's home was on the so-called range. They'd move from building to building, camping out in each for two weeks at a time to inspect and tune it top to bottom before moving on to the next. It would take them five years to tune up all of the buildings on campus, and then they'd start the process all over again. Their tune-ups were making the buildings run more efficiently, saving the company around $250,000 annually—but the new data gold rush will help them save six times that much. The duo now spend most of their time at the ROC, chewing on building data. Though they're no longer camping out together tuning up Microsoft's campus the old-fashioned way, the two have maintained the comfortable rapport and geeky banter they established working in the field.

Facilities engineers like Whitson and Grove think of the buildings they care for as living, breathing things. Just like the human body, buildings have a wealth of indicators that things are going well—or, in some cases, not so well. Also like the human body, small ailments can lead to much larger failures, and an ounce of prevention can lead to many pounds of cure. And now, with the new data-driven software solution that the team built, they can do an even better job of managing the health of Microsoft's buildings—they've gone from country doctors, tapping a patient's knees with a rubber hammer, to specialists with an MRI machine who can examine every layer of the knee inside and out.

Still, Whitson says it can be a tad unnerving to take a building and its network of sensors online and watch as the software immediately discovers a host of inefficiencies. "There is a little bit of a mindset among facilities engineers. Everybody is prideful, and they take such ownership of their buildings, that it's hard for them to find out that there was a lot that they were missing. It was a humbling lesson I had to learn early on while doing this," Whitson says. "We have to get the old-school technicians out there to understand that this is going to help them. This is not to say you've been doing it wrong—you're doing fine. But you can go farther."

Both are quick to tell you that Microsoft's smart buildings solution has revolutionized the way they and their fellow engineers work. "We had the perfect environment and people to put all the pieces together. Our solution is a little unique to Microsoft, but very applicable industry-wide," Grove says.

ENOUGH DATA TO CHANGE THE WORLD

Microsoft's Smart Campus

A bevy of large, wall-mounted displays wrap around the interior of the room, and the same monitors sit atop every desk. The smart buildings tool dashboard is splashed across many of the screens, showing off a colorful collection of maps, dials, lists, and tickers. Engineers can get big-picture information at a glance, like how many kilowatts of energy are being consumed across Microsoft headquarters at any one moment. With a few clicks they can also zoom in on one building, one floor or office in that building, or one piece of equipment.

"Let's see how City Center building is doing today," Whitson says, and within seconds he's clicking through a wealth of information about the building—the number of employees who work there, the outside air temperature, the thermostat, what time the lights come on and go off, even a list of mechanical inefficiencies the software has detected and how much each of those faults is costing the company per year.

Once Whitson sits down in front of one of the terminals, he is suddenly connected to Microsoft's buildings. Those 500 million data transactions per day can be accessed from the dashboard at his desk rather than from crawling through pump rooms or across rooftops to get data. Now that the smart buildings software cooks up a chuck wagon of data every day, what to do with all of that tasty information? The software identifies issues large and small, and even puts them in prioritized order according to how much the problem is costing the company. A majority of problems they can fix right from their desks, and for the rest, the engineers issue work orders (about 32,300 per business quarter).

Apart from efficiency, the surge of data has also made for some eye-popping analytics. These are mechanical engineer Trevor Sodorff's specialty. "We have good people, but without good software there are limits to what you can do," Sodorff says. "Everything lives within the context of the bigger picture." One of Sodorff's party tricks, if you will, is whipping out algo-

rithms to detect new mechanical faults. So during meetings that wander through stretches that don't pertain directly to him? Rather than discreetly checking his email or letting his mind wander, Sodorff writes algorithms. At one such meeting, Sodorff announces that he's just written a new algorithm for detecting when the air in a given building is being overcooled. He projects the algorithm on a screen, and then launches into a deeply technical explanation about when a discharge air pressure set point is something-something, then the air is being overcooled by something-something for a duration of 900,000 milliseconds. "That's 15 minutes," says Grove, his fellow engineer, translating on the fly.

Later in the meeting, Grove is talking about how the smart buildings software helps the engineers measure and validate that the energy reduction they're seeing is due to reduced consumption and not because it was 5 degrees cooler than yesterday. It's an important distinction for companies to make, especially when seeking a utility rebate. "We may do an audit, and find we've done something that saves 200,000 kilowatt hours, which works out to sixteen-thousand dollars."

Darrell Smith beams at his team. "Now you see why developing this software at this scale was a once-in-a-lifetime opportunity," Smith says. "It affords you the ability to work on some very large-scale, world-changing projects with some very smart people. It's Microsoft University." If the smart buildings tool was developed at Microsoft University, today is graduation day. Where much of Smith's time the last few years has been spent developing the software, he now spends hours with visiting business, government and industry leaders offering enthusiastic show-and-tells. He's presented to hospitals, oil companies, automobile manufacturers, cities, and federal government agencies—even at the Pentagon and very soon, this same solution Microsoft has deployed will be available to any business.

"Never in my wildest dreams did I think I'd be presenting at the Pentagon. It was a thrill," Smith says. "It's been interesting, because I don't see myself as a salesperson. I see myself as an evangelist for the smart building industry, and what can be achieved with smarter buildings." Office buildings, hotels, stores, schools, hospitals, malls and other such commercial buildings are responsible for up to 40 percent of the world's total energy consumption. In the U.S. alone, businesses spend about $100 billion on energy every year. "Buildings have been built and run the same way for the last 30 to 50 years," Smith says. "This isn't a Microsoft problem, it's an industry problem."

Microsoft's Smart Campus

125 buildings with...
2,000,000 data points

500,000,000 data transactions every 24 hours

Communicated through an array of different
Protocols, Hardware, and Interfaces

Transforming raw data into
Actionable Information

Assimilating information from
30,000 pieces of equipment across campus

Analyzed and compiled through
Graphics, Charts, and Trending Reports

48% of faults are corrected within
60 Seconds

Provides a forecasted energy savings of
6-10% Per Year
with an implementation payback in less than 18 months

Improves technician efficiency with
32,300 Work Orders Per Quarter

Built using Microsoft technology

Microsoft

A SMARTER FUTURE

In one memorable scene from the 1987 movie Wall Street, Michael Douglas stands on a beach, hair slick and wearing a black bathrobe, barking orders to his protégé via a brick-sized cell phone as the waves crash behind him. It was one of the first times a mobile phone appeared in a movie, and at the time talking on the phone on the beach demonstrated the high-tech fruits of a stock broker's wealth and power. Twenty-six years later, the scene feels dated—solely for technology reasons. That brick with an antenna that Douglas is holding? That is the current state of commercial buildings, Smith says. "At first, all mobile phones were bricks—basically two-way radios. Now, in just a few short years, they have advanced to become a laptop in your hand," Smith says.

"Buildings are still that brick phone. We want to get buildings to where phones are."

Microsoft's campus went from bricks to brains, and Smith believes all commercial buildings can follow suit. Jessica Granderson, a commercial building and lighting research scientist at Lawrence Berkeley National Laboratory, agrees. She says smart buildings are becoming more prevalent, and the commercial real estate industry is going through a "time of rapid change and maturation." Granderson, whose research focuses on intelligent lighting controls and building energy performance monitoring and diagnostics, says the commercial real estate industry is reaching a major tipping point as people realize the power of capturing and analyzing data from buildings.

Granderson says it's still not as easy as it should be to get data out of today's buildings. "There are two sides to the coin really. There's data that is out there that's not put to good use, and businesses that aren't using it. The other side of the coin is that once we do become in the habit of making use of what's there, then you quickly realize the challenges of getting what you really need and want," she says. Another challenge is having the right people in place to analyze and interpret the data and act on it. Granderson contributed to a white paper on Microsoft's new software, "Energy-Smart Buildings: Demonstrating how information technology can cut energy use and costs of real estate portfolios." The paper investigates and evaluates Microsoft's smart building tool, which Granderson says deserves to be held up as a best practice in the industry. "It's one of the more sophisticated implementations that I've seen," she says. "There are a lot of lessons to be learned (in what Microsoft created)."

Probably the most important take-away from Microsoft's smart buildings breakthrough is just how much money and energy businesses can save with relatively little up-front investment, says Jim Young, CEO of the commercial real estate and information technology company Realcomm. Young has traveled all over the globe for more than a decade, visiting and studying all manner of smart buildings and smart campuses. "Companies, even a lot of Fortune 500 companies, have these massive real estate portfolios that they're running with sledgehammers," Young says. "What Darrell and his team can do is watch their buildings at a rate no human has before. When you start connecting to the buildings, generating data, and grouping inefficiencies by cost and priority, all of a sudden you go from the sledgehammer to running buildings with laser precision."

Smith's plan to take Microsoft's smart buildings software worldwide is to adopt the role of matchmak-

er. His team developed the smart buildings software, with the help of vendors, exclusively with off-the-shelf Microsoft software such as Windows Azure, SQL Server and Microsoft Office. Smith says these partners and vendors are eager to help businesses of every size, shape, and need to take their buildings from piles of bricks to data-driven brains. Some companies need only a little push—the know-how to incorporate weather data, or energy meters, or perhaps just the right connections. Others need both hands held, procedurally and technologically. Regardless, Smith and his partners can help. "People may be bumping their heads, but they come here and get a glimpse of the potential," he says.

Young says Smith has been talking about the possibility of making buildings smarter since the two met more than a decade ago at a Realcomm conference on the collision of commercial real estate and information technology. "He'd say, 'I'm going to figure this out.' I'd call him every few months and say, 'How are you doing, Darrell?' and he'd say, 'I'm not ready to talk to you yet.' Then, when he finally had something to show, he was like, 'Here it is,' and it blew us away," Young says. Though Smith and his team were quiet for the years they spent developing the software, their work has already garnered serious attention in the industry.

There's now such an interest in smart buildings in general—and Microsoft's smart buildings software in particular—that Realcomm plans to host a conference on smart buildings in the Seattle area next year. He will hold it there so attendees can visit Microsoft and experience one of Smith's enthusiastic show-and-tells. "Darrell's going to be a rock star. Well, he already is in our world," Young says. "If he's not shaking the president's hand in a year then we've all done something wrong."

Unsurprisingly, Smith is more measured about his part as the leader of a team that developed a game-changing smart buildings tool. Though his team started with little more than a notion that there had to be a better way, they used Microsoft's own software, their own background and expertise, and an unyielding determination to solve an industry-wide problem. Smith played the roles of catalyst and mortar in this process, sparking the revolution but also helping to hold his small team of "ninja innovators" tightly together.

"It's not me that's great, it's our story," Smith says. "Yes I'm passionate, but you could put someone else in my place and it would still be a great story."

Special thanks to Jay Pittenger, Darrell Smith, Tearle Whitson, Jonathan Grove, Trevor Sodorff, Microsoft Real Estate & Facilities, Jessica Granderson, Lawrence Berkeley National Laboratory, Judd Black, City of Redmond, Jim Young, and Realcomm Conference Group, LLC.

Chapter 13

Improving Airport Energy Efficiency through Continuous Commissioning® Process Implementation at Dallas/Fort Worth International Airport*

Bahman Yazdani, PE, Texas A&M University
Guanghua Wei, PE, Bes-Tech, Inc.

Jerry R. Dennis, CEM, CEP
Rusty T. Hodapp, PE, CEM, LEED AP
Dallas/Fort Worth International Airport

ABSTRACT

Large commercial airports consist of many complex systems that require a significant amount of energy to operate. The energy cost is often one of the largest categories of controllable airport operating expense. In the current economic environment for the airline industry, it is increasingly important for airport facility managers to reduce operating expenses for their airline and other tenants. This chapter introduces the Continuous Commissioning® (CC®) process and its application at the Dallas/Fort Worth (DFW) International Airport, one of the busiest airports in the world. Since 2004, several facilities at DFW International Airport have gone through the CC process with excellent results in terms of comfort improvement, energy and cost savings, and carbon emissions reduction. This chapter describes the key steps of the CC process, presents the results at DFW International Airport and notes the potential value of automated diagnostics in maintaining optimized performance over time.

Keywords: airport energy efficiency, Dallas/Fort Worth (DFW) International Airport, Continuous Commissioning® (CC®)

INTRODUCTION

Existing buildings account for about 40% of total energy consumption in the USA and contribute a significant amount of carbon emissions. With over 500 commercial and 2,800 general aviation facilities in the USA, airports are one of the largest public users of energy. Since energy is often one of the largest categories of controllable airport operating expense, airport facility managers must constantly seek to reduce energy consumption to help lower the costs for their airline and other tenants.

One of the most cost-effective ways to improve building energy efficiency without major capital expenses is to perform a process called Continuous Commissioning® (CC®). This process typically achieves 15% whole building energy cost reduction with simple paybacks of less than two years. It has been used in over 450 federal, institutional and commercial buildings and central utility plants with measured savings of over $100m since 1993.

Dallas/Fort Worth (DFW) International Airport and the Energy Systems Laboratory (ESL) at the Texas A&M University started a partnership in 2004 to apply the CC process to DFW International Airport facilities. The application of the process at DFW International Airport has been a great success over the years. The Rental Car Center, which was the first building commissioned, saw its whole building energy consumption reduced by 18% after CC process implementation. A subsequent

*Continuous Commissioning® and CC® are Registered Trademarks of the Texas A&M Engineering Experiment Station, a member of the Texas A&M University System, an agency of the State of Texas.

effort at the Airport's Administration Building reduced whole building energy use by 23%. From 2007 to 2009, the ESL and one of its licensees commissioned the Airport's new Terminal D and the district energy plant. Measured savings during the two-year implementation period were $4.2m. For the time period of October 2004 through to September 2012, the cumulative electricity, hot water and chilled water savings total nearly $9.5m as a result of the CC process. These positive results demonstrate the cost-effectiveness of the CC process.

THE CONTINUOUS COMMISSIONING® PROCESS

CC is a process developed by the ESL at the Texas A&M University. This is an ongoing process that identifies and resolves operating problems, improves comfort, optimizes the system operation and controls based on current building conditions and requirements and reduces energy consumption for commercial and institutional buildings and central utility plants.

The Process

The CC process includes the following key steps.

Step 1: CC Assessment

This involves an on-site visit in which operation of the HVAC systems is evaluated and major CC opportunities are identified. The findings are used to develop a proposal that quantifies potential savings and implementation costs. Any additional energy monitoring that may be needed is also identified.

Step 2: Develop Performance Baselines

Upon approval to proceed, a performance baseline for comfort is developed to document existing comfort conditions. Existing comfort problems, such as areas that are too cold, too hot, too noisy and/or too humid, are noted. Baseline energy consumption models of building energy performance are also developed in order to document energy savings after the building is commissioned. The baseline energy models normally include whole building electricity, cooling energy and heating energy models. The baseline energy models are developed using 1) short-term measured data; 2) long-term hourly or 15-minute whole building energy data; and/or 3) utility bills for electricity, gas and/or chilled or hot water. Techniques used to determine savings should be consistent with American Society of Heating, Refrigerating and Air-Conditioning Engineers (ASHRAE) Guideline 14.

Step 3: Conduct System Measurements and Develop CC Measures

This step involves identification of CC measures needed to optimize the system operation for the actual building use. The CC team, which consists of at least one CC professional, one or more technicians and at least one member of the facility staff, starts the detailed field measurements and inspections. Operating and comfort problems, such as simultaneous heating and cooling, and cold and hot spots, are diagnosed; component failures or degradation, such as leaky control valves, inoperable dampers, disabled mechanical and control systems, are identified. Detailed engineering analysis is conducted based on the measured data to develop solutions for existing problems. Improved operation and control schedules are developed for terminal boxes, air handling units (AHUs), exhaust systems, water and steam distribution systems, heat exchangers, chillers, boilers and other components, as appropriate. Any potential cost-effective energy retrofit measures are identified.

Step 4: Implement CC Measures

The proposed CC measures are presented to the facility owner and building staff. Modifications are made to proposed measures as needed to fit the owner's expectations. It is important that the building staff are comfortable with each CC measure or it will be quickly disabled. The CC professionals then work closely with the building staff to implement the approved measures. Implementation normally starts by solving existing operational and comfort problems, which are the priorities of the building staff. Solving these problems improves occupant comfort and increases cooperation from operating staff.

Step 5: Document changes and Provide Training

When all the CC measures are implemented, it is necessary to document the comfort improvements, energy and cost savings, as well as the changes in operating procedures for the staff. The comfort measurements taken in step 2 should be repeated at the same locations under comparable conditions to determine commissioning impact on room conditions. Post CC process energy performance should be compared with the baselines to determine energy and cost savings using appropriate occupancy and weather normalization. A training workshop is then provided to the building staff to review the new operating sequences and setpoints, as well as trouble-shooting procedures.

Step 6: Keep the Commissioning Continuous

To maintain improved comfort and assure persistence of the savings, follow-up monitoring and fine-tuning of the CC measures are essential. CC professionals can provide follow-up telephone consultation, review the energy data, monitor and analyze system operation, and refine control sequences and setpoints through internet or other remote means as needed. On-site visits will be made if necessary.

Application of the CC Process

Generally, the CC process can be applied in the following applications:

- buildings after energy retrofits
- existing buildings (including new buildings) as a standalone energy conservation project
- as an energy cost reduction measure (ECRM) in an energy retrofit program.

Since 1993, the CC process has been successfully applied in many institutional and commercial buildings and central utility plants, all with similar results, i.e. a 10–25% reduction in building energy usage. The following case studies discuss the application of the process at DFW International Airport.

THE CC PROCESS AT DFW INTERNATIONAL AIRPORT FACILITIES

DFW International Airport was opened in 1974. It covers about 30 square miles and has seven runways and five terminals with a total of 174 gates. In 2008, DFW International Airport served 57,093,187 passengers, making it one of the busiest airports in the country.

In recent years, DFW International Airport has completed several major projects, including the construction of the new International Terminal D, and a $150m upgrade to its central heating and cooling plant. As part of the upgrade project, all chillers and boilers were replaced to increase capacity, improve energy efficiency and reduce air emissions. A six-million-gallon thermal energy storage tank was also added to the chilled water system and a new centralized, preconditioned air system was installed to support heating and cooling of jet bridges and docked aircraft. As a result of these upgrades, NO_x emissions from the facility have been reduced by 91 percent.

Meanwhile, airport management continued to explore other ways to reduce energy further. In 2004, DFW International Airport partnered with ESL to study the energy use at Terminal B and the Rental Car Center. The

standalone, relatively new Rental Car Center was selected to implement the CC process as a pilot project. The result was an impressive 18% whole building energy use reduction. The next use of the process was in the DFW Administration Building, which had just undergone an energy efficiency retrofit following which significant comfort and energy problems were experienced. The CC process reduced the Administration Building's energy use by 23% while improving building comfort dramatically. In 2007, implementation of the CC process in the new Terminal D and central heating and cooling plant began. The CC process was completed in 2009 with cumulative savings of about $4.2m during the two-year period. For the time period October 2004 to December 2009, the cumulative electricity, hot water and chilled water savings totaled nearly $5.0m as a result of the CC process. Implementation of the CC process in the Rental Car Center, Terminal D and the central heating and cooling plant is described below.

CC Process at the Rental Car Center

DFW International Airport's Rental Car Center is a two-story, 130,000 ft² facility that houses all the rental car companies serving the Airport. The building was opened in March 2000. Most of the first floor comprises counter space for the rental companies and an open area for customer circulation. The second floor is mostly office and storage space. Attached to the building is a two-storey, 1.8 million square foot parking garage. Both the building and garage are in continuous use 24 hours a day.

There are six single-duct variable air volume (VAV) AHUs in the building. Each AHU has a supply and a return fan that are both equipped with variable frequency drives (VFDs). There are two 280-ton centrifugal chillers, each has a constant speed primary pump that circulates water through the chiller. Two variable speed secondary pumps supply chilled water to the AHUs. Two constant speed pumps and two cooling towers with variable speed fans provide condenser water for the chillers. The space is served by 133 fan-powered terminal boxes. Boxes serving exterior areas have electric resistance heaters.

During the CC assessment process, many operational issues were observed. For example, higher than necessary outside air intake; excessive duct static pressure setpoints; economizer cycle not fully utilized; excessive heating in the terminal boxes, even during summer time; wide variation of space temperature setpoints, from 65°F to 80°F; over-pressurizing the chilled water loop; and continuous use of the South Garage second level lights.

The CC professionals worked closely with the building staff during the three-month CC process implementation period. CC measures implemented include optimizing the supply air temperature and static pressure reset; improving chiller operations; condenser water temperature reset; modifying the control for the return air fans to optimize minimum outside airflow; utilizing the economizer cycle; improving garage lighting operation, optimizing terminal box minimum airflow setpoints; improving zone temperature control to eliminate simultaneous heating and cooling; and improving secondary pump control.

Figure 13-1 compares the whole building electricity use before and after the CC process. Measured savings were $574,449 since September 2004 with an 18% energy use reduction compared with the baseline.

CC of Terminal D

The new International Terminal D was opened in late 2005. This 2-million ft^2 facility is a world-class international passenger terminal equipped with a three-level roadway system and 29 swing gates. Terminal D also includes a 303,675 ft^2 Grand Hyatt Hotel that is independently operated and not included in the CC project scope.

There are a total of 60 VAV AHUs, 73 constant-volume AHUs and 1,364 fan-powered terminal box units for the terminal. Both constant volume and VAV AHUs have VFDs on the supply and return air fans. Chilled water is supplied from the central heating and cooling plant. There is no chilled water pump in the terminal. Hot water is generated through heat exchangers using steam supplied from the central heating and cooling

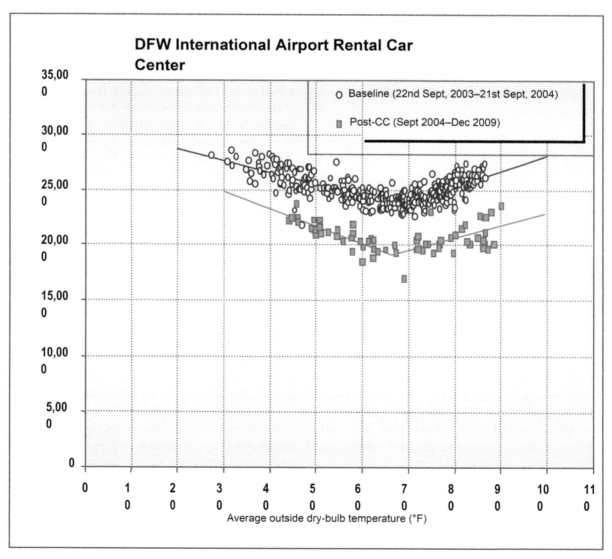

Figure 13-1: Pre and post-CC daily average electricity use versus outside dry-bulb temperature

plant. Two hot water pumps, each equipped with a VFD, circulate hot water throughout the terminal.

The terminal was well maintained and operated. During the CC assessment, however, many opportunities were identified to improve the system's energy efficiency. CC of the terminal started in 2007 and was completed in 2009. Major CC measures implemented include AHU scheduling; terminal box scheduling; AHU supply air temperature and duct static pressure resets; terminal box temperature and flow setpoint adjustments; modulating the VFDs on constant volume AHUs; and turning off return air fans when not needed.

Figures 13-2 and 13-3 compare the actual energy consumption data from October 2007 to December 2009 with the baselines. All consumption data is normalized to daily usage and plotted against the outside air temperature. It can be seen that cooling and heating energy consumption decreased significantly. Average reductions of electricity, chilled water and hot water consumption are 9%, 25% and 43%, respectively.

CC Process in the Central Heating and Cooling Plant

The central heating and cooling plant, or Energy Plaza (EP), consists of six steam boilers with a total capacity of 260,300lbs per hour, six 5,500-ton centrifugal chillers, a 90,000 ton/hour thermal storage tank, five 1,350-ton glycol solution chillers and eight cooling towers with two-speed fans. The glycol solution system provides heating and cooling to jet bridges and docked aircraft via pre-conditioned air (PCA) system equipment throughout the terminals. This system can be switched among cooling, heating and idle modes based on ambient temperature. The steam is supplied to the steam-to-water heat exchangers located near each terminal to produce heating hot water for each terminal.

During the CC assessment process, many opportunities were identified to improve the plant's energy efficiency. CC process implementation in the plant started in 2007 and was completed in 2009. Major CC measures implemented include optimizing the chiller staging and thermal storage tank operation to reduce electrical

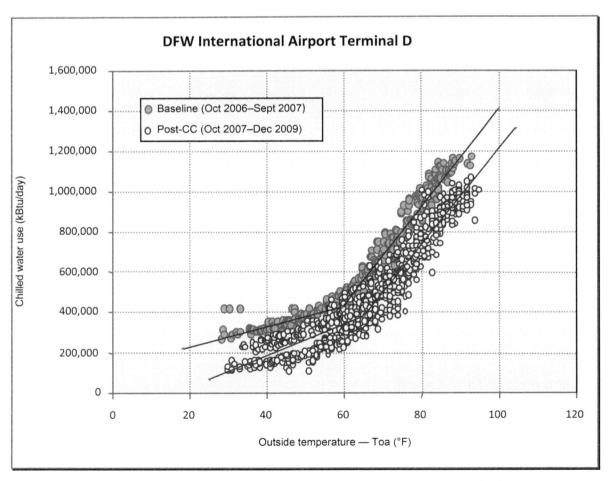

Figure 13-2: Pre and post-CC daily average chilled water use versus outside dry-bulb temperature

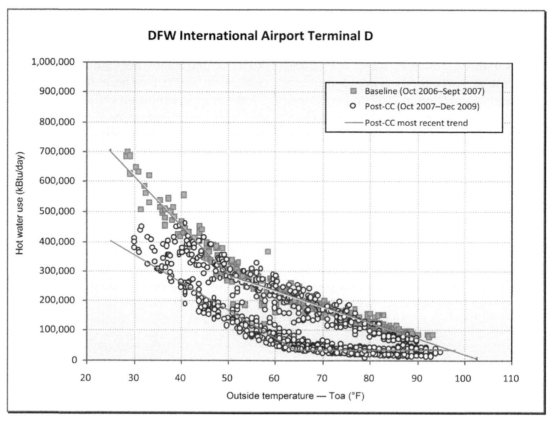

Figure 13-3: Pre and post-CC daily average hot water use versus outside dry-bulb temperature

peak demand; optimizing the cooling tower staging sequence; fine-tuning the boiler steam pressure setpoint; implementing a chilled water supply temperature reset schedule; improving condenser water pump operation; optimizing secondary chilled water system control; optimizing the glycol system supply temperature reset schedule; and improving the PCA system switch mode operation. Figures 4 and 5 compare the actual energy consumption data from April 2008 to December 2009 with the baselines. The measured cumulative electricity and natural gas savings totaled $1,730,000 during that time period.

Summary of CC Process Savings

By commissioning the Rental Car Center, the Administration Building, Terminal D and the central heating and cooling plant, a significant amount of energy reduction was realized at DFW International Airport. Based upon this demonstrated success the process continues to be applied at other DFW Airport facilities with similar results. From September 2004 to September 2012, cumulative cost savings of approximately $11.10 million have been achieved based on monthly actual prices, as shown in Figure 13-6.

CONCLUSIONS

The CC process has been implemented at numerous facilities across the country in the last 15 years. With an average payback of less than two years and typical savings of 10–25% of annual energy costs, it is clearly one of the most cost-effective energy efficiency measures. Other benefits include the emissions reductions that result from the lowering of energy consumption, improved building comfort and enhanced technical skills of the facility staff. The case studies at DFW International Airport demonstrate the effectiveness of the CC process and show that airport facilities can benefit greatly from the CC process.

While not a case study on the use of automated fault detection and diagnostics in itself, this chapter on the implementation of CC at DFW shows what impact can be expected when such tools are used to systematically optimize facility operations. The use of automated systems, techniques and tools is of significant value in the analytic and diagnostic aspects of the CC assessment and measure development. They may be of equal or greater importance in maintaining optimal performance over time. Diagnostic tools focusing on compo-

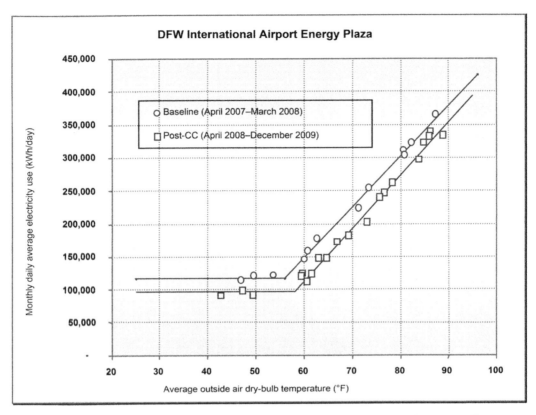

Figure 13-4: Monthly pre and post-CC daily average electricity use versus outside dry-bulb temperature

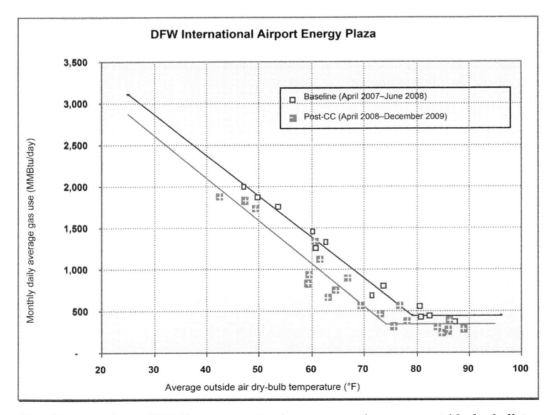

Figure 13-5: Monthly pre and post-CC daily average natural gas consumption versus outside dry-bulb temperature

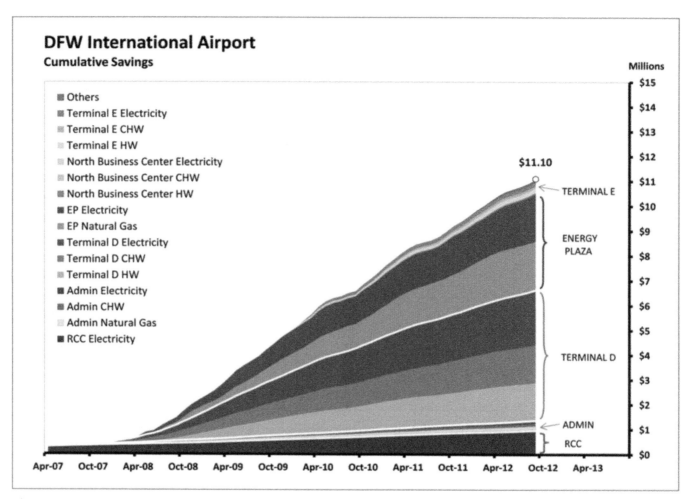

Figure 13-6: Cumulative cost savings from September 2004 to September 2012 (for clarity, savings for the Rental Car Center from September 2004 to April 2007 are truncated in the chart)

nent level deficiencies can identify and highlight leaking control valves, improperly operating economizer dampers, or suboptimal set-points, etc. A building or system level automated diagnostic tool applied after optimization would be of tremendous value in maintaining persistence of savings measures by bringing attention to any significant deviations from the optimized baseline.

Acknowledgments

The case studies presented here would not have been possible without the support from DFW International Airport Board management and Energy, Transportation & Asset Management department staff. The authors would like to thank Larry Kramer, Wayne Suite, Ray Bruce, Dick Chester, Sammy Hicks and Donny Lumpkin for coordinating the fieldwork and gathering

information for the CC process. A number of individuals from ESL and HHS Associates exerted considerable effort to implement the changes described herein. They include Greg Zeig, Steven Esparza, Marvin Zeig, Harold Huff, Daniel Chen, Hui Li, Song Deng, Chen Xu, Steve Gumm, Fred Schroeder and Mustan Shafiq. Savings analysis was provided by Dr. Juan-Carlos Baltazar. The CC process was directed by Dr. Dan Turner formerly of ESL and Dr. David Claridge, Director of the ESL.

An earlier version of the material in this chapter appeared in the article "Improving airport energy efficiency through Continuous Commissioning at Dallas/Fort Worth Airport," by Guanghua Wei, Bahman Yazdani, Jerry R. Dennis and Rusty T. Hodapp, *Journal of Airport Management*, October – December 2010 v5n1.

Chapter 14

ECAM+
A Tool for Analyzing
Building Data to Improve and
Track Performance

Bill Koran, NorthWrite Inc., Lake Oswego, Oregon

INTRODUCTION

ECAM+ facilitates the examination of energy information from buildings or building simulations, reducing the time spent analyzing trend and utility meter data. ECAM+ is an acronym for Energy Charting and Metrics plus Building Re-tuning and Measurement and Verification of Energy Savings.

ECAM+ is a Microsoft Excel® add-in, and hence is very flexible and easy to use. Though some people may initially scoff at the potential of "just a spreadsheet," ECAM+ has nearly 12,000 lines of code and many powerful features. At the time of this writing, many individuals, engineering firms, utilities, and government agencies all over the United States have found it useful. There is an increasing number of international users, although at present many features only work in English versions of Excel®.

This chapter describes some of ECAM's capabilities, provides example charts, and reports on an example project that shows how Measurement and Verification capabilities can support building diagnostics. The goal is not to teach you how to use ECAM—there is a User Guide to help with that—but to demonstrate its features and encourage readers to try this free tool.

CREDITS FOR ECAM+

The following organization have contributed to the development of ECAM+:
- Bonneville Power Administration
- Northwest Energy Efficiency Alliance
- Regional Technical Forum
- California Commissioning Collaborative
- New Buildings Institute
- Pacific Northwest National Laboratory

ECAM+ CAPABILITIES

The broad intent of the tool is to maximize the user's ability to benefit from whatever data is available. Key features include the following:

- Data processing to attach schedule and day-type information to time-series data;

- Filtering by day-type, occupancy schedule, weather data, month/year, project dates, etc.;

- Normalization of data based on user-entered information;

- Creation of standard charts for the points selected by the user;

- Calculation of normalized metrics for the points selected by the user;

- Automated creation of the diagnostic charts supporting the United States Department of Energy (DOE) funded building Re-tuning process developed by the Pacific Northwest National Laboratory (PNNL); and

- Change point energy models and Measurement and Verification (M&V) capability, based on the International Measurement and Verification Protocol (IPMVP) and ASHRAE Guideline 14, Measurement of Energy and Demand Savings.

The tool makes extensive use of Excel® PivotTables to facilitate summarization and filtering of the data. It goes beyond normal PivotTables and PivotCharts, however, by automating the creation of scatter charts based on PivotTable data.

ECAM was developed with partial funding by the Northwest Energy Efficiency Alliance, Bonneville Power Administration, Regional Technical Forum, New Buildings Institute, Pacific Northwest National Laboratory, and the California Commissioning Collaborative.

Most of the early core of ECAM was based on tools that I had developed for my own use, such as timestamp disaggregation, scatter charts based on PivotTables, and Calendar Charts. I was impelled to more formally develop ECAM in recognition of two things:

1. The increasing proliferation of smart meters, and

2. The need to automate tasks that were commonly but tediously performed by energy engineers, or that should be performed but were not because of their difficulty.

The tool takes advantage of earlier Public Interest Energy Research (PIER, sponsored by the California Energy Commission) by using the naming convention and many of the metrics recommended by A Specifications Guide for Performance Monitoring Systems. It can automatically create nearly all of the chart types found in the Web-based Energy Information Systems for Energy Management and Demand Response in Commercial Buildings. The chart types in ECAM include the following:

• Time series

• X-Y (scatter)

• Daily load profile

• 3D (chart of daily load profiles for multiple days) and Surface (3D chart in plan view, of daily load profiles using color as a proxy for value)

• Calendar (chart of daily load profiles laid out as a calendar)

• Load profiles as grouped box plots

• Load-Duration Chart

• X-Y chart of hourly, daily, or monthly data and associated energy models

• Charts of model residuals, including residuals vs. time, residuals vs. independent variable, and a histogram of residuals. A lag chart—a plot of the residuals at time t vs. the residuals at time t-1, is also included.

The data upon which the charts are based can all be instantly filtered by day-type, occupancy schedule, weather data, month/year, pre/post, equipment status, time of day, etc., and the charts will automatically update. Perhaps of greater significance, charts of averages can be created. For example, ECAM can instantly create a chart showing the average load profile by daytype or day-of-week.

ECAM can aggregate data across time (to, get monthly energy use totals, for example) and to sort and bin data for plotting load (or load factor) versus hours at each load.

ECAM also automatically creates the following additional fields to support chart and metrics creation, and additional calculations, based on available point types (Table 14-1).

ECAM supports the DOE-funded PNNL Re-tuning process by automatically creating all of the Re tuning diagnostic charts for which the needed points are available:

• Central Plant analysis
  — CHW Supply, Return, ΔT, OAT vs. Time
  — HW Supply, Return, ΔT, OAT vs. Time
  — CHW Flow and OAT vs. Time

• AHU analysis
  — Outdoor/Return/Mixed/Discharge vs. Time

Table 14-1

Equipment Status	From demand (kW) or current (amps) when a status point is not available.
Demand (kW)	From current (amps) as an approximate calculation when a power point is not available.
Chilled water tons	Whenever a consistent set of flows and temperatures are available.
Watts per square foot	For all electrical demand points that are available, whenever a building square footage is entered.
CFM per square foot	For all airflow points that are available, whenever a building square footage is entered.
kW per ton	For all related points
gpm per ton	For all related points

— Discharge/Discharge Set Point Temperature vs. Time
— Outdoor Air Fraction/Damper Position Signal vs. Time
— Outdoor/Return/Damper Position Signal vs. Time
— Damper/Chilled Water/Hot Water Position Signal vs. Time
— Damper Signal Time-Series
— Discharge Static Pressure vs. Time
— Supply Fan Speed/Status/Static vs. Time
— Return Fan Speed/Status vs. Time
— Discharge Air vs. Discharge Air Set Point Temperature
— Chilled Water vs. Hot Water Signal
— Damper Signal vs. Outdoor Air Temp (Trat>Toat)
— MAT vs OAT (Trat>Toat)

• Zone analysis
— Damper/Reheat Valve/Occ Mode/Zone Temp vs. Time
— Damper/Reheat Valve/Occ Mode/Zone Temp vs. Time repeat for each zone)
— Zone Damper Position vs. Time (All AHUs)

ECAM+ MENU

The list of items on the menu helps show the variety of capabilities included in ECAM+.
Select Data
Definition of Points
Create Schedules
Input Dates for Comparison of Pre and Post
Select Monthly Billing Data

• Time Series Charts
— Point(s) History Chart
— Load Profile by Daytype
— Load Profile by Day of Week
— Load Profile by Month-Year
— Load Profile by Date Range (Pre/Post)
— Load Profile by Year
— Load Profile by Day
— Create 3d Load Profile
— Create Energy Colors (surface chart)
— Load Profile Calendar
— Load Profile as Box Plots

• Scatter Charts
— Scatter Chart by Occupancy

— Scatter Chart by Date Range (Pre/Post)
— Toggle Scatter between all timestamps and aggregated values

Load Duration Chart (Point Frequency Distribution)
Chart to Check Input Schedule
• Matrix Charts
— Information about Matrix Charts
— Matrix Selected Charts
— Matrix All Charts
— Re-Matrix Charts

• Chart Utilities
Set Scales the Same for a Group of Charts

• Metrics and Data Summaries
— Metrics for Points Normalized per Sq. Foot
— Daytype and Occupancy Metrics
— Occupancy and Month-Year Combined Metrics
— Daytype and Month-Year Combined Metrics
— Information about Data Summary
— Summarize Data
— Summarize Data from PivotTable

• PNNL Re-Tuning
— Central Plant Charts
— AHU Charts
— Zone Charts
— AHU Scatter Charts
— Chart Summary

• Measurement and Verification General Inputs
— Input Desired Model and Savings Confidence Level
Input Values for ASHRAE Fractional Savings Uncertainty

• Data-Driven Models and M and V
— Create Load Profile by Day of Week and Evaluate Daytypes
— Evaluate Daytypes (active sheet has LP by Day of Week)
— Create Baseline Models Only
— Create Post Models Only
— Create Baseline and Post Models
— Evaluate Savings for Post Period
— Bring in TMY3 Weather for Annualization
— Annualize Baseline Using TMY3 Weather
— Annualize Post Using TMY3 Weather
— Annualize Pre, Post, and Savings Using TMY3

- Monthly Billing Data Models and M and V
  — Create Baseline Monthly Models Only
  — Create Post Monthly Models Only
  — Create Baseline and Post Monthly Models
  — Evaluate Savings for Post Period Using Monthly Model
  — Bring in Monthly TMY3 Weather for Annualization
  — Annualize Baseline Using Monthly TMY3 Weather
  — Annualize Post Using Monthly TMY3 Weather
  — Annualize Pre, Post, and Savings Using Monthly TMY3

- ECAM Utilities
  — Convert Table format to ECAM List format
  — Create Bin Data from Temperatures
  — "Data" worksheet was changed
     Copy Worksheet and Update ECAM Chart Source

ECAM Help
About ECAM

CHART EXAMPLES

There are two main chart types in ECAM. The first type is a time-series that shows a variable vs. time. The other main chart type is the scatter, or x-y chart. ECAM also provides histograms and a chart to that serves as a visual check on an input occupancy schedule.

In addition to the examples shown here, the chapter titled *Identification of Energy Efficiency Opportunities through Building Data Analysis and Achieving Energy Savings through Improved Controls* has several other examples of ECAM charts.

The simplest time-series chart is the "Point(s) History Chart" that shows one or more variables vs. time, for the full time history of the data. Figure 14-1 shows the basic chart, unfiltered, and Figure 14-2 shows the chart after filtering to show only three days of data.

Note that ECAM+ includes the day of the week associated with the date, to aid in understanding the data. Also note that the y-axis label for the point includes "Avg" which is obviously an abbreviation for average. When ECAM+ aggregates data, it uses the average. Therefore, whenever a PivotTable is created the point name is prefixed with "Avg," even if the most granular, un-aggregated data is used. This will likely change in a future version.

Any particular time period of interest is easily selected using the PivotTable, just as for a regular Excel PivotChart. Here is the same data, but only showing the data for January.

Figure 14-3 shows the chart after 7 days are selected.

The most common scatter charts use outside air temperature as the independent variable. However,

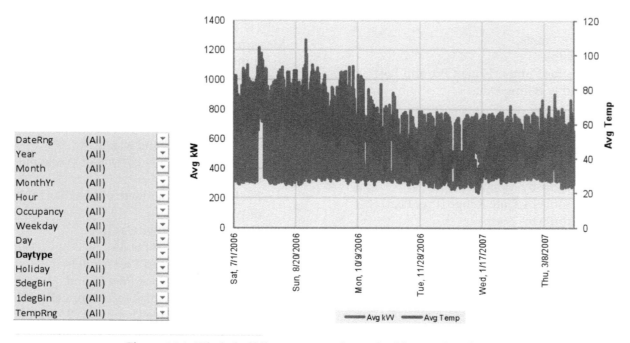

Figure 14-1: Whole building consumption point history chart in ECAM.

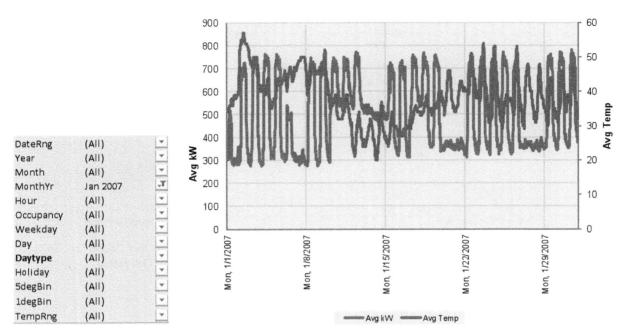

Figure 14-2: Whole building consumption point history chart in ECAM for 1 month.

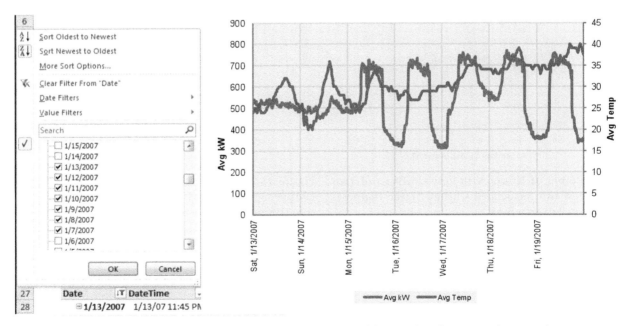

Figure 14-3: Whole building consumption point history chart in ECAM for 1 week.

ECAM can use any point as an independent variable in a scatter chart. Figures 14-4 and 14-5 show examples that illustrate increased demand associated with running more chillers than necessary for the load.

Energy analyses often need the cumulative time at different operating conditions. ECAM makes it easy to summarize the data for this purpose. Here are two histo- grams created using the menu item for "Load Duration Chart (Point Frequency Distribution)."

These charts, like all ECAM charts, can be filtered for different operating conditions. The Load Duration Chart allows you to easily set the bin width as well as the bounds of the independent variable, as can be seen by comparing Figure 14-6 with Figure 14-7.

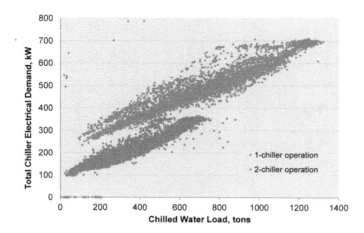

Figure 14-4: Chiller Power vs. Chilled Water Load, by Number of Chillers Operating

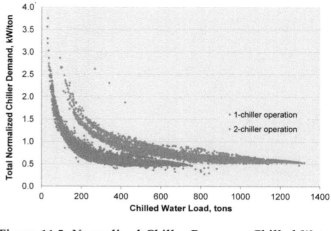

Figure 14-5: Normalized Chiller Power vs. Chilled Water Load, by Number of Chillers Operating

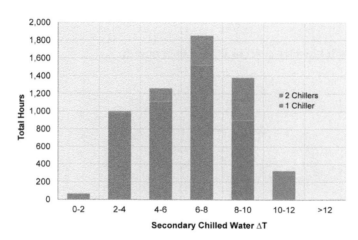

Figure 14-6: Number of Hours with One Chiller and Two Chiller Operation and Associated Chilled Water Temperature Difference in 2 degree Bins

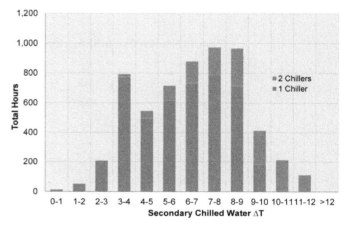

Figure 14-7: Number of Hours with One Chiller and Two Chiller Operation and Associated Chilled Water Temperature Difference in 1 degree Bins

MEASUREMENT AND VERIFICATION

The two most powerful capabilities in ECAM+ are the support of (1) Building Re-tuning (http://buildingretuning.pnnl.gov/) and (2) Measurement and Verification (M&V). The M&V capabilities in ECAM+ are significant yet flexible. This section describes the M&V models, provides an example M&V calculation, and shows how M&V can also be used for building diagnostics.

ECAM+ may have application in a variety of M&V applications. The International Measurement and Verification Protocol (IPMVP) describes four options for M&V:

A. Retrofit Isolation: Key Parameter Measurement
B. Retrofit Isolation: All Parameter Measurement
C. Whole Facility
D. Calibrated Simulation

ECAM+ can provide the analyses for all but Option D. Various ECAM capabilities, such as load profiles by daytype, and evaluation of the relationship of energy use to temperature, can support the simulation calibration required by Option D.

IPMVP also describes two types of savings:

1. Avoided Energy Use, also known as "Reporting Period Basis"

2. Normalized Savings, also known as "Fixed Conditions Basis"

The "Avoided Energy Use" type of savings requires a model of the pre-project energy use (during the baseline period) to forecast what energy use would have been in the Reporting Period if the project had not been done. "Normalized Savings" requires two models, one for energy use during the baseline period and one for energy use during the post period. Both models are then used to forecast what energy use would be under some set of "Fixed Conditions," which most commonly include typical weather.

ECAM+ supports both "Avoided Energy Use" and "Normalized Savings."

A final point about M&V using ECAM+: Monthly billing data, interval meter data, and interval data from systems or equipment are all accepted.

ECAM provides a variety of linear and change-point linear regressions. The linear and change-point linear models, and associated uncertainty, are based on classical statistics and ASHRAE approaches. The ASHRAE approaches were developed and documented through research project 1050-RP, Development of a Toolkit for Calculating Linear, Change-point Linear and Multiple-Linear Inverse Building Energy Analysis Models. Figure 14-8 shows the form that allows users to select the appropriate model type.

2P (two-parameter) models are appropriate for modeling building energy use that varies linearly with another single independent variable. For example, in some buildings, heating and cooling energy use varies linearly with outdoor air temperature.

3P (three-parameter) models are appropriate for modeling building energy use that is varies linearly with an independent variable over part of the range of the independent variable and remains constant over the other part. 3P-heating models, using outside air temperature as the independent variable, are often appropriate for modeling heating energy use in residences with gas or oil heating.

Similarly, 3P-cooling models, using outside air temperature as the independent variable, are often appropriate for modeling whole-building electricity use in residences with electric air conditioning.

Four-parameter (4P) models using outdoor air temperature as the independent variable are appropriate for modeling heating and cooling energy use in variable-air-volume systems and/or in buildings with high latent loads. In addition, these models are sometimes appropriate for describing non-linear heating and

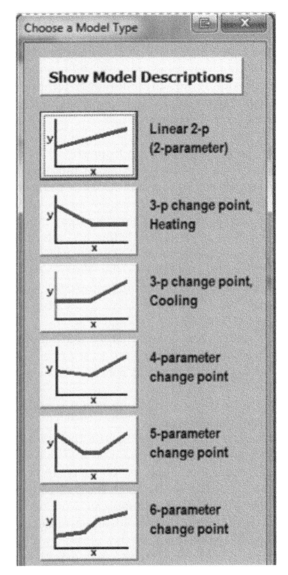

Figure 14-8: Model Selection User Form

cooling consumption associated with hot-deck reset schedules and economizer cycles.

Five-parameter (5P) models using outdoor air temperature as the independent variable are appropriate for modeling energy consumption data that includes both heating and cooling, such as whole-building electricity data from buildings with electric heat-pumps or both electric chillers and electric resistance heating. They are also appropriate for modeling fan electricity consumption in variable-air-volume systems.

6P (six-parameter) models were developed by the author to improve support for diagnostics as well as certain system types and operations. They are an extension of four-parameter models. Whereas a 4P model has only a single change point and two slopes, 6P models have two change points and three slopes. Any of the three slopes can be positive or negative. Six-p models are appropriate when two cooling (positive) slopes are apparent, often as a result of a cooling economizer or insufficient cooling capacity. Six-parameter models can also be valuable for modeling heating or cooling plants as shown later in this chapter.

Measurement and Verification Example

This example shows the use of monthly billing data to estimate savings for a project that reduced natural gas use. When performing regression-based M&V, a good first step is to just plot the data in a scatter chart, with separate chart series for before and after the project. This is easy using the ECAM menu item to create a "Scatter Chart by Date Range (Pre/Post)." The scatter chart for this project is shown in Figure 14-9. Normalizing the data on a "per-day" basis accounts for different billing period lengths.

The scatter chart shows that there is a clear difference between the pre-project and post-project gas use. This gives us confidence that we can use a regression model to estimate savings.

Figure 14-10 shows a four-parameter model of the gas use in the baseline period.

This chart has five series. The first two are the actual gas use per day and the model of that use. The dark points near the bottom of the chart are the model residuals: the difference between the modeled and actual values. This plot of residuals can show us whether we chose a good model form. Ideally, there should not be a shape to the residuals in this plot; they should be randomly scattered about the zero value on the y axis.

The other chart series, shown as the smaller diamonds, are the prediction intervals for the model at the input 80% confidence level. This chapter does not provide further discussion of uncertainty, but ECAM+ provides estimates of prediction, model, and savings uncertainty.

ECAM+ provides four types of residuals charts:

1. Residuals vs. Independent Variable
2. Residuals vs. Time
3. Residuals Lag Chart
4. Histogram of Residuals

This document describes just the first two types. Figure 14-10 included the first type of residuals chart. The second type of residuals chart is very valuable to see if the baseline is changing. The example for this project is shown in Figure 14-11.

Figure 14-11 shows that the residuals show a slight trend over time. They appear to be randomly dispersed around zero, and the trend line indicates that energy use increased about 4.4% over the 2-year baseline period. However, note that last twelve points in the series: They start out low and move up; an indication of decreased energy use. If we had used a 1-year baseline, it would show that energy use decreased about 12% over the year. This shows the importance of picking a relevant baseline, and checking to see whether changes in energy use are solely due to the project being measured, or also due to a general trend in energy use.

After the model is developed, savings can be estimated. In this case, the four parameters describing the model are:

Change point temperature= 55.83°F
Value at Change Point= 95.06 therms per day
Left Slope= -11.306 therms per day per degree
Right Slope= -6.563 therms per day per degree

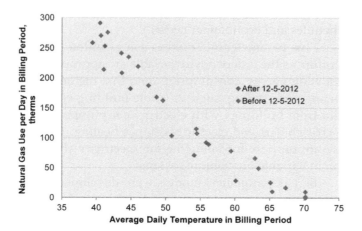

Figure 14-9: Scatter Chart of Monthly Gas Billing Data

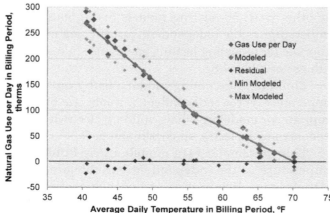

Figure 14-10: Model of Baseline Gas Billing Data

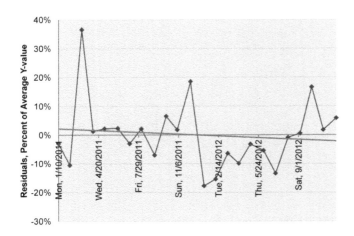

Figure 14-11: Residuals vs. Time

Using these parameters we can forecast what the energy use would have been under the post-project conditions. ECAM+ provides the following savings estimate (Table 14-2).

A time series graphical output is also provided. Note that this chart uses the actual billing period gas use rather than normalized per day.

Building Diagnostics Example

The capability to create accurate models from real data gives us the opportunity to compare operations under different conditions. This can facilitate optimization of things like setpoints, reset schedules, and other controls-related parameters This example uses data from the chilled water plant at Naval District Washington, Bethesda Medical Campus—home of Walter Reed National Military Medical Center. This site is also known as Naval Support Activity Bethesda, or NSA Bethesda, which is

how it is referred to in the remainder of this chapter.

The central plant at NSA Bethesda is undergoing some updates and optimization, with a number of projects underway or planned. One part of this is optimizing the chilled water (CHW) supply temperature setpoint. Finding the optimum chilled water supply temperature setpoint (CHWST) is a multi-dimensional problem, so we can't describe the complete analysis—indeed, it is still underway at this time—but we can demonstrate an approach to using M&V processes for building diagnostics.

The CHW plant at NSA Bethesda has 10,650 tons of chiller capacity. Historically, the plant has distributed CHW to the various buildings and loads at 40°F. Although some newer buildings have controls that are fully DDC, many buildings are not. Hence, it is not possible to confirm that all cooling loads are satisfied except through human feedback. The purpose of this analysis was to see the power and energy impact of resetting

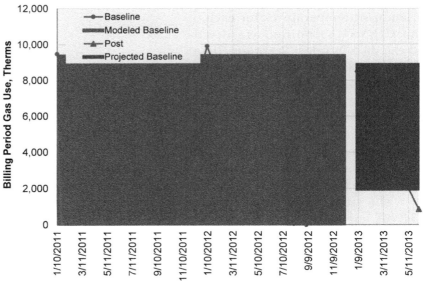

Figure 14-12: Time History of Energy Use and Projected Energy Use After Project

33,241	**Projected Baseline Energy, therms**
27,460	**Measured Energy, therms**
5,781	**Energy Savings, therms**
1,246	**Projected Baseline ±Uncertainty @ 80% Confidence Level**
5,781 ±1,246	**Energy Savings and Uncertainty @ 80% Confidence Level**
3.7%	**Projected Baseline ±Uncertainty @ 80% Confidence Level**
17.4% ±3.7%	**Energy Savings and Uncertainty @ 80% Confidence Level**

Table 14-2

the resetting the CHW temperature set-point. The longer-term goal is to optimize the setpoint to minimize energy, while still ensuring all load—i.e. temperature and humidity requirements—are met.

When ECAM analysis was started, there were about 147 days of data available. After cleaning the data, there were 143 days of data available. Cleaning included eliminating days where the CHWST setpoint changed during the day, and days with incomplete data. Most of the analysis was done after aggregating the as-sampled 5-minute interval data to a daily basis, so any partial days could bias the results. Figure 14-13 shows the history of average daily CHWST.

Since we didn't have a great range of ambient temperatures for each of the CHWST setpoints, we divided the data into just 2 groups: times with CHWST <41.5°F and times with CHWST>=41.5°F. Figure 14-14 shows the number of hours at different setpoints, for three ranges of ambient temperatures and associated loads. The ambient temperature ranges were chosen to fit the ranges in the models, as you will see below.

The site is continuing to vary CHWST, and we will update the results as more warm weather data is available.

The total CHW plant demand, including chillers and pumps, was used for the analysis. At warmer CHW temperatures, the loads should require more CHW flow to provide the same cooling. Therefore, we included the pumps in the analysis.

Study of the data using ECAM showed that there was no clear difference in power levels on different days of the week; ambient temperature seemed to be the primary independent variable. M&V is usually best using daily aggregations. Therefore, we created models of daily average plant demand, in kW, as a function of daily average outside air temperature. A six-parameter model provided a very good fit of the baseline days, as shown in Figure 14-15.

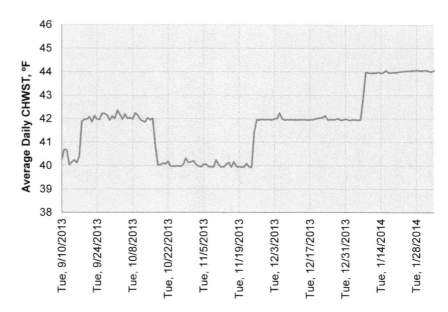

Figure 14-13: Time History of Chilled Water Supply Temperatures

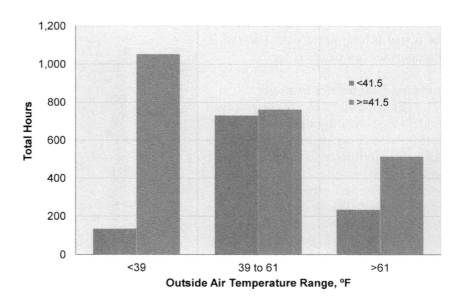

Figure 14-14: Hours of Data at Various Chilled Water Supply Temperature and Ambient Temperature Ranges

However, the savings estimate showed high uncertainty, and an increase in electricity use with the higher CHWST. Here are the results for 116 days in the "post" period, i.e. the times with CHWST setpoint equal to or greater than 41.5°F (Table 14-3).

The uncertainty was greater than the net impact of the change in setpoint. To try and understand the increase in energy use, and the uncertainty, the next step was to model the post period. Again, the fit was very good with a six-parameter model, as shown in Figure 14-16.

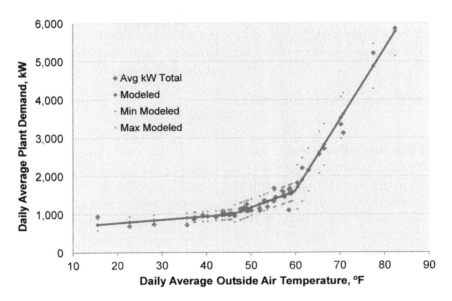

Figure 14-15: Six-Parameter Model for Times with CHWST<41.5°F

4,228,999	Projected Baseline Energy, kWh
4,289,561	Measured Energy, kWh
-60,562	Energy Savings, kWh
210,843	Projected Baseline ±Uncertainty @ 80% Confidence Level
-60,562 ±210,843	Energy Savings and Uncertainty @ 80% Confidence Level
5.0%	Projected Baseline ±Uncertainty @ 80% Confidence Level
-1.4% ±5.0%	Energy Savings and Uncertainty @ 80% Confidence Level

Table 14-3

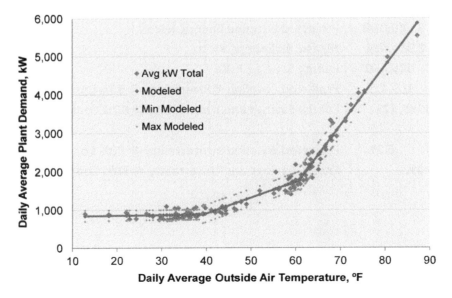

Figure 14-16: Six-Parameter Model for Times with CHWST>=41.5°F

With models of both the baseline and the post period created, they can be directly compared. Figure 14-17 shows the two models. The data points are omitted for clarity.

Figure 14-17: Comparison of Models for Two Chilled Water Temperature Setpoints

Figure 14-17 shows that the higher CHWST saves energy at high loads, at ambient temperatures above about 62°F, but increases energy use at colder ambient temperatures and lower loads. However, there was little data yet at some combinations of ambient temperature and CHWST, as shown previously in Figure 14-14, so these results are not yet conclusive.

The filters in ECAM make it easy to look at specific regimes of operation. Filtering only for temperatures above 65°F—above the upper change point—we can see the savings and that it appears to be well outside the uncertainty in the models (Table 14-4).

So it appears, as expected, that raising the CHWST will reduce plant demand. This savings seems high, and may still be partly attributable to insufficient data. After more data is gathered, the analysis will be refined, and other variables such as chiller staging will be taken into account.

This project shows how ECAM+ and M&V-type regression models can be used to support system optimization. If changes have only a minor effect, improvements may be difficult to detect without robust analyses. The more data that can be incorporated, the more robust an analysis can be. Statistically, more data is almost always better. Using a good regression model can incorporate large quantities of data, and is easier than trying to match identical operating conditions to see the impact of changes using only a few samples of data.

CONCLUSION

ECAM+ is a flexible tool that supports a wide variety of energy analyses. Powerful features like regression-based M&V can be used to aid in building diagnostics and system optimization. Using these models can also support near-real-time fault detection by making unusual operation more obvious. The time-series chart of residuals is a simple yet powerful way of seeing trends in energy use.

References

Taasevigen D.J. and W. Koran. 2013. *User's Guide to ECAM+, Energy Charting And Metrics plus Building Re-tuning and Measurement and Verification*. Northwest Energy Efficiency Alliance, Portland, OR.

Taasevigen D.J., S. Katipamula, and W. Koran. 2011. *Interval Data Analysis with the Energy Charting and Metrics Tool (ECAM)*. PNNL-20495, Pacific Northwest National Laboratory, Richland, WA.

Gillespie, K.L. Jr., P. Haves, R.J. Hitchcock, J. J. Deringer, K. Kinney 2007. *A Specifications Guide for Performance Monitoring Systems*. The Regents of the University of California through Ernest Orlando Lawrence Berkeley National Laboratory, Berkeley, CA.

Motegi, Naoya, M.A. Piette, S. Kinney, K. Herter 2003. *Web-based Energy Information Systems for Energy Management and Demand Response in Commercial Buildings*. Ernest Orlando Lawrence Berkeley National Laboratory, Berkeley, CA.

2,740,014	**Projected Baseline Energy, kWh**
2,314,564	**Measured Energy, kWh**
425,450	**Energy Savings, kWh**
169,123	**Projected Baseline ±Uncertainty @ 80% Confidence Level**
425,450 ±169,123	**Energy Savings and Uncertainty @ 80% Confidence Level**
6.2%	**Projected Baseline ±Uncertainty @ 80% Confidence Level**
15.5% ±6.2%	**Energy Savings and Uncertainty @ 80% Confidence Level**

Table 14-4

Chapter 15

Outside The Box:
Integrating People & Controls
To Create a Unique
Energy Management Program—
A Caltech Case Study

Matthew Berbée, C.E.M.
California Institute of Technology

"An automated whole building diagnostics tools is a software program that takes a top-down approach to diagnostics to detect excess energy consumption of the whole building and its major systems"[1]

At the California Institute of Technology (Caltech), energy projects are financed from a capital revolving fund, the Caltech Energy Conservation Investment Program (CECIP). The industry term for these funds is a Green Revolving Fund (GRF). At Caltech, projects qualifying for CECIP funding must have a return on investment of 15 percent or greater and exhibit verifiable utility savings. The loan fund is replenished quarterly from documented avoided utility cost from completed projects for that period. The Institute's guiding financial mantra is as follows: "The cost to the utility budget during a CECIP project does not change (vs. budget). What does change is that a portion goes to utility bills and a portion to debt service"[2]

From CECIP program inception in 2009 to present, $8M of working capital from the revolving fund has enabled more than $11M of energy efficiency project spend. These investments have generated more than $3.5M in reduced utility cost, capitalized on $2.5M in utility incentives, and is returning approximately $0.5M per quarter in avoided utility costs. The program portfolio is returning at approximately 20 percent.

Managing energy efficiency projects from conceptual design through implementation to measurement and verification (M&V) is well documented throughout the energy industry. Caltech's approach differs in that all project costs must fit within the CECIP payback criteria—soft costs, metering costs, etc.

THE BUSINESS CASE

In 2005, the AHSRAE article "Automated Whole Building Diagnostics" indicated that "The energy-savings potential of retro-commissioning represents the upper boundary of Automated Whole Building Diagnostics (AWBD). Thus AWBD tools could reduce building energy consumption by 5% to 20%."[1] As Figure 15-1 illustrates, energy efficiency drift does occur and increases in severity as time passes from initial project completion. Performance drift presents a significant risk to the value proposition of the CECIP program and is understood to be a challenge for the energy industry—actual savings do not match predicted savings.

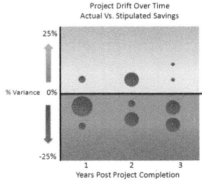

Figure 15-1: This figure represents project performance drift. The blue dots are sized to reflect the number of projects (1-4) of 8 in a sample.

Energy savings don't occur because they are noted in the preliminary project file or associated with a project rebate. Energy efficiency is a result of consistent system operations with design intent. Over time, system parameters may change, settings can be overridden, and actual savings do not match predicted savings. To manage this financial risk, Caltech's Facilities Management Department has developed an Active Energy Management system (AEM), in partnership with the Institute's retro-commissioning provider, controls integrator, and energy retrofit solutions provider.

OPTIMAL PERFORMANCE LEVELS

Standard practice in the energy efficiency industry is to use energy models to evaluate the benefits of various combinations of energy efficiency measures. Outside of the energy performance contracting arena, contract requirements for energy model calibration are the exception, not the norm.

By requiring the use of a detailed, calibrated hourly simulation tool to model existing building energy usage and associated energy savings from select project measures, a high degree of confidence is established upfront. This stringent program requirement satisfies interests for financing and is the first step in the AEM program.

Energy models are developed by inputting building properties such as space geometry, glazing, orientation, mechanical schedule, and regional weather data into modeling software. Normalization equations derived from regression analyses correlate energy usage to the outside air temperature (OAT) (see Figure 15-2).

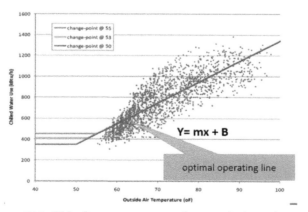

Figure 15-2: This figure represents the correlation of energy use and outside air temperature.

Example Regression Equations:
Chilled Water Regression (kBtu/h):
IF OAT < 53.3, value = 679.6
*IF OAT > 53.3, value = 679.5 + 109.1 * (OAT * 53.3)*
The two IF statements shown above mathematically illustrate the correlation between OAT and chilled water energy usage in kBtu/h.

Energy modeling regression equations are not typically used for any operational purpose. They usually remain buried within the energy modeling or associated companion spreadsheet calculations. However, at Caltech, these regression equations have been programmed in the building management systems (BMS) and are the foundation for an AEM operator interface. The interface shows operations staff the real-time drift, the delta between metered kBtu/h, and calculated kBtu/h representing the energy efficiency of the ideal state (see Figure 15-3).

This interface allows building operators to view the operating efficiency of a building post retrofit. If it is determine a subject building is deviating from optimal operations at any outside air condition, operators can drill down to specific building systems to troubleshoot the problem and minimize efficiency drift.

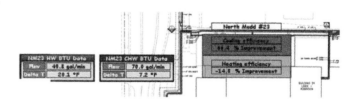

Figure 15-3: Operator Interface

ENABLING ACTION

The BMS at Caltech is an open-protocol architecture. However, the development of the AEM system is not dependent on this feature as the concept and subsequent actionable information can be detailed for any controls specification. Recalling the basic premise for using energy modeling information, operators have an advantage in that information is pushed to them, focusing review of systems on those drifting from optimal efficiency and optimizing time management.

The next step takes the AEM concept to the system level and leverages data from heating, cooling, and electricity meters to highlight devices that can be expected to show efficiency drift, such as HVAC control valves. The AEM program has automated "Valve-Leak" logic to help operators isolate HVAC control valves that are not properly sealing off heating or cooling water. With one click, the operator arrives at a leak-logic graphic that enables real time assessment of the operation of the valve. In addition to this review, valve leak events are automatically emailed to the maintenance dispatch to generate a repair work order.

Example Alarm Logic:

An air-handling unit (AHU) connected to a district heating and cooling plant configured with supply and return fan system with air-side economizer.

If the chilled water valve is closed, the fan is running, and the supply air temperature is 5 degrees F below the mixed-air temperature for 30 minutes, the condition triggers a valve leak notification. The notification is automatically emailed to the maintenance management dispatch and is displayed on the AEM interface. The notification is latched for 2 weeks and is then released. If the valve is still indicating a leak condition after two weeks, the notification is sent again.

A similar logic is applied to heating hot water valves. The system notifies maintenance management of any variable frequency drives (VFD) operating at 100 percent speed. Data from the AEM system is reviewed weekly by the building controls and energy management groups.

THE HUMAN ELEMENT OF AUTOMATED DIAGNOSTICS

Up to this point this paper has discussed energy efficiency performance as it relates to optimal performance values from the energy modeling process to the automated testing of select system components but has not addressed the human element.

When energy retrofits are designed and sequences of operation are engineered, implemented, and tested they capture the operation at that moment, i.e. the setpoints, schedules, and parameters to meet the current customer's requirements. However, over time customer requirements, building energy loads, and space types may change requiring operators to adjust system settings to accommodate such demands. These manual overrides and system changes greatly contribute to efficiency drift from optimal operation. To account for this human element, the AEM program includes override reporting logic that holds operators accountable for changes to the system. Operators must select from a predetermined "pick-list" of reasons overrides are being performed. These overrides are sent to the maintenance supervisor and are discussed as part of routine maintenance management and planning (see Figure 15-4).

A building automation controls "configuration issue" notification system has been provided to Caltech operators. The system takes data entry of observed system configuration issues and immediately emails the issue to the facility service center, and to contracted building automation controls integrators for review and assignment. See Figure 15-5.

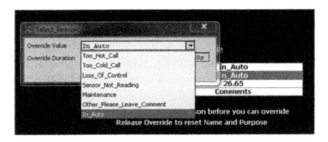

Figure 15-4: System override interface

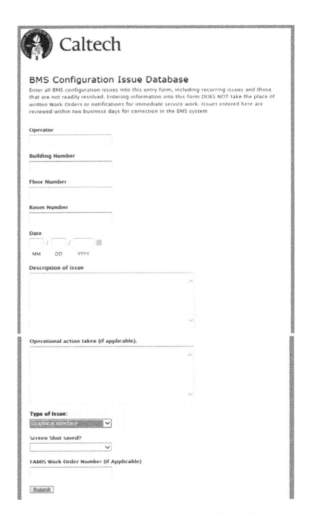

Figure 15-5: Building Automation Configuration Issue Interface

TRANSITION TO OPERATIONS

To further improve operation's capabilities, the Caltech maintenance management group is designing a program called Capital Projects Transition to Operations. Automated diagnostics and the Active Energy Management concept are being further refined. The commissioning agent on a project that installs building automation controls is required to utilize "system tags" a means for easily searching for points within an automation system. These tags are then used to support the functional performance tests that are part of the traditional Cx process. The difference is the hand-off process. At the conclusion of a traditional Cx process, the owner accepts the Cx report and has the expectation that all commissioned systems are functionally to the design intent. The operations group has graphical user interfaces to view the real-time status of a system but not the performance with respect to the functional test rules or the modeled design intent with respect to energy performance. The Caltech transition to operations program requires that the same design intent validation rules that the Cx agent use be programmed into the automation system using the point tags so the operators have a means to evaluate system operating efficiency.

The value to the owner is far more than energy efficiency validation, warranty management is another key focus area. The system design allows system diagnostics to locate sensors that are reading erratically, and programming that requires additional tuning. The value to the owner is improving the use of warranty repair and replace of building automation hardware. A proactive position is taken with the system to itemize warranty items before the warranty period is completed.

AEM FROM THE SOLUTIONS PROVIDER'S PERSPECTIVE

"In our 30 years as an energy solutions contractor, we have seen the evolution of energy efficiency technology, DDC and retro-commissioning. In most of the retrofits and upgrades that we have performed there has been a stipulated savings approach wherein the utility buys into the projected energy savings and the customer finances it based on the projected IRR that will offset the cost of money. Years of research have shown that buildings drift up to 20% or more from the date they are built or retro-commissioned; in dollars and cents this means a building operating at $2.50 a foot could reach $3.00 a foot for electricity alone, in a 5-year period. For example, in one building in Pasadena CA, our customer called us and told us that an energy retrofit we performed was not saving money. Upon investigation, we found that the chief engineer had turned one of the two supply fans on via manual override and had forgotten about it. The fan and the compressors ran 24 hour hours for 6 months until we uncovered the cause of the lack of energy savings. This is a typical situation...The purpose of the FDD is

to alert the DDC operator, Energy Manager or Facility Manager to issues with the sequence or problems with the system that are causing drift in operation form the desired and programmed plan. This can be for lighting, HVAC, transformers, water usage or any measureable system or device that contributes to the overall preferred and commissioned performance for a building. FDD is like looking through the windshield as opposed to the rear view mirror. "[3]

NEXT STEPS FOR THE AEM PROGRAM

The AEM program at Caltech continues to mature as information acquisition is used to improved program operations. The program's strategic next steps include:

- Expanding controls automation to automatically adjust system parameters based on energy efficiency drift, set-point adjustment, load shedding, etc.
- Automate cost avoidance directly linked to AEM efforts—currently the dollars and cents part of the business model is performed after the AEM actions are complete.
- Expand roles and responsibilities throughout the maintenance management department to include participation in the AEM process as it relates to primary job functions.
- Continue to evaluate the viability of turn-key AEM solutions to integrate with building control systems.

CONCLUSIONS

This article is intended to illustrate the numerous options and resources owners have to actively manage the energy and operational performance of their buildings. Building automation controls can be configured to provide a level of feedback focused on operational efficiency, not simply graphical user interfaces. The organizational value add for the long term will manifest itself in lower total cost of ownership, fewer building occupant HVAC related complaints and ultimately lower energy costs. In the context of Caltech's energy investment fund it means tighter building performance and better proof of the energy efficiency performance.

References
[1]   Roth, K., Llana, P., Westphalen, D., & Brodrick, J. (2005). Automated Whole Building Diagnostics. *ASHRAE Journal*, (Vol. 47, No 5, May 2005), page 82.
[2]   Brewer, M. California Institute of Technology, Controller, 2012.
[3]   Fletcher, C. EMCOR Services Mesa Energy Systems, Vice President, 2013. 0

Chapter 16

Value Proposition of FDD Solutions

Sanjyot Bhusari and Michael Watts
Affiliated Engineer, Inc

EXECUTIVE SUMMARY

Building owners, managers and operators continue to face the challenge of "doing more with less". An entire industry has emerged to facilitate this issue. Fault Detection and Diagnostic (FDD) software and technologies are solutions that provide actionable information to better utilize increasingly complicated buildings systems, reduce energy consumption in an environment of, dwindling or stagnating operating budgets.

The value of pursuing an FDD solution should be weighed against a facility's existing investment in Building Automation System and its staff. Only when existing assets have been maximized to their greatest extent should facility management turn to FDD.

FDD solutions come with a range of commitments and investments. Deciding on the best FDD solution can be a daunting challenge. The decision making process will be greatly influenced by age and technological sophistication of the facility, availability of skill sets required to properly manage and execute the application, and investment budget. Above all it will be facility management commitment to the philosophy of FDD that will determine success or failure.

INTRODUCTION—WHY FDD AND WHY NOW?

The concept of FDD has been around for years, however only recently has it become a commonly recognized acronym in the commercial world of facility management.

FDD can be applied retroactively to an existing facility or incorporated into the design of a new facility. FDD typically consist of some sort of vendor specific software, database, or both. FDD solutions could be hosted in the building or applied remotely. FDD algorithms can be open or proprietary. All of these approaches involve some investment of time and capital; however, the return on that investment generally produces very attractive paybacks.

What Are Facility Management Challenges and Why Do They Exist?

Facility management challenges include, but are not limited to issues with *occupant discomfort, increased equipment failure rates, and high energy consumption and overall operational costs.* Key underlying element to these challenges is the ability of O&M staff to operate their facilities in an optimal and effective manner.

O&M Staff Skill Set

Technological advancements continue to infiltrate every part of our lives. Buildings have not been immune to this phenomenon. Advances in building automation continue to provide a wealth of opportunity for the continual optimization of facility performance. Frequently these opportunities are lost due largely to the lack of proper training.

Increasingly Complex Mechanical, Electrical, Plumbing (MEP) Design

Current codes and green building initiatives are resulting in increasingly complicated building design to achieve aggressive energy conservation and sustainability goals. The performance of these buildings in the operations phase rarely matches their promised design. Multiple factors must be contemplated when considering reasons for this performance gap including but not limited to errors in design methodology, improper installation of systems and/or lack of training for operation staff.

Lack of Detailed BAS Specifications

While traditional MEP designs are often complex, the automation software required to manage and control them are technology driven and vary dramatically from vendor to vendor. Traditional design fee structures seldom accommodate the detailed automation engineering required to realize the efficiency of these multifaceted MEP systems. Automation systems are too often specified with little regard for the detail needed to efficiently control the specialized systems and their respective sequence of operation.

Construction Approach

In addition to these complex system designs, construction budgets often demand a low bid solution without regard for qualifications. The disconnect between the design intent and the actual systems installed often result in a less than ideal automation system. Owners are often left with a complex design and an underperforming automation system. To compound this problem, Owners seldom have the skill sets on staff to successfully operate and manage their substantial investments in new systems and technologies designed into their new facility. As a result automation systems are effectively reduced to temperature control and monitoring.

Dwindling operational budgets also contribute to poorly managed facilities. Facility managers and O&M teams continue to react rather than invest. In addition, O&M budgets are seldom increased proportional to the technologies that are intended to minimize labor and energy. This results in a reactive mode of operation that reduces the time required for more productive or proactive tasks. In addition, often the action to resolve occupant complaint or poorly performing system is to "override" an automatic device. Such resolutions satisfy short term problems, but if left in place, can create future complaints and excessive energy consumption.

What is FDD?

In theory, FDD is a way of analyzing building systems data for anomalies (fault detection) and providing supporting information or solutions (diagnostics) to facilitate a corrective action. FDD strategies leverage pattern recognition and data evaluation (not just alarms) to identify opportunities to prevent over-consumption of energy, occupant complaints, improve performance or predict a possible failure. FDD provides higher level of data analytics and graphical outputs that are typically not available in standard building automation systems.

What FDD is Not ...

FDD is not simply the collection of alarm status alerts from the BAS. FDD will not automatically correct faults detected, and it is not the sole solution to an underperforming building.

Why Was There a Need for a Whole New Industry to Emerge?

To solve these facility management challenges an entirely new industry has emerged. Early adopters have included small start ups with a strong software background who have marketed FDD solutions as ret-ro-active strategies for struggling facilities. The void created by mismanaged construction dollar investment (percentage based engineering fees that lead to lack of detailed BAS specifications) and poor implementation driven by low bid approach was filled by these early adopters in the operations arena. The primary driver for FDD solutions have been very attractive returns on investment.

Now that traditional BAS companies are now providing FDD solutions, the question of where in the building's life cycle it makes the most sense to apply these solutions emerges.

ANALYSIS—LEVERAGING VALUE FROM FDD APPROACH

What are the Different FDD Solutions?

There are multiple approaches for the successful implementation of FDD solutions. As technology has advanced, newer solutions are focusing on the continuous detection of anomalies. FDD software monitors building systems via building automation, interpreting the data, indentifying anomalies, and sending the facility owner a monthly report with a list of deficiencies to resolve.

Most FDD solutions follow a cloud-based model which requires an initial first cost for hardware and programming to push building systems data up to cloud based servers. These solutions include a monthly recurring cost for reporting. Another FDD solution is integrated visualization software as a HMI (Human Machine Interface), a solution that typically involves another layer of software integration to an existing BAS platform. In the HMI scenario algorithms and rules to detect faults are loaded on to the facility owner's servers. First costs tend to be higher however there is no recurring fee except for the maintenance costs associated with routine upkeep of the software. Cloud based solutions offer owners the flexibility of ending the service once a building reaches an optimum level of performance or if the solution is no longer effective. However considering that building systems are dynamic, a more continuous approach may have a greater appeal to facility owners.

Irrespective of FDD solutions, one key obstacle that needs to be overcome is the inherent uniqueness of buildings typically designed by architects and engineers. For instance on a single campus, air handlers designed for similar building types could have vastly different sequences, instrumentation and design intent. Some air handlers may use demand control ventilation, while some other may utilize an economizer mode.

Understanding the uniqueness of each building and the equipment serving it, is paramount to evaluation of an FDD solution. Many current solutions provide algorithms that are useful however it should be noted that the more specific the algorithm is to the building and equipment the higher the benefit. Thus, the level of accuracy delivered through fault detection is essentially driven by a business decision (regarding the level of effort and respective budget earmarked for algorithm development).

Utilizing an open source data historian to collect and mine data represents another approach to FDD. Many Building Automation vendors utilize an off the shelf technology database to store building system data. If the facility owner has access to the table structure of the database, simple business intelligence software tools can be utilized to mine the data with the goal of isolating anomalies. Though this approach may be less expensive, the number of faults that can be detected may be dependent on the quality and quantity of FDD "rules" that facility owners and their consultants can provide. In cases where the building automation system does not use off the shelf technology a separate data historian can be specified with open database technology. For example, one healthcare provider in the Southeast is utilizing this approach and after researching and reviewing various FDD Vendors and options chose to implement their own solution. Teaming their energy engineering staff with their IT staff to develop their own data mining "rules" based on their unique requirements turned out to be the best approach for them. See case study for additional details.

A field panel-based approach to FDD is also being implemented by some higher educational campuses. In this case, one university is planning on using faults or rules as part of the equipment sequences. In lieu of adding another layer of software, this solution relies on utilizing the building automation system to report faults. (See Table 16-1.)

What are the Key Elements of FDD?

Regardless of the FDD solution that facility owners may utilize, there are a few key elements that need to be in place to maximize the FDD investment.

FIRST, USE WHAT YOU HAVE

Building Automation System

Building Automation Systems are loaded with tools for energy management. Functions for as reporting, alarms, schedules, trends and graphics offer a tremendous amount of information that can be used to operate systems in an optimum manner. The problem is that most building automation systems are rarely configured for continuous monitoring. Controls contractors bid on design documents that typically focus on control strategies for equipment. Typical control specifications are developed such that the building automation system has a good deal of capability but rarely requires the controls contractor to fully configure it. As a result most facility owners have a very powerful system for which the maximum capability is not fully realized.

Reports

Reports can be configured to identify sensors out of calibration, visualize energy use patterns, confirm that setpoints are within acceptable range, and discover overridden values among other indicators. One college in Florida reviews an overridden values report on a daily basis. They have noted that O&M strategies targeted at overriding values may solve short term complaints but usually result in long term energy nightmares.

Trends

Trends can be used for troubleshooting, reviewing PID loops, and to visualize energy use patterns. For example a university configures its trends for critical values such as chilled water valve position over a long term. When the university reviewed the chilled water valve position trend for a three year period they discovered a distinct jump from year to year. This issue was investigated further to reveal filtration problems that were causing the cooling coil to get increasingly "dirty".

Alarms

Building Automation System alarms can alert building operators when an equipment command does not correspond with feedback. So if AHUs stay online after they are scheduled to be turned off, BAS sends a notification to the operations staff indicating anomalous operation.

Graphics

Key performance indicators can be set up to track building or facility performance. For example energy can be tracked against building area to compare baselines or it can be represented in a calendar format, such that anomalies become readily apparent. One university requires key performance indicators on graphics for all new projects. Key performance indicators include: building square footage per ton of cooling, electrical con-

Table 16-1: Solution Summary

Solution Type	Features	Pros	Cons
HMI	▪ Algorithms are applied on owner's servers. ▪ Continuous reporting of anomalies. ▪ Data remains within the facility	▪ Create accurate fault rules ▪ Visualization of big data	▪ High initial investment cost.
Cloud based	▪ Cloud based solutions facility data is served up through the cloud. ▪ Anomalies are reported once a month or a quarter. ▪ Data leaves the facility.	▪ Rules / Algorithms to find anomalies developed through years of research. ▪ Analyst available to respond to questions.	▪ High first cost and recurring costs per month.
Data historian	▪ An open off the shelf data base technology is utilized to store building systems data. Part of BAS solution. ▪ Continuous reporting or reports emailed through the database at frequency per owner desire. ▪ Data remains within the facility. ▪ No additional software. Business intelligence tools to mine data required.	▪ Lower first costs. ▪ No recurring costs.	▪ Anomalies found are determined by the rules that facility owners and their consultants can come up with. ▪ IT skill set required
Field panel	▪ BAS based solution ▪ Faults served up as alarms ▪ Continuous reporting. ▪ Data remains within the facility ▪ No additional software required.	▪ Lower first costs. ▪ No recurring costs.	▪ Anomalies found are determined by the rules that facility owners and their consultants can come up with. ▪ Dependent on existing BAS controller capability.

sumption per building square footage, heating therms per square footage, and airflow per square footage etc.

Training

In many facilities, operations teams remain in a reactive mode. For example the only time staff investigates building systems or the building automation system is when they receive complaints or need to perform preventative maintenance tasks. Staff can be trained to utilize the building automation system in a proactive manner so as to fully leverage BAS functionality.

Point Naming Convention

One key building block for all FDD solutions is the development of standardized point naming convention. Industry wide efforts are underway to establish such standards so that meaningful information for sensors and instrumentation can be shared across various solutions. One such effort is Project Haystack. Per its website, "Project Haystack is an open source initiative to develop tagging conventions and taxonomies for building equipment and operational data".

Open Protocols

Most FDD solutions require that building systems data remain in an open format. Since some O&M teams use proprietary systems, and integration approach needs to be determined in order to convert proprietary protocols to open sourced protocols thus ensuring that the FDD solutions able to consume data. There are several options available ranging from upgrading controllers to the installation of hardware gateway devices and software drivers.

Staffing Levels

Plan your resources to maximize technology. Many facilities operate in a reactive mode. Staff is overwhelmed responding to occupant complaints and rarely have time to analyze at systems in a more pro-active manner. Before investing in a FDD solution, an assessment of resource bandwidth and in-house technical skill set is necessary. FDD makes more information available for viewing and analyzing so new skills are needed to maintain FDD and optimize buildings system performance.

SOLUTIONS—FINDING THE RIGHT FIT

Selection Criteria

Once a decision is made to pursue a FDD solution, selecting the right approach and vendor can be chal-

lenging. To weed through the options that are available, facility owners evaluate and balance the following criteria: budgets, energy goals, need for analyst support, financial pay back information associated with faults, sensitivity of data and IT support.

Budgets

HMI and cloud-based FDD solutions may require a higher upfront cost investment than Data Historian and Field-Panel based solutions before realizing savings. However if HMI/cloud solutions are desired, facility owners can think of starting small and then scaling their implementation as more funds become available. Both HMI and cloud-based FDD solutions implementation costs are dependent upon the number of points monitored and analyzed. A small start could be initiated which only takes into account big picture energy information.

Energy Goals

HMI and cloud based FDD solutions have years of research behind their rules and algorithms. They may be able to find a few extra anomalies that Data Historian and field panel-based solutions may not find. At the end of the day, the facility owner will have to weigh cost and value to determine the solution that best suits their needs.

Analyst Support

Some cloud based solutions assign an energy analyst to help facility owners understand and prioritize faults discovered and the big picture changes required to transform the facility. This service may be crucial especially in the event that the facility owner does not have such a skill set within their organization.

Financial Payback Information

Most HMI and cloud based solutions come with financial information associated with the faults that they are designed to discover and prioritize. Such information can be a big motivator. For instance, a large university leveraged cost savings information to prioritize faults and further motivate staff to continue these efforts.

Sensitivity of Data

Cloud based solutions require the data to leave the facility. In some cases building systems data is sensitive and in those instances any of the other three solutions should be considered.

Selection Approach

If cloud-based or HMI solutions are determined as the best solutions (for a given portfolio of buildings), naturally the subsequent decision that follows is the selection of a vendor who will provide the best value. Since the quantity and quality of rules are proprietary and not exposed, utilizing them as selection criteria is not possible. Checking up on vendor references can only provide so much value. Most buildings, campuses and facilities are unique in many ways and the same product that worked very well on one campus may not provide optimum results on others.

A major commercial complex came up with a unique way of selecting the right vendor. They used the Request For Qualifications (RFQ) process to shortlist three FDD vendors and assigned each of them a building on campus. This allowed them to see firsthand what each vendor was capable of delivering alongside their cost model/value.

The more information FDD vendors know about the building, equipment and the building automation system during the bidding process, the more precise of a cost proposal can be developed to include approaches to integration. It is advised that building owners/facility managers inventory their buildings as a preparation to developing a RFQ.

The market for FDD implementation is mostly in the post occupancy phase of building operations. Planning for these solutions in the design phase and implementing them in the construction phase will be the most effective approach for facility owners, especially for new buildings. Advantages include reduced hardware costs and having the opportunity to specify the right instrumentation for performance measurement as opposed to merely controls.

CASE STUDY

As energy and operational costs for its three million square foot campus began escalating, a major healthcare provider recognized the need for a more proactive approach. Another motivating call to action was sounded through the "patient experience" feedback mechanism. Management prioritized their commitment to proactively recognize conditions and patterns that could lead to patient room comfort issues. Too often the chosen course of action for the resolution of patient complaints was to "override" a controlled device or setpoint. Such resolutions were observed to satisfy the short term problems, but were typically forgotten about and left in place, thus causing the generation of additional work orders and excessive energy consumption.

This healthcare provider was drawn to the success stories heralded by FDD cloud based solutions and evaluated multiple vendors to implement such solutions. A deeper analysis of their specific requirements, goals, budgets and existing technological conditions lead to the development of a master plan aimed at leveraging the full suite of optimization strategies possible through FDD. A key reason for engaging in deeper analysis was the fear that "faults" generated from FDD software would suffer the same fate as their BAS alarm counterparts: unacknowledged or acknowledged but ignored. It became paramount to assemble and employ a team with the knowledge and commitment to fully embrace FDD as a philosophy as well as technology solution.

To get things started, an assessment of their current BAS investment was employed. Through this effort it was discovered that the BAS had previously been designed and equipped with the technology to implement much of the desired FDD approach. Their BAS included a data historian with an open database to facilitate data mining. BAS Point naming was previously employed and in place, for most of their campus. BAS graphics were already configured to track key performance indicators such as comfort index and energy benchmarks. This detailed and structured approach to the design allowed the healthcare provider to leverage its current investment. Utilizing existing BAS features, multiple anomalies were easily identified and reported with the use of relatively simple inquiries. Through this exercise, a process was developed utilizing a set of rules based on anomalies discovered via ongoing commissioning expertise. These "rules" were then configured as data historian queries such that the process of finding anomalies could be automated.

The approach to leverage the BAS first, along with the data historian is proving to be very effective. Early results have already demonstrated the value of this approach in terms of energy savings and operational gains. Hidden anomalies such as single zone driving a static pressure setpoint higher than required, economizer mode faults and inefficient fan operation have been discovered.

The vision going forward is to combine the analytics generated from the data historian and BAS systems with the work order management system. The healthcare provider intends to use this data to achieve its goals of patient comfort while improving operational and energy efficiency.

CONCLUSIONS

FDD solutions come in variety of different flavors each of which offer some value. Selecting the most appropriate solution can be a daunting task. Building Owners, facility managers, designers and O&M teams spanning across various industries face an increasingly wider array of complications in today's high tech environment. Prioritizing needs and goals to customize a best fit solution is crucial to the success of any FDD implementation program. A well organized, step by step, approach that is required. For this concept to succeed a commitment from the Owner and O&M team to embrace not simply FDD technology but the entire FDD philosophy is a must.

Chapter 17

FDD: How High Are the Market Barriers to Entry?

Gregory Cmar
Interval Data Systems

ABSTRACT

Fault detection and diagnostics (FD&D) is presented as part of the continuum of technology and services associated with Enterprise Energy Management Systems (EEMS). Since the foundation of this information-based technology (computer-based building control) is now over 40 years old and as the concomitant adoption rate EEMS is far from ubiquitous, this chapter examines factors that may hinder owners and operators from taking advantage the latest developments in the field of facility management.

Defining barriers to entry is a complex subject. In the interest of covering as broad a range of factors as possible this chapter uses the 'news style', or Kipling Method to explore a full scope.

- "Who" analyzes barriers in terms of attitudes of the people who will sell, buy, use and support FD&D: Rogers Diffusion of Innovations and the Gartner Hype Cycle

- "What" identifies the various analytics that make this technology useful

- "When" scrutinizes opportunity cost with respect to other building projects and how that affects adoption.

- "Where" assesses the range of buildings and systems that will be affected.

- "Why" looks at expectations of value returned contrasting energy cost and savings, rebates and operating efficiency.

- "How" examines scaling of the technologies and processes that underlies the delivery of FD&D.

The chapter concludes that there is overwhelming value to be gained in FD&D. Equally, it concludes that the real and perceived risks in adopting the technology will make it challenging for companies to enter and build market share.

FAULT DETECTION AND DIAGNOSTICS

Fault detection and diagnostics (FD&D) are two more technologies offered up in the battle for making our building stock energy and cost efficient. They are the logical extension of Enterprise Energy Management, building upon the analysis of utility billing and related smart meter technology. Many start-up as well as old line companies now promote IT technologies and information based services to manage buildings better and optimize the control of equipment that affects energy consumption.

For the record, fault detection as used in this chapter comprises that technology that *automatically* identifies system failure or defective processes that result in energy waste. This is a step beyond simple alarming. For example, a building automation system (BAS) can alarm space temperature that is outside some limit; fault detection can identify whether it is a systemic problem and not just a briefly open window. It can even identify that a heating valve remains 100% is open and space temperature will alarm if the problem is not corrected.

Diagnostics, in contrast, comprises technology that allows the user to explore operating data and find issues that will lead to more effective or efficient operations. It differs from fault detection in that it is not necessarily automatic and often requires an intelligent operator. By extension, as analysis can be made automatic it moves closer into the domain of fault detection.

Like all new technologies, FD&D faces barriers to market acceptance. There are decades, if not centuries of experience running buildings without evaluating the detailed operating data that measures and defines performance. As a result, there is a natural skepticism to any technology that challenges the accepted way of getting the job done. Surmounting this resistance is critical to obtaining the benefits.

WHO

There are three distinct groups involved with the adoption of FD&D, those who sell facility management services, those who buy them and those who use them. Each one of these groups has the ability to stop a successful FD&D implementation; it takes all three working together to create one. According to the theory pioneered by Everett Rogers in his seminal book, *Diffusion of Innovations*, this is called a collective innovation-decision. It is contrasted with other energy efficiency decision processes, such as when an engineer specifies a particular technology will be included (i.e., the growth of the VAV market), or authority innovation-decision. And it differs from choices like a lighting retrofit where a single person, such as a business manager, can make a decision for every one else in the group, or optional innovation-decision.

Rogers characterized how new technology is adopted throughout the population using a normal distribution curve segmented by standard deviation and a process that evolves over time. Diffusion starts when a small segment of customers, innovators and early adopters, examine new technology. Once success has been achieved within these groups the technology will expand to early- then late-majority users. Figure 17-1

should be familiar to most technology marketers as well as their users.

From the standpoint of market barriers, selling into a collective innovation-decision is the most stringent requirement of adoption. Collaboration amongst a variety of user personalities is required. Further, given that the population of expected adopters is coming from the construction and facilities management industry, the task is even more daunting. Consider the technology sector, where product life cycles can be measured in months or a few years at most. Facilities are with us for decades so decisions are much more permanent. When you examine market barriers from this standpoint it seems almost logical that it took 40 years of using computers to control buildings before the industry began to start exploring how information technology might be used to manage those facilities more efficiently.

As attractive as making decisions based upon facts may sound, however, getting it to happen is not a sure thing. FD&D is a service that changes how existing services are delivered. To succeed, according to Rogers, it must have:

1.   relative advantage;

2.   be compatible with existing processes;

3.   be simple to use;

4.   be testable on a limited scale; and,

5.   have results that can be observed by others considering the same technology.

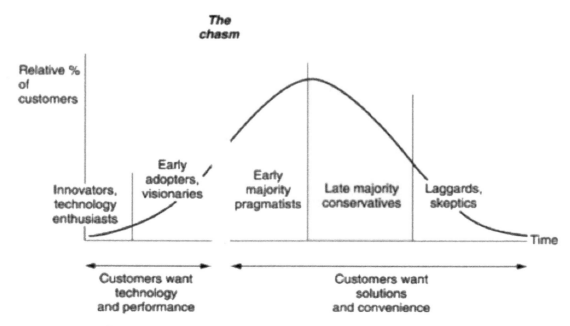

Figure 17-1: Roger's Diffusion of Innovations: Identifying Adopters

On a more individual level there are practical questions that need to be answered. If a company has it's service salesman promoting the offering, does that mean they will cannibalize existing sales? And what of the serviceman who is expected to use this technology? If FD&D identifies that they haven't been properly running the building, what are the implications on his performance? Does surfacing operating flaws put their jobs at risk?

For the buyer there are different questions and risks. Weren't they promised these benefits when they made their initial BAS purchase? Why should they invest more? Finally, as a cap on the barriers to entry, is their recognition of the Hype Cycle (Figure 17-2). Identified by technology research firm Gartner, Inc., the Hype Cycle recognizes that technology gets oversold early on. Most who buy in to the hype are disappointed. At that point, what was once a good idea degenerates into work before the expected advantages can be achieved. This phenomenon leaves the buyer in the position of examining the state of development for the service. In a competitive world, they need to assess whether the FD&D provider is offering something that will be around for the long haul.

WHAT

There are a myriad of technologies that underlie FD&D, each one suited to a particular task or opportunity. This diversity makes it confusing for users to choose the platform best suited for their needs.

The processes used for fault detection are numerous. They can be passive, effectively an observer of the process, or active, where a series of actions are initiated at the behest of the diagnostic in order to precisely measure sequences and results. Some methods are better suited for identifying hardware failures, others are more appropriate for discovering errors in logic, or the actual programming for controlled devices. Model-based fault detection is one commonly used technique. This collection of methodologies, which include causal models, static, dynamic, quantitative, qualitative and engineering based models, identify faults as a difference between what is occurring versus what is expected for the system being analyzed. Pattern recognition is another technique used to identify faults.

Faults, it should be noted, extend beyond identifying mechanical and electrical failures. A significant responsibility of FD&D technology is to identify inefficient operation. These faults can range from improper/incompatible setpoints within a system to changes in the space that redefine design parameters. These, more subtle faults, are typically at the root of inefficient operations

Crossing the boundary from fault detection to automated diagnostics, neural networks are nonlinear, multivariate models that are used to identify improper operations. An engineer trains a model to recognize how a system operates, identify various failure modes and select a diagnosis among a range of possibilities. This technology promises to be a more complete and effective solution, though it currently requires a great deal of technological hand holding early on.

Whether diagnosis is made from neural networks or selected according to some expert system rules, automating diagnoses remains a considerable problem for FD&D. The goal of a having system where decisions are automated will yield a significant reduction in the hours required to keep the HVAC system properly functioning. But the pitfall within the technology is misdiagnosis. By nature, people seem to focus on the shortcomings within a technology as opposed to its benefits. As a marketing hurdle, the issue will be to minimize hype so that disillusionment from overpromising capabilities does not hinder future sales.

WHEN

Another of the decisions operators face with regards to FD&D adoption is when to undertake a project. The major events in the facility life cycle are summarized as:

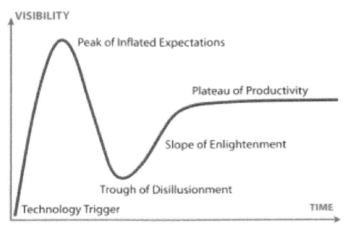

Figure 17-2: Gartner's' Hype Cycle

• Design/Construction

Possible Diagnoses

Alarm Description	Zone temperature sensor drift/failure	Airflow (DP) sensor drift/failure	Discharge temperature sensor drift/failure	Damper stuck or failed	Damper actuator stuck or failed	Reheat coil valve stuck or failed	Reheat coil valve actuator stuck or failed	AHU Supply air too warm	AHU Supply air too cool	Supply air static pressure too low	Scheduling conflict with AHU	Undersized VAV box	Tuning problem with airflow feedback control loop	Tuning problem with zone temperature feedback control loop	Inappropriate zone temperature setpoint	Minimum airflow setpoint too low	Minimum airflow setpoint too high	Maximum airflow setpoint too low	Maximum airflow setpoint too high	Sequencing logic error
High zone temperature alarm	X					X	X	X			X	X	X	X	X			X		X
Low zone temperature alarm	X					X	X		X		X	X	X	X	X		X			X
High airflow alarm		X		X	X							X		X						X
Low airflow alarm		X		X	X					X	X	X	X						X	X
Unstable airflow alarm		X		X	X						X	X	X			X				X
High discharge temperature alarm			X			X	X													
Low discharge temperature alarm			X			X	X													

Figure 17-3: National Institute of Standards and Technology FD&D matrix

- Continuing Operations
- Retrofit/Performance Contract

What separates each of these time periods is how opportunity costs are calculated. By definition, opportunity costs may be assessed in the decision-making process of the facility life cycle to evaluate what investments should be made. If the budget for a facility can purchase either a new FD&D system or redecorate the front lobby, then the opportunity cost of buying FD&D is foregoing a front lobby renovation; good managers implement rational decisions by weighing the advantages of one investment opportunity over another.

Calculating opportunity cost can be a sophisticated process. There are Explicit, Implicit and Non-monetary Opportunity Costs to be considered. Explicit Costs involve direct monetary payment—how much does it cost? Implicit Costs are contained in expenses already made at the facility. They are equivalent to what redeploying costs could earn for the firm in alternative uses. For example, either perform maintenance with their own people or subcontract for the services. The differ-

ence between the two defines money which the facility can use elsewhere.

Opportunity costs are not always calculated in actual monetary units, they can also be unknown. The most common Non-monetary Cost is one that spawns a series of future opportunity costs. For instance, a facility operator could choose to defer maintenance that then will lead to a new analysis of opportunity cost at a later date.

The vagueness encompassing when to implement FD&D at a facility is a significant market barrier to adoption. Not only does a vendor need to understand where the potential customer is in the life cycle of their building but they also need to know the competition for its investment.

WHERE

Consider the differences between residential and commercial landscapes; the needs of each marketplace are radically different. Even when we look within each of these marketplaces there is still great diversity.

For the purposes of this discussion, I will choose the commercial marketplace. Within it, there are offices, schools, hospitals, retail, municipals and more. And within each market segment the requirements for success differ as widely as the differences between segments. For the building manager with widely scattered properties the primary goal might be to minimize "truck rolls," the cost of dispatching a service contractor to the site. For a hospital the goal might be to replace JCAHO mandated preventive maintenance schedules with a more cost effective methodology. And the university building portfolio manager might need tools that help them better manage the processes associated with continually changing space needs and engineering requirements within their building stock.

Further complicating the issues of FD&D is the question of where data processing occurs: is it a function of the control system hardware or is it remote process. Again, needs define what is the appropriate technology. Big box/small box retail stores are characterized by packaged (combined chilling/heating/air handling) HVAC systems; failure modes are limited in scope and consistent in practice making it logical to process the data locally. In contrast, built-up/custom control systems that are typical in the MUSH markets (municipal, university, schools and hospitals) lend themselves to remote processing because failure modes are defined in the design and construction process. This is not because it isn't possible to define hardware-based FD&D logic for a particular controlled device. However, the burden of accommodating the programming needs of a system that might have a myriad of configurations is more cost effectively handled by remote processing.

The market barrier for FD&D with respect to where it is applicable is a practical matter: Once the marketplace is selected, however, FD&D requires a data stream. This makes it easy to determine which buildings are candidates; they are the ones with a computer-based control system.

WHY

People have been managing buildings for millennia; the concept of FD&D is barely a decade old. As a market barrier, in order to gain acceptance FD&D has to have a payback. Fortunately, this payback is there, both from reduced energy costs and reduced maintenance costs.

There is considerable research available that documents how the existing processes for constructing and managing commercial buildings only provide a 70%-85% efficient operation. FD&D, in conjunction with other services such as monitoring based commissioning (MBCx) promises to capture this wasted 15%-30% of operating costs.

Demonstrating the advancing maturity of the marketplace, utilities are also implementing rebate programs to further enhance the payback from improved operations. In Massachusetts, rebates of $0.075/kWh and $0.75/therm are offered for the first year of proven energy savings from MBCx projects. The primary focus is on identification and implementation of low cost/no cost energy efficiency measures (Typically <1-year payback). The rebate assumes vendor and customer enter into first year service contract with commitment from customer to enter into an additional 2-year maintenance savings contract.

Finally, FD&D is a more efficient technology than preventive maintenance. Consider how hospitals are required by JCAHO regulations to schedule inspections of each and every device in order to minimize the probability of catastrophic failure. FD&D allows these plans to be rewritten because it identifies devices as they begin to fail. In turn, the maintenance scheduler can now redeploy his talent. Instead of wasting technicians looking over equipment that might need their skill, handymen can be used to perform routine maintenance while skilled labor is redeployed on more complex work.

HOW

The most significant market barriers to entry for FD&D revolve around how the product works and the service is delivered. What makes it so difficult is the number of steps that need to be mastered in order to be successful. The first hurdle to be cleared is data collection. After that comes a data mapping process where the rules are defined for how all this electronic information is assembled into usable forms. Once the data is in place there needs to be a system to run the required FD&D processes. But FD&D is more than just its information technology. To be successful there needs to be a service organization that can implement the identified repairs and efficiency improvements.

Data collection is not a straightforward process because of several factors. To start, BAS manufacturers do not have a single standard for collecting data. Time periods can vary, or the data is only collected on change of value. Some use UTC, but not all. Neither is there a standard for what data is collected. Can you collect all hardware points? What about setpoints and schedules? Can you get important calculated points or values associated with a particular process algorithm? Just as important, how will you access the original A/E data?

Another issue is that the original design for most installed control systems did not presume significant data collection. As a result, third party devices which poll data can negatively affect operation. A safer alternative is to utilize the specific manufacturers data collection system and access the data after it has been transferred to long-term storage. If the manufacturer cannot reliably collect its own data, a new control system is required.

Once the data have been identified, they still needs to be made useful. This process is known as data mapping or building the data model. Effectively, what this process intends to do is convert the physical structure of the building into digital format so that all the data being collected can be programmatically accessed. It is important to recognize that not only BAS data are considered in the data model. Space inventory data are also required as are architectural, mechanical and electrical specifications.

The FD&D algorithm subsystem becomes a matter of plug and play when data collection and data mapping are properly implemented. The plug and play mechanism is an important consideration for any purchaser of FD&D technology because the technology is still in its nascent stages. No FD&D system is available that can handle every potential fault or diagnoses. Even if one were, building technology is not static. FD&D will continually need to be augmented.

The agile development process is a useful methodology to account for the continual and iterative changes that will be required by FD&D. Each new building added to an FD&D system will inevitably add to a list of desired features. There will be new system types, enhanced requirements and so on. These should be forwarded to the vendor and entered into a feature backlog. From here, which features to build can be prioritized and moved into the development plan. Once in they are coded, tested, fine-tuned and released. If plug and play is part of the FD&D design, the user has nothing more to do in order to obtain the value across their installed base than run an install program and start identifying the benefits. This is critical to avoid the pitfall of a system that requires as much work to maintain as the actual building.

The final market barrier that needs to be overcome in order to have a successful FD&D installation is to have a service organization that is prepared to make the identified changes. For many vendors this may be the gating factor that defines success or failure. Perhaps all that is needed is an adequate customer-training program. Perhaps a network of existing service vendors will provide what it takes. Or perhaps the FD&D vendor needs to be able deliver the service directly.

CONCLUSION

The barriers to entry in the FD&D marketplace are varied and significant. To start, the people who will sell, buy and use the technology are known to adopt change slowly. Compounding this barrier is the complexity of the marketplace; there isn't a one size fits all opportunity. This requires vendors to choose where to make their mark knowing that success in one venue may not transfer to another. Once a market is chosen, vendors are then faced with a multistep, complex process to get their product up and running. Even then, when their product performs, success is not assured. A service delivery mechanism is required to obtain the desired results.

As daunting as these barriers may seem, the benefits from FD&D should prove irresistible. Energy savings are significant. Utilities are sweetening the pot with rebates. Lastly, the potential to use information to reduce energy and maintenance costs while improving building performance will eventually focus management on modernizing building operating systems.

Chapter 18

Diagnostics for Monitoring-based Commissioning*

Michael R. Brambley and Srinivas Katipamula
Pacific Northwest National Laboratory

Patrick O'Neill
NorthWrite Inc.

ABSTRACT

This chapter presents a case for application of automated monitoring, analysis and diagnostic tools for monitoring-based commissioning. Selected examples are presented in which such tools have been used successfully to support commissioning activities in southwestern Canada and the U.S. Pacific Northwest. The first example involves use of spreadsheet-based tools to automatically generate diagnostic plots that are visually examined for specific features that reveal operational problems in space conditioning systems of large commercial buildings. The findings then guide re-tuning actions to increase building energy efficiency. This is followed by application of a tool for continuous monitoring of whole-building energy use to automatically track energy savings resulting from a utility commissioning program. This tool also provides a means by which to detect degradation of savings and performance to guide monitoring-based commissioning actions. The potential use of automated diagnostic tools for chillers and packaged air conditioners is then described for continually commissioning these units. The chapter concludes with a discussion of the impacts of this approach on commissioning, including potential time savings, associated cost savings, and improvements in the quality of commissioning.

INTRODUCTION

Monitoring-based commissioning (MBCx) uses energy consumption and system-performance monitoring to guide the re-and retro-commissioning processes

for existing buildings and to verify the energy savings achieved. Furthermore, monitoring is used to help ensure the persistence of savings by alerting building staff and management to degradation in performance and to detect faults in operation. Monitoring also can help identify improvement opportunities during re-and retro-commissioning and, when implemented to continuously provide data during building operation, can support continual commissioning and renewal of building systems.

A major project that is applying monitoring-based commissioning across a large number of state university campuses in California is showing the value of this approach for existing buildings (Brown and Anderson 2006, Brown et al. 2006). That project involves the installation and "upgrade of permanent energy meters and other instrumentation, augmentation of energy information systems, benchmarking of building energy performance, assistance with initial commissioning efforts, and training of in-house staff." (Brown et al. 2006) The partnership performing this project has defined MBCx as the "adjustment, maintenance or repair of existing equipment as opposed to upgrade of equipment." (Brown and Anderson 2006) In this chapter, we use a broader definition of MBCx, which includes re-and retro-commissioning projects in which retrofits might be included. Monitoring is used to identify opportunities for operational improvements, to verify savings from re-and retro-commissioning, and to provide information to support maintaining building performance after commissioning to ensure the persistence of savings.

Over the last decade or so, diagnostic techniques and automated tools (e.g., PACRAT, the NIST APAR and VAV algorithms, and the WBD; see Friedman and Piette (2001) have been developed that assist in detecting operational faults and degradation in the performance of building systems and diagnosing their causes. These

*Originally presented at and published in the Proceedings of the 17th National Conference on Building Commissioning held June 3-5, 2009, in Seattle, WA.

techniques and tools can significantly reduce the time, effort and level of knowledge required to acquire and analyze data to reveal energy-consuming operational faults, such as failed sensors, inoperable economizer dampers, poorly implemented schedules, and improperly charged direct expansion equipment, to name few. These tools provide information valuable for identifying opportunities for saving energy through improved operations and detecting faults and performance degradation as they occur, enabling their timely correction, thus helping ensure persistent savings.

This chapter presents a case for application of monitoring, analysis and diagnostic tools and techniques through selected examples, where such tools have been used successfully in support of commissioning activities in southwest Canada and the U.S. Pacific Northwest. The examples start with applications across many buildings in two commissioning programs. The next section describes the use of spreadsheet-based tools to format trend-log data from building automation systems and to automatically generate diagnostic plots that are visually examined for specific features that reveal operational problems in space conditioning systems of large commercial buildings. This is followed by a section in which continuous monitoring of whole-building energy use is used to automatically track daily energy savings resulting from a utility commissioning program. This tracking also provides a means by which to detect degradation of savings and performance to guide monitoring-based commissioning actions. The potential use of automated diagnostic tools for chillers and packaged air conditioners is then described for continually commissioning these units based on monitoring. The chapter concludes with a discussion of the impacts of this approach to commissioning, including potential time savings, associated cost savings, and improvements in the quality of commissioning.

GUIDING COMMERCIAL BUILDING RE-TUNING WITH CONTROL SYSTEM DATA

Retro-commissioning studies place the potential energy savings from improved operation and maintenance (O&M) of commercial buildings between 5% and 30%. A pilot program has been initiated in the State of Washington focused on capturing a significant portion of this potential through transformation of building O&M professionals' practices* (Katipamula and Bramb-

ley 2008). One major component of the program focuses on re-tuning large commercial buildings. It is intended to change the way heating, ventilation and air conditioning (HVAC) systems in large commercial buildings are operated, serviced and maintained by targeting high-impact energy efficiency measures that can be delivered immediately, at low or no cost. As part of this effort, companies providing HVAC servicing were trained to provide HVAC and controls re-tuning services. While providing the training, HVAC systems in selected large commercial buildings were "tuned" for efficient operation, and then each trained team re-tuned five additional buildings.

Many large* commercial buildings today use sophisticated energy management and control systems (EMCSs) to manage a wide range of building systems. Although the capabilities of the EMCSs have increased over the last 2 decades, the capabilities of these systems are not fully utilized, and many buildings are not properly commissioned, operated or maintained. Lack of proper commissioning, the inability of the building operators to understand complex controls, and lack of proper maintenance leads to inefficient operations and reduced equipment lifetimes. Tuning building controls using EMCSs helps ensure maximum building energy efficiency and the comfort of building occupants. A poorly tuned system can sometimes maintain comfortable conditions, but at a high energy cost to overcome unrecognized inefficiencies.

Periodic re-tuning of building controls and HVAC systems will enhance building operations and improve building efficiency. Re-tuning, as used in this project, is a systematic, semi-automated process of detecting, diagnosing and correcting operational problems with building systems and their controls. The process can significantly increase energy efficiency at low or no cost—and the impact is immediate. Unlike the traditional retro-commissioning approach, which generally has a broader scope, re-tuning primarily targets HVAC systems and their controls. In addition, re-tuning uses monitored data to assess building operations even before conducting a building walk through. This process is similar to MBCx as implemented in other work cited earlier. However, in contrast to monitoring-based approaches using newly-installed or enhanced permanent meters and energy information systems, our re-tuning approach leverages existing EMCSs to trend data and identify operational faults and opportunities to save energy.

---

*www.retuning.org

---

*For this project, a large commercial building is defined as a building with 100,000 square feet (sf) or more of conditioned space, having an energy management and control system.

Re-tuning Methodology

An early version of the re-tuning methodology was initially developed during the electricity crisis of 2000–2001 for the Federal Energy Management Program (FEMP). The procedures were adopted by FEMP and rolled out as part of the U.S. Department of Energy (DOE) ALERT (Assessment of Load-and Energy-Reduction Techniques) Program for federal facilities. The procedures were further refined and formalized for use in the current project. The re-tuning method consists of six primary steps: 1) initial collection of relevant building information, 2) pre-re-tuning, 3) building walk through, 4) re-tuning, 5) post-tuning and 6) savings analysis. For more details on how to execute each of these primary steps, refer to references 6 and 7. A more detailed description of the pre-re-tuning phase (step 2), which is the focus of this chapter, follows.

The pre-re-tuning phase involves setting up trend logs in the EMCS, collecting trend data for at least 1 week (preferably 2 weeks) for key points in the mechanical system, and analyzing the data to learn more about current building operations. This analysis helps identify operational problems such as: systems running during unoccupied hours, poor economizer operation, outdoor-air ventilation during morning warm-up or cooldown, incorrect "optimal" start and stop of systems, excessive equipment cycling, leaky valves, exhaust fans running continuously, faulty sensors, and high supply-air static pressure, which leads to poor zone control.

Before starting collection of trend data, a monitoring plan is prepared, which is based on the building information gathered in the previous step. To help the service providers, monitoring templates and a list of points to trend on common HVAC systems are provided. The plan identifies trend logs that need to be set up in the EMCS and how the trend data will be analyzed. For each trend log (i.e., sensor or control point), the plan specifies the duration of the logging period (number of days) and the measurement period (time interval between logged values). The monitoring plan depends on the specific HVAC systems in the building.*

The monitoring plan is then implemented, first by creating trend logs in the EMCS. After sufficient data are collected in the logs (as specified by the logging periods in the plan), they are analyzed, using a semi-automated spreadsheet tool and analysis guidelines, to gain insight into current operations and to detect problems with the

building systems and their controls. The spreadsheet reads EMCS trend log files and automatically produces a set of plots that the retuning technicians are taught to use for detecting operational issues.

Analysis of EMCS Trend Log Data

Although most EMCSs can trend and export data to files, the formats of trend logs vary from one EMCS to another. There is no standard for such output, and each EMCS vendor exports the data in a different format. While some vendors provide multiple columns of data, each column for a separate variable, in a single file, others provide values for only one data point per file. Most vendors, however, provide some type of ASCII output, either space or comma delimited.

The spreadsheet developed for use in the project works with many formats but not all. In some cases, some pre-processing is necessary to prepare the inputs so they are compatible with the spreadsheet. The spreadsheet supports both the single-column and the multiple-column data formats, as long as the data columns are separated by commas. When the columns are space or tab delimited, the data can be pre-processed and converted to the compatible format by opening them in Microsoft Excel© and then saving the files in comma separated variable (csv) format.

The spreadsheet tool is tailored to analyze and produce graphs that provide information on the operations of air-handling units, zone variable-air-volume boxes, and chiller and boiler plant operations. As part of training, the service providers are taught how to process the data using the spreadsheet and, more importantly, how to interpret the graphs that the spreadsheet generates. The user enters basic information such as the names of files where the data can be found, the columns where the values of particular data points are located, and the starting row position of the data (see Figure 18-1). Once all relevant information for the various data points is entered in the input sheet, the user clicks on the "Start Analysis" button to generate a set of plots automatically.

The plots can be reviewed visually to detect many common operational problems. Two example plots are shown in Figure 18-2. The first plot (a) shows an improperly working economizer on an air-handling unit (AHU). The mixed-air temperature tracks the outdoor-air temperature, indicating that the outdoor-air damper is continuously fully open while the return-air damper is fully closed. The discharge-air temperature is always greater than the mixed-air temperature, indicating that that the unit is heating the mixed air more than necessary. By modulating the outdoor-air damper

*A sample monitoring plan for a building with variable-air-volume air-handling units with a central plant consisting of chillers and boilers can be found at http://buildingefficiency.labworks.org/media/large_building_trending_requirements_for_retuning.pdf.

and enabling some of the return-air to be recirculated, the desired discharge-air temperature could be reached with no heating or cooling energy required. The second plot (b) shows properly operating chilled-and hot-water valve signals (no simultaneous heating and cooling). When the hot water valve signal is non-zero, the chilled water valve signal is zero and vice versa, as indicated by the data points on the two axes. Data points in the yellow region would indicate poor valve operation causing simultaneous heating and cooling, while points in the red region would indicate even worse control.

DAILY TRACKING OF ENERGY SAVINGS: MODEL AND EXAMPLES

O&M-based energy savings initiatives are gaining more momentum in the marketplace in large part because of the relative ease in implementing them and the commensurately lower cost. Many authors have extolled the virtues of techniques such as tune-ups, retro-commissioning, etc. These methods can yield extremely attractive paybacks (often 1 year or less). Unfortunately, they often suffer from a very practical

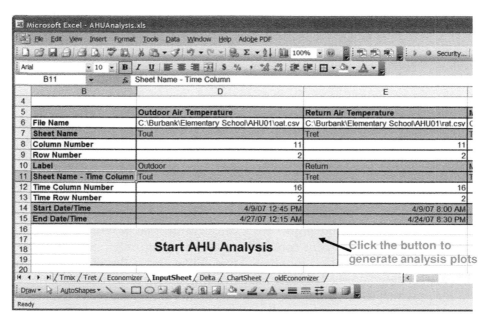

Figure 18-1: Screenshot of the Spreadsheet Analysis Input

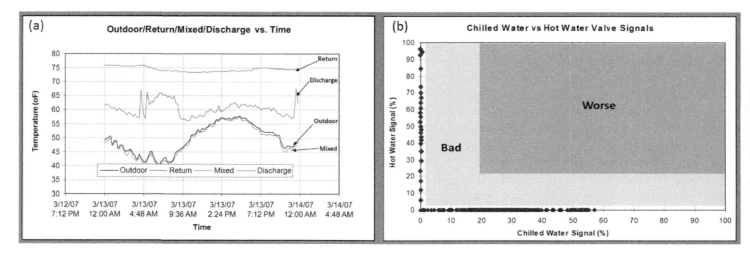

Figure 18-2: (a) Improperly Operating Economizer and (b) Properly Operating Chilled and Hot-Water Valves

problem—mainly, ensuring that the expected benefits are obtained and just as importantly, ensuring that the benefits persist over time. We describe below a modeling technique that is being successfully deployed by a variety of building owners, operators, and utilities to directly address the issues of measuring the savings of O&M-based efficiency improvements and also making it relatively easy to ensure that these savings persist over time.

Technical Approach

We have found the modeling methodology presented below (and commercialized by NorthWrite in their Energy Expert software application) to be effective in establishing performance baselines from which to measure energy savings associated with operational changes in buildings. (Katipamula et al. 2003) This method has the advantage that it can capture both linear and non-linear behavior. The method is based on the concept of data bins borrowed from the field of building energy data analysis. A bin is an interval (bin) of values of an independent variable with which a value of another (dependent) variable is associated. For example, the weather at a location can be characterized by the number of hours per year on average that the outdoor-air temperature falls into 5°F bins between some minimum temperature and some maximum temperature.

When multiple variables are used to explain the variations in energy use, multi-dimensional bins can be used where a multi-dimensional bin is defined as the intersection of one-dimensional bins based on each of the variables. This is shown in Figure 18-3 for three-dimensional bins that characterize a variable such as energy use in terms of three explanatory variables. A representative value of the dependent variable is assigned to

each bin defined by the ranges of values of the independent variables. For an energy use model, the dependent variable is energy consumption.

The model is "trained" by collecting data empirically and assigning it to bins. Given a sample of empirical data with each point of the sample consisting of values for a complete set of N independent explanatory variables $(x_1, x_2, x_3, ..., x_N)$ and the corresponding measured value of the dependent variable, an N-dimensional model is created by assigning each data point in the sample to the bin in which the point defined by the values of its independent variables lies. When a sufficient number of points have been assigned to each bin, the model is considered fully "trained." A representative value of the dependent variable is then assigned to each bin, completing the model. The median of the values of the dependent variable in the bin makes a good representative value for both large and small numbers of points per bin.

The user of this tool defines a "baseline" time period over which they wish to create a model predicting energy use. This baseline can be a period of time prior to retro-commissioning, a preceding year, or other time period of significance for a facility. By using the baseline model with values for the independent variables for times in the post-training period, predictions of what the energy use of the building would have been in the absence of degradation in efficiency or actions taken to improve energy efficiency can be obtained. By comparing the actual energy use to the predictions, energy waste associated with degradation or energy savings from improvements can be determined, while controlling for changes in the independent variables (e.g., outdoor-air temperature).

Examples

Below we provide several examples where facilities have implemented O&M-based energy savings programs and have used the Energy Expert to track their results. One of the analysis features that the Energy Expert provides is called "Cumulative Sum" (CUSUM). CUSUM is simply the integration of the daily differences between the actual and expected energy use for a modeled load. A positive slope on the CUSUM chart indicates that energy is being saved relative to the baseline. A negative slope is an indication of increased consumption relative to the baseline. The days over which the facility experiences the maximum possible positive slope can be considered a "best practice" period and can serve as a model for operating your facility.

The first example is of a large refrigerated dis-

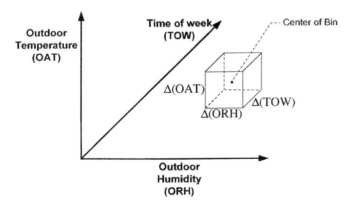

Figure 18-3: An example three-dimensional binning scenario with bins defined by three explanatory variables: outdoor-air temperature, outdoor-air humidity, and time of week.

tribution center. This building underwent a facility tune-up in which approximately 30 energy savings measures were identified as appropriate for the systems and operational needs of the facility. The site visit occurred in mid-October 2008 and approximately half of the measures were implemented while the technical team was onsite. Figure 18-4 indicates that the Energy Expert immediately starts indicating positive savings following the facility tune-up. These savings are relatively consistent for a period of several months. During this time, no additional changes were made to the facility. However, early in 2009, the remaining measures were implemented by the operations staff and the slope of the CUSUM line steepened (indicating a positive step-change in savings).

Another way to see the implementation progress of an energy savings program is by viewing the daily results of the Energy Expert using the Calendar view.

This feature enables you to view the days during the month where your facility uses more (red), less (blue), or about the same (green) amount of energy as the baseline. Figure 18-5 shows a chronological progression for this distribution center from before the tune-up on the left, to after the tune-up on the right. The month in the middle shows the period of time in which the first phase of the tune-up occurred.

The next example (Figure 18-6) is of a retail store that has been receiving significant energy savings as a result of implementing a number of changes to their building control system. However, in early April 2009, the controls vendor upgraded the software and overrode the tuning by resetting the set points and control strategies to an earlier version that had been archived. As a result, the store savings went from approximately $1,000 per month to $0 per month relative to the baseline (notice the decrease and overall flattening of the lines in

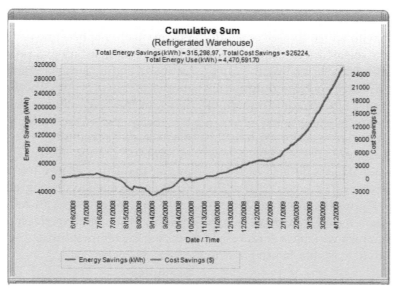

Figure 18-4: CUSUM Graph for Refrigerated Distribution Center

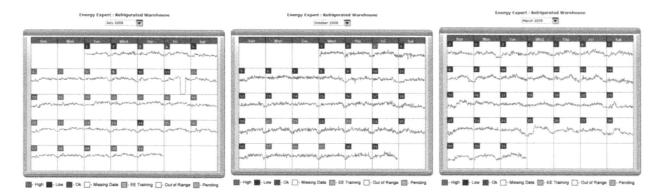

Figure 18-5: Monthly Calendar View of Energy Expert Results

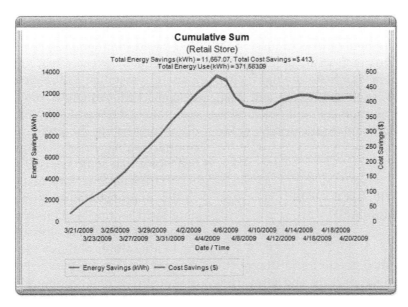

Figure 18-6: CUSUM Graph for Retail Store

Energy Expert Results for: Jan 1, 2009 - Apr 20, 2009

	High Demand ⬍	Actual Consumption ⬍	Expected Consumption ⬍	Consumption Δ ⬍	Savings ($) ⬍	Low ⬍	OK ⬍	High ⬍
Office Bldg. 1	120	178,573	287,911	109,338	8,747	110	0	0
Office Bldg. 2	206	265,015	321,088	56,073	4,486	100	6	4
Office Bldg. 3	770	777,083	860,868	83,785	6,703	73	12	24
Office Bldg. 4	331	376,728	419,051	42,324	3,386	89	13	8
Office Bldg. 5	323	251,769	303,499	51,730	4,138	105	5	0
Office Bldg. 6	294	418,752	454,795	36,043	2,883	86	15	9
Office Bldg. 7	169	280,683	317,090	36,407	2,913	109	1	0
Office Bldg. 8	801	1,083,433	1,023,492	-59,941	-4,795	34	13	63
Office Bldg. 9	303	434,943	477,023	42,081	3,366	79	22	9
Total		4,066,979	4,464,817	397,838	31,827	785	87	117

Figure 18-7: Enterprise Roll-Up Report

Figure 18-6 after about April 5).

The Energy Expert provides two valuable services in this case. First, it notifies the operator that savings the building had been enjoying are no longer accruing. Second, it provides a clear and unambiguous benchmark for returning the facility to its energy saving state.

The final example is for a portfolio of nine office buildings (Figure 18-7). Each building has an Energy Expert with a baseline set to the prior year (2008). The owners of the portfolio tasked the facility manager responsible for each building with "beating" their prior year energy use by as much as possible. The office buildings are between 75,000 and 150,000 ft². As you can see from the Enterprise Roll-Up Report, all the buildings but one made significant progress towards their year-over-year energy savings goals. The owners know exactly which site(s) need more attention by simply scanning down the savings column.

These examples demonstrate how use of models like the one in Energy Expert for monitoring can be used to quantify savings from re-and retro-commissioning and to detect when operation practices are implemented or faults occur that reduce or eliminate savings. When used on a continuous basis, such monitoring and diagnostic tools inform building management and operations staff, enabling them to maintain the persistence of savings over time.

AUTOMATED DIAGNOSTIC TOOLS FOR COMMISSIONING CHILLERS AND PACKAGED UNITS

Other automated fault detection and diagnostic (FDD) tools can be used to support re- and retro commissioning of building equipment and systems. Using data available from a building automation system or from additional sensors installed specifically for this purpose, data can be provided to FDD tools. These tools can be used to identify operational faults that can often be corrected by changes to control parameters or control code, failed components that can be repaired or replaced, and equipment and systems with degraded performance, which generally require further diagnostics or troubleshooting to identify underlying root causes. Although these tools are commonly used to continuously monitor equipment and systems in real time during operation, many can be used to also process data offline. Both approaches can be used to inform re-and

retro-commissioning of existing buildings. To support continued monitoring-based commissioning or condition-based maintenance, these tools should be set up to continually process real-time data feeds from a permanently-installed EMCS or separate data acquisition system. The information from these tools can be used to maintain systems and equipment in peak operating condition preserving the savings achieved from the initial commissioning. Brief descriptions of two examples of FDD tools that could be used to support MBCx follow.

The user interface of an automated centrifugal chiller diagnostician is shown in Figure 18-8. Using data such as the condenser refrigerant pressure, the evaporator refrigerant pressure, the entering and leaving temperatures of the condenser water, and the entering and leaving temperatures of the chilled water, this diagnostician detects operation faults, e.g., condenser fouling, evaporator fouling, system overcharge or undercharge, and high or low condenser or evaporator flow rates.

This diagnostician also tracks the chiller efficiency, providing the ability to detect and quantify degradation in efficiency over time. These are all capabilities that could be used during initial (monitoring-based) commissioning and continuously thereafter to identify operation and maintenance actions that could be taken to improve the system's operating efficiency. The benefit of such a diagnostic tool is that it automatically performs calculations often done manually offline by an analyst or not at all during retro-commissioning.

Another FDD tool currently under development

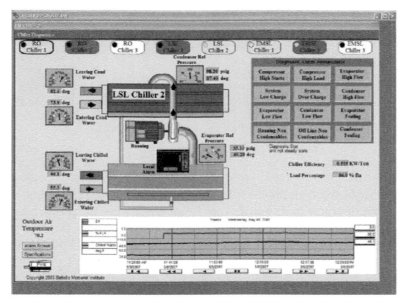

Figure 18-8: User Interface of an Automated Centrifugal Chiller Diagnostician.

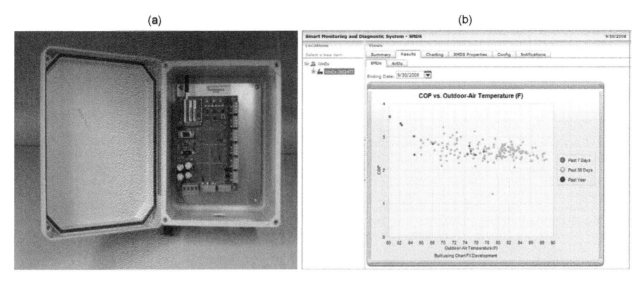

Figure 18-9: SMDS (a) Hardware and (b) User Interface Window Showing Air- Conditioning COP Versus Outdoor-Air Temperature.

and field testing is a Smart Monitoring and Diagnostic System (SMDS) for packaged air conditioners and heat pumps (see Figure 18-9). This tool consists of a hardware package, installed on each packaged HVAC unit, and diagnostic software, which runs locally on the hardware. Diagnostic results and selected data are transmitted wirelessly to a network operations center, where these data are managed and results are made available to users via a web site. Authorized users access results regarding faults detected and changes in the coefficient of performance (COP) of the HVAC unit using a web browser; no installation of special software is required. The system can be used to detect hardware and operation faults during initial retro-commissioning and then over the life of the equipment to guide maintenance and provide information critical for maintaining the savings obtained from retro-commissioning, thus supporting the MBCx process.

Several other diagnostic tools are available for use or are in various stages of development (see, for example, Friedman and Piette 2001). All of these could be used to enhance and streamline MBCx by automating the detection, diagnosis and quantification of impacts of operational faults in building systems and equipment.

POTENTIAL IMPACTS

Potential impacts of monitoring and diagnostic tools on re- and retro-commissioning include:

• time savings in collection and analysis of data compared to temporary monitoring using data loggers, manual performance of functional tests, and manual offline data analysis

• greater consistency across MBCx projects and potentially higher quality commissioning

• better detection of performance degradation and detection and diagnosis of faults, helping ensure the persistence of savings after initial commissioning.

Despite the benefits, use of these tools brings with it the cost of additional instrumentation to supplement the sensing provided by an existing EMCS. Such instrumentation may include end-use energy metering and sensors not part of the existing EMCS. Commissioning providers also require some time to learn to use the diagnostic tools, although many of the tools are user friendly and rather simple to use, possessing user interfaces that are easy to understand. Buildings without EMCS systems, which are generally smaller commercial buildings, would require the installation of sensing and data acquisition systems to provide the required data and are likely not candidates for MBCx. This problem is not unique, however, for MBCx. These same buildings are already not good candidates for retro-commissioning as conventionally done; new approaches for providing commissioning services to these small buildings are needed.

The examples provided in this chapter illustrate how monitoring and diagnostic tools can be used to achieve these benefits and in some cases, the value associated with them (see, for example, Figure 18-6). The re-tuning project described previously is collecting data to quantitatively evaluate the impacts of that monitoring-based process. At the time of writing the original paper, those results were not yet available; they will be presented in future publications.

References

Brown, K. and M. Anderson, "Monitoring-Based Commissioning: Early Results from a Portfolio of University Campus Projects," *Proceedings of the 13th National Conference on Building Commissioning*. Available online: http://www.peci.org/ncbc/proceedings/2006/author.htm, PECI, Portland, Oregon, 2006.

Brown, K., M. Anderson and J. Harris, "How Monitoring-Based Commissioning Contributes to Energy Efficiency for Commercial Buildings," *Proceedings of the 2006 ACEEE Summer Study on the Energy Efficiency of Buildings*, American Council for an Energy Efficient Economy, Washington, DC, 2006.

Friedman, H., and M.A. Piette. 2001. *Comparative Guide to Emerging Diagnostic Tools for Large Commercial Buildings*, LBNL-48629. Lawrence Berkeley National Laboratory, Berkeley, California.

Katipamula, S., and M.R. Brambley. 2008. "Transforming the Practices of Building Operation and Maintenance Professionals: A Washington State Pilot Program." *2008 ACEEE Summer Study on Energy Efficiency in Buildings,* Washington, DC. American Council for an Energy-Efficient Economy, Washington, DC.

Katipamula, S., M.R. Brambley and J. Schein. 2003. Results of Testing WBD Features Under Controlled Conditions. Task Report for the Energy Efficient and Affordable Small Commercial and Residential Buildings Research Program. Project 2.7 – Enabling Tools. Task 2.7.5. Included as part of *Final Report Compilation for Enabling Tools*, P-500-03-096-A7. California Energy Commission, Sacramento, California.

SECTION III

Methodology and Future Technology

This topic of Automated Diagnostics and Analytics for Buildings is one of the most exciting and dynamic areas of technological advancement in our field of energy and facility management today, and is in rapid development and improvement to make it even better and more exciting in the future. The purpose of this section of the book is to share information with a broad audience on the future of automated fault detection and diagnostics and analytics for buildings development and applications, as well as explaining the methodology that is being created and utilized to produce the benefits of these future applications and tools.

Most of the chapters in this section of the book are from authors whose organizations are developing the software tools for AFDD and Analytics for buildings, or are providing the research on which these developments are based. This research and development work also expands to include the areas of basic data collection and big data, as well as the processing of this data to create useful actionable information for their products and services in order to help operate buildings more efficiently and more effectively. They have been gracious enough to share this information with us, and to help other software and service developers, and to help other researchers be more aware of what needs to be done with new research in this area.

The editors hope that this information is helpful to companies that are developing new software and services in these areas to help energy managers, facility managers, building owners and operators, in assisting them in deciding on implementing this new technology to help them operate their buildings; as well as researchers and developers who are interested in what others are doing to produce new methods of creating the operational information needed for Automated Diagnostics and Analytics for Buildings.

Chapter 19

Model-based Real-time
Whole Building Energy Performance
Monitoring and Diagnostics

Zheng O'Neill, University of Alabama, Tuscaloosa, AL, USA

Xiufeng Pang and Philip Haves
Lawrence Berkeley National Laboratory, Berkeley, CA, USA

Madhusudana Shashanka, and Trevor Bailey
United Technologies Research Center, East Hartford, CT, USA Corresponding Author: zoneill@eng.ua.edu

INTRODUCTION

The total energy consumption for US commercial buildings was 17.43 quads (2003 CBECS database), approximately 18% of the total U.S. energy consumption. The Department of Energy (DOE), the International Energy Agency (IEA), Intergovernmental Panel on Climate Change (IPCC) and other agencies have declared a need for commercial buildings to become 70-80% more energy efficient. Although energy-efficient building technologies are emerging, a key challenge is how to effectively maintain building energy performance over the evolving lifecycle of the building. It is well known that most buildings lose a portion of their desired and designed energy efficiency in the years after they are commissioned and recommissioned (Haves 1999; TIAX 2005; Friedman et al. 2010; Mills 2011). Achieving persistent low-energy performance is critical for realizing the energy, environmental, and economic goals expressed in the Energy Policy Act of 2005, Executive Order 13423, and the Energy Independence and Security Act of 2007. Field experience shows that energy savings of 5% to 30% are typically achievable simply by applying FDD (Fault Detection and Diagnostics) and correcting the faults diagnosed in buildings (Liu et al. 2001; Katipamula and Brambley 2005a).

Generally, FDD methods fall into three categories (Katipamula and Brambley 2005b).

- Quantitative model-based methods that include: 1) physical first principles ('textbook') models (Li 2004) and 2) polynomial curve fits of the components and equipment (e.g., fans, pumps, chillers) (Sreedharan and Haves 2001). Faults are detected as the difference between measurements and the model output. Significant differences indicate the presence of a fault somewhere in the part of the system treated by the model.

- Qualitative model-based methods include rule-based systems and qualitative physics (House et al. 2001).

- In contrast to the other groups, process history-based methods (data driven) assume no priori knowledge of the process. In general, these methods are suitable when significant amounts of data are available. The black box models using linear regression models are employed to perform automated fault detection in buildings (Jacob et al. 2010).

Currently, the key barriers and challenges preventing energy diagnostics from being pervasively applied are: 1) an integrated whole building energy FDD system does not exist. Major building subsystems are independently controlled with limited, add-on FDD capability.

Both the controls and FDD do not adequately capture the functional and behavioral interactions between subsystems resulting in sub-optimal building energy performance and increased false alarm rates; 2) existing FDD methods are based on available data and simple, ad-hoc rules that do not adequately capture either the component or system functional and behavioral interactions. This limits the scalability and utility of FDD methods; 3) existing FDD methods, which are currently an "after thought" add-on to building control systems, require manual intervention and labor-intensive analysis. This limits the ability of FDD methods to provide real-time actionable recommendations for ensuring pervasive lower energy building performance; and 4) most of existing FDD systems to perform energy diagnostics are not scalable because they rely on manipulation of data by a limited number of experts which makes the scalability of the existing process to the entire industry infeasible. In addition, significant follow-up and hand-holding to correct faults once they are detected and diagnosed are needed in practice. Unfortunately, knowing a problem in terms of occupant comfort and energy cost impact, which will facilitate the owners and/or facility managers to take correction actions, is not often directly and explicitly available from most existing FDD tools.

Haves et al., 2001 explored the idea of model-based performance assessment at the whole building level and pointed out additional measurements are important to provide the necessary input data. With the increasing need to improve building performance, the use of simulation to assess the actual performance of buildings is starting to gain more attention (Liu et al. 2003; Ramirez et al. 2005). Lee et al. 2007 used an off-line whole building simulation for energy consumption fault detection and concluded that it is important to have a methodology to define an error threshold to differentiate a true system fault from a false alarm caused by an imperfect simulation. A Microsoft Excel based Automated Building Commissioning Analysis Tool (ABCAT) was developed and tested for about 10 buildings on the field (Curtin et al. 2007; Lin and Claridge 2009; Bynum et al. 2010). The ABCAT is a first principles based whole building level top down tool and focuses on detecting faults that have a significant impact if they persist for a long period of time (Bynum et al. 2012). The model used in the ABCAT is a calibrated and simplified first principles based mathematical model, which was developed from ASHRAE's Simplified Energy Analysis Procedure (SEAP) (Knebel 1983). This model based approach is similar to what is proposed in this paper, except for the detailed level of the simulation. However, the scale of building operational data transfer and storage, simulation environment, diagnostic module and visualization module in the proposed system is different from those in the current ABCAT version.

Real-time building simulation, as opposed to off-line building simulation, refers to the use of a building model whose simulation time is synchronized with real time, as represented by the computer clock. Updated values of the input variables are acquired dynamically at each step-time. With the widespread deployment of EMCS (Energy Management and Control Systems) in buildings and the development of open protocols such as BACnet (A Data Communication Protocol for Building Automation and Control Networks), the sensor and control signal information from various components and systems in a building is more accessible (Salsbury et al., 2000). This makes it possible to acquire the dynamic input variables for a real-time model (e.g., an EnergyPlus model) from the EMCS including but not limited to weather data, operation schedules, and control setpoints. However, the EMCS does not normally sense all the necessary model input variables that are needed for real-time simulation. For example, solar radiation, wind speed and direction and additional instrumentation are required to accommodate these needs.

The BCVTB (Building Controls Virtual Test Bed), recently developed by Lawrence Berkeley National Laboratory, provides a platform to synchronize EnergyPlus simulation time to real-time and exchange data with an EMCS in real-time mode (Wetter 2011; Nouidui et al. 2011). It is an extension of Ptolemy II (Eker et al. 2003), a software environment for heterogeneous modeling and simulation. Ptolemy II is a free open-source software developed at the University of California, Berkeley.

An automated, model-based, real-time whole building performance monitoring and energy diagnostics system is presented in this chapter. This system continuously acquires performance measurements of HVAC (Heating, Ventilation and Air-Conditioning) system from the existing EMCS augmented by additional sensors as required. The system compares these measurements in real-time to a reference EnergyPlus simulation model that either represents the design intent for the building or has been calibrated to represent acceptable performance. In current practice, there is a significant time lag that often occurs between identifying and diagnosing a fault and the time it is corrected. If a problem is not dramatic in terms of comfort and/or known energy cost impact, this delay could be months or even years. The proposed approach mainly aims at large problems, e.g., problems that typically lead to increase of 5% or more in energy use (Claridge et al. 1999).

SYSTEM INFRASTRUCTURE

The proposed technology is a dynamic model-based, whole building performance monitoring system that compares measured performance metrics to those generated by a physics-based reference model representing "design intent" or acceptable performance. The system is depicted in Figure 19-1.

The proposed system integrates and compares the output from a building simulation model to measurements to detect deviations from either a design intent model that represents the design intent for the building or a calibrated operation model that represents acceptable performance. The comparison will allow for identification and quantification of sub-optimal performance, identification of the conditions under which sub-optimal performance occurs, a means to compare alternative corrective actions using whole building metrics, and finally a means to validate improved performance once corrective actions have been taken. The six key elements of the system are described as follows:

- Real-time Building Reference Model
- Building Envelope, Systems & Controls
- Extended Energy Management and Control System (EEMCS)
- Data Mining and Anomaly Detection
- Energy Performance Visualization Dashboard
- Integrated Software Environment

The following sections describe these key elements in details.

Real-time Building Reference Model

A whole building EnergyPlus simulation model represents the desired or acceptable performance of the envelope, HVAC, lighting, water, and control systems in a building. EnergyPlus (EnergyPlus 2013) is a whole building simulation program developed by the U.S. De-partment of Energy. It models heating, cooling, lighting, and ventilating processes in buildings and includes many simulation capabilities, such as time steps of less than one hour, modular systems, multizone airflow, thermal comfort, and natural ventilation. The model can also represent "plug" loads including computers and calculates both the direct electrical energy consumption and also the effects of heat gains in the building. The model takes as input a description of the building (e.g., geometry, materials, roof type, window type, shading geometry, location, orientation etc.), its usage and internal heat loads, and the HVAC system description, and then computes the energy flows, zonal temperatures, airflows, and comfort levels on sub hourly intervals for periods of days to years. Conventionally, Energy-Plus is used for off-line building energy simulation, to analyze designs for new construction and retrofit, size HVAC equipment, and model energy and/or water use in buildings. This paper describes an implementation of EnergyPlus in a real-time application, which represents a step towards the development and deployment of model-based building performance assessment techniques.

EnergyPlus model inputs consist of a full description of the building (location, geometry, materials, window type etc.), its usage and the HVAC system description. The weather data is specified separately from the general input file when the simulation is launched. All these input data can be categorized into two types: parameters and variables. Parameters are independent of time and remain constant during the simulation, e.g., building geometry and HVAC equipment nominal capacity etc. Variables may change during the simulation, e.g., weather conditions, control set-points etc. Some of

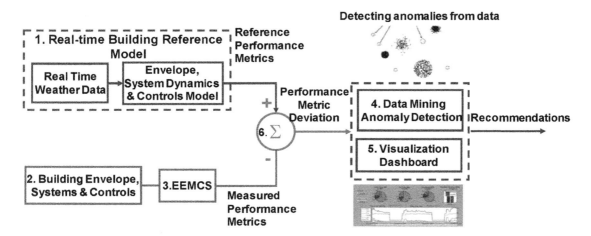

Figure 19-1: Diagram of a real-time energy performance monitoring and diagnostics system

the variables can be specified in the EnergyPlus input file using schedules, e.g., HVAC equipment operational and control set-points, while some of the variables can only be obtained from external data source at each step time, e.g., weather conditions, including outdoor dry bulb temperature and relative humidity, wind speed and direction, and direct normal and diffuse solar radiation. For real-time applications, the external variables need to be updated at each step time. Modern buildings are typically equipped with an EMCS, which can provide a number of these variables. However, except for outdoor dry bulb and relative humidity, weather variables are not usually measured by the EMCS. Supplementary instrumentation, usually hosted by the EMCS, is needed for these non-typical measurements such as solar irradiation.

The design intent baseline model represents the design intent/desired performance of the building. The building descriptions are obtained directly from the design documentation and as-built drawings. In the case where some of the information is not available, an on-site investigation is used to determine these parameters. The HVAC sequence of operations represents the initial design intent or the desired performance that the facility management team is attempting to achieve based on the capability of the installed equipment. The weather data is collected from the on-site weather station if it is available. The lighting and plug load profile in the design intent baseline model signifies an "ideal" performance that has only minimal lighting and plug loads on during unoccupied hours and has lighting and plug loads proportional to the occupancy profile during occupied hours. If the building usage changes (e.g., a conference room is changed to an office room), then the internal load profiles should be updated with the new intent.

The calibrated operation baseline model refers to a whole building EnergyPlus simulation model that represents the current building operational practice. The design intent and operation models share the same model inputs for building information but differ in the description of the HVAC system operation, lighting and plug load profiles. The HVAC system sequence of operation will be obtained by combining the information from the control design documents, existing EMCS programming and interviews with the building operators and control engineers. The weather data, including solar irradiation, outside air temperature, relative humidity, wind speed and direction will be collected from the augmented on-site weather station. The lighting and plug load profiles will be obtained from the additional building level sub-metering. If sub-metering is not available, a onetime measurement along with occupancy profile can be used to approximately determine the lighting and plug load profiles. Real-time dynamic load profiles may also be assessed using a load estimator (Dahilwal and Guay 2012; O'Neill et al. 2008). The estimation is obtained from a reduced-order building model generated from the building thermal network and real-time data (e.g., temperatures, airflow rates) from the EMCS, with considerations for sensor noise and model uncertainties. Figure 19-2 shows the schematics of the real time load estimator (O'Neill et al. 2008).

After the initial model is built, a calibration is required to match the simulation predictions with the

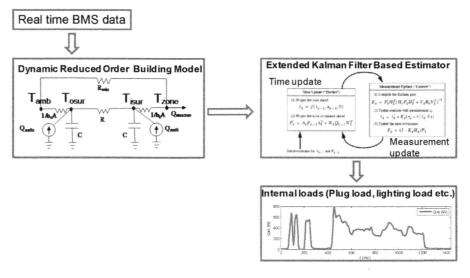

Figure 19-2: Schematics of the real time load estimator

measured data by tuning the model. Figure 19-3 illustrates the usage of a design intent model and a calibrated operation model for building performance monitoring and FDD.

Extended Energy Management and Control System (EEMCS)

The building management and control system, together with additional sensors, are used to measure key building performance metrics. Additional sensors include electrical power sub-metering, fluid flow meters, and temperature sensors to determine thermal energy flow rates. Measurement of electrical input and thermal output, for example, enables the monitoring of chiller efficiency. Installation of permanent instrumentation connected to the existing EMCS ensures that the benefits of the additional performance monitoring capability are available over the long-term. The existing building EMCS is expanded to provide data acquisition for the additional sensors and to interface to a new PC (Personal Computer) where the proposed system will be installed.

Data Mining and Anomaly Detection

Building and HVAC system data represents a hierarchical structure of power usage and the delivered heating/cooling throughout the building. Identifying at which level in this hierarchy a fault-cause occurs is crucial to effectively provide facility management decision support. To perform energy diagnostics, data mining and model-based estimation approaches were used to provide energy anomaly detection. A number of complementary modeling methods can be used to implement energy diagnostic decision support. These include statistical process control algorithms such as T^2 and Q statistics, probabilistic graphical models (Friedman and Koller 2009; Murphy 2001) and expert rule-based threshold methods (House et al 2001). For example, the system level diagnostic model and one of its component diagnostic models for a commercial building with chillers and air handler units are shown in Figure 19-4.

The graphical structure is learned from operational data by discovering the relationships between measured variables. The goal of this step is to learn the nominal behavior of the system. After this step the learned graph-

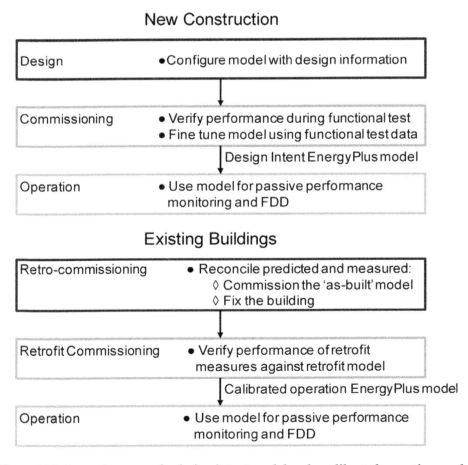

Figure 19-3: Example usage of a design intent model and a calibrated operation model

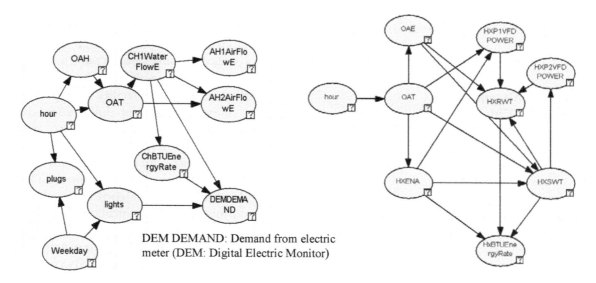

System level graphical network   *Component level heat exchanger graphical network*

Figure 19-4: Graphic diagnostic models

ical structure is validated against domain knowledge and physics based understanding of the system. At this point the graphical network model for FDD is used to analyze new data to generate an anomaly score quantifying the difference between a variable's predicted state and its measured value. Based on the anomaly scores and a suitably chosen threshold, faults are detected & diagnosed by identifying anomalous variables.

In this chapter, we will detail the FDD module that primarily uses algorithms from the statistical process control literature such as T^2 and Q statistics to compute statistics on the deviations of the measured data-points from model predictions for the purpose of fault detection and fault identification. The proposed algorithms take measured and reference data as inputs and process the data to detect outliers or changes, and identify faults. The FDD module utilizes operation data from the EMCS such as temperature, airflows and electricity consumption as well as output data from the EnergyPlus model.

T^2 and Q statistics (Chiang et al. 2000) are used to detect any anomalous events over a substantial period of time. T^2 and Q statistics implicitly use the Principal Component Analysis (PCA) for dimensionality reduction of high dimensional data sets and have been successfully applied in many industrial systems. The main advantage of these techniques is their scalability and their ability to identify anomalous events from multiple data sources without merging data at a central location.

The application of statistical theory to monitor processes relies on the assumption that the characteristics of the data variations are relatively unchanged unless a fault occurs in the system. It implies that the statistical properties of the data, such as the mean and variance, are repeatable for the same operating conditions, although the actual values may not be very predictable. The repeatability of statistical properties allows thresholds for certain measures, effectively defining the out-of-control status, to be determined automatically. This is the essence of the underlying principle used in the FDD module. We describe the method in details below.

Principal Components Analysis (PCA)

PCA is a widely used data driven technique for monitoring industrial systems. It is an optimal dimensionality reduction technique that accounts for correlations among variables. The lower dimensional representations of data produced by PCA can improve the proficiency of detecting and diagnosing faults using multivariate statistics such as the Hotelling T^2 statistic and the Q-statistic (Chiang et al. 2000).

T^2 and Q Statistic

To monitor a univariate variable for out-of-control behavior, we can obtain upper and lower thresholds (either statistically or from domain-expertise) that define boundaries for in-control operation and a violation of these limits would indicate a fault. This is usually achieved by using control charts or *Shewhart* charts as shown in Figure 19-5.

In the case of multivariate data, we could deter-

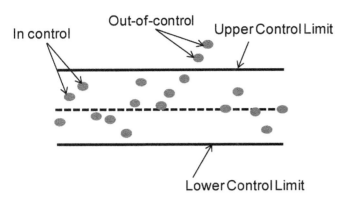

Figure 19-5: Shewhart charts of univariate statistical monitoring

mine thresholds for each observation variable separately. However, analyzing each observation individually will not capture correlations between variables. The multivariate T^2 statistic is a generalization of the above technique to the multivariable case that takes this factor into account.

Let the training data with m variables and n observations for each variable be given by

$$X = \begin{bmatrix} x_{11} & \cdots & x_{1m} \\ \vdots & \ddots & \vdots \\ x_{n1} & \cdots & x_{nm} \end{bmatrix} \quad (1)$$

Then the sample covariance matrix is given by:

$$S = \frac{1}{n-1} X^T X \quad (2)$$

An Eigen value decomposition of the matrix **S**:

$$S = V \wedge V^T \quad (3)$$

reveals the correlation structure of the covariance matrix. The projection **y** of an observation vector **x** onto the orthonormal matrix **V** decouples the observation space into a set of uncorrelated variables corresponding to elements of **y**. Assuming **S** is invertible and using the definition

$$z = \wedge^{-1/2} V^T X \quad (4)$$

the Hotelling T^2 Statistic is given by

$$T^2 = Z^T Z \quad (5)$$

The T^2 statistic is a scaled squared 2-norm of an observation vector **x** from its mean. The scaling on **x** is in the direction of the eigenvectors. Given a level of significance, appropriate threshold values for the T^2 Statistic can be determined automatically (See Chiang et al., 2000 for details).

The Q-statistic is a similar measure and indicates the squared 2-norm of an observation vector from its mean in directions orthogonal to the eigenvectors retained from the PCA decomposition (Chiang et al. 2000). In other words, it is a 2-norm of the residues. T^2 and Q statistics thus are complementary and they together give a good indication of the statistical process going out of the normal operating range.

Along with the raw anomaly scores, we can also identify a list of variables (along with corresponding weights) that are either responsible for the fault and/or are most affected by the fault. Analysis of these variable contributions provides insight into probable causes of a detected change and/or fault.

The FDD module utilizes operational data from the EMCS such as temperature, airflows, and electricity consumption as well as output data from real-time EnergyPlus simulations. T^2 and Q statistics are computed on the deviations of the measured data-points from model predictions for the purpose of fault detection and fault identification. PCA, which underlies T^2 and Q statistics, models the multivariate data as multivariate Gaussian distributions but this assumption may not be true in the cases of the measured data from the EMCS or data from EnergyPlus simulations. However, it is more reasonable to assume that the differences between measured points and corresponding predictions from the EnergyPlus model can be approximated as a multivariate Gaussian distribution. This is the motivation for computing the anomaly scores on the deviations rather than the measured data or model predictions directly.

Energy Performance Visualization Dashboard

The current state-of-the-art energy management and control systems provide facility managers with a rich set of building data. This building data includes system and equipment performance (temperature, pressure, energy consumption, etc.), controller status, and equipment fault status. However, the interconnected complexity and sheer volume of this building data often make facility manager building operation decision-making overwhelming and difficult. Today, facility managers rely on their personal intuition and experience to perform building operation decision-making. An interactive and visual interface for facility managers has been developed to exploit available building data more effectively to improve building operation decision-making. The energy performance visualization dashboard is intended to enable: 1) visualization of energy-related metrics at different building and HVAC systems levels; 2) comparisons between measured quantities and data derived from the integration of both data mining and physics-based modeling methods; 3) energy fault diagnostics to aid in decision support targeting of root cause analysis; and 4) identifying persistent trends in energy usage.

Integrated Software Environment

Represented by the $\sum$ symbol in Figure 19-1, a software environment and supporting signal processing are integrated with the EEMCS and reference Energy-Plus model such that the EnergyPlus simulation outputs can be automatically assimilated with and compared to measurements. This software system is built upon the BCVTB (Wetter 2012), which is an open source software platform for integration of EEMCS data and a range of energy modeling software tools including EnergyPlus. The BCVTB makes use of Ptolemy II (Eker et al. 2003), an open source software environment for combining heterogeneous modeling and simulation tools.

EMCS Integration

Additional sensors are typically required to provide all the variables that are needed by the EnergyPlus model. In order to make EMCS measurements accessible to the EnergyPlus model, a suitable communication protocol is needed. BACnet was used in this study due to its wide acceptance. In cases where the EMCS uses a proprietary protocol, a BACnet gateway is required. The BCVTB contains two actors that can read from, and write to, BACnet devices. The read function acquires the relevant model input variables while the write function provides a hardware-in-the-loop control capability. Both actors use a configuration file to specify the BACnet devices, object types and property identifiers. The detailed procedures to create these configuration files and to use these two actors are described in the BCVTB manual (Wetter 2012).

EnergyPlus Integration

Figure 19-6 shows the connection between the EnergyPlus and the BCVTB. The simulator actor in the BCVTB links to the external interface in the Energy-Plus. In the external interface, the input/output signals that are exchanged between the BCVTB

and the EnergyPlus are mapped to EnergyPlus objects (Wetter 2011). The external interface takes three types of inputs from the BCVTB through three objects:

- ExternalInterface:Schedule,
- ExternalInterface:Actuator,
- ExternalInterface:Variable.

When the BCVTB passes a value to the ExternalInterface:Schedule object, it creates a new schedule. The other two objects are used in the same way as the EnergyPlus Energy Management System (EMS) actuators and variables, except that their numerical values are obtained from the BCVTB at the beginning of each zone time step. The EMS is a newly released feature of EnergyPlus to provide user-defined supervisory control capabilities. It can read a variety of "sensor" data and use this data to direct various types of control actions (EnergyPlus 2013). The actuator objects overwrite various input parameters, such as weather data and setpoints and internal calculated variables, such as fractional window opening. Any EnergyPlus Output:Variable or Energy-ManagementSystem:OutputVariable can be sent to the BCVTB at each zone time step.

Database Integration

PostgreSQL was chosen as the database program in this study. A Java-based API (Application Programming Interface) was built for applications to communicate easily with the database. In order to send data to the database, PtolemyII SystemCommand actors are executed as a wrapper tool around the database API. Another Ptolemy II actor is used to query data from the database.

CASE STUDY

The building used for the case study presented in this chapter is the Atlantic Fleet Drill Hall, at Naval Station Great Lakes, Great Lakes, IL. The implementation of the proposed system greatly depends on the existing

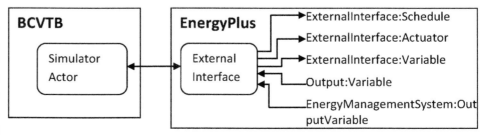

Figure 19-6: EnergyPlus integration in BCVTB

building control system communication capability. It is desirable that the existing EMCS should support open communication protocols such as BACnet, LonWorks, or Modbus.

Building Characteristics

Drill Hall is a two-story facility with a gymnasium-like drill deck, office, classroom, and administrative rooms. The gross area of this building is 69,218 ft² (6,431 m²). The construction was finished in October 2007. This building is LEED® Gold certificated. Figure 19-7 shows the exterior view for this building.

Figure 19-7: Drill Hall building at Naval Station Great Lakes

The Drill Hall HVAC system consists of four airside subsystems and two separate waterside subsystems. The drill deck is served by two VAV (variable air volume) AHUs (air handling unit) with heating and cooling capability. A classroom on the second floor is served by one VAV AHU. Unit operation depends on the occupancy of the drill deck space. Double-walled sheet metal ductwork with a perforated liner and drum louvers distribute the air throughout the space. The office and administrative area is served by one VAV air handling unit with VAV terminal units (with hot water reheat). The chilled water system consists of two 110-ton (386.85 kW) air-cooled rotary-screw type chillers with fixed-speed primary pumping and variable-speed secondary pumping. Heating is supplied from the existing campus-wide steam system through a steam-to-water heat exchanger. The hot water serves unit heaters, VAV box reheating coils, and air handling unit heating coils. There is an instantaneous steam-to-domestic hot water generator for domestic hot water service. The server room and communication service room are served by dedicated duct free split systems. A DDC (Distributed Direct Digital Control) system is installed in this build-

ing to monitor all major environmental systems. Building electric meters are also read by the DDC system. Operator workstations provide graphics with real-time status for all DDC input and output connections.

Additional meters and sensors are required to calibrate models and accurately measure energy consumption to validate results. An on-site weather station, including a pyranometer, aspirated wet and dry bulb temperature sensors, and wind speed and direction sensors, was installed on the roof. BTU meters (a matched pair of supply and return water temperature sensors, water flow meters) were installed for the chillers, secondary chilled water loop, and hot water loop. Lighting load power, plug load power and individual chiller power were also monitored through sub-meters. These sensors and meters were integrated into the existing building EMCS.

EnergyPlus Reference Model

The EnergyPlus model used in this study is version 5.0 (build 5.0.0.035). The structure of the HVAC system in the EnergyPlus model is a series of modules connected by air and water fluid loops that are divided into a supply and a demand side. In order to keep the size of the model and computation time manageable, zoning simplifications were made when entering the building geometry. All the rooms serving by the same VAV box were integrated into one thermal zone. The building model consists of 30 conditioned zones (12, 12, and 6 zones for the drill deck, first, and second floors respectively). Some zones represent a physical room in the building while other zones represent adjacent multiple rooms operating under similar energy usage/requirements. Each zone includes an "internal mass" that represents the thermal storage capacity of the room(s) (e.g., interior walls, furnishings, books, etc.).

Both an extensive sensitivity analysis and an uncertainty analysis were performed to understand and calibrate the EnergyPlus model behaviors (Eisenhower et al. 2012; O'Neill et al. 2011). Originally, the EnergyPlus model was created and selection of its input parameters was performed using the best information that was available at the time. For example, the plug and lighting load profiles were generated by using real time sub-metered data for two months (April and May) in 2010. HVAC system operation schedules and setpoints were taken from actual building EMCS data. During model calibration, a smaller subset of the input parameters that are most critical was identified using sensitivity analysis and subsequently tuned so that the model better matches measured data. This model calibration process relies heavily on characterizing parametric influences on the outputs of the model. This analysis is performed by sampling all parameters of the model around their nominal value to create a database of output data which is used to calculate the sensitivity of these outputs to parameter variation as well as to derive an analytic meta-model based on this model data (Eisenhower et al. 2012). Once the most influential parameters (on the order of 10 to 20) of the model are identified, an optimization can be performed (using the meta-model) in order to identify parameter combinations that produce the best fit to measured data. Only 10-20 of the most influential input parameters instead of thousands were optimized during the optimization/calibration process to avoid the issue of over fitting the model.

Figure 19-8 shows the sensitivity indices of facility electricity consumption (annual total and peak demand) to the 1009 input parameters of EnergyPlus model (O'Neill et al. 2011). The top three input parameters, which have the greatest influence on the facility's annual total electricity consumption in the drill hall, are the AHUs (serving the drill deck) supply air temperature setpoint, chiller reference COP (Coefficient of Performance) and drill deck lighting schedule. The top three input parameters with significant impact on facility electricity peak demand are chiller optimum part load ratio, chiller reference COP and the AHUs supply air temperature set-point. This demonstration building is a military drill hall and occupant schedules are well defined. Therefore, the impact from occupant stochastic behavior on the model prediction accuracy is limited, which may not be the case for other building types. A data-driven approach (Meyn et al. 2010; Liao and Barooah 2010) may also be applied for real-time occupancy estimation in those buildings. This approach is capable of combining information from various sources and providing a consistent and accurate estimate of occupancy in buildings. Advances in sensing, embedded systems and wireless technology enable commercial buildings today to have unprecedented access to data, from temperature sensors, CO_2 sensors, CCTV video cameras, and other access control devices.

Real-time Performance Monitoring and Energy Diagnostics System

The overall system schematic diagram implemented in this case study is shown in Figure 19-9. The PC server running the proposed system is located in the same building location as the PC running the EMCS. The required building performance data are collected through the existing EMCS and then made accessible to the energy diagnostics system through a BACnet gateway. Since the EMCS uses a proprietary communication protocol, a BACnet gateway was installed on top of the EMCS so that the BACnet interface included in the BCVTB can communicate with the EMCS. The BCVTB was installed on a separate computer than the one running EMCS. The real-time simulation is launched by starting the BCVTB through the GUI (Graphical User Interface) or through the console. The latter allows use of the BCVTB in an automated workflow or in a window-less system. The BCVTB version 0.6.0 was used in the case study presented in this chapter.

Within the BCVTB, there are two modules necessary to achieve the proposed functional requirements. The BACnet module is used to acquire the relevant building performance data from the EMCS BACnet gateway through an Ethernet connection. The sampling interval is 5 minutes. The data then is transferred to the PostgreSQL database through the Database (DB) Connector tool. The EnergyPlus (E+) module establishes the communication between the BCVTB and an external pre-built calibrated EnergyPlus model that represents the acceptable building performance. The EnergyPlus simulation time-step is 15 minutes. The EnergyPlus module receives the relevant data (e.g., weather data)

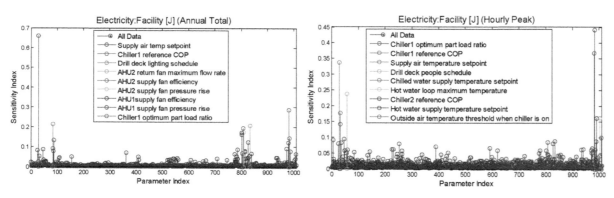

Figure 19-8: Sensitivity indices of facility electricity to 1009 input parameters

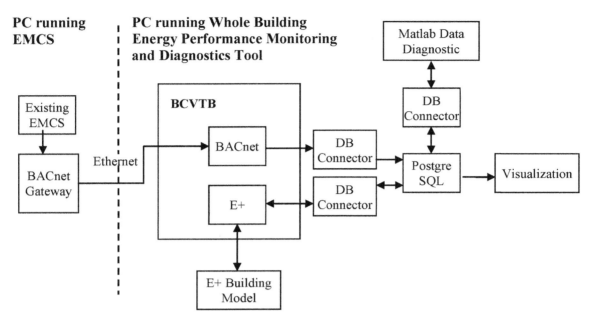

Figure 19-9: Energy diagnostics system schematic diagrams

and executes the external EnergyPlus reference model. The EnergyPlus simulated results then are passed back to the PostgreSQL database through its dedicated database connector tool.

The Data Diagnostic tool runs in the Matlab (Matlab 2012) environment. It communicates with the Database Software through its dedicated DB connector tool. The Data Diagnostic tool applies data mining and anomaly detection methods to identify building faults using building measurements and building reference model predictions data stored in the DB. This tool executes in an automated fashion once an hour. In each instance, it reads the building EMCS data and EnergyPlus model data for the past hour from the DB, performs computations, and archives the results back in the DB.

The Visualization is the user interface to demonstrate the results as well as to display the real-time building performance data. The visualization module is implemented as a stand-alone module and is initiated by the user. The user selects the time period that he/she wishes to explore after which the module reads corresponding data from the database and displays them to the user. Schematics of data diagnostics and visualization modules are shown in Figure 19-10.

It should be noticed that the BCVTB, the EnergyPlus reference model, the Matlab Data Diagnostic tool and database software are running in the background and not visible to the user.

Energy Performance Monitoring and iagnostics Results

The proposed energy performance monitoring and diagnostic tool was installed in the drill hall in April 2010. The facility was well maintained and was well designed from an energy perspective. However, the tool did identify a series of improvement opportunities that include changes to lighting, controls and other further optimizations in the drill hall. The implementation of these recommendations for a retrofit was out of the scope of this study. Currently, anomaly scores and thresholds are automatically computed by analyzing data form the previous 30 days. In other words, data used for analysis comes from a 30-day sliding window and thus the thresholds can vary with time.

Energy Performance Monitoring

The real-time whole building energy performance is compared to the model prediction in Figure 19-11. The differences between the simulated and measured performance are highlighted. The highlighted areas enclosed by the oval dash lines represent the differences that occurred during occupied hours while those enclosed by the rectangles represent the differences occurred during unoccupied hours. Further analysis of the electric power data indicated that different chiller operation strategies were the cause of the considerable performance deviations during occupied hours, as shown in Figure 19-12.

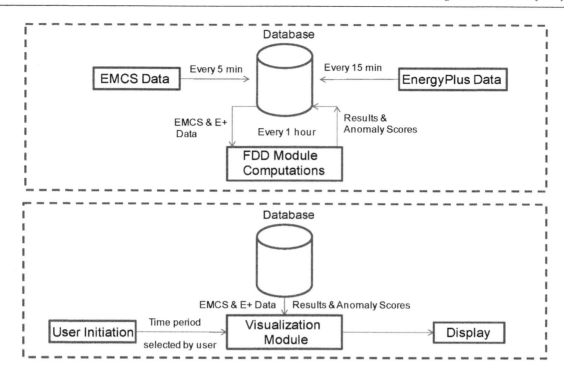

Figure 19-10: Schematics for FDD module and visualization module

It was found that differences arose when the model used free cooling while active mechanical cooling was used in the real building. This missed opportunity for free cooling amounted to 5.3% potential energy savings at the building level for that week. The lights left on overnight was the cause of the deviations during unoccupied hours, as shown in Figure 19-13.

Potential Sensor Bias

Figure 19-14 shows an anomaly in an AHU displayed in the visualization dashboard (discussed in the next section). The biggest contribution to this anomaly comes from a difference between the simulated and measured air temperature exiting the pre-heating coil (the #5 variable). The relative contributions (covariance) are given in top right of Figure 19-14. The anomaly corresponds to potential sensor bias for the temperature sensor located immediately downstream of the pre-heating coil. As the time series plot shows, differences between the mixed air temperatures and the air temperatures after the pre-heating coil were up to 10°F even when the preheating coil valve was closed. By analyzing more historical data, it was found that the air temperature sensor after the pre-heating coil was drifting.

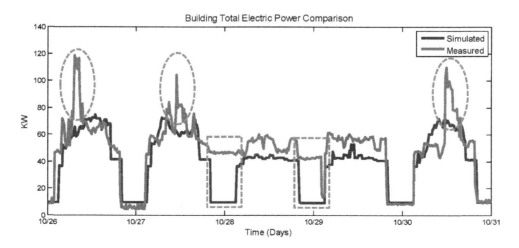

Figure 19-11: Building total electric power comparison between predictions from a real-time

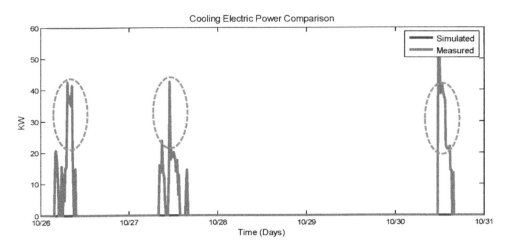

Figure 19-12: Cooling electric power comparison between predictions from a real-time EnergyPlus model and actual measurements (from midnight October 26, 2010 to October 31, 2010). The simulation values are zeros.

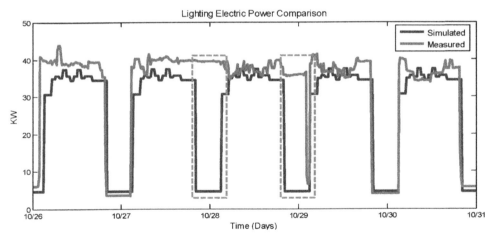

Figure 19-13: Lighting power comparison between predictions from a real-time Energy-Plus model and actual measurements (from midnight October 26, 2010 to October 31, 2010)

Economizer Fault

The upper plot in Figure 19-15 compares the outside air fraction for an AHU on May 4th, 2010 in the actual operation with that calculated from the reference EnergyPlus model. The anomaly scores (blue line) based on T^2 statistics are plotted in the lower part. Whenever the anomaly score is above the threshold (red dash line), a potential fault is indicated. Since only one variable (outside air fraction) was used to compute the anomaly score, there is no contribution weights plot. In non-economizer operation mode on May 4th, 2010, the outside air intake is often more than 50% of total supply airflow, which is 8,000 CFM. According to the design intent, the building needs 4,800 CFM to make up the exhaust and ensure a slightly positive building pressure. Therefore,

there is a potential to further reduce the outside air intake under non-economizer mode, which will save both cooling and heating energy.

The EnergyPlus model was used to offline calculate annual energy impact caused by this economizer fault. A simulation with the design intent was used to calculate energy consumption of the proposed operation case, where the outside air flow rate was set to be 4,800 CFM (i.e., outside air flow fraction was 30%) in non-economizer mode. The current operation of AHU1/2 was simulated in the EnergyPlus as well. For the purpose of simplicity, the outside air faction was set to be 50% for the current operation case.

Table 19-1 shows that annual energy end use breakdown comparisons between the current operation

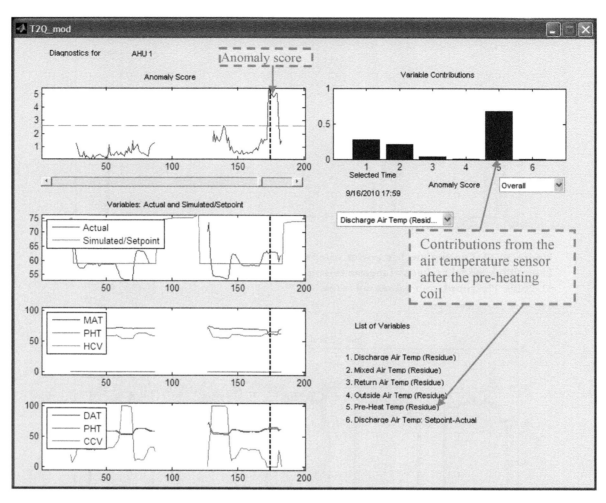

Figure 19-14: Potential sensor bias diagnostics (MAT: Mixed Air Temperature; PHT: Air Temperature After the Pre-heating coil; HCV: Heating Coil Valve Opening Position; CCV: Cooling Coil Valve Opening Position)

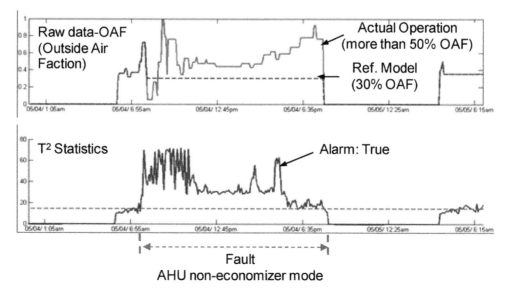

Figure 19-15: Economizer fault

Table 19-1. Annual energy (electricity and steam) end use breakdown comparisons between the current operation and the proposed operation (reduce outside air intake in non-economizer mode)

kWh*1000	Interior Equipment	Interior Lights	Cooling	Pumps	Fans
Current Operation	38.90	227.90	70.11	32.40	45.57
Proposed Operation	38.90	227.90	67.33	31.08	44.49
Difference (%)			-4.13%	-4.24%	-2.43%

MMBTU	Heating
Current Operation	1761.83
Proposed Operation	1254.03
Difference (%)	**-28.82%**

and the proposed operation (reduce outside air intake in non-economizer mode) based on the reference EnergyPlus model prediction. Figures 16 and 17 show the monthly breakdown for total cooling and heating energy consumptions in the drill hall for both cases. For Great Lakes weather (cold winter and cool summer), there are significant heating savings (28.82%) if the outside air intake is reduced to the design intent. This results in an annual savings of $4,418 with an assumption of $8.7 per MMBtu for the steam.

Lighting Issues

Figure 19-18 shows the identified faults due to lights being on during unoccupied hours from November 1st to November 15th, 2010. Lighting sub-metering

data from June 2010 was used as training data. The top plot shows the anomaly score. The middle plot shows the actual lighting electricity consumption. The bottom plot shows the hour of day. The periods marked with red line correspond to the hour when lights on during unoccupied hours. The periods that lights were incorrectly off when supposed to be on is marked with the green line.

The real time data (the middle plot in Figure 19-18) shows the drill hall lighting demand from November 1st to November 15th, 2010. The lighting demand was dominated by the drill deck lights. 64 regular 400W lamps (total 25.6 kW) in the drill deck were turned on from 5:30am to 10:00pm every day. For the period of this study, there was no optimal lighting control in the drill

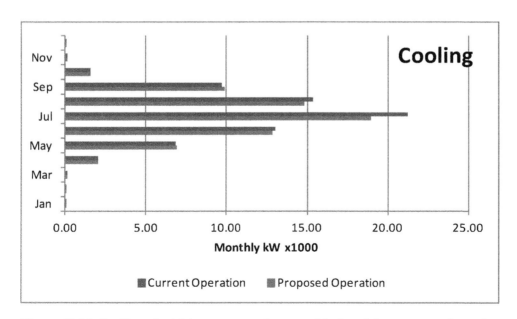

Figure 19-16: Cooling electricity consumption monthly breakdown comparisons between the current operation and the proposed operation (reduce outside air intake in non-economizer mode)

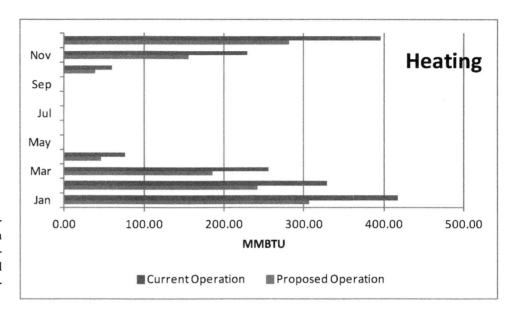

Figure 19-17: Heating steam consumption monthly breakdown comparisons between the current operation and the proposed operation (reduce outside air intake in non-economizer mode)

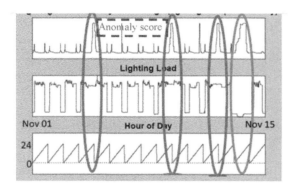

Figure 19-18: Lighting issues

hall. During multiple site visits, it was observed that most of time there were no any activities in the drill deck while all lights were on. It was recommended to install occupancy sensors that will shut lights off when no motion was detected in the drill deck.

The EnergyPlus model was used to evaluate the energy impact from the proposed operation case. The EnergyPlus model with the design intent was used for current operation case, where the lights in the drill deck were turned on from 5:30am to 10:00pm every day. A proposed operation with the assumption that lights in the drill deck will be turned off 50% of the current operation time (5:30am to 10:00pm everyday) was simulated in the EnergyPlus too.

Figure 19-19 shows annual electricity end use breakdown comparisons between the current operation and the proposed operation (occupancy based lighting control). There are significant lighting savings due to

less operated hours from lights in the drill deck. Due to less internal heat gains, cooling (chillers) and fans energy consumption decreased too. Total electricity savings of 23.14% could be achieved through this simple light-reschedule. This optimal lighting control results in an annual savings of $6,542 with an assumption of $0.069 per kW for the electricity. In the winter, the steam consumption is only increased by 2.3% due to less internal heat gains from lights. Figure 19-20 below shows the monthly breakdown for total electricity consumption in the drill hall for both cases.

Visualization Dashboard

Figure 19-21 shows a snapshot of the interactive user-interface. The interface is divided in three panes—(a) loading data (shown in red box in Figure 19-21); (b) energy usage (shown in green box); and (c) system health—anomalies (shown in blue box).

The top part of the user interface is for visualizing energy usage data. There are five visualizations that display various aspects of how energy usage is distributed across different modalities (lights, plug loads, cooling, fans etc.) in the selected time period.

- The first pie-chart displays energy breakdown at any given time instant,

- The second pie-chart displays energy breakdown at the time-step corresponding to peak overall power consumption during the selected time period,

- The third pie-chart displays energy breakdown of the total energy usage over the selected time period,

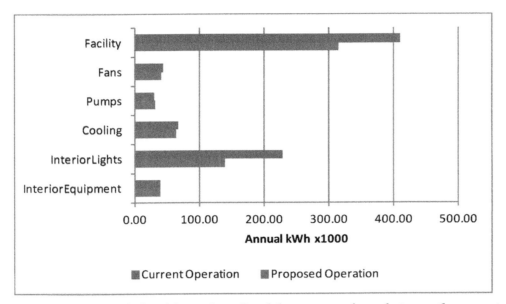

Figure 19-19: Annual electricity end use breakdown comparisons between the current operation and the proposed operation (occupancy based lighting control)

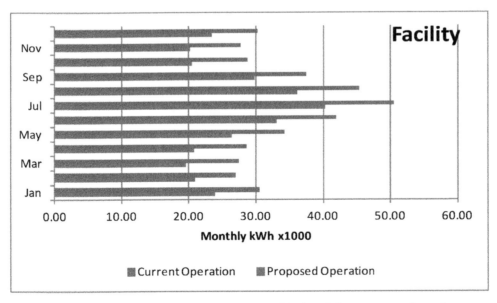

Figure 19-20: Total electricity consumption monthly breakdown comparisons between current operation and proposed operation (occupancy based lighting control)

- The line plot describes the power breakdown over the entire history of the selected time-period.

- The bar chart displays total energy consumed on the HVAC Hot Water side for the selected time period.

There are two kinds of data that can be explored: (a) real-time data from the EMCS and (b) data from the EnergyPlus simulation model. There is a pull-down menu from which user can select either the EMCS data or the EnergyPlus model data to visualize. User also can select a modality (lights, plug loads, cooling, fans, total) and visualize a comparison between the EnergyPlus model predictions and measured data from the EMCS. Once the selection is made, a new plot opens up that displays the comparison for the selected attribute (shown in Figure 19-22).

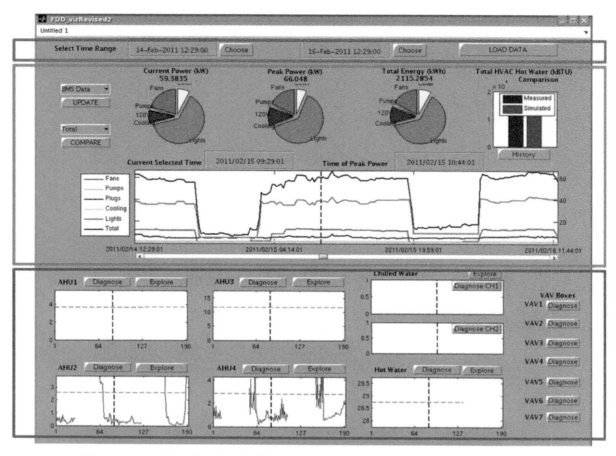

Figure 19-21: Visualization dashboard energy usage and performance monitoring

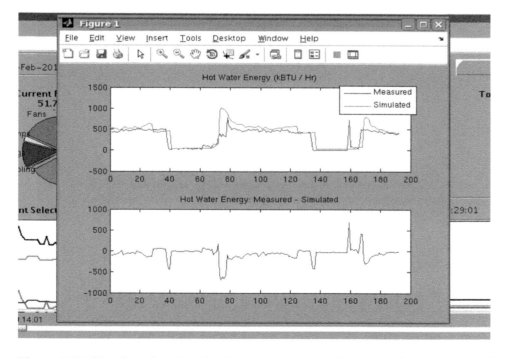

Figure 19-22: User interface showing hot water consumption comparisons between the model predictions and measurements (noon February 14, 2011 to noon February 16, 2011)

System Health—Anomaly Scores

The bottom part of the user interface (Figure 19-21) is dedicated to Anomaly Scores and monitoring the health of each subsystem (Chilled Water System, Hot Water System, Air Handling Units and Variable Air Volume Boxes). Each subsystem (AHU, Chillers, and the Hot Water System) has a graph associated with it indicating the anomaly score (in blue) corresponding to the system health. Also shown in red is a threshold calculated mathematically. If the anomaly score exceeds the threshold at any time instant, an anomalous event is triggered and recorded. The anomaly score is computed only when the system is in operation. No anomaly score is displayed when the system is not running.

Subsystem Drilldown—Diagnose

The user interface displays the anomaly score and the threshold. In addition, the display also plots the "contributions" of individual variables that were used in computing the anomaly score (see an example in Figure 19-14). This gives the user an idea of the significance of different variables in causing an anomaly. A slider functionality is provided where the user can explore a time-instant of his/her choice to understand the variable contributions. The user interface also allows the user to select any of the variables via a pull-down menu and view time-history of the EMCS data corresponding to that sensor, data from the EnergyPlus model and the difference between the measured data and the Energy-Plus model predictions.

CONCLUSIONS AND LESSONS LEARNED

A first-generation, real-time, whole-building energy performance monitoring and diagnostics tool has been developed and demonstrated in a real building. The tool was deployed in April 2010 and has been continuously operated for two years. The system continuously acquires performance measurements of HVAC and lighting usage from the existing EMCS augmented by additional sensors. The system compares these measurements in real-time to a reference EnergyPlus model that has been calibrated to represent acceptable performance. The comparison enables identification and quantification of sub-optimal performance and identification of the conditions under which sub-optimal performance occurs. Opportunities for about 30% energy savings were identified during the first 6-month deployment and actionable information was well received by the facility manager.

There are a few lessons and observations from the case study.

- A real-time whole building performance monitoring and energy diagnostics tool using EnergyPlus has been developed and demonstrated in proof-of-concept form. The EnergyPlus model is a dynamic representation of expected building performance, and it does not represent the real conditions in buildings. Real buildings often don't perform as expected by their designers due to 1) faulty construction, 2) malfunctioning equipment, 3) incorrectly configured control systems, and 4) inappropriate operating procedures.

- A framework of whole building model-based FDD has been established. FDD algorithms based on statistical process control method such as T^2 and Q statistics have been tested.
  - The quality and availability for both nominal and faulty data are important to establish ground truth to test and validate data mining based FDD algorithms.
  - Variable contributions to the anomaly scores provide a good insight into probable causes of a detected change and/or fault.
  - Transient periods including system start-up and shut-off need to be excluded to avoid some false alarms.

- A visualization dashboard for building performance energy monitoring and energy diagnostics has been developed and deployed in a real building. This dashboard provides an effective way for building facility manager to perform building performance decision-making.

- Electrical sub-meters and thermal energy meters (BTU meters) are important for FDD. The lighting faults could be not detected without the lighting sub-meters.

This proof-of-concept case study also suggests the following aspects for the future work.

- A deployment concern about this technology is the skill level required to install and maintain the system. Another challenge is the efficient generation of simulation models of existing buildings from limited, often paper-based, design and as-built documentation. The current development of a comprehensive GUI (Graphical User Interface) for EnergyPlus (See et al., 2011) will make a number of different aspects of modeling buildings, including existing buildings, simpler, faster and less prone to error. However, there are a number of aspects of modeling existing buildings that would be made more efficient by specific enhancements to this GUI.

• Currently, even experienced EnergyPlus model developers may find that the effort required to calibrate the model is significant. The model calibration may be insufficient to discern differences between actual and desired building performance. Model calibration is important and potentially can be handled well by using auto-tuning tools (New et al., 2012, O'Neill and Eisenhower 2013).

• Since use schedule in buildings is often changed according to user's desire, the information about use schedule such as AHU discharge air temperature setpoints and room temperature setpoints need to be read directly from EMCS and sent to the EnergyPlus model in real-time. This could be done by extending the BACnet module in BCVTB.

• For a real-time implementation, real time building operational data was collected through a BACnet gateway by using the open source software BCVTB. A database was used to store both building static data (e.g., model parameters, HVAC configuration, etc.) and building dynamic operational data (e.g., temperature, energy). All the mapping was done manually which increased the implementation cost. It is recommended that the following activities should occur to facilitate the implementation in the field:

— Extend a BACnet compatible data acquisition system to cover the other industry standard communication protocols;

— Develop a database structure that enables rapid mapping and use of both static building information and real time dynamic operational data during the design and operational phases of a building lifecycle. This structure should be tested in a variety of buildings with different types and sizes;

— Develop a services-based architecture to support the data exchange Application Programming Interface (API) and computational services.

ACKNOWLEDGEMENTS

This work was performed under the project EW09-29 administered by ESTCP (Environmental Security Technology Certification Program) technology program of the Department of Defense. We would like to thank Dr. James Galvin, the ESTCP program manager, and Mr. Peter Behrens, the energy manager at Great Lakes, for their support. Views, opinions, and/or findings contained in this paper are those of the authors and should not be construed as an official Department of Defense position or decision unless so designated by other official documentation.

References

Bynum, J.D., Claridge, D.E., and Curtin, J.M. 2010. Development and testing of an Automated Building Commissioning Analysis Tool (ABCAT). Proc. of ASME 2010 4th Int. Conf. on Energy Sustainability ES2010, Phoenix, AZ, May 17-22, 2010, Paper ES2010-90389, CD.

Bynum, J.D., Claridge, D.E., and Curtin, J.M. 2012. Development and testing of an Automated Building Commissioning Analysis Tool (ABCAT). *Energy and Buildings* 55 (2012) 607–617.

Chiang, L., Russell, E. and Braatz, R. 2000. *Fault Detection and Diagnosis in Industrial Systems*, Springer Verlag, London.

Claridge, D., Liu, M. and Turner, W.D. 1999. Whole Building Diagnostics. Proceedings of Diagnostics for Commercial Buildings: From Research to Practice. San Francisco, CA.

Curtin, J.M., Claridge, D.E., Painter, F., Verdict, M., Wang, G. and Liu, M. 2007. Development and Testing of an Automated Building Commissioning Analysis Tool. California Energy Commission, PIER Energy-Related Environmental Research Program. CEC-500-01-044-2007.

Dahilwal, S, and Guay, M. 2012. A Set-based Estimation of Heat Loads for Energy Management in Building Systems. To appear in The 51st IEEE Conference on Decision and Control.

Eisenhower, B., O'Neill, Z.D., Fonoberov, V. and Mezic´, I. 2012. Uncertainty and Sensitivity Decomposition of Building Energy Models. *Journal of Building Performance Simulation*. Vol. 5, n 3. Pp171-184.

Eker, J., Janneck, J., Lee, E.A., Liu, J., Liu, X., Ludvig, J., Sachs, S and Xiong, Y. 2003. Taming Heterogeneity-The Ptolemy Approach, Proceedings of the IEEE, 91(1):127-144.

EnergyPlus. 2013. http://apps1.eere.energy.gov/buildings/energyplus/

Friedman, H., Claridge, D. Choinière, D. and Milesi-Ferretti, N. 2010. Annex 47 Report 3: Commissioning Cost-Benefit and Persistence of Savings. a Report of Cost-Effective Commissioning of Existing and Low Energy Buildings. Directed by the Energy Conservation in Buildings and Community Systems (ECBCS) Program. Available from http://www.ecbcs.org/annexes/annex47.htm#p

Haves, P. 1999. Overview of Diagnostics Methods. Proceedings of Diagnostics for Commercial Buildings: From Research to Practice. San Francisco, CA.

Haves, P., Salsbury, T., Claridge, D. and Liu M. 2001. Use of Whole Building Simulation in On-Line Performance Assessment: Modeling and Implementation Issues. Proceedings of 7th International IBPSA Conference Building Simulation 2001, Aug 13-15, 2001, Rio de Janeiro.

House, J.M., Vaezi-Nejad, H. and Whitcomb, J.M. 2001. An Expert Rule Set for Fault Detection in Air-handling Units. ASHRAE

Transactions. 107(1).

Jacob, D., Dietz, S., Komhard, S., Neumann, C. and Herkel, S. 2010. Black box Models for Fault Detection and Performance Monitoring of Buildings. *Journal of Building Performance Simulation*. Vol.3, n1. pp 53-62.

Katipamula, S. and Brambley, M. R. 2005a. Methods for Fault Detection, Diagnostics, and Prognostics for Building Systems—A Review Part I, *HVAC&R Research*, 2005, vol. 11, n 1.

Katipamula, S. and Brambley, M. R. 2005b. Methods for Fault Detection, Diagnostics, and Prognostics for Building Systems—A Review Part II, *HVAC&R Research*, 2005, vol. 11, n 2.

Knebel, D. E. 1983. Simplified Energy Analysis Using the Modified Bin Method. Atlanta, American Society of Heating, Refrigerating, and Air-Conditioning Engineers, Inc.

Koller, D. and Friedman, N. 2009. Probabilistic Graphical Models: Principles and Techniques. Edited by MIT Press.

Lee, S. Uk, Painter, F.L. and Claridge, D.E. 2007. Whole Building Commercial HVAC System Simulation for Use in Energy Consumption Fault Detection. ASHRAE Transactions, v113, part2: 52-61.

Li, Haorong. 2004. A decoupling-based Unified Fault Detection and Diagnosis Approach for Packaged Air Conditioners. Ph.D. Thesis. Purdue University.

Liao, C. and Barooah, P. 2010. An Integrated Approach to Occupancy Modeling and Estimation in Commercial Buildings. American Control Conference (AACC), 2010.

Lin, G. and Claridge, D.E. 2009. Retrospective Testing of an Automated Building Commissioning Analysis Tool (ABCAT). Proc. 3rd Int. Conf. on Energy Sustainability, ASME, July 19-23, 2009, San Francisco, CA, USA, CD.

Liu, M., Song, L. and Claridge, D.E. 2001. Development of Whole Building Fault Detection Methods. High Performance Commercial Building Systems. California Energy Commission. Public Interest Energy Research Program.

Matlab. 2012. http://www.mathworks.com/products/matlab/

Meyn, S., Surana, A., Lin, Y., Oggianu, S., Narayanan, S. and Frewen, T. 2010. A Sensor-Utility-Network Method for Estimation of Occupancy Distribution in Buildings. The 49th IEEE Conference on Decision and Control.

Mills, E. 2011. Building Commissioning: A Golden Opportunity for Reducing Energy Costs and Greenhouse Gas Emissions in the United States. *Energy Efficiency*, 4(2):145-173.

Murphy, K. 2001. An Introduction to Graphical Models. Technical Report. http://www.cs.ubc.ca/~murphyk/Papers/intro_gm.pdf

New, J., Sanyal, J. Bhandari, M and Shrestha, S. 2012. Autotune E+

Building Energy Models. The Fifth National Conference of IBPSA-USA: SimBuild2012. Madison, MI. August 1–3, 2012.

Nouidui, T.S., Wetter, M., Li, Z., Pang, X., Bhattacharya, P. and Haves, P. 2011. BACnet and Analog/Digital Interfaces of the Building Controls Virtual Test Bed. Proceedings of 12th International IBPSA Conference Building Simulation 2011, Nov 14-17, 2011, Sydney, Australia.

O'Neill, Z.D., Narayanan, S. and Brahme, R. 2010. Model-Based Load Estimation for Buildings. The Fourth National Conference of IBPSA-USA: SimBuild2010. New York City, NY. August 11–13, 2010.

O'Neill, Z.D., Eisenhower, B., Yuan, S., Bailey, T., Narayanan, S. and Fonoberov, V. 2011. Modeling and Calibration of Energy Models for a DoD Building. ASHRAE Transactions, 117(2): 358-365. ASHRAE Annual Meeting. Montreal, Québec, Canada. June 25–29, 2011.

O'Neill, Z.D. and Eisenhower, B. 2013. Leveraging the Analysis of Parametric Uncertainty for Building Energy Model Calibration. Accepted by *Building Simulation: An International Journal*.

Ramirez, R., Sebold, F., Mayer, T., Ciminelli, M., and Abrishami, M. 2005. A Building Simulation Palooza: The California CEUS Project and DrCEUS. Proceedings of Building Simulation 2005, Montreal, Canada. IBPSA, 2005.

Salsbury, T. and Diamond, R. 2000. Performance validation and energy analysis of HVAC systems using simulation. *Energy and Buildings*, 32:5-17.

See, R., Haves, P., Sreekanthan, P., Basarkar, M. and Settlemyre, K. Development of a Comprehensive User Interface for the EnergyPlus Whole Building Energy Simulation Program. Proceedings of Building Simulation 2011, Sydney, Australia, November 2011.

Sreedharan, P. and Haves, P. 2001. Comparison of Chiller Models for Use in Model-based Fault Detection. International Conference for Enhancing Building Operations, TX.

TIAX. 2005. Energy Impact of Commercial Building Controls and Performance Diagnostics: Market Characterization, Energy Impact of Building Faults and Energy Savings Potential. Final Report to U.S. Department of Energy.

Wetter, M. 2012. Building Control Virtual Test bed Manual.http://simulationresearch.lbl.gov/bcvtb/releases/1.0.0/doc/manual/bcvtb-manual.pdf

Wetter, M. 2011. Co-simulation of Building Energy and Control Systems with the Building Controls Virtual Test Bed. *Journal of Building Performance Simulation*, Vol. 4, n 3. pp 185-203.

Chapter 20

Advanced Methods for
Whole Building Energy Analysis*

Jessica Granderson, Mary Ann Piette, Ben Rosenbloom and Lily Hu
Lawrence Berkeley National Laboratory

INTRODUCTION

There is a wealth of methods and tools to monitor and measure building energy use for both the long haul and in real time, and to identify where best to focus facility energy efficiency efforts. Information to plan an energy management strategy that works for your building or facility to make it more energy efficient is presented in the Energy Information Handbook: Applications for Energy Efficient Building Operations {1}. The primary audience for this Handbook is commercial building owners, energy and facility managers, financial managers, and building or facility operators with little or no experience in data analysis and performance monitoring. The secondary audience is software developers and energy service providers in the commercial buildings industry, as well as more experienced owners and managers who wish to improve how they visualize, analyze, and manage their building's energy use.

This chapter is a section of the Energy Information Handbook: Applications for Energy Efficient Building Operations, called Advanced Methods. As classified and defined in the Energy Information Handbook, each of the Advanced Methods rely upon underlying baselines. A baseline is a representation of "standard" or typical energy performance, and used for comparative purposes. Baselines may be expressed according to a variety of metrics, and may account for weather or other independent variables that influence energy consumption. Simple Baselines and Model Baselines are discussed and presented in Fundamental Methods in the Energy Information Handbook. In the best-practices case, the baselines used in the Advanced Methods will be model

based, as reflected in the associated method summaries for each.

As illustrated in Figure 20-1, projected load is determined by inputting measured conditions into a baseline model.

- **Energy Savings** defines energy savings as the difference between the projected and metered load, after efficiency improvements have been made.

- **Cumulative Sum** represents the accumulation of the difference between metered and projected load over time, effectively expressing a running total.

- **Anomaly Detection** compares the difference between metered and projected load to a threshold value.

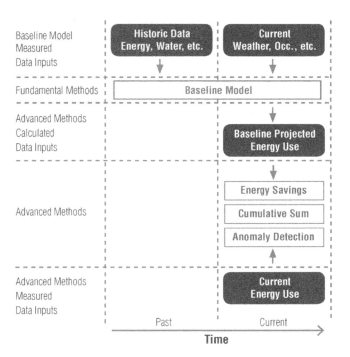

Figure 20-1

---

*This chapter is from Advanced Methods, in the Energy Information Handbook: Applications for Energy Efficient Building Operations, prepared by Jessica Granderson, Mary Ann Piette, Ben Rosenbloom and Lily Hu, Lawrence Berkeley National Laboratory, LBNL-5272E, 2011.

The energy savings method allows building owners, energy service companies, and financiers of energy-efficiency projects to quantify and verify the energy-savings performance of energy conservation measures (ECMs) or efficiency programs. In contrast to previously presented methods that can be used to estimate energy savings, this approach makes use of baseline models, with regression being the most common approach. See Figure 20-2.

Applicable Systems				
Whole Building	Heating	Cooling	Lighting	Plug Loads
●	●	●	●	●

Interpretation		Frequency of Use		
Requires Minimum Expertise	Requires Domain Expertise	Continuous	Monthly	Annual
▰			●	●

Figure 20-2

TECHNICAL APPROACH

Collect metered energy use before and after an improvement has been made, or a tracking period is initiated. Then develop a baseline model that accounts for key energy drivers, i.e., "independent variables" such as outside air temperature, using metered data before the improvement period. Project the baseline model into the tracking/reporting period, to quantify the energy use that would have resulted had no improvements been made. Finally, subtract the energy use from the improvement period from the baseline projected energy use to quantify energy savings. See Figure 20-3.

RELATED METHODS

The relationship between each of the Advanced Methods is summarized in the introduction to the Advanced Methods chapter. See Figure 20-4.

CALCULATION AND PROGRAMMING

State of Commercialization: Energy savings calculations are commonly automated in measurement and verification and continuous commissioning software

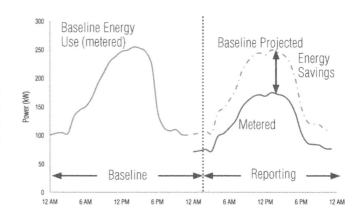

Figure 20-3
As detailed in the Appendix, this approach is largely commensurate with the overall principles in the international protocol for measurement and verification (IPMVP).

tools used internally by providers. They may also be supported in commercial EIS, provided that baselining capability is sufficiently robust. However, there may be adjustments beyond baseline projections that are necessary to quantify savings, yet not automated in commercial software tools.

Step 1: Gather Input Data
Data Resolution

Short interval data is frequently rolled up to daily or monthly increments for use.

1 Hr, 15 Min   Monthly   Annual
◉   ◉   ○

Data Inputs

High meter accuracy (0.5% error or less) is recommended in cases of performance contracting. (Metered Data)

Export energy data from a meter acquisition system, and collect independent variables, such as OAT, required for the baseline model projection. (Baseline Projected)
See: Fundamental Methods: Model Baselines

Step 2: Define the Baseline Period and Reporting Period and Calculate Energy Savings. (See Figure 20-5)

Step 3: Plot Metered Use for the Baseline and Reporting Periods
Plot energy on the y-axis and the reporting interval (day, month, year) on the x-axis. You can also overlay the baseline projected use on this plot to visualize the size of energy savings.

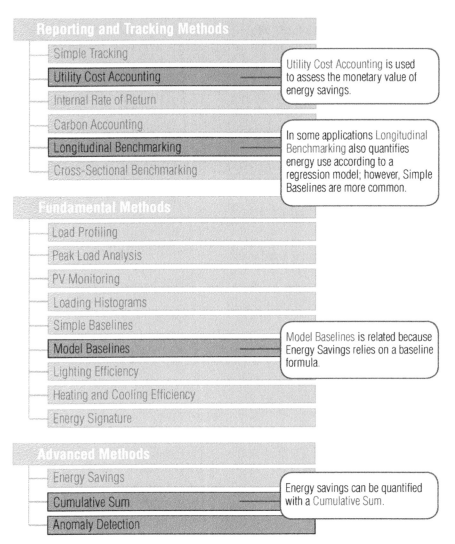

Figure 20-4

Baseline Projected - Metered = Energy Savings

Step 1		Step 2		
Metered Energy Use Baseline Period (kWh)	Reporting Period Baseline Parameters (avg OAT)	Baseline Projected (kWh)	Metered Energy Use Reporting Period (kWh)	Energy Savings (kWh)
Month 1 60,000	58	61,000	58,100	= 61,000 - 58,100 = 2,900
Month 2 58,885	57	59,075	56,250	= 59,075 - 56,250 = 2,825
....	....	....	....	....
Month 12 62,590	52		61,025	....
Total = 700,000	....	= 715,000	= 625,000	= 715,000 - 625,000 = 90,000 or 13% relative to baseline period

Figure 20-5

APPLICATION EXAMPLES

Interpretation: Interpretation of the output of the energy savings analyses is straightforward. However, constructing the underlying baseline model used to quantify those savings may require significant expertise and interpretation of site-specific system and control parameters. Similarly, significant expertise may be required to resolve cases in which actual savings are significantly less than expected, based on the particular efficiency measure.

Example 1: HVAC System Energy Savings

A   Monthly energy use and mean daily temperature are plotted for FY2011 (red).

B   The baseline model (blue) included a base load and weather-sensitive components.

C   Energy use is lower than baseline for each month except July and August. FY11 savings were 10.8%, relative to the baseline period. Figure 20-6.

Example 2: Whole-building Gas Savings

A   Monthly billed use and HDD are tabulated for one year following a retrofit.

B   The baseline model included a base load and a weather-sensitive component.

C   The baseline was used to determine what use would have been without the retrofit.

D   This baseline projection is also tabulated.

E   Energy savings are equal to the difference between measured and baseline use. Savings for 10 months totaled 511k units of gas and $3.2M. Figure 20-7.

Example 3: HVAC System Energy Savings

A   Metered daily HVAC energy use in a 109ksf building is plotted in bars.

B   A retrofit was conducted to improve the control system, and consumption decreased.

C   A regression baseline characterized energy use based on outside air temperature.

D   The baseline was used to determine what use would have been without the retrofit.

E   This baseline projection is shown in blue.

F   Energy savings are equal to the difference between measured and baseline use. A rough estimate indicates approximately 1300-1500 kWh/day energy savings. Figure 20-8.

PURPOSE

The Cumulative Sum (CUSUM) is used to quantify total accrued energy savings or losses over time and to detect energy waste or performance relative to operational changes. CUSUM analysis requires a baseline model, and is applicable to all building types and all building systems. Figure 20-9.

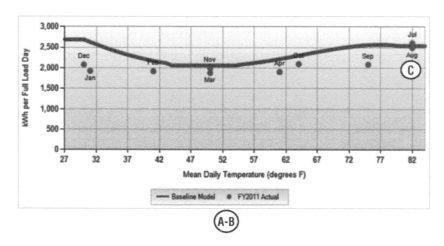

Figure 20-6. *Source: Interval Data Systems*

Meter Reading Date	Actual Post-Retrofit Data		Projected Baseline			Savings		
			Baseload	Weather Sensitive	Total	Gas (units)	Value	
	Consumption	HDD 65	Factors				Price =	
	Units		111,358	173.27			$	6.232
March 6, 2009	151,008	601	111,358	104,135	215,493	64,485	$	401,871
April 4, 2009	122,111	420	?	?	?	?		?
May 6, 2009	102,694	188	111,358	32,575	143,933	41,239	$	257,001
June 5, 2009	111,211	250	111,358	43,318	154,676	43,465	$	270,874
July 5, 2009	80,222	41	111,358	7,104	118,462	38,240	$	238,312
August 6, 2009	71,023	15	111,358	2,599	113,957	42,934	$	267,565
September 8, 2009	65,534	5	111,358	866	112,224	46,690	$	290,972
October 9, 2009	77,354	12	?	?	?	?		?
November 4, 2009	103,000	190	111,358	32,921	144,279	41,279	$	257,251
December 10, 2009	115,112	300	111,358	51,981	163,339	48,227	$	300,551
January 7, 2010	160,002	700	111,358	121,289	232,647	72,645	$	452,724
February 4, 2010	145,111	612	111,358	106,041	217,399	72,288	$	450,499

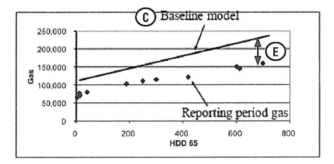

Figure 20-7.
Source: Energy Valuation Organization (EVO)

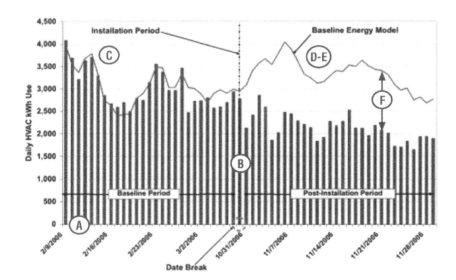

Figure 20-8. *Source: QuEST Engineering*

Applicable Systems				
Whole Building	Heating	Cooling	Lighting	Plug Loads
●	●	●	●	●

Interpretation		Frequency of Use		
Requires Minimum Expertise	Requires Domain Expertise	Continuous	Monthly	Annual
▓		●	●	

The need for interval data depends on the form of the baseline and the system of focus.

Energy Manager

Financial Manager

Figure 20-9.

TECHNICAL APPROACH

Subtract actual metered energy use from the energy use projected by a baseline model to quantify a difference. Aggregate those differences over-time to determine the cumulative sum of difference relative to the baseline, or standard operations. Plot time on the x-axis, and plot CUSUM on the y-axis. Figure 20-10.

RELATED METHODS

The relationship between each of the Advanced Methods, as well as to baselining, is summarized in the introduction to the Advanced Methods chapter. Figure 20-11.

CALCULATION AND PROGRAMMING

State of Commercialization: CUSUM is offered pre-programmed in advanced energy information systems, and may be used in fault detection and diagnostic (FDD) routines. The vendor automates the calculation of the cumulative sum, as well as the underlying baseline.

Computation: You can also use stand-alone data analysis or spreadsheet tools to compute and plot cumulative sums.

Step 1: Gather input data.
Data Resolution

The interval depends on how often CUSUM is calculated. Smaller intervals allow for more granular calculations.

1 Hr, 15 Min   Monthly   Annual

Data Inputs

High accuracy for metered data is not required, but fill in data gaps before computing. Most critical is the accuracy of data inputs to the baseline model used to predict energy use, such as weather variables. (Metered Data)

Export building or system-level electric or gas consumption interval data from a BAS or meter acquisition system. (Baseline Projected)

Compute the associated baseline projected energy use. See: Fundamental Methods: Model Baselines

Step 2: Calculate the difference between metered data and baseline.
Subtract the baseline projected energy use from the metered energy use.

Step 3: Calculate the CUSUM.
Figure 20-12.

Step 4: Plot the CUSUM. The x-axis is Time and the y-axis is CUSUM.

APPLICATION EXAMPLES

Interpretation: A y-value of zero indicates no energy savings, a negative y-value indicates savings, and a pos-

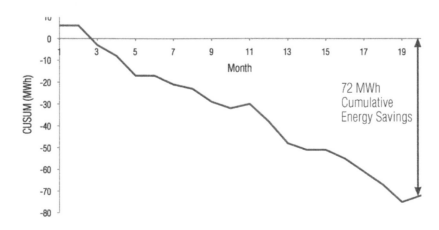

Figure 20-10.

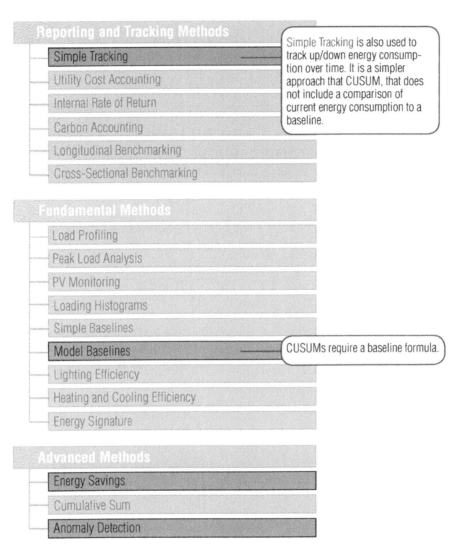

Figure 20-11.

	Step 1			Step 2	Step 3
Interval	Time	Metered Use	Baseline Projected Use	Difference	**CUSUM**
1	1:00pm	34	36	= 34-36 = -2	**= -2**
2	2:00pm	28	29	= 2-29 = -1	**= -2 + -1 = -3**
3	3:00pm	30	28	= 30-28 = 2	**= -2 + -1 + 2 = -1**

Figure 20-12.

itive y-value indicates usage in excess of the baseline. A flat slope marks a period of no change relative to the baseline, a negative slope marks a period of decreased energy use, and a positive slope marks a period of increased energy use.

Example 1: Verification of Energy Efficiency Measures

A   The CUSUM dips, marking a period of energy waste.

B   Efficiency measures are implemented. CUSUM rises to 70,000 kWh savings.

C   More measurements are carried out. The slope steepens, showing additional savings.

D   Five-month total cumulative savings reach 320,000 kWh.

     Figure 20-13.

Example 2: Ensuring Persistence in Savings

A   Months 1-12 form the base year of measurement, and the CUSUM is zero.

B   The slope goes negative, reflecting 8,000 gigajoule (GJ) of efficiency savings.

C   The CUSUM slope goes positive, indicating lost savings.

D   Losses are traced to a missing part, which was replaced, and savings resume.

     Figure 20-14.

Example 3: Quantifying the Effect of Lost Savings

A   A new baseline was computed, following efficiency improvements at Month 13 in Example 2.

B   The CUSUM increased due to the lost savings associated with the missing part.

C   The cumulative lost savings were ~2,000 GJ, with an associated cost of $20K.

     Figure 20-15.

Example 4: Detecting Waste, and Measurement and Verification

A   The CUSUM indicates 15,000 cubic meters (m^3) in total savings.

B   After one month the CUSUM indicates 30,000 m^3 in savings.

C   The slope changed, indicating waste, and an automated alert was generated.

D   A leaking valve was identified and repaired, leading to a new period of savings. Figure 20-16.

PURPOSE

Energy **anomaly detection** is used to automatically identify abnormal energy consumption; it may be paired with alarming and used as part of monitoring-based commissioning routines. It is applicable to all building types and systems. Abnormal energy use can be isolated to a specific system or zone based on a combination of the user's knowledge of the building and supplementary data such as submetered loads, equipment schedules and setpoints, and outside air temperature. Figure 20-17.

TECHNICAL APPROACH

Compare metered use to the use predicted with a baseline model. If metered use surpasses the prediction by a certain threshold value, you have identified an energy anomaly. Figure 20-18.

RELATED METHODS

The relationship between each of the Advanced Methods, as well as to Baselining, is summarized in the introduction to the Advanced Methods chapter. Figure 20-19.

CALCULATION AND PROGRAMMING

State of Commercialization: Anomaly detection may be offered preprogrammed in advanced energy information systems, and is part of some FDD routines.

Computation: You can also use stand-alone data analysis or spreadsheet tools to perform anomaly detection, as described in the steps below.

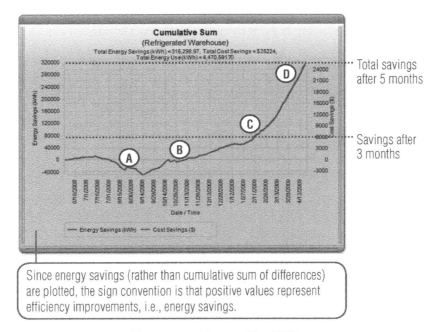

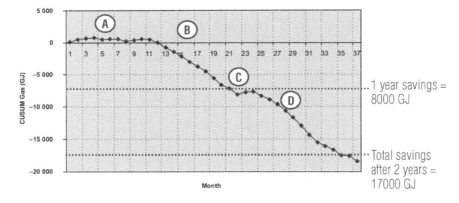

Figure 20-13. *Source: NorthWrite*

Figure 20-14. *Source: Office of Energy Efficiency, Natural Resources Canada, Monitoring and targeting techniques in buildings. Cat. No. M144-144/2007E ISBN 978-0-662-45265-2, 2007.*

Figure 20-15. *Source: Office of Energy Efficiency, Natural Resources Canada, Monitoring and targeting techniques in buildings. Cat. No. M144-144/2007E ISBN 978-0-662-45265-2, 2007.*

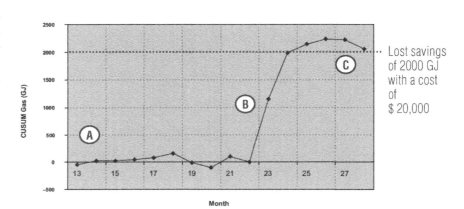

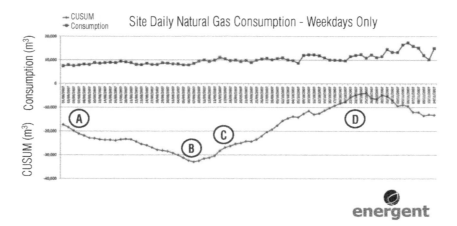

Figure 20-16. *Source: Energent*

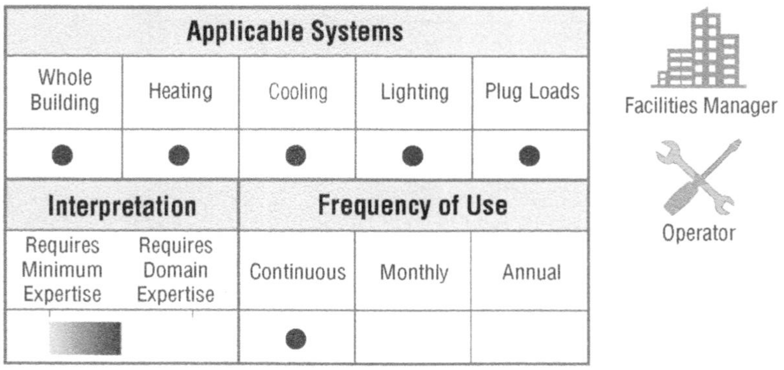

Applicable Systems				
Whole Building	Heating	Cooling	Lighting	Plug Loads
●	●	●	●	●

Interpretation		Frequency of Use		
Requires Minimum Expertise	Requires Domain Expertise	Continuous	Monthly	Annual
�▮		●		

Facilities Manager

Operator

Figure 20-17.

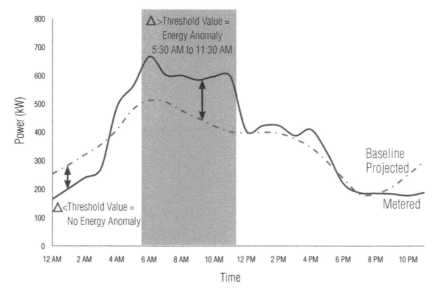

Figure 20-18.

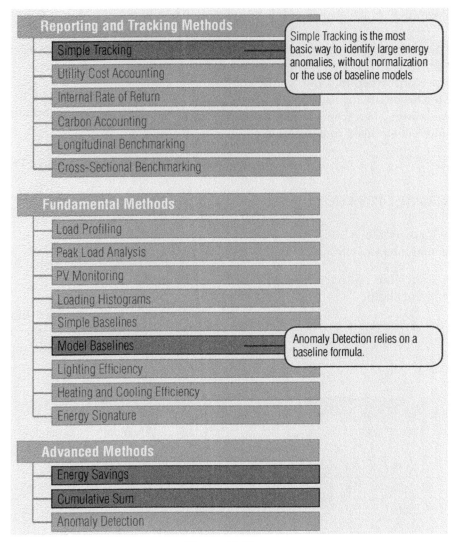

Figure 20-19.

Step 1: Gather input data.
Data Resolution

Anomaly detection requires interval electric or gas data at the whole-building or system level.

1 Hr, 15 Min   Monthly   Annual
    ●   ○   ○

Data Inputs

High accuracy for metered data is not required. Fill in data gaps before computing. Ensure an accurate baseline. (Metered Data)

Export building or system-level electric or gas consumption interval data from a BAS or meter acquisition system. (Baseline Projected)

Compute the associated baseline projected energy use. See: Fundamental Methods: Model Baselines

Step 2: Calculate the difference between metered data and baseline.
Subtract the baseline projected energy use from the metered energy use.

Step 3: Compute the threshold value for each baseline-projected value.
If the difference is greater than the threshold, an anomaly is detected. Figure 20-20.

Step 4: Plot time on the x-axis and metered use on the y-axis. Flag periods for which an anomaly is detected.

APPLICATION EXAMPLES

Interpretation: The threshold is not typically determined analytically; rather it is set according to a default value, such as percent different from predicted, which might be adjusted based on the user's experience. Anomaly detection is distinguished from simple alarming, in that baseline models are used to determine projected consumption.

EXAMPLE 1:
IDENTIFYING ABNORMAL OPERATIONS

A   A 24-hr Sunday load profile is shown for a retail store, with the actual load in yellow.

B   The green band shows the projected load +/- the anomaly detection threshold.

C   Energy use below that projected lies within the blue area.

D   Most of the day the load remains within the green band, but at 7PM it does not fall. Energy use is in the red area, above the projected load, and waste is detected.

The problem was traced to a controls programming error that prevented initiation of nighttime setbacks. Figure 20-21.

Example 2: Avoiding Excessive Peak Demand Charges

A   A hospital experienced high energy consumption on a Monday in late April during the daily peak demand period.

B   The spike in energy use led to a $300 peak demand charge.

Hospital staff discovered that the demand charge resulted from chiller performance testing in preparation for the summer cooling period.

	Step 1		Step 2	Step 3	
Time	Metered Use	Baseline Projected Use	Difference	**Threshold***	**Anomaly?**
12:00pm	34	36	= 34-36 = -2	.10(36)=3.6	-2>3.6= NO
12:15pm	28	29	= 2-29 = -1	.10(29)=2.9	-1>2.9= NO
12:30pm	32	28	= 32-28 = 4	.10(28)=2.8	4>2.8= YES

*In this example the threshold is 10%.

Figure 20-20.

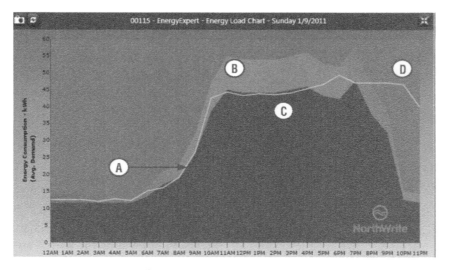

Figure 20-21. *Source: NorthWrite*

Future chiller testing was rescheduled for non-peak periods to avoid future demand charges. Figure 20-22.

Example 3: Anomaly Detection of Daily Peak Loads

A   A baseline (blue line) was constructed from three months of load data (thin line).

B   The baseline was used to determine projected loads subsequent to the first three months.

C   The actual load consistently exceeds the baseline after weekday 100. Energy anomalies are detected, and it is concluded that the building is faulting. Figure 20-23.

Example 4: Identifying After-Hours System Overrides

A   24-hr profiles of OAT (yellow line), and projected load (green band), and actual load are shown.

B   Actual load is color coded red when above, and olive when within the expected range.

C   Excessive after-hours use on Day 1 was traced to a cleaning crew HVAC override.

D   On Day 2 the crew was notified, and reduced the load some, but not enough.

E   On Days 3 and 4 the expected after-hours load increased due to a repeating monthly event.

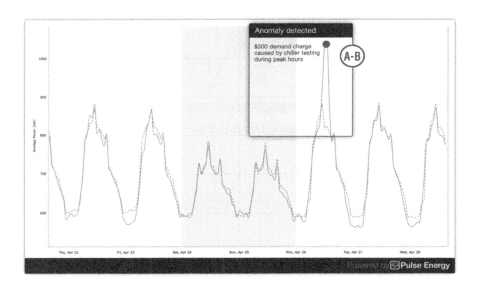

Figure 20-22. Source: Pulse Energy

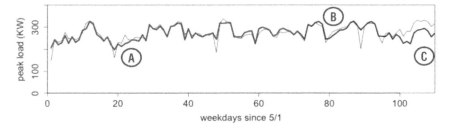

Figure 20-23. *Source: Price, P., Methods for analyzing electric load shape and its variability. LBNL#3713E, 2010.*

F   By Day 4, the BAS was programmed to limit HVAC override times, and no waste occurred. Figure 20-24.

CONCLUSION

Three advanced methods of whole building energy analysis have been presented in this chapter. The three methods were: Energy Savings; Cumulative Sum; and Anomaly Detection. Examples of each of these methods were also presented, which should help readers determine the potential applications and types of savings from these methods in their own buildings and facilities.

References

1.   Energy Information Handbook: Applications for Energy Efficient Building Operations prepared by Jessica Granderson, Mary Ann Piette, Ben Rosenbloom and Lily Hu, Lawrence Berkeley National Laboratory, LBNL-5272E, 2011.

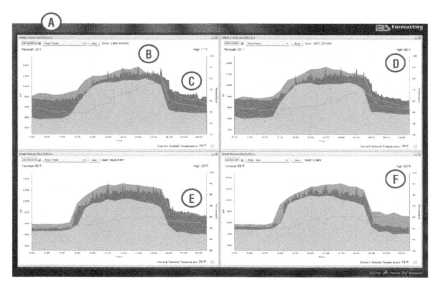

Figure 20-24. Source: Integrated Building Solutions

Chapter 21

Identification of Energy Efficiency Opportunities Through Building Data Analysis and Achieving Energy Savings through Improved Controls

Srinivas Katipamula
Danny Taasevigen
*Pacific Northwest National Laboratory**

Bill Koran
NorthWrite Inc., Lake Oswego, Oregon

INTRODUCTION

A number of studies have shown that commercial buildings in the United States (U.S.) waste as much as 15% to 30% of the energy they use (Katipamula and Brambley 2005). Analysis of whole-building utility data, sub-metered end-use data, and data from the building automation system (BAS) can help identify opportunities to improve building operations and efficiency, and ultimately reduce operating cost.

To ensure efficient energy use and to reduce the cost of electricity, natural gas, and steam, legislation has been published requiring advanced metering across the federal buildings sector in the United States. The Energy Policy Act of 2005 (EPAct) required installation of advanced electric meters by October 1, 2012. These meters must provide data at least daily and measure the consumption of electricity at least hourly. In addition, the Energy Independence and Security Act of 2007 (EISA) requires installation of advanced natural gas and steam meters by October 1, 2015 (US DOE 2011).

While electric meter data are being collected at many federal buildings across the U.S., the challenge is analyzing and identifying opportunities to improve operations and control, to achieve the legislative goals of reducing energy use and operating buildings in an efficient manner. There are commercial software tools available to analyze the data, but there are costs associated with obtaining the software and required training to effectively use the software.

This chapter will highlight analysis techniques to identify energy efficiency opportunities to improve operations and controls. A free tool, Energy Charting and Metrics (ECAM), will be used to assist in the analysis of whole-building, sub-metered data, and/or data from the BAS. All figures used in the chapter were generated in ECAM. Appendix A describes the features of ECAM in more depth, and also provides instructions for downloading ECAM and all resources pertaining to using ECAM.

ANALYZING INTERVAL METER DATA

In addition to legislation requiring installation of advanced electric meters by October 1, 2012 on federal buildings, many utilities in the U.S. and around the world are investing in advanced metering devices that can record electricity consumption by time of use, typically at 15-minute intervals*. These meters are generally referred to as "smart" meters because in addition to recording consumption by time of use, they also provide one- or two-way communication between the utility and the meter, and the data are commonly referred to as interval data. In some cases, the communication is in real time. In addition to the ability of the meter to communicate with the utility in real time, some meters also have the ability to communicate with the various devices in a home or a building. Although these new meters provide the capability for "smarter" building operations, the in-

---

Operated for the U.S. Department of Energy by Battelle Memorial Institute under contract DE-AC05-76RL01830

*Interval data provided by utilities are typically at 15-minute to 60-minute intervals.

formation that these meters provide is seldom used to improve or identify opportunities for improvement in buildings. The high resolution metered data that these meters collect has rich information that can be converted to actionable information either by fully automated processes or by systematic visual inspection of the data in a structured way.

In addition to the interval metered data from "smart" meters, most large commercial buildings (>100,000 sf) typically have BASs to manage operations and control various building systems. The data from the BAS can be leveraged to identify operational problems, and will be discussed further in this chapter. However, over 90% of commercial buildings lack BASs. For these buildings, analysis of whole-building interval data can be an effective way to identify inefficiencies and improve building operations, and ultimately reduce operating cost. In addition to the interval data, if outdoor-air temperature (OAT) is available, additional analysis of the data is possible (Taasevigen et al. 2011).

Interval Data Format

Interval data can provide useful and actionable information through careful review and analysis. In the U.S., as mentioned previously, interval data are typically collected at 15-minute intervals, but some utilities may collect at different intervals (30-minute or 60-minute). The interval data can also be collected using portable data loggers. The data that utilities provide can either be in a row or a column format, while most data loggers provide data in a column format. Figure 21-1 and Figure 21-2 show examples of two different formats of data provided by a utility.

Figure 21-1 shows the data organized in a row-format. Here, the date is generally provided in the first column and the electricity consumption information at each interval for the day is provided in subsequent columns.

Although this format is very convenient for storage, it is not convenient for analysis. It must be converted to a column format, as in Figure 21-2. In the column-format, the data are organized continuously with the date and time in one column and the consumption data in the next column. Note that some utilities and data loggers may provide the date and time in separate columns. Also, note that the consumption is typically reported in kWh for the time interval (i.e., kWh consumed during the 15-minute interval).

Interval Data Visualization

The key to analyzing interval meter data is understanding how to extract patterns that are hidden within large data sets. Time-series data are rich with hidden patterns. The challenge to extracting these hidden patterns requires basic understanding of how the particular building operates, as well as the ability to create relevant graphs from which patterns can be extracted. Figure 21-3 shows four different time series graphs of whole-building electricity consumption from the same building at different time intervals.

Figure 21-3(a) shows an entire year of time-series electricity consumption. Although there are some patterns that you can discern from the data, it is not an ideal plot for a number of reasons. In addition to seasonal variations that are evident, other patterns include: 1) the peaks and valleys have some season dependence, i.e. they change seasonally; 2) the minimum consumption for certain periods even during the same season appear to be different; and 3) the change in consumption from peaks to valleys is significant (from 50% to 80%). Figure 21-3(b) shows a monthly view of the same data, but for just 1 month within the data set. This view shows diurnal variation of consumption and also shows significant changes between weekday and weekend consumption. Figure 21-3(c) shows a weekly view of the data from the

	A	B	C	D	E	F	G	H	I	J	K
1		0:00	1:00	2:00	3:00	4:00	5:00	6:00	7:00	8:00	9:00
2	5/1/2008	521.64	527.16	532.68	532.68	527.16	549.24	596.16	676.2	659.64	643.08
3	5/2/2008	499.56	516.12	521.64	494.04	513.36	496.8	560.28	568.56	579.6	579.6
4	5/3/2008	557.52	538.2	524.4	521.64	532.68	582.36	582.36	604.44	612.72	604.44
5	5/4/2008	549.24	546.48	518.88	510.6	585.12	590.64	607.2	651.36	552	665.16
6	5/5/2008	546.48	549.24	527.16	524.4	491.28	510.6	552	609.96	604.44	604.44
7	5/6/2008	587.88	574.08	560.28	554.76	538.2	538.2	596.16	632.04	565.8	645.84
8	5/7/2008	609.96	585.12	571.32	552	560.28	549.24	598.92	643.08	563.04	604.44
9	5/8/2008	582.36	554.76	535.44	535.44	538.2	532.68	596.16	676.2	678.96	626.52
10	5/9/2008	637.56	604.44	596.16	604.44	618.24	587.88	656.88	678.96	670.68	662.4
11	5/10/2008	549.24	554.76	565.8	549.24	532.68	524.4	571.32	576.84	582.36	612.72

Figure 21-1: Example of interval meter data in a row-format.

⁄	A	D
1	Date and Time Stamp	kWh
2	12/1/2010 13:00	44.00999928
3	12/1/2010 14:00	46.30999947
4	12/1/2010 15:00	48.96000004
5	12/1/2010 16:00	44.25
6	12/1/2010 17:00	40.53999996
7	12/1/2010 18:00	19.0999999
8	12/1/2010 19:00	17.57999992
9	12/1/2010 20:00	17.84000015
10	12/1/2010 21:00	22.44999981
11	12/1/2010 22:00	19.94000006
12	12/1/2010 23:00	22.11000013
13	12/10/2010 0:00	17.37999964
14	12/10/2010 10:00	41.06999969
15	12/10/2010 11:00	38.81000042
16	12/10/2010 12:00	38.50000095

Figure 21-2: Example of interval meter data in a column-format.

same building. This provides even more information than the monthly view because it refines the diurnal variations and also shows that the pattern of consumption for day-to-day, at least for this week, is significantly different on weekends versus weekdays. Figure 21-3(d)

shows a 24-hour profile of the whole-building electric consumption from the same building. Although it provides an even more refined view of the daily profile, the consumption does not look reflective of the entire year of data. Therefore, it is very important not to make general conclusions on the consumption patterns by just looking at a 1-month, 1-week or 1-day profile. There are other ways to look at the profiles to make general conclusions on the building operations, which will be discussed in the following sections.

Weekday/Weekend/Holiday Load Profiles

Most commercial buildings, especially office buildings, have regular schedules and a distinct occupied and unoccupied period. Therefore, the consumption pattern during the occupied and unoccupied periods should also be distinctly different. To analyze and compare the consumption patterns during occupied and unoccupied periods, average load profiles for weekday, weekend, and holidays can be used. The load profiles show average consumption at each hour of the day (from midnight to midnight) by averaging the consumption data at each hour of the day over a number of days (e.g., entire year, summer period, or winter period).

Figure 21-4 shows an average load profile for weekdays, Saturdays, Sundays, and holidays for an of-

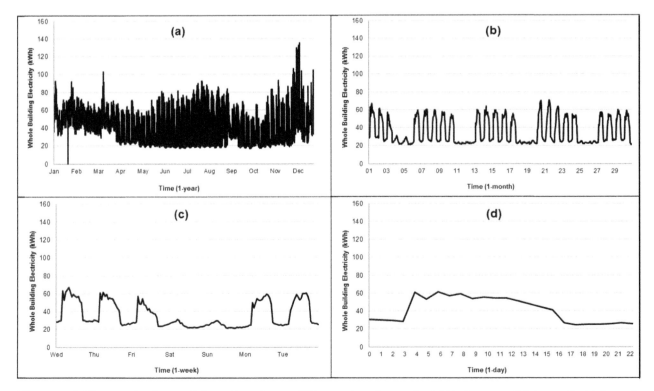

Figure 21-3: Time-series interval data plotted over different time intervals: (a) Entire year, (b) 1 month, (c) 1 week, and (d) 1 day of data.

fice building based on 1 year of interval data. There are a number of conclusions that can be drawn from visually inspecting this graph: 1) the building seems to be in the pre-occupancy mode between 4:00AM and 6:00AM on weekdays based on the pattern of consumption during that period, possibly turning on heating, ventilation and air-conditioning (HVAC) systems to bring the space temperatures to occupied values; 2) the building appears to be occupied sometime between 6:00AM and 8:00AM, when other building systems (lighting, plug loads, etc.) are turned ON; 3) the building seems to be unoccupied between 6:00PM and 4:00AM for weekdays, and completely unoccupied during the weekends; 4) the holiday consumption is significantly different from the weekday and weekend consumption patterns; 5) the night-time consumption (or base load) for the building is significant compared to day-time consumption (53%); and 6) the peak consumption occurs around 1:00 PM, which is right in the middle of the occupied hours. It would have been difficult to make the above conclusions by looking at the yearly, monthly, weekly and daily plots shown in Figure 21-3.

The profiles can be further refined to show average profiles for each day of the week over the data collection period (e.g., entire year, winter season, summer season, shoulder months). This is illustrated in Figure 21-5 for an entire year of electricity data.

For this building, it is clear that the average daily consumption profile does not vary much from Monday through Thursday (with the exception of Thursday mornings from midnight to 4 AM). There is a slight decrease in consumption during occupied hours on Fridays, and then the building appears to be unoccupied on Saturdays and Sundays, with the Sunday profile just slightly higher than the Saturday profile.

When looking at profiles to identify operational problems or opportunities to tighten schedules look at the following:

1. When does the consumption begin to increase in the morning (start-up activity)?

2. Does the consumption start to increase 1 or 2 hours before the building is occupied or does it increase 3 to 4 hours before the building is occupied? If the building consumption shows significant increases 3 to 4 hours before the building is occupied, the building could reduce energy consumption by tightening the morning start-up schedule. During peak summer or peak winter seasons, you may need to start

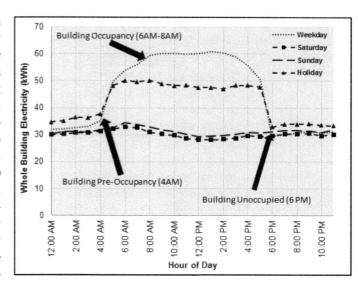

Figure 21-4: An example average weekday, Saturday, Sunday, and holiday profile for a commercial office building.

the HVAC systems 3 to 4 hours before occupancy; however, for most of the year, especially in mild climates, HVAC systems can be started 1 or 2 hours before occupancy. Some thermostats and BASs have optimal start/stop schedules. Alternately, a simple reset schedule based on outdoor-air temperature can be used to start and stop the HVAC systems to "optimize" the start time.

3. When does the building consumption start to decrease in the evening (shut-down activity)?

4. Does the consumption start to decrease 1 or 2 hours after the building is unoccupied or is there signif-

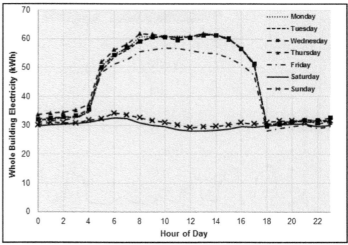

Figure 21-5: An example load profile by day of week for a commercial office building.

icant consumption 3 to 4 hours after the building is unoccupied? If there is significant consumption 3 to 4 hours after the building is unoccupied, the building could reduce energy consumption by tightening the shut-down schedule.

5. What is the base load consumption of the building (i.e., the nighttime consumption when the building is unoccupied)? This consumption should be fairly constant, but is it a significant fraction of the day time peak consumption (30% or higher)? If yes, it is likely that equipment is being left ON and can potentially be turned OFF to save energy. Although the base load consumption will be different for each building, typically for a well operated office building, the nighttime consumption should be around 20% of the daytime peak consumption. Figure 21-5 shows the morning base load consumption on Thursdays being higher than all other days of the week. This should be investigated to identify the reason for the higher base load consumption. Although opportunities exist year-round for base load reduction, this should be evaluated during shoulder months and not during peak heating or cooling seasons. During these times, it is possible that the HVAC equipment operates to maintain setback temperatures.

6. Is there significant consumption during the weekends? If the building is not typically occupied during weekends (like the building depicted in Figure 21-5), the weekend consumption should be close to the nighttime constant load. If the building is partially occupied on either Saturday or Sunday, the consumption during the partial occupancy may be slightly higher than the base load.

Seasonal Variation of Profiles

As a result of the potential occupancy patterns and/or operational changes that can occur during different seasons, it is useful to plot the load profiles by summer, winter, and shoulder periods to identify opportunities to tighten schedules and look for potential operational problems. Figure 21-6 shows the average load profiles for each month in the data set. As seen from the graph, the average load profiles vary significantly from month to month. In this case, it is useful to separate the graphs into seasons to look for trends and opportunities to improve building controls and/or tighten schedules. Figure 21-7 shows only December through March (winter season), Figure 21-8 shows only June through

September (summer season), and Figure 21-9 shows the shoulder months (November, December, March, and April). Note that the seasons will vary depending on building type and location, and this is only an example for one specific commercial office building. Once separated into the seasons, some trends can be spotted, and potential opportunities can be identified. When looking at seasonal profiles to identify operational problems or opportunities to tighten schedules look at the following:

1. Does the consumption pattern change from month to month, or is it relatively stable? In Figure 21-7, it is very close for January through March, but very different for December. During the summer months in Figure 21-8, however, the consumption pattern is fairly consistent from month to month. This indicates that, for this commercial office building, that the cooling (summer) operations are much more consistent than the winter (heating) operations.

2. Is the base load a significant percentage of the peak load from month to month? In Figure 21-7, during the winter months, the base load and the peak load are very similar. This could be the result of very cold weather, requiring heating in the building during all hours of the day, or changes in occupancy patterns. If this occurs, and occupancy has not changed, it may be possible to reduce the nighttime load by investigating the temperature setback controls during nighttime operations (if any). Also check to see if equipment can be shut down at night when it is not needed (i.e., shutting down lights, chillers, boilers, etc.).

3 Are there significant shoulder months at the building location (i.e., do the outdoor temperatures change significantly from summer and winter temperatures between summer and winter seasons)? If so, then there will be a month or two at the beginning and end of the winter and/or summer in which the temperature will change slowly from day-to-day but drastically over the month (Figure 21-9). There may also be several days during shoulder months where the building is operating in "floating" mode (i.e., no heating or cooling is required, so the equipment fans run to ventilate the building with fresh air and use economize cooling as needed). Shoulder months are difficult to target for energy savings because conditions change rapidly. Therefore, months during which only one season is present should be targeted first for energy savings measures.

4. When does the peak demand occur? Is it consistent during occupied hours, or does it peak at certain hours of the day? During winter months, when outdoor conditions are the coldest in the mornings, peak energy/electricity demand will occur in the morning (as seen in Figure 21-7). During summer months, when outdoor conditions are the warmest in the early-to-late afternoons, peak demand will occur in the afternoons (as seen in Figure 21-8). If the demand is consistently high and peaking across all occupied hours, there may be opportunities to improve control strategies (i.e., economizer control, static pressure reset, discharge-air temperature reset, etc.).

To verify findings from seasonal graphs, the load profiles during these seasons can be plotted as box plots to identify the average profile during the time period, and also to identify any outliers. Figure 21-10 shows the winter season data plotted as box plots for weekdays and weekends, Figure 21-11 shows the summer season data plotted as box plots for weekdays and weekends, and Figure 21-12 shows the shoulder months plotted as box plots for weekdays and weekends, respectively.

Comparing the box plot graphs during the winter months (Figure 21-10) and the summer months (Figure 21-11), the look at the following:

1. Is the building operated "tighter" (less variation at each hour) during summer than the winter, or vice versa? If so, then what can be done during the winter/summer to tighten up operation? Can heating/cooling setpoints and controls be tightened to decrease the variability in consumption from 1 hour/day/month to the next?

2. Is the occupancy period more apparent during the summer than the winter, or vice versa? And if so, why? Do occupancy patterns change? Do systems have to stay on because of very cold/hot outdoor conditions?

3. In Figures 21-10 and 21-11, the base load increases from roughly 20 kWh during the summer to roughly 40 kWh during the win-

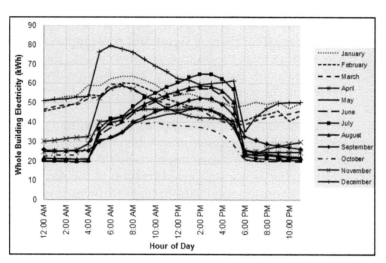

Figure 21-6: Example load profile by month for a commercial office building.

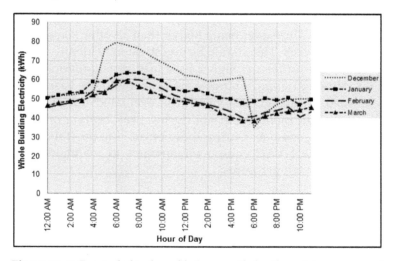

Figure 21-7: Example load profile by month for the winter season of a commercial office building.

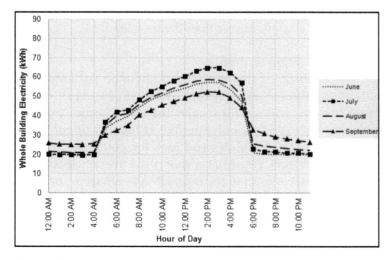

Figure 21-8: Example load profile by month for the summer season of a commercial office building.

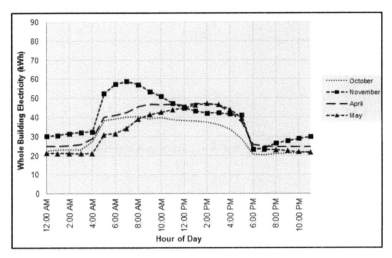

Figure 21-9: Example load profile by month for the shoulder months of a commercial office building.

ter. Why is this occurring? Can some plug loads and equipment be shut down in the winter to mitigate this excess energy consumption during unoccupied hours?

In Figures 10 and 11, the peak demand during the different seasons is close to the same at 60 kWh. The time of day for this peak demand shifts during the seasons, however, with it occurring around 7:00 AM during the winter and 2:00 PM during the summer. This would indicate the dependence on weather on the load inside the building, with much colder outdoor conditions in the morning during the winter months and much hotter outdoor conditions in the afternoon during the summer months. Climate can play a major role in peak demand during summer/winter seasons, along with occupancy patterns.

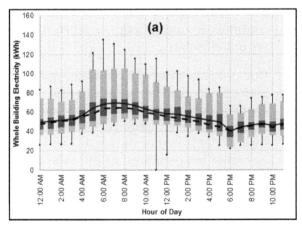

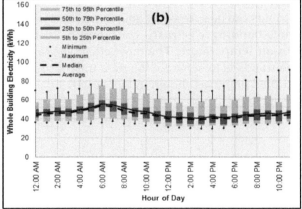

Figure 21-10: Example load profile as box plots for the winter season for (a) weekdays and (b) weekends.

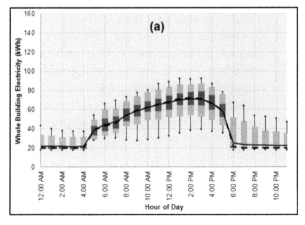

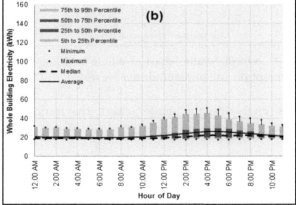

Figure 21-11: Example load profile as box plots for the summer season for (a) weekdays and (b) weekends.

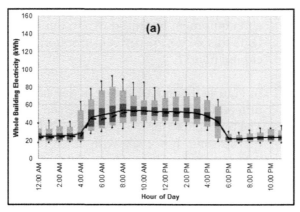
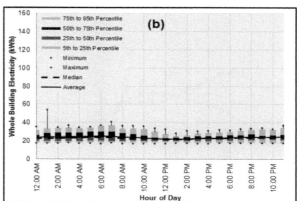

Figure 21-12: Example load profile as box plots for the shoulder months for (a) weekdays and (b) weekends.

Variation of Electricity Consumption with Outdoor-air Temperature

If the outdoor-air temperature (OAT) is available, in addition to the electricity consumption, for the building, other trends can be spotted by analyzing charts such as the whole-building electricity consumption versus the outdoor-air temperature. The most effective way to analyze the trends is to separate the data into occupied and unoccupied hours, as shown in Figure 21-13.

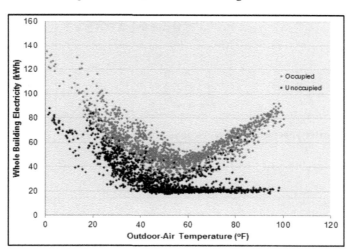

Figure 21-13: Example scatter chart of whole-building electricity consumption and outdoor-air temperature.

When analyzing the scatter chart, the trends to look for include:

1. Is the consumption much lower during unoccupied periods for all range of temperatures? If not, this would indicate that the equipment schedule and the occupancy schedule could be different from one another, and they should be checked for

consistency. If the occupancy schedule indicates occupancy starting at 8:00 AM, but the equipment starts up at 4:00 AM, there will be 4 hours every day where the unoccupied hours can potentially consume as much energy as when the occupants are in the building.

2. Does the cold weather have more impact on energy performance than hot weather, or vice versa? There are two ways to answer this question: 1) Is the consumption during occupied hours (gray points in Figure 21-13) higher during lower temperatures, or higher temperatures, or 2) Is the consumption during unoccupied hours (black points in Figure 21-13) higher during lower temperatures, or higher temperatures? In Figure 21-13, the unoccupied points are much higher during colder weather, but the occupied points are about the same for cold and hot weather. This indicates that, for this commercial office building, cold weather has a greater impact on energy performance than hot weather. If it is discovered that the electricity consumption in one season is greater than the other, the building managers should focus their efforts on building operation during the high consumption season to help reduce the energy use. Things to investigate include insulation, seals on windows and doors, broken economizers, control setpoints in summer versus winter seasons, etc.).

Analyzing Economizer Operation with Interval Meter Data

In some cases, the whole-building electricity consumption data can also be used to diagnose economizer operations. If the whole-building data includes substantial refrigeration consumption in addition to the HVAC, the data may be blurred, thus making it more difficult to find trends. In addition, if there are rooftop units (RTUs) in the building, and the majority of them are working properly while a few are having issues, identifying economizer problems with whole-building data may be difficult.

When analyzing whole-building electricity consumption to identify economizer problems, the evaluations to focus on are:

1. Economizer temperature range and lockout temperature

2. The presence of mechanical cooling at outdoor-air temperatures colder than the design supply-air temperature (if the RTUs are heat pumps or if the RTUs use electric reheat, there could be significant electricity consumption at low temperatures)

3. Inconsistent controls (indicated by a high degree of scatter in the plots)

4. Poor outdoor-air temperature sensor location (i.e., solar impacts on sensor) and

5. Efficiency of mechanical cooling at low loads.

Economizer Temperature Range and Lockout Temperature

To identify the economizer temperature range and lockout temperature, a simple (physics-based) model of whole-building electricity consumption during occupied periods can be generated including an energy balance for a dry-bulb economizer with a lockout temperature at 70°F (as appropriate for many climate zones in the western United States). A plot of the power consumption as a function of the outdoor-air temperature can be seen in Figure 21-14. Note that this model only includes hours when the building is occupied.

In the model, assuming the 100% cooling needs will be met by the outdoor air, mechanical cooling is not needed below 55°F. Because there is no electrical heating, the demand below 55°F is constant. Between 55°F and 70°F, the economizer is assisting the mechanical cooling (assuming an integrated economizer, i.e., the economizer can run at the same time mechanical cooling is being provided). Above 70°F, the economizer is locked out, and the building is cooled by mechanical means. Now, consider the

effects on the model if the economizer were locked out at 60°F, as shown in Figure 21-15. The consumption is now higher in the range of 60°F to 70°F, where the economizer operation is not able to meet the entire cooling load.

Now, to compare the theoretical model to actual building data, consider data from a commercial office building with built up air-handling units and chillers, as seen in Figure 21-16. After fitting a model to the raw data, and adjusting the physical model to match the same loads and economizer lockout temperature, the similarities are apparent, and shown in Figure 21-17.

To extend the analysis more generally, the same can be applied to data for a small building served by a packaged variable-air-volume (VAV) rooftop unit (RTU) in Figure 21-18. Again, these data are only for occupied hours.

Note the higher degree of scatter in the economizer region, and the gap between modes of operation with and without mechanical cooling (in the temperatures between roughly 50°F and 60°F, and about 150 kW). There appears to be mechanical cooling down to 50°F and colder for some points. Figure 21-19 shows data for the same building after some economizer control strategy changes, with regression applied to each mode of operation (i.e., 100% economizing, integrated economizing and mechanical cooling, and 100% mechanical cooling). Note the greater continuity between the modes of operation with and without mechanical cooling compared to that in Figure 21-18. Also note that there is now no mechanical cooling below 55°F.

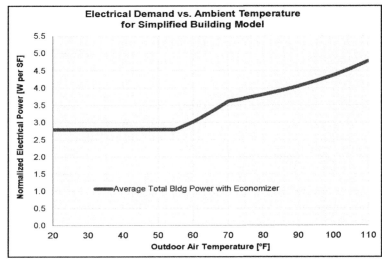

Figure 21-14: Example model of normalized whole-building consumption versus outdoor-air temperature for occupied hours, to show dry-bulb economizer impacts.

Another point to notice is the percentage change in demand from the point with no mechanical cooling (~55°F), to the point with mechanical cooling and no economizer (~70°F) in Figure 21-19 and Figure 21-17(b). In Figure 21-19, the load changes from roughly 116 kW at 57°F to 162 kW at 67°F, a 40% increase. As shown in Figure 21-17(b), the load changes from roughly 727 kW at 55°F to 918 kW at 66°F, a 26% increase. This could be the result of poor turndown on the mechanical cooling, causing significant power at lower loads, or other factors.

When analyzing economizer operation from whole-building data, situations where economizers are most beneficial are the best for analysis, due to the greater need for economizer cooling, and the reduction in power consumption from the use of the economizer. Therefore, the impacts show up the most on the plot of the power versus outdoor-air temperature. These situations include:

a.   Units with relatively low minimum outdoor-air requirements
b.   Units with capacity for 100% outdoor air
c.   Relatively dry climates
d.   Units serving interior zones, and
e.   Units serving zones with high internal gains.

If the impact does not show up, or is less than expected, there may be opportunities for improvement. To finish the discussion from the list of things to evaluate at the beginning of this section:

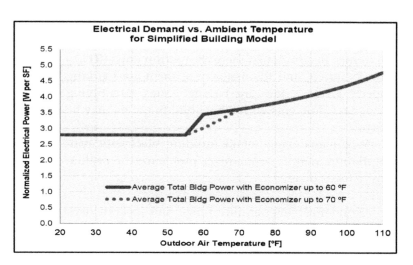

Figure 21-15: Example model of normalized whole-building consumption versus outdoor-air temperature for occupied hours, to show dry-bulb economizer impacts with different lockout temperatures.

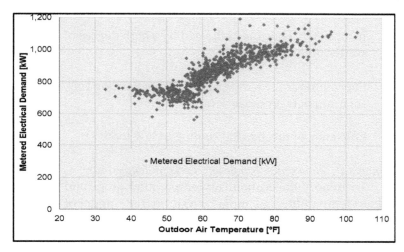

Figure 21-16: Example whole-building electricity data plotted against outdoor-air temperature from a commercial office building.

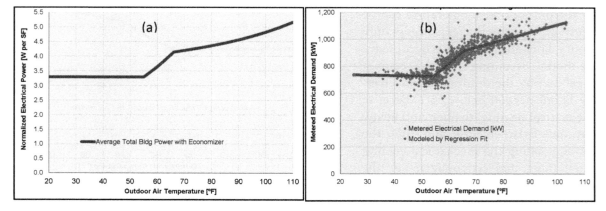

Figure 21-17: Example models for (a) theoretical commercial office building and (b) actual commercial office building.

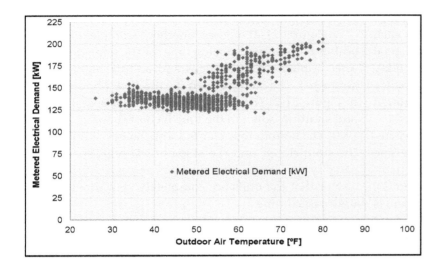

Figure 21-18: Example whole-building electricity consumption data versus outdoor-air temperature for a municipal building served by a VAV RTU.

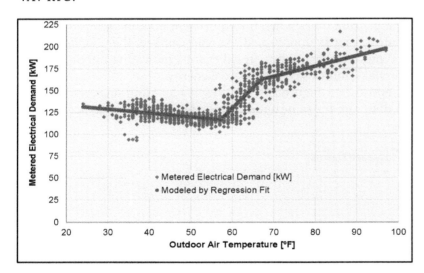

Figure 21-19: Example whole-building electricity consumption data versus outdoor-air temperature for a municipal building served by a VAV RTU after changing some economizer control strategies.

The Presence of Mechanical Cooling at Outdoor-air Temperatures Colder Than the Design Supply-air Temperature

To identify if mechanical cooling occurs if outdoor-air temperatures colder than the supply-air temperature, look to see if the data shows a positive slope at temperatures colder than the design supply-air temperature. If this is occurring, it is an indication that mechanical cooling is active, and improved economizer control and operation should eliminate this.

Inconsistent Controls (Indicated by a High Degree of Scatter in the Plots)

Inconsistent controls can be spotted in the data by the amount of scatter in the data set. If there is significant scatter, this could indicate that the controls are inconsistent.

Poor Outdoor-air Temperature Sensor Location (i.e., Solar Impacts on Sensor)

Poor outdoor-air temperature sensor location can indicate a narrow economizer range, or apparent low lockout temperature. Therefore, before changing any control strategies or setpoints, it is a good idea to verify that the outdoor-air temperature sensor is mounted in a location where solar gains are not affecting the temperature measurement.

Efficiency of Mechanical Cooling at Low Loads

Efficiency of mechanical cooling at low loads can be analyzed by looking at the "height" of the economizer range of the data set (i.e., the power consumption value when the outdoor-air temperatures are favorable for 100% economizing). If this range is high, it may be an indicator of poor HVAC turndown capability. Conversely, a very low height could indicate high outdoor-air quantities during mechanical cooling (so that they don't increase much with economizer operation), or good HVAC turndown but no increasing fan speed as temperatures drop to take full advantage of the economizer.

Evaluating Night Setback or HVAC Shutdown with Interval Meter Data

Night setback adjusts the thermostat setting, or zone temperature setpoint, higher for cooling and lower for heating. Evaluating the presence of night setback is determined by answering the question: Is there heating or cooling in the time period shortly after the end of occupancy?

Just after occupancy ends, the zone temperature(s) will be within the setback heating or cooling setpoints. The zone will be considered satisfied, and neither heating nor cooling will be required. The zone will remain satisfied until heat gains or losses affect the zone temperature sufficiently for the zone to reach the setback temperature. Depending on the setback temperature, the rate of heat gains or losses, and zone heat capacitance, this may take several hours. Figure 21-20 shows

24 scatter charts of whole-building electricity consumption versus outdoor-air temperature, one chart for each hour of the day. The data are for a retail commercial building that is open from 9:00AM to 9:00 PM, Monday through Saturday. The building has reduced occupancy on Sunday, so the charts only reflect Monday through Saturday operation.

The last full hour of occupancy is hour 20 (corresponding to 9:00 PM). Note that the slope of the line on the right side of the plot, which indicates mechanical cooling. The charts indicate that even after hour 20, and up until hour 9 (corresponding to 9:00 AM), there is still mechanical cooling being provided, and there is not much difference in the charts between hour 22 and hour 7. Therefore, it can be concluded that there is no setback for the building, and that some units run all night. The shift in power consumption, however, during unoccupied hours indicates that some units may be turning off, or that lights and other equipment are turning off. In this case, there are multiple single-zone RTUs heating and cooling this building. Further investigation indicated that the thermostats for each RTU were residential-type and were running on battery power, and that a few units had lost power. Some had not been programmed and called for heating and cooling to be available all day

with constant setpoints.

Figure 21-21 shows another example of hourly scatter plots around closing time for the same retail store after implementing changes and adding night setback strategies. The store is open for the first 3 hours shown in Figure 21-21 (hour 18 to hour 20), and begins closing and shutting down in the fourth chart (hour 21, or 10:00 PM). The bottom four charts show the next 4 hours after the store closes, and the slope of the regression line is roughly zero, indicating the site has good setback strategy or that the units are completely shut down during unoccupied times.

Scatter plots of whole-building electricity consumption versus outdoor-air temperature for the entire data set can also indicate the use of night setback if analyzed for occupied and unoccupied hours. Figure 21-22 shows 1 year of data for a small commercial office building that did not utilize night setback. The occupancy schedule was verified, so all unoccupied (black) points are times when the equipment is running but the building is unoccupied. As you can see, there is significant scatter of unoccupied points, indicating opportunity to improve control to save energy.

Figure 21-23 represents an additional year of data for the same building after night setback was implement-

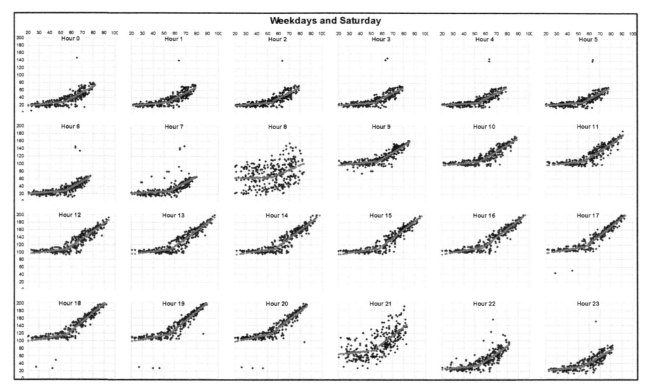

Figure 21-20: Example scatter charts (one per hour of operation) of whole-building electricity consumption versus outdoor-air temperature for a commercial retail space.

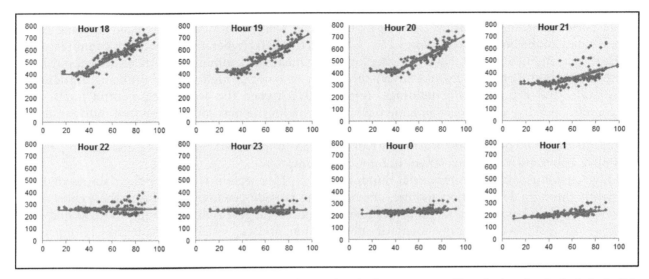

Figure 21-21: Example scatter charts (one per hour of operation for hours surrounding closing time) of whole-building electricity consumption versus outdoor-air temperature for a commercial retail space, after implementing night setback strategies.

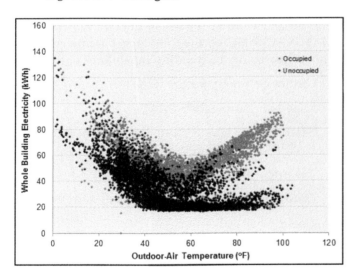

Figure 21-22: Scatter chart of whole-building electricity consumption versus outdoor-air temperature for a small commercial office building without implementing night setback.

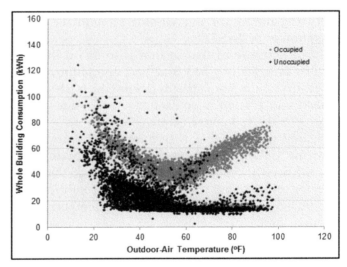

Figure 21-23: Scatter chart of whole-building electricity consumption versus outdoor-air temperature for a small commercial office building after implementing night setback.

ed (as a control improvement for the building). As you can see, the frequency of times black points are above the base load consumption (especially during summer operation) is minimal. As it gets very warm (OAT>80°F), you start to see more scatter in unoccupied hours because of setback limits being reached quicker. In the winter, however, you see more scatter, but in comparison to Figure 21-22, this is drastically improved. For this example building, colder temperatures seem to impact the building more from an energy consumption perspective, so the building owner may want to target building and/or equipment improvements during winter operation as

a starting point, and then start to identify opportunities during summer operation.

ANALYZING BUILDING AUTOMATION SYSTEM DATA

Large commercial buildings (>100,000 sf) typically have a BAS. These buildings include, but are not limited to, office buildings, malls, schools, etc. However, the analysis that follows can be applied to any type and size of facility that has a BAS. While managing various building systems, the automation system ensures the operational performance of the facility as well as the comfort and safety of the building occupants.

While the capabilities of BASs have increased over time, many buildings still do not fully take advantage of these capabilities, and as a result, the buildings are not properly commissioned, operated or maintained, which may lead to increased energy use, reduced lifetime of equipment, and inefficient operations. The Pacific Northwest National Laboratory has developed training to help building operators manage commercial buildings more efficiently through identification and implementation of low-cost and no-cost operational improvements. Most of these low-cost/no-cost improvements are done by making adjustments to controls in the BAS.

The focus on this section will be how to analyze the data that is trended and stored within the BAS, and strategies to start identifying low-cost/no-cost operational improvements.

Analyzing and Improving Air-side Economizer Operation with BAS Data

The purpose of this section is to show, through use of examples of good and bad operation of processed BAS data, how air-side economizers should be utilized and efficiently controlled. An air-side economizer is a duct/damper arrangement in the air-handling unit (AHU) along with automatic controls that allow an AHU to use outdoor air to reduce or eliminate the need for mechanical cooling. When there is a need for cooling, and if the outdoor-air conditions are favorable for economizing (outdoor-air temperature is less than return-air temperature), unconditioned outdoor air can be used to meet all of the cooling energy needs or supplement mechanical cooling. In a properly configured economizer control sequence, the outdoor, return and exhaust-air dampers sequence together to mix and balance the air-flow streams to meet the AHU discharge-air temperature setpoint.

In humid climates, the use of dry-bulb temperature based economizers is not recommended. However, if used, the outdoor-air temperature should be 5°F to 10°F lower than the return-air temperature. There are times when economizing should not be used. This includes during building warm-up periods and cool-down periods, when the outdoor conditions are not favorable for economizing, or during unoccupied periods, when the supply fan is operating.

When an economizer is not controlled correctly, it may go unnoticed because mechanical cooling will compensate to maintain the discharge air at the desired discharge-air setpoint. This may include periods of time where too much outdoor air is being introduced to the AHU (when the economizer control is attempting to maintain a minimum outdoor-air setpoint), or when there is not enough outdoor air being introduced to the AHU (when the economizer control is attempting to bring in the maximum amount of outdoor air). Failure to correct/mitigate this situation, in all likelihood, will lead to increased fan, cooling and heating energy consumption.

This section will cover the data needed to verify the air-side economizer control, the trends to look for and investigate (with examples of good and bad operation, and suggested actions to take to improve operations), and a discussion of various economizer sequencing options and their limitations.

Data Needed from the BAS to Verify the Air-side Economizer Control

To analyze and detect deficiencies in economizer operation and control, for single-duct variable-air-volume (SDVAV) AHU(s), the following parameters must be monitored using the trending capabilities of the BAS:

- Outdoor-air temperature (OAT)
- Mixed-air temperature (MAT)
- Discharge-air temperature and setpoint (DAT and DATSP)
- Return-air temperature (RAT)
- Outdoor-air damper position signal (OAD)
- Cooling-coil-valve signal (CCV)
- Outdoor-air fraction (OAF)

In most BASs, the outdoor-air fraction is probably not computed and trended. If OAF is not recorded in the BAS, it can be computed externally using the outdoor-, return- and mixed-air temperatures:

$$OAF = [(MAT - RAT)/(OAT - RAT)]$$

The recommended frequency of data collection is between 5 and 30 minutes.

Trends to Look for When Analyzing the Data

1. How close is the outdoor-air fraction compared to the outdoor-air damper position signal?

2. Is the minimum outdoor-air damper position reasonable (between 10% and 20%)?

3. Is the outdoor-air damper open above its minimum occupied position when outdoor conditions are not favorable (outdoor-air temperature > return-air temperature)?

4. Is the outdoor-air damper closed or at minimum position when outdoor conditions are favorable for economizing and the AHU is in cooling mode?

5. Does the cooling coil operate during economizer mode?

6. Does the cooling coil operate when the outdoor-air temperature is lower than the discharge-air temperature setpoint?

7. When the cooling coil is open, is the outdoor-air damper fully open, if the conditions are favorable for economizing?

8. Do outdoor-air dampers close to minimum position for freeze protection?

9. Are the mixed-air temperatures between the outdoor-air and return-air temperatures?
   a. When conditions are not favorable for economizing, is the mixed-air temperature closer to return-air or outdoor-air?
   b. When conditions are favorable for economizing, is the mixed-air temperature closer to outdoor-air or return-air?

1. How close is the outdoor-air fraction compared to the outdoor-air damper position signal?

In most cases, there will be significant differences between the OAF and OAD, especially at low OAD values. Therefore, using OAD values can be misleading because it does not reflect the true outdoor-air intake. The calculated OAF should only be used to investigate the true percent of outdoor-air entering the AHU when the outdoor-air temperature is significantly ($\pm$ 5°F) different than the return-air temperature. As these two temperatures get close to each other, the OAF calculation may not be accurate. Review the plot of outdoor-, return-, and mixed-air temperatures, outdoor-air damper position signal, and outdoor-air fraction vs. time to determine the right time to compare the calculated OAF to the outdoor-air damper position signal. Figure 21-24 shows this plot for 1 day.

From Figure 21-24, the best time to compare the OAF to the OAD is between 10:00 AM and 4:00 PM. You can see that the OAF covers the OAD quite accurately at about 20% outdoor air. This is good operation for this case because the AHU is bringing in the minimum amount of outdoor air when the OAT is greater than the RAT. However, if the calculated OAF deviates from the actual outdoor-air damper position by more than 50%, it may be an opportunity to alert someone to investigate the economizer system and its working components.

Suggested Actions

If the calculated OAF isn't automatically calculated in the BAS, it should be added and tracked for all AHU(s).

2. Is the minimum outdoor-air damper position signal reasonable (between 10% and 20%)?

To meet the ventilation requirements, the AHU must provide a certain amount of fresh air when the building is occupied. The ventilation requirements are

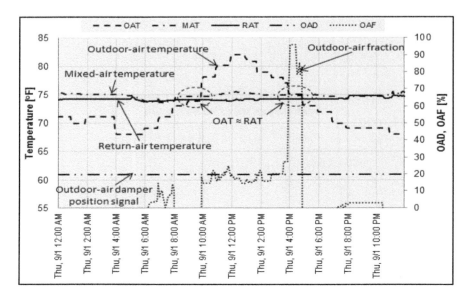

Figure 21-24: Determining the accuracy of outdoor-air damper position signal by comparing it to the outdoor-air fraction during times when the outdoor-air temperature is significantly different than the return-air temperature. In this case, the OAF calculation becomes unrealistic around 10:00 AM and 4:00 PM.

determined at the design stage based on the zone occupancy and other parameters. The ventilation requirements are then translated into the outdoor-air damper opening, which is generally referred to as the minimum damper position to meet ventilation requirements. Typically, the minimum damper position is between 10% and 20%. Check the minimum damper position by reviewing the plot of outdoor-air temperature and outdoor-air damper position signal versus time. Figure 21-25 shows these trends for a 2-day period in which the outdoor-air temperature varies between 65°F and 80°F. Once the OAF has been added to the BAS, it should be added to the graph to provide additional insights. You can see that the OAF matches the OAD for a period of time at 30%, indicating that 30% outdoor-air is being introduced into the AHU. During this period, the OAT is greater than 70°F, so the AHU should not be in economizing mode. Additionally, the OAD is always 30% open, indicating the possibility of 30% outdoor air entering the AHU at all times. If the building is not occupied 24/7, the outdoor-air dampers should be closed during unoccupied times and building warm-up or cool-down periods.

Suggested Actions

The economizer should always operate during favorable conditions (outdoor-air temperatures between 40°F and 60°F in humid zones and whenever the outdoor-air temperature is lower than the return-air temperature in dry climates) to supplement or eliminate the use of mechanical cooling. Outside of this temperature range, the economizer control should set the outdoor-air damper to the minimum position required to satisfy ventilation requirements. The outdoor-air damper should be locked out to the minimum position for freeze prevention when the outdoor-air temperature is below 40°F, and the outdoor-air dampers should be completely closed if the mixed-air temperature becomes lower than 40°F. The building operator should set up alarms in the BAS to monitor these extreme conditions.

If the minimum damper position is greater than 20%, more outdoor air may be entering the building than required, and significant additional heating or cooling may be occurring at the AHU to maintain the discharge-air temperature setpoint. Check ventilation requirements for the spaces served to ensure that the required minimum

damper position is properly configured.

Also, check to make sure the damper is fully closed during unoccupied hours and during the building warm-up/cool-down periods (unless the outdoor-air temperature is conducive to economizer operations during the cool-down period). The damper should open 30-minutes prior to occupancy to flush the building. In addition to checking the damper signal, the outdoor-air fraction should always indicate less than 10% outdoor air during unoccupied hours and building warm-up/cool-down periods (unless the outdoor-air temperature is conducive to economizing).

If it is greater than 10% and the outdoor-damper signal indicates that it is closed, this could be a sign of poor seals or a stuck damper.

3. Is the outdoor-air damper open above its minimum occupied position when outdoor conditions are not favorable (outdoor-air temperature > return-air temperature)?

If the outdoor-air damper is open above the minimum outdoor-air damper setpoint to satisfy the ventilation requirements, at temperatures considered unfavorable for economizing, this can be an indicator of poor economizer control. Figure 21-26 shows the graph of the outdoor-air damper position signal, outdoor-air temperature and return-air temperature versus time for a 2-day period. The outdoor-air temperature varies between 60°F and 85°F. From the graph, the outdoor-air damper opens up to 100% at 6:00 AM both days, and stays open at 100% until 11:00 PM. During this time, the outdoor-air temperature increases to over 80°F (which

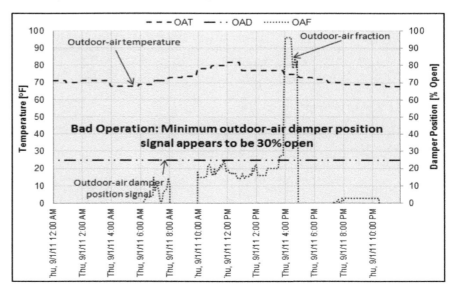

Figure 21-25: Minimum outdoor-air damper position signal too high (30%).

is greater than the return air) for the majority of the afternoon. This could be an indicator that the economizer controls have failed, or that the damper is overridden to be fully open during the occupied period. To understand the impacts of a damper open 100% during high outdoor-air temperature conditions, review the plot of outdoor-air damper position signal and cooling coil valve signal versus time. Figure 21-27 shows the same 2-day period, and you can see that as the outdoor-air damper goes to 100% open, the cooling-coil-valve opens between 40% and 60% to maintain the discharge-air

temperature setpoint. If the outdoor-air damper position signal closed to the minimum when the outdoor-air temperature became greater than the return-air temperature, the cooling-coil would only need to run to maintain the load and the small amount of outdoor air needed to satisfy ventilation requirements.

<u>Suggested Actions</u>

If the outdoor-air damper position signal is greater than the minimum outdoor-air damper position signal specified in the control logic during periods when the out-

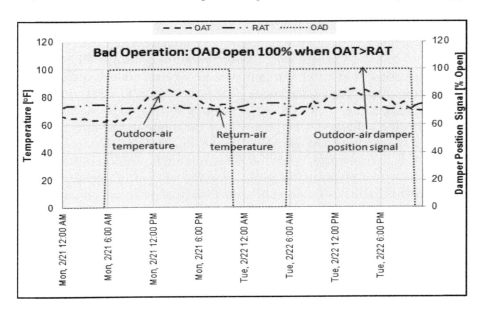

Figure 21-26: Outdoor-air damper letting in too much outdoor-air when conditions are not favorable for economizing.

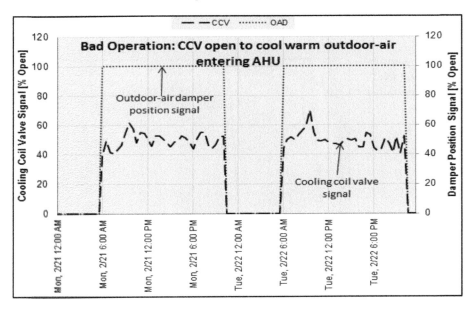

Figure 21-27: Comparison of cooling coil valve and outdoor-air damper position signals.

door conditions are not favorable for economizing, check to make sure that the outdoor-air damper is commanded to the minimum position. If the outdoor-air damper position is set to the minimum, but the OAF is 100%, check to see if something is obstructing the outdoor-air damper from closing or if the damper actuator has failed. Remove anything obstructing the outdoor-air damper, or have any broken parts replaced as soon as possible. When finished, command the outdoor-air damper fully closed, fully open, and to the minimum position, and make sure the outdoor-air damper is responding correctly at each position specified in the control sequences by examining the effect on the calculated OAF, or manually inspect the damper if the OAF is not calculated.

4. Is the outdoor-air damper closed or at minimum position when outdoor conditions are favorable for economizing and the AHU is in cooling mode?

For most climates, when the outdoor-air temperature is less than the return-air temperature and the AHU is in cooling mode, it is favorable to economize. When the outdoor conditions are favorable to economize, the outdoor-air damper should always open first for cooling before mechanical cooling is activated. If the mixed-air temperature is higher than the discharge-air temperature setpoint with the outdoor-air damper fully open, mechanical cooling can be used to supplement free cooling. Figure 21-28 shows the plot of the outdoor-air damper position signal and outdoor-air temperature versus time. This plot is for a Friday, where the outdoor-air temperature varies between 38°F and 50°F. The outdoor conditions are favorable for economizing, and it appears that the outdoor-air damper position signal is only open to the minimum of

15%. This cannot be classified as bad operation, however, until further investigation is done. It is possible that the cooling load is being fully satisfied by the outdoor air with the damper at 15% open.

To determine if the economizer is working properly in this situation, review the chart of outdoor-, return-, mixed-, and discharge-air temperatures versus time (Figure 21-29). For the Friday highlighted in Figure 21-28, the mixed-air temperature should nearly equal the discharge-air temperature. If the mixed-air temperature is higher than the discharge-air temperature, additional (mechanical) cooling is occurring at the AHU to meet the discharge-air temperature setpoint.

The chart in Figure 21-29 reveals that the mixed-air temperature is always 3°F to 5°F warmer than the discharge-air temperature, indicating that mechanical cooling is occurring to meet the discharge-air temperature setpoint for this AHU. Because the outdoor-air damper is at the minimum position, energy is being wasted. It is possible that the outdoor-air damper is overridden to the minimum position in the BAS, or that there is poor economizer control operation. If this economizer were working properly, the outdoor-air damper would modulate accordingly to ensure that the discharge-air temperature is meeting setpoint. Then, if full economizing were required (outdoor-air damper 100% open), and the discharge-air temperature was still warmer than the discharge-air temperature setpoint, mechanical cooling would be enabled.

Suggested Actions

If the outdoor-air damper is closed or set to the minimum position when the outdoor conditions are fa-

Figure 21-28: Outdoor-air damper position signal at minimum position during favorable economizing conditions.

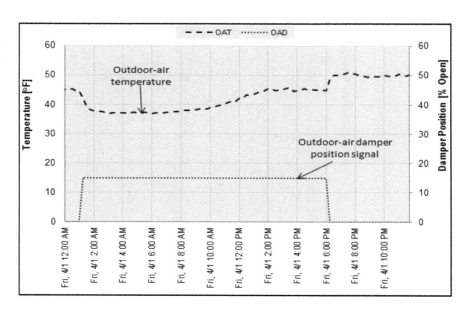

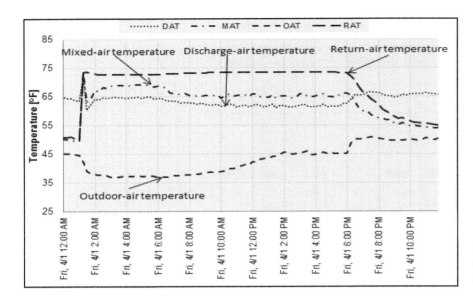

Figure 21-29: Mixed-air temperature is not cool enough to meet discharge setpoint from economizing only.

vorable for economizing, check to see if the mixed-air temperature is greater than or equal to the discharge-air temperature in the BAS or review the plot of mixed-, discharge-, return-, and outdoor-air temperatures versus. time. If the mixed-air temperature is warmer, check to see if the outdoor-air damper is overridden to the minimum position or if the actuator is broken. If the outdoor-air damper is overridden, release the override to allow the outdoor-air damper to operate properly. If broken, fix or replace any necessary parts for the outdoor-air damper system and test the modulating ability by commanding the outdoor-air damper fully open, fully closed, and to the minimum position. Make sure the outdoor-air damper is responding correctly at each position specified in the control sequences. If the outdoor-air damper was already working fine, check the controls to make sure that economizing is allowed, and that it is properly implemented.

5. Does the cooling coil operate during economizer mode?

A cooling coil operating during economizer mode can mean two things: the first is that the economizer is not capable of satisfying the load, so mechanical cooling is required to supplement the free cooling; the second is that the economizer is not functioning correctly. If the outdoor-air damper was stuck open during weather unsuitable for economizing (OAT>RAT), the cooling coil must work harder to maintain the discharge-air temperature setpoint. If the outdoor-air damper is only partially open when outdoor conditions are favorable for economizing, yet the mixed-air temperature is still warmer than the discharge-air temperature, then mechanical

cooling will be required to meet the discharge-air temperature setpoint. See the following section regarding cooling coil operation when the outdoor-air temperature is lower than the discharge-air temperature setpoint.

Suggested Actions

For outdoor-conditions favorable for economizing, require full economizing (outdoor-air damper open 100%) before mechanical cooling can occur. Make sure the outdoor-air damper is being properly controlled, and responds correctly to commands given in the control.

6. Does the cooling coil operate when the outdoor-air temperature is lower than the discharge-air temperature setpoint?

During times when the outdoor-air temperature is less than the discharge-air temperature setpoint, the economizer should be able to meet the setpoint without any additional mechanical cooling. The cooling coil should never open under these conditions, and can be a major waste of cooling energy if it does. Figure 21-30 is a plot of the cooling coil valve signal, outdoor-air damper position signal, discharge-air temperature setpoint, and outdoor-air temperature versus time for a 2-day period. The outdoor-air temperature ranges from 40 to 50°F, while the discharge-air setpoint is always 55°F. The cooling coil operates at times up to 20% open, while the outdoor-air damper position is only open 30%, which is indicative of poor operation. Figure 21-31 shows a similar 2-day period in which the cooling coil valve is closed and the economizer satisfies the discharge-air temperature setpoint.

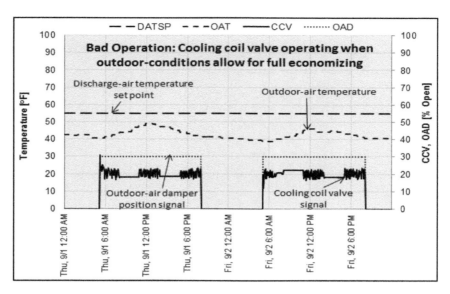

Figure 21-30: Cooling coil valve opening when outdoor conditions allow for full economizing, wasting cooling energy.

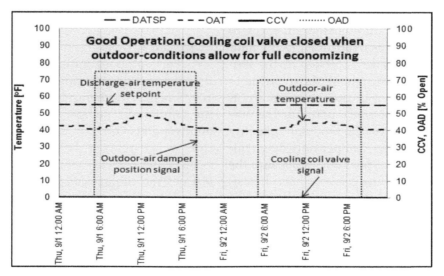

Figure 21-31: Cooling coil valve closed when outdoor conditions allow for full economizing.

Suggested Actions

For outdoor-conditions favorable for economizing, require full economizing (outdoor-air damper open 100%) before mechanical cooling can occur. Make sure the outdoor-air damper is being properly controlled, and responds correctly to commands given in the control. When the outdoor-air temperature is lower than the discharge-air temperature setpoint, the cooling coil should remain closed. Economizing can satisfy the load under these conditions.

7. When the cooling coil valve is open, is the outdoor-air damper fully open, if the conditions are favorable for economizing?

When conditions are favorable for economizing the outdoor-air damper should always be 100% open before mechanical cooling is enabled. This ensures that free cooling is fully utilized before mechanical cooling is utilized. Figure 21-32 shows a graph of the cooling coil valve signal, outdoor-air damper position signal, and outdoor-air temperature versus time for another

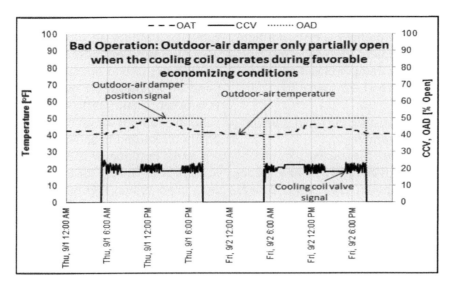

Figure 21-32: Outdoor conditions favorable for full economizing, yet the cooling coil operates because the outdoor-air damper is not fully open.

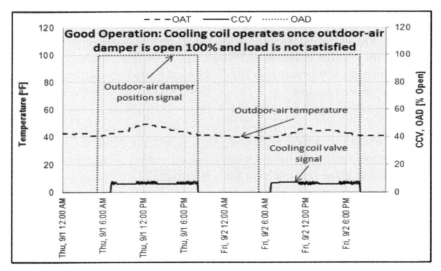

Figure 21-33: Full economizing enabled before mechanical cooling is enabled.

2-day period in which the outdoor-air temperature is between 40°F and 50°F. The outdoor conditions are great for economizing, yet the outdoor-air damper is only 50% open, while the cooling coil operates at 20%. Figure 21-33 shows an example of good operation; where the cooling coil is open roughly 10% when the outdoor-air damper is open 100%. This indicates that economizing cannot satisfy the load fully, so some mechanical cooling is required.

Suggested Actions

For outdoor conditions favorable for economizing, require full economizing (outdoor-air damper open

100%) before mechanical cooling can occur. Make sure the outdoor-air damper is being properly controlled, and responds correctly to commands given in the control.

8. Do outdoor-air dampers close to the minimum position for freeze protection?

The control logic in the BAS should have the outdoor-air damper locked out to the minimum position when outdoor-air temperatures fall below 40°F. Additionally, the outdoor-air damper should be fully closed if the mixed-air temperature drops below 40°F. If this control logic is not implemented, the coils can freeze, causing the air-handler to fail. At outdoor-air temperatures below 40°F, make sure that the damper is responding to the minimum position requirement in the control logic, and make sure the outdoor-air damper closes completely if mixed-air temperatures drop below 40°F.

9. Are the mixed-air temperatures between the outdoor-air and return-air temperatures?

The relationship between the mixed-, outdoor-, and return-air temperatures gives the outdoor-air fraction, which is discussed and shown in equation form at the beginning of this section. By reviewing the graph of discharge-, mixed-, return-, and outdoor-air temperatures versus time, the building operator can get an idea of outdoor-air operations.

The mixed-air temperature must always be between the return-air and outdoor-air temperatures when the AHU is properly working. If the outdoor-air damper is closed, the mixed-air temperature should be equal to the return-air temperature, and if the outdoor-air damper is fully open the mixed-air temperature should be equal to the outdoor-air temperature. For modulating damper position, the mixed-air temperature will be somewhere between the outdoor-air and return-air temperatures. This relationship indicates the outdoor-air fraction.

Suggested Actions

If the mixed-air temperature is not between the outdoor and return-air temperatures, there may be a sensor error. To determine which of these three sensors may be reading incorrectly, override the outdoor-air damper to be fully closed and wait 10 minutes. Make sure the outdoor-air fraction corresponds to the damper being closed, and if the return air and mixed-air temperature sensor readings are nearly equal, the outdoor-air temperature sensor may be faulty or installed in a place where it is receiving solar gain. If the mixed-air and return-air temperature sensor readings differ significantly, command the outdoor-air damper to 100% and wait 10 minutes. Again ensure that the outdoor-air fraction corresponds to the damper being open 100%, and if the outdoor-air and mixed-air readings are nearly equal, the return-air temperature sensor may be faulty. If the readings again differ significantly, the mixed-air temperature sensor is faulty. Replace any sensors that aren't working properly.

9a. When conditions are not favorable for economizing, is the mixed-air temperature closer to the return-air or outdoor–air temperature?

When outdoor conditions are not favorable for economizing, the mixed-air temperature should always

be closer to the return-air than the outdoor-air temperature. The outdoor-air damper should always be at the minimum position for this scenario. Figure 21-34 shows an example of bad operation, when the mixed-air temperature is closer to the outdoor-air temperature when it is between 25 and 35°F. When the outdoor air gets this cold, the outdoor-air damper should be at the minimum position required to satisfy ventilation. In this example, too much outdoor air is being brought in, and the AHU must heat the mixed air to satisfy the discharge-air temperature setpoint. Figure 21-35 shows correct operation for this example, where the mixed-air temperature is nearly equal to the discharge-air temperature just by bringing in the minimum amount of outdoor air required to satisfy ventilation requirements. It is closer to the return-air temperature than the outdoor-air temperature in this case.

Suggested Actions

When conditions are not favorable for economizing, the outdoor-air damper should be at the minimum position to satisfy ventilation requirements. If it is open more than this, check to see if the damper is working properly, and check the minimum position setpoint in the BAS to make sure it is less than 20% open at its minimum position.

9b. When conditions are favorable for economizing, is the mixed-air temperature closer to the outdoor air or return air?

When outdoor conditions are favorable for economizing, the mixed-air temperature should be closer to the outdoor-air temperature than the return-air temperature to utilize as much free cooling as necessary to meet the discharge-air temperature setpoint. Figure 21-36 is an example of bad operation, when the outdoor-air temperature is between 40 and 50°F and the mixed-air temperature is closer to the return-air than the outdoor-air temperature. Cooling energy will be wasted satisfying the discharge-air temperature setpoint when the load could be satisfied by the economizer. Figure 21-37 shows an example of good operation, with the outdoor temperatures favorable for economizing and the mixed-air temperature equaling the discharge-air temperature. This indicates that the economizer is satisfying the load completely, and the cooling coil is closed.

Suggested Actions

When conditions are favorable for economizing, the outdoor-air damper should be modulating between the minimum position (to satisfy ventilation requirements),

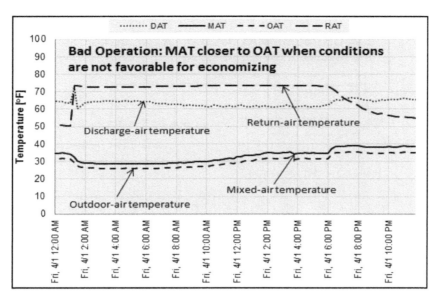

Figure 21-34: Mixed-air temperature closer to outdoor-air temperature when outdoor conditions are not favorable for economizing.

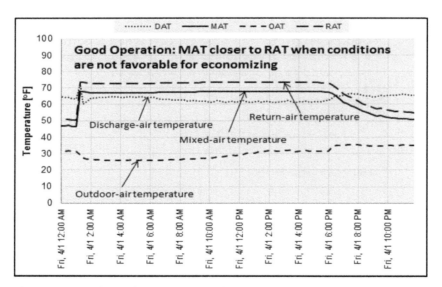

Figure 21-35: Mixed-air temperature closer to return-air temperature when outdoor conditions are not favorable for economizing.

and fully open to satisfy the discharge-air temperature setpoint. If the damper is not modulating, check to see if it is working properly by commanding it fully open, fully closed, and to the minimum position and make sure it responds to the command. Make sure that the cooling coil is opening only when the economizer cannot satisfy the load.

Economizer Controls

There are several economizer control sequences that may be implemented in your BAS. These can include outdoor-air dry bulb, outdoor-air enthalpy, differential dry

bulb, and differential enthalpy. Outdoor-air dry bulb is set up to economize when the outdoor air drops below a fixed outdoor-air temperature setpoint; outdoor-air enthalpy is set up to economize when the calculated outdoor-air enthalpy drops below a fixed outdoor-air enthalpy setpoint; differential dry bulb is set up to economize when the outdoor-air temperature drops below the return-air temperature by some temperature differential, usually 1°F to 2°F; and differential enthalpy is set up to economize when the calculated outdoor-air enthalpy drops below the calculated return-air enthalpy by some differential, usually 2 Btu/lb. Enthalpy controls are typ-

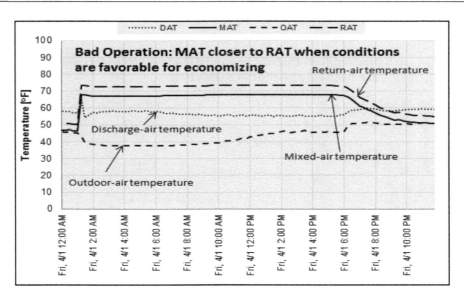

Figure 21-36: Mixed-air temperature closer to return-air temperature when conditions are favorable for economizing.

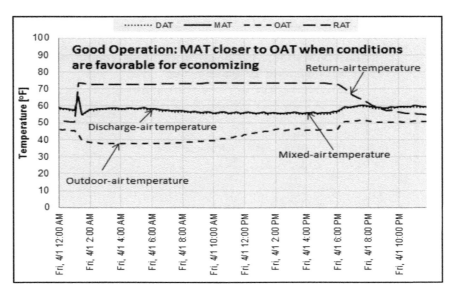

Figure 21-37: Mixed-air temperature closer to outdoor-air temperature when conditions are favorable for economizing.

ically utilized in humid climate zones (Houston, Miami, etc.). There are several economizer sequencing options; three commonly used ones listed below in detail.

Economizer Sequence Option A

When the economizer is enabled, the economizer dampers will modulate open from the minimum position to that required to maintain the mixed-air temperature setpoint. The default mixed-air temperature setpoint is 55°F, but can be user-adjustable. When the fan is off, the outdoor-air damper will be closed. When the fan is running, the outdoor-air damper will be maintained

at the minimum position unless there is a call for more outdoor-air. If the mixed-air temperature drops below the mixed-air temperature low limit setpoint, the outdoor air dampers will close. The default mixed air low limit setpoint is 45°F.

This option has no relation or linkage to mechanical cooling and can still be found on large AHUs. This means that there is a separate discharge-air setpoint for mechanical cooling, and the mixed-air setpoint can be set too high to optimize free cooling, or too low to cause heating to always occur.

Economizer Sequence Option B

When the economizer is enabled, the economizer dampers will modulate open from the minimum position to that required to maintain the discharge-air temperature setpoint. The default discharge-air temperature setpoint is 55°F, but can be user-adjustable. When the outdoor-air damper is greater than 75% to 100% open (usually set in the control code) for more than a user-specified time delay (usually 5 to 15 minutes, but can be as great as 30 minutes), mechanical cooling will be activated (as long as the outdoor-air temperature is above the outdoor-air temperature cooling lockout setpoint; usually 50°F to 60°F). If the mixed-air temperature drops below the mixed-air low limit setpoint, the outdoor-air damper will close.

The default mixed air low limit setpoint is 45°F. When the fan is off, the damper will be closed. When the fan is running, the damper will be maintained at the minimum position unless there is a call for more outdoor air.

This option is a more intelligent than sequence Option A because it links the economizer to the mechanical cooling system. However, it is highly possible that the description above for Option B is not fully implemented, i.e., outdoor lockouts may or may not exist, mixed air low limits may or may not exist, etc. To fully optimize this sequence option, all lockouts and limits should be properly set in the BAS control logic.

Economizer Sequence Option C

When the economizer is enabled, the economizer damper will modulate open from the minimum position to that required to maintain the zone temperature at the zone temperature setpoint. The default zone temperature setpoint is 72°F, but can be user-adjustable. When the outdoor-air damper is greater than 75% to 100% open (usually a setting in the control code) for more than a user-specified time delay (usually 5 to 15 minutes, but can be as great as 30 minutes), mechanical cooling will be activated (as long as the outdoor-air temperature is above the outdoor-air temperature cooling lockout setpoint; usually 50°F to 60°F). If the mixed-air temperature drops below the mixed-air temperature low limit setpoint, the outdoor-air damper will close. When the fan is off, the damper will be closed. When the fan is running, the damper will be maintained at the minimum position.

This option is typical for RTUs or smaller AHU(s), and is also more intelligent than sequence Option A because mechanical cooling and economizing are linked to operate at maximum efficiency.

Evaluating Optimal Start and Night Setback Strategies with BAS Data

The purpose of this section is to show, through use of examples of good and bad operation of processed BAS data, how optimal start and night setback can be utilized in BASs and programmable thermostats to minimize air-handling unit runtime.

Optimal start programs automatically calculate the earliest AHU start times necessary to bring space temperatures up to (heating) or down to (cooling) setpoint, so the space temperatures are within some acceptable value prior to building occupancy periods (usually within 1°F to 2°F of setpoint). Optimal start attempts to start equipment at the latest time possible, rather than starting at some default time. The control processes allow the building owner/manager/operator to configure how the OPTIMAL START will respond to one or more monitored conditions. This usually includes indoor space temperature(s) and the outdoor-air temperature, although the OAT is optional for programmable thermostats. Optimal start may also allow for configuring several parameters related to AHU design/recovery rates, scheduled occupancy time, earliest start time period and desired target temperatures at the scheduled occupancy time. Based on these (and possibly other) parameters, the optimal start program will continually "learn" or adjust the start time to varying conditions to try and "optimize" the start time each day to achieve the target temperature at the occupancy time.

Night setback is an energy-saving aspect related to the occupancy scheduling feature of most programmable thermostats and BASs. When the AHU is turned off, it is placed into a "Night Setback" mode. During this mode, the temperatures in the space(s) served by the AHU will start to rise or fall (depending on several variables such as outdoor-air temperature, building envelope design, internal heat gains, etc.). As the temperatures begin to rise or fall, it is important that the thermostat or BAS be configured to respond to those temperature excursions. If unchecked, the space temperatures may fall or rise to some value unmanageable from the standpoint of recovery or safety. This means that the AHU, even if started at midnight, may not be designed to recover the temperatures to some acceptable value prior to the next occupancy period.

Optimal start programs and night setback directly affect energy consumption by continuously monitoring and changing equipment runtime, while minimizing negative impacts to comfort. Optimal start, if implemented properly, can reduce equipment runtime significantly, resulting in energy savings. Night setback, if implemented properly, allows the zone temperatures in the building to float without starting up the AHU, but does not allow the space temperatures to rise or fall out of some acceptable value for recovery or safety. Both energy savings measures (ESM), although not required, can reduce energy consumption and decrease daily equipment runtime.

Data Needed from the BAS to Verify
Optimal Start and Night Setback

To verify that optimal start has been implemented correctly, for single-duct variable-air-volume (SDVAV) AHU(s), the following parameters should be monitored using the trending capabilities of the building automation system:

- Duct static pressure
- AHU status.

To verify that night setback has been implemented correctly (for VAV boxes only), the following parameters should be monitored using the trending capabilities of the building automation system:

- Zone temperature
- Zone damper position signal.

The recommended frequency of data collection is between 5 and 30 minutes.

Trends to Look for When Analyzing the Data
1. Is optimal start being utilized in the building, and if not, has this opportunity been investigated?

2. Are the zone temperatures being set back at night and on weekends?

1. Is optimal start being utilized in the building, and if not, has this opportunity been investigated?

In absence of optimal start, AHU default start times are originally determined by the building owner or engineer based on the worst case scenario required to bring the space(s) served by the AHU to acceptable comfort levels prior to occupancy. This default time, however, is commonly adjusted based on historical events (equipment failure, arctic cold fronts or extreme hot spells, etc.) and/or occupant complaints. By adjusting the scheduled equipment start time to an earlier time, the operation and maintenance (O&M) staff eliminate additional complaints, and forget about the energy impacts associated with the adjustments. It is not uncommon to find scheduled start times for many AHUs to be configured as early as 1:00 AM, when the actual occupied period for the

building or space served in the building is 6:00 AM or later. This increased HVAC runtime often leads to energy waste. The problem can be even further exaggerated if economizers have failed, resulting in excess outdoor air being introduced into the building, which requires additional energy to heat (winter) or cool (summer).

To determine if optimal start has been implemented, and is working, chart the AHU duct static pressure versus time. If AHU status is available, that could be charted instead of the duct static pressure. Figure 21-38 shows an example of an AHU that is most likely operating without an optimal start program. The graph shows 1 week of data in which the AHU start time (as indicated by the duct static pressure changing from 0 to a value of about 1.5 inches) is 4:00 AM Monday through Saturday, off on Sunday. The scheduled occupancy start time for this building is 8:00 AM, Monday through Friday. Although it is possible that optimal start could have the system starting up at 4:00 AM every day, it is highly unlikely. A good optimal start program will have the AHU start time vary based on several variables such as outdoor-air temperature and indoor space temperatures. Because of the dynamic nature of these variables, it is very likely that the system will start at different times each day.

Figure 21-39, on the other hand, shows an example of an AHU that could be using an optimal start program to start the unit each morning. During this 1 week period, the duct static pressure indicates that the unit starts up at 5:00 AM Monday, 6:30 AM Tuesday and Wednesday, 6:00 AM on Thursday, and 7:15 AM on Friday. The AHU appears to be off Saturdays and Sundays, which indicates no occupancy during the weekends. The scheduled start time for this building is 8:00 AM, Monday through Friday.

When analyzing the varying start times compared to the occupancy start time (8:00 AM), the times for each day make sense. If the building is unoccupied on the weekends, it is very likely that the start time for Monday morning will be earlier than any other weekday because of a longer unoccupied period over the weekend when the building's AHU(s) are not operating. The variation from 1 weekday to the next makes sense because of dynamic outdoor-air temperatures and internal gains in the building. It is possible that the building operator has programmed this schedule manually, but that is unlikely. It is more likely that optimal start has been implemented in this AHU.

Comparing Figure 21-38 to Figure 21-39, with optimal start implemented, the number of operating hours has been reduced by 19 hours during this week of operation. 8 of these hours come from shutting the AHU

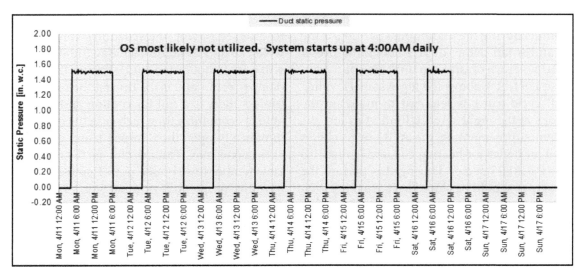

Figure 21-38: 7-day period of duct static pressure versus time to check if optimal start is being utilized. This example shows a consistent start time of 4:00 AM each day except Sunday, when the unit is off. Therefore, optimal start is most likely not implemented for this AHU.

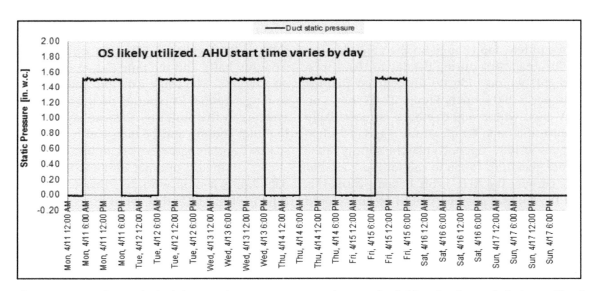

Figure 21-39: 7-day period of duct static pressure versus time to check if optimal start is being utilized. This example shows a varying start time each day (with exception of Tuesday and Wednesday). This could indicate the use of optimal start programming.

down on Saturday, and the remaining 11 hours result in the optimal start program optimizing the AHU start time each morning.

Suggested Actions

If optimal start is not programmed into the programmable thermostats and/or BAS, then this opportunity should be investigated. When implementing an optimal start program, some variables to account for include the temperatures in the zones that the AHUs serve, and the outdoor-air temperature. If in-house capability exists, it should be utilized in programming optimal start; otherwise outsourcing to a controls company is a possibility but should include training so the building maintenance staff can refine (re-tune) the optimal start program or implement the program on other AHUs as needed or required.

2. Are the zone temperatures being set back at night and on weekends?

Night setback mitigates excess temperature issues in the building, and helps mitigate recovery problems.

This is done by setting the minimum/maximum space temperature limits during the unoccupied period. As the space temperatures fall below or rise above the preset limits, the AHU will turn on and run until the space temperature(s) rise 1°F to 2°F above the night setback value. The AHU will cycle (run) as required to maintain the space conditions at the night setback limit. As such, the ancillary system supplementing the AHU (e.g., chiller for cooling coil, or boiler for heating coil, etc.) must also operate when the AHU cycles. During this period, the outdoor-air damper should be closed because there is no requirement to introduce outdoor air (unless there is a call for cooling and the outdoor air is able to economize, in which case the controls should automatically open the dampers).

To monitor the night and weekend setback control scheme, review the chart of the zone temperature and zone damper position signal versus time. For fan powered boxes, this information is not sufficient to know occupied and unoccupied periods, and more information should be investigated (AHU duct static pressure). For VAV boxes, a sudden step change on the zone damper position to 100% (e.g., 40% during the day to 100% open during the night) would indicate unoccupied periods when the damper is 100% open. Figure 21-40 shows an example VAV box, where night setback is most likely utilized because the zone temperature responds when the damper goes to 100% open. (When the supply fan is off, the damper goes 100% open to call for more air, but receives none because the AHU is off. In some cases, it is possible that the damper position would be fully closed during AHU off times instead of fully open.) This means that the zone temperature is allowed to float at night when the AHU is off. (This AHU is scheduled to operate on Saturday, but not on Sunday.) At night, the zone temperature is roughly 80°F, and then decreases to 75°F during occupied hours. You can see that the night setback limit was not reached in this example, or the AHU would have cycled on (along with the chiller supplying the AHU cooling coil) and reduced the zone temperature by 1-3°F below the setback limit, and then shut down. The AHU will cycle when needed to maintain this "buffer" until morning warm up/cool down the following morning.

Suggested Actions

If night setback limits are not implemented at the zone level, consider this an opportunity to investigate because warmer and

colder climate zones without these limits have probably resorted to having the schedules configured to run between 16 and 20 hours each day (otherwise the building will get too hot or too cold). Ensure that the outdoor-air damper command is 0% during the unoccupied periods (unless there is a call for cooling and the outdoor air is able to economize, in which case the controls should automatically open the dampers), and setback limits exist at the zone level. To start, change the night (unoccupied) zone temperature limits by ±5°F, and see how this affects morning recovery. If optimal start is used, a significant change in the setback limits will most likely cause the optimal start program to have to relearn the optimum time to start the recovery, so bear this in mind as well.

Prior to the start of the occupancy period, the further from setpoint the space temperature is, the longer it will take to recover to the desired setpoint. For example, if the occupied heating setpoint is 72°F and the night setback limit value is 60°F, the AHU must be capable of providing enough heating energy to increase the space temperature (air and mass) by 12°F. If the night setback limit value is 65°F, the difference is only 7°F (roughly half), and the time required to recover should also be less. The same example can be applied to the cooling setpoint. Depending on the cooling and heating systems in the building, this can significantly increase the total cooling and heating coil loads (depending on season), and impact recovery times.

In theory, optimal start and night/weekend setback limits should work together to optimize the runtime for the AHUs that serve the space(s) in the building.

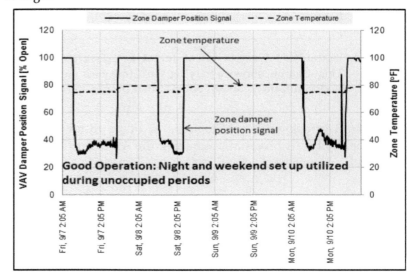

Figure 21-40: Example in which night and weekend setback limits are most likely in place.

CONCLUSIONS/DISCUSSION

A major challenge in meeting legislative goals to achieve reductions in energy consumption across the commercial buildings sector lie not in collecting data, but rather analyzing the data and identifying opportunities to improve operations and control strategies that result in energy savings. This chapter focused on ways to analyze the data with one common tool, ECAM, in hopes of getting building owners and operators familiar with ways to generate and analyze charts to help achieve their goals of energy savings across their buildings portfolio. While interval meter data are useful to identify overall building consumption trends, there are some specific things that can be diagnosed (i.e., economizer operation and night setback) if the data are processed correctly. Building automation system data, on the other hand, allows "dialed"-in analysis of specific operations in the building, again highlighting economizer operation, optimal start and night setback in this chapter. In addition to this analysis, there are several documents and guides published by Pacific Northwest National Laboratory (www.pnnl.gov/buildingretuning) to help building owners and operators become familiar with ways to trend the data and identify corrective actions to operations and controls to achieve energy savings. For more information on ECAM, see Appendix A.

References

Katipamula, S. and M.R. Brambley, 2005. "Methods for Fault Detection, Diagnostics, and Prognostics for Building Systems—A Review, Part II." International Journal of HVAC&R Research, Vol. 11(2):169-187.

Taasevigen D.J. and W. Koran. 2012. User's Guide to the Energy Charting and Metrics Tool (ECAM). PNNL-21160, Pacific Northwest National Laboratory, Richland, WA.

Taasevigen D.J., S. Katipamula, and W. Koran. 2011. Interval Data Analysis with the Energy Charting and Metrics Tool (ECAM). PNNL-20495, Pacific Northwest National Laboratory, Richland, WA.

United States Department of Energy, 2011. Federal operations and maintenance requirements. Washington, D.C. Retrieved from http://www1.eere.energy.gov/femp/program/om_requirements.html. Accessed 10/2012.

APPENDIX A: ECAM

The Energy Charting and Metrics tool is a free, downloadable Microsoft Excel™ add-on intended to facilitate the examination of interval meter and building automation system data. In addition to being easy-to-use, it is also flexible. The key features include:

- Pre-processing of data to attach schedule and day-type information to time-series data;

- Filtering data by day-type, occupancy schedule, binned weather data, month/year, pre/post, etc.;

- Normalization of data based on user-entered information;

- Creation of standard charts for the points selected by the user;

- Calculation of normalized metrics for the points selected by the user;

- Creation of data-driven models and measurement and verification;

- Creation of standard building retuning charts using trend data from the BAS.

The user's original data are not modified in the process of using ECAM, but are copied into a new workbook automatically. The tool makes extensive use of Excel_ PivotTables to facilitate summarization and filtering of the data. It goes beyond normal PivotTables and PivotCharts, however, by automating the creation of scatter charts based on PivotTable data. The ECAM tool, user's guide, interactive webinar videos and practice data files, are all available at (www.pnnl.gov/buildingretuning/webseries.stm). The webinar series includes an introduction on how to install ECAM, pre-processing some sample data in the proper ECAM format, a step-by-step walkthrough of generating charts with interval meter data and building automation system data, and some analysis of the charts (Taasevigen and Koran 2012).

Chapter 22

High-resolution Data Collection for Automated Fault Diagnostics

Tim Middelkoop, University of Missouri

MOTIVATION

The collection and analysis of data is becoming more important for well running buildings. Without data, automated fault diagnostics is difficult or impossible. As fault-diagnostic and energy management tools increase in sophistication there will be an increasing need for high resolution data both in the number of points per building and samples taken. As the amount of data increases we must intelligently manage both the data and the data about the data, the so-called meta-data. It is important to collect as much of this data sooner than later, however, there is a common misconception that storing the data is expensive. This need not be true if the problem is approached correctly. A building that samples 100,000 points every ten seconds without any compression can be done for about $100 in storage and $100 in hardware. However, it is true that it is expensive to analyze data at this scale—this is where the misconception comes from. It is important to remember if you want to analyze building data sometime in the future it is possible to purchase the hardware and software but it is impossible to re-create the data once it is discarded. This chapter is about the infrastructure required to collect, manage, and preserve data in a stable and scalable manner. It is also about how to preserve the data and meta-data at this scale and how to processes it such that it can be used by a wide range of analysis tools both in real time and as time-series data. This chapter should be of interest to both developers of fault diagnostic tools and end users as it provides a road-map for developing a system and the reasoning behind many of the decisions. For end users this chapter provides a set of technologies and capabilities to look for when implementing a data collection program.

Although much of the technology and techniques presented in this chapter can be implemented in isolation it is important that it be a part of an overall data collection program. This process must be initiated before collection and continue along with collection. It is not sufficient to throw hardware and software at the problem. In the next section we will give an overview of this process and in the remaining sections provide a road-map for the hardware and software needed to implement such a process along with an example. The work in this paper is based on a system developed at the University of Florida for the collection of high resolution data to be used for research purposes, the result of which is a living laboratory that combines research and implementation [LMB12, HMBM12]. As such, this chapter is a practical guide based on the research and experience that went into the development of this system.

DATA COLLECTION PROGRAM

Since the goal of long term high resolution data collection is to do it in a stable and usable manner the data collection program should adhere to the design guidelines presented in this section. These guidelines are based on experience in dealing with data at this scale.

Detailed and Structured Logs Must be Made

Although this sounds obvious it is often harder to accomplish in practice. A hard bound lab notebook is ideal for recording what happens in the system. For example if a controller is replaced it should be noted (along with the date, time, building, and device name) and if any changes occur. The details of the changes should be in the meta-database, the log only indicates that these changes are made. This makes it easy to go back and see what has happened to the system at a later date and use it as a resource to find the actual details in the database. If a work-order system is used the work-order or other tracking information should also be indicated in the log. The reason that a physical logbook is important is that it provides a high level overview for all change data even though it may be implemented on various systems. Digital notes are not stable over the life of a building given how fast technology changes. The log system also provides a central location to keep track of changes in the building.

Points Must be Properly Named and Tagged

Most points in a system have both a numerical representation (device, type and instance in BACnet) as well as a textual representation (B27.VAV.221.room_temp). This allows the operators to see what is going on in the building. However numerical representations and text names changes and it is important that these names remain accurate and up to date. Any changes must be logged and stored in a database to track the changes over time. In addition to the names meta-data should be associated with each point that is both human readable and machine readable as automated fault diagnostic information must be able to "reason" about the building. It is not practical or possible to store all the information in name of the point so this information can be stored as tags. Tags are tag-value pairs associated with each point over a period of time (and possibly the equipment as a whole) to describe the point. The tag names must follow a convention so that constant and comprehensive results can obtained when using these tags. Since the tags are associated with points the meta-data can change over time as the system does. However, if a point changes behavior it needs to have a separate set of tags. Tagging allows powerful search capability for both operator and automated systems alike. Table 22-1 is an example set of tags for the room temperature sensor in a thermostat*.

From this we can find, for example, all the rooms (corridors only have a space tag) in zone 1. We can also find that this sensor is associated with a VAV unit either indirectly (unit='VAV' and num=129) or directly (vav=129). Most of this information can be extracted from the name and for this particular application a script was written to create the tags in a few hours populating tens of thousands of points.

The ever changing nature of tags associated with a point means that they must be stored as "tag=value" pairs in a relational database (adding columns is not easy to do in a traditional relational database) to be extracted and used directly by the software. This in-

formation is difficult to use directly as can be seen by the SQL query needed to generate this information. The tag-value structure lends itself to a column-oriented format using database extensions (hstore in Postgres) or newer document based/schema-less database systems commonly referred as NoSQL. Although this format is easier to use* there must be a balance between using newer and changing technology with a stable long term solution. Since meta-data changes slowly one option is to generate newer representations from the stable relational form maintaining the benefits of both.

Data Should be Push Collected Based on Change of Value (CoV)

Pull or polling based data collection arguable works for long interval (greater than 15-minute) collection. However, as the number of points and/or collection interval increases it puts stress on both the network and systems as requests will "spike" at the predetermined interval time. Surges in queries can mean that responses can get queued and come back at different times. Change of value- or push-based collection means that data are only sent when there is a measurable change in the data, which can mean that less data are needed to be transferred. In addition values are sent when the the change occurs, not at the next interval. The downside is that this type of collection can dramatically increase the amount of data and increase the complexity of the analysis since it is now a non-uniform time-series. Excessive data can be reduced through proper configuration (see the next point) and interval data can be collapsed/binned after the fact into interval data. This follows the core idea of this chapter, which is collect data at high rates locally to inexpensive storage then reduce to centralized high performance storage and compute capability for analysis.

Points Must be Properly Configured (Names, Units, CoV)

Field level devices must be configured properly to ensure that data are collected accurately. This includes ensuring that devices are configured for proper scales and units. More importantly Change of Value reporting should be configured to ensure the proper balance between sensor/actuator accuracy and responsiveness. Many devices can be configured how and when the

---

*SELECT name, Tags.tag, Tags.value FROM Tags JOIN (SELECT name, objectID FROM Tags JOIN (SELECT objectID FROM Tags WHERE tag='room' AND value='241') AS t2 USING(objectID) JOIN Objects USING (objectID) WHERE last IS NULL AND tag='attr' AND value='room_temp') AS t3 USING(objectID);

*SELECT * FROM hTags WHERE tag -> 'room' = '241' AND tag -> 'attr' = 'room_temp';

name	tag	value
B0072.RM241.VAV129:ROOM TEMP	address	4
B0072.RM241.VAV129:ROOM TEMP	attr	room_temp
B0072.RM241.VAV129:ROOM TEMP	building	72
B0072.RM241.VAV129:ROOM TEMP	campus	UF
B0072.RM241.VAV129:ROOM TEMP	device	9043
B0072.RM241.VAV129:ROOM TEMP	module	142
B0072.RM241.VAV129:ROOM TEMP	name	B0072.RM241.VAV129:ROOM_TEMP
B0072.RM241.VAV129:ROOM TEMP	num	129
B0072.RM241.VAV129:ROOM TEMP	property	room_temp
B0072.RM241.VAV129:ROOM TEMP	room	241
B0072.RM241.VAV129:ROOM TEMP	space	241
B0072.RM241.VAV129:ROOM TEMP	unit	VAV
B0072.RM241.VAV129:ROOM TEMP	vav	129
B0072.RM241.VAV129:ROOM TEMP	zone	1

Table 22-1

Change of Value event occurs, however, often this also is the rate that the value can trigger updates to other values (PID loops for example). If the Change of Value update interval (the amount of change that triggers an event) is set to low the device will produce excessive amounts of data. This data may not be any more accurate as the interval may be below the accuracy of the sensor usable information and may be just noise in the system. For example if a room temperature sensor update interval is set to 0.01°F it will most likely produce events at the maximum rate as most temperature sensors are not accurate at this level and minor fluctuations will trigger events. As well sensor accuracy at this level is not useful, update intervals of 0.1°F to 0.5°F and even high as 1°F are more reasonable depending on the sensor and application. Setting the value too high can impact operation if internal calculations are also impacted (in-field testing is needed to verify this) by creating larger or artificial dead-bands in controls.

Points Must be Verified

Although somewhat obvious, ensuring that point names and descriptions match what is on the ground is important. In a new installation this may be a daunting task, however since meta-data can retroactively change this can be done over time. However since most modern controls are based on modules simply verifying a modules location can be sufficient to catch most mistakes. Even though this task should be done during commissioning errors may slip through (for example rooms may be improperly named or changed after commissioning). As a part of an overall data collection plan any errors should be logged and the point names and databases should be updated. Leaving a point incorrectly named can cause trouble in the future from a maintenance perspective.

Sensors Must be Calibrated

Again, this sounds obvious but in practice this is not always the case. Automated fault diagnostics requires an accurate view of the equipment in order to properly do the analysis. Some fault diagnostics system may be able to detect sensor calibration issues making the calibration process easier or even auto-calibrate.

Changes In Hardware/Configuration/Calibrated Must be Logged and Have Immediate Meta-Data Updates

Any changes in calibration at the hardware level should be included in the meta-data. Software calibration outside the building automation system should be held in the meta-data database. A proper data management process will ensure that updates happen in a timely manner. Without proper updates fault diagnostic analysis may come to incorrect conclusions.

Meta-data Updates Must be Additive and Not Replace Old Data

In computing terms this means that the data are log based. This means that is any changes should be recorded as changes not altering the data itself. Data are never destroyed, replaced or updated; new and updated information is added and supersedes any previous data. This has a number of benefits the first being that all the analysis can be rolled back based to a snapshot of the system (meta-data is preserved), the second, is that updates can be undone or reverted (through a reapplication of the original information). For example calibration data can be retroactively applied. The meta-data database is a snapshot of how the system is viewed at any one time, and this is logged. In most cases storage will not be an issue due to the small size of the meta-data database.

The System Must Run in Parallel and Independent of Operations so that it is a Non-Critical Device

Making the data collection independent and non-critical facilitates the use of low-cost hardware. This is important due to the cost of high performance computing and storage capability. Running in parallel independent of operations ensures that the data collection system can be installed and then run in the background. Updates to BAS systems, operating systems and other building operations software will not negatively impact the data collection process.

The System Must Run in a Real-Time Streaming Mode

Streaming data allows fault diagnostics systems to be run in real time alerting operators of potential problems as they happen. Additionally, the system can be used to monitor equipment in real time for diagnostics and troubleshooting.

Data Must be Replicated and Backed up

Due to the distributed and low-cost nature of the system data must be backed up to ensure that data are not lost. The recovery of data should

be tested periodically as data backup is actually two operations, backup and restore. If the restore portion is not tested then this may only be discovered when it is too late. Backup systems should also protect against propagating corruption or failures which may destroy backups shortly before they are needed. Database replication allows for analysis to be conducted on high performance systems as well as performing a backup roll. Replication can also be done to the cloud as a low cost solution and give additional analysis capacity. More detail on how this can be accomplished is presented in the next section.

SYSTEM ARCHITECTURE

In this chapter we take the approach that the data collection and analysis should be done in a hierarchal but distributed manner. An overview of such a system can be seen in Figure 22-1. The figure shows that as the data are reduced from left to right the amount of data becomes smaller and the power and cost of the system increases. If this balance is not engineered correctly the entire system will quickly become prohibitively expensive. It costs about 100 times more per node if analysis class machines are used for sensor storage. Briefly, the system is organized as follows: in the left tier of the system the data are collected from floor level equipment by the low cost collection hardware. Then the system replicates this data from the collection points to the preprocessing portion in the middle tier. Finally, after the data

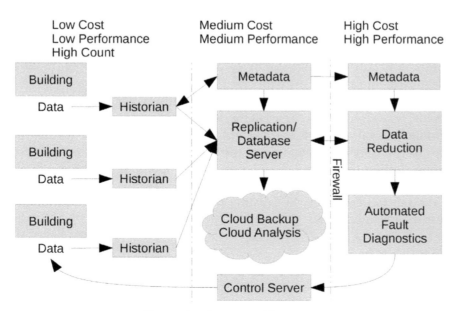

Figure 22-1: System architecture.

are preprocessed it is processed by the fault diagnostics software in the analysis tier of the system. Each tier is protected by a firewall giving multiple layers of isolation to the physical hardware and is described in the remainder of the section.

Data Collection Tier

Given the large data rates collection and storage should be distributed at the building, sub-meter, or line level. This ensures that there is a level of fault tolerance at the building/line level and ensures that collection occurs in close proximity (physically and network wise) to the source of the data. This allows data to be collected in large amounts at relatively low cost. As data are needed they can be processed and pushed up the hierarchy. Since meta-data is both important and small it should be managed centrally but replicated at the data collection points. It is important that data collection is not tied directly to the meta-data so it does not depend on this service being available or accurate. It is very hard to correct meta-data errors embedded in the data after the fact if data are tied directly to the meta-data (for example, point names are used). Instead it is important to store the raw address information along with the data. The data also should be collected in their raw form and any post-processing (conversion of 4-20ma to Amps for example) should be connected with the meta-data database. This ensures that if any errors are made they can be corrected at a later date. During commissioning the data management process should ensure that this information is available and correct.

Data Processing Tier

The data from the data collection tier are replicated to the data processing tier. Replication provides a layer of security between the building systems and the software. Replication also provides backup and a host with more computational power to pre-process the data into a usable form. If complete replication is desired the processing tier can utilize low performance storage for raw sensor data and higher speed storage for the processing functions. This is possible because sequential access of data on rotational media (hard drives) is very fast and can be copied to faster media (SSD) for analysis, which requires random-access. More sophisticated partial replication, where the data collection nodes pre-process the data before replication, can reduce storage and network requirements. However the risk in additional complexity, development and management costs again may not be justified due to the low cost of today's disk and networking hardware.

The replication should be streamed not batched, which means that the replicated data are up-to-date in near real time. Taking a streaming approach architecturally is often quite different than a batched approach and taking a streaming approach from the beginning will allow the use of online (real-time) automated fault diagnostic tools and alarms.

In addition to replication and backup, the primary function of the data processing tier is to pre-process the data for the automated fault diagnostics tools. This level of isolation means that as the fault diagnostics tools evolve and change the data processing tier can be adjusted to meet the needs of the new tools preventing lock in to a single tool. As tools change over time the data processing tier can supply data to multiple tools simultaneously easing the transition from one to another, or supporting multiple tools simultaneously.

Fault Diagnostics Tier

The processing of building data into actionable information is the function of the automated fault diagnostics tool in the fault diagnostics tier. It is here that the automated fault diagnostics tools run supplied by the rest of the system. The data are supplied by the data processing tier as either a stream of processed live building data or in a reduced form. As the needs of the automated fault diagnostics systems change it can be matched by the data processing tier. This tier is separated from the data processing tier by a firewall to add an additional level of isolation and security. Since this tier may also require the interaction of end-users it will be exposed to security risks to a greater degree. The isolation of all three systems means that in order for an attacker or exploit to reach building hardware systems the attack must be both very sophisticated and targeted. Enhanced security comes from the design as well as the ability to severely limit access to the data collection and data processing tiers and to deploy it on hardened platforms.

DATA MANAGEMENT

In this chapter we propose the best way to collect data is to do it independent of, and at a lower level than, the Building Automation System (BAS). Although these systems can "trend" data there are a large number of reasons why this is not a good idea for long term data collection. Ideally this system should be a secure standalone server-class operating system (such as Linux) or embedded system to ensure that data can be collected over the long term. By keeping the function of the system simple it reduces the amount of things that can go wrong over the long term. When looking at building or purchasing a data collection system the following principals should be kept in mind:

Separate Data Collection and Archiving from Analysis

The underlying idea of this chapter is to store data locally (data historian) and to do the analysis remotely. By separating these functions, as stated earlier, the entire system can be built cost-effectively.

Use Open Systems

The system architecture section lays out a general design of a data collection system without specifying a specific implementation. Although this chapter was written based on an open source implementation (BacLog) the concepts apply to propitiatory systems as well. The term "open" has different meanings in different contexts, however, when it comes to managing data the term "open" is an important distinction. An open system means, in the authors opinion, that data must be accessible by third party tools and must be able to be exported, including meta-data, in a usable format for use by future systems. This prevents vendor lock-in ensuring that other automated fault diagnostic systems can use the data and that valuable historical data are not lost.

Don't Try to Save too Much Data Space, ou Will Regret It

It may be tempting use smaller data sizes on columns, not to store full date and time information, normalize device information (originating IP address and port) etc., however this may complicate analysis in the future as well as put the data at risk if somehow the normalized (reducing IP, port etc. to an "ID") information is lost or changed. At such a low cost for hardware even the development time cannot be justified. Just write the Change of Value information to the database.

Use a Structured and Native Data Storage

Text files, although tempting, become unmanageable at large sizes and have a number of other issues when they need to be parsed and processed. A relational database (PostgreSQL, MariaDB) or a fixed format file such as Hierarchical Data Format (HDF) is critical since the data are stored in a structured format and provides ways of extracting the data. Care must be taken to ensure the data are stored in native format not in text form so that there is no loss in fidelity (conversion from text formats is error prone and potentially lossy) and that

data can be extracted efficiently (for example asking for time slice of data). Native data formats can also have significant space savings. Additional space savings and query performance can be achieved by using a database or format that supports column oriented storage (MariaDB, Cassandra).

Store Complete Date Information

A full date and time stamp with millisecond accuracy in native format should be used. It is tempting to store time information in a custom or text-based format, however, this can lead to troubles later on due to parsing and the intricacies of time zones, leap years, leap seconds, and week numbers. It is recommended to utilize the ISO definitions for week numbers and week days due to their clear and standardized definitions.

Index Time, Optionally Index Point

For database access create an index on time. This allows most queries to run orders of magnitude faster when the database grows. For example, even reporting the last 10 minutes of data for an indexed database can take hours since the database has to scan the entire history. Optionally indexing the points can allow fast queries for time and point, however, the amount of storage required to hold the index may not be worth a small speedup depending on the way in which the data are accessed for analysis. Regardless of indexing retrieving data by point may be slow due to the way the data are laid out on disk. Even though the number of points may be large across the entire campus the distributed nature of the system means that the number of points will be smaller at each collection location. For other storage formats using native data formats will allow proper access (time ranges) of data.

META-DATA

One problem in the building HVAC industry is the lack of a formalized standard way of numbering points although there are efforts in this area including project haystack (Haystack). In general point names are static and changing them is neither not very feasible nor desired. In addition, points cannot contain all the information that may be required to do a high level energy optimization or may have errors. To solve these issues only use point names in the initial population of the meta-data database. In order to discuss the meta-data further we present it in the context of the example implementation (BacLog).

In the system an *element* is considered by its end function and it does not change over time. An *element* is considered new if it changes function (if it is moved, or the building configuration changes). For example a temperature sensor in room 240 would be considered an element. Elements are assigned internal numbers so that they can be associated with an object. An object connects a physical point on a device (a BACnet *type* and *instance* on a *device*) for a period of time. The intermediary object is needed as connections can move around physically or virtually over time (most information in the database has an explicit time period for which the information is valid). For example, when a building level controller is replaced all the VAV boxes on the FLN may receive a new address (BACnet object identifiers) and descriptions.

For the above reasons the meta-data is associated with an *object*. The meta-data consists of list of *tags* or key/value pairs. Data includes zone, VAV, AHU, function, room, and building data. In addition, it can include room construction, location and neighbors, or any other data. Given the small size of the meta-data it is expected that an application would load a subset of the meta-data and utilize it directly (as a map or an associative array) as the database structure makes it difficult to easily construct complex queries.

APPLICATION EXAMPLE: PROCESSING NON-UNIFORM TIME SERIES DATA

Processing data for use for automated fault diagnostic systems can take various forms depending on the analysis code. One common form is uniform time series or binned data. The process of down sampling and binning takes high resolution change of value data and creates interval data at various time scales. Time interval data are a lot easier to analyze, however, information is lost during the conversion process. Since the exact format needed is highly dependent on the software used recommendations are present in the context of a small scale example (BacBin) to frame the discussion.

Example Data Collection System

The data collected for this example use an open source BACnet data collection system (BacLog). For the most part we will ignore the implementation except for the main database table for data collection. We will also ignore the meta-data associated with the points, however, a real system would have to map a physical point to a specific functional element, for example a temperature sensor in a room. The main database table where the building data are stored can be found in Table 22-2. The schema only stores raw data collected, it is the job of the data processing tier to map physical measurements to meaningful data via the meta-data when the data are processed. The index on time allows the table to be easily sorted by time. Without sorting the rows/events returned by the database are not guaranteed to be monotonic (ordered). Monotonicity allows a large number of code simplifications and optimizations to be made.

One interesting observation during the development of this particular system was that updating meta-data after-the-fact in the log did not improve performance as modifying rows destroyed the natural time-based on-disk layout of the data and was extremely time intensive. This contributed to the design of splitting data collection with data processing.

Stream Sampling

When data are read from a device the value is most likely just the current value of a device. The trouble with this method is best explained with an example. If the outside temperature where read every day at midnight

```
CREATE TABLE Log (
  time timestamp with time zone,  -- time measurement occurred.
  IP inet,                        -- remote IP
  port integer,                   -- remote port
  type integer,                   -- remote object type
  instance integer,               -- remote object instance
  status integer,                 -- point status
  value real                      -- recorded value
);
CREATE INDEX i_Log_time ON Log (time);
```

Table 22-2: Schema for logging data.

it would give a very different picture than if it were read at noon. The advantage of using Change of Value is that it reports not only the new value but when it changes. This is important even at the somewhat smaller size of 15-minute interval samples as values can change drastically over this period of time as well. In reality Change of Value is a simple form of compression. By changing the reporting interval (a dead-band where no changes are reported) the compression becomes lossy and can remove physical, analog, or digital noise.

The conversion from Change of Value to time-interval bins was done using a streaming approach, which processes sensor events one at a time and immediately discards it. The reason for using the seemingly complex streaming based sampling approach is that there are a lot of barriers to doing the analysis using a traditional array or batch based approach, the primary being that there simply is not enough memory to hold all the samples. If a well-designed streaming approach is used and embraced at all levels then the complexity becomes manageable.

By combining two Change of Value events a vector can be created that shows how long a value was held and what was the resulting change. It is important to think about the change vector as if looking backwards, that is, compute the change in time and value by comparing it to the previous value* and immediately report it. If the new value is held until the next value is known then there will be a delay in the reporting the value adding latency, which increases the time in which actionable information is reported. Figure 22-2 illustrates this representation, where the circles represent Change of Value events (C_1 and C_2). When event C_2 occurs the two component vectors Δt_1 and Δx_1 are computed and reported and are used for on-line fault diagnostics or to fill the

bins. However, we must first discuss the creation of the bins before they can be filled.

Creating Bins

For the data collection tier it was recommended to use a comprehensive native date function due to the one-way nature of data collection. However, since in the data processing tier the information can be re-created it is not critical ensure the correctness of the data. Because of this a custom time format can be used to enhance the usability of the data by breaking time down into separate time scales. This is beneficial because automated fault diagnostics codes will often compare like time periods (intra-hour, intra-day, intra-month, etc.) during analysis, also working with dates directly can be difficult in some analysis software. In our example we choose to align with the commonly used 15-minute interval for a longitudinal study, however, most fault diagnostics software will need a smaller interval period. The remaining intervals are along day, weekday, week, year and daylight savings. Depending on the design daylight savings can either be a separate column or rolled into the year column dividing the year into three parts (spring standard time, daylight savings, and fall standard time). Other qualifiers such as on-peak, off-peak, holiday, workday, etc. can also be added depending on the needs of the analysis. Since there are many possible definitions for year, week, and weekday the ISO definition was chosen as seen in Table 22-3.

The table is also indexed the individual year, week, weekday, and interval columns. The number of items in a bin (n) and the total weight of the bin (w) is also stored with the weighted mean value of the bin. The weight is the amount of sample time covered in the bin. The index provides an efficient way to access a particular bin or point, without indexes the entire dataset would need to be scanned. The combined slot index allows (depending on the type of index used by the underlying database) the efficient retrieval and sorting of events over time.

Filling Bins

The bins can be filled with a number of statistics about the behavior of the equipment during that time period. One of the simplest is the mean value. With Change of Value reporting taking the mean of the samples does not give a correct value since it is a non-uniform time series. One solution is to take a weighted mean with the weighs for the values being the time in which the value was held (Δt in Figure 22-2). The weight (w) value in the example is also stored in the bin is the sum of the time that a value is known, which in the case of 15-minute in-

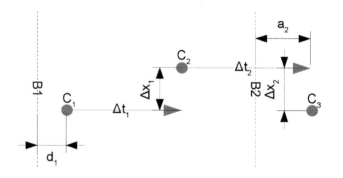

Figure 22-2: Representing Change of Value events as vectors.

---

*This is a trivial process when reported values are monotonic; otherwise it becomes a challenging task.

```
CREATE TABLE Bins (
  isoyear smallint, dst smallint,
  isoweek smallint, isoweekday smallint, interval integer,
  device integer, itype smallint, instance smallint,
  n integer, w real, mean real,
  PRIMARY KEY
(isoyear,dst,isoweek,isoweekday,interval,device,itype,instance)
);
CREATE INDEX i_Bins_slot ON Bins
(isoyear,dst,isoweek,isoweekday,interval);
CREATE INDEX i_Bins_point ON Bins (device,itype,instance);
```

Table 22-3: Schema for representing time-series events created from non-uniform time-series

terval is 900 seconds. This is used for verification of data consistency since unconfirmed messages are used by the data collection module and there is a possibility that that some Change of Value events are lost. To mitigate this problem and to put an upper-bound on the bin filling algorithm (among other things) the data collection system periodically polls the devices for values and re-registers for Change of Value reporting events.

Filling bins at first seems like a straight forward task until the edge cases are considered, which are all the ways in which a vector can cross a bin boundary or boundaries. Figure 22-2 shows the most common case where a vector is contained within a bin (C_1 to C_2) and the case where a vector spans a bin boundary (C_2 to C_3 crossing B_2). When a vector crosses a boundary it is split between bins, the difficulty occurs when a vector spans multiple boundaries or a device no longer reports CoV values. This is the reason that the component vectors should be considered separately so that the duration of the previous event can be determined and the value of the current event is known.

The simplest implementation, the one used in the example for this chapter, is to assume the value remains at the same value until a new event is detected (hence the horizontal arrows for the Δt vectors). In Figure 22-2 the value of the equipment at C_1 is considered to be C_1 up until it changes at C_2 to C_2. There are more sophisticated ways of computing the mean value of a bin (interpolation, etc.) but if the reporting interval for the Change of Value is configured properly then the differences in the methods should be minimal and overall should not be an issue given the length of an interval and the number of values within.

In addition to the mean other values can be computed and stored based on the analysis. Again the advantage of splitting the data collection and the data pro-

cessing into two tiers is that the summary data can be recomputed. Although this cannot be done on a real-time bases it is quick enough to be done fairly often and on a subset of data. For the example problem, just under 24 months of data containing 2.4 billion records and over six thousand points were processed in a little under 4 days on 2009-era workstation hardware (2.3GHz 8 core, 8 GiB RAM and 800 MB of SATA storage). The code was not optimized but based on preliminary indications this number could be at least reduced in half.

Analyzing Bins

When dealing with large amounts of data one should resist using spreadsheet based approaches to analyzing graphing data. For prototype and quick analysis it may be an acceptable tool, however, at this scale the sheer volume of data can overwhelm the spreadsheet software. In addition the effort required to reproduce the analysis and graphs with similar data (say the next period) is high as a lot of effort is required to automate the entire process. The upfront cost of learning a proper analysis and visualization environment is quickly amortized by the ease in which future visualization and analysis can be done.

There are a large number of analysis and visualization software packages in existence, and one should be chosen that works well with the underlying environment that the automated fault diagnostics tool uses. Keeping with the open nature of the example used in this chapter, the open source statistical analysis software "The R Project for Statistical Computing" (R) was used. R has a large statistical library, graphing tools, and database connectivity tools and runs on a large number of platforms.

The binned format makes analysis a lot easier since the data are now both uniform and, for the most part complete. When change of value is used constant values can span the analysis window making querying difficult and is one reason why the software polls all points roughly every hour, making the amount of lead-in time needed to get all the data points finite. Comparing points is also difficult with non-uniform time series since sample times will rarely match (never on a analysis server), which requires tools and analysis techniques that explicitly handle non-uniform time series. The examples in this chapter use 15-minute intervals but this could easily be changed depending on the amount of granularity needed. Since most low hanging opportunities will be obvious over 15-minute intervals (stuck dampers etc.) it is best to start with this granularity as the amount of data per point is low (96 per day, 35,040 per year).

Retrieving Bins

The first step is to read in the data from the database to the analysis package and get it in a form that can be used. This involves taking the row based information about attributes and making them columns, a pivot table the spreadsheet world. This can be either done in the database (specialized queries can make this easier) or in the analysis package (again libraries are available for this). The choice depends on ease of use and performance and how the database is indexed. This form of analysis only works for small subsets of the data (both points and time) as the data can quickly fill all available memory. Table 22-4 is an example of how the data are retrieved and structured for easy analysis using Postgres and R, note the use of the Postgres specific extension hstore that allows easy querying points based on their tags.

To examine the properties of rooms we can now do some quick analysis, for example compare the commanded flow of the VAV box (FLOW_CTRL) vs. the actual flow rate (FLOW) in units of percentage. Figure 22-3 shows that the box is commanded to around 90% and the actual flow ranges between 50% and 100%.

Interestingly enough, the flow range depends if it is in cooling or heating mode, or if the limits are manually changed. Even though these values should not change in a commissioned building they may be altered for a number of reasons. This re-enforces the need to collect all data, even if it seems that it will not change or is not interesting, it also illustrates the need to report values in standard units and in standardized ways (for example air flow in CFM).

Value Standardization, Calibration, and Post Processing

Since we are collecting the data directly provided by the data collection tier values must often be corrected or reported as standardized values for later consumption. It is important that any calibration and exception information used be stored in the meta-data database as well. Again standardization, calibration, and other post-processing should be done in the data reduction part of the system. This choice is strategic; by placing the analysis here corrections must be re-applied during each data reduction forcing it to be both automatic and stored in the meta-data store. If this is not done it may be corrected once in the main database and either the reasoning is lost or more troublesome done incorrectly and not fixable.

Standardization is an important part of the post-processing phase and is a part of the overall data collection program. What values are collected and what are their units. Although this sounds straight forward, value reporting changes from vendor to vendor, from application to application (configuration), and even

```
d <- sqlQuery(odbcConnect('PostgreSQL'),"
SELECT device,itype,instance,
  isoyear,dst,isoweek,isoweekday,interval,
  tag->'room' AS room, tag->'attr' AS attr,
  n,w,mean as value
FROM Devices JOIN Objects USING (deviceID) JOIN hTags USING (objectID)
JOIN bins USING (device,itype,instance)
WHERE tag -> 'room' IN ('120', '241', '240')
ORDER BY isoyear,dst,isoweek,isoweekday,interval")
p <- cast(d, isoyear+dst+isoweek+isoweekday+interval ~ attr + room)
xyplot(flow_stpt_120~flow_120,p)
xyplot(flow_stpt_241~flow_241,p)
```

Table 22-4: R commands to retrieve, reorganize, and plot binned data.

mode to mode (cooling vs. heating). In the case of the example used in this chapter the air flow rates are reported as a percentage of a number of (changeable) control set-point depending if it is in cooling or heating mode. For analysis the use of a standardized value, such as CFM, may be much more appropriate. The standardization of reported values and units should be a part of the overall data management program.

Post processing can remove or mark identified anomalies (not done in this example) so they can be handled appropriately by later analysis (equipment failure, power failure, etc.) if desired. Depending on the type of information this can be either stored as a part of the meta-data or as a part of the fault diagnostics system.

Visualization

An entire chapter or book could be written on visualizing data but since this chapter is about the data collection and analysis infrastructure we will only mention it briefly. Bin data are fairly easy to manage and visualization is easy for overall analysis and developing fault diagnostic information. Simple examples of this type of visualization using R can be seen in the section on Retrieving Bins the section on Analysis. Although spreadsheets and custom BAS software can be used, a direct interface to a statistical and visualization system can be much more powerful and flexible after the software is mastered. In the case of R, the same tools used for visualization can also be used in the fault diagnostics software as we will see later.

The uniform nature and the reduced size of the bins makes a software system such as R suitable for medium scale visualization and analysis. After a certain point the size of the data will overwhelm traditional systems. If large scale analysis is needed specialized tools are needed such as ROOT by CERN [ROOT]. The recommended approach to scaling is to start off small and continue to build larger and larger system. Even though much of the technology will be discarded along the

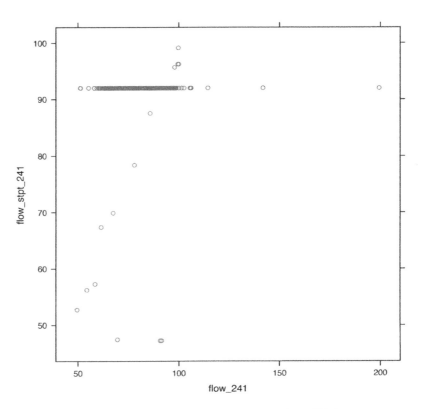

Figure 22-3: VAV flow setpoint (%) v.s. actual flow (%) for room 241.

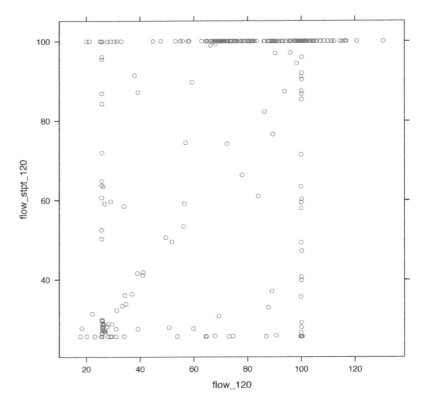

Figure 22-4: VAV flow setpoint (%) v.s. actual flow (%) for room 120.

way important lessons will be learned in the process. By following the recommendations in this chapter the system will be built on a reliable (both in terms of technology and the reliability of the data) data collection and archival system. As the fault diagnostics system grows the tools can evolve with it and the pain of migrating data to newer systems will be lessened by the tiered nature of the system.

Even though R, or other similar tools, may not suitable for visualization and analysis at full scale the data reduction combined with a database make analysis possible. In the following section we demonstrate an example analyses scenario.

Room Temperature Heatmap

In this example we hope to show the simplicity and expressiveness of using a command based (as opposed to a spreadsheet or a BAS) tool. To increase readability the output formatting tools have been replaced with ellipses (...), however the full code is publicly available (BacBin). This example continues with the data loaded in Table 22-5.

In this example we are simply looking at overall usage patterns and what nominal operation should look like (only by knowing what nominal is can we tell what is not nominal). Table 22-5 shows how this is done. We first arrange (cast) the data into 15-min intervals* by (~) the day of the week separated by (|) the attribute of interest and (+) the room. Since we are only looking at the day of the week and interval the other dimensions (week, year, etc.) must be aggregated (the last argument to cast). We can look at both mean (mean) and standard deviation (sd). The dollar ($) symbol in R is used as an index, since we give both the attribute and room we are left with an interval v.s. isoweekday block of data, which the as.matrix function converts to the proper form needed by the heatmap function. The ISO weekday starts on Monday with an index of 1, which conveniently shows the weekends together at the end of the week. The resulting heatmaps are shown in Fig-

ure 22-5 and Figure 22-6 for mean and standard deviation respectively.

Figure 22-5 shows that the room remains constant throughout the week except for three periods. This result shows two things, one that the room is mostly unoccupied and that there is no night setback. Looking at the standard deviation confirms this but also shows that there is a high variance on Thursday (4). Investigating further we drop the week from one axis to allow us to look over the ISO week number (isoweek). Since these are individual values we no longer need to aggregate over some other axis. Figure 22-7 shows the resulting heatmap plot and that our guess was correct, that the event only occurred during ISO week 6 of the year and Figure 22-7 shows a regularly occurring event.

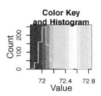

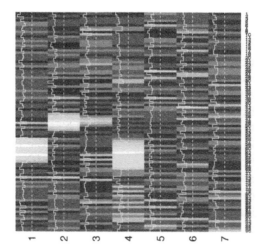

Figure 22-5: A heatmap showing the mean room temperature (ctl_temp) in °F broken down by time of day (interval) and the day of week for room 240.

```
pm <- cast(d,interval ~ isoweekday | attr + room, mean)
ps <- cast(d,interval ~ isoweekday | attr + room, sd)
heatmap.2(as.matrix(pm$ctl_temp$`240`), …)
heatmap.2(as.matrix(ps$ctl_temp$`240`), …)
```

Table 22-5: R commands to create interval/weekday heatmap charts.

*Simply divide the interval by 4 to get the 24 hour time. After a short time thinking in intervals is almost natural.

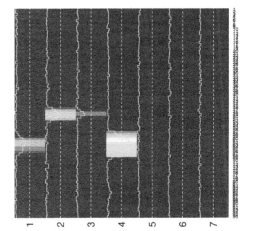

Figure 22-6: A heatmap showing the standard deviation of room temperature (ctl_temp) in °F broken down by time of day (interval) and the day of week for room 240.

This type of visual analysis can be done for a large number of parameters but not for a large number of rooms, or this we need an automated analysis and for this we turn to regression. For the next set of analysis we are looking at all the rooms but are now looking at the measured room air flow as a percentage and the that the damper is commanded as a percentage of maximum air flow (tag -> 'attr' in ('flow', 'flow_ctrl')) instead of temperature.

Analysis and Automated Fault Detection

One approach to automated fault diagnostics is to build the library of analysis based on past events. Once a problem is identified either by user complaints or by exploratory visualization an analysis is performed to detect this condition. Only after the condition can be detected and understood, is it corrected. Once the condition is resolved the before, fault, and after conditions are collected and analysis and packaged as a test case (a unit test in computer science language) that can be used to test other detection methods or to identify other faults in the system. It also forces a "root cause" type of analysis

```
p <- cast(d,interval ~ isoweek | isoweekday + attr + room)
heatmap.2(as.matrix(p$`4`$ctl_temp$`240`), …)
```

Table 22-6: R commands to create interval/week heatmap charts for a single weekday.

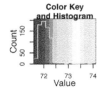

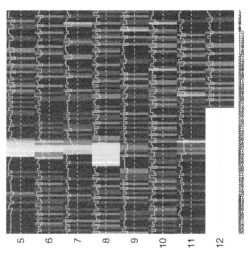

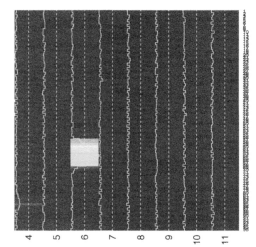

Figure 22-7: A heatmap showing the 15-minute interval (rows) temperatures (°F) for Thursdays (4) from week 5 to week 12 (columns).

Figure 22-8: A heatmap showing the 15-minute interval (rows) temperatures (°F) for Mondays (1) from week 3 to week 7 (columns).

that should lead to a proper solution instead of a Band-Aid fix. To illustrate this type of analysis we present an obvious case conducted early in the example used in this chapter.

Early analysis of the building indicated that there were some issues related to the VAV air flow. The first step was simply to ask if the VAV devices where controlling air flow correctly. Instead of looking for obstructions, broken arms, and other physical indications of failure the question was asked "how should the dampers operate?" For the answer we looked at the fundamentals of VAV operation and at the controller design (all this is obvious but we were developing a structured approach to automated fault diagnostics—learn when it is easy). At the highest level, the controller commands the VAV to a percentage of maximum flow (determined by setpoints) and controls the damper to reach the desired airflow in the room. At this point we do not care what that position is but under ideal conditions the actual flow should meet the commanded flow as shown in Figure 22-9. Unfortunately this is not always the case as seen previously in Figure 22-7 and Figure 22-10. The question is what other VAV units are not operating correctly? Since R is a powerful language we could simply put the visualization in a loop and look at all the room in the building (this was actually done to validate the analysis), however, since we want to automate the detection we need to perform an analysis.

The analysis consists of determining what the expected behavior is, what the actual nominal behavior is, and what failure looks like and then developing a mathematical model of this. As stated earlier, we expect that in a nominally operating VAV box that the actual flow rate (flow) will be the same as the commanded flow rate (flow_stpt). Since the controller does all the flow normalization and abstracts the damper position for us we know that this will be a simple linear model with a slop of 1 and a intercept of 0. By looking at Figure 22-9 we can see that this is indeed the case and that in non-nominal (note this is not necessarily failure) shown previously is not a linear model, which gives us a way of detecting nominal operation.

To accomplish this, as shown in Table 22-7, the data was arranged by time and attribute for each room (year and daylight savings was dropped since the sample was taken from the same year without a change in daylight savings) and the subsequent linear model (lm) was calculated for each room (r).

The summary of the model for room 230 (nominal) and 240 (no nominal) is shown in Table 22-8:

The high R-squared value (0.96, closer to 1 is usually better) for the first model (a slope of 0.94994 and an intercept of 1.81994) indicate a good fit along with it matching the expected model (slope of 1 and intercept of 0). The second model shows a poor fit (R² of 0.1474) indicating that it does not follow the model. Of the 57 rooms analyzed the ones with an R-squared statistic greater than 0.85 all visually showed nominal operation. All the rooms with sever issues had either very low R-squared value or failed to produce a model. Model failure was due to rooms having only a single commanded value (Figure 22-10), which produces model with a vertical line (infinite slope). After further investigation this behavior was found to be due to a failed room sensor reporting a constant 80°F.

AUTOMATED FAULT DIAGNOSTICS

The goal of this chapter was to present a methodology and architecture for the long term collection and analysis of high resolution non-uniform time-series building data for automated fault diagnostics. We showed the use of such a system through a number of example implementations. We also showed how this data can be visualized and analyzed for faults. The source code for the BACnet data collection system and the post processing system is publicly available (BacLog, BacBin). With any computational automated analysis system the real val-

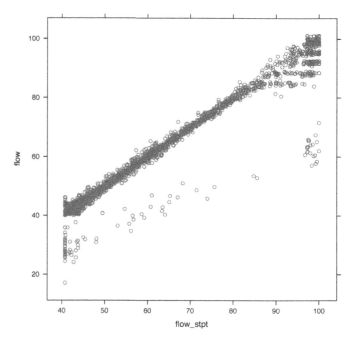

Figure 22-9: Expected commanded flow rate vs. actual flow rate in percent.

```
b <- cast(d, isoweek + isoweekday + interval ~ attr | room)
for(r in names(b)){ m <- lm(flow~flow_stpt,b[[r]]) }
```

Table 22-7: R commands to perform a regression on flow v.s. flow setpoint

```
> lm(flow~flow_stpt,b[['230']])

Call:
lm(formula = flow ~ flow_stpt, data = b[[r]])

Residuals:
    Min      1Q  Median      3Q     Max
-38.363   0.017   0.257   0.767   7.189

Coefficients:
            Estimate Std. Error t value Pr(>|t|)
(Intercept)  1.81994    0.13995    13.0   <2e-16 ***
flow_stpt    0.94994    0.00258   368.2   <2e-16 ***
---
Signif. codes:  0 '***' 0.001 '**' 0.01 '*' 0.05 '.' 0.1 ' ' 1

Residual standard error: 3.214 on 5170 degrees of freedom
  (5 observations deleted due to missingness)
Multiple R-squared: 0.9633,     Adjusted R-squared: 0.9633
F-statistic: 1.356e+05 on 1 and 5170 DF,  p-value: < 2.2e-16

> lm(flow~flow_stpt,b[['240']])
Call:
lm(formula = flow ~ flow_stpt, data = b[[r]])

Residuals:
    Min      1Q  Median      3Q     Max
-50.204   0.108   0.469   0.887  23.464

Coefficients:
            Estimate Std. Error t value Pr(>|t|)
(Intercept)   6.4555     1.8116   3.563 0.000369 ***
flow_stpt     0.8877     0.0297  29.890  < 2e-16 ***
---
Signif. codes:  0 '***' 0.001 '**' 0.01 '*' 0.05 '.' 0.1 ' ' 1

Residual standard error: 3.919 on 5169 degrees of freedom
  (6 observations deleted due to missingness)
Multiple R-squared: 0.1474,     Adjusted R-squared: 0.1472
F-statistic: 893.4 on 1 and 5169 DF,  p-value: < 2.2e-16
```

Table 22-8: Summary of results produced by Listing~22-16.

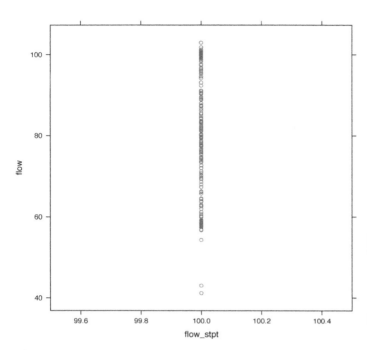

Figure 22-10: Unexpected commanded flow rate vs. actual flow rate in percent.

ue is in its potential to easily and inexpensively scale to large numbers. It is our hope that by following the proposed architecture and data management program presented here that such scalability can be achieved.

Acknowledgements

The author would like to thank the University of Florida Physical Plant Department for their support of this project and Prabir Barooah for spearheading this project.

Acronyms, Software, and Web Links

1. (BacBin) Timothy Middelkoop, Building Automation System Binning and Analysis, Python and R, Apache 2.0, https://github.com/mtim/BacBin.
2. (BacLog) Timothy Middelkoop, An asynchronous BACnet logger written in Python, Python, Apache 2.0, https://github.com/mtim/BacLog.
3. (BAS) Building Automation System.
4. (Cassandra) Cassandra NoSQL database, http://cassandra.apache.org/.
5. (Haystack) Project Haystack, http://project-haystack.org/.
6. (HDF) Hierarchical Data Format, http://www.hdfgroup.org/.
7. (HBase) HBase database, http://opentsdb.net/.
8. (MariaDB) MariaDB database server, http://mariadb.org/.
9. Open Historian, http://openhistorian.codeplex.com/.
10. (oBIX), Open Building Information Exchange, http://www.obix.org/.
11. PI Server, http://www.osisoft.com/.
12. (PosgreSQL) Postgres database server, http://postgresql.org/.
13. (R) The R Project for Statistical Computing, http://www.r-project.org/
14. (SSD) Solid State Drive.

References

Yashen Lin, Timothy Middelkoop, and Prabir Barooah. Issues in identification of control-oriented thermal models of zones in multizone buildings. In IEEE 51st Annual Conference on Decision and Control, pages 6932-6937, Maui, HI, USA, December 2012. [10.1109/CDC.2012.6425958]

He Hao, Timothy Middelkoop, Prabir Barooah, and Sean Meyn. How demand response from commercial buildings can provide the regulation needs of the grid. In 50th Annual Allerton Conference on Communication, Control, and Computing, October 2012. invited paper.

Timothy Middelkoop and Herbert Ingley. Emodel: A new energy optimization modeling language. In Barney L. Capehart and Timothy Middelkoop, editors, Handbook of Web Based Energy Information and Control systems, chapter 23, pages 325-337. Fairmont Press Inc., 2011.

Rene Brun and Fons Rademakers, ROOT—An Object Oriented Data Analysis Framework,

Proceedings AIHENP'96 Workshop, Lausanne, Sep. 1996, Nucl. Inst. & Meth. in Phys. Res. A 389 (1997) 81-86. [http://root.cern.ch/]

Appendix:
Example Implementation Details

The system presented in this appendix and referenced in the chapter was developed to support both research and production (building operation) HVAC-R systems on a campus-wide scale using open source tools and technologies built on open standards. The system combines both data collection and controls and the source is publicly available (BacLog). The goals of the system influenced the design and implementation of the resulting in the following design choices.

• The system utilizes only open standards and open building automation protocols (BACnet). This prevents vendor lock-in and allows for the development and use of open communication stacks.

• The system only uses open-source libraries, frameworks, and systems. It is based around the open source database system Postgres (PostgreSQL).

- The system is developed using open source languages and tools (Python and Eclipse). This will allow contributors to participate in development without having to invest in expensive development tools.

- The system is licensed entirely under an open source license, the GNU GPL compatible Apache 2.0 license. This ensures that contributors and users have perpetual use of the system, and we believe also encourages participation.

The system is built around a Building Data Server (BDS). The BDS hosts a building database to collect, store, and pre-process data from sensors in the building that are already integrated with the building's BMS. In the system we have implemented at Pugh Hall, the data server is physically located in the building to take advantage of the close proximity to the data sources. This reduces the reliance on the campus network, reduces campus network traffic, and increases control reliability as the collection and control occurs at the LAN (Local Area Network) level. However, in principle the BDS can be physically located anywhere as long as it can meet the communication delay requirements of the control algorithms in the EMS.

The relationship between the various sub-systems is shown in Figure 22-11. The BDS connects to the building level controllers on the LAN via BACnet over IP. Building level controllers then, in turn, communicate to floor level equipment and controllers—such as air handlers, VAV (Variable Air Volume) boxes, pump and fan drives—over the FLN. Data are collected from the floor level equipment and controllers. A scheduler updates operational control parameters/points to change operational characteristics based on the data in the schedule table in the control database. The system runs independent of the BMS, which also connects with the building level controllers, and in most cases co-exist without conflicts*.

Data retrieved from the FLN (sensor data) are stored in database table on the sensor database cluster. The database cluster is replicated to a central database server, called the Application Data Server (ADS), which is used by the EMS. A database cluster contains multiple databases and the servers contain multiple clusters. Replication allows a live read-only copy of a database cluster to run on another server and service requests for

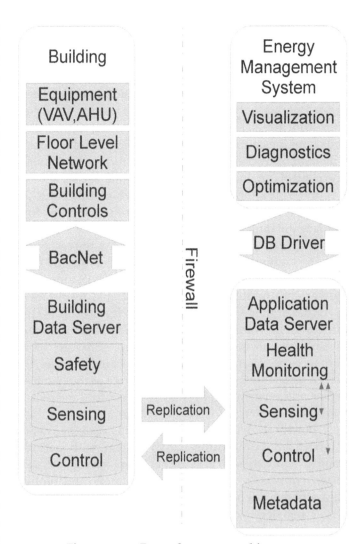

Figure 22-11: Example system architecture

information (queries). In this way the ADS allows clients to query the sensor database cluster on the ADS without accessing the building server. The spl it allows access to the data to be from outside the campus building network or virtual network (VLAN) increasing security. The split also allows the BDS to run on less expensive hardware, relegating the expensive queries to the more powerful ADS. The EMS connects to the ADS server exclusively over a database connection increasing security further.

The ADS server also houses meta-data (information about the FLN), a control database, and a health monitoring system. The health monitoring system ensures that the building remains in a operating safe condition protecting the system from failures in the energy management system. If a failure is detected the system resets the building to a traditional operating mode and locks out any further control. How comprehensive the health monitoring system is depends on the user devel-

---

*Care should be taken when interfacing with physical systems, see the Disclaimer section for more details.

oping it and the building equipment used.

The EMS is a collection of applications that includes (production and research) visualization, diagnostics, control, demand response, etc. that run independently from the building data server and database server. These codes interact with the hardware by reading building data from the sensor database on the database server and changing control parameters in the control database, which are in turned relayed to the building equipment. The control database resides on the database server and is replicated to the building server, which translates the control schedule to BACnet commands, which in turn changes the building environment.

The separation between the EMS and the rest of the system is intentional. This separation allows algorithms of arbitrary complexity to be used in the EMS, since now they are not limited by the computational resources of the either of the two database servers. It also allows easy integration of third party algorithms/applications, such as fault diagnostics by other vendors' software, or by human experts who are physically located off site. The design also allows for the eventual move of the database server and energy management system to the cloud as the system grows.

For these reasons the system was designed using a push Change of Value (CoV) system with periodic polling. More specifically, rather than having the data server connect to the building equipment and request data whether it changed or not (a pull model) the data server registers with the building equipment and the equipment sends updates to the data server when, and only when, a value changes (push model). In this way high resolution data are obtained without the need to constantly poll for data that does not change. Since data are transmitted when the value changes updates will be randomized over time reducing spikes in network traffic. Periodic polling is used to ensure that the equipment is still operational and to reduce the maximum amount of time a point goes without being reported. This enables a number of optimizations when retrieving data from the database.

SERVER ARCHITECTURE

The BDS (Building Data Server) is built using Python and utilizing asynchronous I/O on Linux (Ubuntu 12.04). An asynchronous design was chosen as it is well suited to this type of application, and is easy to implement. In simple terms, the server never waits for an I/O operation to complete (reading network data or writing

to the database); it simply continues processing other data and when the operation completes, takes appropriate action. A blocking approach requires the use of multiple threads (and more difficult parallel programming) to achieve the same high throughput and low latency operation.

BACnet STACK

The building data server utilizes the open standard BACnet protocol. The standard is available for a reasonable price and is royalty-free with a number of open source implementations. The asynchronous design of the server necessitated a custom solution for the BACnet stack and a new stack was written entirely in Python for this purpose with only the Energy Management System in mind. The tight integration between the stack and the application facilitated a simplified design.

META-DATA DATABASE

The meta-data consists of list of tags or key/value pairs. Data includes zone, VAV, AHU, function, room, and building data. In addition, it can include room construction, location and neighbors, or any other data. The schema for the tag database is shown in Table 22-9.

SENSOR DATABASE

The sensor database contains information about the sensors and the measured data. The database consists of point, object, and device tables to hold meta-data and a log table to store the data. The system is bootstraped with an external list of BACnet devices to be monitored. Upon startup this list is used during a discovery phase to collect the list of points and store metadata collected from the controllers, which is used until the topology changes. Since buildings environments change over time point, object, and device information is associated with a time range where a undefined finish time (represented by the SQL NULL) indicates that an item is active (see the previous section for a more detail discussion). The schema for point, object, and device tables can be found in Table 22-10 and the sensor data can be found in Table 22-11. IP (address) and port are the Internet address of the BACnet equipment and network and device are the BACnet addresses of the equipment. There is some schema denormalization to improve query performance.

```
CREATE TABLE Tags (
  objectID integer,
  tag char(8),
  value varchar,
  CONSTRAINT Tags_PK PRIMARY KEY (objectID,tag)
);
```

Table 22-9: Meta-data table

```
CREATE TABLE Devices (
  deviceID SERIAL, -- Internal ID IP inet,
  port integer,
  network integer,
  device integer,
  name varchar,
  first timestamp with time zone, last timestamp  with time zone,
  CONSTRAINT Devices_PK PRIMARY KEY (deviceID)
);
CREATE TABLE Points (
  pointID SERIAL, -- Internal ID campus char(8), building integer,
  first timestamp with time zone,
  last timestamp  with time zone,
  CONSTRAINT Points_PK PRIMARY KEY (pointID)
);
CREATE TABLE Objects (
  objectID SERIAL, -- Internal ID deviceID integer,
  pointID integer,
  itype integer, -- objectType
  instance integer, -- objectInstance
  name varchar, -- objectName
  description varchar,
  first timestamp with time zone,
   last timestamp  with time zone,
  CONSTRAINT Objects_PK PRIMARY KEY (objectID)
);
```

Table 22-10: Sensor point, object, and device tables

```
CREATE TABLE Log (
  time timestamp with time zone,
  IP inet, -- remote IP
  port integer, -- remote port
  objectID integer,
  itype integer,
  instance integer,
  status integer,
  value real - recorded value
);
CREATE INDEX i_Log_time ON Log (time);
```

Table 22-11: Sensor data table.

Currently the database contains over two years' worth of data and over two billion samples. The CoV approach allows sensor resolution around a second. At this rate an equivalent pull approach would require over 150 billion records.

CONTROL DATABASE

Since the control database is replicated from the main database server to the building server the scheduler was designed to have two tables, one on each server. The points that need to be scheduled are inserted as requests to the scheduler table. The SERIAL keyword creates a monotonically increasing sequence of integers and schedules with higher values take precedence over lower values. Schedules that are inserted into the table at a later time take precedence over all schedules even after the schedule value expires (until column), schedules that are overridden are simply discarded. This design was selected so that an EMS application could plan using a rolling horizon with future schedules being constantly updated. If connection to the database server/optimization system is lost then the building will simply continue to use the schedule. This also allows the control computation routine some flexibility in scheduling as some solutions may take longer to compute, there will always be a solution present as backup. Newly inserted schedules, with updated information, override any schedules after the start of the first new schedule. The schema for the scheduler is shown in Table 22-12.

The building server contains a similar control table which is a copy of the scheduler table with additional rows to implement the scheduling precedence logic in an efficient manner. In addition there is a commands table that logs all commands sent to the building controls, which is written to before and after a command is sent and is used for auditing and debugging purposes. Finally, there is a Watches table to limit control and sensor values, which is part of the external health monitoring system that ensures that the physical system operates safely and within expected values, any deviation results in the system being switched to an autonomous fail safe mode. This is an important capability to include when developing sophisticated EMS applications.

APPLICATION INTERFACE

The system is designed simultaneously to support a number of applications by interacting with the various databases. Control commands from the application are communicated to the system by appending a row to the schedule table in the ADS server. The scheduler, which checks for such updates every second, then communicates these new commands to the building actuators/equipment controllers through BACnet. This provides a consistent interface for any third party application in the EMS, since the format of the commands is the same irrespective of the application.

Disclaimer

Since this software interacts directly with building equipment care should be taken before a project of this scope is undertaken. The appropriate administrative, building, equipment, network, security, safety, and any other related personal should be consulted first. No warranty of the concepts or software is expressed or implied. Safety First.

```
CREATE TABLE Schedule (
  scheduleID SERIAL, -- order
  objectID integer,
  active timestamp with time zone,
  until timestamp with time zone,
  value real, -- NULL to clear
  CONSTRAINT Schedule_PK PRIMARY KEY (scheduleID)
);
```

Table 22-12: Scheduler table

Chapter 23

Localized Uncertainty Quantification for Baseline Building Energy Modeling

Abhishek Srivastav, Ashutosh Tewari, and Soumik Sarkar,
United Technologies Research Center

Bing Dong, Department of Mechanical Engineering, University of Texas
Mikhail B. Gorbounov, Corresponding author

ABSTRACT

Prediction of building energy usage and its uncertainty analysis are critical to characterize the baseline performance of any building for impact analysis of energy saving schemes such as fault detection and diagnosis (FDD), control strategies and retrofits. Recently several methodologies have been proposed for localized confidence estimation for baseline building performance modeling, that include methods based on Nearest-neighbors, Gaussian processes and Gaussian Mixture Regression (GMR) among others. This chapter presents the approach based on Gaussian Mixture Regression and a comparative review of different building energy modeling methods. The Comparative evaluation of reviewed methodologies is carried out using one year worth of simulated data for a supermarket in Chicago climate. We show that, GMR approach resulted in a baseline cooling model that was comparable with the other models in terms of prediction accuracy. However, the uncertainty in the estimated energy derived from GMR approach was found to be much more consistent with the observed data compared to other competing approaches.

INTRODUCTION

Building sector uses around 40% of the total energy in the U.S. [EPA, 2008] and the statistics is very similar in the EU countries as well [EUCOM, 2013]. Therefore, the Department of Energy (DoE), the International Energy Agency (IEA), the Intergovernmental Panel on Climate Change (IPCC) and other agencies have declared a need for commercial buildings to become 70-80% more energy efficient. Although energy-efficient building control technologies are emerging, a key challenge is how to effectively maintain building energy performance over the life-cycle of the building. To this end, there is an urgent need for fault detection and diagnostics (FDD), retrofit and advanced control technologies. These technologies can potentially reduce significant energy inefficiencies resulting from faults, degradation of building equipments & materials, and errors in operating schedules and critical design & planning flaws.

Before any advanced control, fault detection and diagnostics (FDD) and expensive retrofit technologies can be applied to improve energy efficiency, a high-fidelity baseline energy performance model is often needed to help understand building operation. A baseline model developed using pre-retrofit data can be used to estimate expected pre-retrofit energy use for the post-retrofit conditions. Thereafter, the difference between the estimated and the actual energy use in the post-retrofit period can be attributed to the retrofit changes. One of the popular methods of baseline modeling is the univariate change-point method [Reddy et al., 1997] and multi-variate linear models [Wu et al., 1992, Katipamula and Claridge, 1993]. The primary reason behind the practical success of these models is the empirical observation that total building energy usage as a function of external conditions such as dry-bulb temperature and humidity have been adequately captured by linear or piecewise linear data-driven models. In a change-point model, the building performance is partitioned into different operating conditions and a linear model is fit to each of the operating modes [Reddy et al., 1997]. However, identification of these distinct operating modes is based on domain knowledge and building control operation. Non-linear models such as Artificial Neural Networks (ANN) [Yang et al., 2005, Wong et al., 2010] have become common place in data-driven building energy modeling literature to model non-linear transitions between linear regimes using a *single* overall model. Increasingly complex models have been proposed that capture this non-linear behavior but it comes at the price of computational complexity and lack of physical understanding of the model itself.

Many other approaches can be found in literature for building baseline energy modeling—Fels et al. [Fels, 1986] utilized variable-base degree-day method to estimate residential retrofitting energy use. Kissock et al. [Kissock, 1993] developed a regression methodology to measure retrofitting energy use in commercial buildings. Krarti et al. [Krarti et al., 1998] utilized neural networks to estimate energy and demand savings from retrofits of commercial buildings. Dhar et al. [Dhar et al., 1999] generalized the Fourier series approach to model hourly energy use in commercial buildings. In addition, in most practical cases, utility bill data are used because they are widely available and inexpensive to obtain and process [Reddy et al., 1997]. Reddy et al. [Reddy et al., 1997] presented a formal baselining methodology at the whole building level based on monthly utility bills and took outdoor dry-bulb temperature as the only model regressor. Internal Performance Measurement and Verification Protocol (IPMVP) [EVO, 2012] and [ASHRAE, 2002], provide some other approaches to develop baseline models for estimating energy savings after retrofits.

Due to multiple sources of uncertainty, it is practically impossible to develop deterministic estimates of building energy performance [Heo, 2011]. These uncertainties can arise from sources such as:

- Physical or operational uncertainty—equipment health status, occupancy or usage patterns and weather conditions,

- Modeling inaccuracies—assumptions and simplifications, and

- Sensing errors—sensor degradation, error in sensor placement and data collection issues

In addition to these uncertainties, risk analysis procedures need to consider economic uncertainties,

such as costs of equipment and labor. For this purpose, several approaches have been adopted in literature for the purpose of uncertainty analysis in building energy modeling. For example, multiple efforts used elaborate simulation models along with statistical sampling techniques to analyze sensitivity and quantify uncertainty [deWitt, 2001, Macdonald, 2002, Moon, 2005, Hu, 2009]. Among many sampling-based techniques, Morris method is recognized to be particularly effective for building applications as it is computationally efficient and is able to capture nonlinear effects. Similarly, Monte Carlo techniques have also been applied for uncertainty quantification [Jackson, 2010, Mills et al., 2006, Soratana and Marriott, 2010]. In [Soratana and Marriott, 2010], retrofit strategy evaluation and payback period computation have been performed for residential buildings under uncertainty using Monte Carlo simulation.

The second generic approach of uncertainty quantification in building energy modeling involves statistical methods, such as Bayesian techniques [Heo et al., 2011, Heo et al., 2012]. Heo et. al. [Heo et al., 2012] present a scalable, probabilistic methodology for analyzing energy retrofit strategies for buildings while considering various uncertainty factors. The methodology is based on Bayesian calibration of normative energy models that are lightweight, quasi-steady state formulations of thermodynamic equations, suitable for modeling large groups of buildings efficiently. Calibration of these models enables improved representation of the real buildings and quantification of uncertainties associated with modeling inaccuracies.

One of the shortcomings of prevalent modeling approaches for building energy is their limited ability to quantify uncertainty in predictions. Usually, error estimates are made for the overall model at the *global* level. This approach not only leads to a conservative estimate of modeling errors, but also smears the uncertainty across the data range. Therefore, locally a model maybe much less (or much more) accurate than global error estimate. In essence, this amounts to assuming an identical distribution for prediction error that is independent of the input conditions, which is inconsistent with most real scenarios. A localized quantification of uncertainties that captures the dependence of error on external conditions is necessary to accurately compute the impact of retrofits or new control and operation strategies.

Only recently, Subbarao et al. [Subbarao et al., 2011] have correctly pointed out the risks of using global confidence intervals for model predictions and proposed a nearest neighborhood method to compute *local* uncertainty estimates. For a given input condition, a set of similar conditions are selected using a distance metric and a chosen cut-off radius. Thereafter, error statistics are computed based only on this local set of information. This is a model-less approach to the local uncertainty quantification using the *k*-nearest neighbors method. In another work, Heo and Zavala [Heo and Zavala, 2012] have recently presented a model-based approach using Gaussian Processes (GP) for building energy prediction and localized uncertainty quantification. It is argued that this statistical method leads to better uncertainty estimates using lesser amount of data compared with those of standard regression methods. While model-less approaches have the advantage of not making any restricting assumptions about the structure of local uncertainty, they work well when local data densities are sufficiently high and estimates made in the local neighborhood of the input conditions are statistically significant. On the other hand, depending on the model choice, model-based approaches might require a smaller set of parameters to be learned and might perform better with sparse data densities in local neighborhoods. The choice of the model, of course, depends on the complexity of the system at hand.

In summary, the prevalent practical techniques in building energy prediction suffer from the following drawbacks (1) high complexity in an attempt to create a single model for the overall system and (2) Global uncertainty quantification, leading to conservative and unrealistic error estimates. In this chapter we describe a regression approach based on Gaussian Mixture Models (GMM) for building energy prediction and uncertainty quantification and perform a comparative evaluation against recently proposed approaches for localized uncertainty quantification, as well as against multivariate linear regression (MLR) with global confidence estimation. We believe the GMM-based regression for building energy prediction and localized uncertainty quantification has the following advantages:

1. *Integrated response surface modeling and local uncertainty quantification*: Conditional probability density of prediction errors is used for the estimation of both the response variable and the associated uncertainty; model parameters are learned from the entire data in the maximum likelihood sense; and a secondary process of localized or global confidence estimation is not needed

2. *Low impact of correlated regressors*: Correlated regressors, such as dry-bulb temperature and humidity, lead to ill-conditioned models that results in model parameter estimates that are highly sensitive to noise and changes in the data. Models that do not take into account the dependence between explanatory variables might be less generalizable and the significance of regressors can be grossly erroneous.

3. *Less sensitivity to data sparsity*: Being a parametric approach to modeling the building performance and uncertainty quantification, the GMR approach is less sensitive to low density of data.

4. *Formal model structure selection*: Identification of the number of modes in the Gaussian Mixture Regression is done using a information criteria (Bayesian Information Criteria) that is counter-weighted by the model complexity to choose the optimal number of modes; modes of the GMR model correspond to unique operational patterns of the building under consideration.

THEORY: REGRESSION & UNCERTAINTY QUANTIFICATION

In regression analysis, we seek a functional map from a set of inputs (regressors), X, to a set of outputs (response variables, Y). For multiple-input-single-output system, such as a building's baseline energy model, the output is a scalar function of inputs, $y = g(x; \beta)$, parameterized by β. Typically, the parameters are calibrated on a training or in-sample dataset. Let, $y_{\text{train}}(i) \in$ R and $x_{\text{train}}(i) \in \text{R}^{n_X}$, represent the ith response variable and regressor vector respectively. Similarly, $y_{\text{test}}(j)$ and $x_{\text{test}}(j)$ are the jth out of sample points of the same dimensions. We denote a set of n_{train} in-sample regressors as $X_{\text{train}} \in \text{R}^{n_{\text{train}} \times n_X}$ such that $X_{\text{train}}(i, \bullet) = x_{\text{train}}(i)$. Similarly, the corresponding n_{train} response values can be arranged in a vector $y_{\text{train}} \in \text{R}n_{\text{train}}^{\times 1}$. The parameters, β, are obtained by solving an optimization (minimization/maximization) problem of form

$$\beta_{\text{train}} = \arg \operatorname{opt}_{\beta} \Omega \left(y_{\text{train}}, g(X_{\text{train}}; \beta) \right) \qquad (23\text{-}1)$$

where the subscript of β_{train} indicate that the parameters are derived from the in-sample dataset and $\Omega (\bullet)$ denotes a suitably formulated objective function. Thereafter, the response value on an unseen regressor set, x_{test} can be obtained as $\hat{y}_{\text{test}} = g(x_{\text{test}}, \beta_{\text{train}})$. Some functional maps also explicitly compute the uncertainty in the estimated value, $\Sigma(\hat{y}_{\text{test}})$, while others require a separate process for uncertainty quantification on top of response variable estimation. This scenario of a two-step process of response modeling as the first step and uncertainty quantification later on, is more common in literature. We provide the expressions for $\hat{y}_{\text{test}}$ and $\Sigma(\hat{y}_{\text{test}})$ for all the modeling approaches described in the paper.

We present three choices for localized uncertainty quantification for baseline building energy modeling— (1) Gaussian Mixture Regression (GMR), (2) Gaussian Processes (GP) [Heo and Zavala, 2012], and (3) k-Nearest Neighbors (kNN) [Subbarrao et al., 2011]. Of these three, in GMR and GP approaches the task of response surface modeling and uncertainty quantification is integrated while the kNN-based approach can be used on top of any response surface modeling method such as Artificial Neural Networks (ANN) or Multiple Linear Regression (MLR). Following Subbarao et al. [Subbarrao et al., 2011], we use a ANN-based response variable prediction model to demonstrate the performance of the kNN-based method. As a comparison point, we also present the popular multiple linear regression model with global uncertainty estimation.

GAUSSIAN MIXTURE REGRESSION

Unlike conventional regression approaches, GMR doesn't directly obtain an optimal map from regressors to the response variable. Instead it solves a larger problem of estimating the joint probability distribution $P(y, x; \theta)$. Once an accurate joint distribution is learnt from the in-sample data, the conditional mean value, $\hat{y}_{\text{test}} = E[y|x_{\text{test}}; \theta_{\text{train}}]$, can be used as the response value on an out-of-sample point. Clearly, the choice of joint distribution model governs the response value. If $P(y, x)$ is unimodal Gaussian distribution, the conditional mean remains linear in θ and coincides with the linear regression estimate. However, for real world scenarios, where the data are typically generated under different operating conditions, a unimodal Gaussian assumption can be quite restrictive. As argued in the previous section 1, a piecewise linear model has been shown to be an adequate and robust model for building energy performance predicted based on measured external conditions. Therefore, we propose using a mixture of Gaussian distribution functions to represent the joint probability distribution as shown in Equa-

tion 23-2.

$$\psi(z;\theta) = \sum_{k=1}^{n_\phi} \frac{\lambda^k}{(2\pi)^{(n_X+1)/2}|\Sigma|^{1/2}} \exp$$

$$\left(-\frac{(z-\mu^k)\left[\Sigma^k\right]^{-1}(z-\mu^k)^T}{2} \right)$$

(23-2)

For brevity we represent the row vector $[y \ x]$ as z. The parameter set θ of a GMM is comprised of $\{\lambda^k, \mu^k, \Sigma^k\}$, with $1 \le k \le n_\phi$, where n_ϕ is the number of Gaussian components. Given a set of n_{train} in-sample data points, these parameters are obtained by maximizing the log likelihood function as shown in Equation 23-3.

$$\theta^*_{\text{train}} = \arg\max_{\theta} \sum_{i=1}^{n_{\text{train}}} \log\left(\psi(z_{\text{train}}(i); \theta)\right) \quad (23\text{-}3)$$

The set of optimal GMM parameters $\theta*_{\text{train}} = \{\lambda^{k,*}_{\text{train}}, \mu^{k,*}_{\text{train}}, \Sigma^{k,*}_{\text{train}}\}$ will be written as $\theta = \{\lambda^k, \mu^k, \Sigma^k\}$ hereon for the sake of notational simplicity. Unless otherwise stated, it should be understood that all GMM parameters were optimally estimated using Equation 23-3 and *only* in-sample or training data-set.

Here we provide equations to compute response value (and the associated variance) for an out-of-sample regressor vector. A more detailed treatment on the use of GMMs for regression can be found in the appendix A and elsewhere in literature [Sung, 2004]. Given an out of sample regressor vector, ξ_{test}, the response value $\hat{y}_{\text{test}}$ and its variance $\Sigma(\hat{y}_{\text{test}})$ can be obtained using Equations 23-4 and 23- respectively.

$$\hat{y}_{\text{test}} = \mathbb{E}[y|x_{\text{test}}] = \sum_{k=1}^{n_\phi} \lambda^k(x_{\text{test}})\mu^k_{Y|X}(x_{\text{test}}) \quad (23\text{-}4)$$

$$\Sigma(\hat{y}_{\text{test}}) = \sum_{k=1}^{n_\phi} \lambda^k(x_{\text{test}})\left(\sigma^k_{Y|X}(x_{\text{test}})^2 + \mu^k_{Y|X}(x_{\text{test}})^2\right) -$$

$$\left(\sum_{k=1}^{n_\phi} \lambda^k(x_{\text{test}})\mu^k_{Y|X}(x_{\text{test}})\right)^2 \quad (23\text{-}5)$$

The expressions for $\lambda^k(\xi_{\text{test}})$, $\mu^k Y|X(x_{\text{test}})$ and $\sigma^k Y|X(x_{\text{test}})$ are provided in Equations 23-6 through 23-88.

$$\lambda^k(\boldsymbol{x}_{\text{test}}) = \frac{\lambda^k \times \phi_X^k\left(\boldsymbol{x}_{\text{test}}; \boldsymbol{\mu}_X^k, \boldsymbol{\Sigma}_{XX}^k\right)}{\sum_{i=1}^{n_\phi} \lambda^i \times \phi_X^i\left(\boldsymbol{x}_{\text{test}}; \boldsymbol{\mu}_X^i, \boldsymbol{\Sigma}_{XX}^i\right)} \qquad (23\text{-}6)$$

$$\mu_{Y|X}^k(\boldsymbol{x}_{\text{test}}) = \mu_Y^k + \boldsymbol{\Sigma}_{YX}^k\left(\boldsymbol{\Sigma}_{XX}^k\right)^{-1}\left(\boldsymbol{x}_{\text{test}} - \boldsymbol{\mu}_X^k\right)^T$$

$$(23\text{-}7)$$

$$(\sigma_{Y|X}^k(\boldsymbol{x}_{\text{test}}))^2 = \boldsymbol{\Sigma}_{YY}^k - \boldsymbol{\Sigma}_{YX}^k\left(\boldsymbol{\Sigma}_{XX}^k\right)^{-1}\boldsymbol{\Sigma}_{XY}^k$$

$$(23\text{-}8)$$

The terms μ_Y^k, μ_X^k, $\boldsymbol{\Sigma}_{XX}^k$ and $\boldsymbol{\Sigma}_{YX}^k$ are obtained by decomposing the parameter set $\{\boldsymbol{\mu}^k, \boldsymbol{\Sigma}^k\}_{k=1}^{n_\phi}$, which were earlier estimated using in-sample data points. This decomposition is illustrated in Equations 23-9 and 23-10.

$$\boldsymbol{\mu}^k = \begin{bmatrix} \mu_Y^k & \boldsymbol{\mu}_X^k \end{bmatrix} \quad \text{with sizes} \quad \begin{bmatrix} 1 \times 1 & 1 \times n_X \end{bmatrix}$$

$$(23\text{-}9)$$

$$\boldsymbol{\Sigma}^k = \begin{bmatrix} \boldsymbol{\Sigma}_{YY}^k & \boldsymbol{\Sigma}_{YX}^k \\ \boldsymbol{\Sigma}_{XY}^k & \boldsymbol{\Sigma}_{XX}^k \end{bmatrix} \quad \text{with sizes}$$

$$\begin{bmatrix} 1 \times 1 & 1 \times n_X \\ n_X \times 1 & n_X \times n_X \end{bmatrix} \qquad (23\text{-}10)$$

GMR Confidence Intervals

For a specified significance level, α, the lower and the upper limits of the $100(1 - \alpha)\%$ CI can be obtained by solving equations 11 and 12 respectively with respect to y. Since, the error function, erf(), has a continuous derivative, we can use Newton-Raphson method to efficiently solve these equations in an iterative fashion. The error bars reported in the experimental section are obtained using this approach.

where

$\hat{y}_{\text{LL}}(\boldsymbol{x})$ and $\hat{y}_{\text{UL}}(\boldsymbol{x})$ are the lower and upper confidence limits of the uncertainty bound at $\boldsymbol{x}$.

In summary, the joint distribution of the response variable, Y, and regressors, X is modeled as a mixture of Gaussians. During the training phase, the model parameters $\{\lambda^k, \boldsymbol{\mu}^k, \boldsymbol{\Sigma}^k\}_{k=1}^{n_\phi}$ are learned form an in-sample (or training) dataset. For an out-of-sample regressor vector, $\boldsymbol{x}_{\text{test}}$, the response value is estimated using equation 4 whose parameters are derived from θ using efficient algebraic operations given in Equations 23-6 through 23-8. The error bands around the estimated response value is obtained by solving Equations 23-11 and 23-12.

Gaussian Process

This subsection introduces another building energy modeling approach that was recently developed by Heo and Zavala [Heo and Zavala, 2012] using Gaussian Processes. Instead of an algebraic relationship structure between the regressors and the response variable, this approach specifies the covariance structure for a finite set of response and regressor values. Hence, this approach becomes rather flexible and can inherently incorporate heterogeneous sources of uncertainties. Following the notations from previous subsection, let $\boldsymbol{X}_{\text{train}}$ be the matrix of n_{train} data samples of the regressor X and $\boldsymbol{y}_{\text{train}}$ be the vector of n_{train} response data samples. Let the covariance matrix be denoted by $\boldsymbol{\Sigma}(X, X, \eta)$, where η is the set of real covariance function hyper-parameters. Gaussian processes make the assumption that the response variable can be expressed as $Y \sim \text{N}(\boldsymbol{m}(\gamma), \boldsymbol{\Sigma}(X, X, \eta)$ where $\boldsymbol{m}(\gamma)$ is the mean model vector. The mean model vector is often assumed to be zero in the absence of any further information. For the the training data set $\boldsymbol{X}_{\text{train}}$, the covariance matrix $\boldsymbol{\Sigma}(\boldsymbol{X}_{\text{train}}, \boldsymbol{X}_{\text{train}}, \eta) \in \text{R}^{n_{\text{train}} \times n_{\text{train}}}$ is modeled using a kernel function $\text{K} \geq 0$

$$\boldsymbol{\Sigma}(\boldsymbol{X}_{\text{train}}, \boldsymbol{X}_{\text{train}}, \eta)[k, l] = \text{K}(\boldsymbol{x}(k), \boldsymbol{x}(l), \eta) \qquad (23\text{-}13)$$

$$\hat{y}_{\text{LL}}(\boldsymbol{x}) = \arg\left[\frac{\alpha}{2} - \sum_{k=1}^{n_\phi} \frac{\lambda^k(\boldsymbol{x})}{2}\left(1 + \text{erf}\left(\frac{y - \mu_{Y|X}^k(\boldsymbol{x})}{\sigma_{Y|X}^k(\boldsymbol{x})\sqrt{2}}\right)\right) = 0\right] \qquad (23\text{-}11)$$

$$\hat{y}_{\text{UL}}(\boldsymbol{x}) = \arg\left[1 - \frac{\alpha}{2} - \sum_{k=1}^{n_\phi} \frac{\lambda^k(\boldsymbol{x})}{2}\left(1 + \text{erf}\left(\frac{y - \mu_{Y|X}^k(\boldsymbol{x})}{\sigma_{Y|X}^k(\boldsymbol{x})\sqrt{2}}\right)\right) = 0\right] \qquad (23\text{-}12)$$

a popular choice for the kernel K is the exponential function. In that case, the $[k, l]^{th}$ term of the covariance matrix can be expressed as:

$$\Sigma\left(X_{\text{train}}, X_{\text{train}}, \eta\right)[k, l] = \eta_0 + \eta_1 exp$$

$$\left(-\frac{1}{\eta_2}\|x(l) - x(k)\|^2\right) \tag{23-14}$$

where, $\eta = [\eta_0, \eta_1, \eta_2]$ and $\| \cdot \|$ is the Euclidean norm. Optimal η is estimated by maximizing the log-likelihood function of the training data set, i.e.,

$$\eta^* = \arg \max_{\eta} \mathcal{L}(Y|\eta, X) \tag{23-15}$$

where,

$$\mathcal{L}(Y|\eta, X) = -\frac{n}{2}\log(2\pi) - \frac{1}{2}(y - m(\gamma))^{\text{T}}\Sigma^{-1}$$

$$\tag{23-16}$$

$$^{\text{T}}\Sigma^{-1}\left(X_{\text{train}}, X_{\text{train}}, \eta\right)(y - m(\gamma)) \cdots$$

$$-\frac{1}{2}\log(\det\left(\Sigma\left(X_{\text{train}}, X_{\text{train}}, \eta\right)\right))$$

At the testing phase, given an out-of-sample data point x_{test} of regressors and $\hat{y}_{\text{test}}$ of response variable. Using the trained Gaussian Process, the predicted mean $\mu\left(y_{\text{test}}\right)$ and variance $\Sigma\left(y_{\text{test}}\right)$ of the response variable can be computed as the moments of the posterior distribution $P(y_{\text{test}}|y_{\text{train}}, X_{\text{train}}, x_{\text{test}}, \eta*)$. The expressions of these statistical moments are:

$$\hat{y}_{\text{test}} = \mu\left(y_{\text{test}}\right) = \Sigma\left(x_{\text{test}}, X_{\text{train}}, \eta^*\right)\Sigma^{-1}$$

$$\left(X_{\text{train}}, X_{\text{train}}, \eta^*\right)y_{\text{train}} \tag{23-17}$$

and

$$\Sigma\left(y_{\text{test}}\right) = \Sigma\left(x_{\text{test}}, x_{\text{test}}, \eta*\right)\ldots$$
$$- \Sigma\left(x_{\text{test}}, X_{\text{train}}, \eta*\right)\Sigma^{-1}\left(X_{\text{train}}, X_{\text{train}}, \eta*\right)$$
$$\left(X_{\text{train}}, x_{\text{test}}, \eta*\right) \tag{23-18}$$

k-Nearest Neighbors

Recently, Subbarao et al. [Subbarrao et al., 2011] have proposed a model-less approach to localized un-

certainty quantification for baseline building energy modeling. The key idea of this approach is as follows: for any given point in the space of regressors, uncertainty bounds around predicted value of response variable is computed using *only* a set of k-nearest neighbors of the given regressor data-point. Instead of using the entire observed data set, which produces a fixed global estimate of uncertainty, a local set of points around the datapoint in question are used. Subbarao et al. have proposed a weighted Euclidean distance for computing the nearest neighbors

$$\nu_{ij} = \left(\frac{1}{n_X}\sum_{k=1}^{n_X}w_k^2\left(x_k(i) - x_k(j)\right)^2\right)^{\frac{1}{2}} \tag{23-19}$$

where weights wk are defined as the partial derivatives of the response variable y with respect to the regressor xk and quantify its sensitivity in predicting y.

$$w_k = \frac{\partial y}{\partial x_k} \tag{23-20}$$

The partial derivatives are proposed to be determined numerically by perturbing the inputs by a small amount (say 1-4%) and recalculating the energy use. Different amounts of perturbations must be evaluated to compute a more reliable estimate of wk. The weight used in equation 20 are *global*, that is they remain constant for all $x(i)$.

To ascertain the uncertainty bounds around a prediction $\hat{y}(i)$ for a given value of the regressors $x(i)$, k nearest neighbors $\{x(i_1), x(i_2), \ldots, x(i_k)\}$ of the the point $x(i)$ will be found using the distance given by equation 19. Nearest neighbors are selected from the space of the training data-set. The predicted set of values $\{\hat{y}(i_1), \hat{y}(i_2), \ldots, \hat{y}(i_k)\}$ for this set of nearest neighbors can now be used to determine the uncertainty bound around the value $\hat{y}(i)$. For example, to compute a the 90% confidence bound from a neighborhood size of 20, points with the lowest and largest residual errors $(y - \hat{y})$ are dropped, and the residual values of the 2nd and 19th residuals are selected as the uncertainty bounds.

DATA SET

The proposed methodology for building energy use modeling and localized confidence estimation was applied to one year simulation data generated using DOE reference model for a supermarket in Chicago climate. The simulated data used in this study is gen-

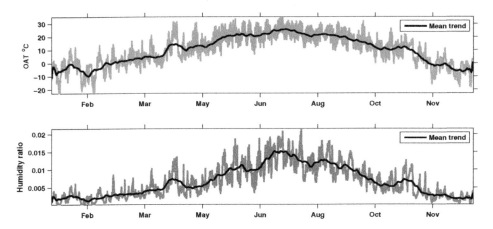

Figure 23-1: Yearly temperature and humidity ratio profile for Chicago area

erated from DoE super market reference model for ASHRAE 90.1-2004 [Deru et al., 2010]. It is simulated as a single-story, six-zone building with total area of 4,181 m^2. Building envelop thermal properties vary with the climate according to ASHRAE 90.1-2004. In this study, Chicago was selected as the climate location that is representative of a cold ASHRAE climate zone. Simulations were carried out for cooling electrical energy use only. The weather information is from Typical Meteorological Year 3 (TMY3) data, the yearly weather profile for Chicago is shown in Figure 23-1.

For the simulated study a retrofit energy savings analysis was also performed. A retrofit was simulated to occur mid-year on July 1st, where the windows glazing type and internal equipments were retrofitted to ASHRAE 90.1-2010 standard. This generated six months worth of both pre- and post-retrofit data from the simulations. Baseline model developed on pre-retrofit data was used to quantify the energy savings and confidence bound around these savings due the retrofit effort. Energy savings, using both the MLR and GMR based baseline models, were computed and compared.

APPLICATION: REGRESSION & UNCERTAINTY QUANTIFICATION

Multiple Linear Regression

A Multiple Linear Regression (MLR) model was developed and a global estimate of uncertainty was computed for comparative evaluation of localized uncertainty estimation methods presented earlier. For all approaches presented in the application section the following are chosen as regressors—(1) daily mean outside-air dry-bulb temperature (OAT), (2) daily mean outside-air humidity ratio (OAHR), and (3) daily mean direct solar radiation. The total daily electrical energy use for cooling was chosen to be the response variable. To ensure a principled approach to model order selection for MLR, Bayesian Information Criteria is used, see Appendix A.1 for details of model selection procedure. For the MLR model, the term nMLR in the BIC Equation 23-27 corresponds to the number of coefficients in the model and the log-likelihood of the observed data was computed as follows:

$$\mathcal{L}_{\text{MLR}}\left(\boldsymbol{\theta}|\{\boldsymbol{z}(i)\}_1^{n_{\text{train}}}\right) = \sum_{i=1}^{N} \log\left(\mathcal{N}(y(i); \boldsymbol{\theta}(i))\right) \quad (23\text{-}21)$$

where, N represents a univariate normal distribution with parameters $\boldsymbol{\theta}(i) = \{\mu(i), \sigma^2\}$, consisting of the means and the variance. The mean is nothing but the estimated response value by the MLR model and hence is indexed by i. The variance is computed globally by Equation 23-22, and is identical for all data samples.

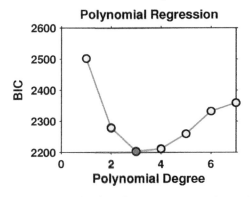

Figure 23-2: Model selection for MLR (3rd order polynomial model selected)

$$\sigma^2 = \frac{1}{n_{\text{train}} - n_{\text{MLR}} + 1} \sum_{i=1}^{n_{\text{train}}} \left(y(i) - \hat{y(i)}\right)^2 \quad (23\text{-}22)$$

The plot of BIC versus model order is shown in figure 2. Model order for the MLR model is the degree of the polynomial used. It can be seen that for the MLR model, BIC is minimized by a 3rd degree polynomial.

To compute the global uncertainty bounds of the MLR model, the estimation error is assumed to be i.i.d with a zero mean normal distribution, then the 95% CI consists of the range $\pm 1.954\sigma^2$ around the estimated value $\hat{y}$; the error variance σ^2 is computed using Equation 23-22.

Figure 23-3 shows the measured (squares) and estimated daily energy use (blue line) for the MLR model for one year of simulated data. Data from first that half of the year (Jan-Jun) was used for training, while predictions were made for the entire data set (Jan-Dec). The bottom plot shows the estimated energy savings $y_{save} = \hat{y}_{pre} - y_{post}$ and uncertainty band of the MLR-based baseline model. It can be seen that the global error band *smears* the uncertainty across the entire data domain i.e. the months with smaller perdition errors do not have tighter uncertainty bands as compared to months with larger prediction errors. Note: the prediction error can be judged from the top plot that compares baseline prediction with measured data.

Uncertainty quantification is necessary to find confidence bounds around estimated energy savings predicted using the baseline model. The energy savings $y_{save} = \hat{y}_{pre} - y_{post}$ is calculated by comparing the MLR baseline model estimate for the pre-retrofit case $\hat{y}_{pre}$ with the measured data y_{post} for the post-retrofit case, and the uncertainty around it are shown in Figure 23-3 bottom plot. It can be seen that the energy savings show a sharp increase right after the retrofit in July. For the MLR baseline models, the total energy savings for post-retrofit time period were estimated to be in the range -2,345 to +27,823 MJ based on global CI estimate.

Gaussian Mixture Regression

The choice of the number of components for GMM was based on the BIC criteria (Eq 23-27) as discussed in section A.1; for the GMR model the minimum BIC occurs for a 4-mode mixture model as shown in Figure 23-4.

As mentioned in section 2.1, the GMR approach solves the larger problem of estimating the joint probability density $P(Y, X; \theta)$ given the training data in the maximum-likelihood setting. Therefore the task of uncertainty estimation and model prediction, as the conditional mean value $E[Y|X; \theta]$, are integrated together as one modeling task. Figure 23-5 shows a contour plot of the joint probability density of the

data estimated by the GMR modeling process. Since GMR model is 4-dimensional with 3 regressors and one response variable, 2D contour plots are shown for 6-unique pairs of variables. It can be seen that learned GMM was able to adequately capture the distribution of the training data. Moreover, the correlation among regressors is naturally captured by the GMM model; this is evident from GMM density contours over the scatter plot of OAHR vs. OAT, in Figure 23-5.

Figure 23-6 shows the measured (squares) and predicted daily energy use (blue line) for the GMR model for one year respectively. Similar to the MLR model, data from first that half of the year (Jan-Jun) was used for training, while prediction were made for the entire data set (Jan-Dec). Note that the GMR model has locally adaptive confidence intervals as is shown by the uncertainty bands. The bottom plot shows the energy savings $ysave = \hat{y}_{pre} - y_{post}$ and uncertainty for GMR baseline model. Confidence intervals for this case are tighter for regions of good prediction (e.g. March) while they adapt locally for different conditions based on the uncertainty present in training data. For the GMR model the energy savings were in a much tighter range -1,544 and +22,567 MJ.

Gaussian Process

As noted in the section 2.2 that a key task in Gaussian Processes Modeling is the selection of covariance structure. The covariance matrix given by equation 14 can be synthesized by aggregating different similarity measures between two vectors, as long as the resulting matrix remains positive definite. To clarify this point, we rewrite an expanded version of equation 23-14 as follows,

$$\Sigma(X, X, \eta)[k, l] = \underbrace{\eta_0}_{\text{Const.}} + \underbrace{\eta_1 \exp\left(-\frac{1}{2}\sum_{i=1}^{n_X} \frac{(X(l,i) - X(k,i))^2}{\eta_{2,i}}\right)}_{\text{SE}_{\text{ard}}} \cdots$$

$$+ \underbrace{\eta_3 \sum_{i=1}^{n_X} \frac{X(l,i)X(k,i)}{\eta_{4,i}}}_{\text{IP}_{\text{ard}}} + \underbrace{\eta_5 \delta(l, k)}_{\text{Noise}}$$

$$(23\text{-}23)$$

Equation 23-23 shows that the $[k, l]$th entry of the covariance matrix is obtaining by aggregating four terms. The first and the fourth terms correspond to the constant, and the noise component of the covariance respectively. The second term is weighted squared exponential (SE) and the third term is weighted inner product (IP). Both these terms ensure that the nearby input vectors are more correlated to each

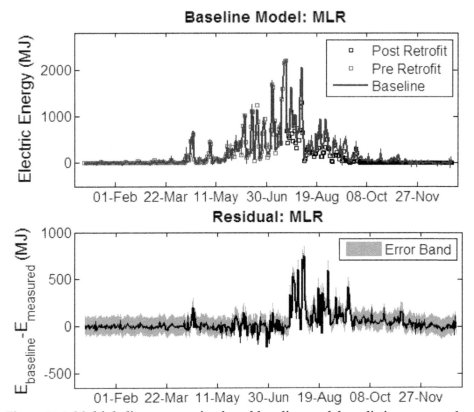

Figure 23-3: Multiple linear regression based baseline-model prediction compared with measured data pre- and post-retrofit (top); and estimated energy savings based on the global error estimation (bottom)

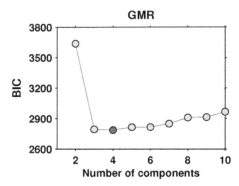

Figure 23-4: Model selection for Gaussian Mixture Regression (4-component Gaussian mixture selected)

other that far away regressor vectors. In addition, the subscript 'ard' in SE and IP stands for *Automatic Relevance Determination*, signify that the weight parameters $\{\eta_{2,i}\}_{i=1}^{nx}$ and $\{\eta_{4,i}\}_{i=1}^{nx}$ hold the information about which regressors are more pertinent in terms of response variable prediction. The covariance selection task involves identifying which combination of these components yield the best model.

Table 23-1 illustrates the task of covariance kernel selection. The BIC metric, on the simulated training dataset, is listed for all the different ways the covariance matrix can be computed using the four components and Equation 23-23. The combination of SE_{ard}, IP_{ard} and $Noise$ yielded a GP model with lowest BIC. Also, combinations IP_{ard} + Const., IP_{ard} and Const. didn't produce a BIC because the resulting covariance matrices weren't positive definite. Table 23-2 lists the ARD weights for SE_{ard} and IP_{ard} covariance functions. Both these set of weights indicate that the outside-air humidity ratio is the most relevant indicator of the building energy use, followed by outside-air dry-bulb temperature and solar radiation.

Figure 23-7 shows a plot of a comparison of measured (squares) and GP-based predicted daily electric energy use (solid line) in the top plot. The energy savings $ysave = \hat{y}_{pre} - y_{post}$ and uncertainty for the GP baseline model are shown in the bottom plot. Although locally adaptive, the performance of the GP model for local uncertainty quantification is very close to the global uncertainty estimation presented earlier (also see Figure 23-9). For the Gaussian Process approach the net energy savings were estimated to be in the range -5,886 to 28,014 MJ.

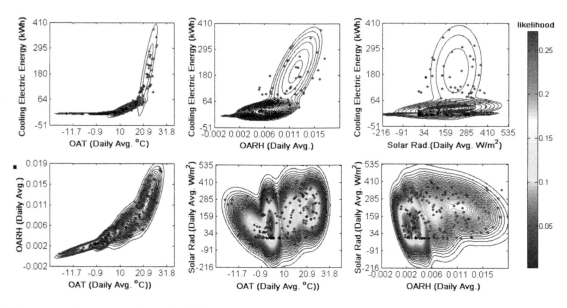

Figure 23-5: Contours of marginal distributions of model variables taken two at a time and scatter plots of data (red points). It can be seen that dependence of the predicted variable (energy) on regressors and correlation between regressors is well captured by the GMR model of 4-components

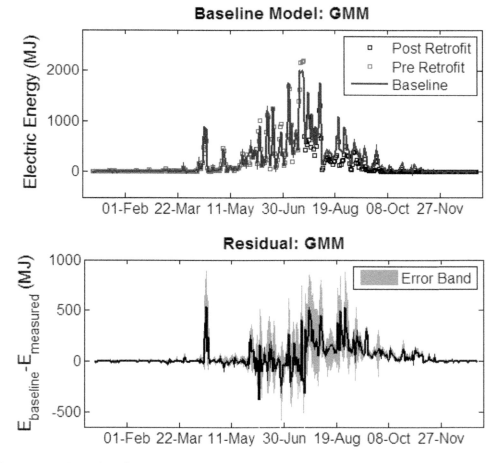

Figure 23-6: GMR baseline-model prediction compared with measured data pre- and post-retrofit (top); and localized uncertainty quantification based on the GMR baseline model

Table 23-1: The performance comparison of different GP models with different covariance structures (first column). The second, third and fourth columns list the number of parameter, log-Likelihood and Bayesian Information Criterion (BIC) of each GP model.

Comb.	Params	LogLike	BIC
$SE_{\text{ard}} + IP_{\text{ard}} + \text{Noise}$	8	29.96	-41.79
$SE_{\text{ard}} + IP_{\text{ard}}$	7	25.72	-35.58
$SE_{\text{ard}} + IP_{\text{ard}} + \text{Noise} + \text{Const.}$	9	27.57	-34.73
$SE_{\text{ard}} + IP_{\text{ard}} + \text{Const.}$	8	24.11	-30.07
SE_{ard}	4	15.30	-21.54
$SE_{\text{ard}} + \text{Noise}$	5	16.24	-21.14
$SE_{\text{ard}} + \text{Const.}$	5	13.85	-16.36
$SE_{\text{ard}} + \text{Noise} + \text{Const.}$	6	2.81	7.98
$IP_{\text{ard}} + \text{Noise}$	4	-159.56	328.19
$IP_{\text{ard}} + \text{Noise} + \text{Const.}$	5	-161.04	333.41
Noise	1	-263.48	529.22
$\text{Noise} + \text{Const.}$	2	-264.96	534.45
$IP_{\text{ard}} + \text{Const.}$	3	–	–
IP_{ard}	1	–	–
Const.	1	–	–

Table 23-2: The ARD weights obtained for SEard and IPard covariance functions

Regressor	$\eta_{2,i}$	$\eta_{4,i}$
OAT	0.38	4.89
OARH	0.11	1.51
Solar Rad.	1.86	7.83

k-Nearest Neighbors

The response variable modeling required for the k-NN method was done using a Artificial Neural Network (ANN) model. A Neural Network of two hidden layers with different combinations of neuron numbers in each hidden layer was tested on the data. The neural network used in this study is a fully connected feed-forward network. A fully-connected has each node connected to all other nodes in the adjacent layers, while in a feed-forward network information is passed in a single direction—from input nodes to output nodes.

The learning algorithm employed is the back-propagation, generalized delta method. In this algorithm, the value of the output of the Neural Network is compared to the target (observed) value to determine the error. The weights associated with the connection between nodes are then adjusted in a backward direction from the output layer to the input layer in order to minimize this error.

The ANN model was implemented using the Matlab Neural Network toolbox. The input layer consists of temperature, humidity and global solar radiation. The LogSigmoid function is used as the transfer function in all hidden layers and a linear function is used in the output layer. Because neural networks are not guaranteed to reach a global solution, training is repeated 10 times, and the output results are averaged.

Using the method described in section 4.4, a 95% uncertainty bound is computed at each point by finding 40 nearest neighbors of any regressor data-point in question and removing the smallest and largest residuals. Unlike the proposed approach in [Subbarrao et al., 2011], weights in the equation 19 are fixed at unity. This is done as partial derivatives are noisy and numerically

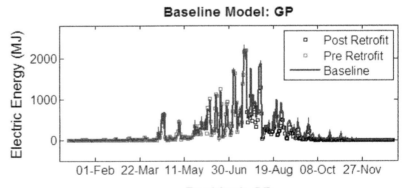

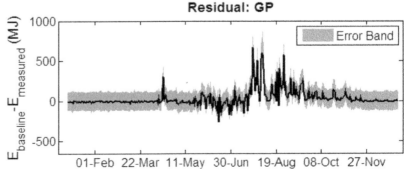

Figure 23-7: GP-based baseline-model prediction compared with measured data pre- and post-retrofit (top); and localized uncertainty quantification based on the GP baseline model

unstable. We believe, this simplification does not affect the essence of the proposed approach.

Figure 23-8 shows a comparison of measured (squares) and ANN-based predicted daily electric energy use (solid line) in the top plot. The energy savings $ysave = \hat{y}_{pre} - y_{post}$ and uncertainty using the kNN based approach on top of the ANN-baseline predictions are shown in the bottom plot. The kNN based approach is locally adaptive but can be seen to be very noisy. This is due to poor data density leading to unstable estimation of uncertainty bands. The net energy savings for the ANN + kNN approach was estimated to be in the range -1,577 to + 24,903 MJ.

DISCUSSION

This chapter presented three approaches to localized uncertainty quantification for baseline building energy modeling using (1) Gaussian Mixture Regression (GMR), (2) Gaussian Process (GP), and (3) k-nearest neighbors (k-NN). Gaussian Mixture Regression (GMR) is a novel approach for modeling building energy use with parameterized and locally adaptive uncertainty quantification. All three approaches were applied to simulated data set generated by DoE reference model for a supermarket. The results from all three approaches were presented and localized confidence estimation was demonstrated. A comparison of energy savings computed based on the three approaches in the post retrofit period is given in Table 23-3. Figure 23-9, shows a comparison of the localized confidence estimation of the three approaches. For an easy comparison of local adaptability of uncertainty bands, global uncertainty estimate computed using the MLR model is shown in the background for all three

As discussed earlier, the k-NN based method makes no assumption about the error statistics (e.g. normal distribution) and therefore maybe more generally applied to situations where error statistics are complex. However, this model-less approach to local uncertainty quantification might be inaccurate and

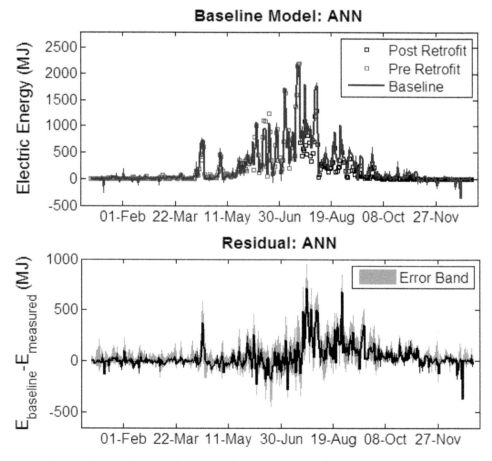

Figure 23-8: ANN baseline-model prediction compared with measured data pre- and post-retrofit (top); and localized uncertainty quantification based kNN-based approach

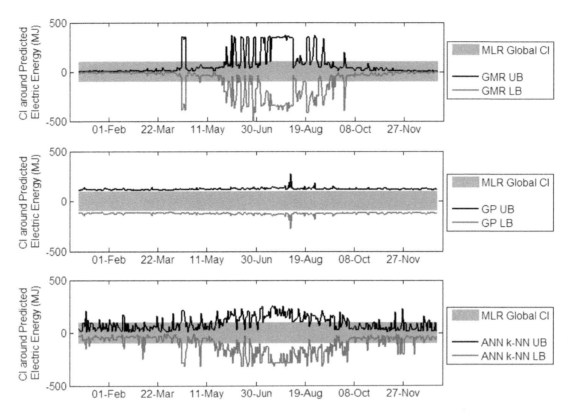

Figure 23-9: Uncertainty quantification comparison for three cases (1) GMR based localized CI (top), (2) GP-based localized CI, and (3) k-NN based localized CI (bottom). Localized CI upper bound (CI-UB) and lower bound (CI-LB) are shown as black and red solid lines respectivelysubplots.

Table 3: Energy savings computed using different baseline modeling approaches

Modeling approach	Lower Bound	Upper Bound
Global CI	-2,345 MJ	+27,823 MJ
GMR based local CI	-1,544 MJ	+22,567 MJ
GP based local CI	-5,886 MJ	+28,014 MJ
kNN+ANN based local CI	-1,577 MJ	+24,903 MJ

unstable due to sparse population of data in local neighborhoods. Also, the degree of similarity of a set of k nearest-neighbors may not be comparable for different input conditions, leading to inconsistent error statistics. While increasing the neighborhood size might resolve the stability and accuracy issue, a larger local neighborhood defeats the purpose of localized uncertainty quantification unless local neighborhoods are sufficiently dense in data. Heuristics have been proposed in [Subbarrao et al., 2011], however, there is no formal and rigorous way to balance this trade-off and select an optimal neighborhood size.

The problem of confidence interval (CI) estimation in local neighborhoods with poor data density can be tackled using a model-based approach so as to limit the number of parameters required to be learned. GMR and GP fall under this category of model-based approaches. The choice of the model depends on the expected characteristics of the dataset, desired training and run-time complexity. The model-based approach is a good choice when the biases incurred by fixing a model structure are expected to be outweighed by the accuracy gains in error quantification. This seems to be the case for building energy modeling and local confidence estimation. A model-based approach is justified here as within an operating mode and given key regressors such as outside temperature and humidity, the errors in predicting the building energy use can be adequately modeled by a parameterized Gaussian density function. These parameters depend on the current conditions (value of regressors), thereby giving a localized estimate of uncertainty.

The GP approach can capture complex input-output dependencies without the use of elaborate domain knowledge [Rasmussen and Williams, 2006]. With a suitable choice of a kernel for the data covariance matrix in a GP, non-linear response surfaces can be modeled with few parameters. However, both the learning and run-time complexity of the GP based models scales cubically with the number of training samples used; which can quickly become impractical as the size of the data-set becomes larger. On the other hand, a Gaussian mixture model for regression can be easily learned using efficient algorithms such as Expectation Maximization [Neal and Hinton, 1998, Xu and Jordan, 1995] and its run-time complexity scales with the dimension of the data-space and not the number of data-samples used for learning.

For the current example, the GMR approach had comparable performance in baseline prediction accuracy and while performed better for local confidence estimation. It can be seen that the GMR confidence intervals adapt smoothly to changes in outside conditions, as compared to the nearest neighbor based and GP based methods. In particular, the GMR approach has the following key advantages (1) integrated response surface modeling and local uncertainty quantification (2) low impact of correlated regressors, (3) less sensitivity to data sparsity, and (4) formal model structure selection. Mixture modeling is not restricted to using Gaussian densities for modeling the distribution in data, for certain scenarios other constituting densities, such the log-normal, might be more relevant. Also, linear dependence structure assumed in this work can be relaxed by dependence modeling approaches such as using Copula functions. These extensions of the current work are proposed to be explored in the domain of building energy performance modeling.

References

[ASHRAE, 2002] ASHRAE (2002). ASHRAE guideline 14-2002 for measurement of energy and demand savings.

[Deru et al., 2010] Deru, M., Field, K., Studer, D., Benne, K., Griffith, B., Torcellini, P., Halverson, M., Winiarski, D., Liu, B., Rosenberg, M., Huang, J., Yazdanian, M., and Crawley, D. (2010). U.S. department of energy commercial reference building models of the national building stock. *U.S. Department of Energy, Energy Efficiency and Renewable Energy, Office of Building Technologies*.

[deWitt, 2001] deWitt, S. (2001). *Uncertainty in predictions of thermal comfort in buildings*. PhD thesis, Delft University of Technology.

[Dhar et al., 1999] Dhar, A., Reddy, T., and Claridge, D. (1999). A Fourier series model to predict hourly heating and cooling energy use in commercial buildings with outdoor temperature as the only weather variable. *Journal of Solar Energy Engineering*, 121:47–53.

[EPA, 2008] EPA (2008). EPA report on the environment.Final Report EPA/600/R-07/045F. Technical report, United States Environmental Protection Agency.

[EUCOM, 2013] EUCOM (2013). Report from the commission to the european parliament and the council. *Brussels, COM(2013) 225*.

[EVO, 2012] EVO (2012). IPMVP Volume I: Concepts and options for determining energy and water savings.

[Fels, 1986] Fels, M. (1986). Special issue devoted to measuring energy savings, the princeton scorekeeping method (PRISM). *Energy and Buildings*, 9.

[Heo, 2011] Heo, Y. (2011). *Bayesian Calibration Of Building Energy Models For Energy Retrofit Decision-Making Under Uncertainty*. PhD thesis, Georgia Institute of Technology.

[Heo et al., 2011] Heo, Y., Augenbroe, G., and Choudhary, R. (2011). Risk analysis of energy-efficiency projects based on bayesian calibration of building energy models. *Building Simulation*, pages 2579–2586.

[Heo et al., 2012] Heo, Y., Choudhary, R., and Augenbroe, G. (2012). Calibration of building energy models for retrofit analysis under uncertainty. *Energy and Buildings*, 47:550–560.

[Heo and Zavala, 2012] Heo, Y. and Zavala, V. M. (2012). Gaussian process modeling for measurement and verification of building energy savings. *Energy and Buildings*, 53(0):7–18.

[Hu, 2009] Hu, H. (2009). *Risk-conscious design of off-grid solar energy houses*. PhD thesis, Georgia Institute of Technology.

[Jackson, 2010] Jackson, J. (2010). Promoting energy efficiency investments with risk management decision tools. *Energy Policy*, 38:3865–3873.

[Katipamula and Claridge, 1993] Katipamula, S. and Claridge, D. (1993). Use of simplified system models to measure retrofit energy savings. *Journal of Solar Energy Engineering*.

[Kissock, 1993] Kissock, J. K. (1993). *A methodology to measure retrofit energy savings in commercial buildings*. PhD thesis, Texas A&M University, Department of Mechanical Engineering.

[Krarti et al., 1998] Krarti, M., Kreider, J., Cohen, D., and Curtiss, P. (1998). Prediction of energy saving for building retrofits using neural networks. *Journal of Solar Energy Engineering*, 120(3):47–53.

[Macdonald, 2002] Macdonald, I. A. (2002). *Quantifying the effects of uncertainty in building simulation*. PhD thesis, University of Strathclyde.

[Mills et al., 2006] Mills, E., Kromer, S., Weiss, G., and Mathew, P. (2006). From volatility to value: analyzing and managing financial and performance risk in energy savings projects. *Energy Policy*, 34(2):188–199.

[Moon, 2005] Moon, H. J. (2005). *Assessing mold risks in buildings under uncertainty*. PhD thesis, Georgia Institute of Technology.

[Neal and Hinton, 1998] Neal, R. and Hinton, G. E. (1998). A view of the em algorithm that justifies incremental, sparse, and other variants. In *Learning in Graphical Models*, pages 355–368. Kluwer Academic Publishers.

[Rasmussen and Williams, 2006] Rasmussen, C. and Williams, C. (2006). *Gaussian Processes for Machine Learning*. MIT Press, Cambridge, MA.

[Reddy et al., 1997] Reddy, T. A., Saman, N. F., Claridge, D. E., Haberl, J. S., Turner, W. D., and Chalifoux, A. T. (1997). Baselining methodology for facility-level monthly energy use–Part 1: Theoretical aspects. *ASHRAE Transactions*, 103(2).

[Soratana and Marriott, 2010] Soratana, K. and Marriott, J. (2010). Increasing innovation in home energy efficiency: Monte carlo simulation of potential improvements. *Energy and Buildings*, 42:828–833.

[Steele and Raftery, 2009] Steele, R. J. and Raftery, A. E. (2009). Performance of bayesian model selection criteria for gaussian mixture models. Technical Report 559, University of Washington, Dept. of Statistics.

[Subbarrao et al., 2011] Subbarrao, K., Lei, Y., and Reddy, T. A. (2011). The nearest neighborhood mthod to improve uncertainty estimates in statistical building models. *ASHRAE Transactions*, 117(2).

[Sung, 2004] Sung, H. G. (2004). *Gaussian mixture regression and classification*. PhD thesis, Rice University, Houston, Texas.

[Wong et al., 2010] Wong, S., Wan, K. K., and Lam, T. N. (2010). Artificial neural networks for energy analysis of office buildings with daylighting. *Applied Energy*, 87(2):551–557.

[Wu et al., 1992] Wu, J., Reddy, T. A., and Claridge, D. (1992). Statistical modeling of daily energy consumption in commercial buildings using multiple regression and principal component analysis. In *Eight symposium of improving building systems in hot and humid climate*, pages 155–164, Dallas, TX.

[Xu and Jordan, 1995] Xu, L. and Jordan, M. I. (1995). On convergence properties of the EM algorithm for gaussian mixtures. *Neural Computation*, 8:129–151.

[Yang et al., 2005] Yang, J., Rivard, H., and Zmeureanu, R. (2005). Online building energy prediction using adaptive artificial neural networks. *Energy and Buildings*, 37(12):1250–1259.

GAUSSIAN MIXTURE MODEL (GMM)

Let (Y, X) be represented as a multivariate random variable Z for brevity. Then $z = [y, x_1, ..., x_d]^T \in R^{(n_X +1)}$ is a column vector of length $(n_X + 1)$, where n_X is the number of regressors predicting a scalar value y. A Gaussian mixture density $\psi(z; \theta)$, with n_ϕ components, that describes the probability distribution of Z with parameters θ can be written as follows

$$\psi(z; \theta) = \sum_{k=1}^{n_\phi} \lambda^k \phi^k \left(z; \mu^k, \Sigma^k \right) \qquad (23\text{-}24)$$

where each Gaussian component ϕ^k is parameterized by the mean vector μ^k of the same length as z and a $(n_X + 1) \times (n_X + 1)$ positive definite covariance matrix Σ^k. The expression of a Gaussian density function is given in Equation 23-25.

$$\phi^k(z; \mu^k, \Sigma^k) = \frac{1}{(2\pi)^{n_X/2}|\Sigma^k|^{1/2}} \exp$$
$$\left(-\frac{1}{2}(z - \mu^k)^{\mathrm{T}} \left[\Sigma^k \right]^{-1} (z - \mu^k) \right) \qquad (23\text{-}25)$$

The scalar λ^k is the non-negative mixing proportion of the kth component such that $\sum_{k=1}^{n_\phi} \lambda^k = 1$. Thus, the parameter set θ of a GMM is comprised of $\{\lambda^k, \mu^k, \Sigma^k\}$, with $1 \le k \le n_\phi$. The estimation of GMM parameters is done in the maximum likelihood setting. The goal in maximum likelihood estimation is to obtain parameter values that maximize the likelihood of observing a given data-set. Given N i.i.d samples $\{z(i)\}_1^{n_{\text{train}}}$, in the training data set, the logarithm of the likelihood function is defined as,

$$\mathcal{L}\left(\theta | \{z(i)\}_1^{n_{\text{train}}} \right) = \sum_{i=1}^{n_{\text{train}}} \log \left(\psi(z(i); \theta \right) \qquad (23\text{-}26)$$

Equation 23-26 when maximized under the afore-mentioned constraints on the covariance matrices (Σ^k) and the mixing proportions (λ^k), yields the desired solution. This optimization is typically carried out using Expectation-Maximization (EM) algorithms, which is a widely studied area, with numerous commercial software implementation available. For this work, we used MATLAB's Statistical toolbox to learn the Gaussian Mixture Models. Also, good reviews on EM algorithm and its variants can be found in [Neal and Hinton, 1998, Xu and Jordan, 1995].

Model Selection

A key task involved in the GMM approach to regression is the identification of the optimal number of components (n_ϕ) to be included in the model. To this end, we propose using Bayesian Information Criterion (BIC) (given in Equation 23-27) as a metric for selecting the number of components. In a rigorous study [Steele and Raftery, 2009], BIC was shown to outperform other methods such as DIC (Deviance Information Criterion), ICL (Integrated Completed Likelihood) and AIC (Akaike Information Criterion) for GMMs in a wide array of application domains.

$$\mathrm{BIC} = -\mathcal{L}\left(\theta | \{z(i)\}_1^{n_{\text{train}}} \right) + \frac{n_\theta}{2} \log(n_{\text{trai}} \quad (23\text{-}27)$$

In Equation 23-27, n_θ represents the total number of free parameters, which for a GMM with n_ϕ components can be obtained as

$$n_\theta = \underbrace{n_\phi(n_X + 1)}_{\text{mean}} + \underbrace{\frac{1}{2}n_\phi(n_X + 1)(n_X + 2)}_{\text{covariance}} + \underbrace{(n_\phi - 1)}_{\text{mixing}}.$$
$$(23\text{-}28)$$

he first term in Equation 23-27 is the logarithm of observed data likelihood, the expression of which is given by Equation 23-26. The value of $n\phi$ that minimizes BIC is chosen as the number of components for the Gaussian Mixture Model.

Nomenclature

Throughout the manuscript, upper case letters (e.g. X) are used for random variables, bold faced lower case letters (e.g. x) for a vector, and normal face lower case letters x for a scalar variable. Bold faced capital letters (e.g. A) are used for matrices. The symbol $\square$ in the nomenclature below is a place holder, therefore $\square_{\text{train}}$ should be read as *any* symbol with subscript "train."

X = random variable (r.v.) for regressors, $X \in R^{n_X}$

x = instantiation of X

$x(i)$ = i^{th} sample of X

x = scalar component of x

X = matrix of regressor data samples $x(i)$ such that each row $X(i, \bullet) = x(i)$

Y = random variable for response variable, $Y \in R$

y = instantiation of Y

$y(i)$ = i^{th} sample of Y

y = vector of response data samples y such that each row $y(i) = y(i)$

$\hat{y}$ = estimated value of y

$\hat{y}LL$ = lower limit of the confidence interval around $\hat{y}$

$\hat{y}UL$ = upper limit of the confidence interval $\hat{y}$

Z = joint variable for the concatenation $[X, Y]$, $Z \in R^{(n_X +1)}$

z = instantiation of Z

$z(i)$ = ith sample of z

g = map of regressors to response variable

β = parameters of g

$P(Y, X; \theta)$ = joint probability density function of (Y, X) with parameters θ

$P(Y, X; \theta)$ = joint cumulative distribution function of (Y, X) with parameters θ

$P(Y|X; \theta)$ = probability density function of Y given X for parameters θ

$E[Y|X; \theta]$ = expected value of Y given X with parameters θ

$\psi(z; \theta)$ = Gaussian mixture probability density with parameters θ

θ = parameters of the GMR model

λ = mixing parameter for Gaussian mixture components

ϕ = mixture component probability density

μ = mean vector

$\mu(y)$ = mean of vector y

Σ = covariance matrix

$\Sigma(X; Y)$ = covariance of r.v. X and Y

$\Sigma(X)$ = variance of r.v. X = $(X;X)$

σ = standard deviation of a scalar r.v.

n_X = number of regressor

n_σ = number of Gaussian mixture components

n_{MLR} = number of parameters for the MLR model

n_{train} = number of data samples in training phase

n_{test} = number of data samples in testing phase

n_θ = number of free parameters of the GMM model

α = condence value 2 $[0; 1]$

erf = error function

N = Normal (Gaussian) probability distribution

L = log-likelihood function

$\det(A)$ = determinant of matrix A

$\dim(v)$ = dimension of a vector v

v^T = transpose of a vector v

$\|\bullet\|$ = Euclidean norm of

1_m = m m identity matrix

η = hyper-parameters = $[0; 1;::::]$ of a Gaussian Process (GP)

K = kernel function of a Gaussian Process (GP)

v = distance function for nearest-neighbor selection

w = weights used in the distance function

Superscripts

$\square^k$ = $\square$ corresponding to k^{th} component of Gaussian mixture model

Subscripts

$\square_{Y|X}$ = corresponding to the joint probability density $P(Y|X;)$

v_X = part of vector v corresponding to r.v. X

A_{XX} = part of matrix A corresponding to r.v. X in both rows and columns

v_Y = part of vector v corresponding to r.v. Y

A_{YY} = part of matrix A corresponding to r.v. Y in both rows and columns

A_{XY} = part of matrix A corresponding to r.v. X in rows and r.v. Y in columns

A_{YX} = part of matrix A corresponding to r.v. Y in rows and r.v. X in columns

$\square_{pre}$ = $\square$ corresponding to the pre-retrot period

$\square_{post}$ = $\square$ corresponding to the post-retrot period

$\square_{save}$ = $\square$ corresponding to energy savings

$\square_{train}$ = $\square$ corresponding to training or in-sample data set

$\square_{test}$ = $\square$ corresponding to testing or out-of-sample data set

$\square_{ard}$ = corresponding to Automatic Relevance $\square$ Determination

Chapter 24

Pushing the Envelope
Building Analytics beyond HVAC

Jim Sinopoli, Smart Buildings, LLC

INTRODUCTION

Propelled by growing energy concerns and technology advancements, the building industry has made several strides in building controls and automation using analytics for buildings. However, despite the progress, we're not even close to utilizing the full potential of analytics for buildings. Automated, real-time data, much more than anything currently deployed, is not only possible but would provide the performance we seek and need in our buildings to better manage operations and energy. Analytics support facility management personnel who are challenged with progressively more complex building systems and the constantly changing skill sets and knowledge required to operate them.

Over the recent past, the best use of analytic software applications for building systems has been fault detection and diagnostics (FDD) specific to HVAC systems. There is research and a number of case studies with verified results showing analytic software reduces energy consumption, improves the efficiency and effectiveness of building operation, and reduces building operating expenses (OPEX). Once used, FDD becomes a core operational tool for many facility managers.

Despite the impressive progress with FDD, the industry is in its infancy of utilizing data analytic applications in buildings. If analytics for the HVAC system has provided outstanding outcomes, we need to take that template to other building systems.

Analytic applications are based on "rules" of how the system should optimally operate, generally obtained from the original design documents and monitoring key data points in near real-time. Essentially you compare the real-time data with the rules and if the data adheres to the rule, the system is fine; if not, the system is not running optimally and has a fault. For those systems that are not process based, applying analytics generally uses statistical monitoring of key performance indicators (KPIs) to monitor outliers. This may not provide the diagnosis of an issue, but it can identify faulty equipment for maintenance. What if we had an application that not only could automatically detect faults in a building, but could automatically diagnose those faults, and maybe even tell you how to fix or correct those faults? This would be total automation, with fault detection and diagnosis, and analytics to determine the corrective action needed for the building and its systems. You could consider it to be something similar to an "autopilot" for an airplane.

Not only can buildings have "autopilots" but they should, and we have the technology and the capability to do this. Buildings are not airplanes but the traits of the aviation inventor, boldness, innovation and vision will be needed to increase automation in our buildings, and we have the technology to accomplish this goal.

WHAT OTHER BUILDING SYSTEMS COULD BE AUTOMATED WITH ANALYTICS AND FDD?

Most building systems will benefit from analytics. Below are examples, illustrations and recommendations for a few key systems and how they can benefit from analytics.

Lighting Systems

The key to analytics for lighting systems is addressable networkable light fixtures communicating with an open standard protocol such as DALI® or IEC 62386. This allows the acquisition of data and command points which can develop into "value Information." This value information for standard lamp fixtures can be as simple as:

- Lamp failure
- LED driver/Lamp Ballast failure
- LED overvoltage
- Mains Failure

- DALI network failure

The more complex reporting would be:

- Run time
- Calculated power per fitting
- Dirty lamp fitting versus lamp fitting failing
- General lamp fixture failure
- Lamp change versus energy consumption cost analysis

There are also analytic opportunities with Emergency lamp fixtures. These fixtures typically are a lamp and rechargeable battery pack unit and need to be tested regularly by law in most jurisdictions. While these tests are carried out mostly manually, networkable emergency fixtures can be fully automated from initiating the tests to recording results and sending alarms when failures occur. The data that can be acquired includes:

- Battery status
- Battery charge level
- Lamp status
- Emergency fixture availability
- Runtime
- Mains Failure
- Network failure

In addition, lighting control systems often include fault and system alarms for the devices and controllers. They rarely have algorithms defined to measure how well the sequences are performing and if the overrides are impacting the overall energy performance of the building. Similar to HVAC, lighting sequences and control strategies are customized for the building and its occupants. It is simply not possible for the manufacturer to pre-determine what analytics or fault detection algorithms to include in the system before the sequences are commissioned. Once the sequences are understood, rules can be created that will measure the effectiveness of the sequence as compared to occupancy, HVAC schedules, energy goals, safety and other related and non-related data sets.

Water Usage Data

Analytic rules applied to water usage data can identify elevated water use during known unoccupied periods, which could be attributed to a leak. Analytic rules may also be used to find periods of time when peak usage was too high. In addition, rules can be written to calculate the cost of the water usage during peri-ods of leaks and the cost associated with usage above expected consumption levels.

Another useful analytic for water distribution or irrigation is the monitoring of power usage and gallons per minute (gpm) of the pumps. If the kW/gallon increases in comparison to the historical data for the same gpm (i.e. not just a performance curve variance), you may assume that the pump (while operational) needs servicing because it has become inefficient.

Energy Data

Analytics can be used in the utilization of raw energy consumption data supplied by a utility company. On a daily basis interval meter data (KW and KWh) is supplied in a CSV file format. Using analytics it is possible to identify buildings operating outside of their occupancy schedule simply by analyzing the "signature" of KW demand. Buildings that follow an occupancy schedule see a notable increase (and decrease) in KW within a short window of time and this represents occupancy transition times. Buildings where systems operate continuously do not see that step change. Adjusting for the magnitude of the change at the occupancy transition provides "filtering" to detect lack of schedule control even in cases where there are systems that operate continuously.

Correlation of Equipment Service Data

Analytics can be used to correlate service/repair events with equipment and identify the equipment that was responsible for high levels of service calls and related costs. Fairly simple data such as service dates, equipment subject to service and service cost or hours allow for initial targeting of costly units. Additional information such as type of repair (based on a defined set of classification names) allows for understanding of the types, frequency and costs of different issues.

Conveyance Equipment

For conveyance equipment, monitoring of the weight load sensor, drive power consumption, and travel distance can determine the status of the motors and whether they are working beyond capacity. For instance, if the kW/ft.*lb increase in comparison to historical data, you may assume the motor (again, while operational) requires service because it is becoming inefficient.

Power Management Systems

Power management is another area to consider for fault detection. Similar to lighting control, the power management system rules cannot be built until the building sequences for HVAC and lighting are implemented and time has passed to build a baseline of information. Energy managers largely look at the energy consumption in reports, either through dashboards or spreadsheets, to determine if they are meeting their energy goals. Analytics and fault detection can be applied to look for the same anomalies the energy managers look for, but they do it near real time and inform the appropriate energy manager of the situation. If the proper systems are in place, additional rules can be written to automatically take corrective action and avoid new demand peaks or energy records.

IT Infrastructure

Many organizations have an IT infrastructure that is both centralized (data centers) and distributed (telecommunications/server rooms). Although energy efficiency is sometimes considered, especially during construction, the real value propositions that matter to the IT department are reliability and risk mitigation (ensuring uptime). The reliability of the infrastructure depends on a number of factors including environmental conditions, power quality, and IT equipment performance. Fault detection and diagnostics (FDD) can be applied to the reliability problem, but it requires the fusion of data from disparate sources such as HVAC, power metering, electrical switchgear, uninterruptable power supplies, backup generators, servers, routers and switches. Data is collected via SNMP, Modbus, BACnet, and proprietary protocols. Today's servers can provide management data, including internal temperatures and computational workload metrics, which are useful inputs to a FDD system. For IT, the benefits of FDD are to predict equipment failures before they occur and to provide insight into the cause of failures that do occur. This requires that FDD analytics operate at the system level as well as at the equipment level.

Demand Response and Refrigeration

One of the benefits of near real time analytics is the ability to determine if a demand reduction program has met its bid/goals. If the customer finds themselves in a situation where they are not meeting bid/goals, analytics will be able to tell why goals were not met in order to make changes for the next DR event to be successful. For this to happen, all building systems that consume energy need to be monitored for correct operation and coordinated in a central location. For instance monitoring the operating modes of a lighting control system in a retail establishment will determine if the system

is "ready" for an automated DR event such as shutting off certain lighting tracks in order to reduce load. This automated load reduction cannot be achieved if the lighting circuit HOA switch has been overridden to the hand position. Therefore the analytics will send a notice to the facility manager to make the correction so that the system is "ready."

Another benefit of analytics is the ability to monitor systems that are an integral part of the client delivering their product or service to their end customers. For instance, monitoring the refrigeration system in a grocery/convenience store will determine if the system is working at peak efficiency and effectiveness thereby preventing product loss due to spoilage. The criteria for peak efficiency and effectiveness are typically defined by the refrigeration case manufacturer. A typical requirement is to achieve a minimum number of defrost cycles per day to determine if the evaporator coil can achieve maximum flow, thereby protecting product from frost buildup.

A ROADMAP TO IMPLEMENTATION OF ANALYTICS AND FAULT DETECTION AND DIAGNOSTICS

The roadmap to advanced automated buildings involves several key issues for industry and building designers, contractors, managers and owners to address:

Granular Data

Building-wide or system-wide data will not be sufficient for a highly automated building. The metrics are too broad and vague. To really manage a building we need to get down to the details. The spaces within most buildings are too different regarding their orientation, use, occupancy, needs, etc. Granular data provides for more precision in properly managing specific spaces within a building, potentially resulting in squeezing out the smallest amount of excess energy consumption and improving occupant satisfaction. Going "granular" will mean more sensors, tailored controls for individual spaces and a bit more investment.

Detailed Policies and Logic

For a building to be fully automated it will require the "logic" or the "policies" of the automation be fully developed. These are pre-determined rules using an array of data sources and data. The building senses real time conditions and then automatically responds or adjusts, much like Sperry's gyroscope stabilizer.

The development of this logic will not necessarily

be easy; as buildings become increasingly complex the decisions regarding their performance become more complex as there are many more variables in the decision making process. Defining the logic or policies will take extensive planning which is sometimes a pitfall of typical facility management; an example being a dearth of detailed written alarm management plans, reflecting the lack of planning and forethought. The policies will need to touch on every significant building situation or scenario affecting energy, operational costs, life safety and tenant comfort. Planning will involve diverse groups within the building's ownership and management. This is really an extensive exercise to develop the brains of the automation systems and in the process, decide exactly how the building should adapt to changes and how it should perform.

Much of the data used as the basis for "policies" will be near real-time data from the building systems however critical data and system-to-system communications are needed with the facility management systems, business systems, the utility grid and other external systems, such as weather or energy markets. A highly automated building will require numerous policies, control logic and sequences of operations taking into account a great number of variables.

A major development in preparing policies and logic will be the evolution of facility management from a rather reactive to an assertive proactive orientation and operation. Yes, things break, alarms and emergencies happen and FM will always react to those events, but FM must embrace planning and become more proactive.

Data Analytics

If you are buying books or music from an internet site it's likely that the company analyzes your purchases, creates a profile of what type of books or music, authors or performers you like and then sends you email regarding other books or music they think you may be interested in purchasing. This is an example of "mining data" to improve business performance. Generally facility management has not traditionally used these techniques. We've focused on analyzing energy consumption data and have analytic tools to optimize HVAC but there's a lot more data out there to be generated and analyzed.

A critical component in building automation is data, because it's the data that will be the foundation for the development and revisions to the logic or policies of the automation. Call it data mining, business intelligence or predictive analytics; it comes down to analyzing the building data, finding trends in how the

building is performing or being used, inferring relationships between variables and creating rules; then using that information to predict how the building performs under different scenarios. This progression is likely to bring new perspectives to the building operation and new ideas for how to operate the building. Finally, the need for data analysis is one rationale for more integrated building management systems which can provide for a unified database of building system data and facilitate the integration to many other data sources.

Vast Amounts of Sensors

Highly automated buildings will need many additional sensors and metering; some for energy systems (plug load, lighting, HVAC), others for air quality, building occupancy, external lighting conditions, water consumption, security, etc. A key building metric is occupancy and it may be the most challenging building metric to obtain. It's not because there is no technical solution to measure or sense occupancy because in fact, a number of solutions exist, each with advantages and disadvantages. Most lighting control systems incorporate an occupancy sensor into their systems; some can even track the path the occupant is taking, others use the lighting control occupancy sensor for control of the plug load within the room or space. However, occupancy sensors attached to lighting control systems alone may not be enough.

Video cameras, access control systems, infrared sensors on door frames, RFID tags, monitoring whether the spaces' IT equipment is on, etc. are all ways to determine occupancy. Some systems are able to not only sense occupancy but count people. Each building owner will have to sort through options on the market for the best solution for their building.

Understanding the Larger Context of ICT

We can't be constructing highly automated buildings in isolation. All around us is a society and world where people are connected in oftentimes a pervasive and hyperactive manner to other people and objects. Everyone occupying, managing and owning buildings is part of this community. In addition, we also have concepts such as the "Internet of Things" and "ambient intelligence" on the horizon, indicating the trends of technology and connectivity will not only continue to evolve but most likely accelerate.

We've seen the relentless penetration of IT into the "traditional" building control model. We expect information and communication technology will play a very large role in increasing and improving building automation. Sometimes IT and FM organizations seem like two sides of the same coin, both involved with networks and systems, albeit different systems. Within each company or organization it will require greater accommodations and a stronger relationship between IT and FM to facilitate increased automation; possibly organizing both under a System Engineering banner.

This level of building automation is not illusory. You see the first steps of heightened automation in smaller and medium size companies creating new BMS platforms that will be required for this level of automation and new analytic tools such as fault detection and diagnostics. You also see it in ICT companies increased interest in buildings, energy and analytics. Enhanced automation is a device to eventually get to the nirvana of minimal energy consumption and improved performance of buildings and it's achievable without any of us sitting out on the wings of a plane in flight to do so.

CONCLUSION

Analytic software is a relatively new class of tools for building owners and managers, with the ability to provide real-time analysis and diagnostics of multiple systems and add "smarts" to the building. As this field continues to expand, we will find new and innovative tools that expand beyond HVAC and into other building systems with built-in intelligence for facility management.

Contributors to This Chapter
Jim Lee of *Cimetrics*
John Petze of *Skyfoundry*
Andres Szmulewicz of *Smart Buildings*
Brian Thompson and Chuck Sloup of *Ezenics*
Brian Turner, *Controlco*
Mike Welch of *Control Network*

Chapter 25

Data to Information for Intelligent Buildings

Nirosha Munasanghe, Consultant

ABSTRACT

In today's data-rich environment the ability to extract meaningful information at the right time to the right audience is becoming a challenge. Building Management Systems (BMS) has evolved over the last decade with open protocols and web connectivity increasing the volume and different types of data that is describing the behavior of an intelligent building. The current systems extracts the data from sensor networks into an embedded device which takes actions predefined by the user and export the information for dashboard reporting to an audience. Are we doing enough with the data? Clearly, No. The data further needs to be fed into models to make proactive decisions at the local device level and further historical analytics on unstructured data handler to uncover patterns and behavior to further improve efficiency of the building. The chapter outlines underlying issues with current state of building management systems that can be resolved by data enhanced models and examines an automated diagnostic architecture using machine learning techniques at the building management device level and introduces the new big data phenomenon to provide historical data analytics to inform meaningful information and knowledge to stakeholders of the building.

INTRODUCTION

The drive for smarter buildings is generating a torrent flow of data from various sources from a building. In the data rich environment today's corporations operate in, the capacity to understand what is happening around them is an important weapon. The growth of open standards in Building Management Systems (BMS) is exposing a plethora of data measuring various metrics of a building to manage its energy.

"Each DDC in the system shall contain a trend that will store samples of all the of the data points."

"The logged data must be sampled every 15 minutes and archived over 2 years."

"Provide trend logging, event logging and hours run on selected points with selectable time intervals."

These are typical phrases from specifications. What does these mean? Lots of data. The reduction in the price of computer memory has allowed BMS to store unlimited data metrics that describe the behavior of the building. What do we do with the data? At present a BMS presents the data in a reporting tool in graphical format for the user to interpret comparisons and report what has happened in the past and what is happening at present. Is this enough? No. We need transform the data to information to automatically diagnose the system, make decisions and predict future behavior to make it a truly a smart building.

It can be argued that the word "Smart" has driven the BMS Industry over the last decade. It has become the fundamental buzzword for product marketing teams. But what really is smart? The oxford dictionary defines smart as "informal having or showing a quick-witted intelligence." Are the devices managing current buildings really intelligent compared to 20 year ago? No, the BMS device functionality set has not changed much in the implementation. The devices are capturing data via input and output signals according to predefined logic. This is not intelligence. To achieve a true automated diagnostic system, it is time device vendors begin to implement well known artificial intelligence techniques such as machine learning at the local device level to provide fast proactive decisions to the current building conditions without user input. The data further needs to be transferred to an unstructured data framework platform for historical analysis. Finally the information needs to be reported to the correct audience at the right time. The chapter examines the concept of intelligence, the current issues of the building management systems and outlines a true automated diagnostic architecture for intelligent buildings using machine learning and big data architecture to use information to proactively make decisions.

The decision support system can be used for tactical and strategic decisions for the facility department. Figure 25-1 illustrates the data hierarchy and types of decisions that can be made with the respective data types.

WHAT IS INTELLIGENCE?

A large collection of data is useful to the daily operations of a building. A temperature graph over a few days can illustrate room temperature behaviors to a plant technician. The data assists the facility department operating the building on a daily basis. Therefore it is only operational data. To further enhance the decision making process for the facility managers, the data needs to be fed into models to discover further intelligence. Intelligence is the ability to learn or understand from experience and use reasoning to solve problems. Therefore past data can be used to learn and solve future problems. It involves acquiring data and information from a variety of sources and modeling them for decision making. The modeled data from the BMS can then be a decision support system for the facility department.

FUNDAMENTAL BUILDING MANAGEMENT ISSUES

Before delving into the possible automated diagnostic architecture, let's examine some fundamental issues in the current state of BMS. Web technology, cloud computing, virtual servers and open protocol standards are some common drivers of the BMS industry today and for tomorrow. These technologies define the core elements of BMS and the future will define further enhancements to these technologies by natural technological revolution. However, we must remember that technology is just a strategic enabler to complete a specific task in a better and improved method, to gain competitive advantage. The technology has advanced the BMS paradigm to for the end user to have more options and cost effective solutions. However the technology has not

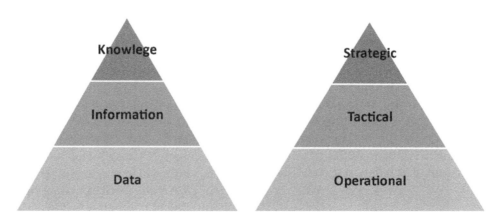

Figure 1: Data hierarchy

Knowledge, Strategic	Support top management decisions. The knowledge can provide cost benefit data of the current system and can assist decision making process to further install the system at another location
Information, Tactical	Support middle management to allocate resources. The information can assist the middle management to further improve the facility.
Data Operational	Support daily activities. The facility technicians can use the data to fault find and fix daily issues.

Figure 25-1: Data hierarchy

addressed some of the core issues. Let's examine three fundamental issues in BMS; lack of commissioning, user error in control algorithms, and lack of visibility that can be improved by better usage of data.

Commissioning

One of the fundamental problems in the BMS industry that does not have a clear solution is commissioning. A building can have state of the art controls products, but without proper commissioning the full functionality will never be realized. In the building construction life cycle, the BMS is generally the last piece of work to be completed and combined with reduced construction programs, limited commissioning durations and a lack of expert personnel, collectively leads to poorly commissioned systems. There are many cases where the system integrator is installing the BMS as tenants are moving into the building. Therefore in most cases at the end of the commissioning phase, the end result does not comply with the original specification.

User Error

There are many specifications with many optimizing control algorithms, which do not behave in reality as in theory, or the engineer programming the system lacks the knowledge to program the system preventing simple control strategies from performing. Also if optimizing algorithms are implemented, it has to be carefully tested and retuned throughout the lifecycle of the building. The expectation of continual tuning of a system is not realistic due to time pressures and lack of understanding. Surveys conducted on a group of senior project engineers, building managers and building service technicians on their understanding of the Proportional, Integral, Derivate (PID) algorithm was a failure. The questions were related to the underlying theory of the PID algorithm and many were only able to answer 50% of the questions correctly. It was clear from the questionnaire that the group was able to tune a control system by trial and error by manipulating the algorithm parameters but when it came to explaining the meaning of each parameter, it was a clear failure.

Visibility

There have being enormous resources allocated to reduce energy consumption of buildings. Government initiatives, green councils and many other bodies have been setup worldwide to promote the "Green" message. BMS manufactures have joined on this green technology road to improve their product range, from improved energy calculation formulas to dynamic reporting tools. Consultants are specifying new methodologies to controls contractors to improve energy performance of building and these are been implemented in new and retrofitted buildings. But the question is, do the end customers who use the building see the end results of energy management actions? In majority of the cases, the answer to this question is No. At present, the facility manger monitors the results but there is lack of feedback to the end customer who utilizes the building. The facility manger generally receives feedback from the building users when the temperature is either too cold or hot. There is no feedback between facility managers and building users about how much energy did they use today and how it can be reduced. The facility manager has this information from the BMS, but in most cases the final link in the chain to improve energy performance of building is missing due to this broken feedback loop. The web connectivity has been on the market for over decade. Almost every BMS specifications defines it but when it comes to day to day usage not many building fully utilize it. In most cases there is local connection to a limited number of machines for facility managers. The power of the web is not fully utilized and the visibility is not exposed to the right audience at the right time.

CURRENT BUILDING MANAGEMENT DEVICE

Let's examine the core functionality of a so-called smart device in a building management system (Figure 25-2). A device consists of inputs which collates data from sensor networks, wires and wireless and provides a set of outputs driving peripheral mechanical devices to control to optimal value. The outputs are triggered by a logic pre-defined by the user using a simple logic using vendor dependent programming language, which also triggers the fundamental theory in control systems, the Proportional, Derivative, Integral (PID) loop. The core function of the device is based on the logic driving the outputs with a PID loop algorithm assisting to control with minimal error. The device also contains peripheral features such as scheduling to schedule an event on a calendar, and a data logging mechanism to log critical data and IP base connectivity to access via web. The data are transferred to a database for further analysis by a software tool and then reported to an audience.

Table 25-1 gives the description of each component.

What is the fundamental issue with the above device? The core functionality of a BMS device has not changed since its inception apart from connectivity to

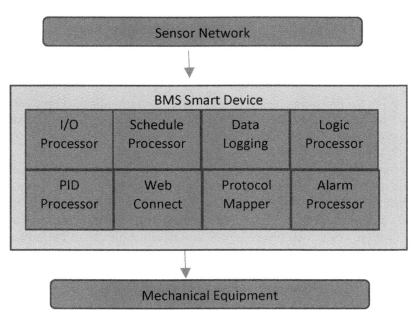

Figure 25-2: Typical components of building management device.

Table 25-1

Component	Description
Sensor Network	Data as input from sensors measuring temperature, humidity, CO2, feedback status of mechanical equipment
I/O Processor	Engine processing the raw inputs to appropriate units
Schedule Processor	Scheduling algorithm which executes 365/24/7 to trigger data points
Data Logging	Process which logs required data into flash memory and export to external interface if required
PID Processor	Proportional, Integral, Derivative control algorithm to drive an output to desired setpoint with minimum error
Web Connectivity	IP interface for connectivity to outside world
Protocol Mapper	Multiple protocol supported in the device which will map the data
Alarm Processor	Handle configured alarms and send it to selected destinations
Mechanical Equipment	The outputs drive mechanical equipment to reach desired conditions.

web. The procedure to setup the device is mundane and has many user inputs leading to user errors. The typical setup procedure for device as follows:

1. Discover the device into a software over a network
2. Configure the data points to requirement of mechanical equipment under control
3. Define the logic using a programming language
4. Define peripheral functionality such as schedules and logs
5. Define the PID loop and estimate the tuning parameters
6. Commission the device and re-tune the loop parameters to achieve minimal error.

A BMS engineer completes the above typical work flow for each equipment under control in a building. What is the fundamental problem? The lack of self-learning from data at the device level to correct user errors and more importantly automatically diagnose to make proactive decisions throughout the life cycle of the building.

AUTOMATED DIAGNOSTIC
ARCHITECTURE FOR BMS

To address the above core issues, BMS require an architecture where the smart device itself learns and optimizes its operation from the raw data via the sensor networks and historical data needs to be fed into a big data platform for advance analytics to report to the right audience at the right time.

The architecture depicted in Figure 25-3 illustrates the smart device executing machine learning techniques to automated decisions locally at the building to provide proactive actions to alter behavior of the building without user intervention. Secondly, the historical data from the device is exported into a big data architecture for further analytics. The analytics is dissected to provide the right information to the various stakeholders of the building.

WHAT IS MACHINE LEARNING?

Machine learning is a core sub area of artificial intelligence that studies computer algorithms for learning to complete a specific task without user intervention. For instance a device can be interested in learning to complete a task, or to make accurate predictions, or to behave intelligently. The learning that is being done is always based on observations or data, such as examples, direct experience, or instruction. In summary, machine learning is the concept of learning to do better in the future based on what was experienced in the past. The emphasis of machine learning is on automatic methods. In other words, the goal is to implement learning algorithms that do the learning automatically without human intervention or assistance. The machine learning paradigm can be viewed as "programming by example." Often there is a specific task in mind, such as finding a pattern. But rather than program the computer to solve the task directly, in machine learning, it seeks methods by which the computer will come up with its own program based on examples that are provided by the user. It is very unlikely that a system will be able to build any kind of intelligent system capable of any of the facilities that we associate with intelligence, such as language or vision, without using learning to get there. These tasks are otherwise simply too difficult to solve. Further, a system cannot be considered to be truly intelligent if it were incapable of learning since learning is at the core of intelligence.

There are many examples of machine learning problems in real world today. Typical problems are shown in Table 25-2.

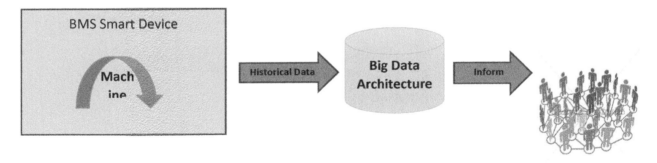

Figure 25-3: Data learning architecture for automated diagnostics for buildings

Table 25-2

Application	Description
Face detection	Find faces in images or indicate if a face is present
Optical character recognition	Categorize images of handwritten characters by the letters represented
Voice Recognition	Within the context of a limited domain, determine the meaning of something uttered by a speaker to the extent that it can be classified into one of a fixed set of categories
Fraud detection	Identify credit card transactions (for instance) which may be fraud- ulent in nature
Medical diagnosis	Diagnose a patient as a sufferer or non-sufferer of some disease
Customer segmentation	Predict, for instance, which customers will respond to a particular promotion
Weather prediction	Predict, for instance, whether or not it will rain tomorrow

HOW DOES MACHINE LEARNING WORK?

Machine learning techniques involves various complex algorithms. A machine learning system can be described as taking a set of input data to process and providing output as a decision.

The type of technique used to learn depends on the desired outcome of the algorithm or the type of input data available during training of the device. Typical techniques include:

• **Unsupervised Learning**: The training is based on unlabeled examples where there are no previously defined outputs with an objective of discovering structures in data. In such cases the system mines for new and interesting patterns that may be useful. The clear advantage of the approach is that the system has the potential to discover new patterns that could benefit. However it can also output patterns that are redundant.

• **Supervised Learning**: The training is based on labeled examples where the desired output is known. It attempts to generalize a function from inputs to outputs which can be used to generate an output from previously unseen inputs. The advantage of the supervised learning is that it can learn complex useful patterns with good performance. The drawback is that the user must feed in examples of useful outputs for it to learn.

• **Semi-supervised Learning**: The approach combines labeled and unlabeled examples to generate a function to classify outputs.

Typical learning algorithms for the above types are shown in Table 25-3.

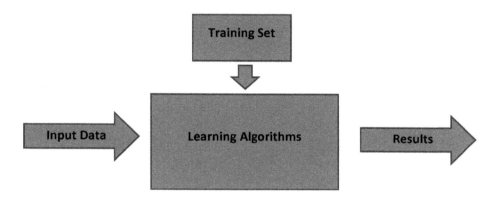

Figure 25-4: Machine Learning

Table 25-3.

Algorithm	Description	Advantage/Disadvantage
Nearest Neighbor	Find the nearest labeled example. For example "This article is about baseball. The last baseball article I saw was about sports". Therefore this article is about sports.	It is easy to implement and can generate complex outputs. The drawback is that it requires every example to be stored making classification slow.
Decision Tree	Construct a tree of decisions to follow where each leaf applies a label. For example this document says baseball therefore its sport. If another word check the next leaf.	Very easy to understand and good for identifying few critical features, however slow to train
Artificial Neural networks	Construct a graph of neural networks by defining input and output nodes. The internal nodes are discovered dynamically.	Can learn non-linear functions but not easily interpretable.
Clustering	Find set of clusters that best describe the data.	Applicable to any attribute type and ease of handling any forms of similarity.

BUILDING MANAGEMENT DEVICES

The fundamental theory for machine learning is a powerful and attractive proposition. However how does it fit the BMS? Let's examine typical examples where machine learning can be applied to BMS device to improve its performance.

PID Algorithm

The fundamental goal is to control an input at a desired set point by manipulating an output. We must not forget this fact. This primary goal is starting to get lost behind the great technological advancements such as open protocols, web and IP integration. "Control" must be the fundamental goal. Without control, technological advancements are useless. The history of control theory goes back over 100 years to Proportional, Integral, Derivative (PID) control theory. The PID theory is the fundamental control strategy used in the BMS industry and we are not putting in enough time to understand the concepts and to apply theory to reality. Incorrect PID settings can cause a system to hunt if set points are exceeded by outputs reacting too aggressively, causing increase in energy consumption and mechanical equipment failures.

The PID controller is a feedback controller which calculates error, which is the difference between the input and the desired set point. The goal is to minimize the error by adjusting an output. For example in building automation as depicted in Figure 25-5, the input is the temperature sensor, the set point is the desired value of the room temperature and the output is a damper which allows air flow. Therefore to minimize the error between the actual temperature and set point, the amount of air flow is adjusted with a damper.

How do we maintain this dynamic system? It is achieved by setting the correct factors for the proportional (P), Integral (I) and Derivative (D) coefficients.

Proportional (P): The proportional term makes a change to the output that is proportional to the current error value. Therefore a large proportional increase to the output by that factor causes instability, if only proportional is used.

Integral(I): The integral term is proportional to both the magnitude of the error and duration of the error. Therefore summing the instantaneous error over time gives the accumulated error that should have been corrected previously. This sum of error is multiplied by the integral factor to obtain the output. It accelerates the input towards the set point and correct for steady state error using only proportional control.

Derivative (D): The derivative term calculates the rate of change of error and this is multiplied by the derivative factor to the output. The derivative term slows the rate of change of the controller output. It is used to reduce the overshoot caused by the integral component and improve stability of the output.

The goal of control is to tune the PID parameters to correct values to control the dynamic system in a stable and accurate manner.

We have a mathematical formula, but why do we keep failing to obtain optimal control parameters?

Trial and Error: Tuning PID parameters is a trial and error procedure. To truly tune a system, the engineer must tune the parameters with minimum, maximum and normal system operation and observe the output over a time period. It is not as simple as entering a value. This process is neglected by many engineers due to time pressures in the project.

Vendor Definition: PID is a generic algorithm for control, and the P, I and D factors are just coefficients. Therefore different vendors have different meanings for these

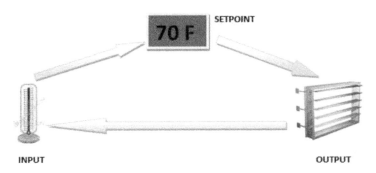

Figure 25-5: PID example

factors. Some treat it just as a factor; others treat it in time scale. Therefore if a system is tuned using certain PID parameters in one BMS system, the same parameters most likely fail in another system. PID parameters used in one BMS system cannot be used in another system.

Continuous Tuning: Tuning a PID to correct values at the start of a project is not enough. During building life cycle maintenance, it must be reviewed as the dynamics of the system can change. However, this is not a standard practice in our industry. Generally the PID values are set by the engineers at the start of the project and those values remain there for life. We are continuously deriving strategies to save energy in buildings. Many dollars are spent on new research to develop theories and to retrofit buildings. It can be argued that with simple retuning of PID values it can save substantial amounts of energy without the need to spend millions of dollars. For example; a typical solution to improve energy performance of an existing building is to change of pulsed based heating valves to analog based. However in most cases the root cause of the energy increase was not the pulse based valves, but the tuning of the PID algorithm. It was originally tuned based on the analog valves causing the system to only perform full heating or no heating. Therefore with a simple change of PID values, system stability was achieved with substantial energy savings.

Copy and Paste Technology: A common engineering principle is to reuse good solutions. Do not reinvent the wheel. In order to use this principle the user must understand how and why the original solution is a good solution; not just copy and paste the solution. This copy and paste phenomenon is used commonly in the BMS industry but not correctly. Many BMS solutions are copied from past projects without really understanding what is being copied. Five years ago many copied solutions went unnoticed as the equipment being automated was the standard fan coil units, air handlers, chillers and boilers. Even a slight variance was unnoticed as it did not affect the end customer. However, now with the evolution of energy management concepts many buildings are using new strategies such as chilled beams to cool buildings, weather stations, window control, rain water harvesting etc... Therefore the older solutions developed 5-10 years ago are out of date and will not operate efficiently in this environment. The new specifications are more stringent in terms of error to reduce energy consumption, therefore even a PID parameter to control a basic fan coil unit

must be reviewed. Therefore do not just copy and paste solutions without knowing what you are copying. Time is precious, however with a wrong solution at the start of the project, you will spend lot more time throughout the life cycle of the building fixing the defects.

Tuning the PID parameters is a classic application where machine learning can be applied on a BMS device. An adaptive method of tuning the parameters eliminates tedious techniques of tuning and more importantly user error. Also it will adjust to the conditions for the life time of the system.

Optimal Schedule Operation

Typically in a building the heating and cooling equipment are turned on and off according to a pre-defined 24/7 schedule based on occupancy time of the tenants. This method does not take into account the current thermal dynamics of the building and past temperature. With simple machine learning techniques equipment can be operated optimally to reflect the current building conditions. For example if it has been a cool summer night, why turn on the cooling at a fixed time. Using historical data the BMS device can calculate the optimal time to turn on the cooling.

Optimal Energy Consumption

Machine learning can be also used on a BMS device to calculate consumption of energy on mechanical equipment. For example let's examine operation of a chiller with the data set shown in Table 25-4.

Feeding the above data into a learning system, the conclusions shown in Table 25-5 can be deduced about the operation of the building.

The example illustrates simple association rules that can be used by a facility department to assist the decision making process. The example association rules just touches the surface of machine learning. Also, the predictive nature can assist in reducing energy in a building as the models can predict how much energy a building will use for any given weather condition or day.

WHAT IS BIG DATA?

Machine learning on the BMS device allows proactive decisions locally at the field level provides fast response to building conditions without user intervention. As the size of the data set grows it will improve the learning patterns of the device to provide further

Table 25-4.

Time	Temp	Chiller A	Chiller B	KWH
12:00	69.8	OFF	OFF	1.2
12:30	71.24	ON	OFF	2.4
13:00	71.78	ON	ON	3.5
13:30	70.34	ON	OFF	2.6
14:00	69.8	OFF	OFF	1.4
14:30	69.98	OFF	OFF	1.3
15:00	71.42	ON	ON	3.8
15:30	70.7	ON	OFF	2.4
16:00	71.42	ON	ON	3.9
16:30	70.34	ON	OFF	2.6
17:00	71.42	ON	ON	4.1
17:30	73.4	OFF	OFF	0.5
18:00	75.2	OFF	OFF	0.4
18:30	73.76	OFF	OFF	0.4
19:00	73.58	OFF	OFF	0.3

Big Data is driven by significant growth in data sets over the last few years due to growth in unstructured data from mobile devices, sensing technology, wireless sensors, and social networks, satellite images, photo/video and speech, along with the reduction in cost of storage. The data growth has put limitations on the current rational database technology in obtaining the right data at the right time, which has introduced a new wave of database architecture such as MapReduce and Hadoop. The fundamental advantage of current big data architectures is the ability to process unstructured data. The traditional database tools required structured datasets with relationships to process the data. However architecture such as Hadoop and MapReduce allow processing of unstructured data at high speed. Hadoop allows big problems to be decomposed into smaller elements so that analysis can be done quickly and cost effectively. It is a versatile, resilient, clustered approach to managing files in big data environment. This architecture has been a key breakthrough in data management as unstructured data is ever increasing in our day to day life.

Table 25-5

Rule	Explanation
IF Temp >= 71.9 the Chiller A and Chiller B ON	It indicates that if temperature is above 71.42 F then both chillers are always running. This is an indication to the operator that either Chiller A is not functioning properly or chiller logic is incorrect. It could also be indication the management of Chiller A capacity is not enough for the building.
IF Chiller A and Chiller B is ON Then KWH > 2.5	It indicates that when both chillers are on then the KWH is always greater than 3.5. It indicates to the operator the cost of running both chillers
IF Time > 17:00 Then KWH >= 0.5	It indicates that after 17:00 the KWH is less than 0.5. This indicates to the operator that the mechanical plat of the building is not operating.

accurate predictions. The concept of machine learning at the BMS device level is to take immediate actions to building reactions in the depicted architecture. The data from the sensor network further needs to be fed into an unstructured data architecture for further analysis to provide historical metrics of the building behavior and make long term predictions.

Big data describes a set of complex data that is difficult to process using traditional database management tools. The processing includes capturing, storing, searching, sharing, transferring, analyzing and visualization.

Figure 25-6 outlines the three key characteristics of big data architecture:

Volume: Applying Moore's law, what is considered to be large today of 10 terabytes, in 12 months it could be 50 terabytes. Therefore a technology is required which can store vast amount of data in a scalable architecture and provide distributed approach to querying the data.

Variety: A proliferation of data types no longer fits into neat, easy to consume structures. Big data architecture deals with unstructured data including data mining, text analytics, and noisy text analytics.

Velocity: Velocity describes the frequency at which data is generated, captured, and shared. The velocity of large data affects the ability to parse text, detect sentiment, and identify new patterns. Big data addresses velocity include streaming processing and complex event processing. In addition, the use of in-memory data bases, columnar databases, and key value stores help improve retrieval of pre-calculated data.

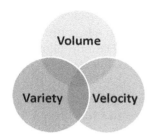

Figure 25-6: Characteristics of Big Data

BIG DATA FOR BUILDING MANAGEMENT SYSTEM

The new big data architecture has a significant advantage for the BMS industry. The fundamental feature of a smart building is the ability to capture endless data from key performing building resources using sensors, wireless and wired networks. The data is transferred to a central processing unit to analyze to take proactive action and then visualize the data for an audience. The rapid growth of data in a building has driven the dashboard industry over the last two years with many players entering the market with various data visualizing panels. As the growth of data increases the big data architecture is a must for the BMS industry to process unstructured data to meaningful information. Current systems report raw data from the building and display in elegant dashboards. The systems lack advance analytics to combine various data types to output data that discovers new patterns to improve the performance of the building.

The challenge is not only to process the data to meaningful information but also to report the information to the correct audience at the right time. A big data architecture allows easier integration to corporate enterprise systems to transfer the information flow to the correct audience. For example: a 20 story building with 4 clients leasing 5 floors each. At present in such a scenario, the facility manager acts on the data from the build-

ing to maintain the energy performance. It is a reactive approach. With a big data architecture encapsulating historical data and providing advanced and predictive analytics the information can be easily integrated to each of the client's corporate IT network. With a web based system the client in respective floors can access their respective daily energy profiles using a web browser via the BMS from their desktop PC. Therefore the building users and management of each floor have a better understanding of how much energy they are using and can accurately allocate costs and assign responsibility to meet their energy key performance indicators. Also, say for example, one client leasing the building is not comfortable using a BMS reporting tool, then with web integration the energy data for the user can be imported to a social networking tool such as Twitter for easy access. Similarly, most Enterprise Resource Planning (ERP) vendors are introducing energy reporting modules as part of the accounting system, therefore the data from the BAS can be imported to ERP system to fit requirements of the client. Therefore it be seen that using a big data architecture opens up many options to share data and report to the correct audience in the required format to assist in improving energy performance of buildings.

CONCLUSION

The ICT industry has already mapped their big data architecture to handle the perfect storm of business needs and evolving technology to allow organizations to leverage data to accelerate the business. The building automation industry must follow and plan their data architecture to transform data to information to open up new opportunities. The BMS needs to segregate its data analytics both at the device and software application level. Most BMS vendors have proven smart devices on their platform which can be used to perform convert data to information locally using machine learning techniques to provide proactive feedback to optimally monitor and control energy resources. Imagine the power of advance analytics using a big data architecture to visualize the long term perspectives of building behavior and true smart devices performing proactive day to day decisions.

References

Franklin G, Powell J, Emami-Naeini A, *Feedback control of dynamic systems*, 4th Edition, 2002, Prentice-Hall, Inc

Wang Ray, Beyond Three V's of Big Data, http://www.forbes.com/sites/raywang/2012/02/27/mondays-musings-beyond-the-three-vs-of-big-data-viscosity-and-virality/

Kearns M, Vazirani U, *An introduction to computational learning theory*, 1997, The MIT Press

Chapter 26

The 12 Things You Need to Know about Monitoring-Based Commissioning (MBCx)

Jim Butler and Jim Lee, Cimetrics
Craig Engelbrecht, Smart Services and Technology—Siemens
Jim Sinopoli, Smart Buildings, LLC

From the moment a building is constructed and starts consuming resources, it starts deteriorating. Mechanical, electrical, automation, and all building systems naturally decline over time in terms of both performance and efficiency. The results can be the same; operations and maintenance costs rise, energy consumption rises, and occupant comfort may decline.

Building owners and managers know they have options to prevent this natural progression, but are often overwhelmed by a glut of contradicting information, misinformation, and misconceptions in the marketplace. Different vendors different use terms like commissioning, existing building commissioning, fault detection and diagnostics, smart buildings… the list is nearly endless.

Although the terms are different, the message is the same: "Let us analyze and optimize your building and we'll save you money." It's certainly an attractive proposition—spend a little money now, save a lot of money later. But because of the terminology confusion, abundance of market options, and lack of standardization in approaches to dealing with this issue, building owners and managers have become understandably leery of sales pitches. In the midst of the confusion, though, lies a truth; what the industry calls continuous or monitoring-based commissioning does, in fact, deliver improved building performance and energy efficiency.

1. It goes beyond energy savings.

2. It's a team effort.

3. Process, process, process.

4. Implementation requires full buy-in.

5. It's more than just a rule developed in a lab.

6. You might not have all the data you want and need.

7. When it comes to data, it's all hands on deck.

8. MBCx doesn't work on its own.

9. The software doesn't "drop in" and get to work.

10. Don't turn it on all at once.

11. If you're not committed to fixing issues, don't waste your money on MBCx.

12. Optimization strategies go hand-in-hand with analytics.

The real long-term value of commissioning is in the building performance improvements and compliance efforts. That value is possible through a comprehensive combination of the right people, processes, and technologies.

DEFINING MONITORING-BASED COMMISSIONING

According to a report by the Lawrence-Berkeley National Laboratory, "Monitoring based commissioning (MBCx) combines ongoing building energy system monitoring with standard retro-commissioning (RCx) practices with the aim of providing substantial, persistent, energy savings*." What we are really talking about is a sophisticated package of software applications that combines building data from a wide variety of sources to better manage building performance and efficiency. MBCx involves the implementation of improvement measures along with ongoing service and insights necessary for full transparency, measurement, and reporting. That is, what facility engineers have done manually for decades can now be completed more efficiently, more comprehensively, and more accurately by combining building and energy system data with an engineering team's expertise through the MBCx process.

When MBCx is built into a continuous building improvement process, it allows combined technologies involved in data mining to identify faults or issues in building systems with the necessary human analysis to determine how to address those faults or issues. Truly advanced MBCx solutions will also help identify and prioritize resolution paths; for example, if there is simultaneous heating and cooling in air handler 5, facility engineers should investigate a leaking chilled water valve to avoid a potential costly expenditure.

Defining MBCx is the first step toward understanding it; we will now review the 12 things you need to know about monitoring-based commissioning.

It's Not Just about the Energy Savings

The same Lawrence-Berkeley National Laboratory report asserts that MBCx is a highly cost-effective strategy. The buildings involved in this benchmarking analysis realized average energy savings of 10%, with some ranging as high as 25%; and the investment had a simple payback period of 2.5 years, according to the report. Those are impressive results to be sure.

Many vendors today offer metering and commissioning services and promise energy cost savings. While that's an important goal that often gets building owners interested in a service package, what's equally as important is the improved building performance and compliance that can result from a truly comprehensive MBCx strategy.

Through our experience over the years, we have come to understand that building maintenance is not always thoroughly completed for a variety of reasons, including, without limitation: the mechanical, electrical, and HVAC systems are considerably more complex than they once were; maintenance issues are resolved when the building is no longer comfortable and a tenant complains, or when a piece of equipment breaks and must be repaired or replaced; and because regular maintenance schedules (for example, filter replacements every six months) are not sufficient to support the building's most effective performance. MBCx is a comprehensive business process to improve the way buildings are maintained by using technology as an enabler.

The results can include a lower likelihood of premature equipment failure; assistance with federal, state, local, and corporate requirements and regulations; energy and operational cost savings; and, finally, sustaining and perpetuating the improvements you have made.

It's a Team Effort

While MBCx involves highly sophisticated technologies, even the best software applications do not always perform exactly as you need or want them to without

the right people working with them. Monitoring-based commissioning means assembling a high performance team comprised of building engineers, controls contractors, mechanical contractors, commissioning resources, and others. The objective is to assemble subject matter experts from every aspect of the building's performance to most effectively review and analyze the mined data. This team will understand the building's unique operations and how to interpret the issues that might arise on a dashboard.

Process, Process, Process

The technology and the people are there. Having the right processes to support their efforts is critical to maximizing their effectiveness.

Fault detection and diagnostics (FDD) and similar solutions identify out-of-range states for various systems or equipment within a building. A fault is not an alarm; a fault indicates that the system is not running optimally. FDD analysis continually monitors building performance and can identify many faults each month, depending on the complexity of the systems and the size of the building. The technology can give you great insight and information, but you need the people available to finish the interpretation and analysis. Those teams require effective processes to facilitate that regular analysis and proactively optimize building performance and efficiency. Building owners who engage in MBCx may need to develop new business processes to support the teams and technologies in new ways.

It requires a new approach that is no longer simply reactive to break/fix scenarios but now can be proactive based on data, analysis, and operational processes built into the business. The upfront investment in proactive services will ultimately reduce operating costs by enabling your team to work more efficiently and, more importantly, improve building performance.

Implementation Requires Full Buy-in.

The term "building owner" refers to more than a single person or entity who owns the building. "Building owner buy-in" means getting all facility engineers, technicians, and information technology specialists on board to plan for and participate in an MBCx implementation. And because it requires continued, long-term engagement and participation, all stakeholders should be aware of and committed to implementing and sustaining the process.

It's More than Just a Rule Developed in a Lab

MBCx is no off-the-shelf, lab-developed process.

That is, every building has unique controls, unique configurations, and unique data. The many diagnostic algorithms must be continually tested and improved upon, rather than developed and tested in sterile lab environments. Preparing these algorithms means getting a picture of the building's systems and creating a model that considers the original control drawings and design intent. No rules can come out of the box and apply to every building, every climate, every situation.

MBCx can provide a series of tried-and-true algorithms that support the rule development process in the field. The mechanical, electrical, and energy systems produce vast amounts of data that can be measured and compared against a set of expected values. When the actual values do not match the expected values, the algorithms will generate a fault, and in the cases of evolved MBCx, a root cause. Engineers get a wide-ranging picture of what is going on in the building, and MBCx can be customized to the characteristics of a particular building.

You Might Not Have All the Data
You Want and Need

Most of today's building automation systems are designed specifically to control the building's systems; not as tools to analyze and trend data. IT professionals can extract data, but it is not necessarily going to be comprehensive enough to provide the analysis you want as quickly as you need it. Insufficient data leads to fewer insights and more tentative conclusions, and so effective commissioning may require a building automation system upgrade or hardware improvements, for example, in order to uncover and leverage the trending data that's buried in the building automation system.

When It Comes to Data, It's All Hands On Deck.

A complete picture of building performance is only available when you have all data hands on deck. Automation system data, energy data, equipment data, service work orders—all of it must work together. Each set of data provides valuable information, but is not effective as a comprehensive strategy on its own. Looking at utility usage as an MBCx strategy is akin to diagnosing cancer by listening to a stethoscope. Many of today's "commissioning" vendors may position their work as comprehensive, but if they're examining just one set of data, you're getting a partial picture at best.

MBCx Doesn't Work on its Own

MBCx can provide persistent performance improvements and operational savings, but without a comprehensive commissioning strategy, it's not going to deliver all the results building owners want, need, and demand from commissioning vendors and partners. A comprehensive commissioning strategy must include existing building commissioning (EBCx) to help ensure that a building is performing optimally according to current needs, while MBCx can not only ensure savings, but also identify advanced measures for performance improvement. This comprehensive commissioning strategy is only effective when combined with the traditional diagnostic activities of skilled personnel in the building. Consider one example where a building's HVAC system was overtaxed in trying to cool a building. Engineers could determine, through MBCx, that one of the temperature sensors was reading temperatures significantly

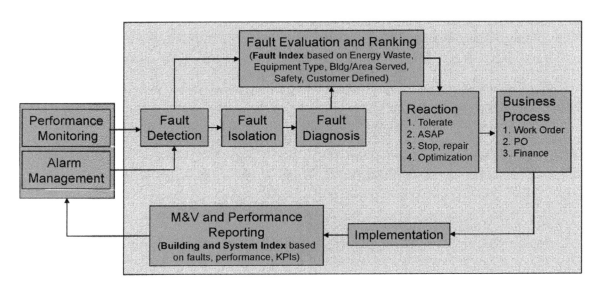

Figure 26-1

higher than in other areas of the building; but it took the intervention of a person walking the building to realize that the sensor was located behind the soda machine to realize the full value of that data. The EBCx was a critical aspect of that full diagnosis and correction.

For further reading on a comprehensive commissioning strategy, please read *Commissioning: An Essential Part of a Comprehensive Energy Strategy*, by Michael Chimack.

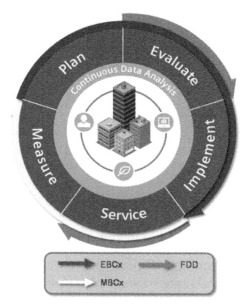

Figure 26-2

The Software Doesn't "Drop in" and Get to Work

In our fifth point, we learned that no off-the-shelf MBCx package is available to work in any building in any situation. Likewise, no software package can simply be "dropped in" and get to work for the building. Configuration requires a team-wide effort to gather data, configure rules, and generate usable information. Regardless of what commissioning vendors' marketing departments might promote, no solution can be turned on and immediately generate savings. Often, because not all data is available and because of a building's preexisting conditions, engineers will have to rely on imperfect data in order to arrive at solutions. Again, MBCx is one part of a comprehensive energy strategy that involves the right people, processes, and technologies to be effective.

Don't Turn it on All at Once

Most organizations don't have business processes built around ongoing building commissioning. If you turn on MBCx across an entire portfolio of buildings, you may identify so many issues that building engineers will become overwhelmed. It can be difficult to establish

effective and reliable response processes. Rather, it's best to start slow and with a smaller workload so you can develop a process that will allow you to be successful. As you get more comfortable with your strategy, then it's possible to take on more and more responsibility and to bring data into the process as it's refined.

If You're Not Committed to Fixing Issues, Don't Waste Your Money On MBCx

MBCx will undoubtedly identify issues in a building to be resolved. And though you may have dedicated budget to the MBCx implementation, if you're not prepared to dedicate an ongoing operational budget to correct faults or implement facility improvement measures, you're wasting your money. Monitoring-based commissioning is a comprehensive process that involves the right people and technologies; use the technologies to change your processes and the way you manage a building

Optimization Strategies Go Hand-in-hand with Analytics.

Optimization strategies can work well on their own, but buildings need to combine those strategies with fault detection and diagnostics and building analytics. Chilled water optimization solutions like Siemens Demand Flow™, demand response for load optimization, and Aircuity with demand based ventilation all provide automated optimization but make sure proper fault detection rules are implemented to support and keep the optimization strategy on track.

SUCCESSFUL MBCX DELIVERS VALUE

As the Lawrence-Berkeley National Laboratory report confirmed, MBCx does deliver energy savings and a return on investment, but the real long-term value of commissioning is in the building performance improvements and compliance efforts. That value, however, is only possible through a comprehensive combination of the right people, processes, and technologies to continuously optimize a building's performance and efficiency.

ABBREVIATIONS

EBCx   Existing Building Commissioning
FDD   Fault Detection and Diagnostics
MBCx   Monitoring-based Commissioning
RCx   Retro Commissioning

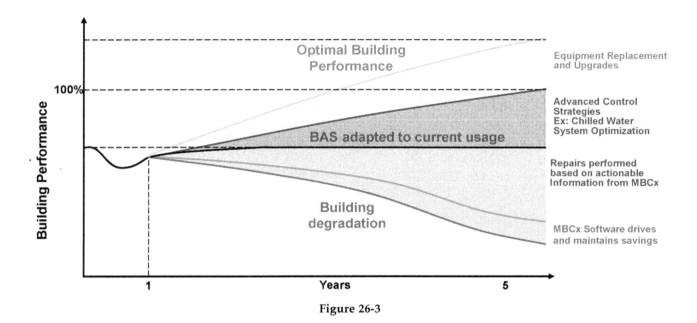

Figure 26-3

Chapter 27

The Building of the Future

John Greenwell, CEPort

INTRODUCTION

Firstly, the building of the future has already been built. In today's economy, new construction has slowed to an all-time low and corporate budgets being strained. Companies will seek to leverage their existing real estate holdings through the use of retrofits and re-commissioning.

In order to create intelligent buildings from existing ones, the built environment must be updated to provide energy efficient, healthy and productive workspaces for a diverse and increasingly mobile workforce. Companies will seek to maximize their ROI, by leveraging their existing building technologies rather than replace them. This can be challenging based on the age of the various systems in the building, some older systems may be obsolete or may not be able to share its data. Any retrofit replacements must promise a reasonable ROI to be considered. That said, traditionally ROI was linked exclusively to energy savings, but today the ROI may be expanded to look at the enablement of new use cases that drive space management and productivity gains. As an example, once LED light fixtures become the norm, it will be hard if not impossible to justify the additional cost of the lighting controls. However, newer lighting control systems can produce a wealth of data that enables new business processes to be defined. These new business processes can be much more valuable than just energy saving. The ability for companies to gain insight into how their buildings are performing, not only from an energy prospective but more importantly from a utilization perspective, will allow them to manage risk and save significant amounts of money.

UNDERSTANDING THE VALUE PROPOSITION

In order for companies to move forward with updating their buildings, they must first gain insight into the value proposition. This can be accomplished by providing two pieces of information, one is the building capability portfolio and the other is a use case analysis.

THE BUILDING CAPABILITY PORTFOLIO

The building capability portfolio is essentially a study of the existing building systems and their current capabilities with regard to granularity of control and ability to share data. The portfolio study should not be limited to only building systems but should also include business data systems and external systems. Systems such as Active Directory and Human Resources can allow you to obtain information about occupancy. Card Access systems, Cameras and others can also provide valuable data for understanding use patterns. Once integrated with building systems this data can be used to write new operational rules and provide valuable context when you're looking at energy efficiency.

One of the goals of this study is to define energy zones. Energy zones are the overlapping functional areas of each building system. These zones will typically be defined by the least granular system. For instance, the HVAC system is usually the least granular and so it becomes the energy zone and all other systems will fit into this zone. How many offices and lighting zones are feed by a common HVAC system? This sets the limits of control for the integrated area.

USE CASE ANALYSIS

The use case analysis uses the information learned from the portfolio capabilities study and from interviewing key stakeholders to define the value proposition of the integration of systems. When and where does it make sense to integrate systems in order to facilitate new polices that drive energy efficiency and productivity? Each use case should have its own ROI model and risk assessment so the owner can pick and choose which ones they want to implement.

PRESENCE BASED AUTOMATION

Presence based automation will drive the building of the future. The mantra for this is, "if you don't need it, don't use it." For most buildings, people have the largest impact on energy consumption and therefore buildings must be responsive to its occupants. Through the use of data produced from many different sources, facility managers can start to understand use patterns throughout the enterprise and make better decisions about their operations. It may seem far-fetched today, but the ability to track where people are within facilities will become commonplace. The value of knowing this information far outweighs the perceived invasion of privacy. Everything from providing just in time HVAC to informing first responders of where potential victims are in an emergency will provide the needed value proposition. Geo-fencing, utilizing cell phone GPS and Radios will allow buildings to understand when occupants will be arriving and set their spaces to the desired setting just in time.

INTELLIGENT BUILDINGS AND THE BOS

Intelligent buildings will require an orchestrated response to constantly changing conditions. Occupancy, Scheduling, Emergencies and other situations will require normally siloed systems to work together in harmony. System controls must first and foremost defend their mission but be programmed to accept signals from other external systems as required. This orchestrated response is not easy to accomplish, systems must be loosely connected, such that a failure of one or more external systems doesn't cause a cascading failure across multiple building systems. This is why it will be important to have a single system responsible to organize and control when and how systems are required to deviate from its normal operation in order to provide a higher level of control. A new term for this type of system is "the building operating system" or "BOS," the analogy to the computer operating system is appropriate. Just as a computer operating system orchestrates the function of computers, the BOS will orchestrate the function of systems in buildings. The BOS will allow new operational rules or policies to be defined and published as a service for individual systems to use.

Applications where system operation is tied to higher level systems will enable Enterprises to put their energy use into context with their work product. Simple energy use intensity per square foot is rather meaningless until it's tied to the business. As an example, an airport might look at kW/sf/takeoffs and landings. A hospital would want to look at kW/sf/census as the true measure of efficiency. Mixed use commercial real estate might want to look at kW/sf/leased % and foot fall. All of these KPIs require information from multiple systems to function.

So, if the key to intelligent buildings is data, then where are the data going to come from and who's going to manage them? In Figure 27-1, we can take a look at some of the many data sources in today's buildings.

THE BDMS

All of these systems produce data for their own internal use, but when integrated, they now represent a rather large volume of disparate set based and runtime data. All of these data are connected to systems via wired or wireless networks. These networks may be daisy chained serial communication for device to network controller or they could be native IP level communications riding directly on the building IP network. In any case, the value comes from integrating, normalizing and contextualizing some of this data into a single repository that turns it into actionable information. This system will be the "Building Data Management System" or "BDMS," you can think of it as an ETL system for buildings. It will provide some key benefits over an *ad hoc* integration approach. One, is that moves, adds and changes are handled in a single system. The BDMS will also be a key component to the success of the Internet of Things and Big Data as they apply to buildings. It will provide Enterprise wide taxonomies for data so that buildings and their data are represented in a unified manner. Imagine being able to Google® your building to find anything from published reports, inspection certificates all the way down to individual sensor values. It will also allow lower level systems to be replaced at will without being locked in to a single vendor. For that matter entire buildings can be added or deleted as the business needs change without the risk of having a stranded asset.

APPLICATIONS

The BDMS will enable cloud based applications for maintenance management, automatic fault detection, constant commissioning and energy analysis to connect to a more holistic and preformatted data source. This will lower the cost of implementation of these systems and provide consistent data across all systems. A more holistic data set will enable these vendors to create new application features that add even more value. Data

LOW LEVEL BUILDING SYSTEMS
- Building Automation Systems
- Lighting Control Systems
- POE/IP based multi-sensor devices
- Fire and Life Safety
- Elevators
- Card Access Systems
- Electrical Distribution,
  Meters, UPS, Generators
  and Renewables,
  EV Charging Stations

MID-LEVEL IT SYSTEMS
- VoIP Telephony
- Video Conferencing
- IP Cameras
- Audio/Visual

HIGH LEVEL IT SYSTEMS
- Human Resources
- Enterprise Asset Management
- Business Process Management
- Customer Relations Management
- Active Directory
- Email
- Calendaring
- Accounting

EXTERNAL SYSTEMS
- Building Information Modeling (BIM)
- Weather Stations
- DRAS Servers (VTN)
- Transactive Energy Markets
- Utility Bill Integration (Green Button)
- Community Alerts

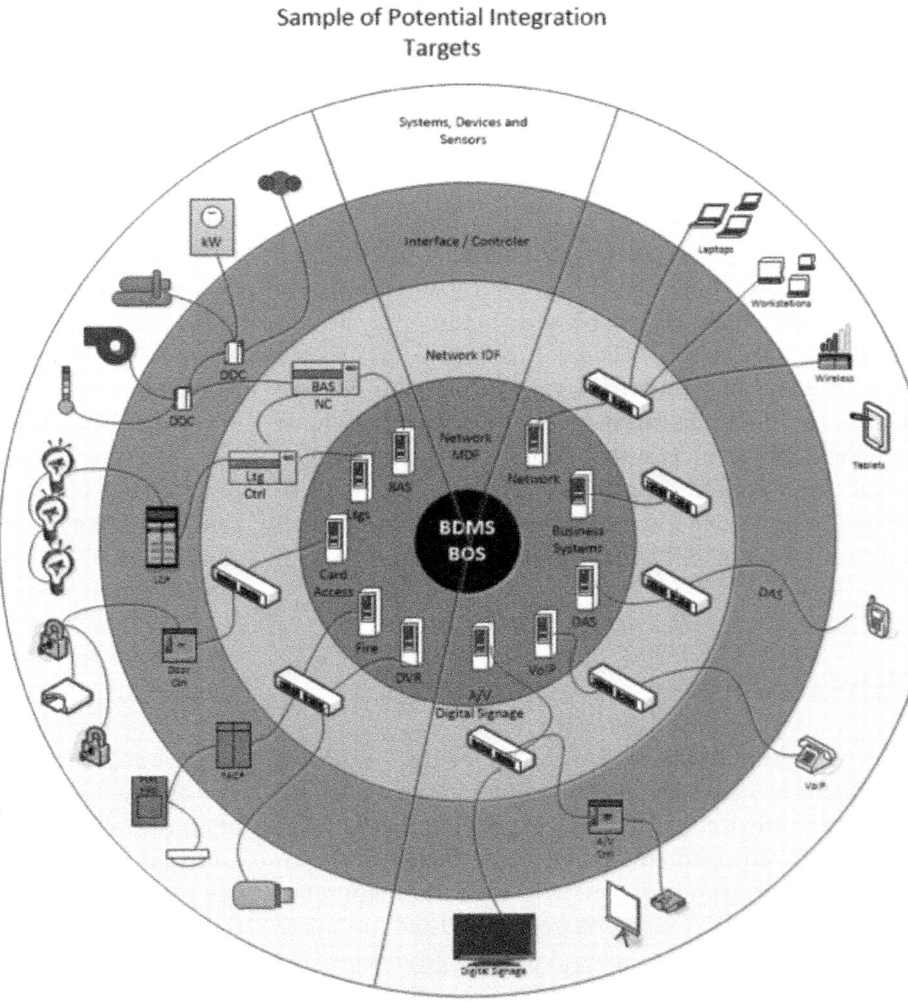

Figure 27-1. *Original concept drawing from ESD, Chicago*

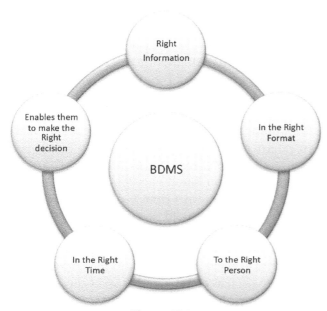

Figure 27-2

sharing across these systems will enable even more capabilities. Over time a convergence of these systems will take place and vendors will offer all of these applications as modules of a single platform. The question is which vendors, traditional BAS vendors or will it be new vendors from the Enterprise IT space?

FAULT DETECTION

With more holistic data sets, predictive fault detection systems can implement more sophisticated algorithms to provide more accurate results. Fault algorithms tied to the work product of a facility will provide businesses with a true measure of the cost of these faults in near real time. The ability to proactively manage risk will be the key value proposition for these systems. These systems will also have to incorporate fault data generated independently by equipment and devices.

As hardware costs drop more and more manufacturers will build in compete fault analysis into their equipment controllers.

CONSTANT COMMISSIONING

By enabling these systems to connect in real time to building systems from multiple vendors and multiple facilities, they will be able to provide better analysis and control. Once these systems are exposed to occupancy data they can start to proactively adjust the control setpoints to fine tune systems. Having access to holistic data will enable artificial intelligence to become a reality. Control points based on number of occupants, type of work being performed, evaporation rate of moisture off the skin and others will allow for very fine tuned control of facilities.

ENERGY ANALYSIS

As mentioned before, the ability for these systems to apply context to their results will be very important. Better and more accurate KPIs will provide the detailed information required to run efficient facilities. Even more important, these systems will predict energy consumption so facility managers can proactively manage costs. Being able to know several days in advance when new demand peak will occur allows time to change operations to avoid it.

MAINTENANCE MANAGEMENT

It's common today for Computerized Maintenance Management Systems to be integrated to the BAS in order to use actual runtime of the equipment as the preventive maintenance triggers. However, as mentioned above, there's now a multitude of systems that have the potential of providing valuable data. CMMSs that are capable of integrating data from these disparate systems will lead the way. New and valuable use cases will be enabled through this integration. Use cases such as using lighting system data to understand use intensity patterns of areas within facilities will provide new metrics for scheduling maintenance. When integrated with elevators and escalators, data could be complied to show when the equipment is least used. For example every third Wednesday there is a lull in use and work could be done with little or no impact on occupants.

The (IoT) Internet of Things will impact CMMSs by providing not only more data but also more complete data sets. In the future it will be common for the equipment to directly notify supervisory systems over wireless networks. You can image equipment nodes that not only provide runtime data but also manufacturer's data like wiring schematics, parts lists and O&M information. Maintenance mechanics will know what parts they need and the system will check the inventory on their trucks to make sure they have everything they need. Using GIS systems for information on the location of maintenance personnel, will allow the system to automatically route work tickets to the nearest resource.

There will be a blurring of the lines between these previously specialized systems once data is easily shared. FDD systems and CMMSs are a prime target for convergence. Their missions are closely related and their target audiences are the same.

NEW USE CASES AND KPIS

Today the key performance indicators between management groups aren't always the same. Converging building and IT data allows all groups to see data in a format that's appropriate for them but still relative to the core mission. Sending building energy use intensity data to a CFO has little meaning, but sending that same data with its impact in dollars does. An added benefit of getting this information to key executives is that it can create a fundamental shift in responsibilities of the other groups below them. As an example, most building engineers are keenly focused on occupant comfort because they get direct feedback from the occupants -- once the building energy costs are normalized and put in context for executives they will provide direct feedback as well. This causes a fundamental shift in the way building engineers think about their responsibilities and how they handle comfort issues in the future.

Buildings will be required to respond automatically to external signals from, community alerts, demand response and transactive energy markets. These alerts will be subscription based encrypted messaging transmitted across the Internet from multiple sources. Buildings that have the ability to respond to these signals can provide a safer and more efficient work environment. One roadblock to wide spread adoption of this is that the majority of the building stock in the United States is 100,000 square feet or smaller. Many of these buildings do not have sophisticated control systems for HVAC, lighting and others. This represents a huge market opportunity for technology vendors in that space. In the past few years there have been some companies starting to focus on this underserved market. New, lower cost building control systems utilizing wireless communication for sensors lowers the installed cost, thus making it cost effective for these smaller facilities to implement. With computing power dropping in cost, it now feasible to have POE/IP based multi-sensor devices deployed throughout facilities. These low cost sensors will enable applications that were not even conceivable just a few years ago. These lower cost systems and sensors will also make centralized management of geographically dispersed facilities possible at a reasonable ROI.

CONCLUSION

In summary, the building of the future will be dynamically responsive to its work product through the use of data acquired from many different internal and external systems. Buildings will be required to respond in near real time to signals generated from these systems. Data platforms that can manage and organize the vast amount of converged data related to buildings will provide companies a way to become technology agnostic. Building Operating systems will provide new applications to manage risk, publish new operational policies, provide predictive analysis for building performance and new productivity tools to all stakeholders. Information about these buildings will be online and readily available via well-established web technologies. Security of these converged data sets will be handled just as online banking. As has been the case thus far in the building technology market, IT technology will lead the way to the enablement of the online intelligent buildings of the future.

Chapter 28

Application of Automated Price and Demand Response for Large Customers in New York*

Joyce J. Kim, Rongxin Yin, and Sila Kiliccote, Member, IEEE

ABSTRACT

Open Automated Demand Response (OpenADR), an XML-based information exchange model, is used to facilitate continuous price-responsive operation and demand response participation for large commercial buildings in New York who are subject to the default day-ahead hourly pricing. We summarize the existing demand response programs in New York and discuss OpenADR communication, prioritization of demand response signals, and control methods. Building energy simulation models are developed and field tests are conducted to evaluate continuous energy management and demand response capabilities of two commercial buildings in New York City. Preliminary results reveal that providing machine-readable prices to commercial buildings can facilitate both demand response participation and continuous energy cost savings. Hence, efforts should be made to develop more sophisticated algorithms for building control systems to minimize customer's utility bill based on price and reliability information from the electricity grid.

Index Terms—Price response, demand response, dynamic pricing, real-time pricing, automated control, energy management, load management, load shedding, load forecasting, dynamic response.

*The work described in this report was conducted by the Lawrence Berkeley National Laboratory and funded by the New York State Energy Research and Development Authority under the Agreement No. 20723. This work was supported in part by the California Energy Commission (CEC) under Contract No. 500-03-026 and by the U.S. Department of Energy (DOE) under Contract No. DE-AC02-05CH11231.

Joyce Jihyun Kim is with the Department of Architecture, Building Science Program, University of California, Berkeley, CA 94720 USA and also with the Environmental Energy Technologies Division, Lawrence Berkeley National Laboratory, CA 94720 USA (e-mail: joycekim@lbl.gov).

Rongxin Yin is with the Department of Architecture, Building Science Program, University of California, Berkeley, CA 94720 USA and also with the Environmental Energy Technologies Division, Lawrence Berkeley National Laboratory, CA 94720 USA (e-mail: ryin@lbl.gov).

Sila Kiliccote is with the Environmental Energy Technologies Division, Lawrence Berkeley National Laboratory, CA 94720 USA (e-mail: skiliccote@lbl.gov).

INTRODUCTION

To ensure reliable and affordable electricity, the flexibility of demand-side resources to respond to the grid reliability requests and wholesale market conditions is required [1-2]. Large customers are often the immediate target for demand response (DR) because they are major contributors to peak demand for electricity and they are equipped with centralized energy management control systems (EMCS) to adjust electric loads. However, much of DR is still manual, which makes providing frequent DR a daunting task. This undermines the full potential of demand-side management among large customers. The customer's ability to perform DR can significantly improve by enabling automated demand response (Auto-DR) [3]. By eliminating human in the loop, Auto-DR eases the operational burden to provide price response in real-time and reduces the cost associated with monitoring and responding.

It has been argued that Auto-DR and enabling technologies would play a critical role in creating price-responsive load [4]. The application of Auto-DR to dynamic pricing has attracted attention since several states and utilities deployed full-scale dynamic pricing programs. To facilitate the communication of price and reliability signals, Open Automated Demand Response (OpenADR), an XML (eXtensible Markup Language) based information exchange model, was developed [5]. Strategies were presented to operationalize dynamic pricing signals into load control modes using OpenADR communication protocols [6]. This chapter reports on the latest efforts to automate customer response to reliability and pricing signals for large commercial buildings in New York. It is significant in two ways. First, the chapter raises the awareness to key cost challenges in commercial buildings who are subject to the default day-ahead hourly pricing and provides a practical solution that the facility can adopt for continuous energy management. Second, it paves the way to develop and test more sophisticated control algorithms in commercial buildings.

A note on terminology: dynamic pricing is referred to energy prices that are available to customers in regular intervals no more than a day in advance. In New York, dynamic pricing is set day-ahead, hour-ahead or in real-time by the New York Independent System Operator (NYISO) wholesale markets. In this chapter, we focus on the day-ahead hourly pricing, which is the default tariff for large customers in New York.

In the rest of this chapter we summarize the existing demand response programs in New York, and we discuss OpenADR communication, prioritization of demand response signals, and control methods for large commercial buildings in New York. Then the application of Auto-DR under the day-ahead hourly pricing is explored through energy simulation and field tests of two commercial buildings in New York City (NYC) that participated in a demonstration project. Then preliminary findings from the demonstration buildings are discussed. We conclude with suggestions for future research directions.

DEMAND RESPONSE IN NEW YORK

In New York, DR is mainly promoted through reliability-based and price-based programs. There are a number of reliability-based programs offered to customers by NYISO and utilities, commonly referred to as DR programs. Since the initial offering in 2001, NYISO's DR program registration has grown steadily. In 2001, there were approximately 300 participants enrolled in reliability-based programs such as Special Case Resource/ Emergency Demand Response Program (SCR/EDRP) with the total participating load of 750 MW. By 2011, NYISO had a total of 5,807 participants for the SCR/ EDRP program providing 2,173 MW of curtailable load [7]. In New York, customers are enrolled in DR programs through Curtailment Service Providers (CSPs). CSPs manage a portfolio of demand resources and aggregate demand reduction to maximize DR compensation. They help customers assess the DR potential and develop load curtailment strategies. Contracting a CSP typically means that customers meet the minimum shed requirements during the DR test/event and receive DR compensation in return.

Price-based programs are offered to flatten system demand by applying high prices during peak periods and low prices during off-peak periods. Pacific Gas and Electric (PG&E) Critical Peak Pricing and Southern California Edison's (SCE) Real-Time Pricing are examples of price-based programs. In 2005, the State of New York Public Service Commission ordered utilities to provide day-ahead hourly pricing as the default tariff to non-residential customers [8]. This tariff is also known as Mandatory Hourly Pricing (MHP). Since then, utilities in New York applied MHP to commercial and industrial customers as the default electric service. However, instead of staying on MHP, many customers ended up choosing alternative rates through retail access. As of 2011, only 15% of the MHP-eligible customers were enrolled in MHP and the rest (85%) were retail access customers [9]. The problem is that existing retail rates (i.e., fixed rates) do a poor job of reflecting wholesale market prices [4] and they tend to be expensive due to the inherent risk of offering a less variable rate. Therefore, switching from MHP to a retail rate can hamper the development of price responsive demand.

The primary barriers to the adoption of MHP are identified as the insufficient resources to monitor hourly prices and inflexible labor schedule [9]. This is not surprising since most customers rely on manual approach to provide DR. Providing DR manually is a resource-intensive process. If customers are not capable of monitoring hourly price variations and managing their loads in real-time, they are likely to choose a more conventional rate such as a fixed rate. Moreover, customers have not yet found a compelling business case to stay with MHP. Many customers think that the DR monitoring cost outweighs the savings. Even if the savings exist under MHP, they are not as obvious as the DR compensation because the savings are embedded in the total electric bill. Therefore, in order to increase the adoption of MHP, we not only need to automate DR but also need to clearly communicate resulted savings to the customers.

In NYC, MHP is billed under Rider M: Day-Ahead Hourly Pricing from Con Edison where the cost of energy is calculated based on the customer's actual hourly energy usage multiplied by NYISO's day-ahead zonal locational based marginal price (LBMP) [10]. In addition, customers pay demand charge imposed on the maximum demand of each billing cycle. The demand charge varies depending on the Time-of-Day (TOD) and season [11]. Based on our billing analysis, the demand charge accounts for 19%-55% of the customer's electric bill. To reduce the total electric bill, customers need to control their electric consumption according to the hourly price variations and limit the building's electric demand during expensive hours.

APPROACH

Since October 2011, Lawrence Berkeley National Laboratory (LBNL) researchers have worked with New York State Energy Research and Development Authority (NYSERDA) to demonstrate the capability of OpenADR to facilitate response to reliability and price signals in New York electricity markets. The recruitment efforts were focused on large commercial buildings in NYC. Preferences were given to the buildings that represented the typical construction of commercial buildings in NYC and previously participated in DR programs. Four facilities were recruited for the demonstration project. The participation was driven by the motivation to automate DR strategies and improve the DR capabilities. All facilities are on a retail rate and are not enrolled in MHP. In this chapter, we set out to investigate the hypothetical scenario wherein the customer purchases electricity under the MHP tariff and therefore has to respond to the variability of wholesale market prices.

OpenADR Communication Model

To automate price and demand response using OpenADR, three basic technologies are required: a Demand Response Automation Server (DRAS) to receive reliability and price signals; a communications device at each facility to receive DR signals; and an EMCS to control customer loads [12]. We used OpenADR version 1.0 for the demonstration project. OpenADR version 2.0, available currently, was not released at the time of the project implementation. Figure 28-1 shows the OpenADR communication architecture for the New York demonstration project. Day-ahead hourly prices are obtained from NYISO's website and DR test/event notifications are received from the customer's CSP. Based on the reliability and pricing signals, operation mode is determined for each hour of the following day. Once the signals are processed, DRAS sends twenty-four hourly

prices and corresponding operation modes to the facility to activate preprogrammed DR strategies for next day. Throughout the day, DRAS monitors the building's electric demand and logs the load data at 15-minute intervals via kyz pulses. All information exchange is accomplished through a secure Internet connection with 128-bit Secure Sockets Layer (SSL) encryption. The participants have an option to opt-out if they do not wish to participate in price and/or demand response. The opt-out can be scheduled in advance for a specified period which can be a few hours or days depending on the facility's operational needs.

Prioritization of DR signals

Three types of DR signals are issued: 1) reliability, 2) demand limiting, and 3) day-ahead hourly price signals. These signals are prioritized differently depending on the next day's DR test/event status as described in Figure 28-2. For non-DR test/event days, customers respond to price signals until the building's electric demand exceeds a pre-set threshold, in which case, DRAS would switch the signal type from price to demand limiting. When a DR test/event is issued, customers only respond to reliability signals during the DR test/event period. If demand exceeds the pre-set threshold before or after the DR test/event, demand limiting signals would be sent to the facility to reduce the load. We decided to turn off price signals during DR test/event days to prevent curtailment activities affecting the customer baseline. This is particularly necessary if the customer uses the NYISO's Weather-Sensitive Customer Baseline which uses 2-4 hours of load data prior to the DR test/event to adjust the Average Customer Baseline (CBL) [13].

The reliability, demand limiting, and price signals are mapped into simple operation modes that can activate preprogrammed DR strategies via EMCS. OpenADR version 1.0 supports four levels of operation

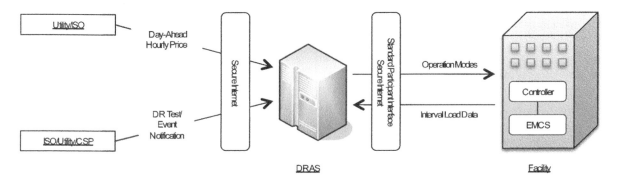

Figure 28-1. OpenADR communication architecture for the New York demonstration project.

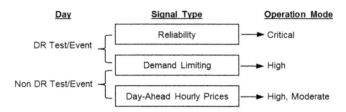

Figure 28-2. OpenADR signal prioritization.

modes: *Normal, Moderate, High,* and *Special* (which we call *Critical* hereafter).

- *Normal* indicates that the energy price is acceptable and there is no DR test/event issued. The facility would operate at normal system settings.

- *Moderate* indicates that energy price is moderately high and electric loads need to be curtailed at the moderate reduction level.

- *High* indicates that energy price is high and electric loads need to be curtailed at the high reduction level. High is also issued when electric demand exceeds the demand limit.

- *Critical* indicates that a DR test/event is issued and electric loads need to be curtailed at the maximum reduction level.

Control Logic

Using OpenADR communication protocols, customers can control electricity cost by responding to both price and demand limiting signals. The control logic can reside 1) within the facility or 2) in the cloud (i.e., DRAS). While the first option has the advantage of unrestricted building data retrieval and direct control over building system, it requires additional control algorithm and data processing capabilities. Locating the control logic within DRAS has the advantage of leveraging the existing server system to program control algorithm and process data. However, the building data retrieval and control over building system are restricted due to security reasons. As such, DRAS can only send simple operation mode with limited ability to control the building operation.

Using the first option, the customer's energy cost for a given day can be minimized through load optimization in response to NYISO's day-ahead zonal LBMP (C_t), as expressed in (28-1).

$$\min \sum_{t=1}^{k} C_t \cdot g\left(u_t, x_t, w_t\right) \qquad (28\text{-}1)$$

Optimal electricity usage (kWh) is determined by

the objective function (g) based on following variables: is the input constraints for load control strategies; x is the building system states (i.e., HVAC setpoints, operation schedules); and w is the weather (i.e., outside air temperature, relative humidity). t represents the time interval and k indicates the total number of time intervals in a day. The demand charge can be minimized by reducing peak demand during a billing period, as expressed in (28-2).

$$\min \left(\max_{t \in 1, \ldots N} h\left(u_i, x_i, w_i\right) \right) \qquad (28\text{-}2)$$

represents the electric load (kW) at a time interval (i) and N indicates the total number of time intervals in a billing period. For the current phase of the New York demonstration project, we established the OpenADR connectivity between DRAS and facilities but the programming and data processing capabilities will not be installed in the facilities until the next phase. Hence, this chapter focuses on the implementation of the control logic within DRAS.

Open-Loop and Closed-Loop Control

There are two types of controls used in DRAS: open-loop control and closed-loop control [14]. In open-loop control, DRAS sends DR signals to the facility but does not use feedback to determine if the building has achieved the performance targets. The performance targets can be defined by load reduction amount, a demand limit, or a thermal comfort limit. Closed-loop control, on the other hand, uses feedback to control DR outcomes. Closed-loop control is more advantageous if the DR performance has to be guaranteed. However, it requires more granularity of control over the building systems and real-time data processing capabilities. For the New York demonstration project, open-loop control is used to respond to price and reliability signals and closed-loop control is used to provide demand limiting. The feedback is provided via interval meter readings, which is used to generate demand limiting signals and to calculate load reduction in near real-time. To predict the DR performance under different operation mode, we conducted whole building energy simulations using EnergyPlus. EnergyPlus is an energy analysis and thermal load simulation software which allows calculating heating and cooling loads based on building geometry, building envelope, internal loads, HVAC systems, and weather [15]. Based on the energy simulation results, we selected control strategies and inputs for each operation mode that would produce the target load reduction and thermal comfort level.

APPLICATION

Developing DR strategies is a multi-step process. First, we need to understand the building's current and historic electric use patterns and evaluate building systems, control capabilities, and operational constraints [16]. Then, we identify DR opportunities and develop DR strategies for each facility. Finally, proposed DR strategies need to be tested and modified to improve the DR outcome. In this section, we explain the process of developing DR strategies for two of the participating buildings from our demonstration project.

Site Description

The first building, located in NYC, is a 32-storey office building with a glass curtain-wall extending the full height of the building (here in called "office building"). The office building has a total conditioned floor area of 130,000 m^2 (1.4 Million ft^2). The building's HVAC consists of multiple-zone reheat systems with constant air volume and air-handling units (AHUs) controlled by variable frequency drive (VFD). There are three 1,350-ton centrifugal chillers with constant speed and one 900-ton centrifugal chiller with variable speed that supplies chilled water to AHUs. Each zone temperature is controlled via direct digital control (DDC). Currently, the office building does not have the Global Temperature Adjustment (GTA) capabilities to change zone temperature setpoints for the entire facility [17]. The facility is heated via Con Edison steam. The building is equipped with Honeywell's EMCS for HVAC control. Multi-zone control is available for lighting through relays but it is not connected to the EMCS. The facility is in operation from 6am to 6pm during weekdays and closes during weekends.

The second building is a 14-storey university building also located in NYC (herein called "campus building"). The campus building recently went through a complete renovation and system upgrades and was recently occupied in September 2011. The newly renovated building has the total floor space of 11,330 m^2 (122,000 ft^2) containing classrooms, computer labs, offices, and conference rooms. There are eleven AHUs, each equipped with VFDs. The building is equipped with a 400-ton chiller supplying chilled water to AHUs. Heating is provided with steam, which is used for AHU reheat, unit heaters, and stairwell heating. The campus building has an Automated Logic Control (ALC) system used for HVAC control. The indoor space is largely lit by T5 fluorescent fixtures located within hallways, offices, and the lobby. Office lighting is on motion sensors. The campus building is equipped with the NexLight two-way digital lighting control system but this system was not used for DR in the past. There are three elevators in the campus building: two passenger elevator and one passenger/freight elevator. Previously, one of the three elevators was shut off during DR events. The facility is open from 7am to 11pm for seven days a week.

Load Characteristics

Approximately two years of 15-minute whole building electric load data was made available to the project team for the office building and the campus building. Table 28-1 summarizes the data over one year period (Sep 2011-Aug 2012). To characterize the behavior of building energy use, we plotted the load profile against different time scales. First, weekly electric demand and consumption was plotted from January 2011 to August 2012 in Figure 28-3. Examining these plots revealed following findings: 1) both the office and campus buildings had relatively constant minimum demand throughout the year; 2) the maximum demand was higher in summer than in winter for both buildings; and 3) maximum demand (kW) varied more significantly from season to season than electric consumption (kWh).

Table 28-1. Load Summary*

Facility	Peak Load (kW)	Peak Load Intensity (W/m^2)	Load Factor	Annual Consumption (kWh)
Office Bldg	6,192	47.6	0.51	27,611,976
Campus Bldg	605	53.4	0.40	2,149,722

*Computed for Sep 2011 - Aug 2012, with 15-minute interval data.

Next, the buildings' interval load was plotted over a one-week period for summer months (May to Aug 2012) in Figure 28-4 and for winter months (Nov 2011 to Feb 2012) in Figure 28-5. The scatter plots reveal following things. 1) The office building was in use during weekdays while the campus building was in use for seven days a week, confirming the operation schedule of the two buildings provided to the project team. 2) In both facilities, the spikes shown at the beginning of each weekday during summer months indicated precooling activities and the system overload. For the office building, precooling typically started at midnight and for the campus building, it started at 7am. The campus building had a start-up electric surge during the first hour of the building operation which marked the highest demand of the day. In summer, starting precooling at 7am would add more loads to the morning ramp-up and increase the demand even higher. 3) Both buildings showed a

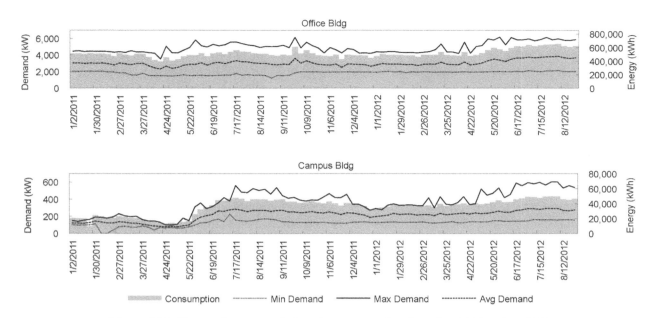

Figure 28-3. Demand usage and electric consumption from Jan 2011 to Aug 2012.

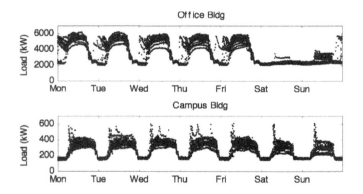

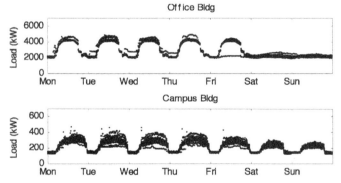

Figure 28-4. Scatter plot: time-of-week from May to Aug 2012 excluding holidays (Memorial Day and Independence Day).

Figure 28-5. Scatter plot: time-of-week from Nov 2011 to Feb 2012 excluding holidays (Veterans Day, Thanksgiving Day, Christmas Day, New Year's Day, Birthday of Martin Luther King, Jr., and Washington's Birthday).

wide range of daily loads during summer months versus winter months while the base load stayed relatively constant throughout the year. This was more prevalent in the office building than the campus building. Since both buildings were heated with steam, the difference in summer and winter loads was likely to be influenced by the amount of cooling loads.

To understand the dependence of the building load on outside weather, we plotted the electric data for occupied hours during weekdays against outdoor air temperature and relative humidity as shown in Figure 28-6. From the National Climatic Data Center, we acquired hourly outdoor air temperature data for each facility from the nearest weather station [18]. Some of the missing data were filled in by linear interpolation. As seen in Figure 28-6, both of the office building and the campus

building electric load were highly sensitive to the outside air temperature. However, the upper range of the campus building load was more influenced by the classroom schedule than outside weather. Both buildings did not show a significant relationship between building load and relative humidity.

Demand Limiting and Price Thresholds

For DRAS to determine operation mode for each hour of the day, customers need to establish the demand and price thresholds to which the selection of a particular operation mode can be based upon. These thresholds can be updated as frequently as required (i.e., weekly, quarterly, or yearly). To help customers choose the right

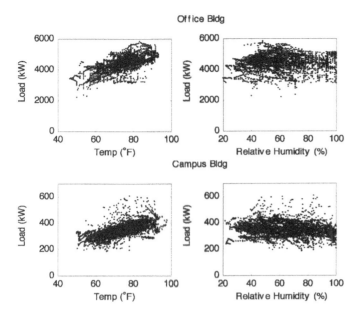

Figure 28-6. Scatter plot of load versus temperature and humidity. Data shown are from May to Aug 2012.

demand and price thresholds for their facility, we first studied the buildings' load duration curves to evaluate demand reduction opportunities.

Figure 28-7 shows the one-year load data (from September 2011 to August 2012) plotted in descending order over the proportion of time. For the office building, the weekday load duration curve descended at a gradual slope and there was no unusual peaks observed in the plot. The weekend/holiday curve was much lower than the weekday's since the office building was not in service during weekend/holidays. However, the weekend/holiday load during the top one percent was "peakier" than the rest. This was probably caused

by night flushing and precooling of thermal mass performed during Sunday evenings in preparation for the next business day or occasional use of the facility over the weekends. For the campus building, the difference between the weekday and the weekend/holiday load duration curves was small since the building was in operation for seven days a week. Both curves showed a significant increase in load during the top one percent of the time. This behavior was probably caused by the system overload experienced during the first hour of the building operation. This issue can be resolved by shifting some loads to earlier times in the morning or later during the day and limiting demand below the level corresponding to the top one-percent of the time.

Similarly, price thresholds can be established by analyzing hourly price distribution over time. Figure 28-8 displays price duration curves over the time period of September 2011-August 2012. We used NYISO's day-ahead LBMP for Zone J: NYC since both the office building and the campus building were located in NYC [19]. Day-ahead LBMP did not vary significantly between weekdays and weekend/holiday and most of the time the price stayed below $100 per MWh. Only significant deviation was seen during the top one percent of the time where the price increased up to $363 per MWh. The loads corresponding to the top one percent of the time are concentrated in summer and winter months. When plotted against the time of day (Figure 28-9), it was clear that the expensive hours were either cooling hours (mid-day) or heating hours (morning and evening). Therefore, limiting the electricity consumption during the top one percent of the time via Auto-DR can help customers minimize their energy cost.

DR Strategies

Both the office building and the campus building currently participate in NYISO's SCR/EDRP through separate CSPs. For the NYISO initiated DR test/event, the office and campus buildings have a minimum shed

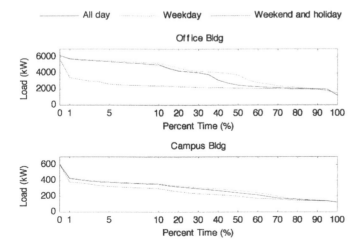

Figure 28-7. Load duration curves. Data shown are from Sep 2011 to Aug 2012.

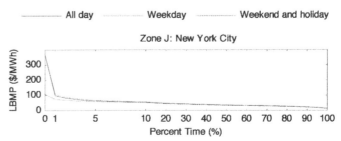

Figure 28-8. Price duration curves. Data shown are from Sep 2011 to Aug 2012.

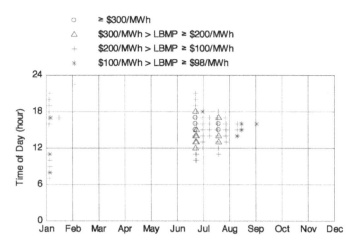

Figure 28-9. LBMP distribution against month and time-of-day during the top one percent of the time from Sep 2011 to Aug 2012.

Table 28-2. DR Strategies and Operation Modes

Facility	Operation Mode	Global temperature adjustment	Passive thermal mass storage	Supply fan speed reduction	Exhaust fan quantity reduction	Chilled water temperature increase	Chilled water pump speed reduction	Shutting off chilled water pumps	Chiller quantity reduction	Condenser water temperature increase	Shutting off condenser water pumps	Turning off lighting in auxiliary space	Slow recovery	Sequential equipment recovery	Extended DR control Period
Office Bldg	Critical	x	x	x	x	x	x	x	x	x	x		x	x	x
	High	x	x	x	x	x	x			x	x		x	x	x
	Moderate	x		x				x		x	x		x	x	x
Campus Bldg	Critical	x	x	x	x						x	x	x	x	x
	High	x	x	x	x						x	x	x	x	x
	Moderate	x		x							x	x	x	x	x

requirement of 2,000 kW. The shed requirement of the campus building has not yet been established. To help the facilities meet their DR targets, CSPs developed DR strategies for their clients that were used for previous DR test/events. Based on the customers' existing DR strategies, we selected the ones that can be automated through EMCS and grouped them into Critical, High, and Moderate operation mode, as shown in Table 28-2. The project team added GTA capabilities to the office building and automated lighting control to the campus building to enhance DR control. The lighting control is restricted to auxiliary space such as hallways and lobby. As for elevators, we recommended that the facilities maintain manual control over their elevators for both DR and non-DR days. To minimize the post-DR rebound effects, Normal operation mode returns slowly with sequential equipment recovery. If there is less than one hour left until the end of occupancy period, DR is extended to the end of the occupancy period and then the building returns to Normal operation mode.

EVALUATING DR PERFORMANCE

In this section, we show how Auto-DR can be performed on a non-DR event day and on a DR event day through field-test results and energy simulation. First, we examined the load data taken from the actual DR event day on June 20, 2012 that the office building participated, as illustrated in Figure 28-10. The DR event was called between 2pm and 6pm, during which the minimum 2,000 kW reduction was expected in reference to NYISO's Average Coincident Load (ACL) baseline

[20].* The office building achieved the reduction target only during the last two hours of the event period by activating all DR strategies listed under Critical operation mode. It experienced a post-DR rebound for a few hours after the DR event. Next, we compared the load reduction with two different baselines to evaluate customer's DR performance: 1) NYISO's Average Customer Baseline (CBL) and 2) the weather regression baseline developed by LBNL [21].† NYISO's CBL has a tendency to underestimate or overestimate the building's power usage for the days with unusual weather conditions. In general, the weather regression baseline provides a more accurate prediction of weather-sensitive loads than NYISO's CBL. As seen in Figure 28-10, NYISO's CBL underestimated the baseline load because the DR event day was warmer than previous days. As such, DR payments would have been smaller if the compensation was calculated based on NYISO's CBL instead of the weather regression baseline.

Figure 28-11 illustrates the office building's response to price signals on a non-DR event day. The load data were taken from August 9, 2012, representing a typical weekday. The building underwent three hours of Moderate operation mode from 2pm to 5pm based on the price thresholds set at LBMP ≥ $98 for Moderate operation mode and LBMP ≥ $200 for High operation mode. We used EnergyPlus simulation to predict the ef-

---

*NYISO's ACL baseline averages customer's 20 highest loads of 40 highest system load hours excluding hours in which DR events were previously activated.
†NYISO's CBL averages customer's five highest of the previous ten weekdays excluding holidays and previous DR event days.

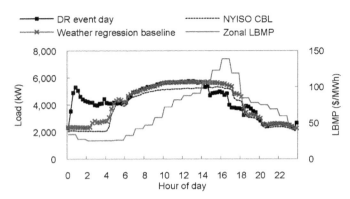

Figure 28-10. Load and price data of the sample DR event day.

fects of DR strategies for Moderate operation mode and compared the simulated load to the actual load which was unaffected by Auto-DR. According to the simulation results, the office building can reduce demand up to 700 kWh by implementing DR strategies listed under Moderate operation mode for this day.

It is noted that continuous energy management in response to hourly prices can impact the customer's DR baseline, potentially reducing DR payments due to lowered baseline usage. This can make DR programs less attractive to energy efficient customers under the day-ahead hourly pricing. However, DR program events are called only a few days a year and the incentives collected from DR programs are likely to be small compared to the utility cost savings achieved under the day-ahead hourly pricing due to continuous energy management. Hence, as the commercial buildings move towards more dynamic response to prices, the applicability of baseline-based DR payments should be evaluated.

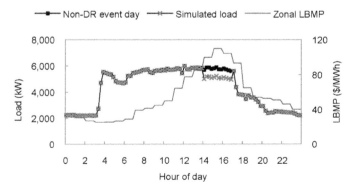

Figure 28-11. Load and price data of the sample non-DR event day.

CONCLUSIONS AND FUTURE STUDIES

We presented the process of automating continuous energy management with day-ahead hourly prices and demand response for large commercial buildings in New York who were subject to the default MHP tariff. OpenADR version 1.0 was used to facilitate the communication of price and reliability signals. Based on the preliminary findings from the New York demonstration project, we concluded that: 1) understanding customer's financial goals, such as reduction in utility bills including demand charges, and curtailment requirements by CSPs was critical in establishing Auto-DR goals and performance targets; 2) price and demand response opportunities were unique to customer's electric load characteristics, control capabilities, and operational constraints and; 3) energy cost savings would vary depending on how the buildings were operated in response to prices.

Future studies include: 1) creating dynamic optimization capabilities in buildings given the availability of price and demand response signals; 2) monitoring and evaluating the effects of DR strategies on load and occupant comfort during operations; 3) increasing the customer's ability to modify and disable individual DR strategies within the facility; and 4) evaluating benefits and drawbacks of having control logic in the cloud versus inside the facility. Finally, we recommend a comparative study on customer economics between MHP and retail rates to be conducted and the role of Auto-DR in cost savings to be further explored.

Acknowledgment
The authors give special thanks to Duncan Callaway for great advice and feedback. We also thank Con Edison Company for the electric load data.

References
[1] S. Borenstein, M. Jaske, and A. Rosenfeld, "Dynamic Pricing, Advanced Metering and Demand Response in Electricity Markets," University of California Energy Institute, Berkeley, CA, Rep. CSEM WP 105, 2002.
[2] E. Hirst and B. Kirby, "Retail Load Participation in Competitive Wholesale Electricity Markets," Edison Electric Institute, Washington, DC, Rep. Jan. 2001.
[3] M. Piette, O. Sezgen, D. Watson, N. Motegi, and C. Shockman, "Development and Evaluation of Fully Automated Demand Response in Large Facilities," Lawrence Berkeley National Lab., Berkeley, CA, Rep. CEC-500-2005-013. Jan. 2005.
[4] C. Goldman, M. Kintner-Meyer, and G. Heffner. "Do enabling technologies affect customer performance in price-responsive load programs?" Lawrence Berkeley National Lab., Berkeley, CA, Rep. LBNL-50328, Aug. 2002.
[5] M. Piette, G. Ghatikar, S. Kiliccote, E. Koch, D. Hennage, P. Palensky, and C. McParland, "Open Automated Demand Response Communications Specification (Version 1.0)," Cali-

fornia Energy Commission, Rep. CEC-500-2009-063 and LB-NL-1779E, 2009.

[6] G. Ghatikar, J. Mathieu, M. Piette, and S. Kiliccote, "Open Automated Demand Response Technologies for Dynamic Pricing and Smart Grid," in Grid-Interop Conference, Chicago, IL, 2010.

[7] D. Patton, P. LeeVanSchaick, and J. Chen, "2011 State of the Market Report for the New York ISO Markets," Potomac Economics, Rep. Apr. 2012.

[8] NYPSC, "The State of New York Public Service Commission 03-E-0641: Mandatory Hourly Pricing - September 23, 2005 order," [Online]. Available:http://documents.dps.ny.gov/public/Common/ViewDoc.aspx?DocRefId={0DDFAD32-C84A-4DDD-BF18-924E6C7DE954}

[9] KEMA, "Mandatory Hourly Pricing Program Evaluation Report," prepared for Consolidated Edison Company of New York. Rep. 2012.

[10] Con Edison, Electricity Service Rules (Riders) [Online]. Available: http://www.coned.com/documents/elecPSC10/GR24.pdf

[11] Con Edison, Electricity Service Classifications [Online]. Available: http://www.coned.com/documents/elecPSC10/SCs.pdf

[12] G. Wikler, A. Chiu, M. Piette, S. Kiliccote, D. Hennage, and C. Thomas, "Enhancing Price Response Programs through Auto-DR: California's 2007 Implementation Experience," Proc. AESP 18th National Energy Services Conference & Exposition, Clearwater Beach, FL, 2008.

[13] NYISO, "NYISO manual 7: Emergency demand response program manual," [Online]. Available: http://www.nyiso.com/public/webdocs/products/demand_response/emergency_demand_response/edrp_mnl.pdf

[14] S. Kiliccote, M. Piette, D. Watson, and G. Hughes, "Dynamic controls for energy efficiency and demand response: Framework concepts and a new construction study case in New York," in Proc. ACEEE Summer Study on Energy Efficiency in Buildings, Pacific Grove, CA, Aug. 2006.

[15] DOE, EnergyPlus Energy Simulation Software [Online]. Available: http://apps1.eere.energy.gov/buildings/energyplus/

[16] J. Mathieu, P. Price, S. Kiliccote, and M. Piette, "Quantifying changes in building electricity use with application to demand response," IEEE Trans. Smart Grid, vol. 2, no. 3, pp. 507–518, Sep. 2011.

[17] N. Motegi, M. Piette, D. Watson, S. Kiliccote, and P. Xu, "Introduction to commercial building control strategies and techniques for demand response," Lawrence Berkeley National Lab., Berkeley, CA, Tech. Rep. LBNL-59975, 2007.

[18] NOAA, NNDC Climatic Data [Online]. Available: http://www.ncdc.noaa.gov/

[19] NYISO, Day-ahead market LBMP zonal [Online]. Available: http://www.nyiso.com/public/markets_operations/market_data/custom_report/index.jsp?report=dam_lbmp_zonal

[20] NYISO, "NYISO manual 4: Installed Capacity Manual," [Online]. Available:http://www.nyiso.com/public/webdocs/products/icap/icap_manual/icap_mnl.pdf

[21] K. Coughlin, M. Piette, C. Goldman, and S. Kiliccote, "Statistical analysis of baseline load models for non-residential buildings," Energy Buildings, vol. 41, no. 4, pp. 374-381, Apr. 2009.

Chapter 29

Understanding Microgrids as the Essential Architecture of Smart Energy

Toby Considine, TC9 Incorporated
William Cox, Cox Software Architects
Edward G. Cazalet, PhD CEO, TeMix Inc.

ABSTRACT

This chapter describes microgrids in the smart grid architecture, autonomous systems interacting through the Energy Services Interface as defined by the OASIS Energy Interoperation [1] specification.

We define for the purposes of system architecture what a microgrid is. The several types of existing microgrids are defined, based on the motivations of those that operate them, the technologies they contain, and the operating characteristics they produce. This chapter includes an analysis today's variety of microgrids and how they are leading us to future

We describe a model that represents component systems in a microgrid as systems able to negotiate optimal outcomes for energy allocation based only on the internal self-knowledge of each system, interacting through the means of software agents. These agents are each able to respond to changes of mission as conveyed by the timely application of abstract sets of priorities [policies].

We term the process whereby these outcomes are developed "micromarkets." We further describe hoe how microgrids themselves can be organized into larger microgrids which are themselves operated by micromarkets.

We discuss the benefits of the micromarket model of integration. Micromarkets provide a simple model that can support each type of existing microgrids, both in technology, and in motivation. Autonomy in microgrids simplifies central control architectures.

We also discuss how the architecture of agents and micromarkets can be used to encapsulate the complexity of legacy systems, providing a path for existing systems into the future even as they reduce the effort required for the incorporation of new technologies.

Keywords: Microgrids, Micromarkets, Integration, Autonomous, Economic

INTRODUCTION

Our smart energy goals demand rapid innovation yet capital assets have long lives. This makes for growing diversity. Smart grid infrastructure must interact not only with technologies extant when that infrastructure is first deployed, but must continue to interoperate with new technologies over its long life. Interactions must specify a result, not a mechanism, an architecture style referred to as Service Orientation. Early deployments must not become a barrier to next generation deployments.

A microgrid is a small grid that can operate as a part of a larger grid or that can operate independently of the larger grid. A stand-alone microgrid never connects to a larger grid. In this chapter, we consider autonomous microgrids, whether attached to a larger grid or not. Because autonomous microgrids operate themselves and hide their internal characteristics from external markets, microgrids are a natural fit with service orientation.

Microgrids can manage their own storage, conversion, and recycling of energy. They can choose to buy when energy is abundant and inexpensive. A microgrid able to do so is inherently adapted for DR events. So long as transactions clear in real time, virtual microgrids share almost all characteristics with actual microgrids.

The salient characteristic of distributed and renewable energy sources is volatility of supply. Current attention focuses on the supplier pain point, when excess supply is gone, when use is at its maximum, and to avoid calling expensive and often dirty sources into production. Long term interests urge us to focus at least as much attention on the surpluses, i.e., when the wind is blowing, the sun is shining, and there is more local power than can be consumed. Even the most successful wind farms do not reliably provide their product to end users. Site-based conventional generation is still subject to fuel availability and costs. Energy storage is only available until you use it.

Smart Grid Architecture addresses this diversity change by limiting direct interactions across each interface between domains. Management of generation, storage, and load is by service request; the resource providing the service may be a device, an aggregation of devices, or a virtual service. The energy services interface accepts requests for load response, for generation, for storage, and manages its internal operations.

BACKGROUND

The essential problems that smart grids are meant to solve are those of smaller operating margins accompanied by a more volatile supply. Operating margin is the excess electrical power over the amount used at any moment. Volatile supply is the result of using more sources such as wind and solar, whose output cannot be precisely predicted or controlled. Together, these changes lead to an over-supply or an undersupply at any given moment.

First generation efforts to compensate for low operating margins and intermittent supply were not satisfactory. Quick compensation for under-supply relied on fast-start technology that was inefficient, expensive, or both. Reserve near-line generation can cost as much, and require nearly as much fuel with its associated carbon costs as would putting the source on-line and re-introducing a higher operating margin. Many generation assets that only enter the market during shortage are not in the normal market because of greater expense or environmental costs.

Demand Response (DR) is the second generation effort to compensate for these issues. Although the term Demand Response technically includes increasing as well as reducing energy demand, in everyday use, it refers to direct curtailment of load through signals sent by the supplier or grid operator.

These signals were often less effective than hoped for. Many residential customers accepted the incentives but opted not to respond to the signals, either through disabling the controller or through ceasing participation

during the critical high demand* months. Commercial and industrial sites have been known to comply with the control signal while ameliorating the effects on their business with other actions that increased overall electrical use.

Newer specifications based on the OASIS Energy Interoperation standard, including OpenADR 2.0, are defining service interactions. The intent is to pay for actual reduction in power consumption, or to meet a particular load curve, rather than for promises to turn a particular device on or off.

This change centers the focus of smart energy firmly on the end node. The end node becomes an entity that negotiates with suppliers, and controls its energy use to make those contracts.

What Motivates the End Node

The End Node† balances two factors, energy surety and economics. It wants to have the power and power quality it needs or wants available when it needs or wants it. The end node wants to acquire access to this power in a cost effective or economic manner. As the end node has little control over the total supply and the demand made by others, it must look inward to what it can manage itself.

To effect its market operations, the end node has a few broad approaches. (1) It can temporally shift its energy use, to use energy at a more economic time. (2) It can temporally shift its purchase of energy, finding a way to acquire energy now while using it later. (3) It can reduce its price risk by making committed purchases of load over time. This buys reduced price risk at the cost of possibly sub-optimal purchases. (4) It can generate its own power internally. Internal power generation is made more valuable by applying the use shifting and buffering as described above.

Energy efficiency is one tool an end node can use to improve energy surety, but it does not address directly the problems of smart energy. A maximally efficient end node may not be able to shift use. Buffering power, either in batteries or by pre-consumption always has some cost in consumption. On the other hand, a 30% reduction in process power requirement may be as effective as a 50% increase in buffering capacity. With enough efficiency, an end node may be able to achieve surety within

---

*Typically summertime cooling, but also the "cold winter mornings" in the Pacific Northwest United States
†In Energy Interoperation [1], Virtual End Node (VEN) as the relationship is recursive in navigating the composition or decomposition of microgrids we describe later in this chapter.

the supply it generates internally.

Each End Node has a different definition of energy surety, and a different value that it achieves through surety. Each End Node may have a different competence in managing its processes and assets. As these cannot be known centrally, this creates what is known in economics as a knowledge problem [2] [3] as to the optimum allocation of power. As a market is the solution to a knowledge problem, market interactions are necessary and sufficient for interactions between a microgrid and its suppliers [4].

What Demotivates the End Node

The owners/operators/participants of the End Node(s) are concerned with the losses that participation in smart grids poses: loss of control, loss of privacy, and loss of autonomy.

The *loss of control* is straightforward. Before smart grids, you turn on what you want, when you want. Under a model based on central control, you can turn things on only if you are permitted to. How costly this is depends upon what are the effects of loss of control. It may be unnoticed. It may cause minor discomfort. It may reduce sales. It may destroy delicate manufacturing processes.

Microgrids limit external visibility and control to only those aspects that the microgrid chooses to expose. Typically, these are at the level of aggregate power use, and not at the level of individual systems. If those individual systems are themselves microgrids, then the containing microgrid itself gets the benefits of simpler operation, and the contained system the benefit of heightened security (see below).

Microgrids lessen the loss of control by localizing decision making.

The *loss of privacy* arises because smart meters are able to become surveillance devices that monitor the behavior of the customers [5]. Government commissions have expressed strong concern about this issue [6]. Published papers have demonstrated the use of simple power observations to infer detailed information about even the most intimate non- powered activities within a home. The traditional counter argument by suppliers is that they don't care, and it would be too expensive to do anything with the data they collect. New techniques, such as those used for clickstream analysis, have reduced the cost and increased the accuracy.

The *loss of autonomy* is more subtle, and includes some aspects of the loss of control and of loss of privacy. Loss of autonomy includes not only the short term detriments, as above, but the longer term ability to change

ones behavior to anticipate these demands, and the freedom to obviate them as one may determine best.

Microgrids are responsible for their own consumption, storage, conversion, and use of energy. Microgrids create autonomy while increasing control and potentially preserving and enhancing privacy. Microgrids are the means to eliminate the de-motivators for smart energy.

Increasing Concerns with Privacy

Privacy concerns are a growing barrier to smart grid deployments... The techniques now known as Big Data are used to glean significant information through aggregating trivial observations. Published papers have demonstrated the use of simple power observations to infer detailed information about even the most intimate non-powered activities within a home. Similar techniques applied to commercial and industrial facilities can degrade physical security and safety, or reveal trade secrets. Secondary application of Big Data across information reveals personal information that is profitable to the party able to sell the information, advantaging to the party able to buy the information, and disquieting to the party observed. These techniques are becoming trivially cheap to apply broadly. Cheap data storage means that information once revealed is never lost, and its use cannot be controlled.

Microgrids, though, can manage their power use, storage, and generation to blur this information so it is never revealed. Microgrids provide a simple boundary at which to manage security and privacy.

On the Language of Putting Things Together

Many of the concerns and contrasts drawn in this article involve the real costs of assembling things and making sure they work together. We have tried to draw consistent distinctions between similar notions by using the following language:

- **Integration** is the cost of making things work together. When several systems are put together into one, the systems need to interact. Traditionally this is done by an engineer defining deterministic interactions between systems. It may also include some sort of system registration with a controller, etc.

- **Configuration** is the periodic re-setting of parameters to as a system changes over time, or as the needs of those using the system change. There is some overlap with integration. Configuration may be a final step of integration. Minor changes, such as adding another instance, another air handler to an integrated facility, may be treated as configuration.

- **Operation** is the ongoing regular changes in priorities or settings on a system. It may be as simple as changing the time of operation or the thermostat setting. Again, acts performed during operation may also occur during configuration.

Clouds

There are many proprietary and semi-proprietary definitions of clouds. We use cloud here in its broadest sense, i.e., as a non-deterministic expression that does specify particular technology or location.

We use the term cloud to represent a multimodal, multi-participant system wherein energy decisions are made, and energy transactions executed. The cloud may or may not be in the building, on the site, or located elsewhere. The cloud is not tied to a particular technology in use today. The cloud is not tied to a specific application contained in any of energy using or producing systems in the microgrid.

A particular microgrid or even a particular micromarket may be implemented through the use of one or more clouds. For brevity and clarity, in this chapter, we write as if each microgrid makes decisions in its own single cloud.

CHARACTERISTICS OF MICROGRIDS

Our vision of the smart grid architecture is recursive; each grid can be composed from a number of microgrids, and each smart microgrid replicates the architecture of the overall smart grid. A customer interface may front a home or commercial building, or an office park or military base. The office park and military base may contain their distribution network, their own generation, and their own customer nodes. There is no architectural limit on this recursion; recent commercial products provide room-level microgrids that support a single service, and manage generation, storage, and distribution internally.

In this section, we get more specific about the microgrids that are the end nodes we name above.

Defining Microgrids

A microgrid is more than islanded power grids and distributed generation.

A microgrid is a self-guided system with a specific mission that acts in such as to preserve its ability to perform that mission. To this end, it acquires and consumes electric power. A microgrid may store electrical power so that it will be able to perform its mission at a future time whether or not power is then available. It may acquire power in advance of need, so it has power to store. It may generate power, and acquire some knowledge of its ability to generate power to improve its planning.

During times of shortage, a microgrid may adjust its internal systems so as to get through the period of shortage. A microgrid may opt to perform some function sub-optimally during shortage so as to preserve energy for other functions more important to its purposes. For example, a microgrid may choose between availability and performance as dictated by its purposes.

Some microgrids may use less linear strategies. A microgrid may be able to recycle the effluent of its energy use to support other energy uses. A microgrid may be able to convert non-electrical power sources into electrical power. A microgrid may pre-consume electrical power into an intermediate form, even finished goods or activities, which may provide simpler storage.

The Fully Integrated Microgrid

Many of us make daily use of microgrids as defined above. The modern portable computer, tablet, or smart phone are each a microgrid.

Each of these devices is sometimes connected to a power source, and sometime disconnected. These devices come with powerful algorithms to manage power use. Screens may dim when not in use. Disk drives may spin down if not accessed for more than a few minutes. These and many other strategies are used to preserve the ability of the system to provide service until it once again is attached to a power source.

These techniques are policy sensitive, that is, they can respond to high-level guidance. Often a simple slider bar will determine how aggressive a system is in managing its energy supply. Software is available to curtail specific functions that demand higher energy. For example, Wi-Fi uses electrical power at a high rate. Often, smart phones only run applications that require Wi-Fi in known locations. Some phones run software that will disable Wi-Fi except in locations where it has been used before. This is an example of policy-based management.

Fully integrated microgrids are available today because they are mass-produced. Because of economies of scale, their power use can be fully integrated into the software that operates them. This enables a competitive market for software that applies different type of policy (No Wi-Fi on the subway) to an existing system.

Self-integrating Microgrids

The small fully integrated microgrids described above are useful for illustrating the effective use of microgrids today. They work because they do not have the challenges of larger microgrids: diversity of components, diversity of purpose and of technology, and few resources for integration.

With enough engineering time, and enough custom integration, we may be able to solve these issues for any single facility. There is neither enough engineering time, nor enough budget for custom integration for every facility. Ideally, systems within an end node would self-organize themselves into a microgrid, optimize the microgrids energy usage, and be able to respond to the market signals as a microgrid.

If my home is treated as a microgrid, it is a unique one. The mix of appliances and equipment in my house is different than in any other house in my neighborhood. One author (Considine) lives in a house that is nearly two centuries old; the structural shell that determines so much of energy use and storage is different than that of the 1970's-era house across the street. The ways in which he uses energy, and the times, are different now, as an adult who travels frequently, than they were when he had children at home.

Even very similar equipment may have quite different energy use profiles. The motor in a top-loading clothes washer has an entirely different temporal pattern of use than that in a front loader. While these differences are trivial at the scale of the grid, within the scale of the microgrid, they may be significant.

Increasingly, these end nodes may have their own energy resources. They may have intermittent power generation from renewable sources. They may buffer energy in batteries, or as hydrogen, or by pre-consumption. A facility that stores energy as hydrogen may opt to use it as hydrogen, to fuel vehicles, or as a battery through a fuel cell, or even to increase effectiveness of generation though blending it with natural gas to use in a traditional generator [7].

The residential end-node has incredible diversity in purpose and in contained technology, even as it appears to be the simplest and most homogenous class of microgrid. There will never be enough time and resources to pay for custom integration of residential end nodes to support fully smart energy. We must look instead at ways for the components of residential end-nodes to assemble themselves into microgrids.

Each system in a home can leave the factory knowing fully only about itself, and how it uses energy. Within the confines of a home, the homeowner can assign, by policy, priorities to different systems in the home. The systems need to discover each other, and to negotiate with each other how to manage energy use within the over policy set at the microgrid [home] level.

Similar challenges face commercial, whether small or large, institutional, industrial, and mixed sites.

Micromarkets and Microgrids

We use the term micromarket to name the inner decision- making process of a microgrid [8]. A market is any structure that allows buyers and sellers to exchange any type of goods, services, or information. Where the exchange is for money, it is termed a transaction. Previously, we have defined market segmentation based on market rules which include definition of the products traded and converging algorithms for clearing that market.

Hayek described markets as the way to solve the local knowledge problem, that while the data required for rational planning are distributed among individual actors, knowledge is unavoidably outside the knowledge of a central authority. Market-based systems for allocating control resources have repeatedly outperformed traditionally operated control systems in studies as early as 1994.*

As described above, the problem of integration and configuration is a knowledge problem. No system knows what the other systems in the microgrid will be. No system knows what the patterns of energy use the other systems exhibit. During operation, no system knows the priority placed, by policy, on each of the other systems in the microgrid.

Each system can, however, know itself. Software on each system can act as an agent able to express its needs and priorities within the micromarket of the end node.

Figure 29-1: A software agent hides a system's complexity while interacting with a market

It is easy to imagine a portable computer or a cell phone supporting an Energy Services (ESI) [1], and able to find and negotiate with the local micromarket. *

Integration is a problem of applying specific knowledge to a set of components as to their optimum interactions. Integration has typically been labor and knowledge intensive. In this model, the micromarket itself becomes the alternate solution to that knowledge problem, one that can adjust itself as new agents arrive or depart.

TODAY'S MICROGRIDS

Today, microgrids are springing up wherever the needs of the local site are not being met by centrally planned and operated electric power and distribution. Early leaders were those with special requirements for high power and availability. We refer to these as *Industrial Microgrids*.

Others are forced to rely on a microgrid due to the expense of bringing power distribution to a remote location. We refer to these as *Isolated Microgrids*.

Microgrids in undeveloped countries are unable to connect to larger grids, but their motives and operation are quite different than those of Isolated Microgrids. The scale is often smaller, even than home-based microgrids in the developed world. We refer to these as *Development Microgrids*.

Military Microgrids are an area of current attention and rapid development [9]. Driven in part by the recent doctrine of energy surety, these microgrids are characterized by diversity of mission, by changing technologies, by the need for Just-In-Tine (JIT) integration, and the need to re-allocate resources rapidly as mission and assets change.

There is a growing adoption of microgrids out of choice. These choices occur in areas well served by existing distribution grids. In many cases, they are in urban re-developments. The power requirements of these sites are often small; in part this because these sites have already made unusual commitments to site-based energy initiatives. We refer to these as Motivational Microgrids.

Industrial Microgrids

Industrial microgrids hearken back to the early days of electrical power, when industrial sites would produce their own power because no other power is available.*

Industrial sites with high power requirements have long relied on site-based generation. For some, such as Aluminum producers, electric power dwarfs all other supplies. If a site has power requirements similar to the capacity of commercial generating plants, in-sourcing this generation is a natural decision.

Some processes, notably in chemical processing and in regulated pharmaceutical environments, are subject to very large costs for power interruption. A small interruption in power may cause very large process costs, in lost product, in equipment degradation, in lost certification, and in high start-up costs afterward.

Other sites use large amount of energy, but in a form other than of electrical power. In particular, some plants rely on thermal energy. These may form wood-based products or be laundry facilities. Steam or hot water may drive a significant part of their activities. Once they have a boiler in place, using excess capacity to generate electrical power is a natural afterthought. This type of microgrid is usually referred to as cogeneration.

District Energy is cogeneration writ large. A District Energy may provide central distribution of steam or of chill water to a business district, hence the name. More often, District Energy is provided across a college campus or a multi-building industrial site. District energy is characterized by a multitude of choices and by substitution.

For example, the same boiler can generate high pressure steam to spin a turbine or low pressure stream for district distribution. Chill water can be produced using electricity in compression chillers or using steam in adsorption chillers; many facilities switch day-to-day based upon weather and upon relative prices of electricity and steam. Hot waste- water from one facility may be the energy source for chilling the next. Modern Combined Heat and Power plants have similar capabilities.

Each of these types of Industrial Microgrids is driven by internal needs and economics. Industrial microgrids pre-date the concerns of Smart Grids. An Industrial Microgrid may learn to interact with smart grid concerns such as Demand Response (DR), but such concerns will never be the primary driver of their energy strategies.

The other side of the coin, the ability to sell surplus, is commonly limited or prohibited by regulatory action.

---

*We showed in *"Energy, Micromarkets, and Microgrids"* [7] that there is advantage from having a logical micromarket attached to each microgrid.

*In the Industrial Revolution, water-driven manufacturing mills or plants were common.

The service interfaces as specified by Energy Interoperation will enable Industrial Microgrids to interact with a smart distribution grid and with other microgrids. Industrial Microgrids will not allow any significant direct control of their internal systems by third parties.

As microgrids of all types gain renewable resources and site-based energy storage, the microgrid model will grows to resemble that of District Energy.

Isolated Microgrids

Isolated microgrids began as soon as wealthy early adopters put in the first light bulbs in the town. Today there are more often in isolated vacation homes, whether in the mountains or on islands, or even on yachts. Traditionally, these microgrids are fueled by fossil fuels transported by road or boat. A few relied on site-based sources, whether coal, or natural gas, or wood. Traditional local generators are the norm.

Many of these isolated sites are owned by economically well-off individuals. It is a rare site that can support sufficient density of power generation to support the level of amenity their owners expect. Today, intermittent generation sources from renewables are an amenity as well as a resource. This means that the normal requirements for economic justification can be reduced.

These owners often occupy these sites intermittently, that is, it is a vacation home or weekend retreat. Intermittent renewable sources sometimes are used solely for maintenance. Intermittent generation manages humidity in summer to prevent mold growth. In winter, intermittent generation can support keeping the pipes from freezing.

New approaches combine intermittent occupancy, intermittent generation, and storage to enable Isolation Microgrids to operate with less and less combustion as an energy source.

Isolation Microgrids serve today as proving grounds for site-based storage and for temporal relocation of energy use.

Development Microgrids

Development microgrids are small commercial operations in areas in which the existing infrastructure and economy are not based on long-standing assumptions of intermittent power. These microgrids actively compete with other energy sources on a day by day basis. These share many characteristics of Isolated Microgrids but are in areas with limited but competitive energy resources and infrastructure. [10] [11] [12] [13] [14] [15]

For example, a small solar generator may provide the primary electric power for a small village in sub-Saharan Africa. Cell phones provide not only the sole communications for commerce, but provide the essential banking services of the local economy as well. Villagers may vie to purchase the power to charge their phones from the limited power generated. If the price gets too high, a bicycle-based generator with a boy riding may provide a competing service.

In Bangladesh, Dean Kamen's slingshot micro-generation systems, a pocket generator the size of a washing machine is paired with a similar-sized water purification system to provide power and clean water in rural villages. [16] These pocket generators work on multiple fuels including cow dung. Because this LED lighting based on a DC infrastructure replaces burning wood for light, these pocket generators reduce deforestation. Cell phone charging is once again a critical service provided by these systems. Slingshots provide "civilization in a box" (light, water, telecommunications), and are the basis of ongoing micro- industrial transactions at the personal level within the village.

Development Microgrids work without the assumptions built into power markets in the industrial world a century ago, before the computing and telecommunications revolutions. As such, they are a proving ground for the new economics of distributed energy.

Military Microgrids

Military microgrids are driven by the developing doctrine of energy surety and its little brother, power surety. Energy is a means to project force, whether through weapons, through intelligence, through command & control. In other words, protecting the energy position of a base is protecting the mission capability of a base.

Military Microgrids have some desiderata that push through the boundaries of traditional integration approaches:

- There should be no central network operating center (NOC) on a base that can be destroyed in order to destroy base-wide energy surety.

- Energy sources, each with unique characteristics arrive on and depart from bases. The base should be able to accept these sources with little or no reconfiguration. (Think everything from idling engines to PV on pup-tents)

- Energy uses change in priority with each change in mission. Advance? Hold? Defend? Withdraw? Redeploy?

- Even fixed base energy assets may be removed w/o planning (mortar shell hits the sub-station)

- Base energy uses include some exotics such as hydrogen cars, PEVs.

In pure Hayekian terms, there is a knowledge problem about energy sources, energy uses, and the best application of same. Each energy source on base has some capabilities that should be used to the fullest. Each energy use on base has a mission, whose import changes over time. Sometimes the import is situational, as the import of food protection grows to the refrigerator that "skipped" its last few cooling cycles.

The priority of each activity is set somewhere between a knowable baseline, situation awareness, and changing orders or "mission." Without too much stretch, claim that each system can know its priority by Policy, that is, through the techniques of policy-based management.

As the activities know their priorities, and the sources know their capabilities, bases need a clearing house that manages the optimum application of energy from minute to minute. If each source and use of energy is represented by an agent, then these agents can negotiate in a market. Systems that have more policy priority operate with larger budgets at certain times. But even the highest priority activity does not want to pay for energy it cannot use, at a time it does not want to use it. This provides the opportunity for the lower priority activities to make winning bids.

Such Microgrids in effect can be operated by micromarkets. Energy surety is an emergent behavior of the participants in the micromarket. Wherever two or more agents can establish communications, a market can exist. New agents are integrated by entering the market. Agents can build reputations through their participation in the market.

Just as in District Energy, there is the issue of optimizing between different results to get the same effects. Bases may store energy as hydrogen as well as in batteries. A given base may have both hydrogen-fueled and PEV vehicles, each requesting charges. These vehicles may get high priority, sometimes, depending on mission and occupant.

Hydrogen can be used to fill a Vehicle, or to generate electricity in a Hydrogen Fuel Cell, or to supercharge natural gas in a conventional generator.

Military microgrids create a premium for energy systems that can reconfigure themselves. Energy assets come and go. Energy using systems are in regular flux.

Priorities for each system can change in a moment. Military needs are best met through solutions that do not require constant re- configuration at a single base and that can be re-used at all bases.

Military Microgrids will pioneer the ability to compose a microgrid from changing elements without continuous intervention. Military Microgrids will be proving grounds for autonomous self-organization and policy-based management.

Motivational Microgrids

Motivational microgrids are sites that choose to operate as microgrids in the absence of the compelling needs itemized for the microgrid types above. Motivational Microgrids may be in the middle of a city, with easy access to traditional distribution. Motivational Microgrids are islanded because they want to be. Motivational Microgrids are driven by valuing one of the aspects of Microgrids far more than does the general public.

For example (based on personal conversations), consider the movement to renovate inner city industrial sites based on a "Green" ethos. The redevelopment minimizes energy use by using LEED approaches. Its tenants are motivated by local use and small environmental footprints. It adds some site- based renewable generation.

The particular site becomes frustrated with the local utility. It does not get the easy deals that it anticipated for its renewables. The distribution entity properly must defend its capabilities and its other customers from the effects of this site. Because of its already low energy use, it sees an easier path to self-sustainment in electrical power than would a traditional commercial/light industrial site.

There are a growing number of efforts that meet this description in older post-industrial settings across the US.

Another example from personal experience is the larger home whose owner places a high premium on privacy. He starts from a high energy profile, and opts for a natural gas fuel cell to opt out of the local smart grid efforts. Because a cooling tower would anger his neighbors, he decided to shed his heat load to support his Jacuzzi and pool. Today he is considering switching to absorption chillers* [ref] to take additional heat. Slowly, he is replicating the district energy model within his home [8].

---

*Heat-operated refrigeration unit that uses an absorbent (lithium bromide) to absorb the primary fluid (water). The evaporative process absorbs heat, thereby cooling the refrigerant (water) which in turn cools the chilled water circulating through the heat exchanger.

Motivational Microgrids show us the future of consumer attitudes toward microgrids and energy. The occupants of Motivational Microgrids are willing to work out the internal models for applying site-based power management in non- traditional situations.

Hidden Microgrids

Many sites manage their own power surety today. The have generators on-site that they use to provide emergency power. Data Centers, Banks, Hospitals, and Emergency Responders are typical examples. The systems on these sites usually include power storage, even if only to support uninterrupted power during switchover from the distribution grid to their internal resources. These sites are often subject to regulatory limits on their site-based generation, particularly in urban environments.

These sites can be considered as microgrids. By adding an Energy Services Interface (ESI) based on Energy Interoperation, they could increase their energy surety through gaining improved situation awareness of the grid.

There are far more extant microgrids than we normally consider.

Summary of Today's Microgrids

Microgrids are much more widely deployed today than generally acknowledged. We must look to them to understand how microgrids will be used in the future. We can generalize the interaction of supplier and microgrid using existing standards.

OASIS Energy Interoperation provides a common means for interacting with each class of microgrid. It makes no assumptions about the technologies or processes within a microgrid. It does not try to directly manipulate processes inside the Industrial or Military Microgrid. It does not limit the technologies or the diversity that can be deployed within the Microgrid. It was in fact designed to work "to, from, inside, and outside microgrids" [1]

Microgrids today support the ability of their inhabitants to manage their own processes and priorities based on superior local knowledge. Microgrids today are proving grounds for site-based storage and for temporal shift of energy use. Microgrids today are proving grounds for the new economics of distributed energy without requiring transformation of the larger distribution or markets.

Military Microgrids require autonomous self-organization and policy-based management. These techniques can be applied within any microgrids where

further segmentation is desired. Autonomous self-organization and policy-based management will scale up to aggregations of microgrids Aggregation of microgrids to provide resilience and reliability is already underway in Southeast Asian markets [17]. These techniques can apply to any microgrid.

Motivational Microgrids show us the future of consumer attitudes toward microgrids and energy, as early adopters pave the way. The occupants of Motivational Microgrids are willing to work out the internal models for applying site- based power management in non-traditional situations. Hidden microgrids, as described above, show that this approach is already taking off.

BUILDING OUT THE MICROGRID

What we have described above is a microgrid created when multiple devices are able to discover a micromarket and interact through an ESI using the Energy Interoperation specification to communicate. The owner/operator of the microgrid can assign different policies to each of the systems controlling how each interacts with the market.

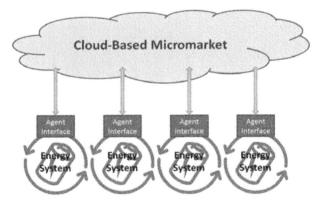

Figure 29-2: Each agent competes in market to optimize its own system performance and mission.

Using Energy Interoperation, each agent can buy or sell power at specific times of delivery. Each agent can establish forward positions based on its own need to fulfill currently applied policy, able to make commitments to deliver or take delivery of power at future points in time. If the policy controlling one of the agents changes, then that agent informs the others by taking different positions within the market.

We do not however, define the form of this market here [8]. There could be a market maker, a single entity responsible for overall market performance. In such a model, all interactions would be transactions with a single market- making entity. Alternately, the agents could make a series of bilateral deals with each other. Successful micromarkets could work under either model.

Legacy, or You Can't Get There from Here

One of the challenges to any model of smart energy is the existing stock of energy using systems. On whatever day we start, those systems installed yesterday will not participate properly.

Many systems can be cost-effectively upgraded to support these agent behaviors. Smart phones and computers are routinely upgraded with new agents. Home networking gear is less routinely upgraded, but still can accept new software with little trouble. Home HVAC systems could be upgraded by adding an agent-capable networked replacement smart thermostat. Even home entertainment systems are now routinely networked. Video players, televisions, and DVRs each routinely receive updated software over the internet.

There are other devices that cannot so easily acquire an agent. These include most traditional appliances and lighting systems. They may be networked using more control-oriented protocols such as SEP (Smart Energy Profile). These systems can participate in markets using Mobile Agent [18] [19] re-location.

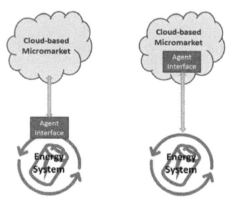

Figure 29-3: Agent functionality can be re-located to support legacy or low-capability systems

Service-oriented architecture has as a key principle the separation of how you do something from what service you request, and is an ideal technology to continue to use legacy systems with more scalable interoperation [20].

For control-oriented systems, the agent itself can be placed in the cloud that supports the micromarket. In this way, existing systems can be encapsulated within the micromarket, and the life of existing assets extended.

In the illustration above, four agents are partici-

pating in the microgrid. One legacy system is unable to participate directly. Each is represented by an ESI. In this model, there may be advantages to a model based on a central market maker.

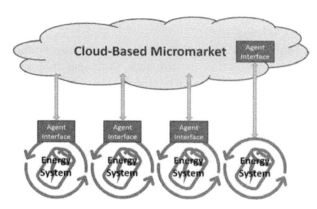

Figure 29-4: Essential operations of the cloud-based market do not change even if some systems are not agent-capable.

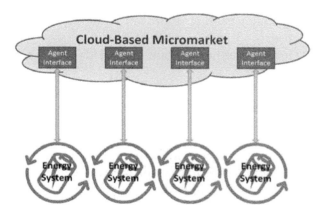

Figure 29-5: A micromarket comprised entirely of low-capability systems can resemble legacy integration.

The model works as well if all existing systems are incapable of supporting agents. In this case, one role of each agent is to act as a traditional "driver," providing an abstract interface to the common operating platform of the end-node. The common operating platform may incorporate higher level models such as that in ASHRAE SPC 201 to understand the other effects on the platform. The platform is the micromarket.

Incorporating Vehicles into the Microgrid

Traditional approaches to the smart grid treat vehicles as special cases, presenting challenges in billing and in integration. This is in part because vehicles are treated primarily as roaming batteries. Vehicles, however, present questions that are both simpler and more complex than these discussions.

Vehicles are run by demanding control systems that are ever-growing in complexity. The primary mission of a vehicle is quite different than that of most grid-attached systems. Within a single household, there may be large differences. A vehicle may be the sports car used on weekends, the delivery van used for short runs around town, the primary transportation for the household or even be driven solely by the teenager who is currently "grounded." In several states, self-driving vehicles have recently been authorized.

Vehicles also move, and must introduce themselves to a number of micromarkets. This puts a premium on security for vehicle interactions. My dishwasher may never see a new micromarket, and my vehicle may do so many times per week.

Under the microgrid model, these interactions are the same as any other in the microgrid. A vehicle must discover the local market, and negotiate its position. As any traveler, the vehicle may find that its currency is not accepted in the local market. A vehicle must assume that its market is untrustworthy, just as the local market may mistrust the vehicle.

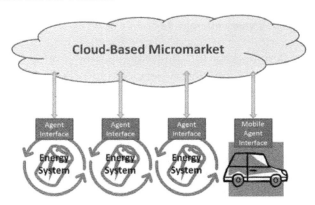

Figure 29-6: While a mobile system such as a vehicle may require some additional services, it does not challenge the model.

Vehicle to microgrid interactions, though, are more complicated than mere power negotiations. Hydrogen vehicles may be negotiating for direct transfer of hydrogen stored in a fuel cell. Natural Gas vehicles may negotiate for a long slow charge.

These and other issues are discussed next.

DIVERSITY OR ENERGY SOURCES AND TYPES

Up until this point, microgrids have been simplified to involve solely electric power. Any microgrid may have multiple ways to store and use energy, and multiple ways to acquire energy.

Traditional smart grid discussions assume that all power comes from the grid. On-site generation is valued for direct sale to the grid. Thermal storage is valued as a pre-purchase from the grid to replace purchases that would be made later in the day. This does not necessarily align with the perspectives of the end node. It also limits the ability of these microgrids to accept new technology in the future.

Assume a small commercial building with several energy collectors. It is normally connected to the grid, and buys its power from the grid. On-site PV cells generate a predictable flow of energy that is stored on-site in hydrogen cells. That energy in hydrogen may be used to improve the site's ability to respond to grid-based (DR) events or to grid failures for energy surety.

This commercial building also uses solar cooling to generate chill-water for a number of internal processes. Whenever the supply is greater than the internal use, that cooling is applied to thermal storage; this storage may be configured later use to support DR just as are systems that use the grid for pre- cooling. It provides exactly the same sort of asset for Demand Response as it would if purchased from the grid.

A commercial building may "host" a hydrogen vehicle that consumes the stored hydrogen. A visiting hydrogen vehicle may wish to fill up. In accord with building policy ("No outside sales unless half full"), and subject to a special market rule ("Sales to strangers are offered at a 25% premium to market") the visiting vehicle may request a purchase. The price offered, though, may be tied to the value of the hydrogen as a battery within the local power market.

The commercial building may have a fixed capacity for receiving natural gas. Some of that natural gas may be used on-site, to back-stop the power markets. It can also support slow filling of a natural gas vehicle. The availability of the natural gas to a vehicle may be limited by prior commitment deriving from the power market.

The micromarket model allows for the fungibility of energy sources. Diverse commodities can coexist in the same market. The complexity of this decisions making is hidden from the suppliers. The end node presents only an aggregate position to each of the markets it participates in.

BUILDING UPWARD FROM MICROGRIDS

Earlier in this chapter, we suggested that a microgrid is the ideal participant in a micromarket. In that case, we used the example of a portable computer as a microgrid pre-adapted for participation in the home-based microgrid. A microgrid knows its energy needs and surpluses. A microgrid is aware of its clearing positions in power. A microgrid is already operating under a policy basis, and is thereby ready to negotiate with other parties. All the microgrid needs is an external ESI that understands Energy Interoperation to be a full participant in the micromarket.

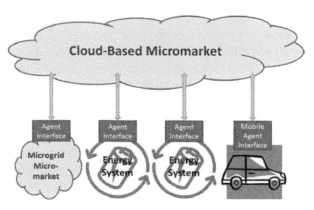

Figure 29-7: The type or complexity of the system represented by an agent does not change the micromarket interaction

From there it is an easy step to building a microgrid entirely upon microgrids. The homes in a neighborhood could participate in the local microgrid. That microgrid's policy limits might include overall capacity of the neighborhood feeder. Microgrids that represent commercial buildings can participate in the office park microgrid. The office park may include local generation, say a wind farm above the common areas. The wind farm, then, is simply an independent participant in the office park micro market.

The composition and decomposition of microgrids is itself a microgrid. [21]

MULTIPLE MICROGRIDS AND MULTIPLE MARKETS

The model described above is consistent with that previously defined as Structured Energy [21]. We have not attempted to show all permutations and exceptions. The discussion above describes each system participating in a single market through an Energy Services Interface (ESI) as described in OASIS Energy Interoperation. That microgrid, in its turn, has a single ESI for communicating with the next level microgrid.

The US grid today has three connected grids, each with many markets and within an overall market that al-

lows transactions among the micro grids with no single agent representing or operating each microgrid. Within those microgrids, functional markets for the same product may be distinct-; within ERCOT, the ISO only operates wholesale spot markets and not the retail and forward markets. Similar market rules can exist within the microgrids described herein.

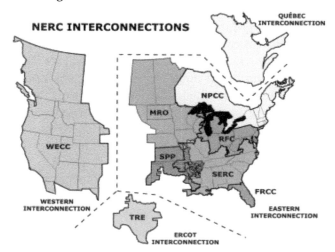

Figure 29-8: A high level look at the North American Grid of today shows many microgrids

Micromarkets for different types of products can coexist as well. Regulatory services markets can coexist with power markets. Microgrids that encompass district energy will have thermal markets as well.

There is an apparent one microgrid, one cloud architecture in the discussion above. Again, this was for brevity and for clarity. A microgrid can be supported by multiple clouds under this model. Multiple microgrids can have their markets in a single cloud.

We can compose up, decompose down. Microgrids simplify the smart energy conversation by defining a scope of concern. A home may contain several microgrids that cooperate in the homes master grid. That home may further participate in a community microgrid that is within a city microgrid. Each microgrid may always or sometimes be disconnected from other grids.

Each grid can be composed from a number of microgrids, and each smart microgrid replicates the architecture of the overall smart grid. An ESI fronts a node that may be a home or commercial building, or office park or military base. The node may contain its own distribution network, its own generation, and its own customer nodes. There is no architectural limit on this recursion; recent commercial products provide room-level microgrids that support a single service, and manage generation, storage, and distribution.

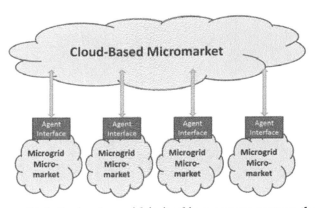

Figure 29-9: Each microgrid is itself an energy system that can interact in larger micro-grids (Recursion)

CONCLUSION

We have shown that microgrids are already much more prevalent today than is generally recognized. Agent based operation of microgrids simplifies the adoption of diverse technologies. Microgrids are locally responsive so they can more easily fulfill their own purposes than can integration based on far-off central offices. Microgrids inherently have more options for balancing intermittent energy generation and intermittent use than do larger grids, because the tradeoffs are visible and local. Microgrids isolate and hide diversity to reduce barriers to innovations.

Microgrids are today the proving grounds for consumer acceptance and site-based management of smart energy. Microgrids today are pioneering consumer-based transactive energy.

We have described a model, of autonomous microgrids operated by agent interactions in a micromarket, that rationalizes the IT architecture of smart grids so that they can self-assemble, minimizing the integration costs that have limited acceptance of microgrids. By creating a common model across many types of microgrids, the model enables techniques and approaches developed in one class of microgrid to later be applied in another.

Smart energy will finally be an emergent behavior of diverse autonomous systems. The architecture described above provides a means for these systems to self-assemble themselves into aggregates [microgrids] that then can aggregate themselves into larger microgrids.

References

[1] OASIS, Energy Interoperation 1.0, 2012.

[2] F.A. Hayek, "The Use of Knowledge in Society," *The American Economic Review*, vol. 35, no. 4, pp. 519-530, 1945.

[3] L. Kiesling, "The Knowledge Problem, Learning, and Regulation: How Regulation Affects Technological Change in the Electric Power Industry," *Studies in Emergent Order*, vol. 3, pp. 149-171, 2010.

[4] P. Centolella, "A Pricing Strategy for a Lean and Agile Electric Power Industry," October 2012. [Online]. Available: http://ElectricityPolicy.com.

[5] D. Carluccio, S. Brinkhaus, D. Löch and C. Wegener, "Smart Hacking for Privacy," in Behind Enemy Lines, Berlin, 2011.

[6] F.o.P. Forum, "SmartPrivacy for the Smart Grid: Embedding Privacy into the Design of Electricity Conservation," Information & Privacy Commissioner, Ontario, Canada, 2009.

[7] W.T. Cox, T. Considine and D. Holmberg, "Energy Ecologies—Models and Applications," in Grid Interop, Dallas, TX, 2012.

[8] W. Cox and T. Considine, "Energy, Micromarkets, and Microgrids," in Grid-Interop 2011, 2011.

[9] T. Podlesak, R. Lasseter, T. Glenwright, T. Abdallah, G. Wetzel and D. Houseman, "Military Microgrids," in Great Lakes Symposium on Smart Grid and the New Energy Economy, Chicago, IL, 2012.

[10] C. Kirubi, A. Jacobson, D. M. Kammen and A. Mills, "Community-Based Electric Micro-Grids Can Contribute to Rural Development: Evidence from Kenya," *World Development*, vol. 37, no. 7, pp. 1208-1221, July 2009.

[11] D.-R. Thiam, "Renewable decentralized in developing countries: Appraisal from microgrids project in Senegal," *Renewable Energy*, vol. 35, no. 8, p. 1615–1623, 2010.

[12] C.G. Kirubi, Expanding access to off-grid rural electrification in Africa: An analysis of community- based micro-grids in Kenya, Berkeley: ProQuest Dissertations And Theses, 2009.

[13] T. Bernard, "Impact Analysis of Rural Electrification Projects in Sub-Saharan Africa," *World Bank Research Observer*, vol. 27, no. 1, pp. 33-51, 2010.

[14] B.K. Blyden and W.-J. Lee, "Modified microgrid concept for rural electrification in Africa," in Power Engineering Society General Meeting, 2006.

[15] G. &. M.C. Venkataramanan, "A larger role for microgrids," *Power and Energy Magazine*, IEEE, vol. 3, no. 78-82, p. 6, 2008.

[16] E. Schonfeld, "Future Energy eNews," Integrity Research Institute, 8 March 2006. [Online]. Available: http://users.erols.com/iri/EnewsMar8,2006.htm. [Accessed September 2012].

[17] T. Mohn, "20/20 Visions for 2030: In the Wider World, Prosperity Requires Sustainable Energy for All," Electricity Policy, 27 February 2012. [Online]. Available: http://www.electricitypolicy.com/archives/4058-20-20- visions-for-2030-in-the-wider-world,-prosperity-requires-sustainable-energy-for-all-20-20-visions-for-2030-in-the-wider-world,-prosperity-requires- sustainable-energy-for-all. [Accessed September 2012].

[18] J. Cao and S.K. Das, Mobile Agents in Networking and Distributed Computing, Wiley, 2012.

[19] R. Gray, "Mobile agents: the next generation in distributed computing," in Parallel Algorithms/Architecture Synthesis, Second Aizu International Symposium, 1997.

[20] OASIS, Reference Model for Service Oriented Architecture 1.0, OASIS, 2006.

[21] W. Cox and T. Considine, "Structured Energy: A Topology of Microgrids (Presentation Only)," in Grid- Interop, 2010.

[22] B. Huberman and S.H. Clearwater, "Thermal markets for controlling building environments," *Energy Engineering*, vol. 91, no. 3, pp. 26-56, January 1994.

[23] B. Huberman and S.H. Clearwater, "A multi-agent system for controlling building environments," in First International Conference on Multiagent Systems, 1995.

Chapter 30

Power Monitoring Solutions through Contextual Data

John Bickel, Schneider Electric

ABSTRACT

Most utility customers do not have the time or expertise to analyze and understand the issues surrounding energy consumption and reliability, much less asset management. Still looking to twentieth century ideas to answer their urgent need for new energy solutions, customers are missing what they really need: real answers. For example, a newly developed non-obtrusive technology provides relevant information that has been unavailable to customers (at least without considerable effort and cost). This technology introduces the fundamental components of artificial intelligence (AI) into the demand-side and utility energy solutions and services market.

INTRODUCTION

Since the late 19th century, when electrical power distribution systems were introduced, energy consumers have needed to monitor various parameters on their electrical systems. Monitoring devices were developed to measure and report such information. They give facility managers and staff the ability to collect, analyze, and respond to information about the electrical power system. This has helped them improve safety, minimize equipment loss, decrease scrap, and save time and money.

Electrical systems are hierarchical in nature, with monitoring devices such as electrical meters installed at various levels in the hierarchy. These devices measure characteristics of the electrical signal as it passes through the conductors (parameters such as voltage, current, and power). Electronics have improved the quality and quantity of data coming from monitoring devices; and communications networks and software have improved the way information is collected, displayed, and stored. Yet, even with the current improvement in metering technologies, effective monitoring of electrical power distribution systems is still cumbersome, expensive, and inefficient.

In its most basic form, a power monitoring system consists of two or more metering devices that use a communications network to provide data to a central software management package. The term 'power monitoring system' is really a misnomer; the components in today's power monitoring systems (meters, software, gateways, etc.) act pseudo-independently from each other. So the facility staff must know how to properly configure hardware, collect and analyze data, and determine the data that is necessary or useful to evaluate potential performance and reliability issues. Two difficulties arise from contemporary monitoring system methodologies: 1) the volume of data to be analyzed, and 2) the data's relationship between each discrete monitoring device. Analyzing large quantities of data generated by a power monitoring system can be automated; however, it necessitates that the data be put into context. Contextual data is the key to understanding how data from one device relates to data from another device, and thus, to leveraging the information accumulated from different points across a power monitoring system.

Today's power monitoring systems often overwhelm system managers with data; and these managers typically have many responsibilities beyond being the "keeper" of the monitoring system. In response, manufacturers of metering products are seeking more efficient methods to process and present information.

EXISTING POWER
MONITORING SYSTEM PARADIGM

In monitoring systems, metering devices are installed at strategic points in an electrical system. The devices determine the system's various characteristics, such as energy consumption patterns or voltage source anomalies. Metering devices collect data at their respective locations in the hierarchy of the electrical system and typically pass the accumulated data to central system management software. The software stores data, helps identify and understand problems, provides statuses, and makes decisions.

In many monitoring systems, data from discrete metering devices have limited relational context with data from other metering devices, leaving it up to the facility staff to "relate" the data. This means that the facility staff must understand the purpose, location, and characteristics of each device as it relates to the other metering devices. For small power monitoring systems that produce minimal data, the facility manager may be able to track the interrelationships of discrete devices and the data they produce. However, as more devices are added to the power monitoring system, not to mention more characteristics to be considered in the analysis, the complexity of the power monitoring system quickly becomes unmanageable and potentially ineffective. Facility managers know that most electrical drawings are out of date soon after they are produced, due to ongoing modifications to the electrical system. These same modifications can change the implications of data coming from the power monitoring system. In short, a facility's electrical system is generally in flux. Staying up to date on these changes and their impact on data being collected across the electrical system may be unachievable.

Power monitoring systems that produce data without relational context are, at the very least, both confusing and burdensome. Figure 30-1 illustrates the indiscriminate perspective of a power monitoring system (as viewed by the software or gateway) without inter-device relational context. Data from multiple points across an electrical system helps determine the system's

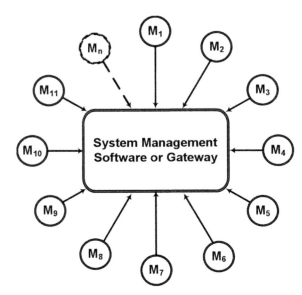

Figure 1. Perspective of Today's Power Monitoring Systems.

overall condition. However, to develop a true "system" methodology to power monitoring, this data must be brought into a reciprocal relationship.

Interrelating data between two devices (hereinafter referred to as contextual data) requires relational knowledge of one or more primary relationships between the devices. The two primary relationships used to provide contextual data between a pair of discrete monitoring devices are their temporal and spatial relationships. It should be noted that other secondary information may be considered as a means to provide additional significance to interrelated data. Table 30-1 introduces several categories of relationships that will be discussed in this chapter.

Understanding the primary (i.e., temporal and spatial) and secondary (i.e., functional, motive, process)

Table 30-1. Examples of Data Relationships.

Data Relationship	Type Relationship	Description
Spatial	Primary	Where a first data point is located with respect to a second data point within the electrical system.
Temporal	Primary	When a first data point was captured in time with respect to a second data point. (quantitatively)
Functional	Secondary	What is metered at a discrete data node* (e.g., main, load type).
Motive	Secondary	Purpose of monitoring at a discrete data node* (i.e., reliability).
Process	Secondary	Whether a first data node* relates to a second data node* by process.

*Node – For the purposes of this chapter, nodes are the building blocks of a hierarchy. In a power system hierarchy, nodes would represent any physical component in the power system, such as switchgear, circuit breakers, conductors, metering. In this case, the nodes all represent physical equipment. However, this is not the only application of nodes. For example, nodes might also represent cost centers and other logical divisions in a billing hierarchy.

relationships between any two devices and their data provides unique insights into the operational parameters of the electrical system. It is a case of the proverbial "the whole is greater than the sum of its parts." In fact, the benefits of true "system" monitoring through the use of contextual data are almost as numerous as the various purposes for energy consumption. Unfortunately, contemporary methods for determining the temporal and spatial relationships of data from multiple devices come with a price, as discussed below.

TEMPORAL RELATIONSHIPS

Understanding how an event propagates through an electrical system is an important aspect in determining the root cause of the event. Knowing the temporal relationship between data from two or more metering devices is critical to determining the sequential order an event's impact as it propagates through the electrical system. To accurately evaluate steady-state data from disparate locations within an electrical system requires a temporally consistent frame of reference.

Difficulty correlating data between monitoring devices results from temporal misalignment of the sampled data. Temporal misalignment occurs for two reasons: 1) communication latencies between monitoring devices, and 2) timekeeping and event time stamping in the individual monitoring device. The facility staff must analyze and interpret pseudo-independently captured data to optimize system performance or to evaluate potential power-quality concerns. Because the captured data is voluminous (both due to the number of devices and the amount of data from each device) and potentially disjointed from one monitoring device to the next, manual analysis can be an enormous effort that requires expert consulting services. Today's sophisticated digital monitoring devices have processing capabilities that derive and accumulate large amounts of complex data from a seemingly simple electrical signal. For facility staff and experts alike, the existing process can be tedious, complex, prone to error and oversight, and time-consuming.

One solution currently used to temporally align metering data is a global positioning system (GPS), which use satellite signals to synchronize the internal clocks of each connected metering device. When the clocks in two devices are synchronized to the GPS signal, time-stamped data from the first device may be suitably analyzed with time-stamped data from the second device, simply by correlating the timestamps.

Global positioning systems require the purchase and installation of additional hardware and data lines to link the metering devices to the GPS signal. This solution still requires the evaluation of large amounts of data. This is because the system is only temporally in context, not spatially in context (the physical location of the monitoring devices in the electrical system is not taken into account). A GPS cannot alleviate the time delay problems created by other hardware in the system. Any alignment of data by a GPS-based system is only as accurate as the propagation delay of the GPS signal to each discrete monitoring device. Generally, the propagation delays are negligible if there is a dedicated link to the devices.

Precision time protocol (PTP) is another technique to align time between two or more devices on a system. When a network uses PTP, one device is established as the master clock, and the other devices are designated as slave clocks. Slave devices determine the time offset between themselves and their master by initiating an exchange of "sync messages" that allows them to determine the network transit times. Once the transit times are quantified, time disparities can be accurately determined between each device's data, and the master and slave clocks can be aligned.

PTP is a viable and proven technology; however, the network must be PTP-compatible and hardware support is required at the device level, which is typically not the case for the components of power monitoring systems. Additionally, assumptions can be made that impact a PTP network's accuracy including:

- equal transit time to and from a master and slave device

- accuracy and precision of the master and slave device clocks

- fast transmission of the sync messages

A popular communication protocol, used by many power monitoring systems to communicate between the devices and the software management system, is the Modbus protocol. Other protocols are used, such as DNP3, BACnet, and LonWorks. Devices may be connected either serially or over Ethernet, depending on the specific devices and their configurations. An inexpensive, and thus prevalent, method of connecting devices over a network is to communicate via Modbus TCP with one master device (a gateway, or even a meter). This "master" device communicates with multiple slave devices (typically up to 32 devices) on a daisy-chained serial network using RS-485 protocol. Although communicating with the master device may have minimal latencies, the latencies on the RS-485 subnet (i.e., sub-network) can be much higher due to other priorities within each slave device. Using the network to provide precision time synchronization across multiple serially connected devices may not be feasible. The latencies can change from instant to instant and device to device. Additionally, as with a GPS, there is no spatial context developed or exchanged with network time technologies.

CHANGING THE PARADIGM

Facility managers rely on power monitoring systems as the "eyes and ears" that identify, comprehend, and respond to energy reliability and usage issues on their electrical systems. Therefore, power monitoring systems must be able to perform these functions. Most power monitoring systems evaluate data from the vantage of each discrete metering device; thus, data is not synchronously related between distinct devices. This is untenable: there is a great deal to gain by generating contextual data and subsequently assessing it. There are many benefits to establishing the spatial and temporal position of data sources (meters, relays, variable speed drives, etc.); and, in fact, the use of contextual data is foundational to the introduction of artificial intelligence (AI) into power monitoring systems.

Some advantages of providing contextual data are:

- Optimization of the cumulative data generated on the monitoring system: Contextual data inherently facilitates superior analysis techniques that would be difficult or impossible to perform without the system electrical data being interrelated.

- Simplification of configuration of hardware and software, which is greatly needed with today's complex power monitoring systems. For example, alarm thresholds may be specified relative to each metering device's position in the electrical system layout (e.g., at the main switchgear, at the load, etc.), the type of load (e.g., motors, variable speed drives, etc.), or the type of energy consumer (e.g., data center, hospital, etc.).

- Ease of embedding system electrical data analysis tools in the software. This reduces the time and money required to compile and analyze data. It can also reduce the need to employ a consultant.

- Reduction of the interface requirements for facility personnel: Additional functionality can incorporated into the system algorithms.

- Production of more useful information from existing monitoring systems: It is fundamentally more sensitive to recurring irregular electrical conditions. This is because analysis of contextual data synchronously evaluates data from each point in the electrical system with respect to other points, and not asynchronously as typical monitoring systems do.

- Justification of capital expenditures based on quantifiable return on investment (ROI). By analyzing contextual data, it is possible to more accurately and rapidly determine the scope of electrical problem(s), and even to notify the facility manager of anomalous electrical conditions that would have gone unnoticed in an asynchronously monitored electrical system. In effect, this gives the facility manager more accurate information to consider when evaluating and prioritizing system investments.

Today's power monitoring system paradigm encourages designers and system integrators to place high-end meters at the service entrance and to reduce meter functionality closer to the load, by installing less expensive meters. These power monitoring systems only evaluate data on a device-by-device basis and, thus, do not take advantage of interrelated data across multiple devices. The functionality of all metering hardware becomes more critical in power monitoring systems that leverage contextual data (explained in more detail when applications are discussed).

Ultimately, the goal of providing context to data is to produce knowledge and intelligence regarding the electrical system being monitored (see Figure 30-2). It is difficult to make important decisions based on reams of data; however, reducing the data to real knowledge and intelligence unlocks opportunities and opens possibilities that may never have been considered. Furthermore, the ability to produce contextual data from metering devices can easily justify the expenses and resources spent to purchase, install, and operate the power monitoring system.

INNOVATIVE TECHNOLOGY

Power monitoring system data often contains subtle clues to its origin that can be overlooked by the casual observer. With this in mind, it is possible to derive both the spatial and temporal position of data and data sources (meters, relays, variable speed drives, etc.) from

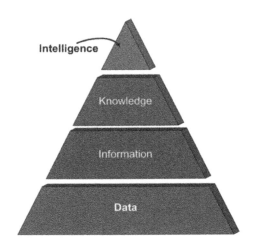

Figure 30-2: Changing Low Value into High Value Intelligence.

the data by evaluating an additional parameter with each data point: time. By aligning "when" data is collected from each discrete data source, with respect to other discrete data sources, the data can be placed into context across the power monitoring system. Contextual data can be leveraged to provide almost unlimited applications.

METHOD TO PROVIDE
DATA WITH SPATIAL CONTEXT

This feature uses a special learned hierarchy algorithm in the system's software that is based on rules and statistical methods. Periodically, the system's software polls each metering point on the system to determine certain characteristics of the electrical system at each device. After the information is collected from each capable metering point on the system, the hierarchy algorithm analyzes the data and traces the relationships or links between the meters. This analysis may be performed periodically to increase the probability that the hierarchy is correct, or to ascertain any changes in the electrical system hierarchy. Once this iterative process reaches a predetermined level of statistical confidence that the determined layout of the electrical system is correct, the hierarchy algorithm ends. The final layout of the electrical system is then presented to the facility management staff for concurrence. Each device's data is evaluated by the hierarchy algorithm during the learning period with respect to all other devices. Then, the basic layout of the electrical system's structure is determined, based on the devices available. Obviously, a more detailed hierarchical structure can be determined when more metering

devices are available for analysis.

The primary benefit of this feature is that it automatically provides a basic hierarchical structure of the power monitoring system with little or no input by the designers, system integrators, or facility staff. The system hierarchy is then used as a tool for other system evaluations.

Samples of specific electrical parameters (power, voltage, current, etc.) are simultaneously taken from each device on the monitoring system. These data are stored and analyzed with respect to:

- the time the sample is taken
- the associated value of the data point
- the type of device providing the data (e.g., meter, relay, variable speed drive)

Data synchronously or pseudo-synchronously taken from each device on the electrical system is compared determine the correlation between the devices. The data is analyzed for statistical trends and correlations, similarities, and differences over a given period of time.

Rules for Auto-Learned Hierarchy in an Electrical Power System

It is necessary to make certain assumptions about the electrical system in order to learn the electrical system's hierarchy. Basic assumptions made by this feature are founded on Ohm's Law, conservation of energy, and working experience with typical electrical power systems.

Universal assumptions made by the hierarchy algorithm include:

1. The electrical system being analyzed is in single or multiple radial feed configurations.

2. The device measuring the highest energy usage (or instantaneous power) is always assumed to be at the top of a hierarchical structure (e.g., Main1).

3. Multiple mains (e.g., Main1, Main2, Main3) may exist in the system.

4. The software housing the hierarchy algorithm collects data through a communications medium from each capable device on the system.

5. The rate of sampling data by the devices is at least greater than the shortest duty cycle of any load (statistically, this becomes less of an issue as more data is collected).

6. Energy is consumed, not generated, on the electrical system during the data collection process.

7. The error due to time offset across the devices on the electrical system is negligible when data is collected by the software.

8. Data is not collected for hierarchical purposes from two devices installed at the same point of an electrical system.

9. Devices with no energy consumption (or energy consumption that changes) are ignored or are only used for voltage information to determine their potential location in the hierarchy.

10. Transformer losses on the electrical system are negligible with respect to the loads downstream from the transformer.

Supplemental considerations that are made by the hierarchy algorithm, to determine the location within the hierarchy of a given device with respect to the other devices, include:

1. Loads that start or stop affect the load profiles for corresponding upstream metered data with a direct or indirect link to that load.

2. Voltage information (fundamental, harmonic, symmetrical components) is relatively consistent for all devices on the same bus.

3. General correlation (over time) of loads between devices indicates either a direct or indirect link.

4. Multiple unmetered loads at the same point in an electrical system are aggregated into a single unknown load (e.g., multiple unmetered feeders are combined into a single unmetered feeder).

Using Data to Create a Power Monitoring System Hierarchy

Electrical data is taken from each device (M_1, M_2, ..., M_k) and compiled into a *data table*, which is essential to determining the hierarchical structure of the electrical system. The *data table* consists of the raw data (e.g., power, voltage magnitude, voltage distortion, current, symmetrical data) taken synchronously (or pseudo-synchronously) at regular intervals (T_1, T_2, ..., T_n) over a given time period. The time period between samples depends on the shortest duty cycle of any load on the system (see assumption 5 above). The maximum time period for collecting data ($T_n - T_1$) is based on the level of variation of each meter's load in the given electrical system. The device with the maximum power in the *data table* is assumed to be a Main (highest hierarchical level in the power monitoring system hierarchy); however, this algorithm considers the possibility of multiple hierarchies (multiple Mains).

Once the data table is developed, a *check matrix* is then developed. The check matrix is a matrix of logical contentions of the *data table* based on conservation of energy principles, and is by nature a symmetric matrix ($A = A^T$) with diagonal values equal to zero. A zero (0) indicates that there can be no direct link between a specific pair of devices (due to a violation of the conservation of energy), and a one (1) indicates that there is no conclusive information regarding the relationship between any specific pair of devices. An exemplary *check matrix* is shown in Table 30-3. In this example, no link exists between Meter 1 and Meter 2. This is because the power measured by Meter 1 exceeds Meter 2 in one entry of the *data table* and the power measured by Meter 2 exceeds Meter 1 in another entry of the data table. Meters always perfectly correlate with themselves, so an NA is placed in that cell of the *check matrix*.

Once the *check matrix* is determined, the data from each meter in the *data table* is used to develop a *correlation matrix*, similar to the example below. In this matrix, a statistical evaluation is performed to determine the linear relationship of each meter in the electrical system with respect to the others. The correlation coefficient between any two meters is determined and placed in the appropriate cell. In the example below, M_{12} is the

Table 30-2: Data Table Example

Time	Meter 1	Meter 2	Meter 3	Meter 4	.....	Meter k
T_1	D_{11}	D_{21}	D_{31}	D_{41}	.....	D_{k1}
T_2	D_{12}	D_{22}	D_{32}	D_{42}	.....	D_{k2}
T_3	D_{13}	D_{23}	D_{33}	D_{43}	.....	D_{k3}
T_4	D_{14}	D_{24}	D_{34}	D_{44}	.....	D_{k4}
$\vdots$	$\vdots$	$\vdots$	$\vdots$	$\vdots$	$\vdots$	$\vdots$
T_n	D_{1n}	D_{2n}	D_{3n}	D_{4n}	.....	D_{kn}

Table 30-3: Check Matrix Example

	Meter 1	Meter 2	Meter 3	Meter 4	.....	Meter k
Meter 1	NA	0	1	1	.....	0
Meter 2	0	NA	1	0	.....	1
Meter 3	1	1	NA	0	.....	1
Meter 4	1	0	0	NA	.....	0
:	:	:	:	:	.....	:
Meter k	0	1	1	0	.....	NA

correlation coefficient of Meter 1 with respect to Meter 2. The higher this number is, the higher the probability that these two meters are either directly or indirectly linked. Conversely, the lower this number is, the lower the probability that these two meters are directly or indirectly linked. The equation used to determine the correlation coefficient between meters is:

$$\rho_{x,y} = \frac{Cov(x,y)}{\sigma_x \sigma_y}$$

where:

$\rho_{x,y}$ is the correlation coefficient and lies in the range of $-1 \le \rho_{x,y} \le 1$,

$Cov(x,y)$ is the covariance of x and y,

σ_x and σ_y are the standard deviations of x and y respectively.

$$Cov(x,y) = \frac{1}{n} \sum_{j=1}^{n} (x_j - \mu_x)(y_j - \mu_y)$$

Where:

n is the number of data elements in x and y, and

μ_x and μ_y are the mean values of x and y respectively.

Again, the *correlation matrix* is by nature a symmetric matrix ($A = A^T$); only half of this matrix is required, due to the redundancy of data ($C_{12} = C_{21}$). Furthermore, the diagonal values are always equal to one (1), since each meter has 100% correlation with itself.

Method to Provide Data with Temporal Context

All real-world electrical signals on power systems experience subtle changes in their frequency over time. The modulation of the signal's frequency is both indeterminate and unique with respect to time. Each device located on the same utility grid will see a nearly simultaneous modulation in the frequency of the electrical signal. Devices that are directly linked to each other spatially in their hierarchy (or even within some proximity) will also see a correlation in their amplitude modulation. The frequency modulations of the signal can then be used to precisely align the data from one device with the data from another device (or all the devices to each other). Figure 30-3 illustrates the frequency and amplitude modulation of a signal.

The data alignment function described below allows all devices on the monitoring system to be aligned with the same zero-crossing of all three phase voltages without the use of additional hardware as described above. Potential phase shifts between discrete devices can also be anticipated through data models. Once the devices are aligned with each other, system data is essentially aligned with respect to the time the data points occurred, allowing more complex analyses to occur without the added expense of supplemental hardware.

Aligning the Device Data

Aligning the data to the same voltage zero-crossing is the principal element of the discussion in this section (although better alignment is feasible). This feature

Table 30-4: Correlation Matrix Example

	Meter 1	Meter 2	Meter 3	Meter 4	.....	Meter k
Meter 1	1	C_{12}	C_{13}	C_{14}	.....	C_{1k}
Meter 2	C_{21}	1	C_{23}	C_{24}	.....	C_{2k}
Meter 3	C_{31}	C_{32}	1	C_{34}	.....	C_{3k}
Meter 4	C_{41}	C_{42}	C_{43}	1	.....	C_{4k}
:	:	:	:	:	1	:
Meter k	C_{k1}	C_{k2}	C_{k3}	C_{k4}	.....	1

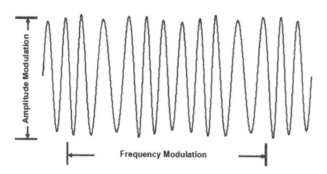

Figure 30-3. Amplitude and Frequency Modulation of a Sine Wave

is facilitated by functionality in both the device and the system management software, and the requirements of each are discussed individually below.

Devices collect and partially analyze frequency data. From the time the device is energized (or reset), a cycle count is performed of the measured voltage signals. In this case, the cycle count is sequentially iterated with each positive voltage zero-crossing. As the device measures the cycle-by-cycle frequency variations of the voltage, a comparison is performed to their respective nominal values. The device firmware tracks frequency variations and the associated cycle count. Note that the associated device clock time at any specified cycle count may also be stored.

The software initiates temporal alignment of the data by sending a global command (broadcast) to all devices on the system to begin storing a predetermined number of consecutive cycle-by-cycle frequency measurements. The cycle count of the first measured cycle is also recorded in each discrete device and is appended to the cycle-by-cycle frequency data. It is also possible to sequentially issue a command to each device (non-broadcast) as long as there is sufficient consecutive cycle-by-cycle data to be logged. The predetermined number of data points to be captured at each device can be established, based on the number of devices on the monitoring system. After the data is collected by each device, the software uploads the data for analysis. The algorithm fully anticipates that there will be a time offset (or more specifically, a cycle count offset) between each device's stored data log (especially for non-broadcast initiations of the algorithms in the devices). This is because the devices on the system may not receive the signal to begin capturing and storing the data simultaneously. The data logs are then analyzed by the alignment algorithm in the software to locate a direct correlation in frequency between all the devices, using the cycle count as a marker for each device's respective data points.

The cycle-by-cycle frequency data from one device is essentially slid across the cycle-by-cycle frequency data from a second device until the frequency variations from the two data sets align with each other as shown in Figure 30-4. A simple analogy is found in turning a puzzle piece until it snaps into place on the puzzle. The cycle count offset between a device pair (M_i, M_j) can be determined directly by the following equations:

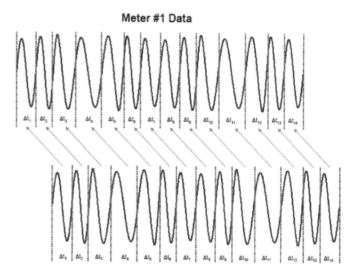

Figure 30-4. Alignment of Data Between Two Devices.

$$M_{ij} = M_i - M_j$$
$$M_{ji} = M_j - M_i$$
$$M_{ij} = -M_{ji}$$

where M_i is the cycle count of a first device and M_j is the cycle count of a second device (taken synchronously). This process is repeated for each device pair on the system until all device pair offsets are determined. The software alignment process then builds a *cycle count offset matrix* of each device's cycle count with respect to every other device's cycle count. Table 30-5 illustrates the construction of the *cycle count offset matrix*, which is a classic form of skew-symmetric matrix.

For example, if M_1 has a cycle count of 20, and M_2 has a cycle count of 25 at the point of alignment for that specific device pair, the cycle count offset between the device pair, M_{12}, equals –5. Conversely, the device pair, M_{21}, will equal +5. When i = j, the cycle count offset is always zero, because the equation is calculating the offset of a device with itself. Hence, the diagonal of cycle count offset matrices is always equal to zero.

Because the cycle counter in both devices iterates sequentially, the cycle count offset between every device pair remains fixed, as does the cycle count offset

Table 30-5. Cycle Count Offset Matrix Construct

	M_1	M_2	M_3	M_4	.....	M_j
M_1	0	M_{12}	M_{13}	M_{14}	(-).....	M_{1j}
M_2	M_{21}	0	M_{23}	M_{24}	(-).....	M_{2j}
M_3	M_{31}	M_{32}	0	M_{34}	(-).....	M_{3j}
M_4	M_{41}	M_{42}	M_{43}	0	(-).....	M_{4j}
:	:	:	:	:	0	:
M_i	M_{i1}	M_{i2}	M_{i3}	M_{i4}	.....	0

matrix; hence, there is no need to align it again unless a device loses the voltage signal or resets. Even in those cases, only the device that loses synchronization must be realigned. Furthermore, it is possible to recover its original alignment through methods not discussed in this chapter. The *cycle count offset matrix* is the solution for providing temporal context to data and deciphering events across multiple devices. When two devices experience an event, the software need only compare the cycle counts between the two devices at the time that the event took place. Using the example above, an event that occurs 5 cycle counts later on M_2 than on M_1 would indicate that the devices experienced the same event.

Figure 30-5 illustrates a small frequency event (Event 'X') captured by six metering devices running the alignment algorithm, possibly caused by a large load starting or a fault occurring on the electrical system. The six devices were installed in two different facilities approximately 500 meters apart from each other. In this case, the cycle count offset from each of the devices is readily apparent. Meter 1 captured the event on its cycle count 13, Meter 2 captured the same event on its cycle count 52, Meter 3 captured the same event on its cycle count 93, and so forth, as shown in the graph. The *cycle count offset matrix* for this metering system is shown in Table 30-6. The reason for the latencies in the cycle counts between each metering device is that the meters received the command to begin logging frequency data at different times (First Meter 1, then Meter 2, then Me-

ter 3, and so forth).

It is interesting to note that there can only be one solution to the cycle count offset matrix (similar to a Sudoku puzzle); and an erroneous solution (even in one cell) is easy to identify, because of the built-in redundancy of the approach. Let's suppose the alignment algorithm determines that the cycle count offset between M_{32} is 40 cycles (not 41 cycles as shown in the example above). It is possible to easily validate this result by using the relationship between M_{31} and M_{21}. Since M_{32} is also equivalent to $M_{31} - M_{21}$, the calculated cycle count offset between M_3 and M_2 is 41 cycles. This can be extended to additional devices:

$$M_{32} = M_{36} - M_{56} - M_{45} - M_{14} - M_{21}$$
$$M_{32} = (-108) - (-36) - (-37) - (-115)$$
$$- (39) = 41 \text{ cycles}$$

As shown, if even one relationship (cell) is incorrect, the error can be identified. More interestingly, it is possible to correct errors (within reason) without obtaining new frequency data logs from the devices.

Synchronizing the Time

After the data are aligned, the time between each device and the system management software can be synchronized. Because human beings prefer to relate temporal events to mean solar time (days, hours, minutes), and not grid frequency (cycle counts), it is easy to use

Table 30-6. Cycle Count Offset Matrix for Frequency Data Shown in Figure 30-6.

	M_1	M_2	M_3	M_4	M_5	M_6
M_1	0	-39	-80	-115	-152	-188
M_2	39	0	-41	-76	-113	-149
M_3	80	41	0	-35	-72	-108
M_4	115	76	35	0	-37	-73
M_5	152	113	72	37	0	-36
M_6	188	149	108	73	36	0

this technology to synchronize on-board device clocks and servers to each other. The master clock (typically, a device or server) can select a specific cycle count in the future to be a specific time, and command all device clocks to reset their time at the chosen cycle count. Because device clocks tend to drift, time resets may be performed as frequently as needed.

The software can read the cycle count in each device and the associated device's on-board clock time (see discussion in pervious section). A device's on-board clock time and cycle count may drift with respect to each other due to the limitation of the on-board clock. Once the data is aligned, the cycle count becomes the absolute reference for a device. Due to the clock drift, it may be necessary to re-read the time associated with a device's cycle count periodically to reestablish the device's time. The software can then update the matrix containing the device time information.

Another capability of this feature is to allow all on-board device clocks to be periodically reset to the same value, to provide a standard time for the entire electric monitoring system. Typically, the time at the system management software server could be set according to some absolute time reference. Once the server time is set, the system management software can then reset the time on all the devices accordingly. In this case, the data and time of each device and the software would be more accurately aligned with the absolute time reference.

Aligning the Data and Time on
Different Points of the Grid

When system devices are located on two different electrical grids, it is still possible to align all the devices together through the use of a GPS interface. In this case, alignment would first be performed between devices within each geographic location. Each geographic location would have at least one master clock device that is synchronized to the GPS via an IRIG-B signal. The software managing the devices across all grids would associate the master clocks from each electrical grid, and thus, associate the cycle count offset matrices as well. Because the frequency variations on separate electrical grids are distinct from each other, grid association is only valid for short periods of time and must be regularly "re-associated."

APPLICATIONS OF CONTEXTUAL DATA

At a high level, metering devices are employed for three general purposes: 1) to quantify energy consumption, 2) to ascertain electrical reliability, and 3) as a means of control (assuming that input/output functionality is integrated into the meter).

Quantifying energy consumption (1) at specific locations around the electrical system helps the facility manager better understand where the energy is being used (location), how the energy is being used (efficiencies), what is using the energy (load or process), and when the energy is being used (profile). At a minimum, the facility manager should be aware of these factors in order to manage or reduce a facility's overall energy consumption and costs.

Poor electrical reliability (2), due to undesirable intermittent or periodic electrical conditions, can adversely influence an electrical system's behavior, culminating in damaged equipment, inefficient operation, safety issues, and increased operating costs. Various standards (e.g., IEEE 1159-1995 and IEC 61000 standards) provide useful definitions, descriptions, and categorizations around the types of electrical system perturbations. Identifying and understanding the effect of power quality-related issues can reduce capital expenditures, increase equipment life, improve product, reduce downtime, and improve the bottom-line growth.

Input/output (3) functionality allows a metering device to interface with external digital or analog signals. The metering device can be the source of the signal (output) or the receiver of the signal (input). Many WAGES (Water, Air, Gas, Electricity, and Steam) devices provide signals that can be aggregated at the metering device and incorporated into the power monitoring system. Other types of data can be received through a metering device's I/O and acted upon accordingly. These include temperature, vibration, status, time, pressure, utility KYZ pulses, etc. It is also possible for the metering device to provide output signals to other equipment, indicating that some event has occurred, or to control some external function. Although applications using contextual data for control are not discussed in this document, many opportunities are available to leverage contextual data for this purpose.

Although placing data into spatial and temporal context are valuable applications in and of themselves (e.g., automated single-line diagrams and time synchronization, respectively), the number of beneficial applications for contextual data are too numerous to list. Beyond the intrinsic applications provided by spatial and temporal context, these two concepts provide the foundation for many other applications related to energy consumption, reliability, and control. A few potential applications will be introduced below in more detail to illustrate the capabilities of contextual data.

Locating Faults Using Contextual Data

Many meters can detect disturbances that occur on the systems they monitor. Metered data typically used to characterize an event may include:

- time of the event
- pickup of the alarm
- dropout of the alarm
- worst-case value
- phase that the event occurred on
- waveform capture of the event

This information, which is very important in determining the cause, severity, and source of an event, has been provided by many meters for years. While the minority of experienced facility personnel can evaluate the data and determine the source of a problem, the typical facility manager or staff may not have this proficiency.

To determine the root cause of the disturbance, a consultant or experienced engineer might be hired to analyze and interpret data captured by the meter. This could be a slow (and expensive) process for the consultant or engineer to acquire the relevant information and come to the correct solution. It is easier (and less expensive) to troubleshoot a power quality problem when it occurs, rather than days or weeks later.

By evaluating the voltages and currents during an event at each capable device, it is possible to determine if a fault's source is located "upstream" or "downstream" from a device. Aggregating the fault source locations (upstream or downstream) from each metering device through the lens of spatial context quickly provides the fault's location within the power monitoring system (or upstream from the utility's meter). Figure 30-6 illustrates a power monitoring system view of a fault located

in a facility's electrical system.

Temporal context is a very important factor when aggregating system information into a single system event (as opposed to numerous discrete device events). If the devices that captured the fault event are in temporal context with each other, it is easy to determine whether each device experienced the same event. If the discrete devices are located on the same electrical grid, their cycle counters will iterate at the same rate. Because of this, the cycle count offset matrix will be static, and the cycle count offset between any discrete pair of devices will be fixed. By comparing the cycle count offset matrix with the cycle counts captured by the discrete metering devices at the time of the fault, it is possible to determine if the event experienced by any two devices was the same event.

Locating Transient Sources Using Contextual Data

High-speed electrical transient events can originate inside a facility or on the utility's grid, and can propagate through various levels of electrical and data systems. Sources of destructive high-speed transient events can range from the obvious—such as a lightning stroke during a thunderstorm—to the subtle—such as static discharge from a human finger. High-speed transient voltages that exceed insulation ratings can stress electrical insulation, leading to gradual breakdown or abrupt failure of the dielectric. Industrial facilities can experience many transient events every hour, with voltage impulses exceeding 5 to 10 times the nominal system voltage. Reducing the magnitude and duration of transient events can extend the life of equipment insulation.

By definition, high-speed transients are very fast events that contain an abundant content of high frequency components. These high frequency components generally attenuate very quickly, due to the inductive characteristics of the electrical system. The brevity and attenuation of transient events make them difficult to capture and even more difficult to locate under normal circumstance. Interestingly, it is easier to capture transient voltage events than it is current events (particularly when a metering device is directly connected to the circuit). This is because the inductive nature of current transformers filters much of the transient signal before it reaches the meter. There are many devices available on the market that can capture high-speed voltage transient events, but locating the source of the transients can be extremely difficult because the current is not available for analysis.

Using spatial and temporal context, voltage transient data that is captured by multiple metering devices can be analyzed to help pinpoint a transient source's

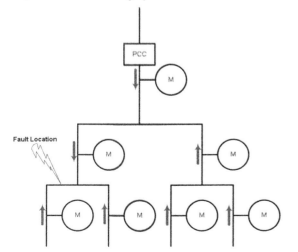

Figure 30-6. Determining the Location of a Fault Using Contextual Data.

location. To illustrate, the single-line diagram in Figure 30-7 shows a power monitoring system model with ten metering devices. A voltage transient source is located upstream from Meter 3 and downstream from a transformer, and is detected by Meters 3, 7, and 8. A transient is the electrical version of a pebble dropped into a pool of water; as the signal moves away from the transient source, it dissipates until it is undetectable (due to the system's natural inductance) by the metering devices. Placing the transient data from each discrete metering device into spatial context allows the facility manager to quickly isolate an "area of interest" for the voltage transient's source. Again, temporal context is needed to accurately aggregate the correct voltage transient data captured by the metering devices for this analysis.

Analyzing Harmonic Distortion Using Contextual Data

It is an established fact that harmonic distortion may result in many potential electrical system issues, including equipment misoperation, degradation, and ultimately, failure. As more and more non-linear loads are connected to the electrical grid, issues associated with harmonic distortion will substantially increase—even in facilities that were previously not susceptible to harmonics.

There are various sources of harmonic distortion, most of which stem from the use of non-linear loads. Harmonic distortion can be caused or exacerbated by a number of sources, including switch mode power sup-

plies, large UPS systems, variable speed drives, high impedance sources, or high impedance electrical wiring.

Because most loads are designed to operate most effectively at or near some designed nominal frequency, it stands to reason that they will not operate as effectively when other frequencies (harmonics and interharmonics) are induced at their terminal. A few problems that can occur as a result of distortion are:

- lack of phase synchronization
- undervoltage circuit activation
- nuisance tripping
- capacitor bank issues
- overheated transformers and motors
- overloaded neutral conductors
- control problems on electronic equipment
- circuit breakers chattering (bouncing)

Employing contextual data, a combination of data from two or more metering devices can alert the facility manager to harmonic distortion concerns that may exist in their electrical power system. Various aspects of an electrical power system's non-fundamental frequency data are evaluated from each device on the monitoring system within the established temporal and spatial context. In this case, temporal context is not as important as spatial context, because harmonic distortion is a steady-state issue. By analyzing aggregated harmonic data from multiple devices, the facility manager obtains higher quality information than available from a single

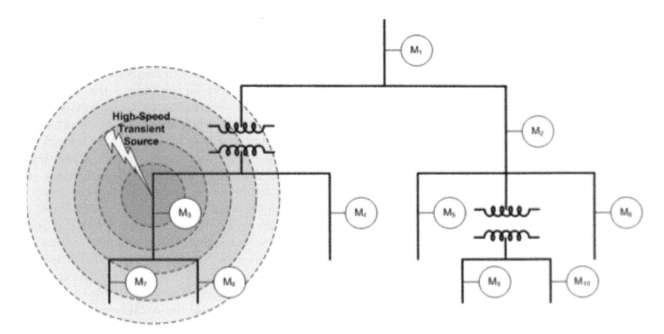

Figure 30-7. Determining the Location of Transient Sources Using Contextual Data.

power monitoring device. Evaluations that may be performed through the contextual data from a multi-meter system include:

- locations of harmonic sources "hot spots" (primary and secondary)

- locations of harmonic sink "hot spots" (where the harmonics flow)

- locations of harmonic current "hot spots" on the neutral conductors

- impact of discrete loads on the entire monitored electrical system, as they relate to harmonic distortion

- impact of processes on the entire monitored electrical system, as they relate to harmonic distortion

- optimal locations for harmonic mitigation devices

- optimal locations for supplemental neutral current conductor runs

- propagation of harmonics through the entire monitored electrical system

- trends in patterns of harmonic component magnitudes throughout the monitored electrical system

- effects of mitigation devices on harmonic component trends throughout the monitored electrical system

- optimal location for harmonic-rated transformers

- location of potentially damaged rectifiers

- locations of adjustable speed drives that inject higher levels of harmonics into the electrical power system

- potential capacitor bank issues

- redundant verification of harmonic power flows to or from the utility

- trending of discrete harmonic components magnitudes and directions from a system perspective

- virtual metering of discrete and total harmonic distortion

Some of the benefits these evaluations can provide for facility managers include:

- simpler analysis of harmonic distortion issues, reducing the complexity of setup and/or required knowledge

- quicker analysis of harmonic distortion issues, resulting in a reduction of the equipment degradation and failures

- easier troubleshooting of harmonic distortion issues, providing quick detection of harmonic "hot spots" that might otherwise not be detected until equipment failed

- more cost effective analysis of harmonics distortion issues, reducing the need for consulting services and associated fees

- decreased downtime due to harmonic distortion related issues

Figure 30-8 illustrates an electrical power system with multiple metering devices installed at various locations. These devices are assumed to be in context, both spatially and pseudo-temporally. They are also assumed to have the capability to measure some level of harmonic characteristics.

After collecting data from each capable metering device, contextual evaluations are automatically performed as they relate to the measured harmonic data. Figure 30-9 illustrates an electrical power system that has information overlaid on it related to harmonic hot spots. This figure illustrates an example of harmonic-producing loads/sources (primary and secondary) and sinks (lowest impedance path for harmonics). Figure 30-10 illustrates the "worst harmonic-producing load." Figure 30-11 illustrates the same electrical power system, with additional information overlaid on it (recommendations for "optimal location of harmonic-related transformers"). Figure 30-12 illustrates the same electrical power system, with additional information overlaid upon it ("optimal location of harmonic mitigation devices" such as line reactors). In each case, the facility manager can

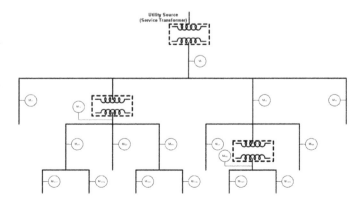

Figure 30-8. Typical Electrical System Single-Line Diagram.

achieve the benefits of contextual data. It provides more intuitive, obvious, and useful information, which will ultimately save time and money.

Data from metering devices are used to determine harmonic distortion issues, and may include voltages, currents, powers, and/or their associated frequency components, or some combination or permutation thereof. It is possible to segment the power monitoring system hierarchy into zones that are determined by the predefined settings or configured by the designer, system integrator, or staff. It is also possible to alarm or notify the facility manager when any predefined threshold has been exceeded. Statistical methods that may be incorporated with this approach to analyze the data include (but are not limited to) correlations, averages, min/max, trending and forecasting, and standard deviations.

Configuring Power Monitoring Systems Using Contextual Data

Power monitoring systems can be an invaluable tool to manage a facility's energy consumption and identify its reliability issues; but they can also be complex and difficult to properly configure. An improperly configured power monitoring system can provide misleading information that wastes time and resources, and even delays corrective actions, resulting in damage equipment and lost production time. While some facilities may be able to configure and manage the daily requirements of their power monitoring systems, most do not. This majority does what they can to get the system working, while relying on factory default settings to save them from disaster.

While minimizing, even eliminating, the facility manager's involvement, it is possible to use each discrete device's location in the monitoring system to efficiently and appropriately configure and commission the power monitoring system. Contextual data also provides the ability to increase the short-term and long-term operational effectiveness of both discrete monitoring devices and the power monitoring system

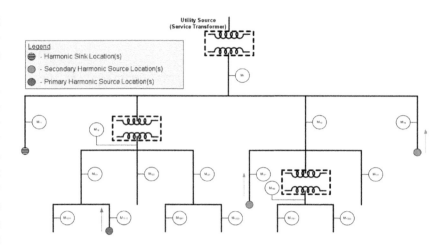

Figure 30-9. Harmonic Distortion "Hot Spots."

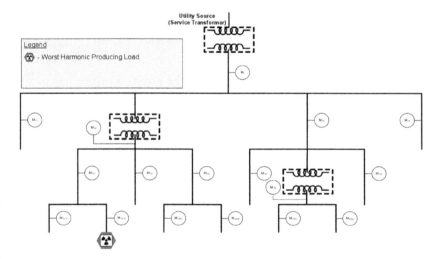

Figure 30-10. Worst Harmonic Producing Load.

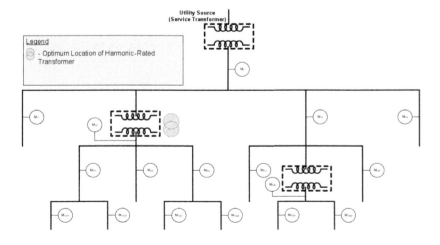

Figure 30-11. Optimum Location for Harmonic Mitigating Transformer

as a whole. For example, device context provides the capability to automatically detect and notify the facility manager of erroneous configurations, potential safety considerations, misapplications of discrete devices (with recommendations), device malfunctions/misoperations, and other issues that might otherwise go unnoticed. Ultimately, contextual data can provide facility managers with a more automated 'plug-and-play' method of configuring and operating their power monitoring system, while saving their time, reducing costs, and improving the validity of information.

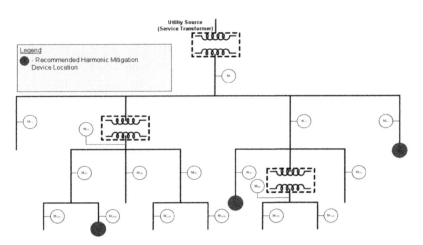

Figure 30-12. Optimum Location for Harmonic Mitigating Devices

Benefits of the Automated Configuration

Improperly configured power monitoring systems can produce errors, resulting in faulty assumptions and costly decisions. However, proper commissioning and maintenance of a power monitoring system is a complex and time-consuming task. Leveraging contextual data can eliminate mistakes that negate the impact a power monitoring system can have on improving a facility's performance. Some of the benefits that contextual data can provide include:

• System installation, configuration, and commissioning:
  — faster commissioning
  — identification of improper discrete device applications
  — proper configuration of discrete devices
  — improved consistency of configuration across devices
  — customized configuration for discrete devices, based on their assigned monitoring task

• System operation:
  — improved confidence in answers and suggestions, due to contextual redundancy and consistency of devices
  — simplified addition, removal, and relocation of devices ('plug/unplug and play')
  — simplified and minimized data generated by the power monitoring system
  — fast notification of malfunctioning, misoperating, or misapplied discrete metering devices

• Safety –
  — faster identification, location, and notification of potential ground fault sources

  — identification of other potential electrical code issues, such as improper neutral-ground bonds
  — quicker identification and notification of nomenclature problems that can lead to electrical hazards and confusion

Installing, Configuring and Maintaining a Power Monitoring System

Installing and configuring a power monitoring system are infrequent events; however, properly re-evaluating and maintaining these systems should be an ongoing activity. Characterizing the electrical system over an extended time period allows a more suitable configuration of the power monitoring system with respect to its loads. It is possible to use contextual data from multiple devices to determine "normal" for a system. This information can then be used to automatically adjust various thresholds and configurations as required. Dramatic changes in the electrical system (or power monitoring system) may be flagged, devices reconfigured accordingly, and the facility manager notified.

Presently, commissioning personnel or third-party system integrators use templates to bring the power monitoring system online. This is often done out of haste to complete the project, to save time, or lack of knowledge. Infrequently, facility staff or system designers evaluate the context of each device with respect to its location in the electrical system, in an attempt properly configure devices. Even with this extra effort, the configuration may not meet the requirements of the system, oversights can occur, or improper devices may be installed in the application. Monitoring devices located on the main switchgear may require that billing features or flicker features be configured, while a monitoring device at a motor load may require that NEMA thresholds or adaptive waveform captures be configured. If EN50160 is required at the main, but the device located at the main is not capable of evaluating data based on EN50160, the power monitoring system will not meet this requirement.

When signals that are present on the electrical system are properly evaluated, a context-based system can determine the effectiveness of its monitoring devices. Because each power monitoring device may be different, contextual data allows the uniqueness of discrete devices at their locations in the system to be considered. For example, if signals are measured at a device with frequency components near the Nyquist frequency (folding frequency), a monitoring device using an anti-aliasing filter or a higher sampling rate might be recommended. Another example is the logging capabilities of discrete devices. A device with on-board logging capabilities could be configured by the software to log data as appropriate. However, to negate this deficiency, discrete devices without on-board logging capabilities could be noted, and the software could self-configure to provide PC-based logging as required.

There are many other aspects of installation, configuration, and commissioning that can be considered when using contextual data. Among them are:

- alarming
- logging
- waveform captures
- I/O, security
- wiring deficiencies and errors
- contextual feature enablement
- safety
- device validation
- firmware validation
- software configuration

System Operation

Electrical systems are rarely static. Without periodic adjustments to accommodate changes, power monitoring systems may not be used to their fullest potential. Furthermore, facility staff may not have the time or knowledge to make adjustments to the monitoring system. Leveraging contextual data in the power monitoring system could either automatically make the appropriate adjustments or notify the facility manager of existing deficiencies.

Permanently installed monitoring systems using contextual data provide the facility staff with many benefits, even after commissioning the monitoring system. By evaluating data from each discrete device in the contextual perspective, confidence in feedback from the monitoring system is greatly increased. This is because data from a single discrete device can be substantiated by one or more additional devices, based on their location.

Modifications to the monitoring system—adding, removing, or moving a discrete device(s)—can easily be determined. Additional monitoring devices can be automatically configured, based on their location with the hierarchy. The removal of a discrete device can be logged and facility staff notified. Moving a discrete device from one point in the hierarchy to another point would be logged, and the meter could be reconfigured contextually. The power monitoring system software can also make recommendations for installing, removing, and moving devices, making the monitoring system more efficient.

Today, facility staff struggle with the amount of data that can be provided by power monitoring systems. If a power monitoring system is evaluated from a contextual perspective, this data can be substantially reduced. Using contextual data, system alarms are easy to incorporate. This can reduce the number of alarms by the same order as the number of meters on the monitoring system. System events that normally would be flagged from multiple devices can be flagged by only the device nearest the occurrence, or by a single alarm of aggregated system data that includes information from multiple devices.

Perhaps the facility manager has certain concerns that outweigh others. For example, energy usage and conservation may be more important than the level of reliability within a specific facility. The facility manager can allow the monitoring system to automatically prioritize feedback of certain electrical information and notify accordingly.

The operational health of the power monitoring system can be evaluated in real time. Discrete devices that are malfunctioning, misoperating, or misapplied can be noted; and the facility manager can be quickly notified. Firmware updates from the manufacturer of each discrete device can also be installed by the software as they become available. This ensures that the latest features and bug corrections are available for each device.

Safety

It is well known that wiring and grounding inadequacies may result in electrical vulnerabilities. Some of these vulnerabilities are:

- fire hazards
- electrocution hazards

- equipment misoperation
- equipment degradation
- equipment failure.

An effective grounding path is defined by the National Electrical Code (NEC) as "an intentionally constructed, permanent, low-impedance electrically conductive path designed and intended to carry current under ground-fault conditions from the point of a ground fault on a wiring system to the electrical supply source and that facilitates the operation of the overcurrent protective device or ground fault detectors on high-impedance grounded systems." An important point of this definition is that the ground path is meant to be used "under ground-fault conditions," not during steady-state operation. A primary reason ground paths are not meant to provide a return path for the current during steady-state conditions is that equipment may become inadvertently energized, posing a safety hazard to facility personnel. Another reason to avoid ground currents is the potential to introduce serious equipment interference problems into the electrical system as a result of steady-state current flows on the grounding system. The NEC allows for "a single neutral-ground bond (connection) at the service entrance or at any separately derived system" to provide a path for ground fault currents to flow.

Individual ground-current measurements from discrete monitoring devices (within a contextual perspective) can help facility managers quickly locate potential issues, minimizing exposure to personnel. Using contextual data, it is possible to determine potential neutral-ground (N-G) bonds that are not allowed, based on electrical code requirements. Figure 30-13 shows how taking measurements, within a spatial context, of N-G ground voltages can indicate the presence of both proper (expected) and improper (unexpected) N-G bonds.

Another safety-related concern, which can lead to confusion when troubleshooting events, is phase nomenclature (labeling of phases) within the power monitoring system. When installing devices on the electrical system, the metering instrumentation (PTs and CTs) may be installed with a positive sequence phase notation (A,B,C) in one discrete device, and installed with a different positive sequence phase rotation (B,C,A) in a second discrete device. Both devices have the same phase rotation, but the phase conductors are labeled differently in the metering devices. This can lead to confusion when determining whether a specific phase is energized, although proper safety guidelines should be followed in either case. Furthermore, troubleshooting issues becomes more confusing when a particular event that appears on one phase in one device appears on a different phase in a second device. Using a subset of the hierarchy learning algorithm allows the power monitoring system to locate and correct phase nomenclature issues. Other safety-related concerns can be identified using contextual data, and the facility manager notified accordingly.

Ancillary Benefits

A unique feature that can be incorporated with contextual data is "distributed features." Features can be automatically enabled or disabled, based on the lo-

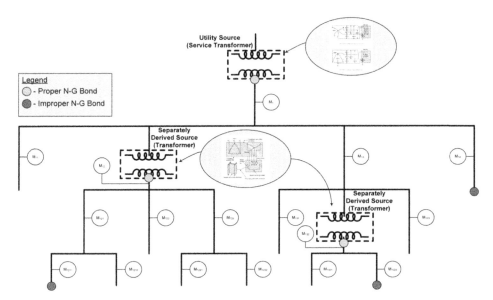

Figure 30-13. Exemplary Electrical System with Proper and Improper Neutral-ground Bonds.

cation of a discrete device within the power monitoring hierarchy. For example, flicker calculations within a monitoring device require an inordinate amount of processing bandwidth, due to the complexity of the flicker algorithms. However, flicker calculations are primarily used at the point of common coupling (PCC) or the main to help quantify the impact of certain load types on the utility system. Flicker-capable meters downstream from the main can have this feature disabled at commissioning to minimize the constraints on the device's processor. Alternatively, asset management algorithms residing in devices would be more beneficial near electrical apparatuses, such as transformers and loads. Ultimately, device context allows each discrete device to be customized for its given application.

CONCLUSIONS

The existing data provided by power monitoring system devices can be cost-effectively converted into data with both spatial and temporal context. Contextual data can then be leveraged to provide useful solutions that improve a facility's efficiency, production, and safety. While some level of expertise will always be necessary to provide more contextual scope to problems and to address them accordingly, the use of contextual data allows a facility manager to put more confidence in the ability of the power monitoring system to identify issues and provide notifications. Ultimately, contextual data is the conduit for bringing the "energy intelligence" of facilities and power monitoring systems into the 21st century.

Bibliography

Jon A. Bickel, Ronald W. Carter, "Automated hierarchy classification in utility monitoring systems," U.S. Patent 7,272,518, September 18, 2007.*

Jon A. Bickel, Ronald W. Carter, Larry E. Curtis, "Automated integration of data in utility monitoring systems," U.S. Patent 7,349,815, March 25, 2008.*

Jon A. Bickel, "Automated system approach to analyzing harmonic distortion in an electric power system," U.S. Patent 7,469,190, December 23, 2008.*

Jon A. Bickel, "Method and apparatus to evaluate transient characteristics in an electrical power system," U.S. Patent 7,526,391, April 28, 2009.*

Jon A. Bickel, Ronald W. Carter, Larry E. Curtis, "Automated configuration of a power monitoring system using hierarchical context," U.S. Patent 7,639,129, December 29, 2009.*

Jon A. Bickel, Ronald W. Carter, Larry E. Curtis, "Automated precision alignment of data in a utility monitoring system," U.S. Patent 7,684,441, March 23, 2010.*

Jon A. Bickel, "Method and system to identify grounding concerns in an electric power system," U.S. Patent 7,684,940, March 23, 2010.*

Jon A. Bickel, Ronald W. Carter, Amjad Hasan, "Virtual Metering," U.S. Patent 7,937,247, May 3, 2011.*

Jon A. Bickel, "Automated data alignment based upon indirect device relationships," U.S. Patent 8,024,390, September 20, 2011.*

National Electrical Code 2011 Handbook, 12th Edition, NFPA, Quincy, MA, 2010.

Precision Time Protocol (PTP) White Paper, EndRun Technologies, Santa Rosa, CA, USA, September 11, 2012.

IEEE Recommended Practices and Requirements for Harmonic Control in Electrical Power Systems, IEEE Standard 519, 1992.

IEEE Recommended Practice for Monitoring Electric Power Quality, IEEE Standard 1159, 1995.

Harvey Wohlwend, Gino Crispieri, Ya-Shian Li, "Factory and Equipment Clock Synchronization and Time-Stamping Guidelines: Version 2.0," Technology Transfer #06094781B-ENG, International SEMATECH Manufacturing Initiative, 2706 Montopolis Drive, Austin, TX 78741, June 30, 2008.

---

*Intellectual property owned by Schneider Electric SA, 35 rue Joseph Monier, 92500 Rueil-Malmaison, France

SECTION IV

Current Technology, Tools, Products, Services, and Applications

More automated fault detection and diagnostic (AFDD) tools have been developed for HVAC systems and equipment than any other specific building systems. A couple key factors likely drive this: 1) HVAC represents nearly 43% of total site energy use and almost 40% of monetary expenditures for energy by commercial buildings in the U.S. and 2) HVAC equipment commonly suffers from a variety of faults that are not repaired, many because they are not recognized as occurring. Application of AFDD can help by increasing awareness of the faults in HVAC system as they occur and by providing information on the impacts faults have on the operating cost of buildings. With the advances in computing, networking technology and communications and drastic decreases in costs, commercial AFDD products and services are now available on the market. The purpose of this section of the book is to provide the reader with information on some of the services and tools currently available.

The chapters in this section are authored by representatives of companies that provide AFDD tools and services for HVAC today, who are leaders in this field. They describe the capabilities of these tools and services, describe examples of their use, and provide information on the enormous value these tools and services based on them bring, implicitly and in some cases explicitly asking the question "Why wouldn't every commercial building owner or operator use this technology given the enormous benefits from using it and the losses from not doing so?"

The editors expect that a wide range of building stakeholders, whether they own, manage, operate, and/or pay the utility bills, will find valuable information that contributes to future decision making. Even researchers and developers in the AFDD field should find value in the information provided by informing them on what is being delivered commercially today and what and how some commercial leaders in the field think.

Chapter 31

Battling Entropy In Commercial Rooftop HVAC Systems with Advanced Fault Detection and Diagnostics

Danny Miller, Transformative Wave Technologies, LLC

INTRODUCTION

We all recognize that there are certain physical laws woven into the universe. Foremost among them is the Law of Gravity. Just as gravity's incessant pull cannot be avoided, so too is the Second Law of Thermodynamics, also known as "entropy." If the concept is unfamiliar, let me spare you the trip to Wikipedia. Entropy is about decay; that it is simply the natural order for things to fall apart. Isaac Asimov (Asimov 1970) explained:

The universe is constantly getting more disorderly! Viewed that way we can see the second law all about us. We have to work hard to straighten a room, but left to itself it becomes a mess again very quickly and very easily. Even if we never enter it, it becomes dusty and musty. How difficult to maintain houses, and machinery, and our own bodies in perfect working order: how easy to let them deteriorate. In fact, all we have to do is nothing, and everything deteriorates, collapses, breaks down, wears out, all by itself - and that is what the second law is all about.

While every orthopedic surgeon specializing in hip replacements probably celebrates this phenomenon, there is an obvious downside for the rest of us. Entropy is a very unpleasant fact of life when it shows up in our daily lives. No area of life is exempt, including my field of commercial heating, ventilation, and air conditioning (HVAC).

I spend a lot of time on rooftops surveying hundreds of HVAC systems every year and entropy beats me there every time. It shows up without fail like some sinister insatiable parasite. It feasts on every aspect of these machines and it all translates into one cumulative outcome: diminished efficiency and performance.

While it occurs quite naturally, I am also witness to the fact that entropy is commonly accelerated by the actions and inaction of humans. Sadly, the very people we trust to maintain and service these systems are often the ones doing the most harm. This is not news to facility operators who pay the utility and HVAC repair bills. These systems are critical to most commercial enterprises and must be kept operational regardless of the cost to do so. Uncomfortable customers, employees, or guests are simply not acceptable regardless of the business.

Facility operators, frustrated with the costs associated with inefficient and inoperable HVAC equipment, have turned to building automation controls as the frontline solution. Unfortunately, after decades of experience, reliance on these systems alone has proven to be inadequate.

My purpose is to provide some evidence that what we are currently doing to confront the impact of entropy on these HVAC systems is not working. I hope to make the case for a more certain solution. One that has not yet been widely embraced but is perhaps our best hope in the battle against entropy: Intelligent Accountability Solutions that act as vigilant watchdogs over these systems.

BACKGROUND

I am the President and Managing Principal of both a 29-year-old commercial HVAC company and a technology company with products that focus on improving the operation and efficiency of commercial HVAC systems. I am a "practitioner" with 37 years of hands-on experience in my field. My partner and I have over 70 employees between these two companies and my personal experience is reinforced by the collective experience of these able and committed people.

We work closely with some of the U.S. Department of Energy (DOE) regional laboratories and other energy efficiency experts that have evaluated and tested our technology. This has allowed me to meet many academics and I have a great deal of respect for them. These are engineers, scientists, utility experts, and consultants who are a lot smarter than I am. My team and I learn a lot from these men and women who dedicate their lives to trying to make things better through their work and research.

I have discovered, however, that academics are at a bit of a disadvantage. They often lack the opportunity for extended exposure to real-life conditions. They find themselves restricted to a world of computer models, test labs, literature, and data. It is difficult to recreate field conditions in a lab, even with the best available resources. This is a challenge that many of them are aware of and seek to overcome with collaborative work in the field. The efforts of these academics often result in reports that advance a theory or share a conclusion based on their research. I am approaching this subject from the perspective of someone who has had a front row seat, observing the field realities of commercial HVAC systems.

A mentor of mine used to say "A man with an experience is never at the mercy of a man with an argument." This doesn't mean that those of us with a lot of practical experience are always right. To the contrary, we all get it wrong at times and misinterpret what we see. Still, I think my mentor had it right. In 1492 the academics of that day argued that the world was flat, Columbus had an experience that said otherwise.

That's the way it works: real life occasionally collides with our theories. I have discovered that it is really tough to predict how commercial buildings and the equipment that serve them will operate and how much energy they will consume. It is the stark realities that we will expose here. The academics and manufacturers are not always the beneficiaries of this kind of information. Hopefully it will shine some light on the cockroaches of inefficiency that continuously undermine our assumptions and challenge our computer models. The truth is, things are not as they should be.

My company, Transformative Wave, provides energy efficiency and accountability solutions for commercial HVAC systems. We manufacture and distribute a product called the CATALYST: Efficiency Enhancing Controller. It is a comprehensive retrofit upgrade for constant volume packaged rooftop HVAC units, commonly referred to as RTUs. The CATALYST upgrade kit includes an advanced intelligent controller, 4-6 new sensors, components to take control of the economizer actuator, a variable frequency drive (VFD), and control of the heat/cool functions. Service technicians of average skill can easily install it.

The packaged RTUs we target with the technology are the most common commercial HVAC systems in the field. The DOE's Commercial Building Energy Consumption Survey (CBECS) database reveals that 69% of the cooled floor space in the United States is served by constant volume RTUs.

A few years ago, the Pacific Northwest National Laboratory (PNNL) approached us about participating in a multi-year analysis of advanced retrofit controls for RTUs. We were flattered that they chose our CATALYST product as the technology they wanted to evaluate. In October, 2013 the results of their research were published in a report entitled "Advanced Rooftop Controls (ARC) Retrofit: Field Test Results" (Wang et al. 2013).

Following is an excerpt from the Executive Summary of that report:

> The multi-year research study was initiated to find solutions to improve packaged heating and cooling equipment operating-efficiency in the field. Pacific Northwest National Laboratory (PNNL), with funding from the U.S. Department of Energy's (DOE's) Building Technologies Office (BTO) and Bonneville Power Administration (BPA) conducted this research, development and demonstration (RD&D) study. Packaged rooftop units (RTUs) are used in 46% (2.1 million) of all commercial buildings, serving over 60% (39 billion square feet) of the commercial building floor space in the U.S. (EIA 2003). The site cooling energy consumption associated with RTUs is about 160 trillion Btus annually. Packaged heat pumps account for an additional 70 trillion Btus annually. The source energy consumption of these units is over 1,000 trillion Btus. Therefore, even a small improvement in part-load operation of these units can lead to significant reductions of energy use and carbon emissions.

A single RTU does not consume an enormous amount of energy, but when you consider the fact that there are millions of these systems on rooftops all across North America, the aggregated energy consumption of these systems is clearly significant.

CURRENT LANDSCAPE

Let's start by saying that these systems are a generally a mess. My staff and I spend a great deal of time on rooftops surveying the existing condition of HVAC units and the systems controlling them. We have touched thousands of RTUs over the last several years across the country, covering a wide array of building types and operations. We do so to document existing conditions and assess the opportunity for energy savings from our technology.

What We Have Learned So Far

#1: Things are worse than we think.

We agree with the statistics that suggest that 65% to 80% of the RTUs are dysfunctional. For this I offer specific evidence from a single client involving ten different retail locations and approximately 60 rooftop HVAC units:

- Economizers operable but harness connectors unplugged on multiple units.

- Economizer dampers seized and inoperable

- Weather station OSA sensor located over the discharge of the condenser section giving automation system false readings.

- Power exhaust sections dismantled with dampers removed allowing unhindered outside air 24/7 creating comfort and humidity issues.

- Dehumidification capability rendered inoperable on multiple units.

- Unoccupied heating and cooling setpoints set the same as occupied period eliminating night setback.

- Control system operating supply fans 24/7 despite store closing at 9 pm

- Poor sensor locations causing excessive heating operation.

- Power exhaust fans operating prematurely with limited economizer damper travel resulting in unnecessary energy use. (Power exhaust fans often consume 75% of the energy used by the main supply fan)

- Dirty evaporator and condenser coils

- Broken belts on supply fan. Motor was still operating but delivering no air.

- Missing or clogged outside air intake screens

- Units operating condenser fans but compressors locked out on multiple units.

- Compressors short cycling on safeties

- Numerous stages of cooling not operating for various reasons

- Runaway operation of inducer draft blowers without a call for heat on multiple units

- Low refrigerant charge due to visible leaks or missing Schrader caps

#2. Much of this is preventable.

There is really no excuse for things to be this bad. The customer referenced above is a well-known national retailer that is committed to sustainability and energy efficiency. They are not reluctant to spend money to properly service and maintain equipment. Like so many other operators, if they knew about these issues, they would address them. There are probably many reasons contributing to this reality. One is that service providers often use their less experienced technicians to service rooftop packaged units and their more experienced technicians are deployed on more complex systems. This is somewhat understandable when you consider how competitive the market is for preventative maintenance contracts and how many HVAC contractors are out there bidding for the same opportunity. The winning contractor was likely awarded the contract based on low price. The contractor must now put pressure on his technicians to minimize the time invested on each unit. Thus, the least costly technicians are deployed and shortcuts ensue.

#3: Reliance on Building Management Control Systems is inadequate.

The majority of the facilities we survey are equipped with a Building Automation System (BAS), also known as an Energy Management System (EMS). However, we find little difference between the facilities with control systems and those running on thermostats. The operator invested in an expensive automation product with the belief that it would eliminate runaway mechanical systems and unnecessary energy use. Yet, the reality on the roof indicates that having a control system is not enough.

#4: Remarkably, even those paying for a facility monitoring service are subject to much of the same.

Most of the issues and deficiencies we find could have and should have been detected by the customer's monitoring service, but were not. A facility manager with one of the nation's largest commercial real estate management firms recently shared an experience with me. He said that he once had of a loss of communication with a critical facility's automation system for over 30 days and was not informed of it by his monitoring service. This is inexplicable and far too common.

#5: The majority of constant volume HVAC equipment deployed in the field lack advanced features even though they are available.

Packaged rooftop units are viewed as a "commodity," purchased in bulk through national account relationships often pitting one supplier against another. There is tremendous competitive pressure on the manufacturers to win big accounts and increase market share. It's all about the numbers and while they would love to sell the more expensive high efficiency models, they know that low price is king. Perhaps 10% of the equipment we see in the field exceeds the minimum code requirements at the time of their installation. Even when advanced features are available from the manufacturers, they are not often selected due to the cost.

#6: Efficiency is regularly subverted in the field.

It is sad to say but technicians are often the enemy of efficiency. It is not that they are intentionally malicious; they just make bad decisions at times. A common flaw is the tendency to err in favor of comfort and make changes in the system configuration without understanding the energy implication of their actions. One of the most common examples is an improper economizer configuration. Technicians will often configure economizer changeover settings to their least efficient position out of a misguided belief that the only way to make sure occupants are comfortable is to use compressors to make cold air.

Even when we do see high-end HVAC equipment with advanced features, this same reality exists. One very well known retailer we work with has invested in a variety of features designed to improve efficiency when they select HVAC equipment. Unfortunately, these capabilities are rarely functioning as intended. I have seen dehumidification capability rendered completely dysfunctional for no apparent reason except perhaps that it was confusing to the service technicians, so they dismantled it completely.

WHAT IS THE SOLUTION?

Utilities and the efficiency industry have a growing interest in finding ways to address the excessive energy use of the RTUs on commercial and industrial facilities. Programs have emerged over the last several years that focus on optimizing these systems using enhanced service procedures and specialized tools. The problem with these programs is that the results tend not to last. The forces of entropy begin to immediately take their toll as soon as the technician climbs off the roof. One Pacific Northwest utility has an RTU Premium Service program that pays contractors to perform advanced service procedures that go beyond traditional preventative maintenance. Oddly, every three years they would pay the contractor to perform the very same procedures again. That is how little confidence the utility community has that these programs will result in permanent change.

While optimization is a critical first step. We believe that technology holds additional answers and promises long-term results.

Beyond optimization, where systems are re-commissioned and restored as much as possible to factory level performance, is the greater challenge of sustaining these results. Here are some of the tools necessary to perpetuate efficiency in these machines. We can characterize these tools as Intuitive Accountability Control Capabilities:

- Automated fault detection
- Remote diagnostics
- Self-correcting controls
- Calibration and testing routines
- Meaningful alarm and alert strategies

When applied, the CATALYST provides a deep reach into the operation of these HVAC systems. We then wirelessly interconnect all of the RTUs and communicate with them for real-time monitoring and control via a web browser over the Internet. We also collect 40 points of data from each RTU and store it in one-minute intervals for analysis. This is far more data than available from the energy management systems typically controlling these simple systems.

We then leverage this data to ensure that the CATALYST is working properly and the energy savings persist. We have embedded each controller with numerous automated fault detection and diagnostic (AFDD) features. The list of controller-level "real-time" faults includes:

- Sensor failures including values out of range and loss of connectivity
- Drive faults involving numerous fault codes generated by the variable frequency drive and communicated via a ModBus IP connection.
- Communication faults
- Inadequate airflow
- Lack of cooling performance
- Lack of heating performance
- Economizer actuator and damper failure
- Fan belt slippage indicating need for replacement or adjustment.
- Proper functionality of the CATALYST and its individual components.

These faults are communicated to the facility operator or the service provider via the Internet in the form of email alarms and alerts. They are also presented visually in the form of a "Health" icon in our web user interface via a browser, tablet, or smart phone. Holding the cursor over the Health icon will enunciate the fault as text. Clicking on the icon will reveal a status list of all potential fault conditions using a green, yellow, red light indication.

In addition to the real-time fault detection and diagnostics from the controller, we have time-based FDD. Many conditions that impact efficiency and performance can only be identified over time using comparative analysis against a baseline. Over the last five years, our cloud-based servers have collected over 6 billion points of data from CATALYST-equipped HVAC systems. This very well may be the world's largest repository of information on how these HVAC systems operate in real life. My team analyzes these histories looking for a different set of problematic issues. These include:

- Improper schedules
- Excessive use of after-hours override functions
- Disproportionate runtime between RTUs
- Changes in the economizer performance
- Degraded cooling output

The single most common failure identified by our AFDD is an economizer damper failure. The New Buildings Institute compiled results from field studies (Cowan 2004) that found economizer problems in 64% of the 500 packaged rooftop HVAC units tested. Other studies claim economizer problems in as many as 80% of RTUs. This is a widespread and serious issue as economizers are the damper arrangement and controls that allow the RTU to utilize outside air for free cooling in commercial buildings. The ability to leverage automated fault detection to identify economizer failures is extremely beneficial as they clearly are not being identified and corrected in the field.

Common sources of economizer dysfunction are sensor failures, controller failures, actuator failures, damper leakage, broken gears or linkage, disabling or improper configuration by field personnel.

The problem is serious enough to warrant the Western HVAC Performance Alliance to assemble a Joint Committee to study the issue and make the case for RTU FDD standards in emerging energy codes such as California Title 24. Utilities in CA have developed financial incentives for products that meet these standards and qualify as Advanced Digital Economizer Controllers (ADEC). The CATALYST is one of three products cur-rently on this list of approved products based on its economizer FDD capabilities.

The second most frequent RTU failure identified by our fault detection is a cooling failure, followed closely by heat failures. These systems primarily exist to heat and cool the buildings they are applied to. The CATALYST fault detection will inform operators whenever the cooling or heating functions have ceased to perform as expected. The CATALYST FDD sequence expects to see acceptable supply temperatures within an allotted time span.

The Appendix following this chapter provides specific details concerning our RTU automated fault detection and diagnostics.

DOES AUTOMATED FAULT DETECTION REALLY MAKE A DIFFERENCE?

We have applied our technology on hundreds of RTUs and have observed the impact that our fault detection and monitoring techniques make. While we cannot easily qualify the financial benefit from AFDD, we have identified thousands of energy and performance issues that would have resulted in discomfort, high repair costs, or excessive energy use. We anticipate longer equipment life and lower overall capital investment as a result of finely tuned fault detection and diagnostic techniques.

OVERCOMING MARKET BARRIERS

It is not surprising that emerging technologies like ours face an uphill battle to gain acceptance in the marketplace. This challenge is not unique to us. Even though our products have been validated by multiple third-party studies and identified as one of the top new energy efficiency technologies, commercial success can still be elusive.

The reality is that while AFDD for RTUs is considered a game-changer, there are obstacles that must be overcome. The challenge is to get the gatekeepers who control the dollars and who oversee these HVAC assets to become receptive to solution providers.

If lasting gains in efficiency via AFDD are to be achieved, we must reexamine the Efficiency Investment Cycle that has become normative in so many organizations. It looks like this:

- Corporate mandates are issued to reduce energy expenditures.

- Efficiency opportunities are identified via audits and proposed.

- Capital budgets are established and projects are authorized

- Efficiency improvements are instituted. (Recommissioning, new controls, equipment replacement, etc)

- Initial positive impact is visible in the energy bills.

- Over time, the effects of entropy eat away at these gains and the investment yields only temporary and partial benefit. The savings erode.

- New mandates are issued and the cycle is repeated.

It occurs to me that one way to break this cycle and gain more support from decision-makers might be to see entropy as an addiction. These machines are "addicted" to inefficiency and bent on returning to an inefficient state regardless of the investment made in them. They are forever demanding more money with the promise of a different future but inevitably return to their old ways.

Those responsible for overseeing facilities with these systems are often "enabling" this cycle of dependency. They are part of the problem and help will only come if they are willing to change their behavior.

I know the first thing a person must do to overcome an addiction is admit they have a problem. I have never been an alcoholic but I have some familiarity with Alcoholics Anonymous (AA). They have established a 12-Step Plan for addiction that has stood the test of time. I don't know that we need twelve steps to get to more permanent efficiency outcomes but I do see some parallels and suggest that we consider five steps, which I have paraphrased, that might serve as a roadmap for getting this situation turned around:

Step One

Admit that we are powerless against the erosive forces of entropy and our HVAC systems are poorly managed despite our best intentions. There is overwhelming evidence that our investments have fallen short. Acknowledge that the reliance on existing building automation systems (BAS) alone is a false hope. Many energy managers and facility operators are heavily invested in the decisions of the past. This makes it difficult to acknowledge the fact that their energy management approach is falling short. Some simply don't know how bad it is and others are in denial.

At Transformative Wave, we regularly approach major facility operators with large portfolios and hundreds or thousands of RTUs. They include: theaters, shopping malls, grocery stores, department stores, fitness centers, manufacturers, office buildings, etc. We introduce our efficiency accountability products and are often rebuffed with the same common response: *"We have a (brand name here) control system and it meets all our needs."* Or, *"We are already doing everything you offer."* Really? If your current systems and strategies are working so well then why do we find so much dysfunction on your roofs? Why do your energy bills fail to show sustained results over the long term? The first step is to "Stop the Denial." What we are doing is not working.

Step Two

Recognize that we need a power greater than our existing building automation systems. Make a decision to trust in something that brings real accountability and the efficiency outcomes we need. The existing control system needs to be supplemented with, or replaced by, an energy accountability technology that is equipped with capabilities that currently are not present. While it is true that a building automation system can be programmed to do almost anything, the fact is they have been applied in a limited manner, offering little more than scheduling and setpoint control. We need specialized technologies that are able to detect, correct, and communicate when energy and performance degradation is occurring.

Step Three

Take an honest and thorough inventory of the true state of our facilities and systems. Unless we stop turning a blind eye and recognize the high cost of our current practices, we won't take concrete action to address them. Face the issue head on by being willing to examine every area of the current operations. There is nothing to be gained by pretending everything is fine or hoping things will get better. As with most detrimental behavior, intervention is almost always required. Be willing to take a fresh look with an open mind and break the cycle of investing in efficiency only to see a return to a lower state, again and again.

Step Four

Envision the cumulative benefit of a more enlightened future. Motivation is critical to any form of recovery. The battle against entropy can be won. Imagining such a future where investments in energy efficiency produce lasting and predictable outcomes will inspire action. Consider the financial capital that will be made available to your organization in years to come if the savings continue to accrue to the bottom line. Consider the human capital that will be preserved if intelligent systems at the machine level are able to anticipate, detect, diagnose, correct, and notify of issues that impact performance and comfort.

Step Five

Be open to outside help. There is no shame in acknowledging that you need help. There is an understandable "vendor fatigue" that plagues the gatekeepers of corporate energy efficiency dollars. Efficiency carpetbaggers and snake oil peddlers abound and have poisoned the well. This leads many decision makers to stiff-arm anyone who approaches them with something new. However, it is detrimental to become resistant to all new products and ideas. It is possible to be both approachable and cautious. There are many third-party organizations that can be trusted to provide unbiased assessments of mature emerging technologies. The DOE regional laboratories and organizations like E Source do great work in this area and have the public interest at the heart of their mission. They can help identify products and vendors who may offer real answers.

CONCLUSION

If we accept as fact that decay is the natural order of everything, we can understand the great challenge faced by those responsible for the efficient performance of all mechanical systems. There is a battle going on against entropy but operators are not without weapons. The day is passed when HVAC systems are out of sight and out of mind, buried in the recesses of mechanical rooms or on difficult to access rooftops. Technology is changing all of that by bringing accountability to every critical function.

AFDD acts as a tireless sentry standing watch over these important assets in the battle against entropy. With it, the invisible becomes visible. Ignorance is revealed and corrective action can then be taken.

Given the widespread presence of Package Rooftop HVAC systems in North America, leveraging AFDD tools to perpetuate efficiency and proper system performance can make a major contribution toward permanent improvement in curbing the excessive energy use of these systems.

References

Asimov, Isaac. 1970. "In the Game of Energy and Thermodynamics You Can't Break Even," *Smithsonian Institute Journal* 1:6.

Cowan, Alan. 2004. *Review of Recent Commercial Roof Top Unit Field Studies in The Pacific Northwest and California.* New Buildings Institute, White Salmon, WA.

Energy Information Administration (EIA). 2003. *Commercial Buildings Energy Consumption Survey 2003.* U.S. Department of Energy, Washington, D.C. Accessed in July 2013 at http://www.eia.doe.gov/emeu/cbecs/contents.html.

Wang, W., S. Katipamula, H. Ngo, R. Underhill, D. Taasevigen and R Lutes. 2013. *Advanced Rooftop Control (ARC) Retrofit: Field-Test Results*, PNNL-22656, Pacific Northwest National Laboratory, Richland, WA.

Appendix 1

Example of a Quick-Serve Restaurant Equipment Survey

Case: A site and equipment survey was performed on a national quick-serve restaurant chain consisting of

- Four restaurants
- Ten total RTUs. Each store had either 2 or 3 RTUs
- Capacity between 5 and 12.5 tons
- Supply fan motors ranged from 1.5 to 5 horsepower

Findings:

Store #1:

- Dead economizer actuator on one unit
- Thermostat setpoints for unoccupied period were 45 degrees cooling and 40 degrees heating
- Thermostat clock had AM and PM reversed
- Extremely loose fan belts on multiple units produced low air flow
- Heat functions had not operated in a considerable amount of time. Heat exchanger had so much dust build up that when gas heat was fired, the system smoke alarm and shutdown was triggered.

Store #2:

- Failed economizer; gears grinding
- RTU serving rear of the store is fighting the RTU serving the front. Front unit is overcooling and driving the other unit into heating.
- Broken fan belt; caused damage to 1/3rd of the evaporator coil surface area

Store #3:

- Broken thermostat
- Two of three new RTUs have non-working economizers
- One of the newer units has never had heat functions enabled

Store #4:

- Main dining room RTU scheduled 24/7
- Second RTU had cooling failure that was identified but not addressed 6 months prior; service technician had permanently jumpered the cooling command in the unit and left the jumper in place. System safeties had locked out the compressors. The unit had been running for months without the ability to provide cooling.

Appendix 2

eIQ Energy Intelligence Platform
Fault Detection Notification System

Transformative Wave's fault detection and diagnostic technology for Rooftop Packaged Units (RTUs) is known as the "eIQ Energy Intelligence Platform." It is both automated and supported by human review. Following is an overview of the wireless communication network that supports the eIQ fault detection process.

RTUs equipped with the CATALYST Efficiency Enhancing Controller are connected via a wireless RTU network. As a retrofit solution, the CATALYST upgrade needs to minimize impact on the client's operations. It is costly to run communication bus wiring in an occupied space or across the roof in conduit. Transformative Wave uses a hub and spoke wireless network with each RTU acting as clients, communicating to a master controller known as the "Access Point" (AP). Additional details are provided in Figure 31-1.

In addition to providing live connectivity for monitoring and control, the eIQ Platform acts as the data collection mechanism for measurement and verification (M&V) as well as the communication vehicle for controller-level fault conditions. Figure 31-2 provides an overview of the eIQ data flow structure.

A screenshot of the eIQ user interface for a single RTU is provided in Figure 31-3.

Figure 31-4 provides another snap shot of the eIQ web interface providing high-level status information for each RTU. Note that if there were a maintenance issue with the RTU alerted by one of the automated FDD algorithms, the "health" icon would indicate a problem and have a notification associated with it providing some insight as to what was the cause.

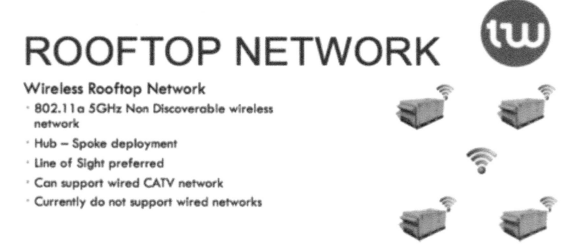

ROOFTOP NETWORK

Wireless Rooftop Network

- 802.11a 5GHz Non Discoverable wireless network
- Hub – Spoke deployment
- Line of Sight preferred
- Can support wired CATV network
- Currently do not support wired networks

Figure 31-1. Rooftop wireless network overview

HOW DOES THE DATA FLOW

About 40 points of information are gathered from each unit

It is transmitted wirelessly back to a central controller

The information goes to a Niagara Supervisor

From the Supervisor, information is sent to a SQL database

The eIQ serves up combined data from the both SQL and Tridium to the end user

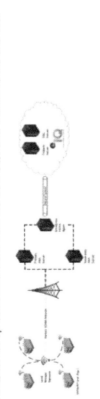

Figure 31-2. The eIQ Platform stores data in one-minute intervals in the cloud

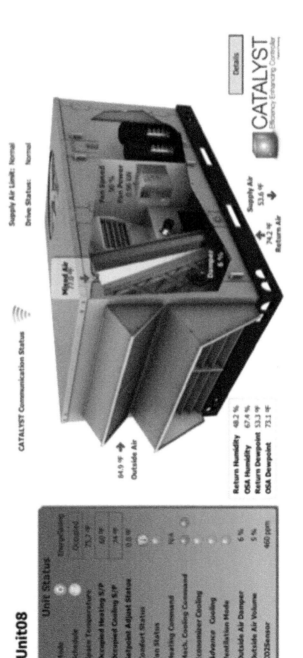

Figure 31-3. eIQ RTU dashboard showing status and basic automated FDD

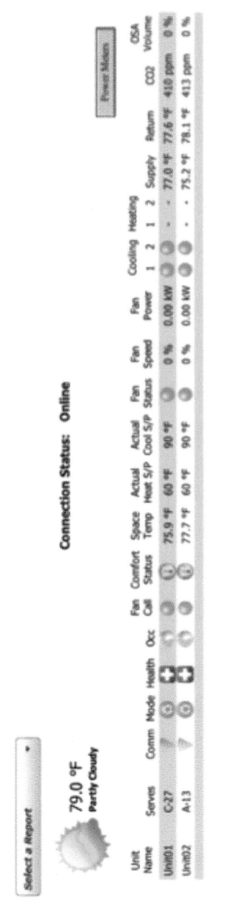

Figure 31-4.

eIQ BMS web interface showing important RTU status and
automasted FDD which is incorporated into the Catalyst defined "health" of the RTU.

Appendix 3

Example of eIQ Fault Detection
Results for 30-day Period

Site	Note
Grocery Store - A	UNIT 1 heating failure. [Tim Snyder - 12/16/2013 10:11 AM]
Grocery Store - B	UNIT 3 fan stopped operating on 12/14. [Tim Snyder - 12/16/2013 9:47 AM]
Manufacturing - B	Unit 7 has a heat fail.HK12/16/13 [Hassan Kahurbet - 12/16/2013 9:44 AM] Needs to be repaired. [Caleb Miller - 1/13/2014 8:00 AM]
Gym - A	UNIT 4 heating and damper failures. [Tim Snyder - 12/16/2013 12:18 PM]
Manufacturing - B	Unit04 economizer damper vanes leaking.HK12/16/13 [Hassan Kahurbet - 12/16/2013 9:47 AM] Needs to be repaired. [Caleb Miller - 1/13/2014 8:01 AM]
Manufacturing - B	Unit 1 sensor mounted on exterior cold wall causing the unit to go on heat all he time. HK12/16/13 [Hassan Kahurbet - 12/16/2013 9:50 AM] Needs to be repaired. [Caleb Miller - 1/13/2014 8:02 AM]
Manufacturing - B	Unit 9 heat fail.HK12/16/13 [Hassan Kahurbet - 12/16/2013 9:53 AM] Cooling only unit. [Caleb Miller - 1/13/2014 8:03 AM]
Manufacturing - B	Unit 3 first stage compressor not working.HK12/16/13 [Hassan Kahurbet - 12/16/2013 10:03 AM] Needs to be repiared. [Caleb Miller - 1/13/2014 8:03 AM]
Manufacturing - B	Unit 9 first stage compressor not working. HK12/16/13 [Hassan Kahurbet - 12/16/2013 10:05 AM] Needs to be repaired. [Caleb Miller - 1/13/2014 8:04 AM]
Drug Store - A	UNIT 6 fan belt alarms. [Tim Snyder - 12/16/2013 12:54 PM]
Manufacturing - A	Unit 12 has 1st and 2nd stage electric Heat tied together. [Daniel Severin - 12/16/2013 7:52 AM]
Retail - A	Changed Unit04 vent speed to 60% 12/6 in order to try and balance Unit01 & Unit04 [Jerry Scott - 12/16/2013 7:48 AM] Changed set point -.5F cooling speeds 90/100% per Ian suggestion [Jerry Scott - 12/16/2013 11:04 PM]
Retail - A	Changed Unit 01 Cool1 speed to 90% in order to try and balance runtime. [Jerry Scott - 12/16/2013 7:49 AM]
Gym - A	UNITS 1, 3 and 5 are getting extended heating runtime. The units are able to effectively heat. [Tim Snyder - 12/19/2013 12:13 PM]
Mall - A	Units 10, 25, and 31 have ECM Motors, no drives [Charles Day - 12/17/2013 8:19 AM]

Grocery Store - B	UNIT 2 has lower setpoints than the other units, causing the unit to go into cooling when other units are heating. Unit is not actually performing a cooling action when in cool call. It's monitor only. [Tim Snyder - 12/18/2013 9:36 AM]
Office - J	Per the model numbers non of the units on this site have heat [Charles Day - 12/17/2013 4:42 AM]
School - A	UNITS 7 and 8 are going into heating with OSA temps of 80+ degrees. [Tim Snyder - 12/18/2013 9:07 AM]
Manufacturing - G	Received a call from JIM the tech. installing the CATALYST on Cedar Graphics, He claimed that unit 13 wasn't responding to the service switch calls. I had him check all the wirings, jumpers, 24volt power and the CT location and installation, all tested OK but unit will not run, I asked Him to call Dermal. HK12/18/13 [Hassan Kahurbet - 12/18/2013 1:11 PM]
Gym - A	No unit location served information has been provided for this site. [Tim Snyder - 12/19/2013 12:07 PM]
Gym - A	All units (except for Unit02) are running almost 24/7 due to optimum start logic. [Jerry Scott - 12/19/2013 2:14 PM]
Drug Store - B	UNIT 1 heat command often doesn't engage until about 30 minutes after heat call begins. [Tim Snyder - 12/19/2013 9:17 AM]
Manufacturing - A	Histories Started 12.18.13, exported to SQL 12.19.13 [Charles Day - 12/19/2013 5:38 AM]
Manufacturing - A	UNIT 12 damper occasionally gets stuck shut, per start-up. [Tim Snyder - 12/19/2013 2:21 PM]
Gym - A	UNIT 2 fan goes to 100% speed for 1 or 2 minutes each time the fan is started from a stop. [Tim Snyder - 12/20/2013 12:08 PM]
Hospital - A	Unit 15 has a bad CTI Board that is not outputting a first or second stage heat call. The unit was tested in the Trane test mode the heat works fine, but the unit does not respond to a W1 or W2 call. The customer has been notified. [Daniel Severin - 12/20/2013 3:47 PM]
Office - D	UNIT 1 is not responding to cool commands. [Tim Snyder - 12/20/2013 3:40 PM]
Office - G	UNIT 20 service switch off since 12/17. [Tim Snyder - 12/20/2013 9:56 AM]
Gym - A	Unit #4 has a damper fail. Connecticut heating and cooling is aware of the issue. CM [Caleb Miller - 12/21/2013 7:02 AM]
Gym - A	Unit #5 has a failed Econ motor. Connecticut heating and cooling is aware of the issue. -CM [Caleb Miller - 12/21/2013 7:03 AM]

Gym - A	Unit #4 has not been fully ducted. It is currently dumping air behind the rock wall. untill the unit is properly ducted the store manager asked me to widen the setpoint range. -CM [Caleb Miller - 12/21/2013 7:05 AM]
Retail - B	Unit #3 has a failed inducer motor on the first heat exchanger. Units 1-6 all have 2 seperate heat exchangers that both come on in low fire for first stage and then high fire for second. [Caleb Miller - 12/21/2013 6:55 AM]
Retail - B	Unit #7's space sensor is in the return duct for the unit. This is where the old t-stat was located as well. The unit serves an upstairs lunch room but the return is not ducted into the lunch room. It is just open to the sales floor, so right now we are not picking up the temp in area served for this unit. This sensor needs to be relocated with new wire ran into the area served. To do this we will need a lift. -CM [Caleb Miller - 12/21/2013 6:59 AM]
Retail - F	No schedule information was provided on the DSI's and the schedules were set one hour before and after the hours of operation. [Dramel Frazier - 12/23/2013 3:18 PM]
Hospital - A	Unit 16 does not appear to be heating 12/23/13 JM [Jerry Miller - 12/23/2013 10:03 AM] Peter has been contacted. [Tim Snyder - 1/7/2014 11:13 AM]
Drug Store - G	UNIT 5 is getting heat failure alarms. Unit is providing adequate heating. [Tim Snyder - 12/23/2013 1:20 PM]
Drug Store - J	UNIT 1 heating failures. [Tim Snyder - 12/23/2013 12:30 PM] Bob has been contacted. [Tim Snyder - 12/27/2013 10:05 AM]
Manufacturing - D	UNIT 19 is getting extended heating runtime. It is able to effectively heat. [Tim Snyder - 12/23/2013 8:33 AM]
Retail - C	UNIT 03 is providing inadequate heating. [Tim Snyder - 12/26/2013 8:50 AM] Mike has been contacted. [Tim Snyder - 1/2/2014 1:24 PM]
Gym - A	UNIT 5 getting fan belt alarms. [Tim Snyder - 12/27/2013 8:31 AM]
Grocery Store - C	Site went offline on 12/26. [Tim Snyder - 12/27/2013 9:35 AM]
Office - D	UNIT 5 heating failures. [Tim Snyder - 12/31/2013 9:47 AM] Janice has been contacted. [Tim Snyder - 1/2/2014 12:15 PM]
Manufacturing - B	Building 2 Unit 2 has 3 compressors and uses 2 compressors in the first stage, but the 2nd compressor has been disabled by the service provider with no notes why. [Daniel Severin - 1/2/2014 8:08 AM] Needs to be repaired. [Caleb Miller - 1/13/2014 8:11 AM]
Manufacturing - B	Building 2 Unit 2 has broken economizer linkage. The actuator opens the outside side air damper, but the connection to the return damper is broken so it never closes. [Daniel Severin - 1/2/2014 8:14 AM] Needs to be repaired. [Caleb Miller - 1/13/2014 8:07 AM]

Manufacturing - B	Building 2 unit 2 has dead power exhaust motor. [Daniel Severin - 1/2/2014 8:15 AM] Needs to be repaired. [Caleb Miller - 1/13/2014 8:12 AM]
Manufacturing - B	Building 2 Unit 3 Has broken powet exhaust wheel housing. [Daniel Severin - 1/2/2014 8:18 AM] Needs to be repaired. [Caleb Miller - 1/13/2014 8:12 AM]
Restaurant - A	UNIT 2 offline since 1/1. [Tim Snyder - 1/2/2014 9:13 AM] Dom has been contacted. [Tim Snyder - 1/13/2014 11:05 AM]
Clubhouse - A	Site offline since 1/1. [Tim Snyder - 1/2/2014 8:56 AM] MacMiller's taking a look. [Tim Snyder - 1/2/2014 12:19 PM]
Retail - D	SQL Exports initiated [Charles Day - 1/3/2014 12:32 PM]
Retail - D	UNIT 2 damper leakage/failure per DSI. [Tim Snyder - 1/3/2014 4:35 PM] [Jerry Scott - 1/15/2014 9:04 AM]
Office - C	Unit #6 has a tripped low voltage transformer due to a shorted connection in one of the commpressor safeties. I rewired the connection and regained communication to the unit. [Caleb Miller - 1/3/2014 8:10 AM]
Gym - A	UNIT 5 getting extended heating runtime. Unit is able to effectively heat (besides damper). [Tim Snyder - 1/3/2014 8:27 AM]
Gym - A	UNIT 2 getting fan belt alarms. [Tim Snyder - 1/3/2014 9:53 AM]
Hospital - A	UNIT 15 getting fan belt alarms [Tim Snyder - 1/3/2014 9:58 AM]
Hospital - A	SQL Histories have been started [Charles Day - 1/3/2014 8:09 AM]
Manufacturing - E	UNIT 12 damper failure. [Tim Snyder - 1/3/2014 8:53 AM]
Manufacturing - E	UNIT 12 compressor failure. [Tim Snyder - 1/3/2014 8:54 AM]
Manufacturing - E	UNIT 46 damper failure. [Tim Snyder - 1/3/2014 9:02 AM]
Manufacturing - E	UNIT 58 heating failure. [Tim Snyder - 1/3/2014 9:10 AM]
Drug Store - D	UNIT 1 damper failure. Unit supplies about 40% OSA when damper is fully open. [Tim Snyder - 1/6/2014 10:43 AM] [Tim Snyder - 1/6/2014 10:44 AM]
Retail - E	UNIT 3 offline since 1/3. [Tim Snyder - 1/6/2014 9:52 AM] Mike has been contacted. [Tim Snyder - 1/7/2014 11:13 AM]
Drug Store - E	UNIT 2 damper failure. Unit supplies about 40% OSA when damper is fully open. [Tim Snyder - 1/6/2014 10:44 AM]
Drug Store - F	UNIT 1 damper failer. Gets about 40% OSA with damper fully open. [Tim Snyder - 1/6/2014 9:14 AM]
Drug Store - F	UNIT 2 damper failer. Gets about 40% OSA with damper fully open. [Tim Snyder - 1/6/2014 9:15 AM]
Drug Store - F	UNIT 3 damper failer. Gets about 40% OSA with damper fully open. [Tim Snyder - 1/6/2014 9:16 AM]
Drug Store - H	UNIT 4 damper failure. [Tim Snyder - 1/6/2014 9:26 AM] Bob has been notified. [Tim Snyder - 1/15/2014 10:57 AM]

Drug Store - C	UNIT 13 damper failure. Unit gets about 40% OSA with damper fully open. [Tim Snyder - 1/6/2014 9:35 AM]
Office - E	All units at site have been having communications issues since 1/5. [Tim Snyder - 1/6/2014 10:04 AM]
Grocery Store - D	UNIT 4 heating failure. Ryan has been notified. [Tim Snyder - 1/6/2014 10:35 AM]
Drug Store - C	UNIT 14 damper failure. Unit gets about 40% OSA with damper fully open. [Tim Snyder - 1/6/2014 9:37 AM]
Drug Store - K	UNIT 2 damper failure. Damper appears to close fine, unknown how it performs when open. [Tim Snyder - 1/6/2014 9:43 AM]
Manufacturing - C	UNIT 4 fan belt faults. [Tim Snyder - 1/6/2014 9:54 AM] Marco has been contacted. [Tim Snyder - 1/7/2014 11:55 AM]
Retail - I	unit 4 service the classrooms went offline on 1/1/14HK1/6/14 [Hassan Kahurbet - 1/6/2014 7:16 PM]
Drug Store - N	UNIT 5 damper failure. [Tim Snyder - 1/6/2014 9:50 AM] Bob has been notified. [Tim Snyder - 1/15/2014 10:58 AM]
Manufacturing - F	UNIT 9 may get extended heating runtime. It is able to effectively heat. [Tim Snyder - 1/6/2014 10:29 AM]
Office - A	UNIT 3 heating failure. [Tim Snyder - 1/6/2014 10:27 AM]
Office - B	UNIT 1 is getting extended heating runtime. Unit is able to effectively heat. [Tim Snyder - 1/7/2014 8:28 AM]
Manufacturing - B	VAVs 201, 202 & 203 are low on airflow due to a failed fire damper in the inlet supply duct. [Ken Hellewell - 1/7/2014 11:44 AM]
Manufacturing - B	Building 2 Unit #4 had a heating failure due to the ingnition module being switched off and the gas valve shut. I turned them both on and regained proper heating functionality. Still unaware of why they were shut off in the first place. [Caleb Miller - 1/7/2014 11:47 AM]
Manufacturing - B	Building 2 Unit #5 has a heating failure due to what apears to be a failing second stage time delay which enables the unit to kick into high fire. There is a low fire for first stage, but it is not strong enough to properly raise the supply air temp. This needs to be repaired. [Caleb Miller - 1/8/2014 8:25 AM] [Caleb Miller - 1/13/2014 8:17 AM]
Manufacturing - B	Building 2 Unit #6 has a heating failure due to what apears to be a failing second stage time delay which enables the unit to kick into high fire. There is a low fire for first stage, but it is not strong enough to properly raise the supply air temp. This needs to be repaired. [Caleb Miller - 1/8/2014 8:26 AM] [Caleb Miller - 1/13/2014 8:17 AM]

Manufacturing - B	Building 2 Unit #7 has a heating failure due to what apears to be a failing second stage time delay which enables the unit to kick into high fire. There is a low fire for first stage, but it is not strong enough to properly raise the supply air temp. This needs to be repaired. [Caleb Miller - 1/8/2014 8:27 AM] [Caleb Miller - 1/13/2014 8:18 AM]
Gym - A	UNIT 2 2nd stage heating failure. [Tim Snyder - 1/7/2014 9:13 AM]
Manufacturing - C	UNIT 4 stage 2 heating is failing. [Tim Snyder - 1/7/2014 8:50 AM] Marco has been contacted. [Tim Snyder - 1/7/2014 11:54 AM]
Office - F	UNIT 2 is getting extended heating runtime. It is able to effectively heat. [Tim Snyder - 1/7/2014 9:23 AM]
Restaurant - A	UNIT 1 getting fan belt alarms. [Tim Snyder - 1/7/2014 9:57 AM]
Mall - B	UNITS 16, 24, 27 and 30 are offline at this time. [Tim Snyder - 1/7/2014 11:07 AM]
Office - G	RTU N3 heat failure since 1/2. [Tim Snyder - 1/7/2014 9:37 AM] Mickey and Marv have been contacted. [Tim Snyder - 1/7/2014 1:41 PM]
Office - I	The site stopped exporting data 1.6.14 @ 1:00 PM. JACE indicates low RAM disk space [Charles Day - 1/7/2014 10:25 AM]
Manufacturing - H	UNIT 4 heating failure. [Tim Snyder - 1/7/2014 8:58 AM] Tom has been contacted. [Tim Snyder - 1/7/2014 1:05 PM]
Manufacturing - B	Unit 7 Bldg 2 Has extremely dirty and corroded damper. Actuator moves damper to about 35% open then gets stuck. I took the shaft off the the actuator and tried to move it and it doesn't budge. Needs to be taken apart cleaned and greased. [Daniel Severin - 1/8/2014 8:16 AM] Needs to be repaired. [Caleb Miller - 1/13/2014 8:10 AM]
Manufacturing - D	UNIT 12 heating failures. [Tim Snyder - 1/8/2014 9:37 AM] Tom has been contacted. [Tim Snyder - 1/9/2014 11:17 AM]
School - B	UNIT 6 getting fan belt alarms. [Tim Snyder - 1/8/2014 11:38 AM]
Manufacturing - D	UNITS 16 and 24 OSA sensor alarms tripping when readings seem to be accurate. [Tim Snyder - 1/8/2014 10:01 AM]
Gym - A	Advance Cool has been disabled due to setpoint requirements by the customer. May be enabled in Spring. [Ken Hellewell - 1/8/2014 10:18 AM]
Library - A	Unit 1 first and second stage compressors are tied together at thermostat and in the unit. ABM is replacing all the thermostats on 1/13/2014 and will seperate at the stat and in the unit. [Daniel Severin - 1/8/2014 8:20 AM]
Library - A	Unit 1 Return damper does not close all the way due to getting hung up on unit side duct connection. [Daniel Severin - 1/8/2014 8:24 AM]
Library - A	Unit 2 has some return leakage when 100% economizing. [Daniel Severin - 1/8/2014 8:26 AM]
Retail - A	UNIT 4 controller down since 1/7. [Tim Snyder - 1/8/2014 10:10 AM]
Retail - E	UNIT 5 heating failure. [Tim Snyder - 1/9/2014 9:47 AM]

Drug Store - L	GP-1 gets about 60% OSA with open damper. [Tim Snyder - 1/9/2014 8:58 AM]
Office - E	UNIT 13 heating failure. [Tim Snyder - 1/9/2014 9:22 AM]
Office - E	UNIT 15 heating failure. [Tim Snyder - 1/9/2014 9:23 AM] Janice has been contacted [Tim Snyder - 1/9/2014 11:38 AM]
Drug Store - M	GP-2 gets about 45% OSA with open damper. [Tim Snyder - 1/9/2014 8:59 AM]
Mall - C	UNIT 9 getting fan belt faults. [Tim Snyder - 1/9/2014 9:28 AM]
Restaurant - A	UNIT 1 offline since 1/9. [Tim Snyder - 1/10/2014 9:09 AM]
Hospital - A	Alternating standard/CAT trending set for 1/11-1/24 per SnoPud grant requirements. [Jerry Scott - 1/10/2014 4:08 PM]
Clubhouse - A	Site went offline Saturday 1/11/13 [Charles Day - 1/13/2014 5:34 AM]
Retail - G	Alternating standard/CATALYST trending set up for SnoPud 1/13 to 1/26 [Jerry Scott - 1/13/2014 1:40 PM]
Manufacturing - B	Building 2 Unit #4 needs a new adjustable sheave to lower the fan amps to keep the drive from over amping. Needs to be repaired. [Caleb Miller - 1/13/2014 1:30 PM]
Manufacturing - B	Building 2 Unit #5 needs a new adjustable sheave to lower the fan amps to keep the drive from over amping. Needs to be repaired. [Caleb Miller - 1/13/2014 1:31 PM]
Manufacturing - B	Unit 7 Bldg 2 Has no power to power exhaust. Needs to be looked at [Daniel Severin - 1/13/2014 7:43 AM]
Manufacturing - B	Building 2 unit #6 apears to have a bad actuator. The damper gets stuck open. Needs to be repaired. [Caleb Miller - 1/13/2014 8:14 AM]
Manufacturing - B	Building 2 Unit #7 has a the second stage commpressor disconnected. [Caleb Miller - 1/13/2014 8:15 AM] Needs to be repaired. [Caleb Miller - 1/13/2014 8:16 AM]
Manufacturing - G	UNIT 14 getting fan belt faults. [Tim Snyder - 1/13/2014 10:09 AM]
Retail - B	Alternating standard/CATALYST trending set up for SnoPud 1/13 to 1/26 [Jerry Scott - 1/13/2014 1:39 PM]
Prison - A	RTU-7 damper failure. [Tim Snyder - 1/13/2014 9:13 AM]
Mall - C	UNIT 9 heating failure. [Tim Snyder - 1/13/2014 9:24 AM]
Retail - H	SQL Exports set up. Histories loaded back to the first full day, 1/11/14. [Charles Day - 1/13/2014 2:14 PM]
Fire Station - A	SQL Exports set up. Histories loaded back to the first full day, 1/11/14. [Charles Day - 1/13/2014 2:14 PM]
Retail - J	SQL Exports set up. Histories loaded back to the first full day, 1/11/14. [Charles Day - 1/13/2014 2:14 PM]
Office - A	UNIT 3 will very rarely get CO2 readings of 0 ppm. [Tim Snyder - 1/13/2014 10:13 AM]
Office - H	Site/Unit offline since 1/6. [Tim Snyder - 1/14/2014 10:06 AM]

Retail - G	Unit02 appears to have damper stuck open. Also OSA sensor is being influence by either return or relief air with significant increase during heating callls. [Jerry Scott - 1/15/2014 9:05 AM]
Drug Store - I	UNIT 3 damper failure. [Tim Snyder - 1/15/2014 8:59 AM] Bob has been notified. [Tim Snyder - 1/15/2014 10:57 AM]
Retail - G	Optimum start disabled for MV due to excessive runtime [Jerry Scott - 1/15/2014 8:32 AM]
Manufacturing - B	A failed fire damper is preventing airflow to VAVs 210, 202 & 203. [Ken Hellewell - 1/15/2014 8:42 AM] Needs Repair [Jerry Miller - 1/15/2014 8:54 AM]
Retail - B	Disabled optimum start during MV due to excessive runtime. [Jerry Scott - 1/15/2014 9:06 AM]
Retail - H	UNIT 1 is in full cooling with 72 degree supply and 80 degree return. [Tim Snyder - 1/15/2014 3:54 PM]
Office - C	UNIT 4 frequently gets drive faults and MODBUS communications alarms. [Tim Snyder - 1/16/2014 1:11 PM] Started on 1/12. [Tim Snyder - 1/16/2014 1:19 PM]
Office - C	UNIT 4 has occasional unexplained spikes in CT reading up to about 20kW. [Tim Snyder - 1/16/2014 2:29 PM]
Fire Station - A	UNIT 3 provides inadequate cooling. [Tim Snyder - 1/16/2014 10:12 AM]

Chapter 32

Clockworks™ Automated HVAC Diagnostics and the Evolution of AFDD Tools

Alex Grace, KGSBuildings

In the last five years, the world of automated fault detection and diagnostics (AFDD) in commercial buildings has grown exponentially. The ability to connect to increasingly open building automation systems has enabled an entire industry to be born around the analysis of existing building automation data to drive operational savings. There is now a rapidly growing competitive marketplace for software that helps building owners and operators find energy waste and mechanical inefficiencies. As the various names for analytics, AFDD, or monitoring based commissioning (MBCx) software demonstrate, the market has now matured to the point where clients are no longer asking "what?" but "how?" How are you different from your competitors? How do you integrate and can you communicate across different building automation systems? Are you a cloud based or an on-site solution? Can you monetize diagnostic results? These and other sophisticated questions from building engineers demonstrate an evolving market. Additionally, consulting firms are playing an important role in helping building owners and managers make sense of the increasingly crowded landscape of technology offers.

With competing technology firms using similar words for different things and different words for similar things, there is an understandable amount of confusion about what the various solutions actually do, and how facilities management teams or service providers can drive value with AFDD. The focus of this chapter will be on the market adoption of AFDD solutions broadly, and Clockworks™ in particular, including: integration to building automation systems, the scalability of cloud technology, a case study of energy savings and mechanical efficiency, organizational factors to success, and stakeholder needs and roles.

BACKGROUND

KGS Buildings began as an engineering team of building experts, troubleshooting HVAC systems, designing control strategies, metering solutions and energy modeling for a range of clients. The three founding team members completed their Ph.D.s in building technology, studying the ways in which HVAC and building system data could be processed to diagnose mechanical inefficiencies, optimize cooling systems and visualize performance data. The Clockworks™ software platform was first commercialized by the company in 2010, incorporating rigorous research and development into an application designed to give facility engineers, maintenance teams and service providers greater visibility and actionable intelligence into the performance of their facilities. An experienced software team builds on the research background to provide the most comprehensive, cost-effective, and cutting-edge performance management software to people who manage and service buildings. Two important cooperative research activities have informed the evolution of Clockworks™ and kept the technology connected with transformative industry research: a Cooperative Research and Development Agreement with Pacific Northwest National Labs funded by the Department of Energy on *Automated Retuning and Fault Detection*, as well as ASHRAE Research Project 1633, *Data and Interfaces for Advanced Building Operations and Maintenance*.

Clockworks™ is a cloud based platform that integrates with building automation systems to provide enterprise-wide automated diagnostics to pinpoint the top energy savings, comfort and maintenance priorities across HVAC systems. Additional software modules provide a range of tools to help building operators identify, investigate and track inefficiencies to prioritize maintenance activities, reduce energy consumption and improve occupant comfort. Clockworks™ currently collects over 2 billion points of data per month and is growing at rate of 50% every 3 months. At this rate, the platform will collect 5 billion data points every month by January 2015. The scalability of the cloud platform has allowed rapid growth not only within the US, but also globally; Clockworks™ is currently driving building performance in Denmark, Finland, Norway, the United Kingdom, the United Arab Emirates, Singapore, and Australia.

SYSTEMS INTEGRATION

There are different approaches in the market today to communicate with building automation systems (BAS) and extract data for fault detection. Clockworks utilizes a customized Data Transfer Service (DTS) to stream data in near-real time into the cloud without the need for additional hardware. The DTS is a Windows service that sits on the same network as the building automation system. It communicates with global/supervisory BAS controllers or databases to request and push temperatures, pressures, flows (inputs and outputs), and setpoints (numeric and virtual points) from the building automation system into the cloud. Clockworks™ DTS supports common open protocols (oBIX, BACnet, OPC, LON, Modbus) and some proprietary protocols as well such as Andover Continuum.

Clockworks™ utilizes the Microsoft Azure cloud which provides many tools for scaling, storing and managing large quantities of data and computations. The ability to push data out one-way to the cloud provides a level of assurance to IT departments. Data are pushed through port 80 or when required, data can be encrypted and pushed https through port 443.

WHY THE CLOUD?

A deliberate decision was made early in the development of Clockworks™ to develop a cloud-based, rather than on-site, software solution. One only needs to open any business section of any newspaper to see the rapidly increasing growth of big data and cloud technologies. Fundamentally, we see two main advantages of the cloud for building owners and operators: scalability and accelerated technological change.

Scalability

For large organizations with hundreds or thousands of sites, scalability across the enterprise is very important. Cloud software technology enables large deployments with an unlimited amount of buildings, equipment and points. Clockworks™ has the ability to process and display millions and even billions of data points quickly through an accessible, intuitive web platform. When data are streaming into Clockworks™ from hundreds of sites, the processing power needed to handle, store, and analyze that data within diagnostics scales to meet demand. As many as one thousand virtual servers are currently used as needed to process all diagnostics on all data globally in under two hours. This number continues to grow as the service grows, with no limit. When many users within an organization log into Clockworks™ at the same time, multiple servers can spin up to handle the requests from those users, eliminating bottlenecks. The bottom line for scalability is that cloud technology eliminates both the data storage and processing power limitations of PC or single-server based applications.

It is worth noting that any software can be "installed on the cloud." One can put a SQL server or any application on a cloud server and access it through the web, but this does not eliminate bottlenecks. There is still a limit to the size of that single server in terms of its ability to process, analyze and store data. If many users try to access the same web server, the system will slow and become unusable. Picture 50 users trying to use your PC remotely at the same time. This is different than software architected around the cloud which allows the use of thousands of servers at the same time to collect and process data, run diagnostics and efficiently serve those results to a website.

Rapid Technological Change

The pace of technology change today is so rapid that one may consider a standard software package loaded on a server or PC outdated within a year of installation. Rather than yearly releases, software teams can now constantly innovate and iterate by building features and functionality continuously and deploying those upgrades dynamically to all users. Customers are no longer asked to upgrade to new versions periodically; upgrades appear through the web interface consistently and seamlessly over a rapid development cycle. This method of software development allows the product team to respond rapidly to customer and market feedback.

As an example, an important Clockworks™ customer wanted one platform to not only provide AFDD but also the visualization and analysis of meter data, and key performance indicator (KPI) tracking as well. There are many platforms that provide this service and few that provide AFDD. However, it was clear that providing both in one unified interface would be valuable to many organizations. A dashboard that allows users to compare multiple facilities to established baselines and normalize energy data for weather and occupancy was a key addition to the Clockworks™ solution. The software team was able to build and seamlessly add this new meter dashboard to the platform in 3 months and they continue to develop it further over time. This type of rapid iteration and response to market conditions has obvious advantages over long product release cycles and the associated costs of upgrades.

PROJECT AT A GLANCE

Location—Massachusetts
Facility—Research lab (450,000 sq. ft.) with a mix of Laboratory, office and educational space
Monitored Systems—AHUs and zone ventilation equipment
Setup Cost—$23,190
Maintenance Cost (annual)—$35,407
Annual Savings—$286,000 (7% of total energy)
Cost to Fix—$150,000

Clockworks™ Goal

Provide facility teams with continuous system fault detection to automatically evaluate and organize the condition of equipment. This information enables proactive building maintenance while ensuring persistent energy efficiency, which may result in significant ongoing energy and cost savings.

Leaking Cooling Coil Valves in Air Handler

Clockworks identified an air handler with leaking cooling coil valves, resulting in a combined loss of approximately $2,200 per week during the heating season due to simultaneous heating and cooling of air supplied to the building.

Solution: The air handlers were repaired and produced annual savings of $61,400.

Malfunctioning Cooling Coil Valve in Air Handler

Challenge: One of the air handlers was found to have a cooling coil butterfly valve that was improperly closing at sporadic intervals, making it easy to over-

look. This valve caused a documented energy waste of $14,500 over a six-month period.

Solution: This valve was repaired, resulting in savings of $29,000 per year.

Preheating Coil Valve in Air Handler

Challenge: One preheating coil valve was causing simultaneous heating/cooling.

Solution: Once the overridden valve was fixed, it resulted in $137,000/year savings.

Leaking Reheat Coil Valves in Terminal Units

Challenge: Nearly 200 of the building's terminal unit reheat valves were found to be leaking, caused by unfiltered hot water during the building startup. Rather than relying on hot/cold calls to identify the problems, Clockworks™ prioritized the most wasteful leaking valves to fix and confirmed the success of these repairs.

Solution: Eighty-four VAV boxes with the highest losses were re-commissioned, resulting in the replacement of 52 reheat valves and 12 actuators as well as some controls changes. The total savings due to repairs was $77,100 per year.

Additional opportunities—including optimizing economizer controls, reducing air handler static pressure setpoints and adjusting energy recovery control sequences—were identified and are under review for implementation. The building is still being monitored by Clockworks™ to discover faults, accelerate retro-commissioning activities and automate verification of energy investments.

THE CLOCKWORKS™ APPROACH TO AFDD

Clockworks™ has a comprehensive library of diagnostics for all HVAC equipment and systems, which is curated and updated continuously. The software has the ability to run in depth diagnostics at scale and to prioritize the results by energy cost, as well as comfort and maintenance on a 0-10 scale (E,C,M below). All results from equipment and system level analyses appear in the punch list that you see below according to the date range selected. This list can be sorted by portfolio or individual building, as well as equipment class and type of analyses. This allows a user to drill into a specific building, piece of equipment, or type of analysis. The below figure illustrates the results of diagnostics* across

---

*"diagnostics" and "analysis" are used interchangeably

a portfolio with a daily interval of 24 hours. Figure 32-1 shows a screenshot from Clockworks™ illustrating the prioritization based on energy cost impact ($ over one day). Each below analyses will generate a report with detailed plain English explanation of findings and graphs for visual representation of fault findings (see Figure 32-2).

DIAGNOSTICS DETAILS

Figure 32-2 is a sample output from an air handler with simultaneous heating and cooling. The diagnostic has identified that both heating and cooling valves are open at the same time, causing the air to overheat and then cool back down to setpoint.

A central challenge for AFDD is the need to deploy diagnostics across many different building profiles, systems and types of equipment. Effective solutions must avoid re-writing diagnostics with every new building, which is cost prohibitive, while at the same time avoid false positives. Clockworks™ addresses both of these

two priorities with a centrally managed code set that is customized to individual facilities through a scalable software configuration architecture that enables rapid deployment. This means that a complete library of diagnostic code can be rapidly applied to individual buildings, entire campuses or even large building portfolios by mapping point, equipment and system parameters directly into an internal online tool that augments the central diagnostic code without re-inventing the wheel.

The KGS building technology team and trained partners mass-customize diagnostics for each building's systems based on the points available on the buildings automation system and system documentation in the form of controls as-built drawings, mechanical schedules, and sequences of operation. Rule thresholds have defaults that can be changed (i.e. short cycling is flagged if a valve is oscillating by 50% or more within 20 minute periods). Horsepowers, rated flows and detailed sequences are all documented and augment the diagnostic library automatically for deep root cause analysis without sacrificing scalability. In addition, users can make adjustments to parameters used for fault detection thresholds and cost

Equipment	Analysis	Start Date	Notes Summary	Cost	E	C	M	Actions
DetCtr_AHU11 (Air Handler)	AHU Coils	12/16/2013	Sensor error. Supply temp higher than setpoint. No supply temp reset. Over-aggressive freeze prevention. Leaking pre-heating valve. Heating or cooling valve issue. Valve short cycling.	$93	10	10	6	▼
STC_BoilerLoop (Heating System)	HW Loop	12/16/2013	Supply temp higher than setpoint. Supply temp reset error. Diff pressure lower than setpoint.	$81	10	0	6	▼
DetCtr_CdWPmp3 (Pump)	Pump	12/16/2013	Status/run error.	$50	10	0	2	▼
MI_HWSys (Heating System)	HW Loop	12/16/2013	Supply temp higher than setpoint.	$26	9	0	5	▼
DetCtr_AHU31 (Air Handler)	AHU Coils	12/16/2013	Leaking cooling valve.	$25	9	0	6	▼
DetCtr_AHU18 (Air Handler)	AHU Economizer	12/16/2013	Flow imbalance. Mixed air temp higher than setpoint. Excess mechanical cooling.	$25	9	0	5	▼
HS_F0BAHU1 (Air Handler)	AHU Heat Recovery	12/16/2013	Heat Wheel off, should be on. Unused free heating available.	$22	8	0	7	▼
PSC_AHU4B (Air Handler)	AHU Fan	12/16/2013	No static pressure reset.	$20	7			▼
MI_AHU_04 (Air Handler)	AHU Economizer	12/16/2013	Flow imbalance. Mixed air temp higher than setpoint.	$19	7	0	5	▼
HS_F0BAHU2 (Air Handler)	AHU Coils	12/16/2013	Supply temp higher than setpoint. Leaking pre-heating valve.	$16	6	10	6	▼
DetCtr_Chiller1Loop (Cooling System)	CHW Loop	12/16/2013	Supply temp lower than setpoint. High supply temp setpoint. Low loop temp difference.	$16	8	0	5	▼

Figure 32-1

Notes: PROBLEM: HEATING OCCURING WHILE COOLING COIL VALVE IS OPEN

- The supply air temperature was more than 3 deg F higher than the mixed air temperature while the cooling coil valve was open, for a total of 24 hrs over the analysis period.

Possible Causes:

- Leaking heating coil valve.
- Stuck cooling coil valve.
- Temperature sensor error.

PROBLEM: HEATING OR PREHEATING AND COOLING VALVE OPERATING SIMULTANEOUS

- The heating or preheating valve is commanded to be open at the same time as the cooling valve is commanded to be open..

Possible Causes:

- Error in control sequences.
- Temperature sensor error causing valve(s) to be controlled incorrectly.

PROBLEM: EXCESS OR SIMULTANEOUS HEATING AND COOLING

- The preheating coil and/or cooling coil are either providing excess heating or cooling or operating simultaneously.
- This may waste around $978 and 79220 kBTUs over 1 day(s).

Possible Causes:

- Valve is not seating properly and is leaking.
- Valve is stuck.
- Temperature sensor error or sensor installation error is causing improper control of the valves.

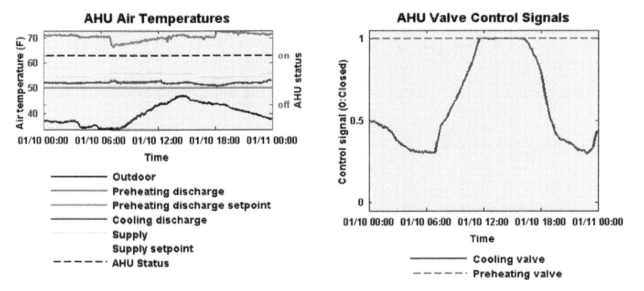

Figure 32-2

calculations on a continuous basis, allowing constant refinement of the diagnostics for each site.

Of course, HVAC system documentation and the sensor data from the BAS that is available differs in every building. Diagnostics are structured to produce results based on available information. The more point data and sequence information, the more precise and informative the output. For example, when only outdoor air temperature and supply air temperature points and setpoints are available on an air handler, the AHU Coil analysis will flag that the supply air temperature is higher than setpoint and quantify the cost of the additional heating energy used based on the excess kBTUs and site utility rate. If valve positions and inter-coil

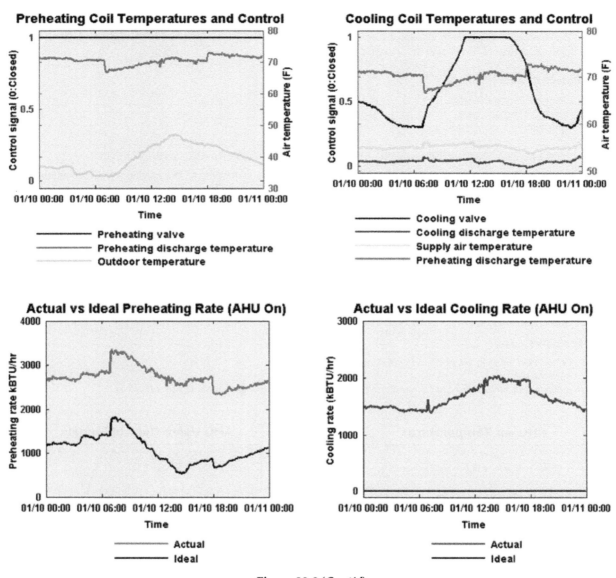

Figure 32-2 (*Cont'd*)

temperatures are available, the analysis may determine that a preheat coil is leaking based on the valve position and preheat discharge temperatures. If there is a flow station, the analysis will use that data point for the energy calculation. If not, the analysis will use fan speed and the rated flow of the unit.

When mechanical information is lacking, Clockworks™ will still identify that a pump is running 24/7 outside of occupancy, or a valve is leaking, but without the horsepowers or rated flow the message will read that "Costs cannot be calculated; missing HP, rated flow." When sequence of operations information is not available, assumptions are made based on the points that are available. For example, when there is no enthalpy point on an air handler, the assumption is made that the econo-

mizer is enabled based on a temperature threshold, when the outdoor air is cooler than the return air or up to a temperature high limit. If an enthalpy point is available, it is assumed that the sequence is for enthalpy control.

Figure 32-3 is a screenshot from Clockworks™ of equipment information, and sequences of operation configured for accurate analyses. These variables (editable by admin users) demonstrate the configuration process that is performed to customize analyses.

Clockworks diagnostics include hierarchical rule-based fault detection and diagnostics, first principles engineering based FDD, as well as model-based FDD. These allow the software to identify root cause, monetize the energy waste/avoidable cost, track maintenance and comfort impacts, and clearly visualize the findings

ElectricityCost:
    Value: .0777
    Default Value: 0
    Description: Cost of electricity

GasCost:
    Value: .5544
    Default Value: 0
    Description: $ per therm

IsEnthalpyEconomizer:
    Value: 1
    Default Value: 0
    Description: True if economizer is controlled using enthalpies, false if controlled using temperatures

ReturnFanRatedFlow:
    Value: 8500
    Default Value: 0
    Description: The rated flow rate in cfm or L/s of a return fan.

ReturnFanRatedPower:
    Value: 5
    Default Value: 0
    Description: Return fan power in HP or kW.

SupplyFanRatedFlow:
    Value: 8500
    Default Value: 0
    Description: The rated flow rate in cfm or L/s of a supply fan.

SupplyFanRatedPower:
    Value: 10
    Default Value: 0
    Description: Supply fan power in rated HP or kW.

WinterControlOutdoorAirTempThreshold:
    Value: 35
    Default Value: 0
    Description: Outdoor air temperature below which winter control sequence is used.

WinterControlPreHeatingValvePositionMin:
    Value: 0.45
    Default Value: 0
    Description: Minimum valve positive used by winter control sequence for AHU preheating coil valve.

Figure 32-3

with graphs and plain English (as shown in Figure 32-2).

Additional Software Modules included in the Clockworks™ platform address different user needs in managing, diagnosing, investigating and reporting on building operations, see Figure 32-4 and section below for descriptions.

BRINGING AFDD INTO FACILITIES MANAGEMENT

Technology is essential, but without a process for managing and responding to diagnostics findings, there is only limited value. Ownership over FDD programs is important and a defined workflow is necessary for scheduling maintenance and responding to diagnostic results. The extent to which work is performed by an in-house facilities engineering team or by outside engi-

neering or services providers is dependent on the organization. Some customers use Clockworks™ to dispatch and plan maintenance every week internally. Others do not have the people or time to manage the platform and prefer to have an energy services company, trusted consulting engineering firm, retro-commissioning firm or controls contractor utilize Clockworks™ to drive energy reduction. Since some diagnostic results require programming changes and sensor replacement and others require mechanical work to be performed, a well-defined process is important for maximizing ROI.

The ways in which different service providers take advantage of Clockworks™ varies:

In House Facilities Teams

When a robust engineering team is capable of directly using Clockworks™ to manage their facilities team, the process for dispatching team members chang-

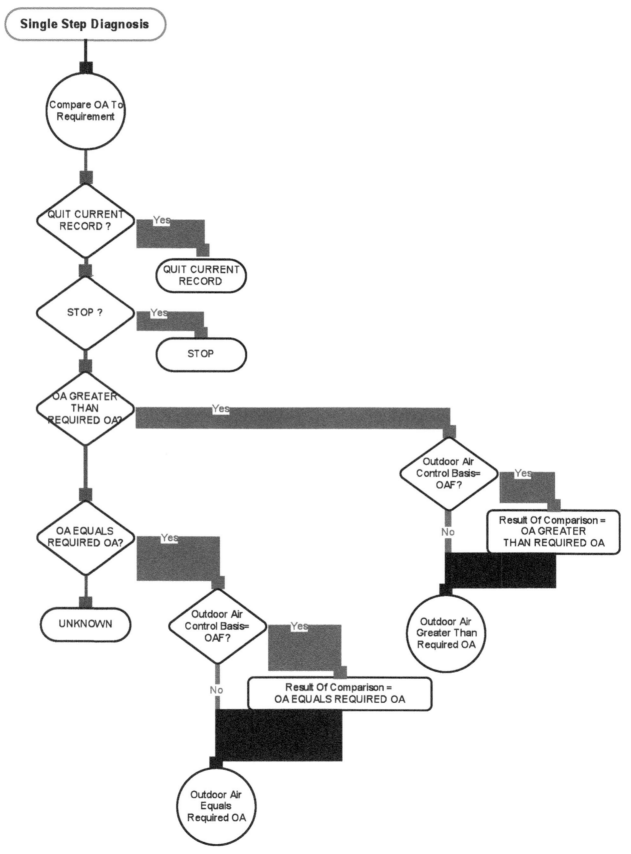

Figure 32-4

es. Facilities teams typically respond to comfort complaints and building automation alarms, and perform time-based maintenance from a scheduled preventative maintenance plan. These 3 categories of work generation are of course essential. Adding Clockworks™ into the equation adds a layer of intelligence to prioritize maintenance based on the financial impact to the facility, as well as comfort and mechanical severity of the problem. Additional problems that may be under the radar are unearthed continuously, such as a leaking valve in an air handler that is still able to meet its supply air temperature, or an under loaded loop or insufficient economizing. In addition, Clockworks provides detailed visibility, with important metrics such as trends in avoidable costs associated with faults alongside trends in energy use intensity, to key decision-makers such as Facility Managers, Energy Managers and financial decision makers to inform funding, staffing, and priorities.

Facility Management Firms

Like proactive in-house facilities teams, forward thinking Facility Management firms can deliver better performing buildings and better service to their customers leveraging AFDD and performance management platforms like Clockworks™. Continuous tracking of equipment performance, combined with records of repair and maintenance activities on a site, readily demonstrate to building management the impact of their services on the financial and overall performance of buildings.

Commissioning Firms

Clockworks™ accelerates the commissioning process significantly. Rather than asking the controls contractor for trends, manipulating the data in Excel™ to identify problems or making sure that specifications and design strategies are being followed, AFDD software will automatically flag anomalies, and prioritize them based on cost. Additionally, software can crunch all HVAC data from the facility across all zones so that manual spot-checking is reduced. AFDD enables the ongoing commissioning process by not only making the initial commissioning (Cx) or existing building commissioning (EBCx) process more efficient but maintaining building performance over time by continuing to flag priority operational inefficiencies.

Design Engineers

Many design firms are increasingly performing measurement and verification (M&V) on new construction projects to make sure that their designs are operating correctly. By incorporating Clockworks™ into their specifications, MEP and performance engineers can ensure that the building is performing consistently with energy models and design. Engineers involved in design and construction may play an expanded role in helping optimize the facility during the first year of occupancy or beyond.

Controls and Mechanical Contractors

Clockworks™ gives a layer of intelligence to the building automation system that enables contractors to provide a significantly higher level of service. In addition to scheduled preventative maintenance, service providers can now arrive on site with a clear checklist of actions items to implement. Controls contractors can focus on energy savings programming changes, faulty or suboptimal sequences, or sensor replacements to implement. Mechanical Contractors can arrive informed about which valves or dampers have failed, which systems are underperforming, and how much could be saved by making repairs.

Energy Services Providers and Utility Companies

For energy services providers and utility companies, such as energy consultants, energy performance contractors, or utility program administrators, AFDD platforms can not only help identify energy saving opportunities and enable monitoring based commissioning programs, but also gather the data and conduct calculations to measure and verify results.

In a rapidly changing and increasingly competitive market for AFDD solutions, software providers must stay on the cutting edge of technological change with a strong connection to industry and academic research. Building owners, operators and service providers are using AFDD tools to enable higher building performance, perform proactive maintenance and drive facilities results with a clear return on investment. Most often, energy savings drives the initial decision to purchase AFDD software. However, maintenance prioritization is the reason for continued utilization of AFDD platforms years after initial savings have been achieved. Building systems are dynamic and equipment performance will continue to degrade over time as new seasons and new problems mechanical anomalies arise and there is therefore a continuous role for AFDD to identify and prioritize energy savings opportunities and the correction of mechanical inefficiencies. At its best, AFDD becomes an integral part of the facilities or service provider workflow by directing the most efficient use of time and money. There are many different roles for service providers and facilities teams to play when implementing AFDD and software needs to provide tools to the appropriate stakeholders in different organizations. It is clear that processes for managing AFDD will continue to evolve with the expansion of the market, as software solutions like Clockworks™ become more and more standard best practice for building management.

ADDITIONAL CLOCKWORKS™
SOFTWARE TOOLS

In addition to the Diagnostics module shown previously, Clockworks includes many other tools to help users understand, track, and take action on AFDD findings and the data in their systems. Below are brief descriptions of each additional module.

Analysis Builder allows users to interrogate and manually analyze all data within Clockworks™ in real time. This is the tool for troubleshooting and investigation.

Analysis Builder

Quickly explore and troubleshoot with all trended data over any date range by selecting data points and generating graphs. Develop your own ad-hoc rules using a user friendly interface where you can build, save, and share graphs with other users.

Figure 32-5: Analysis Builder—Export all raw data from Analysis builder as well into table form and .csv

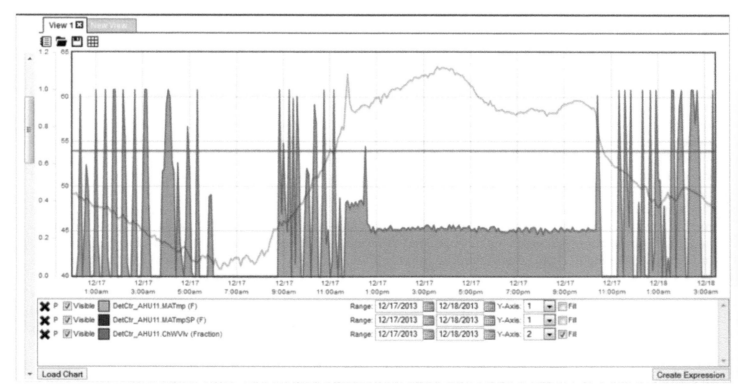

Figure 32-6: Graph Data

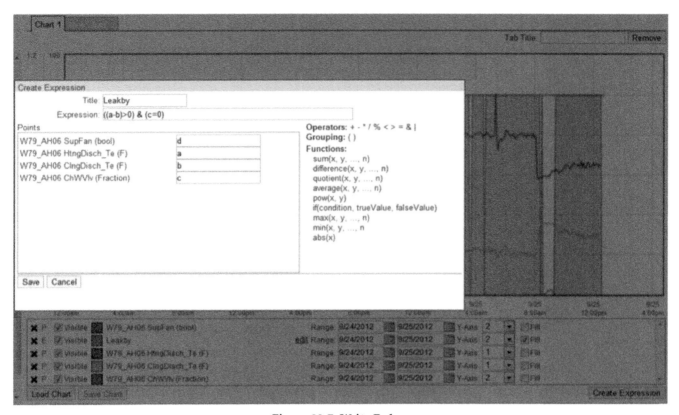

Figure 32-7: Write Rules

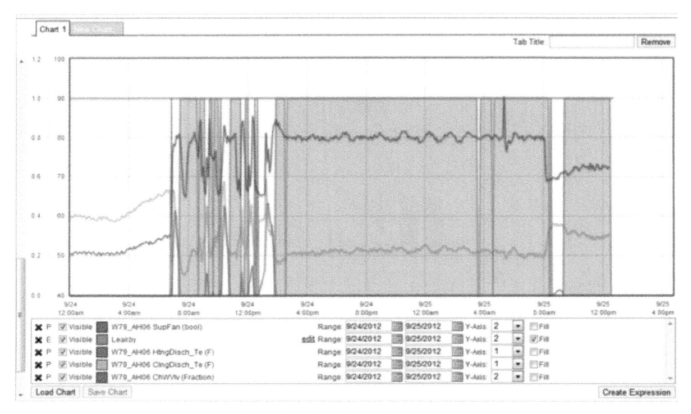

Figure 32-8: Flag where rule is "true"

Profiles

The Profiles pages contain important information about each building, piece of equipment, or system in the building. QR codes deployed in the field provide quick access to building and equipment profiles, from which users can view basic engineering information, link to documents, data and diagnostics, or connect to external tools such as existing document management, maintenance management, GIS, or other tools.

Figure 32-9: Profiles

Commissioning Dashboard

The Commissioning Dashboard provides executive level performance trends illustrating which buildings and equipment have faults or inefficiencies over time, and how much those are worth. Users can view the avoidable costs associated with repair and maintenance needs and operational changes, such as adjustments to control sequences or faulty dampers and valves. The impact of faults and opportunities in terms of cost, energy, comfort and maintenance priority can be viewed over time to observe and analyze trends.

Figure 32-10: Commissioning Dashboard

Figure 32-11: Additional View of Performance Dashboard

ADDITIONAL MODULES
TO BE RELEASED SPRING 2014

Project Tracker

The Projects module enables users to create projects associated with diagnostic results across equipment to track project information, establish costs and payback, set target dates, organize workflows, prioritize projects, assign users, and generate projected cost savings.

Alarm Manage

The Alarms module provides users with information about alarms across their portfolio, including current alarm states, occurrences, percentages, frequencies, durations, thresholds, as well as drill down capabilities to look at statistics on alarms by buildings, equipment, and alarm types.

References

Samouhos, Stephen V. Building Condition Monitoring. Diss. Massachusetts Institute of Technology, 2010. Cambridge, MA: Massachusetts Institute of Technology, Web. <http://hdl.handle.net/1721.1/61611>.

Gayeski, N, and S Katipamula. United States. U.S. Department of Energy. Building Diagnostic Market Deployment—Final Report. Richland, WA: Pacific Northwest National Laboratory, 2012. Web. <http://www.pnnl.gov/main/publications/external/technical_reports/PNNL-21366.pdf>.

Kiosk

The Kiosk is both a community facing portal, and external public touch screen enabled interactive kiosk which allows users to view a real-time graphical overview of your portfolio and building level utility performance, including basic energy performance, breakdowns, and comparisons.

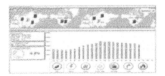

Figure 32-12: Kiosk

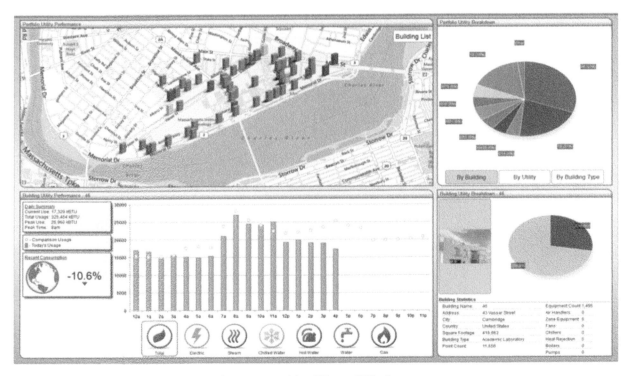

Figure 32-13: Live View of Kiosk:

Gayeski, Nick, Sian Kleindienst, Stephen Samouhos, and John Anastasio. "How I learned to Stop Worrying and Love Building Data." Automated Buildings. 08 2012: n. page. Web. 11 Feb. 2014. <http://www.automatedbuildings.com/news/aug12/articles/kgsbuildings/120726012505kgs.html>.

Mills, Evan. "Monitoring Based Commissioning: Benchmarking Analysis of 24 UC/CSU/IOU Projects." (2009).

Piette, Mary Ann, S. T. Khalsa, and Philip Haves. "Use of an Information Monitoring and Diagnostic System to Improve Building Operations." Proceedings of the 2000 ACEEE Summer Study on Energy Efficiency in Buildings. Vol. 7. 2000.

Meiman, Andrew, K. Brown, and Mike Anderson. "Monitoring-Based Commissioning: Tracking the Evolution and Adoption of a Paradigm-Shifting Approach to Retro-Commissioning." Submitted for publication in Proceedings of the 2012 ACEEE Summer Study of Energy Efficiency in Buildings. Panel 4 (2012).

Reporting

The Reporting module provides the ability to export raw and calculated data from Clockworks and generate reports with graphs and text. Accelerate M&V and share reports to direct the resolution of equipment and system issues.

Figure 32-14: Reporting

Documents

Users can use Documents to store information with specific systems and equipment and share them with service providers, maintenance technicians, engineers, and others. Documents such as testing and balancing reports, calibration notes, maintenance manuals, system diagrams and details, and more can be made available to users of your choice anywhere, anytime.

Figure 32-15: Documents

Utility Billing

Generate tenant utility billing allocation using actual BMS airflow, BTU and available submeter data, and store reports for future reference. Electric, gas, steam, or chilled water meters and submeters can be combined with building management system data from BTU meters, flow stations, and conventional control points to allocate utility costs by tenant or customer consumption measured in multiple systems and formats.

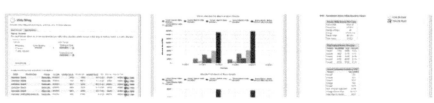

Figure 32-16: Utility Billing

Chapter 33

Automated Diagnostics—
To Pick the Low Hanging Fruits
In Energy Efficiency
In Air Conditioning and Heat Pump Systems

Klas Berglöf
ClimaCheck Sweden AB

ABSTRACT

"To measure is to know" is a valid statement when it comes to air-conditioning and heat pump systems. The optimization of existing systems is the "low hanging fruit" to reduce the energy consumption with 10-40% in commercial and industrial buildings. Considering that two-thirds of the existing buildings will still be in existence in 2050 and that new equipment has seldom proven to reach design performance, the potential energy savings in existing buildings are immense. The International Institute of Refrigeration estimates that 15-20% of the global electrical consumption is consumed by refrigeration and air-conditioning processes.

When systems are measured they are mostly found not to operate at an optimal level and often far from design specifications. Savings of 10-40% are normally possible without significant investments, in a majority of the analyzed systems. In most cases the cost of qualified inspections or monitoring is paid in a few months through energy savings alone, not to mention savings from reduced number of alarms and failures in a well-functioning system.

The challenge is to change "business as usual." Equipment owners expect that the system is performing efficiently as long as the desired temperature is achieved and do not require validation of performance unless there is a very obvious problem. Air conditioning systems and heat pumps are complex systems where efficiency is dependent not only on the design but also on external factors such as air and/or water flows, temperature levels and not to forget the ambient conditions, loads and control system. In reality the air conditioning system is affected by all other systems in the building but often seen as a "Black Box" that is either on or off.

Without a request from equipment owners, few consultants, commissioning agents and contractors invest in tools and training to learn how to measure, analyze and optimize these systems in the field. This is in spite of that they often consume the majority of the electrical energy in a property and in many areas are defining the peak loads in the grid and requiring huge investments to increase the capacity in power supply and grids to deal with increasing loads.

Energy optimization is becoming a major business opportunity with growing awareness among equipment owners of the possibility to reduce cost. For commissioning agents and contractors there is an opportunity to deliver higher quality work. The "black box" can be "opened" in 20-40 minutes and performance measured on heat pump, air-conditioning and refrigeration systems. The "thermodynamic method" is non-invasive and unbiased as it does not require inputs from unit or component manufacturers. The capacity can be established with an accuracy of ±7% and COP with ±5% for almost all compressor-driven heat pump, air-conditioning and refrigeration systems. With systems continuously monitored, it is possible to totally change the way air-conditioning and heat pump systems are maintained and serviced to avoid most of the too frequent and costly problems occurring today.

Performance analyses based on the described method are becoming accepted by leading system and component manufacturers as an important method to reduce warranty discussions and the method is today validated in laboratories of a large number of manufacturers around the globe and by accredited test institutes in several countries in Europe. In the USA Nrel (National Renewable Energy Laboratory) is using the described

ClimaCheck systems in field tests of air-conditioning as well as supermarket refrigeration systems.

Key words: Performance inspection, energy audit, energy optimization, air-conditioning, refrigeration, heat pumps energy efficiency

INTRODUCTION

To measure is to know is a valid statement for air-conditioning and heat pump systems. To combine state-of-the-art analyzing methods with PC or web based technology to optimize existing systems is the "low hanging fruit" to reduce the energy consumption in existing buildings. By collecting the right measurements from air conditioning and refrigeration systems and using powerful PC software or servers to analyze the data, detailed information can be presented in a totally new way both to the technical specialist as well as to the equipment owners. On-line performance can be displayed locally and remotely on any internet-connected computer or smart phone and SMS generated when a system deviates from optimal performance, i.e., lose a few percent in compressor efficiency instead of when compressor fails. Savings of 10-40% are often possible at minimal investments as well as decreased repair cost if systems are analyzed and optimized. In many cases the cost of the inspection and optimization is paid in a few months. It is estimated that 15-20% of the global electricity is consumed by vapor compression systems (IIR, 2002) making the optimization in this sector of global interest. As two-thirds of the existing buildings will still be used in 2050 according to the International Energy Agency (IEA, 2012) the importance of optimizing existing buildings and systems is obvious. Governments around the world are targeting the sector to reduce electrical consumption. In the EU, the Energy Performance in Building Directive (EPBD, 2010) introduced "Performance Inspections" on air-conditioning systems above 12 kW, and in North America incentive programs to re-commission plants were introduced. Few systems show, when measured in the field, the performance they are designed for. The challenge is to change "business as usual"; it is obvious that return of investment of measurements is challenging to present, before the status of a system has been measured and documented, in spite of huge experience that the savings more than justify the relatively small investments required to find out how things actually work.

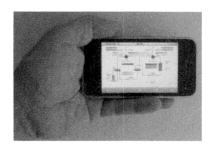

Figure 33-1: Permanently installed performance analyzers give real-time performance online in smart-phone

Figure 33-2: Portable performance analyzers give capacity and COP/EER in 30 minutes

Air-conditioning and heat pump systems are traditionally looked on as a "black box" in a ventilation or plumbing system—good performance is assumed if the temperature in a building is correct. The "Internal Method" based on a thermodynamic analysis of the refrigeration system opens the "black box." Performance and all parameters to evaluate the process can be measured in 20-40 minutes. There is no requirement of pre-installed meters or inputs from unit or component manufacturer. The capacity can be established with an accuracy of ±7% and COP with ±5% (Fahlén, 1989) for most compressor driven heat pump, air-conditioning and refrigeration systems. The method is suitable for inspection and maintenance, but offers additional benefits in monitoring systems with greatly enhanced "early warning" capability. In combination with energy profiling, the detailed performance information makes it possible to tune the plants and then ensure that they stay at optimum performance.

KNOW THE PERFORMANCE—MAXIMIZE THE EFFICIENCY

"If you cannot measure it, you cannot control it" is another valid statement in air-conditioning, heat pump and refrigeration plants. These systems are often operating with efficiency and lifetime far from their design specifications. A historical reason has been the lack of cost-effective methods for measuring performance, allowing many systems to operate with significant problems without anybody being aware of it. Savings of 10-40% and reduced failure rates are possible in many systems at low cost (Prakash, 2006). In ASHRAE's Refrigeration Commissioning Guide for Commercial and Industrial Systems (ASHRAE, 2014) the loss caused by insufficient commissioning is 7-25%. As the inefficiencies are not known to equipment owners, consultants and installers/service companies, they do not see measurements as necessary, even if cost is low compared to the investment, maintenance and failure costs. The prevailing purchasing process does not deliver the expected efficiency, in spite of investments in advanced controls (CABA, 2013). It is not uncommon that prestigious buildings, marketed as sustainable, consume 50-200% more energy than designed to (Roaf, 2013). A major cause besides the lack of awareness and focus from the purchasers is poor coordination between involved contractors/experts, and unclear responsibilities for the building's energy efficiency as a whole. *Without analysis and validation of actual performance at different loads and climate conditions, efficient operation over the year cannot be expected.* When performance and good control is proven, it is also possible to monitor electrical consumption over time and detect unexpected deviations at different conditions.

Methods to cost effectively measure actual operating performance and all information to analyze a system component-by-component are available at reasonable cost which is new and still to most involved unknown. The new technology makes it possible for manufacturers and contractors to deal more efficiently with complaints when they get correct and detailed information from the field and improve the capability of commissioning agents that get access to much more information at lower cost. This gives equipment owners the possibility to, without prohibitive cost, request a full documentation of performance of equipment they buy as well as performance of existing equipment. Detailed information on current performance and what causes decreased performance is a prerequisite to make investment decisions to decrease running cost. With good documentation competent specialists can be consulted at reasonable cost regardless of where in the world they are based.

SUB-METERING OF ELECTRICITY AND TRADITIONAL FLOW BASED MEASUREMENTS DOES NOT GIVE SUFFICIENT ANSWERS

It has historically been rare that electrical consumption of the HVAC systems was measured separately. With an increasing introduction of sub-meters and installation of flow-based energy meters, the challenges to measure and analyze these data are becoming obvious. Electrical measurements are quite straightforward and give valuable information, but with flow-based measurements comes challenges in the field—to measure flows and small temperature differences in dynamic systems with sufficient accuracy to get relevant information is not easy. Due to the complexity of the systems, the variations in loads and climatic conditions, any deviations in performance must be significant before a "warning" can be generated based on such information. Flow-based coefficient of performance (COP/EER) and/or seasonal performance factor (SPF) that show unexpected performance do not give any help to identify if it is a measuring error or a problem in the system. Deviations often cause costly investigations as the data do not contain information for evaluation of the cause. The result is that a lot of data are gathered but seldom turned into useful information.

BUILDING MANAGEMENTS SYSTEMS (BMS) CAN ENHANCE CONTROLS BUT DO NOT ENSURE OPTIMIZED OPERATION OF HVAC SYSTEMS

It is often expected that investment in BMS, energy management systems (EMS) and advanced controls in the units ensures efficiency. But in most cases HVAC systems are treated as a "black box" communicating measured parameters to a central system where they can be presented, but information is seldom upgraded to a level that facilitates the understanding required for optimization. Data normally require a high level of specialist competence and many hours of analyzing to convert into useful information. The obvious challenge is the varying conditions of HVAC systems where they operate over a wide range of loads and climate conditions resulting in that there are no simple limits that can be used to indicate correct/incorrect operation. The result is that even advanced BMS in the context of HVAC equipment tend to be alarm handling systems rather than early warning and optimization tools. Dynamic system performance at different loads and climate conditions, which is almost a prerequisite to optimize performance, is almost never available in BMS systems.

Due to the experienced complexity of BMS and control systems, they can even be counter productive, as they are assumed to ensure efficiency because their configuration requires specialists. But specialists can lack air-conditioning and refrigeration expertise (such as thermodynamic understanding). Based on experience, it is apparent that BMS systems as such are often not effective to provide correct and relevant information to avoid repeated failures without alarms or multiple alarms without faults. The "Internal Method's" thermodynamic evaluation offers a possibility to analyze dynamic systems in a new and cost effective way. A careful commissioning that includes experts on involved sub-systems is important as well as that the presentation of information is adapted to those using it on a daily basis. Thermodynamic performance analyses can be technically integrated in existing BMS/EMS/Scada systems but it has often proven to be more costly than using standardized stand-alone systems due to the cost of integration with suppliers of the general systems that need to develop interphases in areas where their organizations lack sufficient expertise. To identify required calculation and highlight the relevant parameters requires specialist competence.

THE THERMODYNAMIC METHOD FOR IN-THE-FIELD PERFORMANCE ANALYSES

It has been considered challenging and hardly cost-effective to measure performance of HVAC and refrigeration systems outside calorimetric rigs. To cost-effectively establish the dynamic performance of an HVAC system in actual operation is a prerequisite to optimize energy consumption and detect problems before they cause failures. The thermodynamic or "Internal Method" has proven to makes it possible to have all the information required to compare actual performance with design data after 20-30 minutes hook-up or continuously on-line. This provides high level findings on the overall system performance as well as detailed information on a component level required by the specialist. Analysis using a permanent installation provides continuous real time data on an online portal. The method is well proven as hundreds of thousands of analyses have been performed in manufacturer's development and production test rigs as well as in the field. The method presented

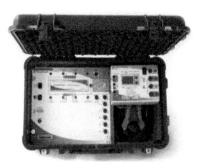

Figure 33-3: Field kit for Performance Analyzer that does not require installation of flow meters to establish the performance and capacity.

below is described in more detail in several published papers i.e. (Berglöf, 2011). The method is unbiased and can be generally applied as it does not require pre-installed sensors or inputs from unit or component suppliers. The "Performance Analyzer" can often be installed without even stopping the compressor. When systems are permanently monitored early warning alarms on a component level can be generated, reducing the risk of failure. There is a close relation between efficient and reliable operation. Optimization of existing equipment is the "low hanging fruit" to reduce energy costs, but equipment owners must realize the benefits of performance analysis, and industry as a whole must increase the competencies for measuring, analyzing and optimization of dynamic systems.

The internal method can be used as portable systems, Figure 33-4 or permanently installed monitoring as in Figure 33-5. It can also be integrated in BMS/EMS to make them more useful for optimization of the HVAC system. Integration requires, as mentioned earlier, that the thermodynamic competence and experience in evaluation is properly integrated in the system which often has proven to be more costly than a stand-alone system.

MEASURE

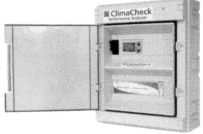

Figure 33-4: Fixed installed stand-alone performance analyzer with energy management capability.

EFFECTIVE AND UNBIASED PERFORMANCE INSPECTION

The "Internal Method" is based on the well-defined thermodynamic properties of the refrigerants, and offers an accuracy that can hardly be achieved by conventional methods in the field at the same time as the analyzed data gives a level of information not available with conventional methods. All relevant data for validation of performance, optimization and trouble-shooting can be presented dynamically in real time, i.e.:

- Coefficient of performance (± 5%)
- Cooling and heating capacity (± 7%)
- Power input (± 2 %)
- Compressor isentropic efficiency at full and part load
- Evaporator heat transfer
- Condenser heat transfer
- Indicators for refrigerant charge for early detection of leaks
- Superheat

A basic system requires ten easy-to-apply sensors that are attached at strategic points around the system shown in Figure 33-5:

- Temperature and pressure at entrance of compressor.
- Temperature and pressure at compressor exit.
- Liquid refrigerant before expansion device.
- Active electrical power.

For reference of operating condition the temperature of air/liquid entering and exiting condenser and heat exchanger are measured.

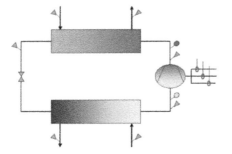

Figure 5, Sensors required and their location to establish performance of a standard refrigeration system.

Pressure sensors are applied on the service ports available on almost all systems larger than refrigerators, and even split air-conditioning systems with only one pressure port can be handled through replacing one pressure sensor with a correctly placed temperature sensor.

Temperature sensors are applied on the outside of the tubes with heat transfer paste, aluminum tape and insulation which gives sufficient accuracy for the thermodynamic evaluation of process as shown in Figure 33-6. Insulation should be adapted to temperature difference between tube and ambient and material in tube.

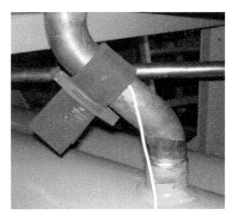

Figure 33-6: Surface applied temperature sensor facilitate application on almost any system

WELL PROVEN METHOD

The method and technology was first developed in Sweden in 1986, and in 1989 validated by SP, The Swedish Technical Research Institute (Fahlén, 1989) and later by many manufacturers. National Renewable Energy Laboratory (Nrel) in the US is using ClimaCheck Performance Analyzers for field evaluation of air conditioning systems and supermarket refrigeration. Several "incentive programs" from utilities in Canada and the US are supporting re- or retro commissioning with ClimaCheck, as 20-30% energy savings are often documented with a systematic approach and proper measuring. Today ClimaCheck Performance Analyzers are used by more than 50 manufacturers and 600 contractors/consultants in more than 20 countries. They are used in development and production tests rigs, validation of prototypes in the field and to improve efficiency of existing systems. Several hundred systems are monitored 24/7 over the internet, to follow system performance and energy consumption in a detail that is new to the industry. Warnings are generated on deviation long before the symptoms are noticed by the users. Equipment owners have access to energy reports with energy profiles, and in combination with underlying information on efficiency of different components systems are optimized. World leading companies in the industry have validated and use the Performance Analyzers to document the efficiency of products in test rigs as well as in the field. More than 20 universities and training centers around the world use Performance Analyzers in education and research work to document and visualize performance.

The internal method was first developed during an early heat pump boom in Sweden to solve the much too frequent problems and disputes occurring on the market when customers made significant investments to save cost but did not experience the promised savings. Portable and fixed equipment is now used widely in all applications of refrigeration processes to optimize air-conditioning, refrigeration and heat pump systems on the global market. The information contains all details a competent person needs to evaluate the plant without even having to go there. The information is as detailed as what has before been only available in manufacturer's test rigs. The opportunity for manufacturers and contractors to measure and validate performance with an unbiased method that does not depend on any component or system manufacturer's data improves customer relations and reduces the risk of expensive, time-consuming disputes that are costly for market reputation regardless of the cause of the problem.

The possibility of connecting (in half an hour) an independent measuring system that instantly gives the performance of any system without use of manufacturing data is relevant both to establish the base line for evaluating return of investment (i.e. of replacing an old chiller with a state of the art chiller) as well as validating performance in connection with commissioning of a

Figure 33-7: Performance Inspections on air-cooled chiller give full information after 20 minutes.

new chiller. It is possible to check performance, component-by-component, creating a new dimension of confidence while avoiding unfair competition. To exaggerate performance or leave poorly adjusted systems behind becomes less attractive when validation of performance is required. The risk of unfair competition decreases when specification requires that performance is documented at commissioning.

EXAMPLE—PERFORMANCE VALIDATION OF CHILLER IN A MANUFACTURING SITE

In Figure 33-8 a protocol done by Carrier Sweden shows that the performance of a chiller in the field corresponds within 2% with the Carrier software performance at the measured conditions—leaving no question that the chiller is performing within specification. The measured performance is unbiased, e.g. totally based on thermodynamics, and does not require any manufacturing data from Carrier.

Example—Combined Chiller and Heat Pump Used for Air Conditioning, Process Cooling, and Heating

This case is from a chiller operating in an industrial plant where they wanted to check performance of the chiller plant as well as to understand if there was a cause to two compressor failures. The chiller was used to cool production as well as office space. The system worked with heat recovery. A performance inspection with portable equipment documented all parameters of the chiller heat pump in less than two hours onsite. The inspection documented that the performance was significantly lower than a state-of-the-art chiller would give, and it explained the cause for the low performance and the two compressor failures.

In Figure 33-9 the detailed information pinpoints several problems in the system:

1. Poor heat transfer in evaporator—resulting in 15% higher energy consumption than even a conventional design and around 25% higher energy consumption than state of the art today due to the high temperature difference—low evaporation.

2. Poor heat transfer in condenser—resulting in 5% higher energy consumption than conventional design and at least 10% higher than state of the art today.

3. High risk operation at part load—causing risk for compressor failures—review of service records

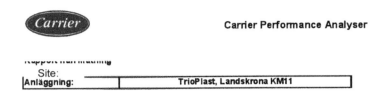

Figure 33-8: Performance validated in the field versus manufacturer data (SI units)

showed two earlier unexplained compressor failures.

4. Poor part-load performance of compressor—displayed by the compressor isentropic efficiency being 47% at part-load but 64% at full-load. The poor compressor performance at part load caused 10% increased energy consumption versus full load in spite of decreased pressure lift at part load. As there were multiple chillers with a total of eight compressors, the part-load valves could be permanently disconnected without negative impact on temperature control. This was done during the measurements resulting in a no-cost annual saving of around 6,000 USD/year.

COMMISSIONING REQUIRES A STRUCTURED APPROACH, GOOD INFORMATION, COMPETENCE AND TIME

There is a growing focus on "Continuous Commissioning," retro-commissioning and re-commissioning due to the increasing number of surveys showing that the energy efficiency of new as well as old buildings is far from the design. That lack of focus on commissioning and maintenance is the main reason for unnecessarily high energy bills, rather than performance of the equipment, and it is generally accepted by most in the industry. But it is difficult to get the appropriate budget to spend the time required. This is in spite of the overwhelming experience that it is extremely cost-effective when done correctly. Contractors experience difficulty "selling" the fact that the buyer must invest extra money to get the fantastic product to work well in their systems. The hours spent on commissioning must be justified in advance, which can be challenging, as no cases are identical. It's obvious that if it were known what would need to be optimized beforehand, it would not be needed to spend time on commissioning. The purchasing process is focused on minimizing the budget, and commissioning is often reduced as it is not believed to be necessary when competent contractors are used and BMS installed.

In reality more sophisticated systems and energy efficient designs require more competence and attention in commissioning as more parameters must be set correctly as savings are achieved by better controlling loads, flows and capacities. Additional savings are often achieved by free cooling and heat recovery. Again, these require careful adaptation to the system, load and climate, to give anticipated savings. These improvements add control functions and require coordination between different systems often installed by different contractors that often operated independently before and did not require as much competence to ensure correct interaction.

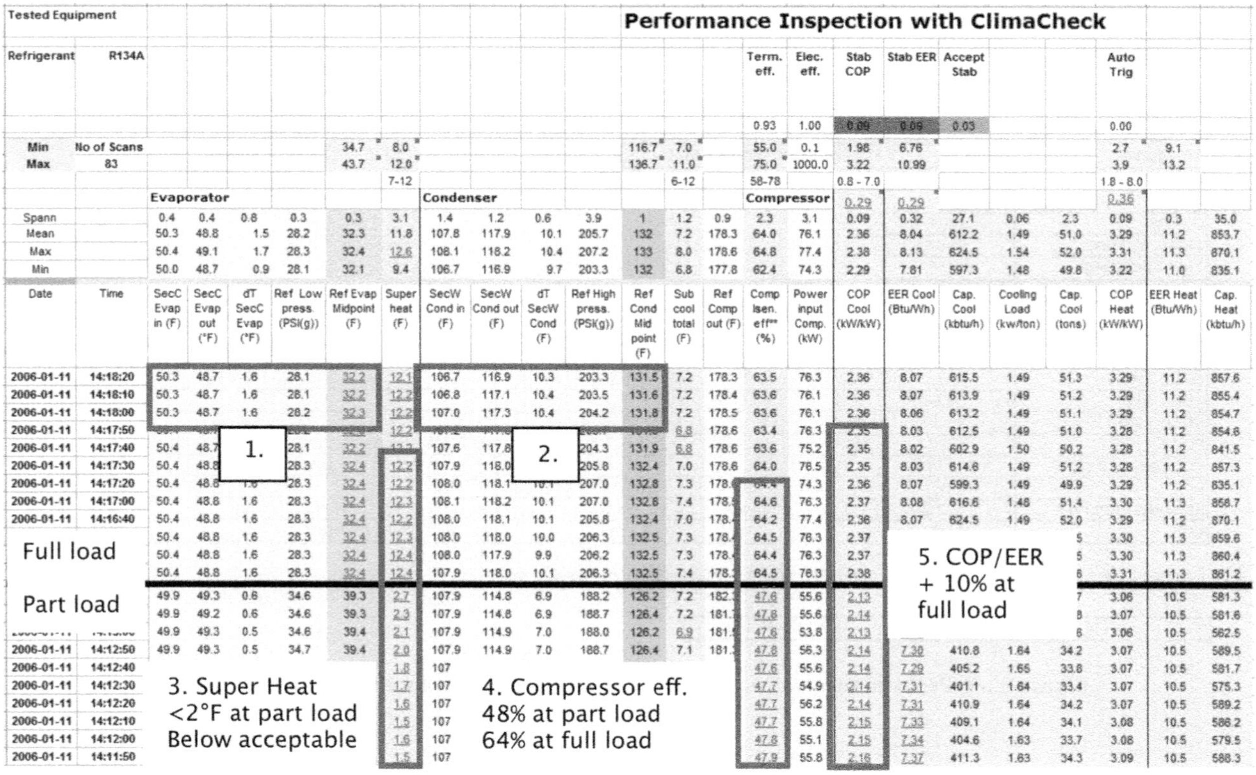

Tested Equipment																**Performance Inspection with ClimaCheck**								
Refrigerant	R134A															Term. eff.	Elec. eff.	Stab COP	Stab EER	Accept Stab			Auto Trig	
																0.93	1.00	0.09	0.09	0.03			0.00	
Min	No of Scans					34.7	8.0				116.7	7.0				55.0	0.1	1.98	6.76				2.7	9.1
Max	83					43.7	12.0				136.7	11.0				75.0	1000.0	3.22	10.99				3.9	13.2
							7-12					6-12				58-78		0.8 - 7.0					1.8 - 8.0	
		Evaporator						Condenser								Compressor		0.29	0.29				0.36	
Spann		0.4	0.4	0.8	0.3	0.3	3.1	1.4	1.2	0.6	3.9	1	1.2	0.9	2.3	3.1	0.09	0.32	27.1	0.06	2.3	0.09	0.3	35.0
Mean		50.3	48.8	1.5	28.2	32.3	11.8	107.8	117.9	10.1	205.7	132	7.2	178.3	64.0	76.1	2.36	8.04	612.2	1.49	51.0	3.29	11.2	853.7
Max		50.4	49.1	1.7	28.3	32.4	12.6	108.1	118.2	10.4	207.2	133	8.0	178.6	64.8	77.4	2.38	8.13	624.5	1.54	52.0	3.31	11.3	870.1
Min		50.0	48.7	0.9	28.1	32.1	9.4	106.7	116.9	9.7	203.3	132	6.8	177.8	62.4	74.3	2.29	7.81	597.3	1.48	49.8	3.22	11.0	835.1
Date	Time	SecC Evap in (F)	SecC Evap out (°F)	dT SecC Evap (°F)	Ref Low press. (PSI(g))	Ref Evap Midpoint (F)	Super heat (F)	SecW Cond in (F)	SecW Cond out (F)	dT SecW Cond (F)	Ref High press. (PSI(g))	Ref Cond Mid point (F)	Sub cool total (F)	Ref Comp out (F)	Comp Isen. eff** (%)	Power input Comp. (kW)	COP Cool (kW/kW)	EER Cool (Btu/Wh)	Cap. Cool (kbtu/h)	Cooling Load (kw/ton)	Cap. Cool (tons)	COP Heat (kW/kW)	EER Heat (Btu/Wh)	Cap. Heat (kbtu/h)
2006-01-11	14:18:20	50.3	48.7	1.6	28.1	32.2	12.1	106.7	116.9	10.3	203.3	131.5	7.2	178.3	63.5	76.3	2.36	8.07	615.5	1.49	51.3	3.29	11.2	857.6
2006-01-11	14:18:10	50.3	48.7	1.6	28.1	32.2	12.2	106.8	117.1	10.4	203.5	131.6	7.2	178.4	63.6	76.1	2.36	8.07	613.9	1.49	51.2	3.29	11.2	855.4
2006-01-11	14:18:00	50.3	48.7	1.6	28.2	32.3	12.6	107.0	117.3	10.4	204.2	131.8	7.2	178.6	63.6	76.1	2.36	8.06	613.2	1.49	51.1	3.29	11.2	854.7
2006-01-11	14:17:50						12.2						6.8	178.6	63.4	76.3	2.35	8.03	612.5	1.49	51.0	3.28	11.2	854.8
2006-01-11	14:17:40	50.4	48.7		28.1	32.2		107.6	117.8		204.3	131.9	6.8	178.6	63.6	75.2	2.35	8.02	602.9	1.50	50.2	3.28	11.2	841.5
2006-01-11	14:17:30	50.4	48.8			28.3	12.2	107.9	118.0		205.8	132.4	7.0	178.6	64.0	76.5	2.35	8.03	614.6	1.49	51.2	3.28	11.2	857.3
2006-01-11	14:17:20	50.4	48.8	1.6	28.3	32.4	12.2	108.0	118.1	10.1	207.0	132.8	7.3	178.	64.1	74.3	2.36	8.07	599.3	1.49	49.9	3.29	11.2	835.1
2006-01-11	14:17:00	50.4	48.8	1.6	28.3	32.4	12.2	108.1	118.2	10.1	207.0	132.8	7.4	178.	64.6	76.3	2.37	8.08	616.6	1.48	51.4	3.30	11.3	858.7
2006-01-11	14:16:40	50.4	48.8	1.6	28.3	32.4	12.2	108.0	118.1	10.1	205.8	132.4	7.0	178.	64.2	77.4	2.36	8.07	624.5	1.49	52.0	3.29	11.2	870.1
		50.4	48.8	1.6	28.3	32.4	12.3	108.0	118.0	10.0	206.3	132.5	7.3	178.	64.5	76.3	2.37					5	3.30	11.3	859.6
		50.4	48.8	1.6	28.3	32.4	12.4	108.0	117.9	9.9	206.2	132.5	7.3	178.	64.4	76.3	2.37					5	3.30	11.3	860.4
		50.4	48.8	1.6	28.3	32.4	12.4	107.9	118.0	10.1	206.3	132.5	7.4	178.	64.5	76.3	2.38					5	3.31	11.3	861.2
		49.9	49.3	0.6	34.6	39.3	2.7	107.9	114.8	6.9	188.2	126.2	7.2	182.	47.6	55.6	2.13					7	3.06	10.5	581.3
		49.9	49.2	0.6	34.6	39.3	2.3	107.9	114.8	6.9	188.7	126.4	7.2	181.	47.6	55.6	2.14					8	3.07	10.5	581.6
		49.9	49.3	0.5	34.6	39.4	2.1	107.9	114.9	7.0	188.0	126.2	6.9	181.	47.6	53.8	2.13					8	3.06	10.5	562.5
2006-01-11	14:12:50	49.9	49.3	0.5	34.7	39.4	2.0	107.9	114.9	7.0	188.7	126.4	7.1	181.	47.8	56.3	2.14	7.38	410.8	1.64	34.2	3.07	10.5	589.5
2006-01-11	14:12:40							107						178.	47.6	55.6	2.14	7.29	405.2	1.65	33.8	3.07	10.5	581.7
2006-01-11	14:12:30						1.7	107							47.7	54.9	2.14	7.31	401.1	1.64	33.4	3.07	10.5	575.3
2006-01-11	14:12:20						1.6	107							47.7	56.2	2.14	7.31	410.9	1.64	34.2	3.07	10.5	589.2
2006-01-11	14:12:10						1.5	107							47.7	55.8	2.15	7.33	409.1	1.64	34.1	3.08	10.5	586.2
2006-01-11	14:12:00						1.6	107							47.8	55.1	2.15	7.34	404.6	1.63	33.7	3.08	10.5	579.5
2006-01-11	14:11:50						1.5	107							47.9	55.8	2.16	7.37	411.3	1.63	34.3	3.09	10.5	588.3

Labels within figure: 1. | 2. | Full load | Part load | 3. Super Heat <2°F at part load Below acceptable | 4. Compressor eff. 48% at part load 64% at full load | 5. COP/EER + 10% at full load

Figure 33-9: Step-by-step documentation of poor chiller operation

Figure 33-10: Detailed information available to global expertise on-line at low cost changes the business.

To achieve low energy consumption it is normally necessary to combine high efficiency equipment with advanced controls, creating challenges as the interaction varies with load and climate conditions. A comfortable indoor climate does in no way indicate that systems are efficient or reliable.

Commissioning requires a structured approach to be meaningful and can rarely be done in an acceptable way on one occasion. In particular it is almost impos-

sible to do it properly before the building is in full use which is often the case at hand-over inspection, when construction is finished and the building is not likely in normal use and the climate might be so that there is no natural load.

To achieve optimized operation it is required to have a systematic approach as often described in re-commissioning programs with:

- **A scoping and data collection phase**—To collect information of system design and log data of performance at different operating conditions.

- **Investigation phase** —Where data are analyzed and measures that improve efficiency and reliability are identified.

- **Implementation phase** —Where identified improvements are done which often require a step-by-step approach as different measures with validation of improvements.

- **Validation phase with hand off reporting**—Where performance and relevant parameters required to keep system operation at optimum conditions are documented.

In phase 1 and 3 it is important to have cost effective measurement and analyzing tools to give sufficiently detailed data to at reasonable cost. The performance at different operating condition as shown in Figure 33-11 and Figure 33-9 is necessary to evaluate a chiller's performance and optimize the operation. At the same time it is important to visualize the result of the optimization over time to ensure that the overall result corresponds to expected results because an efficient chiller in an inefficient system, or with poor controls, will have high energy consumption. Controls not handling varying loads well are a major cause of poor performance. To document and visualize the energy savings is also important to justify the cost. In many HVAC and or refrigeration plants energy profiles are a very powerful tool as the dependence on ambient temperature is high in many plants. Without correlating energy consumption to ambient conditions it can be impossible to detect impact of measures taken. In Figure 33-11 the sequence of scoping, implementation of measures, and validation is clearly visualized.

The lines in the upper part show max, mean and min ambient temperature for each hour or day, depending on selected time range.

Top line at lower part shows the baseline consumption based on measured energy during "scoping period"; e.g., the kWh measured during each hour versus average ambient temperature and the average kWh/h create a profile.

Bottom line shows average kWh/hour during selected period, but typically a rolling 12-month period is useful to see trends and detect if consumption increases or decreases. The deviation between the top bars and the bottom curve indicates savings achieved by implementing results of analysis done during scoping phase and gradually introduced during implementation phase.

In Figure 33-12 it is shown how this information can be used to 1) detect deviations like a leak that increases energy consumption versus optimized operation, 2) generate warnings and alarms if it exceeds limits,

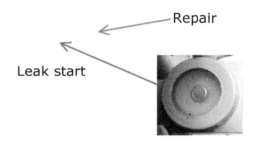

Figure 33-12: Detection of refrigerant leaks through performance monitoring.

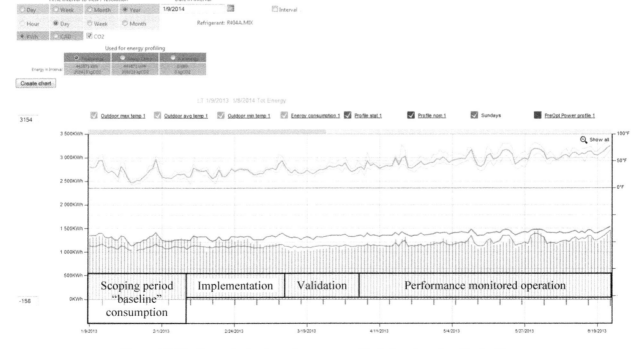

Figure 33-11: Commissioning phases—result of structured optimization

and 3) cost effectively pin-point problems as all details of behavior are available 24/7 through any internet connected device with proper access. *To measure is to know.*

Following are some example pictures of performance analyzing equipment installed in typical projects.

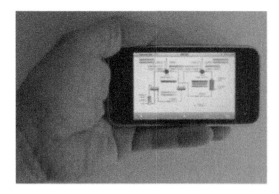

Remote presentation of Performance analyses on smart phone, K:\ClimaCheck Bilder\Bilder till Dropbox\Smartphone_P1010436.JPG

Installing permanent Performance Analyzing on industrial plant

Performance Inspection to validate performance of state of the art heat pump with inverter on compressor
K:\ClimaCheck Bilder\Anläggningar\Schneider_Fregattvägen_2012-01-17\P1020893.JPG

Portable Performance inspection Dubai, K:\ClimaCheck Bilder\Bilder till Dropbox\AC_Dubai_P1000600.JPG

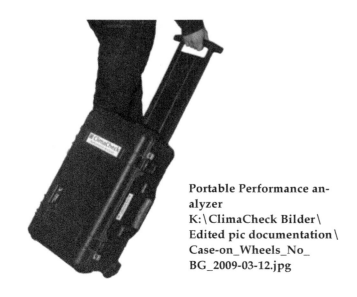

Portable Performance analyzer
K:\ClimaCheck Bilder\ Edited pic documentation\ Case-on_Wheels_No_ BG_2009-03-12.jpg

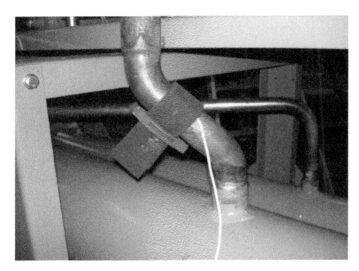

K:\ClimaCheck Bilder\2005-05-16 climacheck web\
DSC05403.JPG

CONCLUSIONS

The heat pump, air-conditioning and refrigeration industry has to phase in a new market situation with more focus on operating efficiency. Experience shows that improvements of 10-40% can be achieved at low investments if plants are operating as intended. This is by far the most "low hanging fruit" to save energy and climate foot print. The saving potential is of strategic interest to businesses as office buildings, hotels and supermarkets often are using 40-60% of their energy to operate air conditioning and refrigeration systems. Increased quality and documentation of commissioning, maintenance and trouble-shooting with portable and permanently installed state-of-the-art performance analyzers decreases energy cost and carbon footprint. Reduced repair cost, down time and minimized bad will

from failures are key driving forces to acceptance of new analyzing methods. The biggest challenge is the necessity to change business as usual in a conservative industry where equipment owners pay energy bills but focus more on initial cost whereas contractors experience problems with getting paid to do proper optimization but not to repair and replace failed components. The price competition on contracts makes compressor replacements and repairs a necessity for many contractors' survival as contracts do not give sufficient margins. The incentives to work proactively are weak, and even if the number of companies categorizing themselves as Energy Service Companies (ESCOs) increases and performance contracting is a buzz word, the HVAC industry with impact of varying loads and climate creates additional legal challenges to make contracts that do not require big legal departments or risk ending up in court.

References

ASHRAE. 2014. Refrigeration Commissioning Guide for Commercial and Industrial Systems. Atlanta : ASHRAE, 2014.

Berglöf, Klas. 2011. PERFORMANCE INSPECTIONS WITH INNOVATIVE ANALYSING EQUIPMENT RESULTS IN SIGNIFICANT ENERGY SAVINGS IN AIR-CONDITIONING AND REFRIGERATION SYSTEMS. Prague : IIR, 2011.

CABA. 2013. http://www.prweb.com/releases/2013/4/prweb10659668.htm. PRWeb. [Online] The Continental Automated Buildings Association, May 2013.

EPBD. 2010. Energy Performance in Building Directive. s.l. : EU-comission, EU-commission, 2010.

Fahlén, P. 1989. Capacity measurements on heat pumps—A simplified measuring method. s.l. : Swedish Counsil for Building Research, 1989. R4:1989.

IEA. 2012. IEA—Energy Technology Perspectives 2012. s.l. : IEA, 2012.

IIR. 2002. Refrigeration Report 2002. Paris : IIR, 2002.

Prakash, John Arul Mike. 2006. Energy Optimization Potential through IMproved Onsite Analysing Methods in Refrigeration. Stockholm : KTH, 2006.

Roaf. 2013. Ozone and Climate-Friendly Buildings. Ohrid : UNEP, 2013.

Chapter 34

Implementing Advanced Diagnostics
In Fleets of Commercial
HVAC Equipment

Dale T. Rossi, Field Diagnostic Services, Inc.

INTRODUCTION

Fault detection and diagnostics technology has the potential to make a positive impact on the cost of operating buildings. That is, to lower utility bills, service and maintenance costs and extend equipment life. Buildings use a significant amount of the energy used in the United States and around the world, and heating, cooling and lighting buildings consume most of the power used in buildings. The roles fault detection and diagnostics play includes finding, analyzing, documenting, and communicating the existence of a fault, degradation or operational anomaly.

Field Diagnostics Services, Inc (Field Diagnostics) has developed three basic types of fault detection and diagnostic technologies.

- The SA Mobile™ application and the optional HVAC Service Assistant™ tool provides in-field diagnostics and is used by technicians to collect and analyze refrigeration cycle data as well as to collect responses to questionnaires used for a variety of purposes including maintenance inspection task reporting. This data is transferred over the internet and reporting is provided from the data to 1) find opportunities with a high potential for energy and non-energy cost savings 2) Improve equipment performance and 3) manage the quality of the service and maintenance work.

- On-board diagnostics is an emerging technology that Field Diagnostics is developing in cooperation with HVAC manufacturers, also known as Original Equipment Manufacturers (OEM's). In addition to the savings mentioned above, this technology has the potential to alert people of operational anomalies prior to loss of comfort or catastrophic equipment failure, either of which can have costly negative impacts on personnel productivity or commerce. Customers don't spend money in over-

ly hot or cold indoor conditions. It is also useful to support and validate service delivery quality.

- Portfolio-level diagnostics uses available data sources including utility bills, smart interval meters, building and energy management systems (BMS/EMS) and NOAA weather data. This analysis can be especially useful to customers with hundreds or thousands of sites because in those types of organizations it is easy for opportunities to be hidden in the large volume of data. It detects, analyzes, documents and communicates the existence of opportunities to save energy and to reduce service costs. These opportunities include continuous lighting operation, sites with abnormally high energy use, and sites and zones that cannot maintain set point under a variety of conditions including heat waves and cold snaps.

TYPES OF FAULT DETECTION AND DIAGNOSTIC (FDD) TECHNOLOGY BEING USED TODAY

In-field Fault Detection and Diagnostics

In-field fault detection and diagnostics refers to computer-based tools and devices that are used by technicians on a job for short periods of time. These devices aggregate and analyze a single data set, or data log for short periods of time ranging from hours or weeks. These devices may also collect additional quantitative and qualitative data useful in managing HVAC maintenance programs, including utility incentive programs.

On-board Fault Detection and Diagnostics

On-board FDD for HVAC units are permanently installed systems that analyze data from sensors or controller signals over a long period of time. There have been attempts to commercialize retrofit/add-on versions of on-board FDD in the past that met with limited success usually because of their cost to purchase and install. OEM-installed FDD is now receiving the most focus and commercialization will occur. OEM-installed FDD is a more attractive option because of many concerns best addressed by the scalability, efficiency and credibility of the information. OEM FDD is just starting to be seen in the marketplace with very limited capacities at the time of this writing (early 2014). It is widely expected that OEM-installed FDD products will increase in both availability and capability in the near future. This development is generally being driven by the reduction in the cost of the components and by developments in code requirements, especially in California.

Portfolio-level Fault Detection and Diagnostics

Portfolio-level FDD currently uses data sources that are not specifically HVAC hardware related. This includes utility bill analysis, service cost analysis as well as trending and analyzing run time and other data available from building and energy management systems (BMS/EMS).

FIELD DIAGNOSTICS PRODUCTS

Field Diagnostics has been developing and delivering FDD products and services that make field data collection easier, more accurate and more analyzable for nearly twenty years. Its primary commercial market focus is end users with hundreds or thousands of rooftop package units and commercial split systems. This includes national retail chains, restaurants and banks along with their service providers.

It also provides FDD products and services for utility quality maintenance (QM) incentive programs around the US. In addition to proven technology, Field Diagnostics recruits customers and service providers for those programs, provides management and training to both and supports utility program implementers.

In-field Fault Detection and Diagnostics

Field Diagnostics developed and produces the HVAC Service Assistant™ tool and SA Mobile™, an application for Smartphones and tablet computers that performs without continuous internet connectivity.

Both platforms assist technicians and others in the chain of responsibility for HVAC performance by collecting the full set of data

required for refrigeration cycle performance analysis. That data is then used to produce diagnostic statements based on the difference between current performance of HVAC systems and the how they should be performing. The definition for how systems should be performing is derived from a model that was developed using manufacturers' published performance and other data. It is called this the "Normal model" or "No-fault model." This type of model is used as a reference to analyze all the different potential equipment designs over the range of driving conditions that equipment is tested under. The SA Mobile software also estimates the current relative efficiency and capacity of the system as compared to what it would be with no-fault performance. The faults detected and diagnosed include the most common refrigeration cycle faults that impact system reliability and energy efficiency:

- Excessive and insufficient low side heat transfer
- Excessive and insufficient high side heat transfer (excessive only during low ambient conditions)
- Excessive and insufficient total charge mass
- Excessive and insufficient (liquid line restriction) refrigerant mass flow
- Reduced compressor pumping capacity
- An advisory to test for non-condensables

SA Mobile™ software is also useful for collecting data with questionnaires, as would be required for equipment maintenance programs. Field Diagnostics

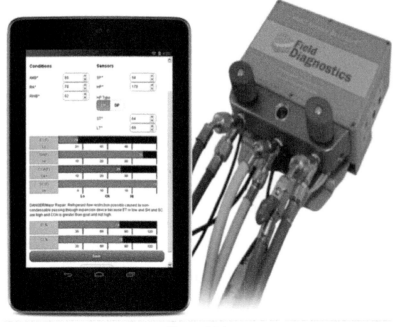

Figure 34-1

provides maintenance-tasking questionnaires that are compliant with the ANSI/ASHRAE/ACCA Standard 180 non-residential and the ACCA 4 residential national maintenance standards. This capability is also used by various utility incentive programs around the United States and for maintenance programs used by national and regional facility managers including retail stores, restaurants and banks.

These tools, used by technicians to collect and analyze data in the field, also easily transmit the data records to a central server using cellular or Wi-Fi connections. That data may then be processed into a variety of reports including those used for technical quality control and return-on-investment analysis for sophisticated financial buyers. These reports can be configured to show unit, site, district and portfolio analysis and evaluations. The opportunities can be monetized and ranked for easy recognition of the opportunities with the most potential.

The Field Diagnostics Mobile Dashboard™ is an application for Smartphones and tablets that allow the mobile service technician to access data and analysis from the BMS/EMS system while they are on-site. They can see twenty-four hour and seven day trends of each zone's set points, the zone temperature the supply air temperature and the calls for cooling or heat. This provides a useful overview of the performance of each zone. This means that a technician can investigate comfort complaints efficiently and pin point the location and the context of a comfort problem quickly and easily. This greatly increases the probability of an effective repair, leading to lower costs and fewer return trips.

The Mobile Dashboard™ also shows the lighting schedules and the lighting operation. Technicians can proactively resolve excessive lighting runtime problems as well as HVAC and zone performance problems while at a site.

The Field Diagnostics Mobile Dashboard™ is the in-field user interface for portfolio level data collection and analysis.

On-board Fault Detection and Diagnostics

Field Diagnostics has developed a software package and sensor specification for Original Equipment Manufacturers (OEM) that is included in control hardware installed in the factory. This software analyzes the performance of the refrigeration cycle and issues advisories over building managements systems and directly to technicians on the job. The faults detected currently mir-

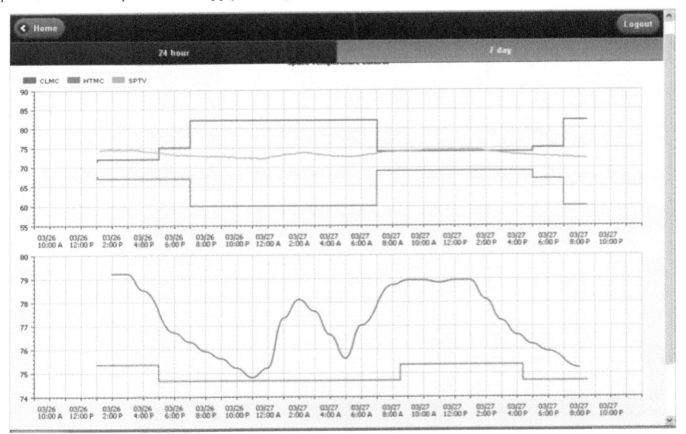

Figure 34-2

ror those detected by the Service Assistant™/SA Mobile ™ in-field FDD technology. On-board FDD collects data continuously. This offers the opportunity to detect faults and degradations that are best observed under specific conditions and to detect subtle faults with much a lower risk of "false-positives," meaning detecting an operational anomaly as a fault when no fault exists, by imposing extended persistence qualifiers prior to issuing a fault notice. Additionally, the embedded diagnostic capability provides a larger continuous and more detailed data set for use with portfolio level diagnostics.

Portfolio-level Fault Detection and Diagnostics

Field Diagnostics provides analytics based on "Big Data" to help define energy and non-energy savings opportunities for large multi-site clients. Recommendations are based on utility bill and service cost analysis as well as Building Management System (BMS/EMS) time-series data. Reports include finding the sites with the greatest savings potential and providing specific and actionable recommendations including HVAC capacity loss, excess ventilation and set point abuse, as well as lighting schedule overrides. Higher level analysis is available to document zones and units that have shown the inability to meet their set point requirements during heat waves and cold snaps. This allows facility managers to target these units to avoid comfort problems and potential facility and equipment damage in the future.

As stated above the Field Diagnostics Mobile Dashboard™ is the in-field user interface for portfolio level data collection and analysis. This capability reduces call center traffic and empowers the field service provider to investigate complaints more efficiently and to be proactive in finding and acting upon problems that cause excess energy use while on site.

When thoughtfully implemented and operationalized, portable, on board, and portfolio-level FDD leverage each other to help maximize energy efficiency (lower utility bills), comfort and reliability (ensure productivity and commerce), and to increase the useful life of HVAC systems in order to reduce and delay capital expenditures.

IMPLEMENTATION OF FAULT DETECTION AND DIAGNOSTIC TECHNOLOGY

In-field Diagnostics

Common initial approach

When applying diagnostic tools in the field, there are two approaches customers commonly choose in non-incentivized environments. One is to authorize a pilot program; often five to ten locations chosen based on some criteria related to energy cost or perceived chronic reliability problems. A second common approach is to add the diagnostic capability to a regularly scheduled visit, usually a maintenance inspection. These attempts at implementing diagnostics have the objective of gathering data, assessing opportunities, effectively addressing issues and reporting achieved benefits.

Another implementation approach to in-field diagnostics has been through utility energy efficiency incentive programs. These programs have the goal of reducing energy use and peak demand by making HVAC systems perform better. Because ratepayer or taxpayer funds are employed, as a rule utility programs cannot make use of proprietary technologies. The unintended consequence of this policy is that patented, market-ready FDD products such as Field Diagnostics' can go unused or unsupported by program implementers for years while less effective technologies and approaches in the public domain are supported.

Despite this barrier, the use of Field Diagnostics' two in-field FDD platforms have been very useful in gathering data about the general condition of equipment in service and the effectiveness of various approaches to diagnostics. They have supported research into the prevalence, the magnitude and the root cause of inefficient system performance.

Field Diagnostics has participated is many utility incentive programs over the years. While meeting the program requirements, Field Diagnostics has delivered analytics and services that exceed most program requirements. The most notable and significant improvement over standard issue publicly-funded programs delivered by Field Diagnostics' approach has been the documentation of both 1) proof that faults have been eliminated by the work done, and 2) the performance gains made by those improvements in terms that make sense to customers The Field Diagnostics approach has been to analyze each refrigeration cycle and apply the services needed to achieve a "Safe and reasonable" diagnosis and at least a 90% efficiency estimate at each unit, based on the Service Assistant™/SA Mobile™ assessment. This methodology is gaining acceptance in the incentive program world and is moving the utility incentive program industry to adopt specifications more along the lines of the Field Diagnostics approach.

Implementation Challenges

HVAC provides several kinds of challenges to facility managers. These include comfort complaints from internal and external clients, perceived and actual high service and energy costs, capitol planning and the complicated decision-making around selecting candidates for replacement, and service provider selection and management. The metrics used to make choices and to judge success are often subjective or poorly focused because of a relative lack of dependable data. There is a need for a simplified and saleable process for interpreting the data that is available. Those in the position to make HVAC maintenance, service and replacement decisions many times are not technical or analytical experts and usually have a different skill set than their "professional energy manager" colleagues where maintenance usually is not considered an energy management measure.

Sometimes, effective implementation of advanced diagnostic techniques are challenged from the onset because objectives are poorly defined, or because the approach requires service providers to behave in ways that unnecessarily increase costs or reduce the value of the outcome.

Some common in-field diagnostic implementation strategies that lead to disappointing results include:

1. Assuming that service providers are already skilled in the new technology and do not require more than a few hours of training.

2. Using an incumbent service provider that is unsuited or not interested in advanced technology or change itself.

3. Selecting a small group of sites based on some criteria other than evidence of a savings opportunity that is capturable through maintenance. These may include selecting sites near their office or sites where the HVAC equipment is beyond reasonable repair.

4. Setting expectations that each unit at a selected site will be working "perfectly" when the work is complete, this means expending valuable resources on units that are performing adequately, merely because they exist at the same site as poor performers and therefore are under added scrutiny during the project.

Lessons Learned

HVAC equipment's current performance, relative to design expectations is highly variable because of a range of factors. These include the equipment's age, how well it is maintained, how much it runs and, how well the system it is a part of was originally designed and constructed. Some fleets of equipment are well

maintained while others less so. Experience has shown that the best maintained equipment may have as few as 10% of the fleet performing poorly. However there are some fleets where essentially every unit has serious performance problems. On average, in a normally maintained fleet of HVAC units, about a quarter to a third of the units will represent 80%-90% of the energy savings and bill reduction opportunity. Finding and documenting the units that are performing poorly, prior to applying a basket of services is a good initial approach. Auditing the entire fleet and producing an accurate current inventory with enough data to detect and rank opportunities is a definite best practice.

Establishing achievable goals for the implementation and then designing an approach that takes into account the condition of the equipment, the budget available for investment in performance improvement and the capabilities of the people and systems involved in implementing the solution greatly increases the probability of success. There are many variables that make defining a cookie-cutter solution that will work everywhere is difficult. However there are some characteristics of a successful plan including:

1. Define the group of units that are performing poorly and have the better opportunities for measurable improvement and return on investment.

2. Target that group in a way that focuses most of the effort on the units with the performance problems. Some examples might be a plan where only the units identified as poor performers are addressed, or a plan where the performance of all the units at a site are averaged and the sites are ranked and a common basket of services are applied to all units at the sites with the most savings opportunity.

3. Understand that customers have budgets and package solutions that do not exceed them. By ranking the units or sites and focusing on the larger opportunities, the budget can be managed to produce the best return with the available investment and the project can pause when the budget is exhausted.

4. Be clear with the service provider about the goals of the program, how success is being measured and how to communicate unexpected information early in the process. Bring the service provider into the planning process and get agreement to the plan from the whole team.

5. Be flexible when unanticipated obstacles arise. Very often the best answer is to temporarily bypass

a problematic site or unit and then re-engage it when the issue is resolved. Some examples of this may include sites with access problems or sites or units that need repairs outside the scope of the program like compressor or fan motor replacement.

6. Train the people that are doing the work. This includes training on the use of the technology, effective testing procedures and equipment performance prior to the audit and then training on effective cleaning and adjustment procedures when remediation work has commenced. Write clear step-by-step instructions and make them available on a single laminated page for use in the field. Do not assume that any technician, regardless of how much experience they have will know what the expectations are and what the definition of success is for a job.

7. Have a plan for reporting results and an expectation that there will be a formal meeting where final results are delivered to the people that will make the judgments about the effectiveness of the program. Work to get explicit agreement about the resources expended and the benefits produced.

Benefits Produced

There are a range of benefits that could be expected from the implementation of in-field diagnostic technology in a HVAC maintenance program. These include:

1. A current and accurate equipment inventory

2. Data about the condition and performance of each unit that can be processed into effective reporting:
   a. Exception reports showing unfulfilled maintenance requirements
   b. List of units that are inoperable and what is required to return them to service
   c. Ranked list of opportunities to save energy
   d. Ranked list of opportunities to resolve problems that lead to service interruptions and premature compressor failures
   e. Ranked list of replacement candidates based of a pre-determined selection formula

On-board Diagnostics

Common Initial Approach

On-board diagnostic capabilities in HVAC equipment is on the cusp of commercialization primarily because, despite barriers, the potential benefits of this technology are high. A university study concluded that automated FDD reduces service costs due to reduced

maintenance inspection (tasks), fault prevention, lower cost fault detection and diagnostics, better scheduling of multiple service activities and better scheduling to the low season. Operating cost savings consist of utility cost and equipment life savings.[1]

Implementation strategies are still being worked out. The current theory is that the diagnostic messaging will be communicated in several ways. Units that exist on a building network will carry the diagnostic messaging, presumably integrated into the BMS/EMS front-end. However, the great majority of HVAC units are "stand alone" meaning no such network connection exists. It is envisioned that these units will integrate with indicators expected to be included on the thermostat interface to alert the building occupants of detected issues. Additionally it is assumed that there will be a way to inform the technician at the unit either through a display in the unit or through a user interface brought to the job. The thought is that it may be a specialized tool that will plug into an access point provided or a general purpose device, like a Smartphone or tablet computer that will communicate wirelessly. Ultimately, the expansion of the wireless Internet is likely to make communication with each unit simple.

It is currently assumed that diagnostic messaging from OEM diagnostic modules will be managed by each OEM's proprietary system. It seems likely based on the history of technology development that an open system will ultimately be the choice of most manufacturers and consumers.

There have been attempts to retrofit diagnostic capabilities into existing HVAC units. The cost of the equipment and especially the installation cost made them unattractive when combined with the immature communication infrastructure and absence of a common system to manage the diagnostic outputs. However, a retrofit solution may make sense for larger units with complicated controls and applications with persistent performance problems.

Implementation Challenges

At the current state of development, the most significant barrier to implementation is the lack of availability of OEM-enabled HVAC units in the marketplace. This is likely to be addressed in the next few years. Product delivery is expected to start in 2014.

When product is available, it is easy to predict that early adopters will embrace the technology, and assuming the rollout is managed effectively, and that the infrastructure to manage the diagnostic messaging is forthcoming, wider adoption will follow.

When that happens, resistance to change and risk aversion are likely to be factors hindering wide scale adoption, while code requirements and perhaps other encouragements and incentives will drive wider adoption.

Portfolio Level Diagnostics

Common Initial Approach

When implementing a "Big Data" analytics strategy, it is easy to see a large variety of opportunities and attempt to "boil the ocean," for example to attempt to integrate several data sources and provide analytics to produce a comprehensive management system for energy opportunities across their portfolio. Such a goal might seem right to ambitious facilities and energy managers, but in reality, setting such an unrealistic goal is likely to lead to frustration and disappointment.

Experience has shown that you can spend a lot of money and time, and discover far more opportunities than the facilities management and mobile workforce systems in place can possibly react to. Successful portfolio analytics implementation programs start by trying to do some simple useful things and then build on small successes incrementally.

One example of a measured approach might include determining the top ten sites in terms of energy use measured in kWh per square foot from analyzing the utility bill and lease data. Then, using the time series data from the BMS/EMS systems, try to determine if the problem is a lighting control problem or a HVAC performance problem. With this information, a technician can be dispatched to investigate further and hopefully find a cause of the problem that can be repaired. The following month, perhaps five more sites could be added to the list. After several months, if things are going well, an impact on energy use will start to be evident in the utility bill analysis. After several more months, some sites may be declared "fixed" and removed from the list. Some sites will probably be "stubborn" and require specialized investigation to determine the cause of the abnormal energy use.

Another example might be to use the BMS/EMS data to detect units and zones that are not performing well. There may be several levels of severity to consider. For example, in order to keep the number of opportunities to a minimum, the goal might be to find zones that the data shows have not maintained the set point only during heat waves and cold snaps in the past. The aim would be to reduce the rush of comfort complaints during extreme weather conditions. After that has been addressed, the goal may switch to finding zones that are currently not maintaining comfort. That will be a larger list and may take a substantial amount of time to clear and have a variety of root causes. A very sensitive version of this concept would be to find the zones that are maintaining the set point but where the units are running nearly continuously to accomplish this.

Another example of a simple approach that pays well in terms of energy saving and cost reductions would be to find the sites where the interior or the exterior lights remain on continuously. Depending on the specifics of the organization and the facilities, addressing this might be as simple as making phone calls to the sites and asking that the hand/off/auto switches be returned to "auto." This works sometimes. However, there are times when the reason for the override becomes clear and a schedule change is required to make the override unnecessary. There will be other times when the people at the site just disagree with the level of lighting during reduced lighting periods. There will be sites where there are equipment failures that must be addressed to get the system back under control.

An example of a relatively sophisticated analysis that can make a huge impact on energy use if acted upon is analyzing the relationship between the outside air temperature, the return air temperature or zone temperature and the supply temperature when the fan is running without heating or cooling. This analysis is used to quantify the amount of air in the supply that came from the outside. Resolving over-ventilation is one of the highest-impact energy savings opportunities available in buildings today. In some cases, something as easy and inexpensive as adjusting a programming variable, like the changeover set point, is something that can be done once and has an impact on the entire portfolio.

Some organizations prefer systems or processes that are self-implementing and self-managing. An example of this would be to provide information about energy use and HVAC performance and lighting operation to site managers or to mobile workforce technicians and creating an incentive for improvement. Incentives are tricky however and it may take several iterations before an incentive scheme is found that produces the desired behavior.

Implementation Challenges

There are challenges at several levels when implementing an analytics strategy that may have to be overcome to produce good data, accurate analysis and action to resolve the operational problems and achieve measurable results. Some of the challenges may include gathering the data required. Some of the obstacles may include the availability of, or access to data, inconsistent naming conventions and inconsistent data collection strategies. Some of these challenges may be resolved relatively easily. Others will not and will limit the analytical possibilities, at least in parts of the portfolio. These obstacles might include permissions to access data, incapable or uncooperative vendors or in some cases, the required data is simply not being collected, or is not collected in a way that can be accessed at scale.

Having accessed the data that is going to be collected, the data will need to be validated and normalized. The data will need to be tested for validity and invalid data removed. It is helpful to create statistics to help judge the quality of the data for analysis. For instance if 95% of a particular data point is invalid, the usefulness of that data may be limited.

If the sites are distributed geographically, it will be necessary to do weather normalization to compare the sites to one another. It is so common as to be considered a universal fact that the cooling capacity per square foot will range widely across a portfolio and the variation is usually not easily explainable. Normalization for equipment density is required if valid runtime analysis is a goal. Other normalizations include variations in operating hours and the activities carried out in the facility as it may cause wide variations in internal heat gains.

Lessons Learned

Experience has shown that overcoming technical challenges are relatively straight forward compared to the challenges introduced by people. Finding opportunities is easier than getting them acted upon. Change management within an organization is an important part of any implementation process involving people. Change management is a field of study in itself.

American John P Kotter (b 1947) is a Harvard Business School professor and leading thinker and author on organizational change management. Kotter's highly regarded books 'Leading Change' (1995) and the follow-up 'The Heart of Change' (2002) describe a helpful model for understanding and managing change. Each stage acknowledges a key principle identified by Kotter relating to people's response and approach to change, in which people see, feel and then change.

Kotter's eight-step change model can be summarized as:

1. Increase urgency—inspire people to move, make objectives real and relevant.

2. Build the guiding team—get the right people in place with the right emotional commitment, and the right mix of skills and levels.

3. Get the vision right—get the team to establish a simple vision and strategy focus on emotional and creative aspects necessary to drive service and efficiency.

4. Communicate for buy-in—Involve as many people as possible, communicate the essentials, simply, and to appeal and respond to people's needs. De-clutter communications—make technology work for you rather than against.

5. Empower action—Remove obstacles, enable constructive feedback and lots of support from leaders—reward and recognize progress and achievements.

6. Create short-term wins—Set aims that are easy to achieve—in bite-size chunks. Manageable numbers of initiatives. Finish current stages before starting new ones.

7. Don't let up—Foster and encourage determination and persistence—ongoing change—encourage ongoing progress reporting—highlight achieved and future milestones.

8. Make change stick—Reinforce the value of successful change via recruitment, promotion, new change leaders. Weave change into culture.

Kotter's eight-step model is explained more fully on his website ww.kotterinternational.com.

Adding analytics to an organization's process is very often not just change, but also a design-as-you-build project. It is not always clear in the beginning which analytical insights will be found to be useful and usable to the team. Attempting small but useful changes over a period of time greatly reduces resistance and increases the probability of success. Contrary to the conventional wisdom, the hardest part of implementing a performance improvement strategy is not getting started, it is getting beyond just getting started and getting the first measureable success. Everything gets easier when people see something working and producing a tangible benefit.

Portfolio level analytics works best as part of an overall analytical strategy that includes in-field diagnostic technology and a management system that drives results. In such a system, portfolio-level diagnostics provides evidence that an opportunity exists at a particular site and sometimes points to a particular system as a suspect. In-field diagnostic technology and data capture may be needed to acquire more granular information to identify specific actions or to capture data about what was found and what was done to resolve the problem. Capturing resolution data can be valuable to inform future investigations and for reporting.

Reporting results to decision makers is essential to getting agreement that benefits are being captured. Getting agreement about the costs and benefits derived from the effort is often required to maintain funding for the project. In a very real way, if success is not reported, it does not exist.

Benefits Produced

The main benefit of analyzing data at the portfolio level is to, in a systematic way, find, define, document and rank the opportunities, including their potential financial value, so that they may be addressed in a logical way and to measure and report improvements. The opportunities found may be systematic, meaning a change made will have an impact across the portfolio, or they may be site or unit specific issues.

Systematic improvements might involve adjustments to control strategies, or changes to policies like the maintenance protocol. Site-specific improvements include finding units bringing in excessive amounts of outside air because of faulty or poorly designed or installed economizers.

The main benefit sought is usually a reduction in costs. These could be energy costs as in the case of finding the sites with the lighting on continuously, or there could be energy reductions combined with comfort improvements as would result for resolving excessive ventilation cases. It also might be reduced service costs by detecting units with diminished capacity and making that information available at the time when a technician is at the site for another reason and can address the issue without a special truck roll. The largest return on investment comes from finding units that are performing poorly and targeting them for visits by technicians, sometimes equipped with in-field diagnostic technology, with the desired result being finding and resolving operational anomalies that would otherwise lead to premature failures of the compressor or gas heat exchanger.

Proactive maintenance has the best return on investment because compressor and gas heat exchanger failures often precipitate an unplanned unit replacement and the net present value of a few years of additional equipment life is often very high compared to the cost of solving the problem prior to failure. This was made clear by a widely distributed study of the economic value of preventive maintenance by one of the largest property management companies in the world. That study shows

that "an investment in preventive maintenance not only pays for itself but also produces a huge return on investment." The study went on to say "At the portfolio level, the analysis indicated… a ROI of 545 percent. The bulk of the return comes from increasing the useful life of equipment."

CONCLUSION

Field Diagnostics Services, Inc. has been developing and deploying HVAC diagnostic technology and using it to create energy savings benefits and non-energy benefits for end users and service providers since the mid 1990's. Nearly twenty years of experience has shown that the common expectation that HVAC equipment performs poorly can be true. There has been some hard won wisdom gained in that time, including these ideas:

- While HVAC equipment performance is too often weak, finding and addressing the specific needs of the weak performers is far more cost effective than applying a common basket of services to every unit in a portfolio.

- Old habits die hard. The customer must drive the adoption of diagnostics and more sophisticated servicing techniques. Contractors do what the customer requires and is willing to pay for; technicians do what the contractors require for them to keep their jobs.

- Contractors will be profitable or they will cease to exist. Pressure from customers to reduce maintenance costs inevitably reduces the time technicians spend with equipment and just as inevitably, HVAC units performing poorly will continue in that state longer and fail earlier. Many fixed price contracts make any activity that doesn't directly impact comfort an uncompensated expense to the service provider. By allowing for extra payments for compressor replacements and no payments for activities that reduce premature compressor failures, customers create an incentive for service providers to neglect their equipment and actively reduce their service life. This is a particularly counter-productive strategy for the end user.

- Technicians need to be trained to use diagnostic technology and they also need to be trained and encouraged to do the tasks that produce good running units. It is not logical to assume that a technician will work harder to solve a problem that does not impact them personally without encouragement and support, and even the requirement from their employer.

- The customer must see and feel the benefits of diagnostics in terms of increased occupant satisfaction and reduced operating costs in order to invest in quality maintenance and better servicing techniques. Data driven reporting is key to customer adoption.

- There are opportunities to find additional savings with very attractive returns on the investment in data analysis by finding easy to understand and easy to repair operational anomalies like the continuous operation of lighting by analyzing utility bills and BMS/EMS time series data.

- There are further savings opportunities provided by more sophisticated analysis of zone temperature performance and runtime analysis.

- Providing access to existing lighting operation, zone temperature and set point data and zone temperature performance to the mobile service technician empowers the technician to be proactive, resolve problems, and produce energy savings with very limited additional cost when they are already at a site.

References

1. Li, H., and J.E. Braun. 2007. Economic Evaluation of Benefits Associated with Automated Fault Detection and Diagnosis in Rooftop Air Conditioners. *ASHRAE publication* LB-07-023
2. Excerpted from http://www.businessballs.com/changemanagement.htm
3. http://maintenance5.com/Determining-the-Economic-Value-of-Preventive-Maintenance-pdf-e585.html

Chapter 35

Energy Savings Benefits of FDD Systems for Light Commercial Unitary HVAC Equipment

Matthew Tyler, P.E. (PECI)

INTRODUCTION

In May 2012, the California Energy Commission approved the proposed revisions to the Building Energy Efficiency Standards (Title 24, Part 6), which take effect in July 2014. One of the significant changes is a new mandatory requirement for Fault Detection and Diagnostics (FDD) on light commercial unitary HVAC equipment. This requirement applies to all newly installed air-cooled unitary direct-expansion units, equipped with an economizer and with mechanical cooling capacity greater than or equal to 54,000 Btu/hr. The FDD system must detect the following faults:

- Air temperature sensor failure/fault
- Not economizing when it should
- Economizing when it should not
- Damper not modulating
- Excess outdoor air

This is an abbreviated list from a longer list that the authors initially considered. The authors investigated numerous HVAC faults while developing these new Title 24 requirements. This chapter describes the energy savings analysis and benefits of FDD systems in detecting these faults on light commercial unitary HVAC equipment.

LIST OF FAULTS

The authors investigated numerous HVAC faults to determine the potential benefit of FDD systems in detecting these faults. These include the following:

1. Air temperature sensor failure/fault
2. Low refrigerant charge
3. High refrigerant charge
4. Compressor short cycling
5. Refrigerant line restrictions/TXV problems
6. Refrigerant line non-condensables
7. Low side HX problem
8. High side HX problem
9. Capacity degradation
10. Efficiency degradation
11. Not economizing when it should
12. Economizing when it should not
13. Damper not modulating
14. Excess outdoor air

As described in more detail later in this chapter, the authors conducted a series of computer simulations to estimate the potential energy savings resulting from FDD systems. Descriptions of the investigated failure modes and the computer simulation assumptions are described below.

1. Air Temperature Sensor Failure/Fault

This failure mode is a malfunctioning air temperature sensor, such as the outside air, discharge air, or return air temperature sensor. This could include miscalibration, complete failure either through damage to the sensor or its wiring, or failure due to disconnected wiring. Calibration issues are more common than sensor failures, thus we modeled this fault as a calibration problem. Temperature sensors are commonly accurate to $\pm 0.35°F$. For a conservative estimate we modeled this fault as $\pm 3°F$ accuracy. Calibration errors greater than this and failed sensors will contribute to an even worse energy impact.

2. Low Refrigerant Charge: 80% of Nominal Charge

Incorrect level of refrigerant charge is represented in this failure mode, designated by a 20% undercharge condition (80% of nominal charge). Refrigerant undercharge may result from improper charging or from a refrigerant leak. While the most common concern about a refrigerant leak is that a greenhouse gas has been released to the atmosphere, a greater impact is caused by the additional CO_2 emissions from fossil fuel power plants due to the lowered efficiency of the HVAC unit.

A typical symptom is low cooling capacity as the evaporator is starved of refrigerant and cannot absorb its rated amount of heat. This causes a high evaporator superheat as the receiver is not getting enough liquid refrigerant from the condenser, which starves the liquid line. The thermal expansion valve (TXV) experiences abnormal pressures and cannot be expected to control evaporator superheat under these conditions. The compressor is pumping only a small amount of refrigerant. Essentially, all the components in the system will be starved of refrigerant.

The computer simulation software does not allow a specific input related to refrigerant charge. However, it does allow specifying the HVAC unit's Energy Efficiency Ratio (EER). Based on laboratory testing, a 20% undercharge condition is equivalent to a 15% reduction in the unit's rated EER.

3. High Refrigerant Charge: 120% of Nominal Charge

Incorrect level of refrigerant charge is represented in this failure mode, designated by a 20% overcharge condition (120% of nominal charge). This fault was added to the list after conducting the energy analysis and therefore is not included in the energy analysis. The energy analysis is thus conservative as it does not include this fault.

4. Compressor Short Cycling

Compressor short cycling means that the compressor is enabled again shortly after being stopped for only a brief period of time. Some manufacturers recommend a minimum runtime of three minutes and minimum off time of two minutes. Thus, short cycling could be considered a runtime shorter than three minutes and off time shorter than two minutes. Short cycling can originate from many sources, for example coil blockage, equipment oversizing and a poor thermostat location (e.g. near a supply air diffuser).

It takes about three minutes of runtime for an HVAC unit to achieve steady state operation and full cooling output. During this time, the unit efficiency is reduced as the refrigerant pressures are established and the evaporator coil cools down. When a unit is short cycling, the startup time becomes a higher fraction of the total runtime. The startup losses thus become a higher fraction of the total cooling output such that the overall efficiency is reduced.

A runtime of three minutes and off time of two minutes corresponds to a runtime fraction of 60% and

an efficiency penalty of 10% according to a design guide for small HVAC systems.[1] The computer simulation software does not allow a specific input related to compressor short cycling. Instead, the simulation used -10% EER (10% reduction in the unit's rated EER), equivalent to a runtime fraction of 60%.

5. Refrigerant Line Restrictions/TXV Problems

Refrigerant line restriction means the refrigerant flow rate is constrained due to a blockage in the refrigerant line. A restriction always causes a pressure drop at the location of the restriction. A suction line restriction will cause low suction pressure and starve the compressor and condenser. This can be caused by restricted and/or dirty suction filters or a bent or crimped refrigerant line from physical damage. A liquid line restriction will cause low pressure and a temperature drop in the liquid line and starve the evaporator, compressor, and condenser. This can be caused by a bent or crimped refrigerant line, a restricted and/or dirty expansion device such as a TXV, a restricted liquid line filter/dryer, or a pipe joint partially filled with solder. In the case of a bent refrigerant line, it acts like an expansion device such that two expansion devices effectively operate in series causing a higher than normal pressure drop. The low evaporator temperature can freeze the evaporator coil and suction line.

The computer simulation software does not allow a specific input related to this fault. Instead, the simulation used -56% EER. This comes from laboratory testing, which reports that reduced mass flow rate caused by a liquid line restriction reduces the EER by 56%.[2] Based on the same lab testing, a restriction in the suction line decreased the EER by 27%. We chose to model the EER penalty as 56% since there is a higher probability of damage to the liquid line as the suction lines are relatively sturdy.

6. Refrigerant Line Non-condensables

This refers to contamination in the refrigerant lines. This is commonly air, water vapor, or nitrogen. They enter the system through leaks or poor service practices, such as not purging refrigerant hoses while working on a unit or not completely evacuating the system after it has been open for repair. The only fluids in a refrigeration system should be refrigerant and oil. Any other fluids within the system can reduce its cooling capacity and lead to premature failure. When air enters a system it will become trapped in the condenser and will not condense. This results in less surface area available for the refrigerant to condense, thus decreasing the capacity of the condenser and increasing its pressure. This causes the compressor to work harder, degrading its efficiency and potentially damaging it by overheating.

The computer simulation software does not allow a specific input related to refrigerant line non-condensables. Instead, the simulation used -8% EER, which comes from laboratory testing.[3]

7. Low Side (Evaporator) Heat Exchange Problem

This failure mode is low airflow through the evaporator coil as measured at the unit's supply air discharge. This could be caused by an evaporator coil blockage for example. When the evaporator coil has a reduced airflow, there is reduced heat load on the coil. This can cause the refrigerant in the coil to remain a liquid and not vaporize. The liquid refrigerant will travel past the evaporator coil and reach the compressor, thus flooding and damaging it. Airflow rates around 400 cfm/ton are desirable, however, a design guide for small HVAC systems reports that 39% of units have airflow less than or equal to 300 cfm/ton.[4]

The computer simulation software does not allow a specific input related to evaporator coil heat exchange problems. Instead, the simulation used -5% EER, which is equivalent to a low airflow of 300 cfm/ton, from the Mowris study.[5]

8. High Side (Condenser) Heat Exchange Problem

This failure mode was modeled as a 50% condenser coil blockage. In this case, the condenser coil fails to properly condense the refrigerant vapor to a liquid.

The computer simulation software does not allow a specific input related to condenser coil heat exchange problems. Instead, the simulation used -9% EER, which is equivalent to a 50% condenser coil blockage, from the Mowris study.[6]

9. Capacity Degradation

This fault was added to the list after conducting the energy analysis and therefore is not included in the energy analysis. The energy analysis is thus conservative as it does not include this fault.

10. Efficiency Degradation

This fault was added to the list after conducting the energy analysis and therefore is not included in the energy analysis. The energy analysis is thus conservative as it does not include this fault.

11. Not Economizing When It Should

This fault was represented as an economizer high limit setpoint of 55°F instead of 75°F. An economizer is equipped with a changeover (high limit) control that returns the outside air damper to a minimum ventilation position when the outside air is too warm to provide cooling. This failure mode is easily modeled in the computer simulation software by changing the high limit setpoint from 75°F (base case) to the failure mode of 55°F. The 55°F setting instead of the 75°F setting results in missed opportunities for free cooling between the range of 55°F and 75°F, thus losing a large number of economizer hours and energy savings potential.

12. Economizing When It Should Not

This is opposite to the previous case of not economizing when it should. In this case, however, conditions are such that the economizer should be at minimum ventilation position but for some reason it is open beyond the correct position. This leads to an unnecessary increase in heating and cooling energy. This could be represented as the economizer high limit setpoint is higher than 75°F or the economizer is stuck open.

13. Damper Not Modulating

This was represented as an economizer stuck closed. When the economizer damper is stuck closed the unit fails to provide any ventilation and is a missed opportunity for free cooling, thus causing an energy penalty during periods when free cooling is available. This was modeled as "no economizer" in the computer simulations.

14. Excess Outdoor Air

This was represented as an economizer stuck fully open. When the economizer damper is stuck open the unit provides an excessive level of ventilation, usually much higher than is needed for design minimum ventilation. It causes an energy penalty during periods when the economizer should not be enabled, that is, during heating and also when outdoor conditions are warmer than the economizer high limit setpoint. During heating mode the stuck open economizer will bring in very cold air and the gas usage will increase significantly. This was modeled as 100% outside air in the computer simulations.

ENERGY SIMULATION

The estimated energy savings resulting from the use of FDD are based on a series of computer simulations. This analysis used EnergyPro (developed by EnergySoft, LLC), which is a building energy simulation software commonly used in California to show compliance with the California building energy efficiency standards (Title 24).

Based on input from stakeholders, the researchers shortened the list of faults under investigation from fourteen to eleven. The simulations included seven of the eleven faults, seven different commercial building types, and sixteen California climate zones.

- 7 building types
- 16 climate zones
- 8 fault scenarios (1 base case, 7 faults)
- 896 total simulations

Figure 35-1 summarizes a number of key inputs used in the energy simulations.

A series of prototype buildings was developed that are based upon actual project designs in terms of building configuration. For example, an actual big box retail store was used for the large retail prototype to provide a realistic approximation of glazing area, number of stories and building geometry. Each prototype building was also configured with Title 24 standard assumptions for insulation levels, glazing type, lighting power density, occupant densities, ventilation rates and other characteristics. The HVAC systems in each case were configured as typical light commercial unitary systems with gas heat, electric air conditioning, and an economizer.

Each prototype building had eight associated EnergyPro simulation models. One model was the baseline case that assumes a perfectly functioning HVAC system. The other seven models included some specific operational fault or degradation of certain portions of the HVAC system. Once each prototype was developed in EnergyPro, the computer simulations commenced for the various climate zones. The energy impact of the faults was then calculated as the difference between the annual energy consumption of the fault models and the baseline models.

EnergyPro is not capable of modeling all the particular faults under investigation. However, these faults can be described by a corresponding EER penalty, which can be easily modeled in EnergyPro. The impact to the unit EER due to each of these faults was available through a literature review.

EnergyPro can thus calculate the annual energy consumption as a function of EER for a given building type and climate zone. Simulations are needed only at three EER values to illustrate the relationship between EER and annual energy consumption. In fact, the energy penalty of a fault is directly proportional to the EER penalty. Thus, the energy impact of any additional faults described by an EER penalty can be derived from these three models via interpolation. Any failure modes not described by an EER penalty will of course still require a unique simulation. This is summarized below in Figure 35-2. An example interpolation is shown in Figure 35-3 for a 5-ton rooftop unit (RTU) at a small office in California climate zone 12, which is a portion of the Central Valley including Sacramento.

PROBABILITY ANALYSIS

As described earlier, the energy impact of the faults was calculated as the difference between the annual energy consumption of the fault models and the baseline models. However, this assumes a 100% failure rate, a 100% chance of the FDD system detecting the fault, and a 0% chance the fault would be detected without an FDD system. In reality, not all units will experience all these faults, the chance of the FDD system detecting the fault is less than 100%, and the chance the fault would be detected without an FDD system is greater than 0%. It is necessary to account for this to avoid overestimating the potential energy savings from implementing an FDD system. This section describes the methodology used to

Prototype Building	Occupancy Type	Area, sf	Number of Stories	# HVAC Systems	Cooling capacity, tons	Avg sf/ton
1	Fast Food	2,099	1	2	11	199
2	Grocery	81,980	1	18	249	329
3	Small Retail	8,149	1	4	25	330
4	Large Retail	137,465	1	22	286	480
5	Small Office	40,410	2	14	113	356
6	Large Office	112,270	2	10	421	267
7	School	44,109	2	39	171	257

Figure 35-1: Summary of Energy Simulation Models for FDD

Failure mode	EER penalty	Energy savings calculation method
Low side HX problem (50% evaporator coil blockage)	5%	Simulation
Refrigerant charge: 80% of nominal charge	15%	Simulation
Performance degradation: 30% cond. block, 300 cfm/ton, -10% charge	21%	Simulation
Refrigerant line non-condensables	8%	Interpolation
High side HX problem (50% condenser coil blockage)	9%	Interpolation
Compressor short cycling	10%	Interpolation
Refrigerant line restrictions/TXV problems	56%	Extrapolation
Economizer high-limit setpoint 55°F instead of 75°F	Unavailable	Simulation
OAT sensor malfunction	Unavailable	Simulation
Damper not modulating (economizer stuck closed)	Unavailable	Simulation
Excess outdoor air (economizer stuck open)	Unavailable	Simulation

Figure 35-2: FDD Failure Modes by EER Penalty

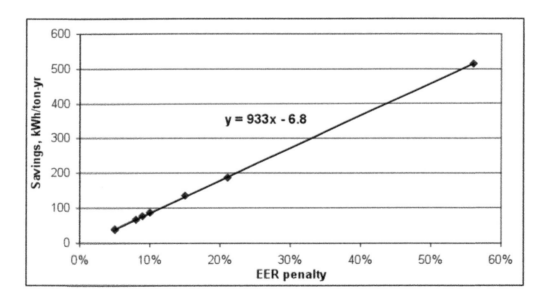

Figure 35-3: Electric Savings as Function of EER Penalty, 5-ton RTU, Small Office

estimate the failure rate and the probability of detecting the faults with and without an FDD system. This method does not account for any interactive effects if multiple failures are encountered, but provides a reasonable distribution of outcome for each test.

This analysis relies on fault incidence. Incidence is the frequency at which a fault occurs in a specific time period or the rate of occurrence of new cases of a fault in the population of interest (e.g., all RTUs in California).

$$\text{Incidence} = \frac{\text{Number of units in a population developing the fault in a time interval (e.g., a year)}}{\text{Total number of units in the population during the time interval of measurement}}$$

This is not to be confused with prevalence, which is the number of cases that exist in the population of interest at a specific point in time, for example, the number of economizer faults in all packaged units in the U.S. presently.

$$\text{Prevalence} = \frac{\text{Number of units in the population with the fault at a specific time}}{\text{Total number of units in the population at a specific time}}$$

For example, with regard to the refrigerant line restriction fault, it is reported as a 60% probability that a filter/dryer restriction fault will occur once during the equipment lifetime.[7] Adding the probability of damage to the liquid line and other restrictions yields an estimated 75% probability for a refrigerant line restriction/TXV fault during the equipment lifetime. Considering the average air conditioner lifespan of 18.4 years as reported by the DOE[8], the annual incidence is 75% ÷ 18.4 = 4.1%. This means 4.1% of RTUs will develop a refrigerant line restriction fault each year. Considering the 15 year nonresidential analysis period, 62% (4.1% x 15) of RTUs will develop a refrigerant line restriction fault within 15 years.

PECI's AirCare Plus (ACP) program provided incidence data for a number of HVAC faults. ACP is a comprehensive diagnosis and tune-up program for light commercial unitary HVAC equipment between 3 and 60 tons cooling capacity. This program has been active throughout the PG&E service territory since 2006 and throughout the Southern California Edison service territory since 2004. It includes inspection of the following HVAC components: thermostat controls, economizers, refrigerant charge and airflow. The ACP program database includes over 17,000 RTUs with documented status of these HVAC components. This massive collection of HVAC data proved useful in identifying the incidence of various HVAC faults.

Figure 35-4 and Figure 35-5 show the number of faults identified by the ACP program as a function of the unit's vintage. The slope of the linear trend lines indicates the number of new faults per year. This is presented for the first five years of a unit's lifetime. In other words, this dataset contains the newest units in the entire ACP dataset. This allows for new equipment design and factory assembly and quality control processes that may affect the incidence of faults, while avoiding most obsolete designs and processes. To convert this data to incidence, these numbers of new faults per year are simply divided by the total number of units in the population during the time interval of measurement (units tested/yr). Figure 35-6 summarizes the results for a variety of faults.

This analysis still assumes a 100% chance of the FDD system detecting the fault, and a 0% chance the fault would be detected without an FDD system. In reality, not all units will experience all these faults. The chance of the FDD system detecting the fault is closer to 75%. The chance the fault would be detected without an FDD system varies depending on typical service and if the fault impacts comfort conditions.

The following fault is quite likely detected by the economizer acceptance test (functional test required in California for new equipment with an economizer) or through regular service such that the fault is 75% likely to be detected:

- Economizer high-limit setpoint 55°F instead of 75°F

The following fault is likely detected through regular service and/or impact comfort conditions such that the fault is 50% likely to be detected:

- Refrigerant charge: 80% of nominal charge

The following list of faults is less likely detected through regular service and does not impact comfort conditions such that these faults are only 25% likely to be detected.

- OAT sensor malfunction
- Compressor short cycling
- Refrigerant line restrictions/TXV problems
- Refrigerant line non-condensables
- Low side HX problem incl. low airflow (50% evaporator coil blockage)
- High side HX problem (50% condenser coil blockage)
- Damper not modulating (economizer stuck closed)
- Excess outdoor air (economizer stuck open)

Figure 35-7 summarizes the results of the probability analysis. The FDD benefit is the difference between the probability of detecting the fault with FDD and the probability of detecting the fault without FDD.

ENERGY SAVINGS

Based on input from stakeholders, the researchers decided to shorten the list of faults proposed for the 2014 Title 24 requirements. The energy savings described in this section are thus based on only a subset of the longer list of faults described earlier. Per the 2014 Title 24 standards, the FDD system is required to detect the following faults:

- Air temperature sensor failure/fault
- Not economizing when it should
- Economizing when it should not
- Damper not modulating
- Excess outside air

For the failure modes that were not simulated, lin-

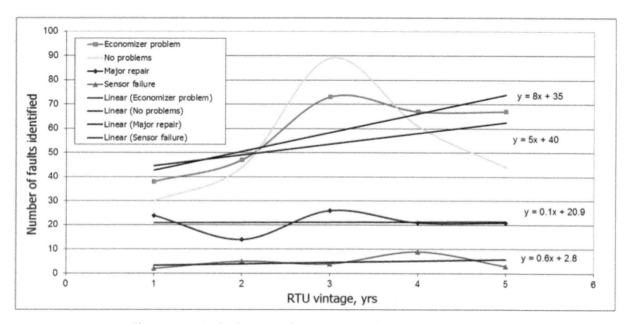

Figure 35-4: Faults by RTU Vintage: Economizer and Sensor Faults

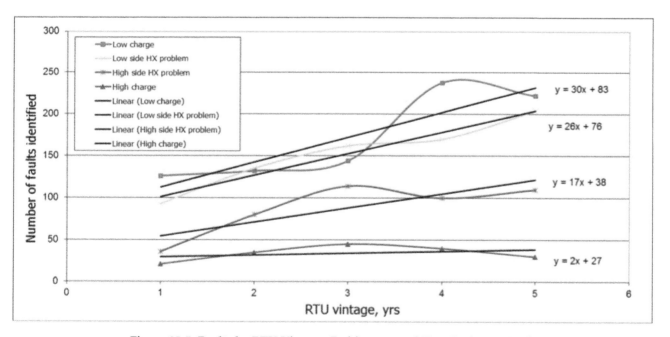

Figure 35-5: Faults by RTU Vintage: Refrigerant and Heat Exchange Faults

	No problems	Major repair	Low charge	High charge	Low side HX problem	High side HX problem	Economizer problem	Sensor failure
Slope (faults/yr)	5	0.1	30	2	26	17	8	0.6
Units tested/yr	527	527	527	527	527	527	251	527
Incidence	0.9%	0.0%	5.7%	0.4%	4.9%	3.2%	3.2%	0.1%
x 15 yrs analysis period	14%	0%	85%	6%	74%	48%	48%	2%

Figure 35-6: Summary of Fault Incidence Analysis

Failure Mode	Fault incidence over 15 years	Prob. of detecting the fault w/FDD	Prob. of detecting the fault w/o FDD	Fault incidence x FDD benefit
Air temperature sensor malfunction	2%	75%	25%	1%
Refrigerant charge: 80% of nominal charge	85%	75%	50%	21%
Compressor short cycling	30%	75%	25%	15%
Refrigerant line restrictions/TXV problems	62%	75%	25%	31%
Refrigerant line non-condensibles	50%	75%	25%	25%
Low side HX problem incl. low airflow (50% evaporator coil blockage; -5% EER)	74%	75%	25%	37%
High side HX problem (50% condenser coil blockage; -9% EER)	48%	75%	25%	24%
Not economizing when it should (high-limit setpoint 55F instead of 75F)	30%	75%	75%	0%
Damper not modulating	24%	75%	25%	12%
Excess outdoor air	24%	75%	25%	12%

Figure 35-7: Summary of FDD Probability Analysis

ear regression was used per climate zone and building type to determine the energy savings as a function of EER penalty. The results of the probability analysis are then applied to the energy savings by multiplying the savings for each failure mode by the last column in Figure 35-7 (Fault incidence x FDD benefit). This yields the benefit of FDD considering the fault incidence and the probability of detecting the faults with and without an FDD system. These savings are then summed by climate zone and building type across all failure modes.

The savings per climate zone and building type were divided by the total RTU cooling capacity (tons of cooling) associated with the simulation building. Then a weighted average was taken based on the 2014 non-residential building construction forecast for the climate zone and building type. The resulting metric is average savings per cooling ton for the first year.

Figure 35-8 summarizes the energy savings results. It shows the annual energy savings for each of these failure modes averaged over the effective useful life (EUL) of the FDD system, assumed to be 15 years. These savings values represent the weighted average by estimated new construction activity for the next 15 years across all climate zones and simulated building types. It also shows the savings values considering the probability analysis and thus the benefit of FDD considering the

Failure Mode	Average kWh/ton-yr savings over EUL	Average therms/ton-yr savings over EUL	Fault incidence x FDD benefit	Average kWh/ton-yr savings over EUL with FDD	Average therms/ton-yr savings over EUL with FDD	kWh/ton savings using $1.86 PV	therm/ton savings using $14.59 PV	PV$ total/ton
Air temperature sensor malfunction	10	0	1%	0.1	0	$ 0.18	$ -	$ 0.18
Not economizing when it should	450	0	0%	0	0	$ -	$ -	$ -
Economizing when it should not	Did not model this failure mode							
Damper not modulating	540	0	12%	65	0	$ 121	$ -	$ 121
Excess outside air	140	70	12%	17	8	$ 31	$ 123	$ 154
Total	1,140	70	25%	82	8	$ 152	$ 123	$ 275

Figure 35-8. Savings by Failure Mode

probability of detecting the faults with and without an FDD system.

The annual energy and gas savings is 82 kWh/ton-yr and 8 therms/ton-yr. For an RTU with cooling capacity of 5 tons (60 kBtu/hr), the annual savings is 410 kWh/yr and 40 therms/yr.

The last three columns consider the Time Dependent Valuation (TDV) of energy, which gives greater weight to energy saved during peak periods—or periods when the generation capacity is at its limit and when the distribution system is near capacity. The weight assigned to energy consumption depends on climate zone, time of use, building type (residential or nonresidential) and fuel type (electricity, natural gas, or propane).

Per the 2014 Title 24 standards analysis methodology, the average present value cost of site energy is $14.59 PV$/therm for natural gas and $1.86 PV$/kWh for electricity for nonresidential measures with 15-year EUL. The Present Value (PV) energy savings over the effective useful life of 15 years is $275/ton. For an RTU with cooling capacity of 5 tons (60 kBtu/hr), the total savings over 15 years is $1,375.

References

1. Integrated Energy Systems: Productivity & Building Science Program, Element 4—Integrated Design of Small Commercial HVAC Systems, Small HVAC Problems and Potential Savings Reports. Submitted to the California Energy Commission. Boulder, CO. Architectural Energy Corporation. 2003. (PIER publication 500-03-082-A-25)
2. O'Neal, D., Haberl, J. Monitoring the Performance of a Residential Central Air Conditioner under Degraded Conditions on a Test Bench. May 1992.
3. Evaluation Measurement And Verification Of Air Conditioner Quality Maintenance Measures, Mowris, October 2010.
4. Integrated Energy Systems: Productivity & Building Science Program, Element 4—Integrated Design of Small Commercial HVAC Systems, Small HVAC Problems and Potential Savings Reports. Submitted to the California Energy Commission. Boulder, CO. Architectural Energy Corporation. 2003. (PIER publication 500-03-082-A-25)
5. Evaluation Measurement And Verification Of Air Conditioner Quality Maintenance Measures, Mowris, October 2010.
6. Evaluation Measurement And Verification Of Air Conditioner Quality Maintenance Measures, Mowris, October 2010.
7. Automated Fault Detection and Diagnosis of Rooftop Air Conditioners for California, Deliverables 2.1.6a & 2.1.6b. Braun, Li, August 2003
8. US DOE, Technical Support Document: Energy Efficiency Standards for Consumer Products, May 2002.

SECTION V

AFDD for HVAC Systems and Equipment— Methodology and Future Technology

The development of building analytics and automated fault detection and diagnostic (AFDD) methodologies and tools continues in earnest. Developments aim to increase the depth and precision at which faults are diagnosed, reduce the number of false alarms, extend the types of systems to which AFDD can be applied, make systems and controls fault-tolerant and self-healing, and accompany FDD results with information on impacts, such as operating cost, that building operators and facility managers can use to prioritize maintenance and repairs.

This section provides readers a glimpse at some of the details of methods underlying analytics and AFDD tools for HVAC and new developments that portend the tools of the future. These methodologies and the tools that implement them will enable maintenance of HVAC equipment and systems to move from primarily preventive to condition-based, enabling buildings and their systems to be maintained much closer to peak operating performance and efficiency continuously. Also included is a chapter on recently initiated efforts on standards and codes for HVAC AFDD, which have the potential to change the marketplace and accelerate the introduction of these new technologies into practice.

The chapters in this section are authored by researchers and developers in universities, industry, and government laboratories. They describe methods for both built up and packaged HVAC systems and equipment, an approach for giving the systems self-healing capabilities, and the role of codes and standards in moving these technologies from laboratories and bench tops to common use on the HVAC systems in commercial buildings. The editors expect that readers who are researchers and developers are likely to have the greatest interest in the chapters in this section, but building managers and operators and other readers with an interest in how AFDD and analytics work will find valuable information in this section.

Chapter 36

Fault Detection and Diagnostics for HVAC&R Equipment*

Jessica Granderson, Mary Ann Piette
Ben Rosenbloom, Lily Hu
Lawrence Berkeley National Laboratory

Shrinivas Katipamula and Michael Brambley
Battelle Pacific Northwest National Laboratory

INTRODUCTION

Poorly maintained, degraded, and improperly controlled equipment wastes an estimated 10% to 30% of the energy used in commercial buildings. Much of this waste could be prevented with widespread adoption of fault detection and diagnostics (FDD), an area of investigation concerned with automating the processes of detecting faults in physical systems and diagnosing their causes.

For many years, FDD has been used in the aerospace, process controls, automotive, manufacturing, nuclear, and national defense fields. Over the last two decades, efforts have been undertaken to bring automated fault detection, diagnosis, and prognosis to the heating, ventilating, air conditioning, and refrigeration (HVAC&R) field. Although FDD is well established in other industries, it is still in its infancy in HVAC&R. Commercial tools using these techniques are only beginning to emerge in this field. Nonetheless, considerable research and development has targeted the development of FDD methods for HVAC&R equipment.

This chapter provides an overview of FDD, including descriptions of fundamental processes, important definitions, and examples that building operators and managers can implement using data collected from the building automation systems or dedicated logging devices.

*This chapter is from Fault Detection and Diagnostics, in the Energy Information Handbook: Applications for Energy Efficient Building Operations, prepared by Jessica Granderson, Mary Ann Piette, Ben Rosenbloom and Lily Hu, Lawrence Berkeley National Laboratory, LBNL-5272E, 2011.

THE GENERIC FDD PROCESS

The primary objectives of an FDD system are to detect faults early and to diagnose their causes, enabling building managers to correct the faults, to prevent energy waste, additional damage to the system, or loss of service. In most cases, fault detection is easier than diagnosing the cause of the fault or evaluating the impacts arising from it. FDD itself is frequently described as consisting of three key processes:

1. Fault detection: Determination that a fault has occurred in the system

2. Fault isolation: Determination of the specific fault that occurred including it's type, location, and time of detection

3. Fault identification: Determination of the size and time-variant behavior of a fault

Together, fault isolation and fault identification are commonly termed fault diagnosis.

APPLICATIONS FOR FDD IN BUILDINGS

Automated FDD shows promise in three basic areas of building engineering:
(1) commissioning/retro-commissioning, (2) operation, and (3) maintenance. However this handbook primarily focuses on operation. During building operation, FDD tools can detect and diagnose performance degradation and faults, many of which go undetected for weeks or months in most commercial buildings.

Many building performance problems are compensated automatically by controllers so occupants experience no discomfort, but the penalty is often increased energy consumption and operating costs. Automated FDD tools could detect these, as well as more obvious problems.

Automated FDD tools not only detect faults and alert building operation staff to them, but also identify causes of those problems so that maintenance efforts can be targeted, ultimately lowering maintenance costs and ensuring good operation. When coupled with knowledge bases on maintenance procedures, other tools can provide guidance on actions to correct the problems identified by FDD tools. By detecting performance degradation rather than just complete failure of a physical component, FDD tools could also help prevent catastrophic system failure by alerting building operation and maintenance staff to impending failures before actual failure occurs. This would enable convenient maintenance scheduling, reduced down time from unexpected faults, and more efficient use of maintenance staff time leading to condition-based maintenance practices.

FDD IMPLEMENTATION

Fault detection and diagnostics can be performed "manually" through visual inspection of charts and trends or can be fully automated. For example, the temperature of the supply air provided by an air-handling unit might be observed to be too high chronically during hot weather. This conclusion can be drawn by visually inspecting a time series plot of the supply- air temperature, for example, within a building automation system. Alternatively, an FDD system could be automated. A computer algorithm could process these data continuously to reach this same conclusion, reporting the condition via an alarm to the operator. Automated diagnostics generally goes a step further, and might conclude for example, that the outside-air damper is stuck fully open. As a result, during hot weather, too much hot and humid outdoor air would be brought in, increasing the mechanical cooling required and exceeding the capacity of the mechanical system for cooling; which would explain the chronically high supply-air temperature. This is a process that can be integrated into a commissioning process.

VISUAL FDD, APPLICATION EXAMPLES

Air-side economizers can obtain free cooling by using cool outdoor air in place of (or to supplement) me-

chanical cooling when outdoor conditions are suitable for doing so. Unfortunately, economizers often do not work properly, causing energy-use penalties rather than savings.

Interpretation

Several common incorrect behaviors of economizers result from: incorrect control strategies, stuck dampers, disconnected or damaged damper linkages, failed damper actuators, disconnected wires, obstructions preventing damper movement, and failed and out-of-calibration sensors.

A number of these incorrect operations can be detected visually by plotting the relevant data. Although there are a number of different ways that economizers can be controlled, in general, when the zone conditions are calling for cooling, and if the return-air temperature (RAT) (or energy content) is greater than outdoor-air temperature (or energy content),
the conditions are favorable for economizing. Sometimes all the cooling needs can be met by outside air. When the outdoor-air temperature (OAT) is less than or equal to the discharge-air temperature (DAT) setpoint, no mechanical cooling is necessary. When the outdoor-air temperature is higher than the discharge-air setpoint, some mechanical
cooling is need to supplement free cooling. By analyzing the data visually you can detect a number of problems with economizer operations. The following three examples illustrate visual FDD for economizers, using plots of outdoor-air, return-air, mixed-air (MAT), and discharge-air temperatures versus time.

AUTOMATED FDD

An automated FDD (AFDD) process uses measured time-series data and set-up data that describe the equipment and system characteristics (such as setpoints and type of control) to create actionable information to help building operations staff make informed decisions, as shown in Figure 36-1.

In addition to the data, the basic building blocks of automated FDD systems are the methods for detecting faults and diagnosing their causes. Approaches to FDD range from methods based on physical and analytical models based entirely on first principles, to those driven by performance data and using artificial intelligence or statistical techniques. Both approaches use models and both use data, but the approach to formulating the diagnostics differs fundamentally.

Example 1: Properly Operated Outside Air Economizer

(A) One day of temperature data is plotted: OAT, return air, mixed air, and discharge air.

The return-air temperature varies between 72°F and 75°F.

(B) OAT is lower than the RAT, and is therefore acceptable for economizing.

(C) DAT closely tracks MAT, indicating no use of mechanical cooling.

(D) Discharge and mixed air trends also indicate proper modulation of outside airflow.

In this example the MAT sensor is located upstream of the supply fan.

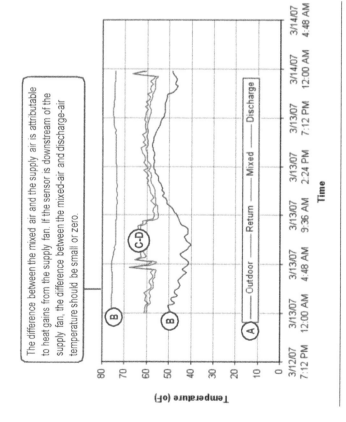

The difference between the mixed air and the supply air is attributable to heat gains from the supply fan. If the sensor is downstream of the supply fan, the difference between the mixed-air and discharge-air temperature should be small or zero.

Source: Pacific Northwest National Laboratory

Example 2: Economizer Fault, Damper Stuck Fully Closed

(A) One day of temperature data is plotted: OAT, return air, mixed air, and discharge air.

(B) OAT is lower than the RAT, and is therefore acceptable for economizing.

(C) MAT tracks RAT, indicating that outdoor air is not entering the mixing box.

(D) The outdoor air damper is not opening, as it should be.

Potential causes are a stuck damper or failed or disconnected actuators or linkages.

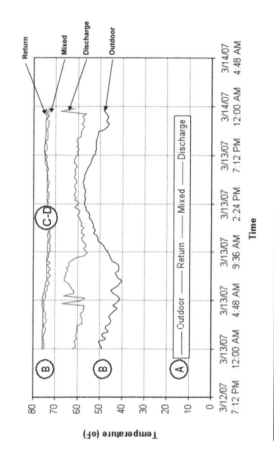

Source: Pacific Northwest National Laboratory

First-principle model-based approaches use a priori knowledge to specify a model that serves as the basis to identify and evaluate differences (residuals) between the actual operating states determined from measurements and the expected operating state and values of characteristics obtained from the model.

Purely process data-driven approaches use no a priori physical knowledge of the process. Instead, they are derived solely from measurement data and therefore may not have any direct physical significance.

Rule-based methods, broadly classified as first-principle qualitative models, are most commonly employed in commercial FDD solutions. (Qualitative relationships or rules derived from knowledge of the underlying system operation.) Strengths of these models are:

1. They are well-suited for data-rich environments and non-critical processes.

2. They are simple to develop and apply.

3. They employ transparent reasoning, and provide the ability to reason even under uncertainty.

4. They possess the ability to provide explanations for the suggested diagnoses because they rely on cause-effect relationships.

5. Some provide the ability to perform FDD without precise knowledge of the system and exact numerical values for inputs and parameters.

Weaknesses of these models include:

1. The methods are specific to a system or a process.

2. Although they are easy to develop, it is difficult to ensure that all rules are always applicable, and to find a complete set of rules, especially when the system is complex.

Example 3: Economizer Fault, Damper Stuck Fully Open

(A) One day of temperature data is plotted: OAT, return air, mixed air, and discharge air.

(B) MAT tracks OAT, indicating that the outdoor air damper is fully open.

(C) Since the discharge air setpoint is higher than OAT, the damper should not be fully open.

(D) The outdoor air damper is not closing, although it should be.

Potential causes are a stuck damper or failed or disconnected actuators or linkages.

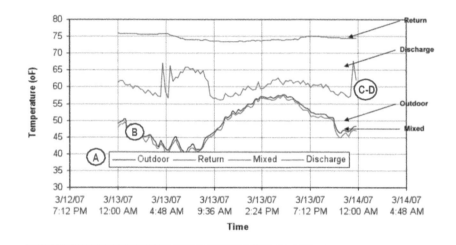

Source: Pacific Northwest National Laboratory

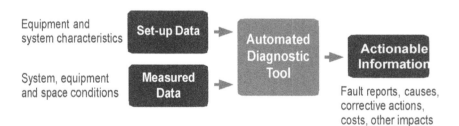

Figure 36-1

3. As new rules are added to extend the existing rules or accommodate special circumstances, the simplicity is lost.

4. To a large extent, they depend on the expertise and knowledge of the developer.

For information on the various methods used for AFDD please refer to the References and Technical Resources.

AFDD OF AIR HANDLER UNIT OPERATIONS

As part of its mission in commercial buildings research and development, the U.S. Department of Energy (DOE) collaborated with industry to develop a tool that automates detection and diagnosis of problems associated with outdoor-air ventilation and economizer operation. The tool, known as the outdoor-air economizer (OAE) diagnostician, monitors the performance of air handler units (AHUs) and detects problems with outside-air control and economizer operation, using sensors that are commonly installed for control purposes.

The tool diagnoses the operating conditions of AHUs using rules derived from engineering models of proper and improper air-handler performance. These rules are implemented in a decision tree structure in software. You can use data collected periodically (such as that from a building automation system) to navigate the decision tree and reach conclusions regarding the AHU's operating state. At each point in the tree, a rule is evaluated based on the data, and the result determines which branch the diagnosis follows. The AHU's current condition is revealed when you reach the end of a branch. The Figure 36-2 illustrates the logic tree used to identify operational states and to build the lists of possible failures.

The boxes represent major sub-processes necessary to determine the operating state of the air-handler, the diamonds represent tests, i.e., decisions, and ovals represent end states that contain brief descriptions of "OK" and "not OK" states. Only selected end states are shown in this overview. The detection and diagnostic imple-

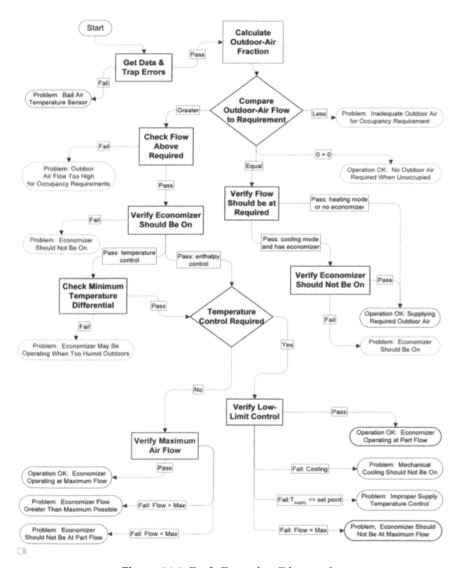

Figure 36-2. Fault Detection Diagnostics

Example 4: OAE Diagnostician, Visual Display

(A) Days of the week are plotted on the x-axis, with hours of the day on the y-axis.

(B) Each hour of the day is color coded according to one of five diagnostic findings.

(C) Though not present in the example, blue denotes low ventilation, and yellow is a catchall for "other" problems.

(D) White indicates fault-free operations; red, a high-energy fault; and gray, no diagnosis.

(E) An object tree allows the user to navigate between sites and air handlers.

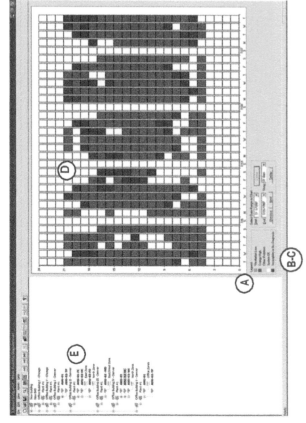

Source: Outside Air Economizer Diagnostician

Example 5: OAE Diagnostician, Fault Descriptions

The OAE Diagnostician has the capability to generate problem summaries

Each summary describes:

(A) The associated equipment, date, and time

(B) Current conditions and cost impacts

(C) Potential causes and suggested corrective actions

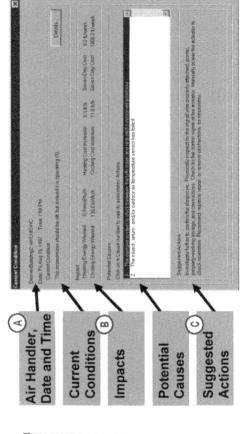

Air Handler, Date and Time

Current Conditions

Impacts

Potential Causes

Suggested Actions

Source: Outside Air Economizer Diagnostician

mentation details are provided in the literature by PECI and Battelle, and Katipamula et al. in the References and Technical Resources.

The outdoor air economizer diagnostician offers a variety of graphical displays; two examples are presented.

This display from a commercial energy information system illustrates another example of rule-based automated fault detection and tools.

CONCLUSION

This chapter has described the generic Fault Detection and Diagnostic (FDD) process, and has discussed applications to the areas of commissioning/retrocommissioning, operation, and maintenance. Then two general methods have been presented and discussed. These two methods were: Visual FDD; and Automated FDD.

Three examples of each of these methods were also presented, which should help readers determine the potential applications and types of savings from these methods in their own buildings and facilities.

References and Technical Resources

(All URLs provided for documents available on the internet were accessed in the summer and fall of 2011.)

Brambley, MR, Pratt, RG, Chassin, DP, and Katipamula, S. Automated diagnostics for outdoor air ventilation and economizers. ASHRAE Journal, Vol. 40, No. 10, pp. 49-55, October 1998. This article discusses deployment of new automated diagnostic tools for building operation.

Friedman, H, Crowe, E, Sibley, E, Effinger, M. The Building performance tracking handbook: Continuous improvement for every building. Prepared by Portland Energy Conservation for the California Energy Commission. California Commissioning Collaborative, 2011. Available from: http://www.cacx.org/PIER/documents/bpt-handbook.pdf.

Handbook presenting basic concepts in tracking the energy performance of commercial buildings, including overall strategies, and FDD and other tool types.

Example 6: AFDD in a Large Commercial Office

The AFDD engine identifies operational inefficiencies in air handler units (AHUs).

(A) A plot of damper position vs. OAT is color coded to show faulty and correct operations.

(B) The light blue points show when the dampers are closed, even though "free cooling" is available.

(C) Green points correspond to fault-free operations.

(D) Yellow points indicate cooling lockouts, and red indicate heating lockouts.

Though not present in this example, purple is designated for scheduling faults.

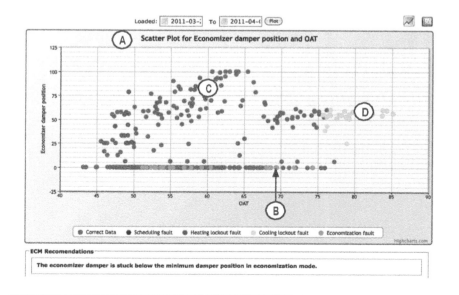

Source: Serious Energy

Katipamula, S and Brambley, MR. Methods for fault detection, diagnostics and prognostics for building systems- A review part I." HVAC & R Research 11(1):3-25. 2005. This paper provides an overview of FDD and prognostics (FDD&P), including definitions and descriptions of the fundamental processes, a review of research in HVAC and refrigeration, and a discussion of the current state of applications in buildings.

Katipamula, S and Brambley, MR. Methods for fault detection, diagnostics and prognostics for building systems - A review part II." HVAC & R Research 11(2):169-187. 2005. This paper provides the second portion of an overview of FDD and prognostics (FDD&P), including definitions and descriptions of the fundamental processes, a review of research in HVAC and refrigeration, and a discussion of the current state of applications in buildings.

Katipamula, S, Pratt, RG, Chassin, DP, Taylor, ZT, Gowri, K, and Brambley, MR. Automated Fault Detection and Diagnostics for Outdoor-Air Ventilation Systems and Economizers: Methodology and Results from Field Testing. ASHRAE Transactions, Vol. 105 Pt. 1. 1999. Article presenting field testing results of the OAE diagnostician.

Ulickey, J, Fackler, T, Koeppel, E, Soper, J. Characterization of fault detection and diagnostic (FDD) and advanced energy information systems (EIS) tools. Report for Project 4, Building Performance Tracking in Large Commercial Buildings: Tools and Strategies. September, 2010. California Commissioning Collaborative; California Energy Commission. Available from http://www.cacx.org/PIER/documents/Subtask_4-3_Report.pdf. Report covering the evaluation and characterization of energy performance tracking tools, with a focus on FDD tools in particular.

Chapter 37

Automatically Detecting Faulty Regulation in HVAC Controls

Daniel A. Veronica, Ph.D., P.E.
National Institute of Standards and Technology

A new method is introduced to automatically detect faulty regulation of temperatures, pressures, and flow rates within the heating, ventilating, and air-conditioning (HVAC) systems and equipment of commercial buildings by using digital data typically available from an existing building automation system. The building automation system passes data by network to a general-purpose microprocessor executing this method. The method computerizes control charts and combines them with expert system logic to identify transients and record excursions of regulated variables beyond allowance bands set by the user. Its three separate functions monitor: (1) variables regulated to a single setpoint value; (2) actuating variables that drive the regulation; and (3), temperatures regulated to duplex setpoints (i.e., thermostats). Faults detected include unstable, excessively oscillatory regulation, and failure of regulated variable to maintain allowed band. A brief background on control charts and expert systems is included.

INTRODUCTION

Properly monitoring and maintaining the complex heating, ventilating, and air-conditioning (HVAC) systems of modern large commercial buildings presents difficulties for any staff of economically practical size and expertise, a fact well established by sources such as Westphalen et al. (2003). Software-based products and services are currently marketed to help address the problem by way of fault detection and diagnostics (FDD) of the HVAC system. The primary information sources for such products and services are digital data sampled from sensors throughout the HVAC system. These sensors are typically—though not always comprehensively—provided by an existing building automation system (BAS) controlling the HVAC system. But, the report by Summers and Hilger (2012) gives evidence that current FDD software can demand uneconomical amounts of expert time and effort from the staff or hired consultants. A remedy is to develop novel *automated*

FDD (AFDD) software "tools" that help the staff ensure everything works well without presenting uneconomical demands for human experts.

A basic task for an AFDD tool is to determine whether closed-loop automatic controls, such as proportional-integral-derivative (PID) compensators, are properly regulating temperatures, pressures, and flow rates. That requires the tool to have an analytical component able to identify autonomously (on its own) unstable, excessively oscillatory behavior ("hunting") in the regulated quantity, as well as in the regulating device (e.g., damper or valve).

Referring at first only to the top plot panel of Figure 37-1, the air temperature T_{az} of a building zone served by a fan-coil unit is shown over about 39 h of a heating season as an irregular black line. The abscissa common to all the panels is time, labeled at the figure bottom as hours after midnight. Straight, horizontal upper and lower gray lines show respectively the cooling and heating setpoint (i.e., intended) temperatures for the zone. It is seen that the unit generally regulates zone temperature acceptably, within 0.5 °C (0.9 °F) of the heating setpoint value. However, there is subtle evidence something is amiss, because the variation of zone temperature between Hours 03 and 18 at the left end of the plot is suspiciously large and rapid compared to that before Hour 03, while between Hours 18 and 04 on the following day the variation is much slower, sluggishly persisting below the heating setpoint for longer periods. From Hour 04 the zone temperature again is regulated well as at the beginning of the plot until a hidden event near Hour 13 causes regulation to jump to the cooling setpoint.

So, the top panel of Figure 37-1 shows temperature regulation of a zone affected, at times adversely, by factors not identifiable in the plotted data alone. It is possible these hidden factors degrade the energy efficiency of the HVAC system or put unnecessary wear on components. Further information is needed to determine if that is true. However, the only information

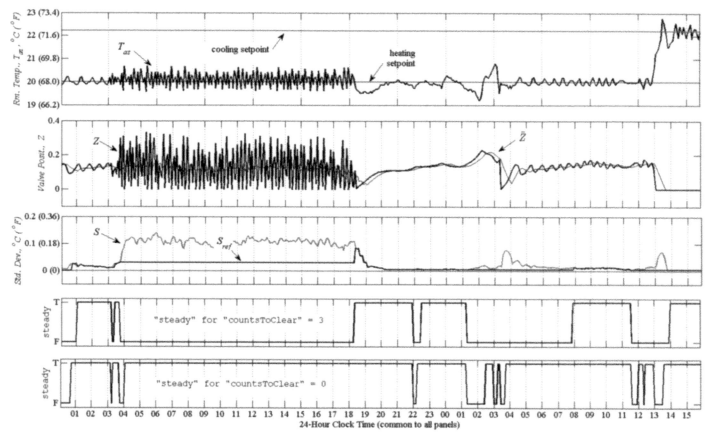

Figure 37-1: Logic Analysis Sequence of Erratic Valve Activity

the building's staff will have are the complaints from occupants during the sluggish excursions of zone temperature below setpoint. In buildings having centralized plants serving tens to hundreds of distributed zones, the staff typically would not have time to create plots such as Figure 37-1 to examine for suspicious features. Even less plausible would be the further time and effort the staff needs to gather evidence about observed features and come to diagnoses resolving whether or not faults exist, and if they do, where they are and what is the cost of their impacts. It is instead the job of an AFDD tool to resolve all those things for the staff as automatically and autonomously as practicable.

PRIMARY CAPABILITIES FOR DETECTING AND DIAGNOSING HVAC CONTROL FAULTS

To fully diagnose situations as that just discussed, the AFDD tool must perform some functions beyond the scope of this article. All those functions rely, however, upon the tool first having at least four primary capabilities:

a) It should reliably discriminate dynamics caused by faults from dynamics reflecting normal, fault-free operation.

b) It should identify excessive variation or drift in a regulated quantity, that is, a dependent quantity such as duct air temperature or pressure having a single setpoint value.

c) Some quantities such as zone temperature at a thermostat are regulated by duplex setpoints (e.g., a higher value activating cooling and a lower one activating heating with an unregulated band between). Hunting and drift should be identified in this case without implicating free dynamics in the unregulated band.

d) Faulty dynamics such as hunting or drift of a regulated variable should be associated to the devices exerting the regulation, such as specific valves, dampers, fan motors, or control modules. Correlations can then infer whether the problem is really a more primary fault in one of those devices.

These four primary capabilities can be accomplished by combining control charts with expert system logic.

A BRIEF BACKGROUND ON CONTROL CHARTS AND EXPERT SYSTEMS

Certain fundamentals of control charts or expert systems are important to AFDD of HVAC systems but are likely unfamiliar to many in the HVAC industry. This section examines those fundamentals as needed to understand the material presented later.

Origin of Control Charts

With the advent in the early twentieth century of large-scale manufacturing involving standardization and interchangeable parts, control charts began as the practice of periodically plotting specific physical, quantifiably measureable attributes of product quality—from here on called "metrics"—as time-series on paper graphs. The earlier practice of simply testing finished parts to discard rejects had become uneconomical, and in control charts producers found a way to track key metrics as data sampled sequentially over time. This allowed a proactive response to avoid waste whenever warning limits on a metric are approached. Not only could the charts track expected devolutions such as tooling wear, but by charting several metrics correlations could be drawn to uncover hidden factors influencing product quality. As mass products such as electrical fuses and light bulbs emerged, quality needed to be quantified consistently, verifiably, and economically for large lots of items whose metrics involved testing individual examples to destruction. Those issues were thoroughly studied in the 1920s by a statistician at Western Electric Company, Walter Shewhart, leading to his seminal book (Shewhart 1931) establishing control charts as analysis tools within a larger, more comprehensive framework now known equivalently as statistical quality control (SQC) or statistical process control (SPC). In this context, "control" has a broader connotation than the closed-loop feedback regulation usually associated with that word in HVAC, where the timescale between measurement and automated compensating action is typically only a few seconds. In SPC the compensating action is generally not automatic, its timescale can range from minutes to days, and it considers both defined and hidden factors, suggesting something more akin to a researcher's notion of "experimental control." Further, SPC takes "process" to encompass all factors in an item's manufacture whether they are physically coupled or not, again a broader notion than usually seen in HVAC literature.

Within SPC many types of control charts have been established for specific industrial uses, and a common practice to aid diagnostics is tracking sample mean and dispersion on separate control charts having distinct warning limits and action procedures (Wise and Fair 1998, Betteley et al. 1994). Two control chart types are considered here for AFDD of HVAC systems, the Shewhart chart and the CUSUM chart.

The Shewhart Chart

Control charts in the general category bearing Shewhart's name are distinguishable by their making a decisive use of the IID (independent, identically distributed) concept from statistics. To be considered IID, the value of any one sample of a particular metric has no relation to the value of any other sample of it, apart from the presumption that all the samples are coming from a common generating process. Further, IID implies the values this process generates under normal (i.e., natural, varied only by an accepted degree of random "chance") conditions can be characterized over time by a single statistical distribution having constant parameters. It is common, but not at all necessary, to model this actual "normal" distribution as a Gaussian normal distribution, so a more distinctive term, "nominal," is used for it here. As long as its samples continue to exhibit the nominal distribution the process is said to be statistically "under control."

Shewhart considered the variation observed over successive samples of the same metric as having two components. One component is the variation assignable to deterministic causes, whether those causes are currently identified or not. This includes all dynamic features seen in the data and understood at least well enough to know they are not occurring randomly. Shewhart called this the *special-cause* component. The second component is all the variation remaining. That is, all variation remaining in the samples, regardless of true origin, is provisionally considered "chance" by default, and called the *common-cause* component. Shewhart's innovation was to improve and control product quality by using the chart to move avoidable variation from the default, common-cause component to the special-cause component and then out of the production process entirely.

Given a series of IID samples, a Shewhart chart reveals when a special-cause component has emerged in their variation, enabling someone to correlate that emergence to current or impending process trouble that can be assigned to causal factors and fixed. Tighter parameters can then be applied to the chart iteratively, revealing successively deeper special-cause factors, the common-cause variation being progressively whittled away as more of it is assigned to fixable causes and eliminated. Upon reaching a practical minimum in common-cause variation, charting of the metric is then continued as surveillance against new special-cause factors.

Setting parameters for the nominal distribution also defines implicitly which factors having influence on the process are its common-cause factors. As an example in parts manufacturing, the imprecision accepted in a caliper measurement becomes a common-cause factor, and wear of a cutting tool detectable as a special-cause factor, depending on how much caliper imprecision the nominal distribution allows. Intended or not, some degree of tool wear ends up as common-cause if the nominal distribution allows for more caliper imprecision than usually occurs. Perhaps two milling stations both feed a single caliper test station, with one milling station consistently having less precision in placing its cutter. That disparity is an assignable special-cause factor in the variance observed at the caliper. But, its detectability might be lost within the caliper's common-cause component, since the disparity between the milling stations would be unknown when the caliper chart parameters are first set. The fault at the errant milling station can be revealed by the caliper's chart, however, if sudden variance changes are caught as special-cause and associated to periods when either mill is offline. The goal of the method presented here is to take that same scheme of control charting followed by logical associations, and automate it for the detection and diagnosis of faults in HVAC systems.

Surveillance of a process by Shewhart charts as they are traditionally used requires that application details be carefully, and often heuristically, addressed for each metric charted. Parameters for the nominal distribution must allow emerging special-cause factors to be identified promptly, while minimizing false warnings on factors properly common-cause. On a paper Shewhart chart these parameters are one or more pairs of parallel lines, commonly called "limits," that are in fact the counting thresholds of sampling bins. Sample values are binned and counted toward the tallies expected by the nominal distribution. By graphs and tables, a paper chart shows the number of samples counted outside specific threshold lines over a moving interval of time. The length of that time interval, the *run length*, results implicitly from the distribution chosen as nominal, as do the location of threshold lines and the counts expect-

ed in the bins they delineate. Any instance of counts exceeding what is nominally expected can be taken as evidence an assignable special-cause factor may have entered the process.

For example, presuming a Gaussian nominal distribution having standard deviation σ, no more than 31 discrete counts are theoretically expected (given the samples are IID) outside threshold lines at $\pm\sigma$ about the mean over any run length of 100 samples. Counting 32 or more samples outside those lines can be evidence a problem has emerged, although that is not a certainty because statistical distributions reflect only probability. It could be asked which of the 32 samples over the run time is to be regarded as the discrepant one. Supporting evidence is gained by binning at other thresholds, for example at $\pm 2\sigma$, $\pm 3\sigma$, or unilaterally. Beyond the $\pm 3\sigma$ threshold only two counts are nominally expected over a 1000 sample run length, an average of one in 500 samples, so a second count in 500 could help isolate the trouble in time, as would observing graphical clustering of the samples. Clearly, automating the traditional use of Shewhart charts, based as it is on statistical expectation, presents some challenges in the case of HVAC processes, notably in the potentially long run lengths necessary, and in replicating human graphical cognition of sample clustering over time.

The CUSUM Chart

The Shewhart chart is suited to detect swift variance changes in a metric, such as caused in manufacturing by a cracked tool bit, or in HVAC by a valve actuator suddenly changing position. Detecting a small or slow drift in the metric mean, like the gradual tool wear in the example above, is problematic for a Shewhart chart because the drift may go undetected a long time if the common-cause variance is relatively wide. A complementary SPC tool for those situations is the CUSUM chart, first introduced by Page (1954). Of the variants now developed, Ryan (2000) defines a CUSUM chart having a *slack value*, k, used in conjunction with the z-score, z (see Equation 37-3), of each sample at time, n. Samples observed either above or below a band of $\pm k$ about the expected mean are accumulated in separate numerical registers:

$$P_n = \max[0, \ ((z_n - k) + P_{n-1})] \qquad (37\text{-}1)$$

$$Q_n = \min[0, \ ((z_n + k) + Q_{n-1})] \qquad (37\text{-}2)$$

Positive z_n greater than k moves sum P further above 0 and sum Q toward 0. Negative z_n less than $-k$

moves sum Q further below 0 and sum P toward 0. Excessive drift in the metric mean is signaled by exceeding alarm thresholds placed upon P and Q.

Automating Control Charts for HVAC Systems

Automating control charts for computer implementation means, among other issues, reducing the need for human graphical cognition as much as is practical. The AFDD method presented here does not exploit sample clustering, although automating its recognition could help diagnostics. Of the traditional paper chart features, only threshold lines are automated here, done by the sample z-score. Parameters defining a normal distribution are the mean, $\bar{x}$, and standard deviation, s, of a sample set. A z-score, z_n, is calculated for each sample, x_n.

$$z_n = \frac{(x_n - \bar{x})}{s} \qquad (37\text{-}3)$$

Common statistical practice is to standardize z-score by sample size, which here is the run length used by any particular threshold line. Shewhart (1931) rigorously considered sample size because a common use of his charts is to infer the quality of a large number of products by testing small sampled sets. However, here we are not estimating statistical parameters of a population, so standardizing by sample size is not necessary. Furthermore, it is not wanted, because Shiffler (1988) showed that if z-score is standardized to sample size, no z-score beyond 3 is even possible unless at least 11 samples are considered. That would be a problem if a long sampling period of several minutes is implemented to minimize traffic in a real-time data network, because it forces longer run lengths that would inhibit detecting faults early.

Control Charts and Autocorrelation

Manufacturing processes typically produce discrete items having physical metrics, such as part dimensions, that are amenable to the IID assumption. In contrast, a HVAC process typically involves quantities such as temperature, pressure, and velocity sampled from moving fluid streams, each creating a continuous time-series of values with physical couplings often existing across successive values. These couplings result from inherent properties such as the specific heats of fluids and heat transfer media, and appear in the sampled data series mathematically as autocorrelation. Time-series from HVAC processes, such as air temperature leaving a cooling coil, also exhibit frequent dynamics that are the normally expected results of exogenous influences like controller action, occupancy, weather, or time of day. If that time-series is viewed in isolation these dynamics also appear as autocorrelation. But, expert knowledge of the system can readily point out such autocorrelation as being, more fundamentally, a cross-correlation to another dynamic variable, typically with an appreciable lag, such as exists at a cooling coil between its chilled water (CHW) control valve position and its leaving air temperature (LAT). Either way, correlation violates the IID premise and thus theoretically undermines control charts as SPC traditionally employs them, which is to isolate instances of special-cause variation explicitly as faults to be fixed. But, correlation does not undermine control charts as the charts are employed here. Instead, correlation helps motivate their use.

The method described here adapts automated versions of the Shewhart and CUSUM charts to be analysis tools that future work will more fully integrate within a larger, more comprehensive AFDD tool of expert system architecture. The Shewhart chart, instead of its traditional use to discriminate faults explicitly, is used here only to mark the beginnings and endings of transients in quantities. Expert logic downstream then checks for any normally expected correlation to changes in other charted quantities, a lag being factored in where appropriate. The CUSUM chart, also, is adapted here for diagnostic roles not inhibited by correlation, by being used in conjunction with complementary expert system logic that the Shewhart charts make possible.

Expert Systems

An "expert system" is a computer program that replicates or enhances the reasoning and judgment a human, having a specific expertise, exercises when given data to be analyzed and acted upon using that expertise. Expert systems are a major field within artificial intelligence (AI), and thus rate exhaustive treatments in foundational AI texts. Expert systems all distinctively have at their core two complementary parts: (1) a "knowledge base" that typically is the body of facts, rules, and relationships (e.g., a model or lookup table) formulating what the expert knows and (2) an "inference engine" that moves the information given to the program through the knowledge base in a manner expressing expert reasoning and judgment. For the purpose of this article it is sufficient to know the following points. The programmed expressions defining rules and relating states such as LATregSteady form some parts of a knowledge base. The inference engine is an algorithm that probabilistically evaluates those states and rules given BAS data and supplemental information the user provides, thus transforming the data into advice the user can act upon. This article only addresses knowledge base parts in an expert system.

Prior Use of Control Charts for HVAC Systems

At least one prior HVAC research effort with control charts (Schein and House 2003) has been implemented by industry. There, CUSUM charting is embedded within the controllers of variable air volume (VAV) units by appending FDD logic to the units' usual control programs, demonstrating on-line (i.e., real-time) detection of faults in the units. To keep that added FDD logic, called VAV Performance Assessment Control Chart (VPACC), small and simple enough for the 1990s-era microcontrollers and interface software it targeted, it does not estimate the two data statistics CUSUM charts typically need: mean and standard deviation. VPACC instead replaces those statistics with exogenous *a priori* parameters obtained offline through controlled testing of similar VAV units.

During operating transients in HVAC systems, autocorrelation normally often appears between samples taken of a single variable over time. Because it can be mistaken as fault evidence, autocorrelation was viewed as an obstacle to the effectiveness of VPACC. Because it is not an expert system, VPACC cannot isolate normal transient behavior from faulty behavior, nor can it attribute faulty behavior to an external cause (e.g., attribute a VAV unit fault to trouble in the air handler upstream). What is needed is presented next: a way for an AFDD tool to expertly and autonomously distinguish changes in a quantity due to faults from changes due to normal dynamic operation—capability (a) in the preceding list.

DISTINGUISHING CONTROLLER FAULTS FROM NORMAL DYNAMICS

A "normal" transient in a variable is one caused by dynamics in an acceptable influence such as a valid control signal, an occupancy or plug load change, weather, or hour of day. Of those four, the first presents a greater challenge because it has the potential to drive the most rapid transients, and a primary concern is a normal transient occurring fast enough to fill control chart registers to an alarm limit when in fact no alarm is warranted. Slower transients are much less problematic, because the charting algorithm can be designed to maintain its own autoregressive mean of the variable as a moving datum to key the chart registers. This mean will follow the dynamics of slow-acting influences and thus

help avert false alarms. While occupancy or plug load changes could also conceivably produce fast transients, most BASs do not evaluate (at least not numerically) those factors, and so data on them is generally sparse or absent.

An Expert System Infrastructure Supports the Task

In many HVAC processes, such as heating and cooling coils, the controlled quantities (e.g., leaving air temperature [LAT]) normally follow regulation (e.g., valve motion) with a significant and varying first-order lag. Such lags make it more difficult to infer causes to the effects observed in data by way of numerical techniques such as correlation, regression, or hypothesis testing. Instead of a numerical technique, correlating logically between binary (i.e., Boolean, true or false) "states" is much simpler, and yet fully sufficient to provide the four capabilities listed earlier. This is illustrated by the following simplified case.

A control chart monitors each instrumented quantity—such as controller setpoint, air flow rate, and chilled water (CHW) valve position—normally driving the values of each regulated quantity, such as the LAT from a cooling coil. A binary variable `ZvcSteady`, one of the many logical states in this expert system, is output by a Shewhart chart function, `ShewChart(...)`, operating on the current CHW valve position, `Zvc`:

$$ZvcSteady = ShewChart(Zvc, \qquad\qquad\qquad\qquad \\ ZvcMean, ZvcStdDev) \qquad\qquad (37\text{-}4)$$

where `ZvcMean` and `ZvcStdDev` are the running mean and standard deviation of the valve position, calculated for the chart from a circular data buffer. After all control charts have operated upon the current round of sampled data, their output states can be used in logical expressions downstream either alone (as "primitive states") or combined to form "derived states," such as

$$LATfactorsSteady = \qquad\qquad\qquad\qquad\qquad \\ setptSteady \ \& \ airflowSteady \ \& \qquad \\ ZvcSteady \qquad\qquad\qquad\qquad (37\text{-}5)$$

The symbol & is the logical "and" operator, and `LATfactorsSteady` is a binary state characterizing the steadiness of the factors an expert knows normally affect a coil's LAT, being in this example respectively, its setpoint, airflow, and valve position. Downstream of derived states, the AFDD tool program steps through arrays of paired, if-then "rule" statements testing that the states correlate in ways reflecting normal HVAC system operation. For example, given a control chart has generated the primitive logical state `LATsteady`, two rule pairs (numbered "13" and "14" in this example) use it with a derived state in:

$$ruleIf(13) = LATfactorsSteady \qquad (37\text{-}6)$$

$$ruleThen(13) = LATsteady \qquad (37\text{-}7)$$

$$ruleIf(14) = not(LATsteady) \qquad (37\text{-}8)$$

$$ruleThen(14) = not(LATfactorsSteady) \qquad (37\text{-}9)$$

where the `not(...)` function negates (i.e., flips to opposite) the binary state in its argument. The `ruleIf(n)` and `ruleThen(n)` terms are not programmer's if-then syntax, but are another form of binary variables derived from previously evaluated states. A key distinction is that while the "true" or "false" value of any state itself does not necessarily connote either "normal" or "faulty," rule statements explicitly use states so that a "false" value for any `ruleThen(n)` when its `ruleIf(n)` is "true" is evidence of a fault in the HVAC system. In the example, it is normally expected that any "true" value for `LATfactorsSteady` correlates to a "true" for `LATsteady`. Conversely, a "false" for `LATsteady` is normally expected only when `LATfactorsSteady` is "false." A fault in the LAT controller is suspected when either is not the case, as tested by:

```
ruleTest(n) =
    (ruleIf(n) == false)|((ruleIf(n)
    == true)&(ruleThen(n) == true))   (37-10)
```

where the operator "==" tests for logical equality, and " | " (i.e., the "verbar" character) is the logical "or" operator. Each rule index, n, where `ruleTest` is "false" causes the AFDD tool to initialize a fault detection "case" for interactive diagnosis later using tool components beyond the scope of this article. Note that the ostensibly simpler test of merely checking for logical equality between `LATfactorsSteady` and `LATsteady` would not express an equally full knowledge about feedback control. So, authoring effective states and rules for an expert system is an expertise in itself.

The essential points for now are that control charts and expert system logic are two complementary parts in an infrastructure of computer programming engineered to produce autonomous capabilities in the AFDD tool, and that those capabilities begin with control charts that generate primitive states sequentially from data they are fed. The remainder of this article focuses on that last point.

DISTINGUISHING CONTROLLER FAULTS FROM NORMAL DYNAMICS

Some prior research on FDD of HVAC equipment, such as Li and Braun (2003), employ *a priori* models of acceptable steady-state process behavior to detect faults. These models can be computational (e.g., first-principles equations), statistical distributions, or tabulated performance data. The AFDD tool then needs a "steady-state detector," such as described by Kim et al. (2008), to distinguish the periods of data compatible with such models from those periods it must ignore. If the tool does not use a steady-state model, though, there is no reason to bar valid dynamic data from it. So, filtering data down to have only steady-state content is potentially unnecessary. What expert logic does need instead is automatic detection of when specified quantities begin and end transient periods. This feeds states and rules logically correlating, for example, that an excursion in duct air temperature is due not to some fault, but to a normal transient in CHW valve position instigated by a scheduled reset of CHW temperature. Transients are identified automatically by adapting the Shewhart chart as follows.

Automating the Shewhart chart involves calculating the z-score, z_n, of each sample, x_n from the sample set mean, $\bar{x}$, and standard deviation, s. Equation 37-3 shows a z-score in general terms, but to accomplish the purpose here two customizations are necessary. Given the HVAC process has continuing duration, $\bar{x}$ and s must be computed from a circular "rainfall" buffer of running sample values extending back a defined time span. That span was forty minutes (eight samples) for all results seen in this article. Also, unlike Equation 37-3, the standard deviation used to calculate the z-score at sample time n, z_n,

$$z_n = \frac{(x_n - \bar{x})}{s_{ref,n}} \qquad (37\text{-}11)$$

is not always from the current contents of the rainfall. Instead, the standard deviation used is that from the instance of the rainfall when the quantity was last automatically classified as being steady: a "running reference" standard deviation, s_{ref}. It can differ from the current value, s, as expressed for sample n by the following pseudocode:

```
Sref, n = (Sn * isTrue?(steady)) +
    (Sref, n-1 * isFalse?(steady))   (37-12)
```

The casting functions `isTrue?(steady)` and `isFalse?(steady)` each test whether a binary logical variable generically named `steady` (such as `LATsteady` in Equation 37-7) is respectively "true" or "false." The casting functions return the number "1" if their test (not the variable) proves true and "0" (zero) if it does not. The numbers returned are multiplied (the "*" operator) by, and thus select between, the standard deviation of all samples currently in the rainfall buffer, s_n, and the reference standard deviation brought forward from the previous iteration, $s_{ref, n-1}$.

Equations (37-11) and (37-12) are computed for each sample, x_n, of each quantity, x (e.g., a temperature, flow rate, or actuator position) monitored by Shewhart chart for transients. The Shewhart chart in automated form is a programmed function, `ShewChart(...)`, that assigns a "true" value to a normally "false" binary logical state, `shewTrip` (meaning a sample has "tripped" a chart threshold), for any z_n outside the three-sigma band defined by $\pm 3s_{ref}$. The chart thus takes that sample as evidence that x is now, as explained earlier, influenced by a special-cause factor.

No more is asked each sample period of a Shewhart chart than to revalue its `shewTrip` output state for the variable it monitors. It is up to expert logic downstream of all the charts to determine what any "true" value for a `shewTrip` means in the context of the prevailing HVAC system operation. For example, when a special-cause factor found by a chart correlates to a valid change in the setpoint of a controller known to affect the charted quantity, it can be inferred that a normal event has occurred. The method is to first pass each `shewTrip` to a state characterizing whether the monitored variable is "steady" (i.e., within the band allowed as steady-state variation):

```
steady=(not(shewTrip) & is
    Equal?(counter, countsToClear))    (37-13)
```

where `counter` and `countsToClear` are both positive-valued integers. The functions `not(...)` and `isEqual?(...)` apply respectively a negation and a numerical equality test. Initially, `shewTrip`, `steady`, and `counter` are respectively "true," "false," and zero. Each sample period that `shewTrip` is revalued "false" increments `counter`, up to a maximum called `countsToClear`. Each period that `shewTrip` revalues as "true" decrements counter, down to a minimum of 0. Changing `steady` from "true" to "false" is thus immediate, but going from "false" to "true" is delayed by `countsToClear`. This reduces chattering and spurious values of

`steady`, as seen by comparing the plot panels of Figure 37-1, collectively called a *logic analysis sequence* because it shows changes over time in specified logic states of the expert system alongside the analog quantities driving those states.

The bottom two panels Figure 37-1 show `steady`, which refers to the behavior of the heating coil water valve position, Z, seen in the second panel from the top. The only two values of `steady` possible are "true" (T) or "false" (F) on the vertical axis. In the bottommost panel `countsToClear` had been set to 0 (i.e., no delay action), resulting in `steady` being incorrectly "true" during an interval of wide, obvious hunting of the valve. In the panel second from bottom, where `countsToClear` is set to 3, the same hunting interval is correctly identified by `steady` being "false." Only when `steady` is "true" is s_{ref} updated to equal s.

The benefit of combining control charts with expert logic is seen by comparing the top four plot panels of Figure 37-1. The second panel from the top shows hunting of the hot water valve position, Z, as the obvious immediate cause of the zone temperature oscillation seen in the top panel and discussed in the Introduction. The expert logic infers the same automatically without human graphical cognition. A crucial factor is proper update of the reference standard deviation, s_{ref}, by the standard deviation, s, from the current 40-minute rainfall. To prevent divide-by-zero errors, rainfalls are binned so 0 values of s do not occur even for constant data.

It is seen that erratic valve activity just before Hour 04 trips the Shewhart chart, assigning "false" to `steady` which in turn saves away the current standard deviation s as s_{ref}. The valve action is not again considered steady until after Hour 18, when it calms back to a variance that repeatedly (recall the role of `countsToClear`) yields "false" for `shewTrip`. At that point the current value, s, of standard deviation becomes the "new" s_{ref} used by the Shewhart chart (via Equation 37-12). This new s_{ref} begins larger than its previous value, the valve activity held in the rainfall buffer only just satisfying the chart, but s_{ref} settles adaptively to lower values thereafter.

Values of `countsToClear` greater than 0 put the switching of `steady` from "false" to "true" later than a human expert reading the valve plot would more accurately put it. But, it is evident that 0 for `countsToClear` yields misidentifications that a non-zero value prevents. In the bottom panel of Figure 37-1, much of the Day 1 period of erratic action by the valve between Hours 04 and 18 is misidentified as being steady, because a single sample met the three-sigma chart criterion (with s_{ref} as "sigma"), switching `steady` to "true," which jumped s_{ref} to a very large value. That subsequently allowed a factually erratic period to be marked as steady, an error the same routine avoids in the above panel where `countsToClear` is non-zero.

The Shewhart chart thus proves useful to identify and demarcate transients in the operation of HVAC equipment. Since dynamic data from the equipment can be due to many factors, some totally acceptable, it is expert logic downstream that determines whether or not any transient so observed indicates a fault.

IDENTIFYING FAULTS IN DEVICES UNDER REGULATION

Besides identifying transients, control charts can detect faults directly. Automated functions based upon the CUSUM chart, described earlier, put out binary logical states characterizing regulation explicitly, such as "tracking too high," "tracking too low," or "hunting." These states are "primitive" in the expert system because analytical (e.g., chart) functions generate them directly from data with no other logic intervening. Those actions constitute items (b) and (c) in the capabilities list.

The example of Figure 37-1 concerns temperature regulation, so discussion continues here in terms of a generic temperature T, although pressure or flow regulation could apply as well. Two incremental quantities based upon the setpoint temperature T_{set} are updated upon each sample T_n of the regulated temperature at time step n:

$$\Delta U = \left(T_n - \left(T_{set} + b\right)\right)\Delta t_{ddc} \qquad (37\text{-}14)$$

$$\Delta L = \left(T_n - \left(T_{set} - b\right)\right)\Delta t_{ddc} \qquad (37\text{-}15)$$

where b is a parameter chosen by the user to reflect the amount of dispersion in T_n the chart algorithm is to allow. The BAS sampling period is t_{ddc}—1/12 hour (5 minutes) was used here—meaning the units of ΔU and ΔL are degree-hours per sample. These increments then tally into two chart registers, A and C, as follows:

$$A_n = A_{n-1} + q\left(\Delta U > 0\right)\Delta U \qquad (37\text{-}16)$$

$$C_n = C_{n-1} + q\left(\Delta L < 0\right)\Delta L \qquad (37\text{-}17)$$

where the function $q\ (...)$ converts the true and false results of logical expressions into the numbers 1 and 0, respectively. Non-zero A is always positive and non-zero C negative, as the case with P and Q used by CUSUM.

Unlike CUSUM, here each data sample affects at most only one register, affording a capability to identify hunting that CUSUM lacks.

Units of A and C are degree-hours, analogous to watt-hours in electric meters. The units are relevant because the user sets a warning parameter K_{warn} in them by which the chart algorithm judges quality of controller regulation via the following:

$$\texttt{tracksHigh} = \left(\left(A_n + C_n\right) > K_{warn}\right) \qquad (37\text{-}18)$$

$$\texttt{tracksLow} = \left(\left(A_n + C_n\right) < -K_{warn}\right) \qquad (37\text{-}19)$$

$$\texttt{hunts} = \left(\left(A_n > K_{warn}\right) \wedge \left(C_n < -K_{warn}\right)\right) \qquad (37\text{-}20)$$

where "$\wedge$" is the logical "and" operator. The three outputs (left of each equal sign) are primitive states which indicate regulation of the quantity is satisfactory when all are valued "false" and faulty when any is valued "true."

In the case of a quantity regulated to duplexed setpoints, such as the zone thermostat of Figure 37-1, the band parameter in Equations (37-14) and (37-15) is here made redundant by the unregulated band between the two setpoints. Band b is thus eliminated, and the increments become

$$\Delta U = \left(T_n - T_{set,cooling}\right)\Delta t_{ddc} \qquad (37\text{-}21)$$

$$\Delta U = \left(T_n - T_{set,cooling}\right)\Delta t_{ddc} \qquad (37\text{-}22)$$

These increments then tally into four chart registers, A through D, as follows:

$$A_n = A_{n-1} + q\left(\Delta U > 0\right)\Delta U \qquad (37\text{-}23)$$

$$B_n = B_{n-1} + q\left(\left(\Delta U < 0\right) \wedge \left(A_n > -B_{n-1}\right)\right)\Delta U \qquad (37\text{-}24)$$

$$C_n = C_{n-1} + q\left(\Delta L < 0\right)\Delta L \qquad (37\text{-}25)$$

$$D_n = D_{n-1} + q\left(\left(\Delta L > 0\right) \wedge \left(C_n > -D_{n-1}\right)\right)\Delta L \qquad (37\text{-}26)$$

Non-zero B is always negative and non-zero D positive, and B and D are limited to no more than the additive inverse of A and C, respectively. Introducing registers B and D is better than treating duplexed setpoints as two separate instances of the single-setpoint case. With separate single setpoints the tool would need logic selecting which register sums are relevant over any given period so other register sums can be disregarded to prevent false alarms. Instead, the three primitive logic states are determined without selective logic:

$$\texttt{tracksHigh} = \left(\left(A_n + B_n\right) > K_{warn}\right) \qquad (37\text{-}27)$$

$$\texttt{tracksLow} = ((C_n + D_n) < -K_{warn}) \qquad (37\text{-}28)$$

$$\texttt{hunts} = ((A_n > K_{warn}) \wedge (B_n < -K_{warn})) \vee$$

$$((C_n < -K_{warn}) \wedge (D_n > K_{warn})) \qquad (37\text{-}29)$$

where "$\vee$" is the logical "or" operator.

The three states defined above explicitly identify faults associated with regulation of a quantity. They also serve expert system logic downstream with evidence to diagnose more primary faults or faults having no explicit test of their own.

IDENTIFYING FAULTS IN DEVICES EXERTING REGULATION

Capability (d) in the list requires identifying faulty dynamics in quantities (e.g., position of valves or dampers, or fan speed) exerting regulation on other quantities. For example, a hot water valve position, Z, regulates heating of the air a fan-coil unit supplies to the zone whose temperature is plotted in Figure 37-1. There is no setpoint for Z to track, thus no "tracks high" or "tracks low" fault states to consider. Instead, Z modulates as needed to maintain a set temperature in the air discharged from the unit. Various other factors external to this modulation loop influence Z, notably the fan, outdoor air damper position, and air temperatures in the zone and outdoors. So, Z cannot be expected to remain steady even if discharge air temperature remains steady. However, those external influences are typically slow relative to hunting, the principal fault mode. Thus, it is suitable to base diagnostics upon the mean of Z from a rainfall of past samples going back a specified number of steps in time. The register increments now become

$$\Delta U = \left(Z_n - \left(\overline{Z}_n + b\right)\right)\Delta t_{ddc} \qquad (37\text{-}30)$$

$$\Delta L = \left(Z_n - \left(\overline{Z}_n - b\right)\right)\Delta t_{ddc} \qquad (37\text{-}31)$$

where $\overline{Z}_n$ is the mean of Z as obtained from the rainfall buffer at the current time step n. As done for regulated variables, these increments tally into two chart registers, A and C, according to Equations (37-13) and (37-14), and the registers then generate two logic states, $\texttt{swing}$ and $\texttt{hunts}$:

$$\texttt{swing} =$$
$$((A_n + C_n) > K_{warn}) \forall ((A_n + C_n) < -K_{warn}) \quad (37\text{-}32)$$

$$\texttt{hunts} = ((A_n > K_{warn}) \wedge (C_n < -K_{warn}))) \qquad (37\text{-}33)$$

where "$\forall$" is the exclusive "or" operator.

A "true" for $\texttt{swing}$ indicates the regulating quantity has changed in the same direction over several samples, something notable but not necessarily a problem. The margin implied by those samples depends upon K_{warn}. Given reliable values in $\texttt{swing}$ and $\texttt{hunts}$, subsequent logic can check whether any "true" value obtained correlates with $\texttt{steady}$ being "false" from any acceptable causal factor such as valid change of setpoint, fan speed, or hot water supply temperature. If that check fails, a fault is suspected, triggering further diagnostic logic including user interaction.

RESULTS FROM LABORATORY TESTING

The top panel of Figure 37-2 repeats that of Figure 37-1 so accumulations in registers A through D, seen in the second panel, can be compared directly to developments driving them. Winter weather causes zone temperature to modulate at the heating setpoint for most of the plotted period, where only registers C and D accumulate value. These registers toggle zone temperature state $\texttt{hunts}$ from "false" to "true" and back near Hour 14, 11 hours after hunting visibly begins. A change to parameter K_{warn} could toggle the state sooner, but it is also desired that $\texttt{hunts}$ not falsely toggle "true" during the acceptable modulation seen between Hours 05 and 12 on the second day. Since acceptable modulation can go on indefinitely, a balanced choice of K_{warn} and a register reset period is needed to prevent false alarms. Such a reset occurs at Hour 12 of Day 2. In the bottom panel, the state $\texttt{tracksLow}$ toggles through "true" during the period of sluggish response between Hour 19 on the first day and Hour 03 on the second day. Subsequent logic can check for any correlation with "true" $\texttt{hunts}$ from charts of the hot water valve, fan, and outdoor damper. A coincidence infers that primary cause lies in a specific control loop, averting redundant alarms. The second panel of Figure 37-1 shows that the intervals of hunting and sluggish regulation of zone temperature correlate with similar behavior of the hot water valve. In fact, during those intervals faulty valve controller gains had been applied deliberately as a test.

Figure 37-3 shows a fault in regulation of mixing box air temperature T_{am}, by the outdoor air damper position Z_{dm}. Gray dashed lines in the top panel show the band b by which the user sets an acceptable variation for T_{am} that here is 0.5 °C (0.9 °F). In this case, regulation is satisfactory between Hours 08 and 12, where Z_{dm} is seen

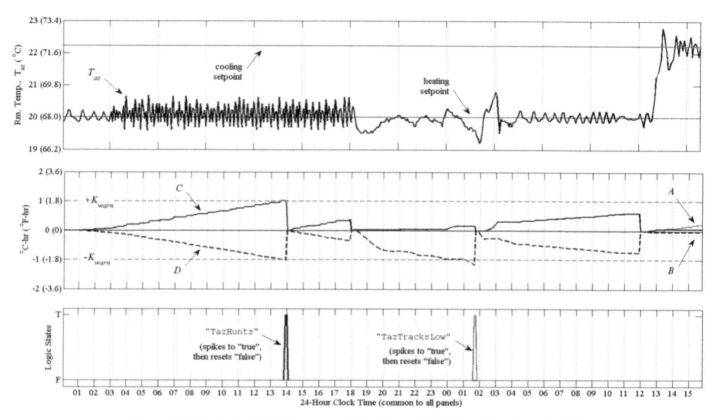

Figure 37-2: Logic Analysis Sequence of Hunting and Drift in a Regulated Variable

in the second panel to slowly trend open (outdoor air is shut off at zero Z_{dm}) as daytime outdoor temperature rises. After Hour 12, however, a change occurs that affects regulation of T_{am} very adversely. The subsequent large swings in T_{am} above setpoint are captured as repeated "true" values of the `TamTracksHigh` state seen in the bottom panel. Coincident hunting of the damper is captured by "true" values for the `ZdmHunts` state. Subsequent logical statements then combine these primitive states with other states, such as `TamSetptSteady`, to localize the cause of the change and determine whether a fault exists.

ADDITIONAL ASPECTS OF THE METHOD

Managing Chart Registers

It is necessary to ensure sums accumulating in chart registers do not result in numerical overflows. Also, fault evidence can pass undetected if sums accumulating in registers grow large and "stale," meaning much of the data contributing to them occurred so long ago as to be irrelevant or even deleterious to current surveillance. Proper selection of allowance bands such

as b in Equation (14) lets fault-free operation make only small additions to chart registers over time, although getting no addition is unlikely. Further, fault-free operation will always add large values to some registers in duplex setpoint charts. One provision is resetting all registers to 0 when any of the three fault states test as "true," a practice used in CUSUM. Unlike CUSUM however, here each data sample outside the accepted band does not tally to both positive and negative registers, making hunting detectable but also removing a feature that intrinsically helps keep registers bounded.

Another way to manage registers is through so-called "forgetting factors" applied to all register equations, such as F_f in this example for A:

$$A_n = A_{n-1} + q(\Delta U > 0)\Delta U - (F_f A_{n-1})q(\Delta U \le 0) \quad (37\text{-}34)$$

Trials showed the value of F_f must be chosen carefully to not inhibit detection of hunting, so its use adds to the parameters needing expert attention. Since satisfactory results were obtained from periodic resets alone, forgetting factors were not tried further.

The solution taken here simply reset all registers to 0 at a specific period regardless of other factors. It was

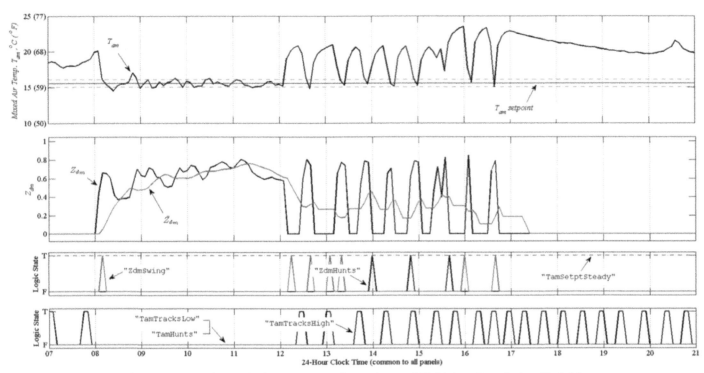

Figure 37-4: Logic Analysis Sequence of Hunting and Drift in a Regulating Variable

found here that reset periods longer than 18 h left the registers too stale to exploit recent data features, while periods below 10 h inhibited detecting slow-evolving faults. The heuristic demand to manually find robust settings for field deployments offers a future opportunity for fuller automation.

Field Implementation

This method is implemented most feasibly in the field as one component in a larger, self-contained AFDD program running on a generic, general-purpose computer having open-protocol real-time network access to present value data from the BAS controllers existing in the building. This is in contrast to prior AFDD approaches such as VPACC, which are to be programmed into existing BAS products alongside the conventional HVAC control logic, using the typically very constrictive user programming package the BAS vendor supplies to its customers. That approach was reasonable in its era, when the hardware supporting the automation of buildings was much more limited and expensive than it is now. It accounted for a field technology lacking the comprehensive open-protocol networking and more capable hardware and programming interfaces that are available today at reasonable cost. Real-time data from the BAS network is not mandatory—the method described here also works on archived "trend logs"—but

a real-time implementation is much more reliable, as no routine and potentially troublesome synchronization of disparate or corrupted data files is required. Real-time implementation also lets the user answer diagnostic queries, asked by a suitably designed AFDD tool, with contemporaneous information not available from the BAS but beneficial to the tool's effectiveness.

Rethinking the AFDD tool as a program networking autonomously from its own hardware and software frees it from the constraints of existing BAS products tailored primarily to enable technicians of little programming skill to set up and maintain simple, conventional HVAC control sequences. That orientation, lacking the usual capabilities of general-purpose computing, fully excludes programs of the sophistication described in this article. In buildings where the BAS can supply an autonomous AFDD tool with sensor data by open-protocol networking, the tool along with its hardware and maintenance could easily be an integrated product offered by a third-party specializing in HVAC AFDD and having the advanced skills needed to create and service such tools for use by building staff. Recognizing that potential, no specific way to implement the chart or expert logic structures is given here. Researchers able to build upon the method already have the skill in general-purpose programming to implement it in laboratories, perhaps better than could be detailed here. The crucial issue instead is developing such methods into user-oriented commercial products of the building automation industry. Future collaborative development and field testing projects with HVAC industry partners are essential to build within that industry the new, cross-discipline competencies they need to make effective AFDD products a marketed reality.

CONCLUSIONS

Control charts combined with expert system logic offer a viable way to automatically detect and diagnose hunting and other faults in HVAC system controllers. Historically involving recordkeeping and calculations done manually on paper, control charts now take computerized form as numerical registers "charting" the behavior of a monitored process by accumulating specified data, particularly by summing excursions of designated quantities beyond allowed bands. Autocorrelation in the data, traditionally a bane to control charts given continuous dynamic processes such as HVAC, becomes a diagnostic asset when expert system logic downstream of the charts is used to test for the cross-correlations normally expected between quantities having known physical relationships.

Equations 37-14 through 37-33 describe the "first generation" of charting functions keeping control loops under surveillance in a far more comprehensive AFDD tool for commercial HVAC systems, called "AFDD Expert Assistant," being developed by the U.S. National Institute of Standards and Technology (NIST) for field testing in 2014. With the viability of control charts proven for that role, further research and field trials will aim for many improvements, for example, to eliminate delayed return of the logic state `steady` to "true" in Figure 37-1, and eliminate the occasional spurious transitions of that state. Broader areas for future work include estimating critical parameters such as the band and alarm limits autonomously, and charting variation of the mean of a quantity as a way to extract frequency-domain information that could improve diagnostics.

References

Betteley, G., and N. Mettrick, E. Sweeney, D. Wilson. 1994. *Using Statistics in Industry.* Hertfordshire, UK: Prentice Hall International.

Kim, M., and S.H. Yoon, P.A. Domanski, W.V. Payne. 2008. Design of a steady-state detector for fault detection and diagnosis of a residential air conditioner. *International Journal of Refrigeration* 31: 790-9.

Li, H., and J.E. Braun. 2003. An improved method for fault detection and diagnosis applied to packaged air conditioners. *ASHRAE Transactions* 109(2):683-92.

Page, E. S. 1954. Continuous inspection schemes. *Biometrika.* 41(1):100-15.

Ryan, T.P. 2000. *Statistical Methods for Quality Improvement, 2nd Edition.* New York, NY: Wiley & Sons.

Schein, J., and J.M. House. 2003. Application of control charts for detecting faults in variable-air-volume boxes. *ASHRAE Transactions* 109(2):671-82.

Shewhart, W.A. 1931. *Economic Control of Quality Manufactured Product.* New York, NY: D. Van Nostrand.

Shiffler, R. E. 1988. Maximum z-scores and outliers. *The American Statistician.* 42(1):79-80.

Summers, H., and C. Hilger. 2012. *Fault Detection and Diagnostic Software.* Emerging Technologies Program ReportET11PGE3131, Pacific Gas and Electric Company, San Francisco, CA.

Westphalen, D., and K.W. Roth, J. Brodrick. 2003. System & component diagnostics. *ASHRAE Journal* 45(4): 58-9.

Wise, S.A., and D.C. Fair. 1998. *Innovative Control Charting.* Milwaukee, WI, USA: ASQ Quality Press.

Chapter 38

AHU AFDD

Jin Wen, Ph.D. and Adam Regnier
College of Engineering, Drexel University

INTRODUCTION

An air handling unit (AHU) connects primary heating and cooling plants with building zones and controls building ventilation air intake. It therefore has a significant impact on the energy consumed by the heating, cooling, and ventilating systems, as well as on the supply air temperature, humidity, and carbon dioxide levels, which further affects the occupants' health and comfort.

Configuration

Many configurations exist for an AHU. An AHU can be categorized as

- single duct or dual duct AHU based on the supply air duct design;

- single fan (without return fan) or dual fan (with return fan) AHU based on the availability of return fan;

- constant air volume (without supply and return fan speed control) or variable air volume AHU (with supply and return fan speed control) based on the fan control type and downstream terminal units;

Moreover, the number, type and location of heating and cooling coils are also diverse. Some AHUs have a preheating coil before the mixing air section, some AHUs do not have heating coils (nor preheating coils) at all. Some AHUs have hot water heating coils and some have electrical heating coils. Some AHUs have direct expansion (DX) cooling coil, and some AHUs have chilled water cooling coil. The availability of a humidifier is also different. More detailed information is provided by ASHRAE (ASHRAEa, 2012). Figure 38-1 demonstrates the configuration of a typical single duct dual fan AHU with a pre-heating coil.

The major components of the AHU, shown in Figure 38-1, are the supply air and return air fans; preheat,

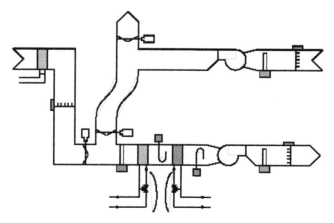

Figure 38-1: Air handling unit (Price and Smith, 2003).

cooling, and heating coils; heating and cooling control valves; recirculated air, exhaust air, and outdoor air (OA) dampers; and the ducts to transfer the air to and from the conditioned spaces.

Commons Sensors

The number and type of sensors equipped in a typical AHU vary greatly. Newer AHUs are normally instrumented more heavily than older ones. The most commonly available measurements include: supply, return, and mixed air temperatures, outdoor air temperature (often is not the immediately air temperature outside of the specific AHU). For a variable air volume AHU, the supply air pressure and fan speed are typically measured. Some AHUs provide outdoor, return and supply air flow rate measurements. Humidity (especially outdoor, supply, and return humidity) is often measured, especially for those AHUs that need to perform humidity control. Some AHUs offer fan power measurements, supply and return (or mixed, if by-pass valve is used) hot water and chilled water temperature measurements. Although highly valuable, hot and chilled water flow rates are not often measured due to the high sensor cost. Control signals, including outdoor, re-circulated, and exhaust air damper signal, coil valve signal, fan on/

off signal, and fan speed signal (for variable speed fan), are typically available, although few AHUs offer real damper and valve positions, as feedback signals from the damper/valve.

Common Control Sequences

An ASHRAE Guideline, SEQUENCES OF OPERATION FOR COMMON HVAC SYSTEMS (ASHRAEb, 2005), provides a fairly detailed and comprehensive description of common AHU control strategies, including CAV and VAV AHUs, as well as single duct and dual duct AHUs. In this section, some commonly used single duct dual fan VAV AHU control strategies are described to provide a basic understanding of AHU control sequences.

System Operation

The AHU is often controlled and operated differently during different system operation modes. Commonly used system operation modes include occupied mode (OCC), unoccupied mode with system off (UNOCC), night set-back mode (SET-BACK), and system start-up mode (START-UP). The modes can be manually selected or can be scheduled with the EMCS. When system operation is in the OCC mode the fans, dampers, and valves are controlled with the normal control sequences discussed in this Section. During the UNOCC mode, the fans are typically turned off, and the dampers and valves are indexed to a fully closed position. Fully closed dampers and valves refers to 100% recirculation air with both the heated and chilled water valves closed.

During a SET-BACK mode, the zone temperature is maintained within a heating and cooling setpoints, which are different from the setpoints during the OCC mode. The heating setpoint is typically lower than that of the OCC mode and the cooling setpoint is typically higher than that of the OCC mode. The systems are typically off, similar to that of the OCC mode, if all of the zone temperatures are within the setpoint. The systems are controlled to provide heating or cooling if the temperature of one of the zones is outside of its setpoints.

During a START-UP mode, setpoints used in OCC mode are typically used. However, special operation, such as 100% recirculation of indoor air without any heating or cooling coil operation is typically used if the zone temperatures fall below the zone heating setpoint. 100% outdoor air purging, when the economizer conditions are satisfied, is often firstly used if the zone temperatures rise above the zone cooling setpoint.

Supply and Return Fan Control Sequence

The speed of the supply fan is modulated to maintain duct static pressure at a static pressure setpoint. A PI control algorithm is typically used to control the speed of the fan, although floating control is often also used for this purpose.

The control sequence used to control the return fan can be in one of four typical modes, namely, fan speed matching, air flow rate matching, air flow rate differential, or return fan off. When using the fan speed matching mode, the speed of the return fan is maintained at an adjustable percentage of the supply fan speed. When using the air flow rate matching mode, the speed of the return fan is maintained so that the return air flow rate is at an adjustable percentage of the supply fan flow rate. When using the fan air flow rate differential mode, the return fan speed is maintained so that the return air flow rate is the same as the supply air flow rate minus an adjustable constant differential.

Supply Air Temperature Control Sequence

Sequencing control, in which, one PI loop is used to control various AHU components to maintain supply air temperature at its setpoint, is very popular and is introduced here. Although relatively simple, sequencing control has inherited stability issues. More detailed discussion and potential solutions can be found in Seem et al, 1999. When using sequencing control, the control sequence used to maintain supply air temperature at its setpoint is divided into different control regions, such as, mechanical cooling, mechanical and economizer cooling, economizer cooling, and mechanical heating, as shown in Figure 38-2. Each region depends on whether or not outdoor air temperature is greater or less than a reference temperature and humidity known as the economizer air temperature and humidity setpoints (typically around 55 to 65°F, and 80% RH), and whether supply air temperature is above or below its setpoint.

The regions correspond to the output from a single PI control algorithm that is split to control the heated water coil valve, the outdoor air damper, and the chilled water coil valve. The PI control algorithm has an output that ranges such as from -100 to 200.

The cooling coil is modulated to maintain supply air temperature at the setpoint when the control sequence is in either the mechanical or mechanical and economizer cooling ranges. During the mechanical and mechanical and economizer cooling modes, the output from the PI control algorithm ranges from 100 to 200, corresponding to a cooling coil position from 0 to 100% open. The outdoor air damper is held in the minimum position (mechanical cooling only) when either outdoor air temperature or humidity are above the economizer setpoints. The outdoor air damper is fully open (mechanical and economizer cooling) when the economizer condition is satisfied and when the output of PI is larger than 100.

As outdoor air temperature drops, the need for mechanical cooling is eliminated (output from the PI control algorithm drops below 100) thereby switching the control sequence to the economizer cooling mode. In this mode, Tsa is maintained by modulating the outdoor air damper. The output from the PI control algorithm ranges from 0 to 100, corresponding to outdoor air damper position of 0 to 100% open. Note that the supply air temperature control sequence has a lower priority than maintaining a minimum outdoor air damper position. Therefore, during the period from an output corresponding to the minimum outdoor air damper position (minimum outdoor air damper position, typically 30% open) to 0, the outdoor air damper stays at minimum outdoor air damper position, there is no control for supply air temperature, and supply air temperature deviates from its setpoint.

If outdoor air temperature continues to decrease, a point is reached where the outdoor air damper is in the minimum position and mechanical heating is required (output from the PI control algorithm drops below 0) causing the control sequence to switch to the mechanical heating mode. During the mechanical heating mode, the valve for the AHU heating coil is modulated to maintain Tsa. The output from the PI control algorithm ranges from 0 to -100, corresponding to a heating coil valve position of 100 to 0% closed.

AHU FAULTS

Reported AHU Faults

A wide variety of faults have been reported in the literature for AHU systems. Most of the studies focus on single duct VAV AHU. Very few discussions about how the fault severities are determined for.

Faults can either be described by their symptoms, such as "supply air temperature is too high," or by their sources (faulty devices), such as "cooling coil valve is stuck closed." Among all available AHU AFDD studies, several studies discussed the AHU faults systematically and in great details and are summarized here.

As part of the IEA ANNEX 25 study (ANNEX 25, 1996), various possible AHU faults are provided. A survey was conducted among AHU designers, construction engineers, and commissioning engineers to rank the various possible AHU faults based on their importance. Seven reasons were used when considering whether a fault is important, which include 1) Environmental degradation and occupant complaints; 2) Increased energy consumption; 3) Serious secondary damage; 4) Frequent occurrence; 5) Difficult detection; 6) Lengthy repair time; and 7) Costly repair. Ten most important faults for a VAV AHU system described by their symptoms with possible sources are summarized by ANNEX 25 (1996). No fault simulation method or experimental data are supplied in this study.

A series of papers (Lee et al., 1996a, Lee et al., 1996b, Lee et al., 1997, and House et al., 1999) summarize up to thirteen faults for an AHU system including: complete failure of the supply and return fans, chilled water pump, supply air temperature sensor, pressure sensor, supply

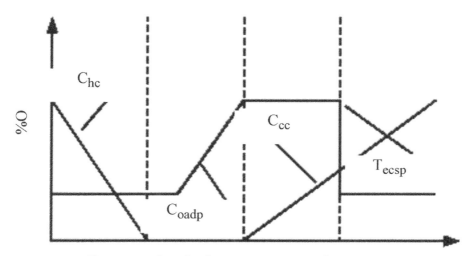

Figure 38-2: Supply air temperature control sequence.

flow rate sensor, return flow rate sensor, and damper linkage; supply air temperature sensor offset; stuck cooling coil control valve; fouled cooling coil; leaking heating coil control valve; and supply fan performance degradation. The faults and the studied AHU are modeled based on the component models reported by Wang (1992).

Besides the above studies, the following section summarizes other available AHU AFDD researches that provide AHU fault description, modeling, and experiments.

Norford and Little (1993) describes four AHU faults symptoms including 1) Failure to maintain supply air temperature setpoint; 2) Failure to maintain supply air pressure setpoint; 3) Increased pressure drop; and 4) Fan motor, coupling to fan, and fan controls. Possible causes for the above four symptoms are also given. Neither simulation nor experimental data for the above faults are discussed. Coil fouling and its modeling are discussed by Dumitru and Marchio (1995). Two mixing box faults, which are 1) reverse acting recirculated damper and 2) leaking dampers, are discussed and modeled using HVACSIM+ by Dexter (1995). Several faults related to heating exchanger and AHU are described and simulated by Fasolo and Seborg (1995). The faults include control valve stiction, water pump failure, and supply air temperature sensor failure. Both experimental and simulation tests are reported for several common AHU faults including sensor offset and damper and valve mechanical blockage (Glass et al., 1995). The experiments were conducted using the Landis & Gyr HVAC laboratory. MATLAB SIMULINK was used to simulate the system and related faults. Cooling coil fouling and valve leakage are simulated using HVAC-SIM+ by Haves et al. (1996). The simulation model was later used by Dexter and Benouarets (1996). Peitsman and Soethout (1997) also reports using HVACSIM+ to simulate cooling coil valve leakage, coil fouling (slowly increasing), stuck outdoor air damper, and degenerating supply air temperature sensor. Details are supplied about how the AHU is modeled using HVACSIM+ but not about how the faults are simulated. Various faults related to economizer control are examined with hourly measurements from a real building by Brambley et al. (1998). No detailed fault description is supplied. DOE-2 is used to simulate various economizer faults including stuck damper and temperature and humidity sensor failures (Katipamula et al., 1999). Faults related to a cooling coil, including leaking and stuck valve, coil fouling, and supply air temperature sensor offset, are simulated (Ngo and Dexter, 1999). A sensitivity test is used to study the effect of fault size on the reported AFDD methods ca-

pability. Nine faults during the heating mode and nine faults during the cooling mode were implemented in an AHU of a real building (Carling, 2002). Limited information is supplied about how the faults are implemented in the test facility.

Based on the literature review and logical analysis, a comprehensive and categorical list of AHU faults are provided by Wen and Li (2011) in Table 38-1. Theoretically, all AHU devices including control software could develop faults. Therefore, the faults listed in Table 1 are categorized based on the specific device corrupted by a fault, with the devices grouped into four categories: sensor, controlled device, equipment, and controller. Such categories are mostly used among control engineers. It is noted that some devices, such as fans and pumps can be either grouped into the controlled device category or the equipment category. For each device, common faults are summarized in Column 3 of Table 38-1. Column 4 provides fault severity magnitudes cited in the literature with a format of: severity [literature]. Studies addressing fault models for each fault are assembled in Column 5. Column 6 specifies the experimental data sets that are associated with different faults which have been conducted at the Iowa Energy Center Energy Resource Station (ERS) test facility. Column 7 lists studies that have cited the associated faults without providing fault severity.

AHU Fault Impacts

It is well acknowledged that the existence of faults in an AHU impacts building energy consumption, occupant comfort, maintenance cost, and equipment life cycle. It is estimated that 15 to 30% of energy are wasted in commercial buildings due to various HVAC equipment and control faults (Katipamula and Brambley, 2005). A detailed discussion about how various AHU faults affect energy, comfort, and equipment is provided by AN-NEX 25 (1996). A survey among a wide range of HVAC professionals was performed as part of the ANNEX 25 study in Japan to rank faults in VAV AHU systems and VAV units based on the seven reasons listed in Section 2.2.1 (ANNEX 25, 1996). From the survey results, it is observed that the most important VAV AHU faults that: 1) cause environmental degradation are: air filter being clogged, and humidifier failure; and 2) cause energy loss are: air filter being clogged, fouling on coil fins and tubes, abnormal OA damper opening, and improper static pressure sensor location.

Lunneberg (1999) discusses how economizer faults would impact building energy consumption. Various case studies show that up to 50% of total building energy could be wasted by mal-functional economizers.

Various field test results are summarized by Katipamula et al (2003) which provides annual energy cost impact of different AHU faults (from $110.00 to $1,2000.00). Economizer faults that cause excessive flow of outdoor air and limit the outdoor air flow rate in economizer mode had the largest impact on energy use.

A recent study (Regnier and Wen, 2013) examines more closely experimental data generated from ASHRAE project RP 1312 (Wen and Li, 2011) to obtain daily energy impact caused by a fault in different seasons. During the RP 1312 study, two identical AHUs which serve two identical sets of building zones, are used. Faults are artificially injected in one AHU and the other AHU serves as the baseline system (more details are discussed in Sec. 3.4.1). By analyzing energy consumption differences between the two systems, it is found that the various tested faults could cause 10 to nearly 150% daily energy consumption difference between the two systems. Table 38-2a lists the energy consumption differences among the two systems under fault free conditions. Notice that all of the data are taken from comparing two AHUs, each serving 4 VAV boxes from 8am to 6pm. The differences found from the two different AHUs span in magnitude from 6% to 20%, and from 5 kWh/day to 19 kWh/day. The average magnitude of difference is 12%, and the average difference is 3% more energy being used by AHU-A than by AHU-B (or 2.2 kWh/day). This provides some context with which to view the remaining results. Table 38-2b illustrates the energy differences between the two systems under different fault conditions in the summer season. Only those faults that exceed the highest fault-free differences from Table 38-2a are included here. Data from other seasons have similar results and are not included here.

AHU FAULT DIAGNOSTICS

Challenges

Compared with other HVAC systems, AHU system present the following unique challenges for its fault detection and diagnostics:

AHU systems have diverse configuration, system parameter, sensor system design, and control strategies. Unlike primary system (such as chillers and boilers), or terminal units, which only have limited configurations and designs, AHU system, as a build-up system, is often customized. The duct system design, major components, component placement, sensor system design, and control strategies are often different from one AHU to another

AHU. For those system that has limited configurations, it is possible to conduct laboratory tests to develop physics based models which are often used in model-based AFDD, or to obtain faulty operational data and to identify fault thresholds (a value used to differentiate fault free behavior and faulty behavior), which are necessary for rule-based AFDD. Because of the diverse configurations, it is very unlikely to obtain universal physics based model or fault threshold. Obtaining these via individual AHU testing is time consuming and expensive.

AHU systems have uniquely different operating modes that are determined by weather conditions and internal conditions. AHU systems need to re-configure themselves to adapt to different weather and internal conditions. Even during a day or within couple hours, an AHU could experience heating mode, economizer mode, and cooling mode. Each operating mode involves different components and control strategies. This feature presents great challenges for pure data driven AFDD methods.

Moreover, AHU systems are connected with both primary systems and terminal units. Faults occur in those systems could have impacts on AHU systems, which further complicate its AFDD process.

Besides to the above challenges, AHU systems, similar to other HVAC systems, also suffer from poor measurement accuracy and data quality. For example, because of the large duct cross section area in most of the AHU systems, it is very difficult to measure air temperature and flow rate accurately, especially the mix air temperature and outdoor air flow rate. AHU controllers, often are secondary controllers, may not trend enough measurements for AFDD purposes. The above discussed challenges make AHU AFDD a difficult task.

Available Methods

Over the past couple of decades, significant research efforts have been undertaken to address the need for reliable AFDD methods for AHU-VAV systems. The research can be grouped into three broad categories of approaches: physical modeling methods, rule-based methods, and data-driven methods (Katipamula & Brambley, 2005a, 2005b). Additionally, a number of researchers have suggested hybrids of these method categories. For the sake of brevity, this section includes a discussion focused on the most recent advances in AFDD with the promise for commercialization. For this reason, methods requiring detailed physical models of the system have been largely excluded and can be found in other publications such as (Katipamula & Brambley, 2005a, 2005b, Li and Wen, 2014).

Table 38-1: AHU fault summary (Wen and Li, 2011)

Category	Device	Type		Literature with Severity	Literature without Severity	Available Exp. Data
Sensor	SA Temp	offset	discrete	0 [4]*; 2.8C [21]; 3 and 5F [22]; 2.5C [25]	[15], [16], [24], [26], [27]	[21], [22]
			drift	5.4F [1]; 2.7F [4] , 1.8F [18]; -4 to 4 °C [23]		
	MA Temp	offset	discrete		[27]	
			drift	-4 to 4 °C [23], 1C [25]		
	OA Temp	offset	discrete		[26], [27]	
			drift	10F [20]; -4 to 4 °C [23]; 1C [25]	[11]	
	RA Temp	offset	discrete		[26], [27]	
			drift	5.4F [1]; -4 to 4 °C [23]; 1C [25]		
	SA Pressure	offset	discrete		[4], [15], [16], [27]	
			drift	[25]		
	Building Pressure	offset	discrete			
			drift			
	SA CFM	offset	discrete	0 [4]*	[15], [16], [27]	
			drift			
	RA CFM	offset	discrete	0 [4]*	[15], [16], [26], [27]	
			drift			
	OA CFM	offset	discrete		[26], [27]	
			drift			
	SA HUMD	offset	discrete			
			drift			
	OA HUMD	offset	discrete		[26]	
			drift			
	RA HUMD	offset	discrete		[26]	
			drift			
Controlled Device	OA Damper	Stuck	fully open	[1], [19], [20]		[22]
			fully closed	[1], [20]		
			partially open	45% [19]; 40% [23]	[20], [24]	
		Leaking		10% 25% 40% [23]	[4]	
		Faulty position indication				
	RA Damper	Stuck	fully open	[19]		[21], [22]
			fully closed	[2], [19], [25]		[22]
			partially open	50% [8];55% [19]; 10% 25% 40% [23]	[10]	
		Leaking		10% 25% 40% [23]	[12]	[22]
	EA Damper	Stuck	fully open			
			fully closed			
			partially open			
		Leaking			[4]	
	Valve of Preheating Coil	Stuck	fully open	[4]		
			fully closed	[4]		
			partially open			
		Leaking				
	Valve of Heating Coil	Stuck	fully open			
			fully closed	[1]		
			partially open			
		Leaking		25% [1]; 10% 25% 40% [23]; 2%, 3% [21]; 5% [25]	[5], [6], [19]	[21], [22]

Table 38-1 (*Cont'd*): AHU fault summary (Wen and Li, 2011)

	Valve of Cooling Coil	Stuck	fully open	[7]; [19]; 10% 25% 40% [23];		[22]
			fully closed	[1], [7]		
			partially open	50% [7]; 50% [19]	[5], [15], [16], [17]; [25]	
		Leaking		25% [1], 10% [7], 20% [8], 3% [13], 3% [14], 10% [18], 10% 25% 40% [23]	[24]	[22]
Equipment	Fan	Increased pressure drop				
		Complete failure of SF and RF		[4], [15], [16]		
		decrease in the motor efficiency		10% [25]	[5]	
		belt slippage			[5]	[22]
	Duct	Leaking				
	Heating Coil	Fouling (fin and tube)				
		Reduced capacity				
	Cooling Coil	Fouling (fin and tube)		1mm [13, 14]; 2mm [18]; air 0.4m^{2}K/kW [5]; water 0.5m^{2}K/kW [5]	[24]; [25]	[22]
		Reduced capacity		70%, 42%, 27% [22]		[22]
	Preheating Coil	Fouling				
		Reduced capacity				
Controller	Mixing Dampers	Unstable				
	Heating Coil	Unstable				
	Cooling Coil	Unstable				
	Sequence of heating and cooling devices	Unstable	open loop [3]			
	Reverse Action				valve actuator [3]; RA damper [12]	
	Reverse Flow					
	Return Fan	stick at a fixed speed			[5]	
	Supply Fan	Unstable				[22]

Note:

1) SA – Supply Air; MA – Mixing Air; OA – Outdoor Air; RA – Return or Recirculated Air; EA – Exhausted Air; Temp – Temperature; CFM – Air flow rate; HUMD – Humidity; SF – Supply Fan; RF – Return Fan;

2) *: faults with a severity of zero means a total failure.

3) The list of AHU AFDD papers:

[1]: House et al. (2001); [2]: Xu et al. (2005); [3]: Kelso and Wright (2005); [4]: Lee et al. (1997); [5]: House et al. (1997); [6]: Glass et al. (1995); [7]: Dexter and Ngo (2001); [8]: Salsbury (2002); [9]: Liu and Dexter (2001); [10]: Brambley (1998); [11]: Fasolo and Seborg (1995); [12]: Dexter (1995); [13]: Dexter and Benouarets (1996); [14]: Haves et al. (1996); [15]: Lee et al. (1996a); [16]: Lee et al. (1996b); [17]: Yoshida et al. (1996); [18]: Ngo and Dexter, 1999; [19]: Carling (2002); [20]: Katipamula (1999); [21]: Castro et al. (2003); [22]: Norford et al. (2000); [23]: Bushby et al. (2001); [24]: Peitsman and Soethout, 1997; [25]: Lee et al. (2004); [26] Wang and Xiao, 2006; [27] Du and Jin, 2007.

Table 38-2a: Energy comparison of A and B systems at the Iowa Energy Center Energy Resource Station under fault free conditions.

	Fault	AHU-A Minus AHU-B (kWh)					
Date	Description	Fans	Cooling	Heat	Reheat	Net	Pct
01-29	Fault Free	-0.9	0.0	-3.6	-3.8	(8)	-6%
08-19	Fault Free	0.2	11.1	0.0	0.0	11	6%
08-25	Fault Free	-0.7	13.9	0.0	-0.2	13	8%
05-05	Fault Free	-0.5	11.6	0.1	0.5	12	10%
05-04	Fault Free	-0.9	10.6	1.1	0.4	11	13%
05-03	Fault Free	-0.9	6.3	0.2	-0.4	5	13%
02-17	Fault Free	-0.9	0.0	-15.1	-1.5	(18)	-16%
02-16	Fault Free	-0.6	0.0	-18.0	-0.3	(19)	-17%
05-02	Fault Free	-1.2	13.2	0.2	-0.1	12	20%

Table 38-2b: Summer fault energy impacts (airstream)

	Fault	AHU-A Minus AHU-B (kWh)					
Date	Description	Fans	Cooling	Heat	Reheat	Net	Pct
09-03	Cooling Coil Valve Reverse Action	-1.1	140.2	156.7	1.0	297	146%
08-31	Cooling Coil Valve Stuck (15%)	-1.7	134.8	134.2	-0.7	267	157%
09-02	Cooling Coil Valve Stuck (65%)	-0.2	121.6	124.0	-0.3	245	126%
08-30	Heating Coil Valve Leaking (2-2.0GPM)	-0.3	109.6	95.3	-0.6	204	135%
08-27	Cooling Coil Valve Stuck (Fully Closed)	23.4	-193.3	-0.2	0.0	(170)	-79%
08-29	Heating Coil Valve Leaking (2-1.0GPM)	0.2	77.2	58.1	-0.6	135	86%
08-28	Heating Coil Valve Leaking (1-0.4GPM)	3.4	50.1	28.6	-0.7	81	40%
08-23	RF complete failure	-1.7	54.6	-0.2	-0.3	52	34%
08-22	RF at Fixed Speed (30%)	-0.4	33.1	-0.5	0.0	32	18%
09-06	OA Damper Leak (55%)	-1.0	28.0	0.1	-0.3	27	17%
08-20	EA Damper Stuck (Fully Open)	1.0	21.3	-0.1	0.0	22	11%
08-24	Cooling Coil Valve Control Unstable	2.9	17.2	-0.1	-0.2	20	15%
09-07	AHU Duct Leaking (after SF)	0.3	17.2	-0.1	-0.2	17	11%

Rule-based Methods

The most straightforward AFDD methods utilize a rule-based approach derived from the physics of the system. The National Institute of Standards and Technology (NIST) generated a useful set of AHU performance assessment rules (APAR, refer to Table 3) that was demonstrated to be effective at detecting and diagnosing faults when the proper fault thresholds were identified for specific applications (House, Vaezi-Nejad, & Whitcomb, 2001; Schein, Bushby, Castro, & House, 2006). To augment the basic rules, it has been suggested to utilize "proactive" commissioning, whereby the detection and/or diagnostics are performed by manipulating the control system for assistance (Brambley et al., 2011; Pakanen & Sundquist, 2003). These proactive methods can also be augmented by a fault correction algorithm (Fernandez, Brambley, & Katipamula, 2009; Fernandez, Brambley, Katipamula, et al., 2009).

The rules are typically based on a priori knowledge of the system dynamics, and these variable interactions can be studied to infer values for variables that are difficult to measure, like flow through a valve, or variables that are known to be unreliable, like air flow rates (Tan & Dexter, 2006).

Some rule-based methods have resulted in an FDD tool for whole-building diagnostics (Song, Akashi, & Yee, 2008), while others have focused on specific components, like sensor faults (H. Yang, Cho, Tae, & Zaheeruddin, 2008). Although detailed modeling has been proven to be intractable for widespread commercial adoption, some methods have effectively combined simpler modeling approaches with rule-based techniques (H. Wang, Chen, Chan, & Qin, 2012; H. Wang, Chen, Chan, Qin, & Wang, 2012). The key to simplifying the modeling required with these methods is the use of a genetic algorithm-based optimization method that reduces the residual between the actual and predicted values to continuously refine the model as the system operates.

Beyond some of the implementation difficulties previously mentioned, there are numerous difficulties with extending test bed and laboratory results to commercial applications. Some of these include unstable control issues, a lack of standard control sequences, and data handling challenges. Seem and House (2009) addressed these problems by recommending a novel AFDD method based on finite-state machine sequencing logic combined with the measurement of residuals versus model-based expected conditions.

Rule-based methods are a straightforward method for detecting and diagnosing faults, but it can be difficult to select fault detection thresholds that maintain fault detection accuracy without generating excessive false alarms. This difficulty is due to the uniqueness of different system installations, the transient and multi-modal nature of the systems, and the noise in the sensor measurements. As a result, many of the rule-based methods can require a significant engineering effort for each installation. In an effort to overcome this obstacle many studies have utilized various statistical process control (SPC) based methods for AHU-VAV AFDD applications.

PCA-based Methods

One of the simplest and most commonly employed SPC methods is the use of principal component analysis (PCA). PCA-based methods have been successfully employed for fault detection in many other industries, and a number of promising approaches for AHU-VAV systems have been suggested in the literature. S. Wang and Xiao (2004) proposed a PCA-based scheme for AHU sensor fault diagnosis, using the squared prediction error (SPE) for fault detection, and a combination of contribution plots and expert rules for fault isolation. This scheme was then updated to include a condition-based adaptive step to allow for seasonal changes (S. Wang & Xiao, 2006). PCA-based methods have also been augmented by rule-based methods for increased diagnostic power _ENREF_12(Qin & Wang, 2005).

The possibility of using different combinations of variables for PCA models to detect different faults has also been explored (Du & Jin, 2007a, 2007b; Du, Jin, & Wu, 2007; Du, Jin, & Yang, 2009). These studies utilized joint-angle-analysis (JAA) methods and expert rules for diagnosis after using PCA for detection. Another method proposed to deal with the wide variety of AHU operating conditions was to use a spatial-temporal partition strategy to isolate different system states prior to the incorporation of PCA techniques (S. Wu & J.-Q. Sun, 2011; S. Wu & J. Q. Sun, 2011).

In general, the PCA-based methods demonstrate promising results, but may suffer from high false alarm rates and many of the methods proposed incorporated sensors that are not typically available in the majority of AHU installations. One suggested approach to overcome these limitations is to pre-treat the data using a wavelet transform method (Li & Wen, 2014). This method was demonstrated to effectively differentiate the impact of weather/load effects from faulty/abnormal system behavior. Another approach to solving the same problem was the use of a pattern-matching algorithm for the selection of training data for the PCA models (Xiao et. al., 2013, Regnier et. al., 2013). By only selecting training data from similar operating conditions, the effects of external load fluctuations and transient system operation are automatically eliminated from the fault detection algorithm.

Neural Network-based Methods

Another common SPC method that has been extensively studied for AHU-VAV AFDD is the use of artificial neural networks (ANN). Many different types of ANN applications have been proposed, including general regression neural networks (Lee, House, & Kyong, 2004) and the use of a recurrent cerebellar model articulation controller to detect faults specifically affecting the coil valves (S. Wang & Jiang, 2004).

A couple of studies investigated the use of wavelet analysis for preprocessing the data, one using an Elman neural network for fault diagnosis (Fan, Du, Jin, Yang, & Guo, 2010), and the other combining the wavelet analysis with fractal methods for sensor fault detection (Zhu, Jin, & Du, 2012).

These neural-network based methods show promising detection capabilities, but the sole use of a classifier-based technique for fault isolation requires faulty data from each system and set of faults. Beyond the PCA- and ANN-based methods, a number of other SPC methods have been suggested for AHU AFDD.

Other Statistical Process Control-based Methods

A number of different methods utilizing control

charts have been proposed. Control chart methods have been widely used in other SPC applications, and Li et al. (2012) demonstrated one potential use by combining with the cumulative sum (CUSUM) method. The predicted variables for VAV boxes can also be estimated using autoregressive time-series models, prior to the implementation of the CUSUM method (H. Wang, Chen, Chan, & Qin, 2011, 2012; H. Wang, Chen, Chan, Qin, et al., 2012). The control chart strategies are typically focused on detection, with the use of a rule-based, or other, framework required for diagnosis (Veronica, 2013).

Another SPC technique is the use of a Bayesian framework, which has the advantage of providing the building operator with probabilistic outputs with regard to the faulty behavior of the system (Najafi, Auslander, Bartlett, Haves, & Sohn, 2012). Bayesian network approaches can be combined with other SPC techniques, like the use of Hidden Markov Models (Wall, Guo, Li, & West, 2011). The probabilistic output of these methods can provide a benefit due to the uncertainties associated with diagnosing some types of faults.

Recent SPC methods have also demonstrated some novel approaches specifically designed for dealing with the dynamic behavior of AHU-VAV systems. One method was to utilize Kalman filters on variables with known interactions in order to generate adaptive control limits (Sun, Luh, Jia, O'Neill, & Song, 2013). Another approach investigated was the use of fractal correlation dimension (FCD) methods to directly counteract the noise found in the direct sensor measurements (X.-B. Yang, Jin, Du, & Zhu, 2011). This FCD method was later updated, and combined with a support vector regression technique (X.-B. Yang, Jin, Du, Zhu, & Guo, 2013).

Literature Discussion

While a number of these proposed methods have made important progress, the lack of widespread industry adoption of AFDD for AHUs is indicative of certain drawbacks or limitations in each of the existing methods. The most common shortcomings of proposed AFDD methods can be summarized as the following:

1. The requirement for the installation of additional sensors not typically found in building AHUs. The installation of additional expensive sensors, like Btu/flow meters on the coil piping or additional air flow stations, is prohibitive in the widespread adoption of a proposed method.

2. The requirement for many costly engineering hours to implement the method. The most common expenses related to additional engineering

include requirements for the generation of faulty data, requirements for fault thresholds to be individually customized for each installation, and requirements for modeling the physical system.

3. Ineffective at maintaining accuracy through the multiple operational modes and transient states experienced during normal AHU operation. Many methods that have been demonstrated to be effective in specific operational conditions are not adaptive for the use in all operating conditions experienced by AHU-VAV systems.

4. High false alarm rates during normal, or fault-free, operation. False alarm rates are rarely reported in the literature, so it is difficult to quantify this factor, but extremely low false alarm rates have been widely reported in industry as a key factor delaying widespread commercial adoption of AFDD for AHU-VAV systems.

A number of the methods discussed above show promise, but accurate evaluation of the efficacy of the contrasting methods would require extended in-situ testing in multiple buildings. The extent to which various methods overcome the key difficulties outlined above requires information about the implementation effort/cost for more than a single building, as well as demonstration of how the different methods maintain high accuracy and low false alarms as buildings shift through different modes of operation. Most of the studies were implemented in controlled environments, or utilized specially selected data that do not reflect the environment found in real buildings. A summary of the testing data used to evaluate the AFDD methods reported in the open literature is provided in Table 38-4. Going forward it will be important to extend the most promising concepts/methods to demonstrate performance of AFDD during typical building operation.

Available Simulation Models

Dynamic simulation models that can simulate fault-free and faulty system behaviors, as well as real faulty system operational data, are very important during the FDD strategy development and evaluation processes. In this section, available models and operational data reported in the open literature are summarized.

AHU and Building Zone Dynamic Simulation

Various building HVAC simulation models have been developed during the past decade for different

Table 38-3: APAR Rule Set (Schein et al., 2006)

Mode	Rule #	Rule Expression (true implies existence of a fault)
Heating (Mode 1)	1	$T_{sa} < T_{ma} + \Delta T_{sf} - \varepsilon_t$
	2	For $|T_{ra} - T_{oa}| \geq \Delta T_{min}: |Q_{oa}/Q_{sa} - (Q_{oa}/Q_{sa})_{min}| > \varepsilon_f$
	3	$|u_{hc} - 1| \leq \varepsilon_{hc}$ and $T_{sa,s} - T_{sa} \geq \varepsilon_t$
	4	$|u_{hc} - 1| \leq \varepsilon_{hc}$
Cooling with Outdoor Air (Mode 2)	5	$T_{oa} > T_{sa,s} - \Delta T_{sf} + \varepsilon_t$
	6	$T_{sa} > T_{ra} - \Delta T_{rf} + \varepsilon_t$
	7	$|T_{sa} - \Delta T_{sf} - T_{ma}| > \varepsilon_t$
Mechanical Cooling with 100% Outdoor Air (Mode 3)	8	$T_{oa} < T_{sa,s} - \Delta T_{sf} - \varepsilon_t$
	9	$T_{oa} > T_{co} + \varepsilon_t$
	10	$|T_{oa} - T_{ma}| > \varepsilon_t$
	11	$T_{sa} > T_{ma} + \Delta T_{sf} + \varepsilon_t$
	12	$T_{sa} > T_{ra} - \Delta T_{rf} + \varepsilon_t$
	13	$|u_{cc} - 1| \leq \varepsilon_{cc}$ and $T_{sa} - T_{sa,s} \geq \varepsilon_t$
	14	$|u_{cc} - 1| \leq \varepsilon_{cc}$
Mechanical Cooling with Minimum Outdoor Air (Mode 4)	15	$T_{oa} < T_{co} - \varepsilon_t$
	16	$T_{sa} > T_{ma} + \Delta T_{sf} + \varepsilon_t$
	17	$T_{sa} > T_{ra} - \Delta T_{rf} + \varepsilon_t$
	18	For $|T_{ra} - T_{oa}| \geq \Delta T_{min}: |Q_{oa}/Q_{sa} - (Q_{oa}/Q_{sa})_{min}| > \varepsilon_f$
	19	$|u_{cc} - 1| \leq \varepsilon_{cc}$ and $T_{sa} - T_{sa,s} \geq \varepsilon_t$
	20	$|u_{cc} - 1| \leq \varepsilon_{cc}$
Unknown Occupied Modes (Mode 5)	21	$u_{cc} > \varepsilon_{cc}$ and $u_{hc} > \varepsilon_{hc}$ and $\varepsilon_d < u_d < 1 - \varepsilon_d$
	22	$u_{hc} > \varepsilon_{hc}$ and $u_{cc} > \varepsilon_{cc}$
	23	$u_{hc} > \varepsilon_{hc}$ and $u_d > \varepsilon_d$
	24	$\varepsilon_d < u_d < 1 - \varepsilon_d$ and $u_{cc} > \varepsilon_{cc}$
All Occupied Modes (Mode 1, 2, 3, 4, or 5)	25	$|T_{sa} - T_{sa,s}| > \varepsilon_t$
	26	$T_{ma} < min(T_{ra}, T_{oa}) - \varepsilon_t$
	27	$T_{ma} > max(T_{ra}, T_{oa}) + \varepsilon_t$
	28	Number of mode transitions per hour $> MT_{max}$

Where

MT_{max}	=	maximum number of mode changes per hour
T_{sa}	=	supply air temperature
T_{ma}	=	mixed air temperature
T_{ra}	=	return air temperature
T_{oa}	=	outdoor air temperature
T_{co}	=	changeover air temperature for switching between Modes 3 and 4
$T_{sa,s}$	=	supply air temperature set point
ΔT_{sf}	=	temperature rise across the supply fan
ΔT_{rf}	=	temperature rise across the return fan
ΔT_{min}	=	threshold on the minimum temperature difference between the return and outdoor air
Q_{oa}/Q_{sa}	=	outdoor air fraction = $(T_{ma} - T_{ra})/(T_{oa} - T_{ra})$
$(Q_{oa}/Q_{sa})_{min}$	=	threshold on the minimum outdoor air fraction
u_{hc}	=	normalized heating coil valve control signal [0,1] where $u_{hc} = 0$ indicates the valve is closed and $u_{hc} = 1$ indicates it is 100 % open
u_{cc}	=	normalized cooling coil valve control signal [0,1] where $u_{cc} = 0$ indicates the valve is closed and $u_{cc} = 1$ indicates it is 100 % open
u_d	=	normalized mixing box damper control signal [0,1] where $u_d = 0$ indicates the outdoor air damper is closed and $u_d = 1$ indicates it is 100 % open
ε_t	=	threshold for errors in temperature measurements
ε_f	=	threshold parameter accounting for errors related to airflows (function of uncertainties in temperature measurements)
ε_{hc}	=	threshold parameter for the heating coil valve control signal
ε_{cc}	=	threshold parameter for the cooling coil valve control signal
ε_d	=	threshold parameter for the mixing box damper control signal

Table 38-4: Testing data used to evaluate AHU AFDD methods.

Author	Source for Test Data	Seasons
Brambley et al. (2011)	1 AHU from PNNL Laboratory	AHU sensor faults tested during heating/transition seasons Rules are agnostic to seasonal changes Threshold refinement required for fault detection
Du & Jin (2007a, 2007b), Du, Jin, Wu (2007)	Simulation Testbed (Quin, 1999)	AHU sensor and hardware faults tested during cooling season
Fan, Du, Jin, Yang, Guo (2010)	Simulation Data	AHU sensor faults tested during cooling season
Fernandez et al. (2009a, 2009b)	1 AHU from PNNL Laboratory	AHU sensor faults tested during transition season Rules are agnostic to seasonal changes
Lee, House, Kyong (2004)	Simulation testbed (IEA reference AHU)	AHU sensor and hardware faults tested during cooling season
Li, Wen (2013)	Data from Iowa Energy Center	AHU hardware faults tested during cooling season Models could be trained for all seasons
Najafi et al. (2012)	Data from Iowa Energy Center	AHU hardware faults tested during heating and economizing seasons Models could be applied for all seasons
Pakanen (2003)	1 AHU from college building in Oulu, Finland	AHU hardware faults tested during heating season
Qin and Wang (2005)	TRNSYS simulation, and data from building in Hong Kong	AHU and VAV hardware and sensor faults tested during unknown seasons
Schein, et al. (2006)	Simulation data, and AHU data from several field sites	AHU and VAV hardware and sensor faults tested during cooling season Rules for all operational modes are provided
Seem & House (2009)	HVACSIM+ based testbed	AHU sensor and hardware faults tested during all seasons
Song, Akashi, Yee (2008)	Simulation study	AHU hardware faults tesed during all seasons
Sun, Luh, Jia, O'Neill, Song (2013)	Simulation study	AHU hardware and sensor faults tested during the cooling season
Tan (2006)	Data from Iowa Energy Center	N/A
Veronica (2013)	NIST Laboratory Testing	N/A
Wall, Guo (2011)	Data from Iowa Energy Center	AHU hardware faults tested during the cooling season
Wang, Chen, Chan, Qin (2011, 2012), Wang, Chen, Chan, Qin, Wang (2012)	Data from office building in Hong Kong	VAV hardware and sensor faults tested during the cooling season
Wang, Jiang (2004)	TRNSYS simulation	VAV reheat valve tested
Wang, Xiao (2004, 2006)	Simulation study, and data from building in Hong Kong	AHU sensor faults were tested during the cooling season
Wu, Sun (2011a, 2011b)	Simulation testbed, and data from a building	AHU sensor and hardware faults tested during the cooling season
Yang, Cho, Tae, Zaheeruddin (2008)	Simulation testbed, and data from a building	AHU sensor faults tested during cooling season Rules for all operational modes are provided
Yang, Jin, Du, Zhu (2011), Yang, Jin, Du, Zhu, Guo (2013)	TRNSYS simulation	AHU sensor faults tested during the cooling season
Zhu, Jin, Du (2012)	AHU simulator	AHU sensor faults tested during the cooling season

purposes (Reddy et al., 2005): 1) Simplified Spreadsheet Programs, such as BEST (Waltz, 2000); 2) Simplified System Simulation Method, such as SEAM and ASEAM (Knebel 1983 and ASEAM 1991); 3) Fixed Schematic Hourly Simulation Program, such as DOE-2 (Winkelmann et al., 1993), and BLAST (BSL, 1999); 4) Modular Variable Time-Step Simulation Program, such as TRNSYS (SEL, 2000), SPARK (SPARK, 2003), ESP (Clarke and McLean, 1998), Energy Plus (Crawley et al., 2004), ASHRAE Primary and Secondary Toolkits (Bourdouxhe et al., 1998 and Brandemuehl, 1993), and Modelica Building library (LBNL, 2011); and 5) Specialized Simulation Program, such as HVACSIM+ (Park et al, 1985), GEMS (Shah, 2001), and other CFD programs (Broderick and Chen, 2001). Detailed building and HVAC simulation model reviews can also be found in Kusuda (1999 and 2001), Bourdouxhe et al. (1998), Shavit (1995), Ayres and Stamper (1995), and Yuill and Wray (1990).

Because the major objective of this project is to adapt existing dynamic AHU simulation model for AFDD simulation and evaluation, only those simulation programs that are component based, with open source, contain sensor, controller, and actuator models, are suitable. Therefore, HVACSIM+, TRNSYS, SPARKS, and EnergyPlus+ are reviewed further below.

HVACSIM+ (Park et al, 1985) developed by National Institute of Standards and Technology (NIST) uses a unique hierarchical variable time step approach in which components are grouped into blocks and blocks into super-blocks. The actual breakdown of the system is left to the user. Each super-block is an independent subsystem, whose time evolution is independent of other super-blocks. The only exception is the building envelope, which uses a fixed, user-specified time step. The time step in a super block is a variable, which is automatically and continuously adjusted by the program to maintain numerical stability. HVACSIM+ is especially appropriate for simulating secondary system and control strategies, and has been undergoing experimental validation and improvements for several years (Dexter et al., 1987).

TRNSYS (SEL, 2000) developed by the Solar Energy Laboratory, University of Wisconsin Madison, uses component based methodology in which: 1) a building is decomposed into components each of which is described by a FORTRAN subroutine, 2) the user assembles the arbitrary system by linking component inputs and outputs and by assigning component performance parameters, and 3) the program solves the resulting non-linear algebraic and differential equations to determine system response at each time step.

ASHRAE 825-RP (Norford and Haves, 1997) extends the ability of HVACSIM+ and TRNSYS in the following areas: 1) new models such as those for controller, sensor, and air flow path are developed; 2) component models of the building fabric and mechanical equipment are enhanced; 3) a real building (building E51 at the MIT campus), including the AHU system, is simulated and documented in detail to demonstrate the use of the models.

SPARK (SPARK, 2003), which is similar to a general differential/algebraic equation solver, is an object-oriented software system that can be used to simulate physical systems described in differential and algebra equations. In SPARK, components and subsystems are modeled as objects that can be interconnected to specify the model of the entire system. Models are expressed as systems of interconnected objects, either created by the user or selected from a library. An HVAC tool kit library is supplied with SPARK. However, SPARK is not supported or further developed by the Lawrence Berkeley National Laboratory (LBNL) any longer.

EnergyPlus (Crawley et al., 2004) is a building energy simulation program developed from BLAST and DOE-2. It includes many innovative simulation capabilities such as time steps of less than an hour, modular systems and plant integrated with heat balance-based zone simulation, multizone air flow, thermal comfort, and photovoltaic systems (EnergyPlus, 2005). However, it is relatively difficult to model only one AHU system and to be combined with other programs.

Modelica (LBNL, 2011), which was not available during the majority time of this project, is the newest dynamic building models library currently being developed and supported by the LBNL. Two versions of the Modelica building library were released by the LBNL in March and May 2011. Modelica allows the separation of the "modeling (i.e., defining the model equations) and simulation (i.e., computing a numerical solution to the equations)" (LBNL, 2011). Such a separation allows (LBNL, 2011): "1) a high degree of model reuse; 2) graphical "plug and play" modeling since modular models can be connected in an arbitrary way; 3) the integration of models from different domains (controls, thermodynamics, heat and mass transfer, fluid flow, electrical systems, etc.); 4) the coupling of models with fast dynamics in the order of seconds (local loop control) and slow dynamics (energy storage); 5) the coupling of models whose evolution is described by continuous time equations (for the physics and local loop control), discrete time equations (for supervisory control) and state events (for control that switches when a threshold is reached); 6) the exchange of models with other simulation platforms; and 7) the use of state-of-the art numerical solvers."

Fault Modeling

In general, models of faulty component and process are used either as part of a AFDD method or used as part of the simulation to develop or evaluate a AFDD method (Haves, 1997). None of the simulation models discussed in Section 3.3.1 directly provides the capability to simulate faulty operation. Although many AFDD studies simulate various faults for their own methodology development, few supplies detailed information about how the faults are modeled. Fewer studies describe how their simulated faulty operation data are validated.

A general discussion about fault modeling methodology is provided by Haves (1997), in which faults are grouped into Design, Installation, Abrupt, and Degradation categories. He suggests that faults be modeled in two different ways, i.e., by 1) changing parameter values in a fault free model, such as reducing UA value to model a fouled coil in a simple coil model; and 2) extending the structure of a fault free model to treat faults explicitly, such as adding a new parameter that specifies the thermal resistance of the deposit when modeling coil fouling fault in a detailed coil model. Furthermore, it is noted that if a fault is such that a basic assumption of the model is no longer valid, a major change in the fault free model is needed, such as poor sensor placement, which invalidates the perfect mixing assumption. Examples of cooling coil and valve faults modeling are also provided.

As part of the scope for ASHRAE Project 1043-RP, a simulation model was developed for a vapor compression centrifugal liquid chiller (Bendapudi and Braun, 2002). The model is based on first principles and is able to capture start-up and other transients caused by changes in steady state operation. Four faults, namely, 1) 20% reduced condenser and evaporator water flow rates; 2) 20% reduced refrigerant charge; 3) 20% refrigerant overcharge; and 4) 45% fouling in condenser, are modeled in the simulation tool. The fault free and four faulty simulation data sets are compared with experimental data under steady state, start-up, and other transient states. System pressure, power, and various temperatures are generally used to compare the simulation model against real system. Large deviation in the model predictions have been observed for evaporator pressure prediction under both fault free and faulty operations. Furthermore, it is hard to judge what are the criteria

used to claim that the model is "validated." Different levels of difference exist between model prediction and real measurements especially under transient states. For example, the model over-predicts the motor power by nearly 30% and over-predicts the sub-cooling by nearly 100% under load change (LC9) for 20% excess refrigerant fault simulation.

Bushby et al. (2001) describes two tools, namely a AFDD test shell and the Virtual Cybernetic Building Test-bed (VCBT), used for AFDD tool development. The VCBT employs HVACSIM+ as the simulation program and is able to emulate the characteristics and performance of a cybernetic building system. Twelve faults associated with a VAV AHU are modeled using VCBT, which include supply, return, mixed, and outdoor air temperature sensor offset faults; stuck open, closed, or partially open outdoor air damper; leaking outdoor air damper; stuck closed cooling coil valve; leaking cooling coil valve; stuck closed heating coil valve; and leaking heating coil valve (refer to Table 1). The fault modeling details are provided in Bushby et al. (2001). Experiments were also conducted at the Iowa Energy Center Energy Resource Station testing facility to examine the simulated faults. However, differences exist between the simulation conditions and testing conditions, such as that during simulation, historic weather data were used which were different from the testing weather conditions. Such differences prevent a rigorous validation comparison. Hence, only the trends between faulty operation and fault free operation displayed in the simulated data are compared with those shown in real measurements. It is noted that during the tests, two identical AHUs were employed at the test facility. One AHU served as fault free AHU while the other served as the faulty AHU.

ASHRAE 1312 Model

In an ASHRAE research project (ASHRAE RP 1312, Wen and Li, 2011), dynamic behaviors of the following systems: an AHU, four building zones that are served by the AHU, and four VAV boxes, are modeled using HVACSIM+ software. The model (called 1312 model hereafter) is developed based on two previous ASHRAE projects (RP 825 and RP 1194). However, significant modifications, including new parameter, control strategies, and component models (a new coil valve model and a new fan energy model), are developed in this study to ensure that the 1312 model simulates the dynamic behavior of the systems in a test facility located in Iowa Energy Center Energy Resource Station. The new coil valve model considers nonlinear behaviors of a three way valve. The new fan energy model outputs fan energy consumption that includes energy consumptions for fan, belt, motor and VFD. Coefficients for the new fan energy model can directly be estimated from the total fan energy measurement.

The developed 1312 AHU model is then systematically validated using experimental data for both fault free and faulty operation under three seasons (winter, summer, and spring).

Common AHU faults, including their features and severities, are modeled in this project and are summarized in Table 5. Existing experimental data from other research projects that can be used to validate the 1312 AHU model under fault free and faulty operation conditions are collected. Additional experiments are performed to thoroughly validate the 1312 AHU model under both fault free and faulty operation conditions. The fault models are able to replicate all major fault symptoms although detailed dynamics between simulated data and measured data do not always overlap.

User-friendly interfaces (Figure 38-3) that enable the 1312 model to work with a third party AFDD tool, and to allow users the maximum flexibility to select various faults and system configurations are also developed.

In a later project (EEB-HUB, 2013), the 1312 model is extended to include the following common VAV faults: 1) Sensor drift and bias faults for both room air temperature and discharge air flow rate sensors; 2) VAV damper stuck (fully open or closed, and partially open) fault and leaking fault; 3) VAV reheat valve stuck (fully open or closed, and partially open) fault and leaking fault; 4) Unstable control fault (for both VAV damper and/or reheat valve); and 5) Reheat coil air and water side fouling fault.

In the 1312 model, the following strategies are used to model common AHU faults:

Sensor Fault

Two types of sensor faults, namely, discrete fault and drift fault were simulated in this project. Discrete faults were modeled by adding a user specified bias to the simulated sensor output, which was achieved by Eq. (38-1)

$$Y_{output} = Y_{input} + Bi \tag{38-1}$$

where Y_{output} and Y_{input} were the output and input of a sensor model, Bi was an user specified bias which was kept as a constant.

Drift faults were modeled by linearly varying a bias, which resulted in a linearly varied simulated sensor bias as described in Eq. (38-2).

$$Y_{output} = Y_{input} + S \times T \tag{38-2}$$

where T was the amount of time after a fault occurred, S was a user specified slope of drifting bias.

Fault Controlled Device Category

Controlled device category included three dampers (recirculate, exhaust, and outdoor air dampers) and two valves (heating and cooling valve). Two types of faults, namely, stuck fault and leaking fault were modeled. Damper or valve stuck faults were modeled by fixing the simulated controlled device position to be a user specified position. Leaking faults were modeled by adding a user specified flow rate when the controlled device was 100% closed.

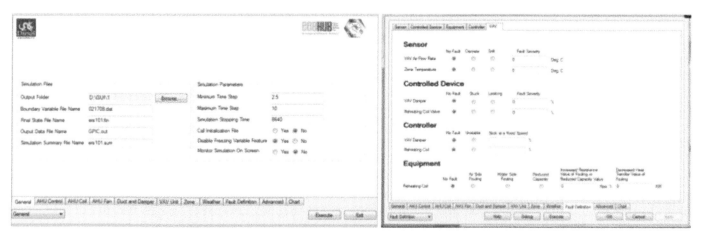

Figure 28-3: The dynamic fault simulation testbed.

Fault Equipment Category

For an AHU, faults could occur in two types of equipment, namely, fans (supply or return) and coils (heating or cooling). For supply and return fans, three types of common faults, namely, decreased pressure rise fault, complete failure fault, and decreasing motor efficiency fault, were modeled.

The decreased pressure rise fault was modeled by adding a user specified bias to the fan pressure rise model:

$$\Delta P = \Delta P_{org} \times bias \qquad (38\text{-}3)$$

where ΔP_{org} was the pressure rise calculated from the original fault free fan pressure rise model. ΔP was the pressure rise output from the faulty fan pressure rise model. bias was a user specified value.

The complete failure fault was modeled by outputting a zero value from both the fan pressure rise model and the fan efficiency model.

The decreasing motor efficiency fault was modeled by using a user specified ratio to decrease the simulated motor efficiency:

$$\eta_t = \eta_{torg} \times ratio \qquad (4)$$

where η_{torg} was the total fan efficiency calculated from the original fault free fan efficiency model, η_t was the total fan efficiency calculated from the faulty fan efficiency model, and ratio was a user specified value ranging from 0 to 1.

Three types of faults, namely, air side fouling fault, water side fouling fault, and reduced capacity fault, were modeled for coils. Several studies in the literature discussed how to model coil faults.

Bendapudi and Braun (2002) suggested that the heat exchanger fouling fault should be modeled by altering the heat-transfer coefficients. For the severest condenser fouling cases tested in their study, the water side heat transfer coefficient was reduced down to 55% of its fault free value. House et al (1999) also suggested that deposition of dirt and scale on a coil surface would increase the heat resistance. Moreover, the thermal resistance can be represented by a fouling factor. Haves (1997) recommended that coil fouling could be treated in a simple coil model by reducing the overall heat transfer UA value.

In the 1312 model, the reduced capacity fault was modeled by increasing the water flow pressure resistance. In this project, the heating coil and cooling coil were simulated using two different models. For heating coil fouling fault, the values of two variables, one repre-

senting thermal conductivity of the fin material and one for the tube material, were decreased to imitate the coil fouling. For cooling coil fouling fault, the water conductivity and aluminum fin conductivity were decreased to simulate coil fouling.

Fault Controller Category

Four types of control faults, namely, unstable control fault, fixed speed fault (for fans only), unstable sequencing fault, and reverse control fault, were modeled in PR 1312. Unstable control was modeled by implementing a user specified proportional band for PID controller. Fan fixed speed fault was modeled by outputting a user specified fixed speed from the fan controller model. Unstable sequencing fault was modeled by implementing a user specified proportional band for the PID controller that controls the AHU supply air temperature. Reverse control fault was modeled by changing the controller output to simulate a reverse control output.

Available Experimental Data

Simulation models can help in the process of developing AFDD methods. However, the developed methods need to be evaluated using fault data, either from laboratory facilities or from real buildings, because 1) real fault symptoms are often rather complicated, which may not be completely replicated by the models; 2) real data quality (due to sensor quality and communication quality) often strongly affects the accuracy and robustness of AFDD methods; and 3) diverse weather conditions, control and operation strategies and methods, system configurations and parameters, exist in real systems, which all have strong impact on the accuracy and robustness of AFDD methods. In this section, AHU fault data that are reported in the open literature are summarized.

AHU Fault Experimental Data

ASHRAE 1020-RP (Norford et al, 2000) selected seven AHU faults to be implemented in the test facility at the Iowa Energy Center Energy Resource Station during different seasons (Table 1). The experimental data were used to compare two AHU AFDD methods. The criteria used to choose the AHU faults and their severities (magnitudes) were based on the ability of the AHU AFDD methods. Those faults that could not be detected by the two AFDD methods were not considered. In general, abrupt faults were implemented for one testing day and degradation faults were implemented for one testing day per severity stage. Because of the fast response of pressure control loop, the supply air pressure sensor degradation faults were implemented in one day

Table 38-5: Summary of the AHU faults modeled in the 1312 model (Wen and Li, 2011).

Category	Device/Process	Fault Type	Subroutine Type
Sensor	Outdoor air temperature	0 – no fault, 1 – discrete offset, 2 – drift offset	311
	Mixed air temperature	0 – no fault, 1 – discrete offset, 2 – drift offset	
	Supply air temperature	0 – no fault, 1 – discrete offset, 2 – drift offset	
	Return air temperature	0 – no fault, 1 – discrete offset, 2 – drift offset	
	Supply air pressure	0 – no fault, 1 – discrete offset, 2 – drift offset	305
	Outdoor air flow rate	0 – no fault, 1 – discrete offset, 2 – drift offset	313
	Supply air flow rate	0 – no fault, 1 – discrete offset, 2 – drift offset	
	Return air flow rate	0 – no fault, 1 – discrete offset, 2 – drift offset	
	Outdoor air Humidity	0 – no fault, 1 – discrete offset, 2 – drift offset	312
	Supply air Humidity	0 – no fault, 1 – discrete offset, 2 – drift offset	
	Return air Humidity	0 – no fault, 1 – discrete offset, 2 – drift offset	
Controlled Device	Outdoor air damper	0 – no fault, 1 – Stuck*, 2 – leaking	325
	Recirculate air damper	0 – no fault, 1 – Stuck, 2 – leaking	
	Exhaust air damper	0 – no fault, 1 – Stuck, 2 – leaking	
	Heating coil valve	0 – no fault, 1 – Stuck, 2 – leaking	533
	Cooling coil valve	0 – no fault, 1 – Stuck, 2 – leaking	534
Equipment	Supply air fan	0 – no fault, 1 – Increased pressure drop, 2 – Complete failure of SF and RF, 3 -- Decrease in the motor efficiency	355
	Return air fan	0 – no fault, 1 – Increased pressure drop, 2 – Complete failure of SF and RF, 3 -- Decrease in the motor efficiency	
	Heating coil fouling	0 – no fault, 1 –Air side fouling, 2 –Water side fouling, 3 –Reduced capacity	533
	Cooling coil fouling	0 – no fault, 1 –Air side fouling, 2 –Water side fouling, 3 –Reduced capacity	534
	Duct leaking	0 – no fault, 1 – Leaking*	325/347*
Controller	Supply fan control	0 – no fault, 1 – unstable	581
	Return fan control	0 – no fault, 1 – unstable	582
	Mixing dampers control	0 – no fault, 1 – unstable	586
	Heating coil control	0 – no fault, 1 – unstable	
	Cooling coil control	0 – no fault, 1 – unstable	
	Sequence of heating and cooling devices	0 – no fault, 1 – unstable	586
	Reverse control action	0 – no fault, 1 – reverse	586
VAV Sensor	Room temperature	0 – no fault, 1 – discrete offset, 2 – drift offset	311
	Airflow rate	0 – no fault, 1 – discrete offset, 2 – drift offset	525/526*
VAV Controlled Device	VAV damper	0 – no fault, 1 – Stuck*, 2 – leaking	490
	VAV reheat coil valve	0 – no fault, 1 – Stuck*, 2 – leaking	490
VAV Controller	VAV Damper control	0 – no fault, 1 – unstable	490
	VAV reheat coil valve control	0 – no fault, 1 – unstable	490
	Reverse Action	0 – no fault, 1 – reverse	490
VAV Equipment	VAV reheat coil fouling	0 – no fault, 1 –Air side fouling, 2 –Water side fouling, 3 –Reduced capacity	522/252,526*

Note:

Stuck: 1-Stuck fully close, 2-Stuck fully open

VAV airflow sensor: 525 and 526 are two different VAV boxes

Duct leaking: 325 Duct leaking before supply fan, 347- Duct leaking after supply fan

VAV reheat coil fouling: 522-foulining thermal effect, 525,526- fouling flow effect for three severity stages. Unlike those faulty operation experiments conducted for chillers in ASHRAE 1043-RP, where the chiller input conditions were systematically and manually varied, the AHU faulty operation experiments were under real weather and building zone load conditions, which were similar to those in a real commercial building. Except weather conditions, other factors that would affect the AHU operation, such as supply air temperature and pressure setpoint, supply chilled and heating water temperatures and pressures, supply and return fan control strategies, and other operation strategies were not varied in ASHRAE 1020-RP.

Three AHU faults, supply air temperature sensor offset fault, stuck open recirculation air damper fault, and a leaking heating coil valve fault were implemented in the test facility at the Iowa Energy Center Energy Resource Station as part of the NISTIR 6964 study described by Castro et al. (2003) (Table 1). The experimental data were used to examine the fault models described by Bushby et al. (2001). There are two major differences between the fault implementation methods in ASHRAE 1020-RP and NISTIR 6964: 1) Longer time period (two weeks) were used for supply air temperature drifting offset fault implementation in NISTIR 6964; and 2) One AHU among two identical AHUs was operated under faulty conditions, while the other AHU was operated under fault free conditions in NISTIR 6964. In ASHRAE 1020-RP, the two AHUs were used for different faults implementation. Similar to ASHRAE 1020-RP, the AHUs were subject to the real weather conditions and building zone load in NISTIR 6964. However, experiments were only conducted in winter season in NISTIR 6964 with fixed AHU operation sequence and condition.

A comprehensive list of faults was artificially implemented in the test facility at the Iowa Energy Center Energy Resource Station during ASHRAE project RP 1312. Similar to the NISTIR 6964 project, both of the two identical systems (including building zones and AHUs) at the facility were used, of which, one was used to implement faults and one was used to serve as a baseline. Three seasons, namely, winter, spring, and summer, were used to test faults. Each test season consisted of two or three weeks of controlled tests, either under normal or faulty test conditions. Table 6 shows the tested faults, their magnitude, their implementation methods and test seasons associated with each fault.

Since it was impractical to test all levels of severity for a degradation fault, in ASHRAE RP 1312, three levels of severity was tested for each degradation fault. Some faults have similar symptoms in different seasons. For such faults, only one season was selected.

Faults were manually introduced into the air-mixing box, coils, and fan sections of AHU-A at the test facility. To help analyzing fault symptoms, the parallel test systems B, was operated under fault free conditions. Although the two test systems were designed and constructed to be identical, there were still unavoidable differences between the two systems. Therefore, several normal operation days were included in each test period to identify the inherent differences in the systems. By comparison of the difference between AHU- A and B systems, the impacts of the fault could be observed and fault symptoms can be identified.

During the experiment, the system operation was scheduled to be occupied from 6:00 to 18:00; and unoccupied from 18:00-6:00 the following day. The system operation schedule was to simulate the schedule of working hours for a commercial building. The minimum outdoor air damper position was set to be 40% open (spring and summer) or 47% open (winter). The positions were determined to satisfy the minimum ventilation requirement for test rooms served by the AHUs. The economizer control was enabled when outdoor temperature was less than 65°F (winter and summer) or when the outdoor air temperature was lower than the return air temperature by at least 3°F (spring). The return fan was operated with a speed tracking control sequence (80% of supply fan speed). The supply air static pressure setpoint was at 1.4 psi. The supply air temperature setpoint was 55°F (spring and summer) or 65°F (winter). A higher supply air temperature was maintained during the winter so that the heating coil will be used.

Notice that during the test for any damper fault, only the specific damper that had a fault was affected. The other two dampers were operated as usual.

The AHU faulty operation experiments were under real weather and building zone load conditions, which were similar to those in a real commercial building.

In ASHRAE RP 1312, fault symptoms for each fault were also analyzed by comparing measurements from the faulty system (A system) and the identical fault free system (B system). The fault symptoms are summarized in Wen and Li, 2011.

Field Data

In a laboratory test facility, such as the Iowa Energy Center Energy Resource Station facility, faults are generally artificially introduced in the facility. Artificial fault implementation may yield different operation data than those from the field, especially for those degradation faults such as coil fouling fault. Field data that contains faults, on the contrary, reflect the behavior of a

Table 38-6: Faults implemented in ASHRAE RP 1312 experiment.

Category	Fault description	Magnitude or location	Summer	Winter	Spring	Implementation
Controlled Device	OA Damper Stuck	Fully Closed	X	X	X	Manually control OA damper at faulty positions
		40% Open		X	X	
		45% Open	X			
	OA Damper Leak	55% Open	X			
		52% Open		X		
		62% Open		X		
	EA Damper Stuck	Fully Open	X	X	X	Manually control EA damper at faulty positions
		Fully Close	X	X	X	
		40% Open			X	
	Cooling Coil Valve Stuck	Fully Closed	X		X	Manually control valve at faulty positions
		Fully Open	X	X	X	
		Partially Open - 15%	X			
		Partially Open - 20%		X		
		Partially Open - 50%			X	
		Partially Open - 65%	X			
	Heating Coil Valve Leaking	Stage 1 - 0.4GPM	X			Manually open heating coil bypass valve
		Stage 2 - 1.0GPM	X			
		Stage 3 - 2.0GPM	X			
Equipment	AHU Duct Leaking	after SF	X			Remove sealing from one access door
		before SF	X			
	Heating Coil Fouling	Stage 1		X		Partially block heating coil using a piece of cardboard
		Stage 2		X		
	Heating Coil Reduced Capacity	Stage 1		X		Manually throttle heating coil balancing valve
		Stage 2		X		
		Stage 3		X		
	Return Fan complete failure		X		X	Manually stop return fan
	Air filter blockage fault	10%				Partially block air filter using a piece of cardboard
	Air filter blockage fault	25%				Partially block air filter using a piece of cardboard
	Return Fan at fixed speed	30%spd	X			Manually maintain return fan speed at faulty speed
		20%spd			X	
		80%spd			X	
Controller	Cooling Coil Valve Control unstable		X			**Reduce the PB value for supply air temperature PI control algorithm to be 50% of its original value**
	Cooling Coil Valve Reverse Action		X			Change cooling coil valve scaling factor
	Mixed air damper unstable				X	Change the PB value for supply air temperature PI control algorithm from -45.7 to -10
	Mixed air damper unstable/Cooling Coil Control Unstable				X	Change the PB value for supply air temperature PI control algorithm from -10 to -5
	Sequence of Heating and cooling unstable				X	Increase the PB value for supply air temperature PI control algorithm
Sensor	OA temperature sensor bias	+3 F			X	Manually change the sensor calibration equation in the control system
	OA temperature sensor bias	-3 F			X	Manually change the sensor calibration equation in the control system

fault in a real building environment. However, field data 1) generally are obtained with an hourly time interval; and 2) may contain various unknown faults mixed together, especially with sensor degradation faults.

Publicly accessible field data that contain AHU faults are briefly reviewed here. Relatively high quality field data exist as part of the ANNEX 34 study (ANNEX 34, 2001). The following ANNEX 34 field studies (indicated by the PI's surname) contain degradation fault data: 1) Grob (Factory building, coil fouling, leaky valves and dampers); and 2) Dexter (Office building, leaky valve, fouled coil). Some other studies also contain degradation data but in which faults were artificially implemented. A large set of field data used to evaluate the Whole Building Diagnostician exist and are well documented (Katipamula et al., 2003). But the field data contain mostly economizer faults.

Yang, et al (2004) investigated the role of filtration in maintaining clean heat exchanger coils and overall performance. Although this project is not an experiment aimed at studying fault symptoms, it provides near real world data for coil fouling studies. Combinations of 6 different levels of filtration (MERV 14, 11, 8, 6, 4, and no filter) and 4 different coils (an eight-row lanced-fin coil, an eight-row wavy-fin coil, a four-row lanced-fin coil, and a two-row lanced-fin coil) were tested at 4 different air velocities (300, 400, 500, 600 ft/min). A standard-size air duct of 24 inch×24 inch was constructed as part of the test facility. The fouled conditions were obtained after injection of 600 grams of ASHRAE standard dust upstream of the filter/coil combination. This magnitude of dust is representative of a year of normal operation for an air conditioning system. Each of these coil-filter combinations was tested under clean and fouled conditions. Measurements taken in this project include: test time, air velocity, coil pressure drop, air inlet temperature, air outlet temperature, water inlet temperature, water outlet temperature, air upstream relative humidity, air downstream relative humidity, water flow rate, upstream filter pressure drop and downstream filter pressure drop. Therefore, data from this experiment provide valuable understanding about the impacts of air side fouling fault in a nearly real world condition.

SUMMARY

The performance of AHU systems significantly affect a building's energy consumption and indoor air quality. A large variety of faults could exist in an AHU system. Over the years, many types of AHU AFDD

methods have been reported in the literature. But due to unique AHU AFDD challenges, such as the variety of AHU systems and control strategies, the dynamic operation modes of AHU systems, and the poor measurement quality, more research is still needed to develop robust, plug-and-play, and low cost AHU AFDD methods. Moreover, it is emphasized here that accurate evaluation of the efficacy of an AFDD method should require extended in-situ testing in multiple buildings. The extent to which various methods overcome the key difficulties outlined above requires information about the implementation effort/cost for more than a single building, as well as demonstration of how the different methods maintain high accuracy and low false alarms as buildings shift through different modes of operation.

References

1. ANNEX 25, Building Optimization and Fault Diagnosis Source Book, ANNEX 25 Real Time Simulation of HVAC Systems for Building Optimization, Fault Detection and Diagnosis, International Energy Agency, Finland, 1996.
2. ANNEX 34, Demonstrating Automated Fault Detection and Diagnosis Methods in Real Buildings, ANNEX 34, International Energy Agency, Finland, 2001.
3. ASEAM 3.0, 1991. A Simplified Energy Analysis Method, U.S. Department of Energy, Washington, D.C.
4. ASHRAEa, ASHRAE Handbook—HVAC Systems and Equipment, ASHRAE, Atlanta, Georgia, 2012.
5. ASHRAEb, TC1.4 Recommended control strategies, 2005.
6. Ayres, J.M., and E. Stamper. 1995. "Historical Development of Building Energy Calculations," ASHRAE Transactions, 101(1), pp. 31-38.
7. Bendapudi, S., and J.E. Braun, 2002. Development and Validation of a Mechanistic, Dynamic Model for a Vapor Compression Centrifugal Liquid Chiller, Report #4036-4, Deliverable for Research Project 1043-RP, ASHRAE, Atlanta, GA.
8. Bourdouxhe, J.P., M. Grodent and J. Lebrun, 1998. Reference Guide for Dynamic Models of HVAC Equipment, ASHRAE, Atlanta, GA.
9. Brambley, M., R. Pratt, D. Chassin, S. Katipamula, and D. Hatley, 1998. "Diagnostics for Outdoor Air Ventilation and Economizers," ASHRAE Journal, October, pp. 49-55.
10. Brambley, M., Fernandez, N., Wang, W., Cort, K., Cho, H., Ngo, H., & Goddard, J. (2011). Self Correcting Controls for VAV Systems Faults.
11. Brandemuehl, M., 1993. HVAC 2: Toolkit for Secondary HVAC System Energy Calculations, ASHRAE, Atlanta, GA.
12. Broderick, C.R., and Q. Chen, 2001. "A Simple Interface to CFD Codes for Building Environment Simulations," Seventh International IBPSA Conference, pp. 577-584, August, Rio de Janeiro, Brazil.
13. BSL, 1999. BLAST 3.0 Users Manual, Building Systems Laboratory, Department of Mechanical and Industrial Engineering, University of Illinois, Urbana-Champaign, IL.
14. Bushby, S. T., N. Castro, M.A. Galler, and C. Park, 2001. "Using the Virtual Cybernetic Building Testbed and AFDD Test Shell for AFDD Tool Development," NISTIR 6818, National Institute of Standards and Technology, Gaithersburg, MD.
15. Carling, P. 2002. "Comparison of Three Fault Detection and Diagnosis Methods Based on Field Data of an Air Handling

Unit," ASHRAE Transactions, 108 (1): pp. 904-921.

16. Castro, N.S., J. Schein, C. Park, M.A. Galler, S.T. Bushby, and J.M. House. 2003. Results from Simulation and Laboratory Testing of Air Handling Unit and Variable Air Volume Box Diagnostic Tools, NISTIR 6964, National Institute of Standards and Technology, Washington, DC.

17. Clarke, J. and D. McLean, 1998. ESP—A Building and Plant Energy Simulation System, version 6, Release 8, University of Strathclyde, Glasgow, UK.

18. Crawley, D.B, L.K. Lawrie, C.O. Pedersen, F. C. Winkelmann, M.J. Witte, R.K. Strand, R.J. Liesen, W.F. Buhl, Y.J. Huang, R.H. Henninger, J.G., D.E. Fisher, D.B. Shirey III, B.T. Griffith, P. G. Ellis, and L. Gu. 2004. "EnergyPlus: New, Capable, and Linked," Journal of Architectural and Planning Research, 21 (4).

19. Dexter, A.L., M.M. Eftekhari, P. Haves, and J. G. Jota, 1987. "The Use of Dynamic Simulation Models to Evaluate Algorithms for Building Energy Control: Experience with HVA-SIM+," Proceedings of ICBEM'87, International Congress on Building Energy Management, October, Lausanne, Switzerland.

20. Dexter, A.L., 1995, "Fuzzy Model Based Fault Diagnosis," IEE Proceedings—Control Theory Application, 142 (6), pp. 545-550.

21. Dexter, A.L., and M. Benouarets, 1996. "A Generic Approach to Identifying Faults in HVAC Plants," ASHRAE Transactions, pp. 550-556.

22. Dexter, A.L., D., Ngo, 2001. "Fault diagnosis in air-conditioning systems: a multi-step fuzzy model-based approach," HVAC & R Research, 7(1), pp. 83-102.

23. Du, Z., & Jin, X. (2007a). Detection and diagnosis for multiple faults in VAV systems. Energy and Buildings, 39, 923-934.

24. Du, Z., & Jin, X. (2007b). Tolerant control for multiple faults of sensors in VAV systems. Energy Conversion and Management, 48(3), 764-777. doi: 10.1016/j.enconman.2006.09.007.

25. Du, Z., Jin, X., & Wu, L. (2007). Fault detection and diagnosis based on improved PCA with JAA method in VAV systems. Building and Environment, 42(9), 3221-3232. doi: 10.1016/j.buildenv.2006.08.011.

26. Du, Z., Jin, X., & Yang, X. (2009). A robot fault diagnostic tool for flow rate sensors in air dampers and VAV terminals. Energy and Buildings, 41(3), 279-286. doi: 10.1016/j.enbuild.2008.09.007.

27. Dumitru, R. and D. Marchio, 1995. "Fault Identification in Air Handling Units Using Physical Models and Neural Networks," Proceedings of Building Simulation 95, International Building Performance Simulation Association, Madison, WI.

28. EnergyPlus, 2005. EnergyPlus Energy Simulation Software, U. S. Department of Energy, http://www.eere.energy.gov/buildings/energyplus/

29. EEB-HUB, Energy Efficient Buildings Hub Budget Period 2 Annual Report, Task 4.3.1, January 2013, Philadelphia, PA.

30. Fan, B., Du, Z., Jin, X., Yang, X., & Guo, Y. (2010). A hybrid FDD strategy for local system of AHU based on artificial neural network and wavelet analysis. Building and Environment, 45(12), 2698-2708. doi: 10.1016/j.buildenv.2010.05.031

31. Fasolo, P. S. D. E. Seborg, 1995. "Monitoring and fault detection for an HVAC control system" HVAC&R Research, 1 (3), pp. 177-193.

32. Fernandez, N., Brambley, M. R., & Katipamula, S. (2009). Self-Correcting HVAC Controls: Algorithms for Sensors and Dampers in Air-Handling Units. (PNNL-19104).

33. Fernandez, N., Brambley, M. R., Katipamula, S., Cho, H., Goddard, J., & Dinh, L. (2009). Self-Correcting HVAC Controls. (PNNL-19074).

34. Glass, A. S., P. Gruber, M. Roos, and J. Todtli, 1995. "Quali-

tative Model-Based Fault Detection in Air-Handling Units," IEEE Control systems, August, pp. 11-22.

35. Haves, P., T.I. Salsbury, and J. A. Wright. 1996. "Condition Monitoring in HVAC Subsystems Using First Principles Models," ASHRAE Transactions, pp. 519-527.

36. Haves, P., 1997. "Fault Modeling in Component-Based HVAC Simulation," Proceedings of Building Simulation 97, IBPSA, pp. 119-126.

37. House, J.M., W.Y. Lee, and D.R. Shin, 1999. "Classification Techniques for Fault Detection and Diagnosis of an Air-Handling Unit," ASHRAE Transactions, 105 (1).

38. House, J.M., Vaezi-Nejad, H.; Whitcomb, J.M. 2001. "An expert rule set for fault detection in air-handling units." ASHRAE Transactions, v 107 PART. 1, 2001, p 858-871

39. House, J.M., K.D. Lee, and L.K. Norford, 2003, "controls and Diagnostics for Air Distribution Systems," Journal of Solar Energy Engineering, Transactions of the ASME, 125, pp. 310-317.

40. Katipamula, S., R. Pratt, D. Chassin, Z.T. Taylor, K. Gowri, and M. Brambley. 1999. "Automated Fault Detection and Diagnostics for Outdoor-Air Ventilation Systems and Economizers: Methodology and Results from Field Testing," ASHRAE Transactions, 105 (1).

41. Katipamula S., M.R. Brambley, N.N. Bauman, and R.G. Pratt. 2003. "Enhancing Building Operations through Automated Diagnostics: Field Test Results." In Proceedings of 2003 International Conference for Enhanced Building Operations. PNNL-39693.
http://www.buildingsystemsprogram.pnl.gov/publications. stm

42. Katipamula, S. and M.R. Brambley, 2005a,b. Methods for fault detection, diagnostics, and prognostics for building systems, Part I and II, HVAC&R Research Journal, Jan, April.

43. Kelso, R.M. and Wright, J.A. 2005. "Application of fault detection and diagnosis techniques to automated functional testing." ASHRAE Transactions, Volume 111, Part 1.

44. Knebel, D. 1983. Simplified Energy Analysis Method, ASHRAE, Atlanta, GA.

45. Kusuda, T., 1999. "Early History and Future Prospects of Building Systems Simulation," Proceedings of the Sixth International IBPSA Conference, Kyoto, Japan.

46. Lee, W. Y., C. Park, and G.E. Kelly, 1996a. "Fault Detection in an Air-Handling Unit Using Residual and Recursive Parameter Identification Methods," ASHRAE Transactions, pp. 528-539.

47. Lee, W.Y., J.M. House, C. Park, and G.E. Kelly. 1996b. "Fault diagnosis of an air-handling unit using artificial neural networks." ASHRAE Transactions 102(1): 540-549.

48. Lee, W.Y., J.M. House, and D.R. Shin. 1997. "Fault diagnosis and temperature sensor recovery for an air-handling unit." ASHRAE Transactions 103(1): 621-633.

49. Lee, W.-Y., House, J. M., & Kyong, N.-H. (2004). Subsystem level fault diagnosis of a building's air-handling unit using general regression neural networks. Applied Energy, 77(2), 153-170. doi: http://dx.doi.org/10.1016/S0306-2619(03)00107-7

50. Li, S., and J. Wen, "A model based Fault Detection and Diagnostic methodology based on PCA method and Wavelet Transform," Vol 68, Part A, pp. 63-71, Energy & Buildings, 2014.

51. Liu, Xiong-Fu and Dexter, Arthur. 2001. "Fault-tolerant supervisory control of VAV air-conditioning systems." Energy and Buildings, Volume 33, Issue 4, pp. 379-389.

52. Lunneberg, T. 1999. "When Good Economizers Go Bad." E Source Report ER-99-14, September.

53. MATLAB, MATLAB Function Reference, the Mathworks Inc., 2006. http://www.mathworks.com/products/matlab/functionlist.html

54. LBNL, 2011, Modelica Library for Building Energy and Control Systems, Lawrence Berkeley National Laboratory, https://gaia.lbl.gov/bir.

55. Najafi, M., Auslander, D.M., Bartlett, P. L., Haves, P., & Sohn, M. D. (2012). Application of machine learning in the fault diagnostics of air handling units. Applied Energy, 96, 347-358. doi: 10.1016/j.apenergy.2012.02.049

56. Ngo, D. and A.L. Dexter, 1999, "A Robust Model-Based Approach to Diagnosing Faults in Air-Handling Units," ASHRAE Transactions, 105 (1).

57. Norford, L.K., and R.D. Little, 1993. "Fault Detection and Load Monitoring in Ventilation Systems," ASHRAE Transactions, pp. 590-602.

58. Norford, L.K. and P. Haves. 1997. A Standard Simulation Testbed for the Evaluation of Control Algorithms and Strategies. Final Report of ASRAE 825-RP. American Society of Heating, Refrigerating, and Air-Conditioning Engineers, Inc.: Atlanta, Georgia.

59. Norford, L.K., Wright, J. A., Buswell, R.A., and Luo, D. 2000. "Demonstration of Fault Detection and Diagnosis Methods in a Real Building," Final Report of ASHRAE 1020-RP, American Society of Heating, Refrigerating, and Air-Conditioning Engineers, Inc.: Atlanta, GA.

60. Park, C., D.R. Clark, and G.E. Kelly, 1985. "An Overview of HVACSIM+, A Dynamic Building/HVAC/Control System Simulation Program," the 1st Annual Building Energy Simulation Conference, Seattle, WA, August 21-22.

61. Pakanen, J.E., & Sundquist, T. (2003). Automation-assisted fault detection of an air-handling unit; implementing the method in a real building. Energy and Buildings, 35, 193-202.

62. Peitsman, H.C. and L.L. Soethout, 1997. "ARX Models and Real-Time Model-Based Diagnosis," ASHRAE Transactions, 103 (1).

63. Price, B.A. and Smith, T.F. 2003. Development and Validation of Optimal Strategies for Building HVAC Systems, Technical Report: ME-TEF-03-001, Department of Mechanical Engineering, The University of Iowa, Iowa City, Iowa.

64. Qin, J., & Wang, S. (2005). A fault detection and diagnosis strategy of VAV air-conditioning systems for improved energy and control performances. Energy and Buildings, 37(10), 1035-1048. doi: 10.1016/j.enbuild.2004.12.011

65. Reddy, T.A. and al. 2005. Literature review on calibration of building energy simulation programs: uses, problems, procedures, uncertainty and tools, accepted for publication by ASHRAE Trans., May.

66. Regnier, A., Yang, X.B., and Wen, J., "Pattern Matching PCA for Fault Detection in Air Handling Units," IEEE CASE, Madison, WI, August 2013.

67. Salsbury, T.I. "A controller for HVAC systems with fault detection capabilities based on simulation models." Proceedings of Building Simulation '99, Japan.

68. Schein, J., Bushby, S.T., Castro, N.S., & House, J.M. (2006). A rule-based fault detection method for air handling units. Energy and Buildings, 38(12), 1485-1492. doi: 10.1016/j.enbuild.2006.04.014

69. Seem, J.E., & House, J.M. (2009). Integrated control and fault detection of air-handling units. HVAC and R Research, 15(1), 25-55.

70. Seem, J., C. Park, J. M. House, "A New Sequencing Control strategy for Air-Handling Units," HVAC&R Research, Volume 5, Issue 1, pp. 35-58, 1999.

71. SEL, 2000. TRNSYS Version 15 User Manual and Documentation, Solar Energy Laboratory, Mechanical Engineering Department, University of Wisconsin, Madison, February.

72. Shah, D.J., 2001. "Generalized Engineering Modeling and Simulation (GEMS)," Seventh International IBPSA Conference,

pp. 723-730, August, Rio de Janeiro, Brazil.

73. Shavit, G., 1995. "Short-time Step Analysis and Simulation of Homes and Buildings during the last 100 years," ASHRAE Transactions, 101 (1).

74. Song, Y.-h., Akashi, Y., & Yee, J.-J. (2008). A development of easy-to-use tool for fault detection and diagnosis in building air-conditioning systems. Energy and Buildings, 40(2), 71-82. doi: 10.1016/j.enbuild.2007.01.011

75. SPARK 2003. Simulation Problem Analysis and Research Kernel. Lawrence Berkeley National Laboratory and Ayres Sowell Associates, Inc.

76. Sun, B., Luh, P.B., Jia, Q.S., O'Neill, Z., & Song, F. (2013). Building Energy Doctors: An SPC and Kalman Filter-Based Method for System-Level Fault Detection in HVAC Systems. Automation Science and Engineering, IEEE Transactions on, PP(99), 1-15. doi: 10.1109/tase.2012.2226155

77. Tan, H., & Dexter, A. (2006). Estimating airflow rates in air-handling units from actuator control signals. Building and Environment, 41(10), 1291-1298. doi: 10.1016/j.buildenv.2005.05.027

78. Wall, J., Guo, Y., Li, J., & West, S. (2011). A Dynamic Machine Learning-based Technique for Automated Fault Detection in HVAC Systems. ASHRAE Transactions.

79. Waltz, J. P., 2000. Computerized Building Energy Simulation Handbook, Fairmont Press, Lilburn, GA.

80. Wang, S. 1992. Modeling and simulation of building and HVAC systems—Building and HVAC system and component models used in emulation exercise C.3. IEA ANNEX 17 working paper.

81. Wang, F., Yoshida, H., and Miyata, M., 2004. "Total Energy Consumption Model of Fan Subsystem Suitable for Continuous Commissioning," ASHRAE Transactions, 110(1).

82. Wang, H., Chen, Y., Chan, C.W.H., & Qin, J. (2011). A robust fault detection and diagnosis strategy for pressure-independent VAV terminals of real office buildings. Energy and Buildings, 43(7), 1774-1783. doi: 10.1016/j.enbuild.2011.03.018

83. Wang, H., Chen, Y., Chan, C.W.H., & Qin, J. (2012a). An online fault diagnosis tool of VAV terminals for building management and control systems. Automation in Construction, 22, 203-211. doi: 10.1016/j.autcon.2011.06.018

84. Wang, H., Chen, Y., Chan, C.W.H., & Qin, J. (2012b). An online fault diagnosis tool of VAV terminals for building management and control systems. Automation in Construction, 22(0), 203-211. doi: http://dx.doi.org/10.1016/j.autcon.2011.06.018

85. Wang, H., Chen, Y., Chan, C.W.H., Qin, J., & Wang, J. (2012). Online model-based fault detection and diagnosis strategy for VAV air handling units. Energy and Buildings, 55, 252-263. doi: 10.1016/j.enbuild.2012.08.016

86. Wang, S., & Jiang, Z. (2004). Valve fault detection and diagnosis based on CMAC neural networks. Energy and Buildings, 36(6), 599-610. doi: 10.1016/j.enbuild.2004.01.037

87. Wang, S., & Xiao, F. (2004). Detection and diagnosis of AHU sensor faults using principal component analysis method. Energy Conversion and Management, 45, 2667-2686.

88. Wang, S., & Xiao, F. (2006). Sensor Fault Detection and Diagnosis of Air-Handling Units Using a Condition-Based Adaptive Statistical Method. HVAC and R Research, 12(1), 127-150.

89. Wen, J., and S. Li, Tools for Evaluating Fault Detection and Diagnostic Methods for Air-Handling Units, Final Report, ASHRAE 1312-RP, ASHRAE, Atlanta, GA, 2011.

90. Winkelmann, F.C., B.E. Birdsall, W.F. Buhl, K.L. Ellington, A.E. Erdem, J.J. Hirsch, and S. Gates, 1993. DOE-2 Supplement Version 2.1 E, Lawrence Berkeley National Laboratory, November.

91. Wu, S., & Sun, J.Q. (2011). Cross-level fault detection and diagnosis of building HVAC systems. Building and Environment,

46(8), 1558-1566. doi: 10.1016/j.buildenv.2011.01.017

92. Wu, S., & Sun, J.Q. (2011). A top-down strategy with temporal and spatial partition for fault detection and diagnosis of building HVAC systems. Energy and Buildings, 43(9), 2134-2139. doi: 10.1016/j.enbuild.2011.04.020

93. Xiao, F., Y. Zhao, J. Wen, and S. W. Wang, "Bayesian network based FDD strategy for variable air volume terminals," Automation in Construction, To be published, Available online November 18th, 2013 (http://www.sciencedirect.com/science/article/pii/S0926580513001878)

94. Xu, P., P. Haves, M. Kim. 2005. "Model-based automated functional testing—methodology and application to air-handling units." ASHRAE Transactions, Volume 111, Part 1.

95. Yang, L, Braun, J.E., Groll, E.A., and Luo, D. 2004. "The Role of Filtration in Maintaining Clean Heat Exchange Coils," Final Report of ARTI-21CR/611-40050-01.

96. Yang, H., Cho, S., Tae, C.-S., & Zaheeruddin, M. (2008). Sequential rule based algorithms for temperature sensor fault detection in air handling units. Energy Conversion and Management, 49(8), 2291-2306. doi: 10.1016/j.enconman.2008.01.029

97. Yang, X.-B., Jin, X.-Q., Du, Z.-M., & Zhu, Y.-H. (2011). A novel model-based fault detection method for temperature sensor using fractal correlation dimension. Building and Environment, 46(4), 970-979. doi: 10.1016/j.buildenv.2010.10.030

98. Yang, X.-B., Jin, X.-Q., Du, Z.-M., Zhu, Y.-H., & Guo, Y.-B. (2013). A hybrid model-based fault detection strategy for air handling unit sensors. Energy and Buildings, 57, 132-143. doi: 10.1016/j.enbuild.2012.10.048Yoshida, H., Yuzawa, H, Iwami, T, Suzuki, M. 1996. "Typical faults of air-conditioning systems and fault detection by ARX model and extended Kalman filter." ASHRAE Transactions, v 102, n 1, 1996, p 557-564

99. Yuill, G.K., and C.P. Wray, 1990. "Overview of ASHRAE TC 4.7 Annotated Guide to Models and Algorithms for Energy Calculations Relating to HVAC Equipment," ASHRAE Transactions, 96 (1), pp. 91-97.

100. Zhu, Y., Jin, X., & Du, Z. (2012). Fault diagnosis for sensors in air handling unit based on neural network pre-processed by wavelet and fractal. Energy and Buildings, 44, 7-16. doi: 10.1016/j.enbuild.2011.09.043

Chapter 39

Methodology for Evaluating FDD Protocols Applied to Unitary Systems

David P. Yuill, PE and James E. Braun, PhD, PE
Ray W. Herrick Laboratories, Purdue University

INTRODUCTION

Author's note: This chapter describes a method for evaluating the performance of FDD methods applied to air-cooled unitary air-conditioning systems at a steady operating condition. The method was published in Yuill & Braun (2013), and the description below is adapted from a report to the program director, New Buildings Institute, for a project funded by the California Energy Commission.

When considering which fault detection and diagnosis (FDD) approach to use, or when considering whether a particular FDD approach meets a code requirement, the obvious question to ask is: how well does it work? Answering this question is not simple. There is currently no standard method of evaluating the performance of FDD applied to unitary equipment, and "there are currently no available military or commercial standards to support a systematic and consistent approach to assessing the performance and effectiveness" of FDD applied to engineered systems in general (Vachtsevanos et al. 2006). This means that an evaluation method can't be adapted from another field or engineering application, in the same way that HVAC FDD itself was adapted from other fields.

There has been some previous research that considered evaluation of FDD for vapor compression air-conditioning equipment. Breuker & Braun (1998) studied the accuracy of a FDD tool developed by Rossi and Braun (1997) when applied to a specific rooftop unit, and methods of tuning parameters within the tool to achieve optimal performance. Reddy (2007) discusses generic evaluation methodologies for assessing different FDD protocols applied to large chillers. None of the previous research proposed a standard method of test or evaluation and rating system.

FDD has the potential to provide significant benefits. Surveys of air-conditioning systems have found a large fraction to be operating with a fault (Rossi 2004; Breuker et al. 2000) that can have significant effects on capacity, efficiency and equipment life. For example, if refrigerant undercharge faults were eliminated from only the existing residential air conditioners in the US, it is estimated that residential cooling energy consumption would be reduced by 0.1 to 0.2 quad per year, i.e. a 5 to 10% reduction (Roth et al. 2006). The evaluation methodology presented below will allow users of FDD—including equipment manufacturers, facilities operators, utility incentive managers or equipment owners—to make informed decisions about whether to use FDD and which protocol will work best for them. It will also aid in the development and improvement of FDD algorithms by providing a measure by which improvements can be tracked.

EVALUATION METHOD

Several approaches to evaluating the effectiveness of FDD protocols have been developed and considered in the current project. There are significant challenges to evaluating FDD because there are so many approaches to conducting FDD, using different inputs, giving different outputs, and having varied objectives. One major division is between protocols intended to be used in maintenance and installation work (typically run on a handheld device), and protocols intended to be used in a permanently-installed onboard application (automated FDD). The focus of the current discussion is the former—handheld devices—but much of the evaluation methodology could be applied to the latter. The methodology also focuses on FDD methods that are based on steady-state measurements from unitary equipment operating in cooling mode.

Another example of a challenge in evaluating FDD is that the benefits and costs associated with applying FDD vary for potential applications of a given FDD tool. For something as complex as FDD, ideally an evaluation provides a simple output, such as typical economic benefit from deploying the FDD. However, this value depends on fault prevalence, which is currently not well understood, so the evaluation method that was chosen is one in which the evaluation provides output based on the performance degradation. This allows flexibility in using the evaluation results for a wide range of expected scenarios. The method is summarized below then described in greater detail within the context of a case study, so that examples of the evaluation calculations are readily available.

Approach Summary

The approach to evaluation of FDD protocols is to feed a set of data to each protocol and observe the responses, collecting and categorizing them to develop summary statistics. The data represent typical conditions that a FDD tool may encounter:

- Several different systems with different properties, such as configuration, refrigerant type, SEER rating, and expansion device type
- A range of ambient and indoor thermal conditions
- Different types of faults, or with no fault
- Different intensities of fault

For each test case (a single combination of the conditions above) the protocol gives a response. These responses are tallied and organized to give statistics that reflect the overall utility of the protocol. The evaluation process is summarized in Figure 39-1.

The following subsections describe the components of the evaluation method in more detail. The library of input data compiled for this project is described in the "Data Library" section of this report. Fault types and their effects on operation of unitary air-conditioners are described in the "Faults" section.

Faulted and Unfaulted Operation

Faults are conditions that affect performance negatively and they have some level of severity. We have developed two ways to characterize this level of severity.

The first is Fault Intensity (FI), which is related to measureable quantities. For example, a 20% undercharge. The second is Fault Impact Ratio (FIR), which is related to equipment performance, and is tied to either capacity or coefficient of performance (COP). For example, when $FIR_{COP} = 95\%$, it says that the equipment is operating at 95% of its maximum efficiency under a given set of driving conditions. Each of these terms—FI and FIR—are formally defined in later sections of this chapter.

There is not a direct relationship between FI and FIR. This means that it is possible to have faults that have some FI, but with no measureable degradation of performance. This raises the question of how do we draw a distinction between faulted and unfaulted operation. For the evaluation method described here, the answer is that we consider FIR, because the equipment performance is generally what equipment operators and users of FDD are concerned with. This leads to another question, which is: how much performance degradation constitutes faulted operation? Our approach is to leave this as a variable quantity, using FIR *thresholds* to draw the distinction between faulted and unfaulted. We evaluate each protocol at several thresholds so that a user of the results can choose the threshold he or she considers appropriate. If the FIR threshold is 99%, it means that test cases with FIR above this threshold are considered to be unfaulted, regardless of the FI. This threshold concept is important in the consideration of False Alarms, described below.

Test Case Outcomes

When FDD is applied, there are five possible outcomes with respect to fault isolation:

1. **No response**—the FDD protocol cannot be applied for a given input scenario, or does not give an output because of excessive uncertainty.

2. **Correct**—the operating condition, whether faulted or unfaulted, is correctly identified

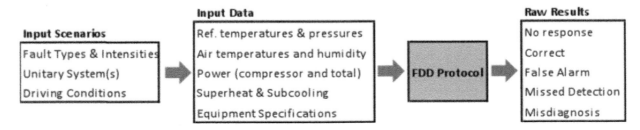

Figure 39-1: Evaluation method

3. **False alarm**—no significant fault is present, but the protocol indicates the presence of a fault. More specifically, a False Alarm is indicated when the protocol gives a response that a fault is present and
   a. the fault impact is below a given threshold, and
   b. the system is not overcharged by 5% or more

The special requirement in bullet b. is included for the following reason. An overcharged system may have a significant fault, but no significant impact on capacity or COP. Consider the example case of a system that is 10% overcharged, but has no significant degradation of capacity or COP. An equipment operator may want to know about the overcharge, since it can be associated with reduction of compressor life, even though it doesn't impact the current performance of the equipment. To address this situation, if the refrigerant is overcharged by more than 5% the system is considered faulted, even if the fault impact is below the given threshold.

4. **Misdiagnosis**—a significant fault is present, but the protocol misdiagnoses what type of fault it is. In the method described here, misdiagnosis rates are presented within specific bands (ranges) of fault impact ratios. Three criteria are met for cases to be categorized as Misdiagnoses:
   a. Fault Impact Ratio (FIR) is within the specified range
   b. Experimenter indicated the presence and intensity of a fault
   c. Protocol indicates that the system has a fault different from the type of fault indicated by the experimenter

5. -—a significant fault is present, but the protocol indicates that no fault is present. Missed Detection rates are presented within specific bands (ranges) of fault impact ratios to better understand where the Missed Detections are most important. Three criteria are met for cases to be categorized as Missed Detections:
   a. Fault Impact Ratio (FIR) is within the specified range
   b. Experimenter indicated the presence and intensity of a fault
   c. Protocol indicates that the system has no fault

To evaluate an FDD protocol, one feeds it multiple input scenarios, each of which gives one of these five test outcomes. Test outcomes 1, and 3 to 5 are gathered and

expressed as rates, using percentages. Test outcome 2 is implied by the other outcomes. The rate calculations are provided here and demonstrated within the description of the Case Study.

Test Case Outcome Rate Calculations

In rate calculations, the numerator is the number of test cases that have a given test outcome (one of the five listed above). The denominator for each test outcome rate is described below. Each denominator is defined based on determining a meaningful rate. The denominators include only the cases that could apply to each type of outcome. For example, a Misdiagnosis can't be made on a test in which no fault is present, so only those cases determined to be faulted are included in the denominator for Misdiagnosis rate. (If a protocol indicates a fault when none is present, this is a False Alarm, not a Misdiagnosis).

No Response

Numerator: number of cases that meet the "No Response" criteria

Denominator: total number of test cases

False Alarm

Numerator: the number of cases that meet the "False Alarm" criteria

Denominator: the number of cases in which the fault impact is below a specified threshold and the refrigerant is not overcharged by more than 5%

Misdiagnosis

Numerator: the number of cases that meet the "Misdiagnosis" criteria

Denominator: the number of cases that meet the following criteria:
• Fault Impact Ratio (FIR) is within the specified range
• Experimenter indicated the presence and intensity of a fault
• Protocol indicates that the system has a fault

Missed Detection

Numerator: number of cases that meet the "Missed Detection" criteria

Denominator: the number of cases in which three criteria are met:
• Fault Impact Ratio (FIR) is within the specified range
• Experimenter indicated the presence and intensity of a fault
• Protocol gives a response

FAULTS

Faults Considered

There are six degradation faults that are included in the evaluation. These are listed in Table 39-1, along with a description of each fault. A diagram showing the components of an air-conditioning system is shown in Figure 39-2, and referred to in the descriptions of faults and how to impose faults in experiments.

Imposing Faults in Laboratory Experiments

To provide measurement data as an evaluator input, faults were simulated in a laboratory. We have developed a term to quantify the severity of the fault— Fault Intensity (FI)—and defined FI for each type of fault. A description of how faults were imposed in the laboratory, and the definition of FI for each fault type is provided below.

nominal mass can be based on the refrigerant mass that provides the maximum capacity or efficiency.

2. **Low-side heat transfer faults**: In a typical laboratory setup the airflow across the evaporator coil can be modulated using a variable speed booster fan or dampers. Reducing the airflow accurately duplicates the effect of most faults in this category: airflow reduction from fan or distribution system design problems, obstructions or filter fouling. The effect of evaporator coil air-side fouling is also assumed to be well represented by reducing airflow, particularly if the fouling is assumed to be evenly distributed across the face of the heat exchanger. The fault intensity is defined analogously to FI_{charge} using either mass flow rate or volumetric airflow rate.

Table 39-1: Faults included in evaluation of FDD protocols

Fault	Abbr.	Description
Under- or overcharge	UC, OC	A mass of refrigerant that is less or more than the manufacturer specification
Low-side heat transfer	EA	Faults in the evaporator coil such as coil fouling or insufficient airflow
High-side heat transfer	CA	Faults in the condenser coil such as coil fouling or insufficient airflow
Liquid line restriction	LL	Flow restrictions such as crimps or fouled filter/drier in the liquid line (Figure 2)
Non-condensables	NC	The presence of gases that do not condense (e.g. air or nitrogen) in the refrigerant
Compressor valve leakage	VL	Leaks in the compressor from high to low pressure regions, reducing mass flow

1. **Charge**: To impose an under- or overcharge fault, charge is simply removed from or added to the system. The fault intensity is:

$$FI_{charge} = \frac{m_{actual} - m_{nominal}}{m_{nominal}}$$

(Eq. 39-1)

where m_{actual} is the measured mass of refrigerant in the system

$m_{nominal}$ is the nominally correct mass of refrigerant (see discussion on charge effect on COP and capacity later on)

Thus a system designed for 5 lb of charge that had 4.5 lb would be referred to as 10% undercharged or having $FI_{charge} = -10\%$. If manufacturer specifications are not available, an alternate definition for

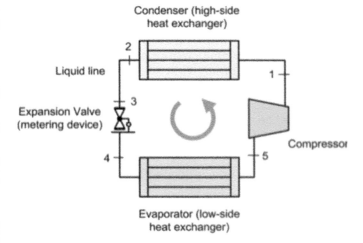

Figure 39-2: Components of a typical vapor-compression air-conditioner

$$FIEA = \frac{V_{actual} - V_{nominal}}{V_{nominal}}$$

(Eq. 39-2)

3. **High-side heat transfer faults:** Similarly to low-side faults, a reduction in airflow is used to implement high-side heat transfer faults. Some experimenters have simulated blockage by large-scale debris, such as leaves, by covering the face of the condenser coil with paper or mesh. Although this may, in some cases, more realistically represent the fault physically, it is not repeatable nor easily quantified as a fault intensity. Furthermore, the general effect—to increase the refrigerant's high-side pressure—is the same as with reduced airflow. Therefore, reduced airflow is proposed as the standard means of imposing this fault in the laboratory. Accordingly, the fault intensity is defined with airflow rates in the same manner as with low-side heat transfer faults.

$$FICA = \frac{V_{actual} - V_{nominal}}{V_{nominal}}$$

(Eq. 39-3)

4. **Liquid line restriction**: A liquid line restriction is implemented by using one or more valves to impose the desired pressure loss. The fault intensity is defined using the ratio of the increase in pressure drop through the liquid line caused by the faulted condition to the liquid line pressure drop under non-faulted operation and at the same operating condition.

$$FILL = \frac{\Delta PLL,faulted - \Delta PLL,unfaulted}{\Delta PLL,unfaulted}$$

(Eq. 39-4)

5. **Non-condensables in the refrigerant**: A non-condensables fault is imposed by introducing nitrogen into the refrigerant line. The maximum amount of non-condensables to be expected is in the case where a system has been open to the atmosphere and not evacuated prior to charging. Therefore, the fault is defined with a mass of nitrogen compared to the mass of nitrogen that would fill the system at atmospheric pressure.

$$FINC = \frac{m_{N2,fault}}{m_{N2,ref}}$$

(Eq. 39-5)

6. **Compressor valve leakage**: Compressor valve leakage is simulated with the use of a hot gas

bypass—a pipe carrying refrigerant from the discharge to the inlet of the compressor (from point 1 to point 5 in Figure 39-2. The fault intensity for this fault is defined as the change in mass flow rate (at a given operating condition) to the original mass flow rate.

$$FIVL = \frac{m_{faulted} - m_{unfaulted}}{m_{unfaulted}}$$

(Eq. 39-6)

Faults Not Considered

There are many possible faults that occur in unitary air-conditioners. The faults that are used in the evaluator are controlled largely by the available experimental data. Previous researchers have considered these six faults important, and have conducted experiments to quantify their effects. The results of these experiments are included in the data library.

Some other faults that are not included in the evaluation method developed here are

- Economizer faults
- Thermostatic Expansion Valve (TXV) faults
- Control faults, such as short cycling, sensor failure or degradation, etc.

These faults can have major impacts on performance, and anecdotal evidence suggests that they may be quite prevalent. The evaluation method presented here can be adapted to include these faults when appropriate input data become available.

Heat Exchanger Fouling Effect

An important question for the evaluation method, and for FDD in unitary equipment, must be considered: is airflow reduction an appropriate proxy for air-side coil fouling? Ali and Ismail (2008) suggest that it is not. However, Bell et al. (2012) and Yang et al. (2007a) both conclude that the effect of fouling on the heat transfer coefficient is small, but the impact on air pressure drop across the coil is large and dominant in the fouling degradation effect for constant speed fan systems. We did a review and analysis of Ali and Ismail and concluded their results aren't applicable to the systems we are considering, and do not affect our assumption that reduced airflow is indeed a good proxy for all air-side heat transfer impediments. One reason is that the system they studied is a window-mounted air-conditioner. Another is that their conclusions are based in part on performance at very high face velocities. Typical face velocities are about 1.5 m/s, whereas Figure 39-3 shows that the face velocities in their experiment range up to 5 m/s. The scatter in the 1 to 2 m/s range is larger than the effects being measured. Furthermore, the data show a trend of reduced COP for a given mass of fouling as airflow rates get higher than 2.5 m/s. This suggests that fan power is included in the COP calculation, which would blur the effect being considered.

Therefore, based on the results of Bell et al. (2012) and Yang et al. (2007a), it is concluded that reduced airflow is an appropriate and reliable proxy for coil fouling faults.

DATA LIBRARY

One of the most significant outputs from this project is the development of a data library. The library supplies input data for evaluation. It consists of both experimental and simulation data for unitary systems operating at steady state over a range of operating conditions and fault conditions.

Summary of Data Library

Table 39-2 shows a summary of the units and numbers of test cases contained in the experimental data library. The number of tests is separated to show the number for each fault type for each system, using the fault type abbreviations given in Table 39-2. There are a total of 607 test cases, gathered from experiments on nine unitary systems. Three of the systems are rooftop units (RTU), but the tests on these three make up 60% of the total test cases.

The rightmost columns show the limits of the range of ambient temperature during testing for each of the test units in the library. The distribution of tests by return air wet-bulb temperature and by ambient temperature (among the entire set of 607) is shown in Table 39-3.

Description of Data

The data in the library represent steady-state cooling operation. Each datum is an average of multiple measurements—from 8 to several hundred—taken while the equipment was operating steadily. The experimenters followed the same standards that equipment performance rating experiments follow, such as AHRI Standard 210/240 (AHRI 2008) and ASHRAE Standard 37 (ASHRAE 2009). More details on the experimental approaches of the data included in the data library can be found in the following references: Breuker (1997), Kim et al. (2006), Shen et al. (2006), Palmiter et al. (2011).

The data library contains measurement data and system information. The types of measurement (and calculated) data are listed in Table 39-4 in IP units. The entire data library also has an SI-unit version so that inputs are readily available for protocols that use these units, such as European protocols.

For each test case there are also pieces of system information about the test unit. The types of system information are described in Table 39-5.

Some of the test units in the data library do not have all of the types of data listed in these tables.

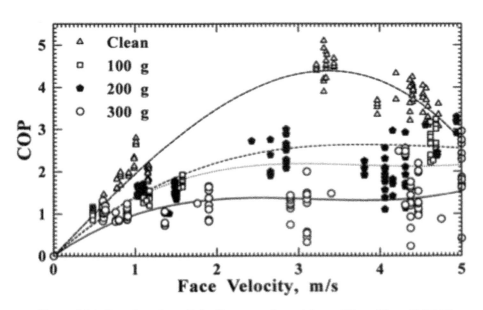

Figure 39-3: Results of a coil fouling experiment from Ali and Ismail (2008)

Table 39-2: Summary of test cases in experimental data library

Note 1: RTU 2 is a split system, but was named using a previous naming convention.

#	ID	Type	Capacity [tons]	Refrig.	Exp. Device	Comp. Type	Number of tests								Temp.	
							No Fault	UC	OC	EA	CA	LL	NC	VL	Min. [°F]	Max. [°F]
1	RTU 3	RTU	3	R410a	FXO	Scroll	24	25	12	21	6	0	0	0	67	125
2	RTU 7	RTU	3	R22	FXO	Recip.	39	34	0	26	36	34	0	33	60	100
3	RTU 4	RTU	5	R407c	FXO	Scroll	17	15	12	19	8	0	0	0	67	116
4	Split 1	Split	3	R410a	FXO	Recip.	1	29	1	0	0	0	0	0	82	127
5	RTU 2[1]	Split	2.5	R410a	TXV	Scroll	16	12	12	21	15	16	15	16	70	100
6	Split 2	Split	3	R410a	TXV	Recip.	2	30	7	0	0	0	0	0	83	127
7	Split 3	Split	3	R410a	TXV	Scroll	4	4	7	0	0	0	0	0	82	125
8	Split 4	Split	3	R22	TXV	Scroll	4	8	0	8	0	0	0	0	82	125
9	Split 5	Split	3	R22	TXV	Scroll	4	4	4	6	0	0	0	0	82	125
						Total:	111	161	55	101	65	50	15	49		

Data Vetting, Uncertainty and Removal of Questionable Data

FDD protocols use different inputs to detect and diagnose faults. To ensure that the evaluations are meaningful they must be fair, which requires consistent data. A great deal of effort has gone into studying the data to look for inconsistencies because of the importance of using reliable input data. This was done manually.

In conducting and reporting on experiments, results can't be removed from the dataset without justification because this could skew the overall results or conclusions of the experiment. However, in vetting the data for the evaluator, removal of a test case doesn't necessarily skew results because it is removed for all protocols that will be evaluated.

Of more than 1000 test cases that were collected for 14 units, about 40% were removed. Some of the reasons for removal:

- *Data that don't follow physical laws*—for example if significant refrigerant pressure increases occur in locations other than the compressor, if energy is not conserved, humidity is generated across the evaporator, etc.

- *Data that show too much scatter, or are obvious outliers when compared to other data within the set.* For unfaulted tests, a normal model (described below)

Table 39-3: Distribution of tests by return air wet-bulb temperature (left) and ambient temperature (right)

Return air wet-bulb		Ambient Temperature	
Range [°F]	Number of occurrences	Range [°F]	Number of occurrences
50-55	16	60-70	53
55-60	260	70-80	58
60-65	85	80-90	223
65-70	240	90-100	186
70-75	6	100-110	26
	607	110-120	39
		>120	22
			607

was an effective tool for assessing outliers. One of the test units was rejected completely because of questionable data.

- *Data that are not self-consistent*—as with fault detection, redundancy in data can be used to detect problems. For example, in cases where an experimenter provided two forms of humidity, such as wet-bulb and relative humidity, each was checked using psychrometric relationships and the associated pressure and dry-bulb temperature. If they didn't agree, the test was further investigated by

Table 39-4: Data library measurement data types

Variable ID	IP Units	Description
T_RA	[°F]	Return Air dry bulb temperature (evaporator inlet)
DP_RA	[°F]	Return Air dewpoint temperature (evaporator inlet)
WB_RA	[°F]	Return Air wet bulb temperature (evaporator inlet)
RH_RA	[%]	Return Air relative humidity (evaporator inlet)
T_SA	[°F]	Supply Air dry bulb temperature (evaporator outlet)
DP_SA	[°F]	Supply Air dewpoint temperature (evaporator outlet)
WB_SA	[°F]	Supply Air wet bulb temperature (evaporator outlet)
RH_SA	[%]	Supply Air relative humidity (evaporator outlet)
T_amb	[°F]	Ambient air dry bulb temperature
P_LL	[psia]	Liquid line pressure
T_LL	[°F]	Liquid line temperature
P_suc	[psia]	Suction pressure
T_suc	[°F]	Suction temperature
P_dischg	[psia]	Compressor discharge pressure
T_dischg	[°F]	Compressor discharge temperature
Power	[W]	Total electrical power of system
T_air_ce	[°F]	Condenser exiting air temperature
T_sat_e	[°F]	Refrigerant saturation temperature in the evaporator
T_sat_c	[°F]	Refrigerant saturation temperature in the condenser
Power_comp	[W]	Compressor power
Fault	[-]	Experimenter's identified fault type (or unfaulted)
Q_ref	[Btu/hr]	Refrigerant side capacity
Q_air	[Btu/hr]	Air-side capacity
SHR	[-]	Sensible Heat Ratio
COP	[-]	Coefficient of performance
SH	[°F]	Suction Superheat
SC	[°F]	Subcooling
m_ref	[lbm/min]	Refrigerant mass flow rate
Chrg	[lbm]	Mass of refrigerant charge
Chrg%	[%]	Charge as a percentage of nominally correct charge
V_i	[CFM]	Indoor coil volumetric airflow rate
V_i_nom	[CFM]	Nominal indoor coil volumetric airflow rate
V_i_%	[%]	Indoor coil volumetric airflow rate as percentage of nominal
V_o	[CFM]	Outdoor coil volumetric airflow rate
V_o_nom	[CFM]	Nominal outdoor coil volumetric airflow rate
V_o_%	[%]	Outdoor coil volumetric airflow rate as a percentage of nominal
Blk%	[%]	Portion of outdoor coil blocked
LL restr.	[psia]	Pressure loss through liquid line restriction
NonCond	[lbm/lbm]	Mass fraction of non-condensables in the refrigerant
NonCond%	[%]	Mass of non-condensables as a percentage of reference mass
VlvLeak	[lbm/min]	Compressor hot-gas bypass mass flow rate
VlvLeak	[%]	Compressor hot-gas bypass mass flow rate as % of total mass flow
$FIR_{capacity}$	[%]	Fault Impact Ratio for capacity
FIR_{COP}	[%]	Fault Impact Ratio for COP

Table 39-5: Data library system information

Variable ID	Description
Expansion Type	Expansion valve type (TXV, FXO or EEV)
Manufacturer	Manufacturer
Model (indoor)	Model of indoor unit (for split systems)
Model (outdoor)	Split system outdoor unit model or RTU model
Nominal Capacity	Nominal Capacity (tons)
Refrigerant	Refrigerant
Operating Mode	Cooling or heating
Compressor Type	Reciprocating, scroll, etc.
Compressor Model	Compressor Model
Target SC	Target subcooling rate (for TXV systems)
EER	Energy efficiency ratio
SEER	Seasonal energy efficiency ratio
C1 to C10	Compressor map coefficients

comparing to air-side capacity or sensible heat ratio if sufficient data were available. If it was unclear which variable was flawed, the test case was removed. Similar approaches were used to check other calculated data, such as capacity and COP.

- *Insufficient data*—some data sets did not contain enough data to determine the impact of a fault. Four of the 14 test units were removed for this reason.

Charge Effect on COP and Capacity

For faults such as the presence of non-condensable gas in the refrigerant, compressor valve leakage, or reduced airflow across the outdoor coil, the unfaulted condition is clear. However, the unfaulted or "correct" mass of refrigerant charge in a system is less clear. The experimenters have charged their experimental units by methods that they may not have detailed within their description of the experiments. Their data sets usually identify which tests they consider to be conducted with correct charge. However, when evaluating FDD protocols that are attempting to diagnose charge faults, it's imperative that the experiments with nominally correct charge truly have correct charge.

An earlier approach to defining the correct charge was to use the experimenters' nominally correct values. However, this was criticized because we can't be certain that the experimenters' values were correct. To provide a consistent approach, we currently define the correct charge as being **the mass of charge that gives the maximum COP at the standard rating condition (95/80/67).** In most cases this approach agrees with the experimenter value. For example, consider Figure 39-4. Although the COP flattens out around 100% of nominal charge for the rating condition (purple line), there is a point at 100% that gives the highest COP (2.5). However, there are four units in the data library for which the experimenter's nominally correct value gave the maximal capacity, but not the maximal COP.

In the four test units for which the maximal COP at the rating condition was not reached at the experimenter's nominal charge, the nominal charge was updated to match the charge for which the maximum COP was achieved. In this report "nominal charge" refers to the maximal-COP charge.

The updated nominal charge for the four units changes the fault category for many of the tests in the affected dataset. One complication of this update is that the other fault test cases became multiple fault cases. For example, cases with evaporator airflow faults imposed became evaporator airflow and over-charge or under-charge fault cases. These tests were removed from the evaluation inputs.

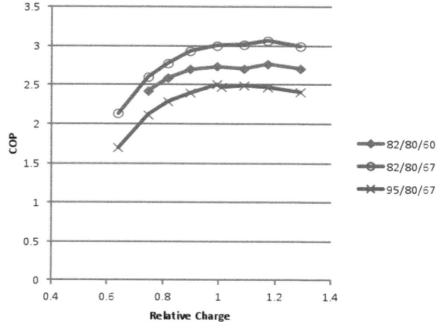

Figure 39-4: Relative COP as a function of charge at three conditions for a FXO RTU

Normal Model & FIR

In six of the nine systems in the data library there was a sufficient set of no-fault tests to enable development of a normal model*. The six modeled systems are numbers 1 to 6 in Table 39-2. A normal model is a multiple linear regression of the driving conditions that predicts capacity or COP, as shown in Eqs. 39-7 and 39-8, where the coefficients a_i and b_i are found using a least squares approach. The normal model is developed using unfaulted tests (those with no faults imposed, and with maximal-COP charge), so that it can be used to assess what the capacity or COP degradation is for faulted tests at any given condition. The normal model approach for determining degradation is preferable to a measurement-based approach for two reasons. The first is that it significantly reduces bias error, because it obviates the problem of trying to exactly match the test conditions for a faulted and an unfaulted test. The second is that it reduces or eliminates one half of the random error associated with a comparison of two test results (faulted and unfaulted tests at the same conditions).

$$Q=\alpha0+\alpha1 \cdot wbra+\alpha2 \cdot wbra2$$
$$+\alpha3 \cdot wbra \cdot Tamb+\alpha4 \cdot Tamb+\alpha5 \cdot Tamb2$$

(Eq. 39-7)

$$COP=\beta0+\beta1 \cdot wbra+\beta2 \cdot wbra2$$
$$+\beta3 \cdot wbra \cdot Tamb+\beta4 \cdot Tamb+\beta5 \cdot Tamb2$$

(Eq. 39-8)

For wet-coil cases, the two external driving conditions are ambient air dry bulb and return air wet bulb temperature. For dry-coil cases, the two driving conditions are ambient dry bulb and return air dry bulb. To use a single two-input model (as shown in Eqs. 39-7 and 39-8) to represent both dry- and wet-coil cases, an approach has been followed in which a fictitious return air wet bulb temperature, $wb_{ra,f}$ is used in place of the actual return air wet bulb temperature, wb_{ra} for all dry-coil

---

*Systems 4 and 6 from Table 39-2 are among the four that had their nominal charge adjusted to give maximal COP at the rating condition. These systems had a large number of tests conducted with the experimenter's original nominal charge, but only one or two tests with the maximal-COP charge. For these two cases the normal model was conducted using the original nominal charge. To adjust the FIR values for the updated nominal charge level, each FIR value was divided by the FIR at the maximal-COP charge. For example, in system #4, the maximal-COP charge gave a FIR_{COP} of 104%. Therefore, all FIR_{COP} values were divided by 104% when the nominally correct charge level was adjusted. Although the resulting FIR values are not exact, this method correctly represents the trends caused by adjusting charge, and the magnitude of the inaccuracies introduced by this method is small. The effect of the inaccuracies on an FDD evaluation is insignificant.

cases (see Brandemuehl (1993) for details). This $wb_{ra,f}$ is calculated using an iterative approach that involves a bypass factor (BF). BF indicates the fraction of air that would need to bypass an ideal coil, $mbyp/mlvg$, to give equivalent performance to the real coil. Using energy and mass balances and psychrometric relationships, BF can also be expressed in terms of specific enthalpies, h, or humidity ratios, w, as shown in Eq. 39-9.

$$BF=mbypmlvg=hlvg-hadphra$$
$$-hadp=\omega lvg-\omega adp\omega ra-\omega adp$$

(Eq. 39-9)

For a wet coil condition, the air leaving an ideal coil will have a dewpoint temperature equal to the surface temperature of the coil—the apparatus dewpoint (*adp*). In the fictitious wet bulb approach, BF is iteratively varied until the enthalpy calculations of Eq. 39-9 give the same result as the humidity ratio calculations with an assumption of 100% relative humidity for the air at the apparatus dewpoint.

The BF values calculated for the wet coil cases are averaged, and this average is then used to calculate sensible heat ratios for each dry coil test using Eq. 39-10. In Eq. 39-10, w_{adp} is calculated using Eq. 39-9, and the fictitious return air enthalpy, $h_{ra,f}$ is varied until SHR converges to 1.0. Finally, the fictitious wet bulb, $wb_{ra,f}$ is calculated from $h_{ra,f}$ and T_{ra} and is used in Eqs. 39-7 and 39-8 for any dry coil cases in the data set.

$$SHR=h(Tra,\omega adp)-hadphra,f-hadp$$

(Eq. 39-10)

This approach is described in more detail by Brandemuehl (1993).

During model validation, the measured unfaulted cases (the basis for the model) are compared with model outputs for the same set of conditions. The capacity and COP are compared, and residuals calculated. For example, Eq. 39-11 shows the calculation for capacity.

$$Residual=capacitymodel-capacitymeasuredcapacitymeasured$$

(Eq. 39-11)

An example plot, showing the residuals for the normal model of capacity for RTU 3, is shown in Figure 39-5. This plot indicates the level of scatter for this unit, which is typical for a laboratory-tested unit. The dry coil and wet coil data are shown separately to illuminate any difference that could be caused by problems associated with the fictitious wet-bulb approach to model generation. The wet and dry coil cases are very similarly distributed, indicating that this modeling approach hasn't introduced any obvious bias or scatter error. The dry coil cases are associated with lower-capacity cases on average, as one would expect because unitary system capacity decreases with decreasing indoor humidity.

An example of a normal model from RTU 3 is shown in Figure 39-6. The mesh surface is the model and the circular markers are the measurement data upon which the model is based. This figure is typical of normal models; it has a fairly planar shape, with a slight increase in COP as return air wet bulb increases, and a strong decrease in COP as ambient temperature increases. If the surface is rotated so that it can be viewed from the side, there is typically a very small amount of twist to the planar shape. This is demonstrated in Figure 39-7. Similar figures showing the normal models of other units are provided in the appendices.

The completed model is used to calculate fault impact ratios (FIR), which form the basis of fault-impact based evaluations, as discussed earlier. FIR are defined as:

$$FIR_{COP}=\frac{COP_{faulted}}{COP_{unfaulted}}$$
$$FIR_{capacity}=\frac{capacity_{faulted}}{capacity_{unfaulted}}$$
(Eq. 39-12)

where the *faulted* COP and capacity values come from measurements, and the *unfaulted* values come from the normal model. The FIR values are included in the data library, and form the basis of the impact-based evaluation approach.

SIMULATION

As the evaluation method's development has progressed, we have become increasingly convinced that the future of FDD evaluation needs to use complete system models to generate input data, rather than relying on measurements. The arguments in favor of using simulation data are summarized below. The main argument against this approach is typically that engineers find it difficult to believe in simulations that they aren't deeply familiar with. To paraphrase William Beveridge: Everybody believes an experiment except the experimenter; nobody believes a model except the modeler. A second argument against the approach is that it is too difficult and time consuming to generate models that can accurately model faulted system operation. This second argument has been addressed recently by developing a new method for rapidly simulating unitary systems, using an inverse modeling approach. This method is described in Cheung and Braun (2013a, 2013b). The outputs from models developed with this approach are being added to the evaluation method at the time of writing. These models are based upon the library of fault data described earlier.

Arguments in Favor of Using Simulation Data for FDD Evaluation
Reliability of Data

The exercise of vetting the data for this project has shown that measurement data have significantly uncertain results. We also are aware through direct experience that obtaining accurate results for air-conditioning systems under varying driving conditions and with faults

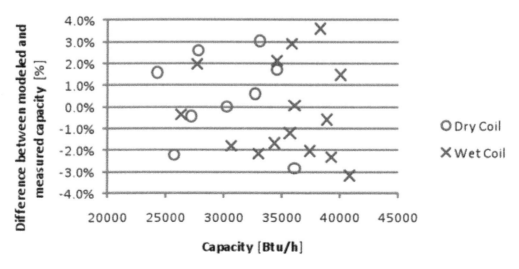

Figure 39-5: Normal model residuals as a function of capacity for RTU 3

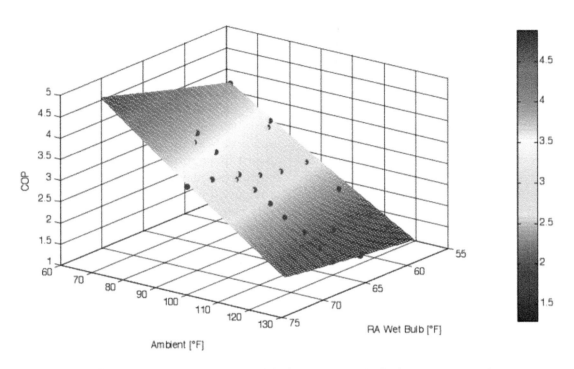

Figure 39-6: RTU 3 normal model of COP and unfaulted measurement data

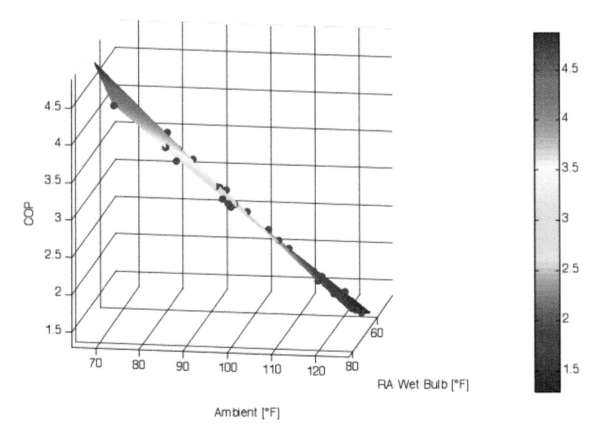

Figure 39-7: RTU 3 normal model of COP and unfaulted measurement data—side view

imposed is extremely difficult. Since errors typically don't affect all variables equally, a protocol that relies on an error-affected variable may perform worse than a protocol that uses a different variable as its input.

Additional Systems

The current database has just nine systems, and represents all of the known data that is sufficiently reliable and detailed. This may not be a large enough sample, since protocols perform better with some systems than others. The cost of conducting additional experiments is prohibitive. Simulation can be conducted with much less expense.

Finer Resolution of Driving Conditions

It is likely that a developer or potential user of a protocol may be interested in an exact condition that hasn't been tested, or may wish to know a more precise fault intensity for which a protocol begins to flag faults than what the data can provide. A simulation can be set to give any reasonable conditions.

Multiple Faults

Multiple faults are known to exist in the field and methods of diagnosing multiple faults are being developed (Li and Braun 2007). Adding combinations of faults at varying intensity drastically increases the number of test cases required for even the coarsest evaluation. However, multiple faults can be simulated quite accurately.

Gaming

The input data for an evaluation are analogous to the answers to a test. A finite set of input data, such as the set contained in the data library, can quite easily be programmed into a protocol so that it recognizes the conditions and gives the correct response. This will render evaluation meaningless, because the ability to get perfect evaluation results isn't related to the ability to detect and diagnose faults in the field. With simulation data, tables of correct answers won't exist.

Case Study

The California Title 24 HVAC Refrigerant Charge and Airflow (RCA) diagnostic protocol was used as an experimental subject during the development of the evaluation method. It was chosen because it is readily available and is in current widespread use. This section describes an evaluation of this protocol based on the performance criteria and data library previously described. In applying the method, data were supplied to

the FDD method from the laboratory measurements. In determining fault impact ratios for any fault, the measurement results were compared to the outputs from the normal model determined from regression as previously described.

RCA Background

The RCA protocol is specified in California's Title 24—2008 building energy code (CEC 2008). It is included in a modified form in the 2013 version of the code that was implemented in 2014. RCA, as its name implies, is intended only to detect and diagnose high or low refrigerant charge and low evaporator airflow. The airflow diagnostic is intended to ensure that the evaporator has sufficient airflow for the charge diagnostics to be applied. It is an available option if direct measurement of the airflow isn't conducted. The RCA protocol is based primarily on manufacturer's installation guidelines.

Title 24 specifies that the RCA protocol is to be applied to residential systems. However, it has been used as the basis for utility-incentivized maintenance programs on residential and commercial unitary systems. For this reason, and because there is no fundamental difference between commercial and residential unitary systems, the input data from both RTU and split systems were used in the evaluation.

The protocol is applied sequentially. The evaporator airflow is checked first. If the airflow is deemed acceptable, then the charge algorithm is applied. The RCA uses the following as its inputs: (1) return air dry bulb and wet bulb; (2) supply air dry bulb; (3) ambient air dry bulb; (4) either evaporator superheat for FXO systems, or subcooling for TXV systems; and (5) the manufacturer's specified target subcooling value (for TXV systems). Some of these inputs are used to gather target temperature split and target superheat values from two lookup tables. The inputs, and the values from lookup tables, are used to determine whether temperature split (the air temperature difference across the indoor unit) and superheat (for FXO systems) or subcooling (for TXV systems) are within acceptable ranges, using a difference (D) between the measured and target values. For example, $D\,SH$ is calculated as: $\Delta\,SH = SH measured - SH target$.

The range of driving conditions for the lookup tables is limited, which means that the protocol can't be applied to some tests in the data library (i.e. gives *No Response* outcomes). A flow diagram of the RCA protocol logic is shown in Figure 39-8. In this figure the inputs listed above are shown in red. The RCA output results are shown in grey boxes. The process starts in the top left corner (if the temperature-split airflow diagnostic is used), with return air dry bulb and wet bulb temperatures as inputs.

RCA Versions: 2008, 2013, Installer & HERS

The RCA protocol has been modified with each new version of Title 24. In the 2008 energy code, a special version of the protocol was given for use by Home Energy Rating System (HERS) raters, who provide field verification and diagnostic testing to demonstrate com-

pliance with the standard. This version was identical except that it included looser tolerances when comparing measured and target values of superheat, subcooling, and temperature split. The standard provides a rationale for the different tolerances:

"In order to allow for inevitable differences in measurements, the Pass/Fail criteria are different for the Installer and the HERS Rater." (RA 3.2.2.6.1, note #5).

For example, the charge diagnostic for FXO systems is:

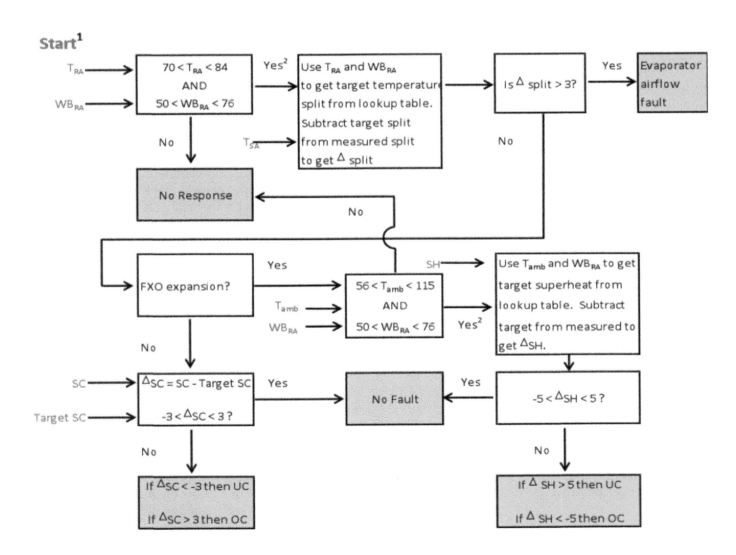

Note 1: The first part of the protocol, intended to determine whether there is sufficient evaporator airflow, is an optional approach that can be used if direct airflow measurement isn't conducted. If the evaporator airflow diagnostic isn't used, the process starts in the box labeled "FXO expansion?"

Note 2: The lookup tables cover the ranges specified above, but there are several cells on each table that contain dash marks, indicating that the protocol should not be applied. In these cases the result is "No Response".

Figure 39-8: Flow diagram of logic for applying the RCA protocol (using 2008 Installer's version)

$-5°F<SHmeasured-SHtarget<5°F→$
No charge fault

(Eq. 39-13)

while for the HERS rater the tolerance is increased 1°F above and below the target:

$-6°F<SHmeasured-SHtarget<6°F→$
No charge fault

(Eq. 39-14)

The 2013 version of Title 24 has removed the temperature-split evaporator airflow diagnostics option. There are other compliance options available to confirm that sufficient airflow is attained prior to diagnosing charge faults. These generally involve showing by direct measurement that the evaporator airflow is above 300 or in some cases 350 CFM per nominal ton of cooling capacity.

The 2013 version also has additional restrictions on the driving conditions under which the protocol can be applied, such as a maximum outdoor (condenser inlet) air temperature of 120°F for TXV-equipped systems, and a minimum return air (indoor) dry-bulb temperature of 70°F (whereas the 2008 protocol had this limitation only for outdoor air temperatures from 55-65°F).

A summary of the differences in tolerances within the four versions of the RCA protocol in the current (2008) version and the future (2013) version is shown in Table 39-6.

In the case of charge, if a fault is detected, it is diagnosed as "undercharged" if the difference in Equation 39-12 is above 5°F and "overcharged" if the difference is below -5°F. This distinction is not specified for the HERS rater; the system simply fails the charge test. However, to present a more meaningful evaluation here, the distinction is taken as implied in the results presented below.

Test Results

Results of the tests are presented for each of the four RCA versions' with respect to the evaluation outcome categories *No Response, False Alarm, Missed Detection* and *Misdiagnosis*.

No Response

The *No Response* rates for the four RCA versions are shown in Table 39-7.

In the RCA protocol, a *No Response* result is generated when the driving conditions—ambient air temperature, indoor wet-bulb temperature, and indoor dry-bulb temperature—are not within the range of the look-up tables that are used to determine target temperature split and superheat values, or when they are outside of the limits discussed above. A higher rate of *No Response* means that the protocol is less useful, particularly for maintenance technicians, as detailed by Temple (2008). However, since the rate is dependent on the conditions of the input data, the rates themselves aren't very meaningful because the distribution of input data conditions

Table 39-6: Tolerances for 2008 and 2013 Installer and HERS versions of the RCA protocol

	2008		2013	
Charge	**Installer**	**HERS**	**Installer**	**HERS**
FXO (Δ superheat)	±5°F	±6°F	±5°F	±8°F
TXV (Δ subcooling)	±3°F	±4°F	±3°F	±6°F
Airflow				
(Δ temperature split)	+3°F	+4°F	-	-

Table 39-7: Total test cases, and No Response results for four RCA versions

	2008		2013	
	Installer	**HERS**	**Installer**	**HERS**
Test Cases	607	607	572	572
No Response	127	128	158	158
No Response Rate	21%	21%	28%	28%

may not exactly represent the typical conditions when a technician might want to deploy the protocol. A comparison of rates from one protocol to the next would be more meaningful.

The number of test cases and responses differ in the versions of the RCA presented in Table 39-7. In the 2013 version, all cases with indoor airflow rates below 300 CFM/nominal ton are assumed to be eliminated by direct measurement of airflow, and so they have been removed from the input data (35 test cases were below this criterion). The number of responses varies in the 2008 version because the temperature split (airflow diagnostic) table has a wider range of acceptable conditions. This means that a test case can be flagged as having an airflow fault under conditions where the charge diagnostic would give *No Response*. Since the protocol is sequential (airflow diagnostic first), the charge diagnostic isn't applied if an airflow fault is flagged. With the looser tolerance of the HERS version, some test cases passed the airflow diagnostic (which hadn't passed for the Installer version) and were then flagged as *No Response* when the charge diagnostic was applied.

False Alarm

The *False Alarm* results for each of the four RCA versions are presented in separate plots in Figure 39-9 to 39-12. Below Figure 39-9, the data that form the basis of the figure are also presented, in Table 39-10, to indicate the sample sizes.

Calculation of *False Alarm* Rate

The *False Alarm* rate is calculated at several Fault Impact Ratio (FIR) thresholds. Test cases with FIR above the threshold are considered unfaulted. The rate calculations follow the procedure described in the section *Test Case Outcome Rate Calculations*. Referring to Figure 39-9, the *False Alarm* rate is 45% for the 95% FIR_{COP} threshold, which refers to all test cases in which COP is degraded by 5% or less. Some of these *False Alarms* are cases where the experimenter had imposed a fault, but it wasn't significant enough to cause a 5% degradation in performance, and the others are cases in which the experimenter did not impose a fault.

There is experimental uncertainty in all measurements, including the measurements used in calculating

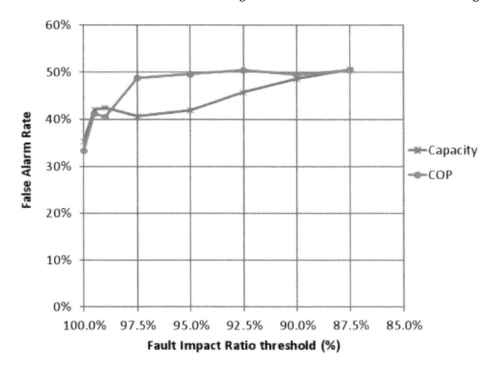

Figure 39-9: RCA 2008 Installer False Alarm rate as a function of FIR threshold

Table 39-8: Numerical results for RCA 2008 Installer False Alarm rate as a function of Fault Impact Ratio thresholds

	FIR threshold:	100.0%	99.5%	99.0%	97.5%	95.0%	92.5%	90.0%	87.5%
Capacity	Responses	68	107	120	165	236	288	335	361
	False Alarms	24	45	51	67	99	132	163	183
		35%	42%	43%	41%	42%	46%	49%	51%
COP	Responses	63	97	116	174	254	311	343	370
	False Alarms	21	40	47	85	126	157	170	187
		33%	41%	41%	49%	50%	50%	50%	51%

capacity and COP. With randomly distributed error, about half of the unfaulted tests will give FIR values above 100%, and half below 100%. All of the data with values above 100% are included in the calculations. Since these cases are overwhelmingly unfaulted cases (as opposed to cases slightly below 100%, many of which have small faults imposed), including them gives lower *False Alarm* rates than if these cases were omitted.

Results and Discussion

The results for the 2008 Installer version, in Figure 39-9, are surprisingly high (the numerical basis of these results is shown in Table 39-8. In a third of the cases in which the system performs at 100% efficiency, the RCA diagnoses a fault. For systems performing at 97.5% or greater efficiency, the RCA diagnoses about half with a fault. When the protocol is applied in the field, each *False Alarm* is associated with costly and unnecessary service (which may degrade performance), so this result suggests a very poorly performing protocol.

The *False Alarm* rate stays fairly constant, which is surprising. We would expect that as we move to the right across the plot (i.e. consider larger and larger fault impact cases to be unfaulted) that the rate of *False Alarm* would increase significantly.

The *False Alarm* rate at the 100% FIR threshold in Figure 39-9 is not 0%. As noted above, the 100% threshold includes all test cases for which the FIR is above 100%. If these test cases were not included, then besides increasing the *False Alarm* rate at lower thresholds (as explained above) the 100% point would be undefined (0/0), since there are no cases with FIR between 100% and 100%.

Comparing Figure 39-9 and 39-10 we see an improvement in performance with respect to *False Alarm* rate—almost 10% in most cases. The only difference between these versions is the tolerances applied, as shown in Table 39-8. This suggests that the protocol is overly sensitive. Looser tolerances will reduce the *False Alarm* rate, although they may also have a detrimental effect on the ability to detect and correctly diagnose faults.

Comparing Figure 39-9 and 39-11 we again see improvement. The only difference between these versions is that the airflow fault diagnosis module is removed in the 2013 protocol (Figure 39-11). This brings reductions in the *False Alarm* rate of 5-10%, in most cases.

Although the removal of the temperature-split airflow diagnostic reduces the *False Alarm* rate, it is not necessarily an overall improvement to the protocol, because it reduces the utility of the protocol. The temperature split method is generally easier to apply than the alternative (directly measuring the airflow). Furthermore, the present evaluation necessarily assumes that the direct airflow measurement approach does not provide any *False Alarms*, but this may not be true in actual application of the protocol.

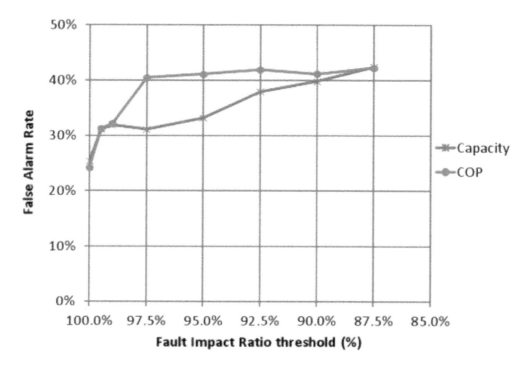

Figure 39-10: RCA 2008 HERS False Alarm rate as a function of FIR threshold

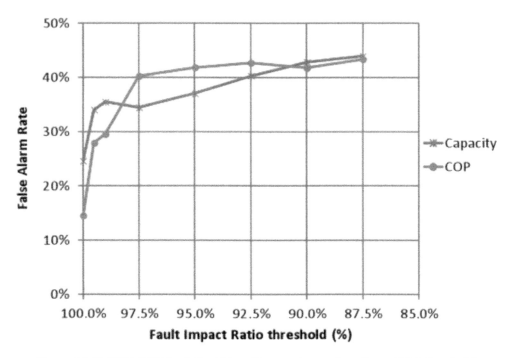

Figure 39-11: RCA 2013 Installer False Alarm rate as a function of FIR threshold

In Figure 39-12 we see the best performance, with respect to *False Alarms,* of the four versions. This can be attributed to the even looser tolerances for the 2013 HERS version. Although these results are improvements over the other versions, they still seem far too high to be able to consider this an effective protocol. Considering the 95% FIR_{COP} threshold, we still have over 1 in 4 cases falsely flagged as having a charge fault (since there is no airflow diagnostic evaluation in this version), meaning that charge would be added to or removed from a system that is operating acceptably.

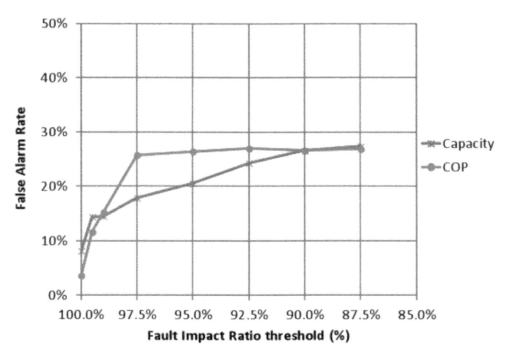

Figure 39-12: RCA 2013 HERS False Alarm rate as a function of FIR threshold

Misdiagnosis

Calculation of *Misdiagnosis* rate

As noted in the evaluation outcomes definitions, a Misdiagnosis is a test case in which the following criteria are true:

• The RCA flags a fault

• The experimenter identifies the system as having a fault (or in the units in which we have determined that the maximum COP at the standard 95/80/67 condition occurs at a charge level different from what the experimenter considered to be 100% charged, we have redefined the "correct" charge to coincide with the charge level at maximum COP. When these systems have a charge level different from this "correct" charge, they are classified as having a charge fault)

• The RCA-flagged fault is not the same as the experimenter-identified fault

The rate is calculated as the number of *Misdiagnoses* divided by the number of tests for which the first two criteria are true. The *Misdiagnosis* rate addresses the question: if the protocol is applied to a system operating with a fault, how often will it diagnose a different fault?

For evaluation of *Misdiagnosis* rates, we group results into five FIR bins: <75%, 75-85%, 85-95%, 95-105%,

and >105%. The *Misdiagnosis* rates for each FIR bin are shown as bars in Figure 39-13 to Figure 39-20 for each of the four different RCA versions. The number of responses (meeting the first two criteria listed above) is shown in the base of each bar. For example, in Figure 39-14, the bin for $FIR_{capacity}$ from 95-105% shows that there are 118 cases in which the RCA flags a fault and the experimenter has indicated the presence of a fault. Of these, 83 were correctly diagnosed and 35 were misdiagnosed, which gives 30%. In the bin for $FIR_{capacity}$ greater than 105% there were four cases, all of which were correctly diagnosed.

Results and Discussion

In Figure 39-13 the Misdiagnosis rate for FIRCOP in the 75-85% bin is very high: 65%. This bin contains six cases of condenser airflow and six cases of compressor valve leakage faults, all of which the RCA diagnoses as overcharged, contributing to this unusually high rate. However, the overall rate for all data is 26%, which is also quite high.

The relaxed criteria for the HERS version of the 2008 protocol means that less cases are flagged as faults—238 versus 266 for the installer protocol. The results, shown in Figure 39-14, are otherwise quite similar to the Installer version results, shown in Figure 39-13. The overall Misdiagnosis rate here is 25%.

In Figure 39-19 the RCA again uses the Installer tolerances, but differs from the evaluation in Figure 39-11

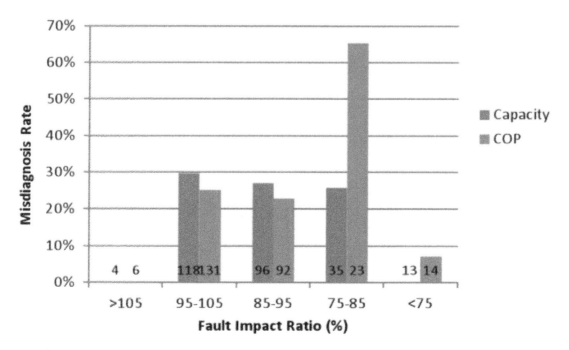

Figure 39-13: RCA 2008 Installer Misdiagnosis rates as a function of Fault Impact Ratio (FIR)

in that there is no longer any airflow diagnostic. Also, all cases with evaporator airflow below 300 CFM/nominal ton are removed from the inputs. The overall Misdiagnosis rate here is 32%. This is slightly higher than the 2008 Installer version's 26% (Figure 39-15) despite the removal of the airflow diagnostic and the removal of input cases. This suggests that the *Misdiagnosis* rate for the airflow diagnostic may be lower than that of the charge diagnostic for the full range of fault types.

In Figure 39-16 the results slightly better than the results in Figure 39-15, suggesting that the looser tolerance provides a small improvement to the Misdiagnosis rate. The overall rate for the 2013 HERS version of the RCA is 29%. The spike in the 75-85% FIR_{COP} bin remains, but is reduced to 60% compared with 64 or 65% for the other versions of the protocol.

In terms of *Misdiagnoses* the RCA performs quite poorly for each of the four versions. The aggregated

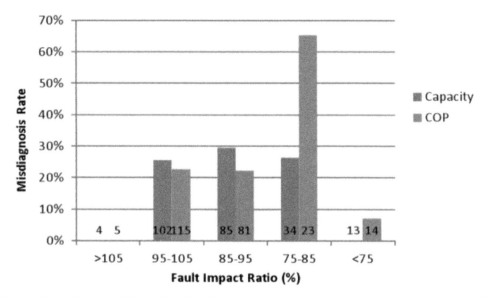

Figure 39-14: RCA 2013 Installer *Misdiagnosis* rate as a function of Fault Impact Ratio (FIR)

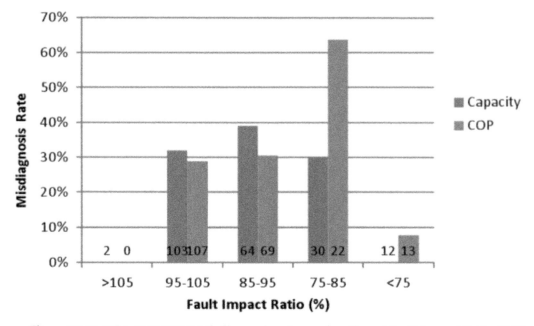

Figure 39-15: RCA 2013 HERS Misdiagnosis rate as a function of Fault Impact Ratio (FIR)

Misdiagnosis rates are summarized in Table 39-11. For the all versions, more than 1/4 of the times it's applied to a system with a fault, it reports that a different fault is present. This will result in a maintenance or installation technician performing the wrong corrective action, which may make the performance worse or at the very least cause the technician to repeat the diagnosis, then apply different corrective actions.

Table 39-9: Summary of aggregated Misdiagnosis rates for the four RCA versions

2008		2013	
Installer	HERS	Installer	HERS
26%	25%	32%	29%

Missed Detections

Calculation of *Missed Detection* rate

Missed Detections are cases in which two criteria are met:

- The experimenter identifies a fault (or a charge fault has been determined based on comparison with the charge giving maximum COP, as discussed in the section "Charge effect on COP and capacity").

- The RCA response is "No Fault"

The test cases are grouped by Fault Impact Ratio into the same bins as *Misdiagnosis*, as described above. The *Missed Detection* rate is calculated by dividing the number of *Missed Detections* by the total number of tests in which the experimenter identifies a fault. It addresses the question: when the protocol is applied to a system with a fault, how often does it miss the fault and report that the system is operating properly?

The *Missed Detection* rates in Figure 39-17 are 30 to 40% for the low impact faults—those with FIR above 95%. These faults are of lesser importance than the higher impact faults, so the high rate is not very alarming. However, the 85-95% FIR bin still has this high rate of *Missed Detections*, and represents a significant missed potential for energy savings. The overall *Missed Detection* rate is 32%.

Comparing the *Missed Detection* rates in Figure 39-18 and with the rates in Figure 39-17 and shows that the looser tolerances of the HERS protocol cause many more faults to be missed, raising the rate in most FIR categories by 5 to 10%. This is a tradeoff with the improved False *Alarm* rate that is associated with looser tolerances. The overall rate of *Missed Detection* for the RCA 2008 HERS protocol is 39%.

In Figure 39-19 we see the effects of removing the airflow diagnostic, by comparing with Figure 39-17 since the 2008 and 2013 Installer protocols use the same tolerances for charge diagnostics. The overall performance is slightly worse for the 2013 protocol (even ignoring the result for FIR > 105%, which are not very meaningful). This could suggest that the airflow diagnostic in the 2008 version is less likely to miss a detection than the charge diagnostics. However, investigation showed that the airflow module was misdiagnosing faults, rather

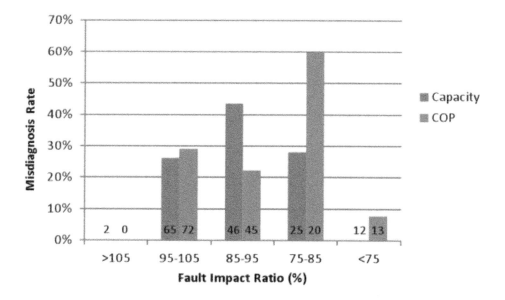

Figure 39-16: RCA 2013 HERS Misdiagnosis rate as a function of Fault Impact Ratio (FIR)

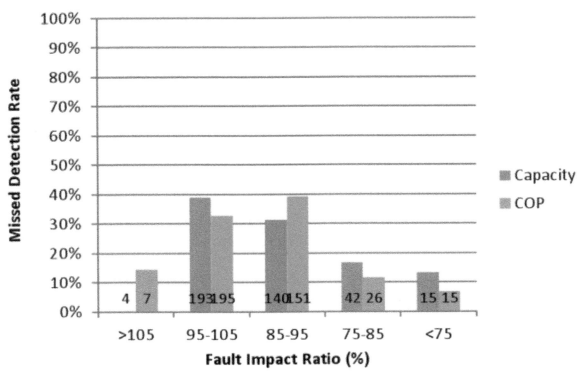

Figure 39-17: RCA 2008 Installer Missed Detection rate as a function of Fault Impact Ratio (FIR)

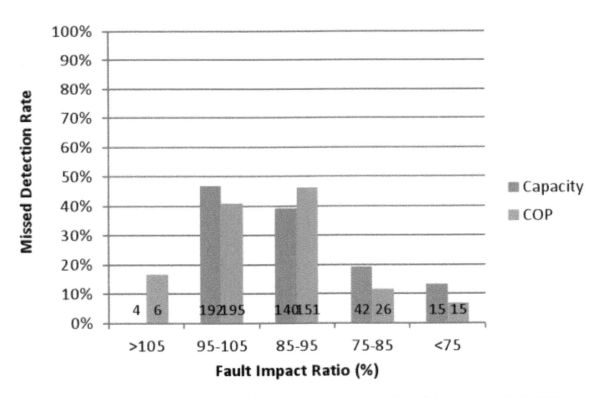

Figure 39-18: RCA 2008 HERS Missed Detection rate as a function of Fault Impact Ratio (FIR)

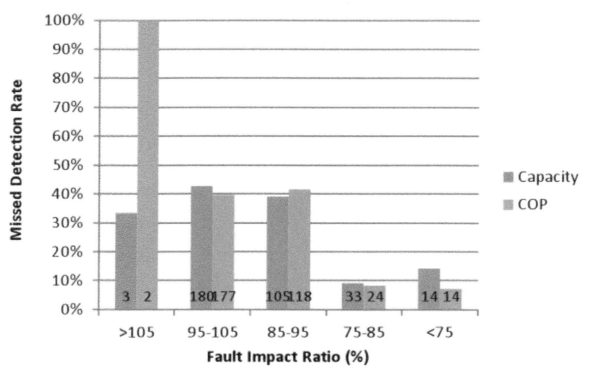

Figure 39-19: RCA 2013 Installer Missed Detection rate as a function of Fault Impact Ratio (FIR)

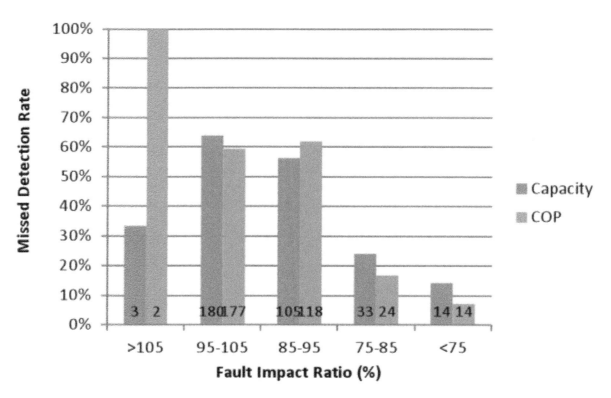

Figure 39-20: RCA 2013 HERS Missed Detection rate as a function of Fault Impact Ratio (FIR)

than missing detections, and this is why the rates are lower in Figure 39-20.

Finally, the plots showing the Missed Detection rates for the 2013 HERS version, which has the loosest tolerances, show quite poor performance. In Figure 39-20 almost two thirds of the faults that cause an 85-95% $FIR_{capacity}$ are missed.

Summarizing the *Missed Detection* evaluation, performance can be characterized as poor, with significant faults being missed in all versions of the protocol. A summary of the aggregated *Missed Detection* rates is given in Table 39-10.

Table 39-10: Summary of aggregated Missed Detection rates for the four RCA versions

2008		2013	
Installer	HERS	Installer	HERS
32%	39%	37%	55%

Conclusions of Case Study

The case study of the RCA protocol has demonstrated that the evaluation method developed in this project can effectively evaluate the strengths and weaknesses of an FDD protocol in a way that reflects the overall usefulness of the protocol to a user. This is done by using a fault-impact based analysis, and by providing results across a range of FIR. A user can select his own sensitivity to fault impact based on either capacity or COP, and read the results in each outcome from the charts. For example, if a user feels that all faults that reduce capacity by less than 10% should be tolerated, she can read the evaluation results from *False Alarm* rate, *Misdiagnosis* rate, and *Missed Detection* rate at those values, and have a reasonable idea of how the protocol will perform. If these results are available for several different protocols they can be compared so that the best protocol for that user can be selected.

One issue that hasn't been discussed is the distribution of faults by type and intensity. The performance of the RCA, and of any protocol, will vary by fault type and by intensity. This means that the results of the evaluation are dependent upon the distribution of faults in the data library. Ultimately, it would be best if this distribution matched the likely distribution of faults that occur in the field. However, this distribution is largely unknown. Even if the distribution were known with some confidence, it would be difficult to get data from the data library to match the expected distribution

in the field because it is a finite set. In the next stage of development of FDD evaluation methods, simulation data will be used, which will allow the fault distribution to be controlled. At that time any fault distribution data that become available can be incorporated into the evaluation.

The fault distribution of the current data library is shown in Table 39-9. This distribution has a heavy concentration of charge faults, with more than 1/3 of all tests, and 44% of faulted tests having charge faults. This distribution may skew some results in favor of the RCA protocol, since charge is a fault that it is intended to diagnose.

Table 39-11: Distribution of fault types in data library as a percentage of all tests

Fault Type	% of library
No Fault	18%
Undercharge	27%
Overcharge	9%
Evaporator Airflow	17%
Condenser Airflow	11%
Liquid Line Restriction	8%
Non-condensables	2%
Valve Leakage	8%

Besides being an example for application of the evaluation method developed in this project, the case study provides evaluation results for the RCA protocol. The protocol's main strength may be that it is simple to apply, not requiring any difficult computations that require a computer. The protocol's weaknesses are many. Although there is no yardstick for FDD performance, it seems safe to conclude that the results are not good. The current (2013) protocol's *False Alarm* rates of 40% or higher for all FIR thresholds below 97.5% are very troubling. It seems unlikely that the net benefit of this protocol could be positive, when it has such a high *False Alarm* rate.

It is tempting to consider using looser tolerances to improve this *False Alarm* rate, but even if we apply the loosest tolerances (from the 2013 HERS version) we get *False Alarm* rates above 15% for all thresholds. (This is an imperfect comparison because the 2013 HERS version also has no airflow diagnostic). This rate, 15%, is still unacceptable. Furthermore, the tradeoff with looser tolerances is that we will have increases in the *Missed Detection* rate. Again using the 2013 HERS tolerances, we have *Missed Detection* rates in that are over 70% in the 85-95% FIR$_{COP}$ bin. Even if we only consider charge faults, in this same bin we see that the RCA misses more than half of the faults. One needs to question whether the cost of having 15% *False Alarm* rate is worth the benefit of catching half of the charge faults.

Given this poor performance, it seems that the potential for salvaging the RCA protocol is not very good, and that consideration should be given to replacing it or removing it from the standard.

CONCLUSIONS

An FDD protocol evaluation method for protocols applied to unitary air-conditioning systems at steady-state operation has been developed and described in this chapter. This includes a database of measurement data from laboratory measurements of systems with and without imposed faults.

Several concepts related to the evaluation of FDD have been defined in this chapter, such as Fault Impact Ratio, Fault Intensity, False Alarms, Misdiagnoses, etc. The evaluation method developed is based on a fault's impact on the performance of a unitary system, and provides outputs over a range of impacts that could be of interest to a potential FDD protocol user. A system has been developed to categorize and enumerate the outcomes of an evaluation.

A case study was conducted in which the RCA protocol was evaluated. It was found to perform poorly, with unacceptable levels of *False Alarms*, regardless of the threshold used to differentiate faulted from unfaulted performance.

Some of the main conclusions of the project:

- The evaluation method developed here should be applied to FDD protocols of interest to determine whether their performance is acceptable to potential users

- Model data should be used instead of measurement data in future evaluations, to:
  — control the distribution of faults
  — remove the uncertainty from experimental error
  — widen the field of potential systems, faults, fault intensities, fault combinations, and driving conditions
  — provide datasets that can't be learned, hence

gamed, by unscrupulous developers

- More understanding of the likely distribution of faults in the field is required

- The RCA protocol performs quite poorly, and consideration should be given to removing it from the standard

Moving forward, we are hopeful that this tool can provide a path to improved performance of FDD. It can do this by illuminating poorly performing FDD, to spur further developments and improvement, and by providing a tool for developers to use as they explore improvements in their protocols. It is important to evaluate FDD, because poorly performing FDD is costly, in that it can cause improper corrective actions to be taken, or wasted maintenance service, and can hinder more widespread future adoption of FDD if users don't find FDD benefiting them. The potential for FDD to reduce energy consumption and peak power, and to improve equipment life is still largely untapped, and this potential should be pursued.

The work described here is ongoing. Some additional steps being carried out are:
(a) evaluation of other protocols, as a way of testing and refining our evaluation approach
(b) evaluations using simulation data
(c) assessment of simulation data as a replacement for experimental data
(d) development of simplified figures of merit for FDD protocols

DEFINITIONS

There are several terms that do not yet have consensus definitions, but which are necessary or convenient when discussing FDD for unitary equipment. The definitions below describe the meaning of some applicable terms as used within this discussion. The definitions for terms in italics are proposed as standard definitions.

Driving Conditions—the dry-bulb temperature of the air entering the condenser, and the dry-bulb temperature and humidity of the air entering the evaporator

Fault—an operating condition in a unitary air conditioner that causes degradations in performance. This may include degradations in efficiency, capacity, equipment life, maintenance costs, or ability to maintain comfort conditions.

Fault detection—determination that a fault is present in

the system

Fault Assessment—A quantification of the severity of the fault. This may be expressed as a fault intensity, a fault impact, or in broader terms, such as "low charge," "very low charge," etc.

Fault Diagnosis—Fault diagnosis consists of two processes: fault isolation and fault assessment

Fault Isolation—determination of the type of fault that is present or the component that is faulted.

Fault Impact—the effect caused by a fault on a variable of interest, such as capacity, EER, subcooling, supply air temperature, cost, thermal comfort, etc.

Fault Impact Ratio (**FIR**)—the ratio of COP or capacity under faulted conditions to COP or capacity under unfaulted operation at the same operating conditions.

Fault Intensity (**FI**)—the level of a fault expressed with reference to physical measurements

Protocol—the algorithm that generates outputs in a FDD tool

Return Air—in the context of this report, return air refers to the air entering the evaporator coil or in cases where the indoor fan is immediately upstream of the indoor coil, return air is the air entering the fan

Test Case—a set of input values, including pressures, temperatures, etc. for a system that is operating at steady-state

Threshold—with respect to Fault Impact Ratio, a threshold is the dividing point between faulted operation and unfaulted operation; the FIR above which a test case should be considered unfaulted

Unitary system—in the context of this report, a unitary system is an air-cooled direct-expansion vapor compression cycle air-conditioner with a single-speed compressor and single-speed fans.

NOMENCLATURE

adp	Apparatus dew point
BF	Bypass factor
CA	Condenser Airflow fault
CFM	Cubic feet per minute
COP	Coefficient of performance
EA	Evaporator Airflow fault
EER	Energy efficiency ratio
FDD	Fault detection and diagnostics
FI	Fault Intensity
FIR	Fault impact ratio
FXO	Fixed orifice expansion valve

HERS	Home Energy Rating System
IP	Inch-pound system of units
LL	Liquid Line fault
NC	Non-condensable gas in the refrigerant fault
OC	Overcharge fault
psia	Pounds per square inch, absolute
RA	Return air
RTU	Rooftop unit
SA	Supply air
SEER	Seasonal energy efficiency ratio
SI	International system of units
T	Dry bulb temperature
T_{amb}	Ambient air dry bulb temperature
TXV	Thermostatic expansion valve
UC	Undercharge fault
VL	Compressor Valve Leakage fault
WB_{RA}	Return air wet bulb temperature

References

AHRI. 2008. *AHRI Standard 210/240: Performance Rating of Unitary Air-Conditioning & Air-Source Heat Pump Equipment.* Air-conditioning, Heating, and Refrigeration Institute.

ASHRAE. 2009. *ANSI/ASHRAE Standard 37-2009: Methods of Testing for Rating Electrically Driven Unitary Air-Conditioning and Heat Pump Equipment.* American Society of Heating, Ventilating, and Air-conditioning Engineers, Inc.

Brandemuehl, M.J. 1993. *HVAC2 Toolkit: Algorithms and Subroutines for Secondary HVAC System Energy Calculations.* ASHRAE, Inc., Atlanta, GA: 402 p.

Bell, I. H., E.A. Groll and H. König. 2012. Experimental Analysis of the Effects of Particulate Fouling on Heat Exchanger and Air-side Pressure Drop for a Hybrid Dry Cooler," *Heat Transfer Engineering*, 32 (3): 264-271

Braun, J.E. 1989, *Methodologies for the Design and Control of Central Cooling Plants*, Ph.D. Thesis. University of Wisconsin-Madison, WI

Breuker, M.S. 1997. *Evaluation of a Statistical, Rule-based Detection and Diagnostics Method for Vapor Compression Air Conditioner*, MSME dissertation, Ray W. Herrick Laboratories, Purdue University, Ind. Report No. 1796-6 HL97-29

Breuker, M.S. and J.E. Braun, 1998. Evaluating the Performance of a Fault Detection and Diagnostic System for Vapor Compression Equipment. *HVAC&R Research* 4(4):401-425.

Breuker, M.S., T.M. Rossi, and J.E. Braun. 2000. Smart Maintenance for Rooftop Units. *ASHRAE Journal* 42 (11):41-46.

California Energy Commission (CEC), 2008. *2008 Building Energy Efficiency Standards for residential and nonresidential buildings.* CEC-400-2008-001-CM. Sacramento: California Energy Commission.

Cheung, H. and J.E. Braun, 2013a. Simulation of Fault Impacts for Vapor Compression Systems by Inverse Modeling Part I: Component Modeling and Validation, HVAC&R Research, Vol. 19, Issue 7.

Cheung, H. and J.E. Braun, 2013b. Simulation of Fault Impacts for Vapor Compression Systems by Inverse Modeling Part I: System Modeling and Validation, HVAC&R Research, Vol. 19, Issue 7.

Kim, M.; Payne, W.V.; Domanski, P.A.; Hermes, C.J.L. 2006. *Performance of a Residential Heat Pump Operating in the Cooling Mode With Single Faults Imposed.* NISTIR 7350; 173 p.

Li, H. and J.E. Braun. 2007. A Methodology for Diagnosing Multiple Simultaneous Faults in Vapor-Compression Air Conditioners. *HVAC&R Research*, 13(2): 369-395.

Palmiter, L., J-H. Kim, B. Larson, P.W. Francisco, EA. Groll and J.E.

Braun. 2011. Measured effect of airflow and refrigerant charge on the seasonal performance of an air-source heat pump using R-410A. *Energy and Buildings*, 43(7):1802–1810.

Reddy, T.A. 2007. Development and evaluation of a simple model-based automated fault detection and diagnosis (FDD) method suitable for process faults of large chillers. *ASHRAE Transactions* 113(2):27-39.

Rossi, T.M. 2004. Unitary Air Conditioner Field Performance. *Proceedings of the 8th International Refrigeration and Air Conditioning Conference at Purdue*, July 2004. Paper 666.

Rossi, T.M., and J.E. Braun. 1997. A Statistical, Rule-Based Fault Detection and Diagnostic Method for Vapor Compression Air Conditioners. *HVAC&R Research*, 3(1):19–37.

Roth, K.W., D. Westphalen, and J. Brodrick. 2006. Residential Central AC Fault Detection & Diagnostics. *ASHRAE Journal* 48(5):96-97.

Shen, B., E.A. Groll and J.E. Braun. 2006. *Improvement and Validation of Unitary Air Conditioner and Heat Pump Simulation Models for R-22 and HFC Alternatives at Off-Design Conditions*. Final Report for RP-1173, ASHRAE.

Vachtsevanos, G.J., F.L. Lewis, M.J. Roemer, A. Hess, and B. Wu. 2006. *Intelligent fault diagnosis and prognosis for engineering systems*. Hoboken, N.J.: Wiley.

Yang, L. Braun, J. E. and Groll, E. A. 2007. The Impact of fouling on the performance of filter-evaporator combinations," *International Journal of Refrigeration*, 30:489-498

Yuill, D.P. and J.E. Braun. 2013. Evaluating the performance of FDD protocols applied to air-cooled unitary air-conditioning equipment. *HVAC&R Research*, 19(5): 882-891.

Chapter 40

Self-correcting HVAC Controls:
They May be in Your Future
Sooner than You Think

Michael R. Brambley,
Pacific Northwest National Laboratory

INTRODUCTION

Automated fault detection and diagnostic (AFDD) tools are beginning to appear on the market embedded in heating, ventilating, air-conditioning (HVAC) equipment and lighting systems, as stand-alone tools, and as remote services provided to buildings. One family of AFDD tools from Field Diagnostics Inc. (FDSI 2014) has even been available commercially for use by HVAC service technicians for more than 15 years. AFDD capabilities have also been deployed within building automation systems (BASs) and as add-on software. Yet the market penetration of these tools is still relatively small. Even when AFDD or manual analyses indicate that faults with equipment condition or operations exist, often building staff or owners do not take corrective action to repair the faults or contact an outside service provider. AFDD tools detect faults, but field experience testing these tools shows that information, although essential, is not sufficient alone to change behavior in maintaining HVAC equipment and systems.

As shown by many studies and commissioning projects (e.g., Cowan 2004, Katipamula et al. 2003, Brambley et al. 2005, and Mills 2011), faults are endemic across the commercial buildings sector. Common faults include, among others, economizers improperly implemented or with stuck (not modulating) dampers, leaking valves, incorrect system operation schedules, and absence of or suboptimally-implemented supply-air temperature and static pressure reset. They can increase the energy consumption of commercial buildings by between 15% and 30% or even more compared to buildings without equipment condition and operation faults. These faults can be detected with labor intensive commissioning, analysis and review by expert teams providing operation-support services, or many can be found using AFDD tools. In all three cases, however, the excess energy use and associated costs are not eliminated unless actions are taken to remedy the faults. All too often, many are not.

That's where self-correcting controls could play a critical role. Any fault associated with an incorrect parameter such as a set point, a systematically erroneous sensor reading (e.g., a sensor bias from drifting), incorrectly implemented control code (e.g., for an economizer), and incorrect scheduling can, in principle, be corrected automatically (or self-healed). Many of the faults that can be automatically corrected are in software or parameters input for use in it; however, some physical faults can also be corrected. A simple example is a faulty sensor that has "drifted" and now provides incorrect readings with a bias of say +5°F. The sensor has physically changed, but its value can be corrected simply by subtracting 5°F from every value reported by it. This, however, first requires detecting that the sensor is giving incorrect readings, characterizing the nature of the errors in those readings (e.g., that the error is constant), and determining the sign and magnitude of the error. Only then can the bias fault in the reading from the sensor be corrected.

Some faults cannot be corrected this way. Many physically damaged devices cannot be corrected through changes in software. For example, when the bearings in a pump are physically damaged, either the bearings in the pump or the entire pump itself must be replaced. The fluid circulation system in which the pump is located, however, could continue to operate while the pump is damaged and even while the pump is being repaired or replaced, if the piping system includes one or more (redundant) parallel pumps that can serve the load while the faulty pump is repaired or replaced. The control system, upon detecting the faulty pump (using AFDD), could automatically control valves to shut down and isolate the faulty pump and transfer the pumping load to the other pumps in the circuit. Such a system would be tolerant to (some) physical faults, i.e., able to continue to operate even in the presence of physical faults, and be known as fault tolerant.

Fault tolerance can extend to controllers themselves, in which case, if a controller fails, its responsibilities can be transferred to one or more other controllers. Fault-tolerant controls were first developed in the aircraft field with development started in earnest in the late 1970s and early 1980s, first for military aircraft and later for civilian. A couple notable examples of fault-tolerance are the ability to land an aircraft safely after loss of a physical control surface such as a wing on one side of the aircraft and landing using only thrust control (e.g., after loss of all hydraulics) (Tucker 1999, Tomayko 2003).

Fault tolerance and self-correction are nearly non-existent commercially in the buildings domain. Certainly some fault tolerance is built into major equipment such as chillers, but only in the most basic ways. Its use is not widespread, although many faults could be addressed with these capabilities, helping keep our buildings operating much more efficiently, effectively, and at lower operating cost. If self-correction (or self-healing) capabilities could be developed and deployed for many common problems in commercial buildings, the excess inventory of deferred maintenance in many buildings could be decreased. The operation and maintenance teams of large commercial buildings could focus on problems that require human intervention, leaving amenable faults for correction or temporary compensation by the control system. Owners and managers of smaller commercial buildings without operation and maintenance teams would have more flexibility in making decisions regarding when to call in service companies and which problems to have them focus on. More degrees of freedom in maintenance management would be enabled for commercial buildings of all sizes by using this technology.

The remainder of this chapter describes a process that can be used to automatically correct faults using algorithms implemented in controls and provides examples of such capabilities that have been developed and demonstrated by a team at the U.S. Department of Energy Pacific Northwest National Laboratory (PNNL). Since 2002, the PNNL team has been developing self-correcting control algorithms for HVAC systems and equipment. The results show that this capability is not science fiction but rather is being realized and could

soon be used to improve the performance and efficiency of the buildings you own, operate or work in.

THE SELF-CORRECTION PROCESS

Self-correction and fault tolerance require redundancy, which can be physical or analytical. Physical redundancy requires duplicate equipment and components so that when one fails or degrades sufficiently, another or a group of others can replace the one that failed. Physical redundancy is sometimes used on larger commercial buildings. An example is two chilled-water pumps of like capacity installed in parallel with one another. One can meet the entire load by itself. When one requires servicing or replacement, it can be valved off and the other used during its repair. Although such switching is usually done manually, it can be done automatically. Analytic redundancy employs models of physical systems, the relationships among them and the sensors used to monitor their states and performance to correct or compensate for a faulty or failed component. A simple example is the use of other sensors to infer the value that should or would be measured by a faulty sensor. This virtual sensor reading can then be used in place of the faulty sensor temporarily until the physical sensor is replaced or in some cases permanently. The rest of this chapter focuses on using analytic redundancy to automatically correct or compensate for faults using control systems or self-correcting controls.

A generic mechanistic self-correction process is shown in Figure 40-1. This process uses automated fault detection during passive observation of an operating system to detect when a fault occurs in the system. After a fault is detected, automated fault diagnosis identifies the specific fault that occurred and its location in the system (i.e., it isolates the fault). Proactive testing can be used to speed and potentially improve the fault diagnosis compared to simply observing operation of the system. The tests can be used to create special situations (e.g., a damper completely closed) where the correct behavior of the system is known better or to create conditions sooner than they would occur during ordinary system operation. Data from the proactive tests to isolate faults can also be used to characterize the fault and additional tests can be run to provide additional data for characterizing faults as shown in the next process. Fault characterization results in a mathematical description of the fault. With that mathematical description, a correction is automatically created and implemented to correct or compensate for the fault.

Fault corrections can be simple subtractive elements or more complex corrections that can depend on time or values of multiple variables. Some types of faults for which deterministic mathematical characterizations cannot be developed are not correctable with such an approach. Consider, for example, an intermittent sensor fault that occurs at random time intervals; such a fault likely could not be corrected because a sufficiently accurate prediction of when the fault would occur could not be developed.

Variations in the process shown in Figure 40-1 can also be used successfully. The first step of detecting the occurrence of faults during passive observation (i.e., data collection) of the operating system might be eliminated. Instead, proactive tests might be triggered periodically (e.g., during times when the building is unoccupied) and the results used for both fault detection and diagnosis. Another potential deviation involves eliminating the explicit characterization of the fault. In this case, an iterative correction process can be used, where incremental corrections are applied until the fault is corrected sufficiently.

A diagram showing schematically how the self-correction process can be implemented is given in Figure 40-2. Critical to the implementation are the virtual sensor, the creation of the mathematical correction, passing the correction to the virtual sensor, and the signals to initiate proactive tests, which are sent to the controller. The virtual sensor is where corrections to sensor faults are implemented with the virtual sensor value, $xV(x)$, provided as an input to the controller in place of the faulty value of the sensor, x. When there is no sensor fault, $xV(x) = x$. The box at the bottom of the diagram represents the computational processes for fault detection, isolation, characterization and correction or variations of these, as described previously. Not shown explicitly, however important, is a virtual control point that can be created in the controller to correct for faults in the controller that are not associated with the control code or values of parameters in it.

EXAMPLES OF SELF-CORRECTION

A number of self-correction algorithms for air-handling units have been developed and tested at PNNL. Testing has been performed on a fully functional air handler installed in a laboratory. The air handler is well instrumented and controlled by a commercial building automation control system with special enhancements to support implementation and testing of self-correcting control processes, as shown in Figure 40-1 and Figure

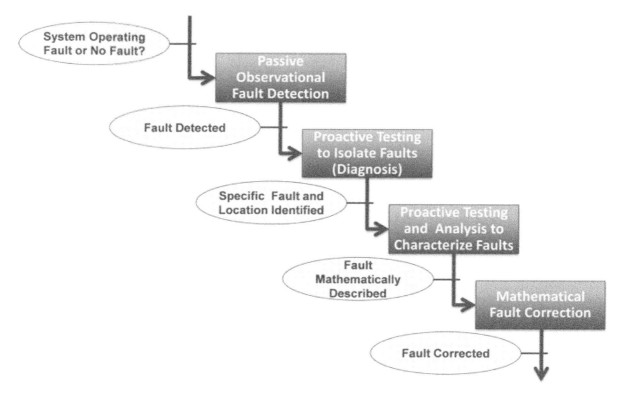

Figure 40-1. Self-correction process

40-2. Especially notable is the ability to implement faults of various types without damaging the physical equipment. Field tests have also been performed on air handlers in two occupied buildings. The representative examples presented in this section are taken from the laboratory tests. A diagram of the mixing section of an air handler is shown for reference in Figure 40-3.

In the first example, the recirculated-air (RA) temperature sensor has a -5°F bias error in the values it reports. Figure 40-4 shows the record of physical sensor measurements and virtual sensor values. The virtual RA temperature is used as an input for controlling the outdoor- and recirculated-air dampers. The measured RA temperature shown is the actual RA temperature from a sensor known not to have a fault. The fault is instigated by giving the virtual RA temperature sensor the -5°F bias relative to the actual RA temperature. The self-correcting algorithm is operating throughout the entire time period of the test. The sequence of events, the values indicated by all virtual and physical sensors, and the outdoor-air damper command, which is controlled by the self-correction algorithm during the proactive tests, are shown in Figure 40-4. Numbered vertical black log lines are used to identify the times of events.

At time zero, the virtual and the actual (measured) RA temperatures are approximately equal (the virtual sensor value is shown as slightly below the actual value for clarity). At 5 minutes into the test (event 1), the virtual RA temperature is given a constant -5°F bias, i.e., a -5°F fault. At event 2, 15 minutes into the test, a fault has been detected, and a proactive test automatically is initiated by moving the outdoor-air (OA) damper position from about 35% open to fully closed. At event 3, a second proactive test is initiated by the algorithm by moving the OA damper to 100% open. By the end of the second test at event 4, the fault has been isolated to the RA temperature sensors, and the OA damper is fully closed again for a final test. The results of the test determine that a correction of +5°F should be added to the faulty RA virtual sensor readings. The correction is implemented at event 5, correcting the faulty virtual RA sensor reading as indicated by the approximate equality of the measured (actual) and the virtual RA temperatures after event 5.

Although this first example illustrates correction of a constant (bias) fault in a temperature sensor, the algorithm can be used to correct drifting sensors by periodically or when a bias exceeds a selected threshold having this algorithm trigger to automatically correct the fault.

The second example (illustrated in Figure 40-5) demonstrates an algorithm for automatically correcting an incorrect minimum outdoor-air damper position.

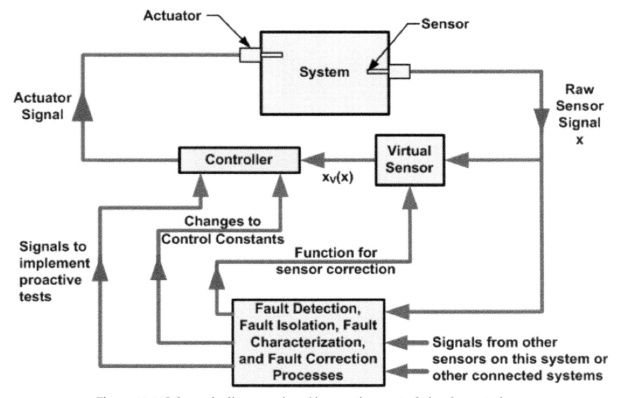

Figure 40-2. Schematic diagram of a self-correcting controls implementation

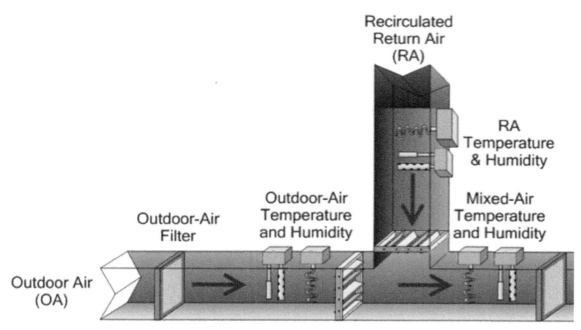

Figure 40-3. Schematic diagram of a generic air-handler mixing box

This position corresponds to the outdoor-air damper set to provide the airflow rate required to meet ventilation requirements; during other times when air-side economizing is used, the damper will ordinarily be open fully. The position is often set visually under the assumptions that the airflow rate is directly proportional to the percentage of open area (which generally is not true) and that the open area can be judged reasonably accurately by visual inspection. The outdoor-air damper position is controlled in this example based on the fraction of outdoor air (OAF) in the mixed air.

An iterative solution approach is used to determine and implement a suitable reset of the damper position, and the process involves detecting a fault in the minimum damper position and then seeking a solution using iterative repositioning of the damper based on the half-interval method. The desired OAF in the test is 30%. The initial OAF determined based on measured temperatures (measured OAF in the figure) is between 8% and 10% when the damper command is set at 20%, indicating a fault. At event 1 (1 minute after start of the test), a process is initiated to empirically determine the limits of OAF when the damper control command is at 0 and 100%. Therefore, at event 1, the algorithm sets the damper command to 0, and the value of the OAF is allowed to reach steady state at about 6%. The damper command is then set at 100% at event 2 and allowed to reach steady state at 85%. The damper command is then set at 60% (the midpoint between 20% and 100%), and

the OAF is allowed to reach steady state, which corresponds approximately to OAF = 40%. The desired OAF is lower than 40%; therefore, the damper command is then set half way between 60% and 20% at 30%, and the OAF is allowed to reach steady state. The process is continued until the OAF is sufficiently close to the desired value of 30% (e.g., within ±2.5%). The corresponding value for the damper command is 35%, which is the value then set for ventilation-only operation. The entire process is automatically executed by the algorithm with no human operator intervention.

The self-correction algorithms for the two examples presented in this chapter, additional test results and algorithms for additional faults can be found in Fernandez et al. (2009a, 2009b) and Brambley et al. (2011).

CONCLUSION

Automated self-correcting controls for an initial set of HVAC systems, equipment and devices have been shown to be achievable. A number of algorithms for this purpose have been developed and are publically available. Extensive additional work is still needed to extend the coverage of self-correcting algorithms to the plethora of equipment and systems in buildings and the faults they experience.

Self-correcting controls technology promises to help ensure that building HVAC and other systems can

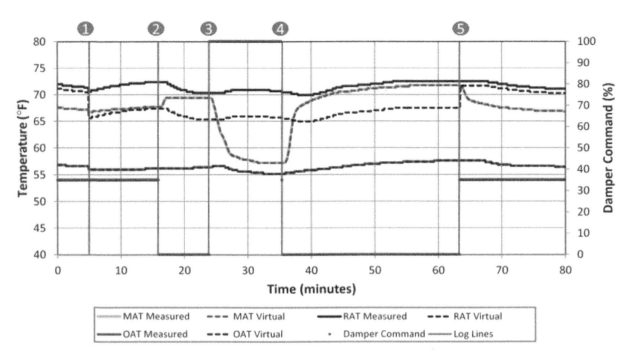

Figure 40-4. Sequence of events and corresponding time series records of temperatures and the outdoor-air damper command for a test of an algorithm to automatically correct bias errors in the return-air temperature sensor.

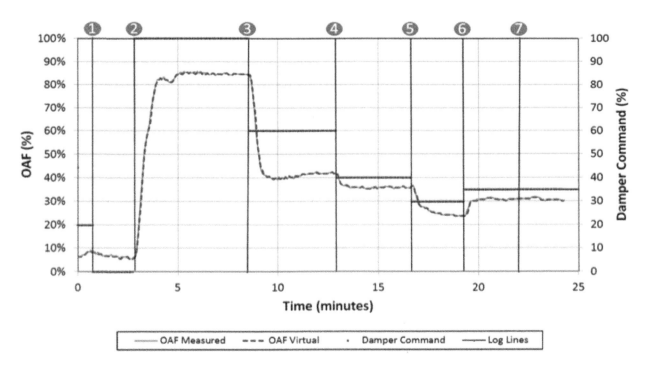

Figure 40-5. Sequence of events and corresponding time series records of outdoor-air damper commands and OAF for a test of the algorithm to automatically correct an incorrectly set minimum outdoor-air damper position for ventilation only.

continue operating efficiently and effectively even as faults occur, meeting occupant needs and minimizing the cost of operation for building owners. Furthermore by correcting many faults automatically, operation, maintenance and service staff will be able to focus their efforts more on faults that require human involvement for physical repair and replacement and allow control systems to fix or compensate for faults amenable to self- correction. Corrections implemented automatically could be reported to building operators, enabling them to monitor the performance of the affected systems and build confidence in the technology. Many problems that arise, limiting the efficiency of building equipment, increasing energy consumption and costs, and negatively affecting the comfort of occupants, that go unaddressed for months or even years could be corrected automatically.

Automated self-correction technology is feasible today and is on the path to a building near you.

Acknowledgements

This chapter would not have been possible without the support of projects on self-correcting building controls over the last decade by: the U.S. Department of Energy Building Technologies Office, Bonneville Power Administration, the California Institute of Energy and Environment (CIEE), and the Air-Conditioning and Refrigeration Technology Institute of the Air-Conditioning and Refrigeration Institute (ARI) through Portland Energy Conservation Inc. (PECI). The author gratefully acknowledges their support; the developments reported in this chapter would not have been possible without it. The author also gratefully acknowledges contributions in all aspects of the self-correcting building controls work by my colleagues at PNNL: Heejin Cho (now at the University of Alabama), Katie Cort, Liem Dihn (student intern), Darrell Hatley, Nick Fernandez, James Goddard, Krishnan Gowri, Srinivas Katipamula, Shun Li (now with Siemens Building Technologies Ltd., China), Guopeng Liu, Andrey Liyu, Hung Ngo, Ron Underhill, and Anne Wagner.

References

Brambley, M.R., N. Fernandez, W. Wang, K.A. Cort, H. Cho, H. Ngo and J.K. Goddard. 2011. *Final Project Report: Self-Correcting Controls for VAV System Faults: Filter/Fan/Coil and VAV Box Sections.* PNNL-20452, Pacific Northwest National Laboratory, Richland, WA.

Brambley M. R., P. Haves, S. C. McDonald, P. Torcellini, D. G. Hansen, D. Holmberg, and K. Roth. 2005. *Advanced Sensors and Controls for Building Applications: Market Assessment and Potential R&D Pathways*, PNNL-15149, Pacific Northwest National Laboratory, Richland, WA.

Cowan, Alan. 2004. *Review of Recent Commercial Roof Top Unit Field Studies in The Pacific Northwest and California.* New Buildings Institute, White Salmon, WA.

Fernandez, N., M.R. Brambley and S. Katipamula. 2009a. *Self-Correcting HVAC Controls: Algorithms for Sensors and Dampers in Air-Handling Units*, PNNL-19104, Pacific Northwest National Laboratory, Richland, WA.

Fernandez, N., M.R. Brambley, S. Katipamula, H. Cho, J. Goddard and L. Dinh. 2009b. *Self-Correcting HVAC Controls Project Final Report*, PNNL-19074, Pacific Northwest National Laboratory, Richland, WA.

Field Diagnostic Services, Inc. (FDSI). 2014. *Grow your bottom line with the HVAC Service Assistant™ from Field Diagnostics*, Field Diagnostics Services, Inc., Langhorne, PA; see also https://www.fielddiagnostics.com/products/hvac-service-assistant, last accessed on March 22, 2014.

Katipamula, S., M.R. Brambley, N.N. Bauman and R.G. Pratt. 2003. Enhancing Building Operations through Automated Diagnostics: Field Test Results, ESL-IC-03-10-16, *Proceedings of the 2003 ICEBO - International Conference for Enhanced Building Operations*, Energy Systems Laboratory, Texas A&M University, College Station, TX. Available at http://repository.tamu.edu/bitstream/handle/1969.1/5209/ESL-IC-03-10-16.pdf?sequence=4; last accessed on March 22, 2014.

Mills, Evan. 2011. Building commissioning: a golden opportunity for reducing energy costs and greenhouse gas emissions in the United States. *Energy Efficiency* 4:145–173.

Tomayko, J.E. 2003. *The Story of Self-Repairing Flight Control Systems.* Dryden Historical Study No. 1 National Aeronautics and Space Administration, Washington, DC.

Tucker, T. 1999. *Touchdown: The Development of Propulsion Controlled Aircraft at NASA Dryden*, Monographs in Aerospace History, Number 16, NASA History Office, Office of Policy and Plans, Washington, DC.

Chapter 41

A Framework to Apply a Structural Pattern Recognition Technique in Automated Fault Detection and Diagnostics of HVAC System and Component Faults

Xiaohui Zhou, Ph.D., P.E., Iowa Energy Center

INTRODUCTION

In commercial buildings, heating, ventilating, and air-conditioning (HVAC) systems often perform below design expectations due to lack of commissioning, component failures, undetected faults, software problems, and human errors. These problems can cause significant waste of energy and increase operation and maintenance costs. Over the past two decades, many automated fault detection and diagnosis (AFDD) methods have been proposed and developed for various types of HVAC systems and their components in academic, industrial, and government laboratory research. In 2005, a group of field experts also assessed the status and needs of AFDD and recommended research focus areas in a U.S. Department of Energy (DOE) report [1]. Katipamula and Brambley also published two review articles summarizing and categorizing these methods and their development [3] [4].

Progress has been made in removing barriers to applying many of these methods and algorithms to the real world since 2005: the emergence of virtual sensor technology, the dramatic drop in the unit price of computer storage in the last 5~7 years, and the emergence of technologies such as cloud computing, all have made it possible to collect and store massive amounts of building system data in the control system server or database. This presents a great opportunity to use advanced algorithms or methods to mine these data for useful information, optimal operations and control, as well as fault detection and diagnostics to improve energy efficient building operations and management. In the meantime, standards for evaluating and quantifying

AFDD methods are being developed, providing potential benchmarks with which to compare different AFDD methods and quantify potential energy savings from deploying AFDD systems. A software tool that is capable of reading and retrieving HVAC data directly from multiple building automation systems and converting these data to a universal format and then processing them with "add-on" third-party AFDD modules in real-time is emerging. All these technology advancements make it much easier to deploy AFDD methods at the system level (rather than at the field panel or controller level). There are now real AFDD applications deployed on large campus-type facilities. For example, the energy management system at the Microsoft campus in Redmond, Washington manages about 125 buildings totaling roughly 15 million square feet and collects HVAC system and equipment operational data from the campus' more than 20,000 VAV boxes, 150 chillers, and other HVAC equipment. The recently deployed "smart solution"—an "add-on" system that uploads all HVAC data from multiple existing building automation systems to a centralized "analytical layer," and then uses rule-based, customized AFDD algorithms to detect various faults. The "smart solution" can even assign each fault an economic value when notifying facility engineers so they can prioritize the maintenance work needed in order to minimize overall energy waste and cost. It's estimated that this fault detection program can save $1.5 million in energy costs for fiscal year 2013 for Microsoft.

However, the commercial AFDD applications or products are still not widely available and/or used in real-world environments. One of the major reasons is the limited universal applicability and usability of these algorithms, which are developed for application to real building systems with so many possible different system designs and configurations, as well as noise in data collected. From quantitative and qualitative model-based diagnostic methods, to process history-based diagnostic methods, almost all algorithms or methods developed so far need some kind of tuning, modeling, or fitting parameters into actual components and systems in the field. These algorithms or methods all have their own pros and cons when used in different applications. This adds complexity when implementing the AFDD applications and often requires a person with special knowledge to do it. Another reason is that these AFDD algorithms in many cases can detect faults but have limited capability to pin-point the cause of the faults, and this decreases its commercial value. The high cost-to-benefit ratio for developing, deploying, and maintaining AFDD technologies seems to prevent its wide adoption by building facility managers and operators.

In this chapter, a framework of using structural pattern recognition technique on automated fault detection and diagnostics of HVAC systems and components is proposed. Pros and cons of this approach are discussed, and some sample chart patterns produced during a recent fan coil unit AFDD experiment are shown.

HOW FACILITY ENGINEERS DETECT AND DIAGNOSE FAULTS

Many experienced facility engineers and commissioning agents detect and diagnose HVAC system or component problems by analyzing graphical patterns of certain trended points or studying the correlations among several related points. Look at the two charts below (Figures 41-1 and 41-2) showing the patterns of several control points recorded from a unit ventilator on a summer day from 00:00 to 23:59.

It can be easily seen that something is wrong during the day: a) the unit ventilator's discharge air (UV-DAT) and the cooling valve (UV-CVLV) both oscillated and appeared in "peaks" until around the 8:00am; b) comparing with the room cooling setpoint (RM-CLG-SPT), the room temperature (RM-TEMP) was out of control during most of the day, and the unit ventilator return air temperature (UV-RAT) followed a similar graphic pattern; c) the cooling valve maxed out at 100% during the day and then dropped, forming a near trapezoid or rectangular shape. Based on a), an experienced engineer could guess there might be a problem with cooling valve control tuning or oversizing of the valve, or some kind of system error may cause the chilled water flow to be available only intermittently. Based on b)

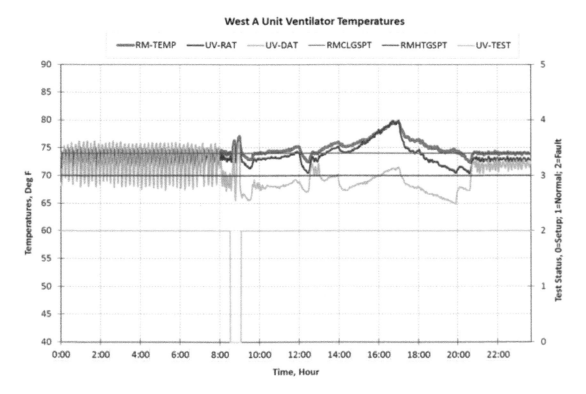

Figure 41-1. An Unit Ventilator Temperature Patterns

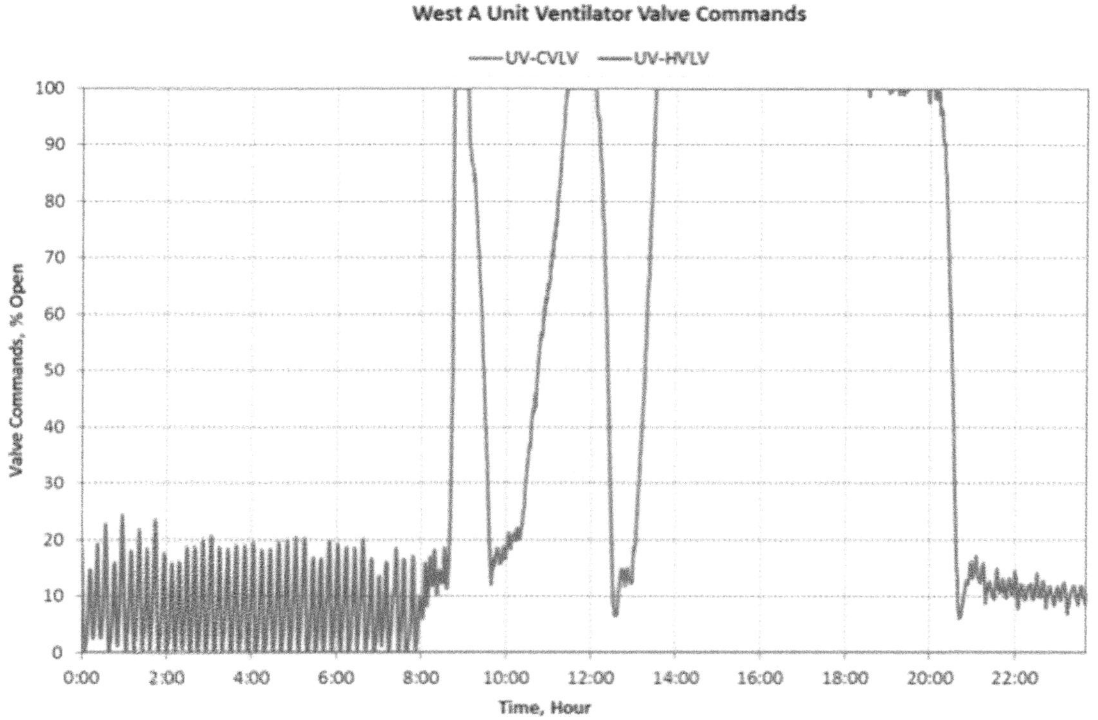

Figure 41-2. An Unit Ventilator Valve Command Patterns

and c), the cooling coil may be under-sized, or there may not have been enough chilled water flow, but control tuning seems ok because after around 9:00pm the room temperature can be controlled stably and the cooling valve commands were relative stable and did not oscillate that much. The most probable fault was that the chilled water system flow may be obstructed or blocked somewhere.

The fault actually was instigated in a recent research project on unit ventilators and fan coil units. For this day, the fault instigated was "restricted water flow"—the manual valves on the chilled water return pipe were closed down from their original balanced positions. Before 8:30am, the maximum flow (when the cooling valve is 100% open) was restricted to 50% of the designed flow rate, and after that time, the maximum flow was restricted to 25% of the design flow rate. The point "UV-TEST" in Figure 41-1 was used to signal the test status—when the test cases were switched among "Normal," "Setup" and "Fault," and how long the setup time was for that day.

Now look at the following two charts (Figures 41-3 and 41-4) that show patterns from another unit ventilator in a different room (East A). Though not exactly the same, they show very similar graphical patterns as those described above for the "West A" unit ventilator.

Now look at yet another set of charts in Figures 41-5 and 41-6. Here FCU-MAT represents the fan coil unit mixing air temperature; FCU-DAT is the unit discharge air temperature; FCU-TEST indicates the test setup status; and FCU-CVLV is the unit cooling valve command.

The two charts above (Figures 41-5 and 41-6) were from a room controlled by a different type of terminal unit—fan coil unit made by a different manufacturer. Again, they show very similar graphical patterns, even though the characteristic parameters of both types of units (equipment dimensions, configurations, heating coil construction, capability, etc.) may differ.

Without needing to know any equipment parameters, an experienced engineer can quickly identify potential problems/faults with this equipment or its control using graphical/structural pattern recognition. The reason is that they have seen similar graphical patterns many times before and have learned from past failures and problems and understand what these patterns mean when they happen again. This approach is more intuitive, straightforward and easier to understand than many existing AFDD methods that use quantitative and qualitative model-based diagnostic methods. It will be worth trying to use artificial intelligence to simulate the human visual pattern recognition capability in identifying graphical patterns and then detecting faults. So far in the field of HVAC AFDD, some process history-based diagnostic methods have been developed using a statistical pattern recognition approach, neuron-network approach, etc. but the author found no study using structural pattern recognition techniques, even though it has been used in fields like process control, speech recognition, electrocardiogram diagnosis, and other detection and identification applications [5] [6] [7] [10] [13].

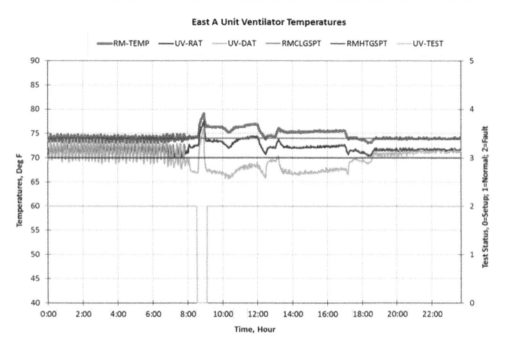

Figure 41-3. Room "East-A" Unit Ventilator's Temperature Patterns

STRUCTURAL PATTERN RECOGNITION TECHNIQUES IN HVAC APPLICATIONS

Structure Pattern Recognition and How it Can be Used in HVAC AFDD

In time series data applications, while statistical pattern recognition is based on the "quantitative features" of the data, the structural approach is based on the arrangement of "morphological" (shape-based) events evident in the underlying waveform—e.g., speech recognition, electrocardiogram diagnosis, seismic activity identification, radar signal detection, and process control. Each pattern is expressed in terms of a composition of its component sub-patterns—or primitives. The primitives are the smallest units or basic patterns in a structural pattern recognition problem, and one of the keys to a successful solution is to select the primitives properly for a specific application.

Any HVAC data series can be illustrated by its graphical representation with time as the x-axis and its

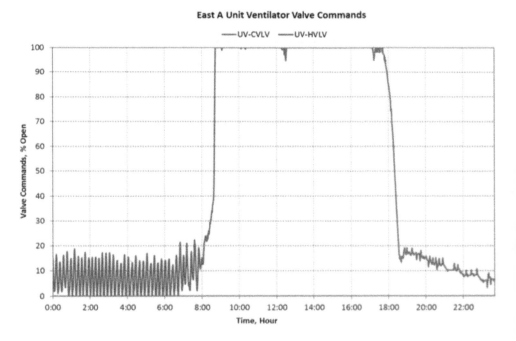

Figure 41-4. Room "East-A" Unit Ventilator Valve Command Patterns

analog value as the y-axis—these are the "charts" we use every day in monitoring and analyzing HVAC systems and component performance. Any of these charts can be considered consisting of a combination of many sub-patterns or primitives. These combinations form a "signature" of a chart. For a specific process variable,

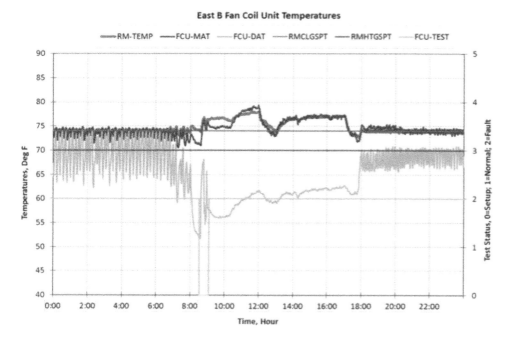

Figure 41-5. Room "East-B" Fan Coil Unit Temperature Patterns

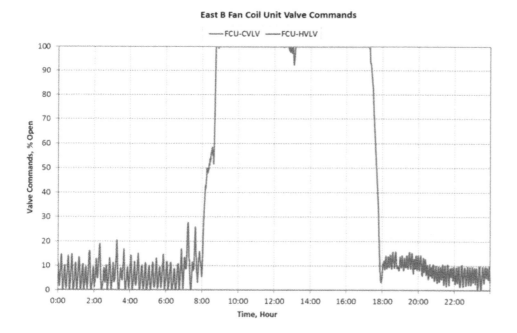

Figure 41-6. Room "East-B" Fan Coil Unit Valve Command Patterns

The overall process of using a structural pattern recognition technique to automatically detect HAVC system and component fault is illustrated in Figure 41-7.

After the data collection process that usually results in time-series data, feature description describes the structural characteristics of the data series. In structural pattern recognition, a set of primitives should be first selected, and then a coding scheme designed for best describing the set of primitives using numeric or alphabetical values. Feature extraction uses algorithms to detect and extract the primitives from the data series and assign values based on the designed coding scheme. The primitive combinations form the feature for the data series. Feature classification assigns the set of features to different groups (e.g., normal, faults, etc.), and finally fault diagnostics tries to identify detailed faults and causes. In the next several sections, each step is discussed in detail.

Feature Description

Proposed Basic Patterns for HVAC AFDD Applications

The first step in feature description is to select a set of basic patterns, or primitives. Good primitive selection should not only catch the essence of the internal struc-

normal operation can be described as a signature or a group of signatures, while a faulty case's signature in the data series may be significantly different than those of normal operation. By differentiating the signatures of normal and faulty data series, faults can be detected. By analyzing a group of related process variables' signatures, detection of specific faults may be achieved.

tural relationship of the pattern for that specific application, but also make it easier to interpret and classify for the later classification phase. The selection of primitives is, however, often highly domain and application dependent [8] [9]. Simple primitives are domain and application independent, but they capture very little relational information and often cause difficulty and complexity

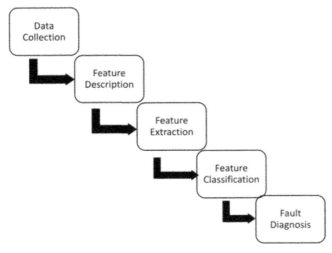

**Figure 41-7. Steps in Using
Structural Pattern Recognition for HVAC AFDD**

in the later classification phase. On the other hand, good or proper selection of primitives for a domain-specific application requires the involvement of a domain expert with deep understanding of the internal relationships among data patterns.

In the past 10+ years, many lab and field tests of HVAC AFDD algorithms and methods have been conducted at the Iowa Energy Center Energy Resource Station facility [2]. Data from these tests with normal operations and instigated HVAC system and component faults were collected and analyzed. From the experiences and observations from these tests, it is proposed that the following "shapes" be used as the primitives for HVAC AFDD algorithms when using structural pattern recognition techniques:

1) Linear;
2) Exponential;
3) Step;
4) Peaks;
5) Sinusoid;
6) Outdoor air temperature.

Linear Pattern

The linear pattern shows no change (constant) or increase/decrease at certain steady rate for process variables recorded in time-series format. The variables could be any measured HVAC process point like temperature, flow rate, pressure, or commanded points like an analog valve or damper position command. This pattern normally does not indicate any errors or faults, unless the absolute values are at maximum or minimum positions (e.g., 100% damper open position) or are way off their setpoints (e.g. air handling unit supply air temperature far exceeds or is below its setpoint).

Exponential Pattern

The exponential pattern is also very common in time-series data collected from various HVAC process variables (e.g., AHU supply air temperature) during the transition periods. Any step change in control setpoint could cause the control variable to exhibit an near exponential pattern during the transition period to a new steady state. The pattern could appear in normal operations or could be caused by faults. For example, in normal operation, an air handling unit supply air temperature could change exponentially with a temperature setpoint change. In stuck damper/valve situations, the air/water temperature(s) after the damper/valve could exhibit no change (linear pattern), while an exponential pattern may be expected when the damper/valve is used to control some process variables to maintain their setpoints.

Step Pattern

The step (up or down) pattern describes sudden changes or disturbances occurring in the system. This pattern usually gives a warning sign that something in the HVAC system is not working properly, since usually the HVAC control process is relatively slow, stable, and smooth (except for the duct pressure control). The exception is when the HVAC system control mode changes (e.g. economizer enable/disable).

Peak Pattern

The peak pattern consists of a sharp increase and then a sharp decrease (and vice versa) in a process variable or command signal in a very short period of time, and it may repeat many times. This pattern normally happens when some kind of binary event or problem occurs like frequent start/stop of fans or pumps, or unstable control causing a damper or valve to oscillate rapidly. Some sudden disturbance by internal/external load changes or solar/weather condition could also produce this sort of pattern in some of the process variables or command signals.

Sinusoid Pattern

Many process variables would show a continuous sinusoidal pattern when controls are oscillating. To determine the frequency of oscillation, the time series data in the time domain can often be transformed via a Fourier transform to frequency domain. This is very useful in detecting faults and finding correlations with other process variable patterns.

Outdoor Air Temperature Pattern

The outdoor air temperature greatly affects HVAC system performance in buildings. Many process variable patterns have high correlations with the weather condition, especially the outdoor air temperature, when outside air damper is used to provide fresh outside air to a building or in air-side economizer control to minimize cooling energy. To detect outside air damper or economizer related fault, it is important to determine if a process variable or command signal has any correlation with the outdoor air temperature.

The selection of these primitives is based on current experience, intuition, analysis of many cases of testing patterns, and a "trial and error process." Further changes or additions may be needed with the progress of future research.

Coding Scheme

The second step in feature description is to design a coding scheme. A coding scheme is a way to describe the pattern so that it is easy to classify later in the feature classification step. The three-digit number system described in Tables 41-1 through 41-3 is proposed for use as the coding scheme for HVAC AFDD.

The first digit is used to identify the data pattern using primitives described in Section 3.2.1.

The second digit is used to describe the data pattern's shape direction (if applicable) as shown in Table 41-2.

Table 41-3. Third-digit Code

Coding Scheme	Third Digit - Goodness-of-fit
Numbers	
1	Excellent
2	Good
3	Ok
9	No fit / Not applicable

The third digit is used to describe the data pattern's goodness-of-fit to a primitive at three levels: "excellent" fit, "good" fit, or just an "ok" fit. These numbers will be evaluated quantitatively based on optimal parameters determined during curve fitting or using other methods.

Figure 41-8 illustrates graphical patterns of three digit codes and representing primitives used in the framework of HVAC fault detection and diagnostics using structural pattern recognition.

Feature Extraction

Feature extraction in structural pattern recognition for time series data includes curve fitting, piecewise-linear regression, and chain codes to generate basic shapes (primitives) that encode sequential, time-ordered relationships. Stockman and Kanal [12] used curve fitting to identify instances of parabolic and straight line primitives in developing a waveform parsing system called WAPSYS. Rengaswamy and Venkatasubramanian [10] also selected parabolic and straight lines as primitives in monitoring sensor data for process control and fault diagnosis. The structural approach is also commonly used for electrocardiogram (ECG) diagnosis. Straight lines, parabolic curves, and peaks are picked to be primitives in Trahanias and Skordalakis [13] for ECG data recognition. Olszewski [9] developed a domain-independent structural pattern recognition system that is "capable of acting as a black box to extract primitives and perform classification without the need for domain knowledge." The generalized approach uses the following six simple shapes as

Table 41-1. First-digit Code

Coding Scheme Number	First Digit - Pattern Type
1	Line
2	Exponential
3	Step
4	Peak
5	Sinusoid
6	Outdoor Air Temperature
9	None of the above

Table 41-2. Second-digit Code

Coding Scheme Number	Second Digit - Pattern Direction				
	Line	Exponential	Step	Peak	Sinusoid
1	Constant	Exponential Growth Increase	Up	Up	Up
2	Up	Exponential Decay Increase	Down	Down	Down
3	Down	Exponential Growth Decrease	-	-	-
4	-	Exponential Decay Decrease	-	-	-
9	-	-	-	-	-

primitives: a) constant; b) straight line; c) exponential; d) sinusoidal; e) triangular; and f) rectangular. These are simple patterns that are commonly used in waveform or time-series data. The main feature extraction techniques used in Olszewski's approach (structure detector) is the linear regression for linear structures (line, constant) and simplex method (commonly used in optimizing non-linear functions) for non-linear structures (exponential, sinusoidal, triangular, and rectangle). Zhou and Nelson [14] [15] also utilized the linear regression, piecewise linear approximation, and curve fitting in developing a plug and play framework for an HVAC air handling unit and temperature sensor auto recognition technique.

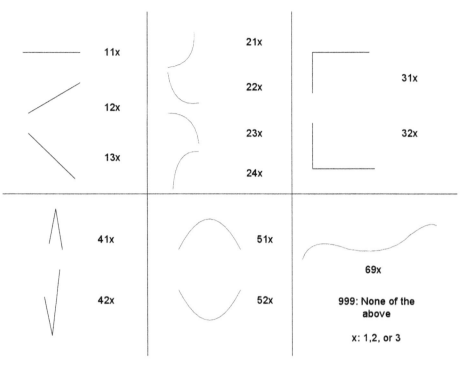

Figure 41-8. Illustrations of Primitives and Coding Scheme

Feature Classification using Template Matching

A template is a set of predetermined primitive code combinations describing normal or faulty operations of an HVAC system or component's monitoring or control point. The assumption is that the chart patterns of normal operations and faulty operations will be different for a process variable (or a group of related variables). For a specific fault or fault type, the chart patterns may be unique so that the fault detection and diagnostics are possible and accurate. The structural patterns or "shapes" of time-series data collected for the same monitoring/control point in different HVAC systems with identical system designs should be similar under the same operating conditions, as long as the sequence of operations are standardized—even in cases where HVAC component/equipment sizes, models, and capacities could be different. For example, a normally controlled AHU supply air temperature should be nearly constant at its setpoint, so a chart pattern with an average value near setpoint and detected pattern code as "111," "112," or "113" (see Figure 41-8) representing "perfect-fit linear," "good-fit linear" and "ok-fit linear" chart patterns can be considered normal operation. A sinusoid pattern such as "511," "521," "512" or "522" (see Figure 41-8) or a peak pattern such as "411" or "412" (see Figure 41-8) often mean there is oscillation in the temperature control loop and it could

be caused by bad loop tuning or other reasons.

Template matching is the process of comparing structural pattern codes identified in the field with a group of known templates to determine whether there is a match. In HVAC fault detection and diagnostics, once time-series data for HVAC process variables are collected (in real-time or from historical data) and structural pattern features extracted into codes, the codes can be compared to a group of known templates representing normal or faulty operations to see if there is a match. If a match with one of the "faulty operation" templates is found, a fault is detected.

The creation of the templates is based on data obtained in experimental testing. Normal operations and various instigated known faults were implemented and data recorded in these experimental testing. Many experiments may be needed to discover as much possible faulty patterns for monitoring and control points in an HVAC system or its components. The consequence of missing a template for a fault can occur is missed opportunities in detecting this fault.

A possible sample template for a fault case on an HVAC process variable could look like as in Table 41-4.

Table 41-4. A Sample Template for an HAVC Fault

Scenario 1:	411				
Scenario 2:	111	312	322		
Scenario 3:	691				

TEST CASES: NORMAL AND FAULTY PATTERNS FOR FAN COIL UNIT OPERATIONS

In this section, sample chart patterns are given for fan coil units under normal and various faulty operations to illustrate how these chart patterns may be related to the primitives proposed.

Test System and Setup

Normal and multiple faulty test cases were implemented in a facility [2] with two independent HVAC systems: systems A and B. System A is equipped with four four-pipe unit ventilators controlling four A rooms (East-A, South-A, West-A, and Interior-A), and system B is equipped with four four-pipe hydronic fan coil units for four B rooms (East-B, South-B, West-B, and Interior-B). Both systems are served by the central heating and cooling plants, which supply hot and chilled water to the facility. The fan coil units are floor mounted vertical cabinet models with outside air intake capability. These units have outside air openings on the back with a louver on the exterior of the building. A modulating outside air damper is equipped on each of these four fan coil units. Figure 41-9 gives the schematics of these fan coil units used in this research. During any testing day, South-B fan coil unit always ran in normal operating mode, while same fault was instigated in East-B and West-B room fan coil units. The Interior-B room unit was not used during the experiment.

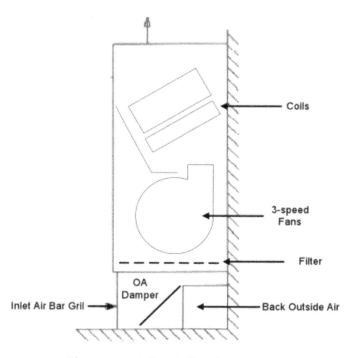

Figure 41-9. A Fan Coil Unit Schematic

These fan coil units have a 3-speed (high, medium and low) fan (2-speed for the interior room unit), which has several modes of operation: "Automatic ON/OFF" with three fan speeds automatically switched based on load, "Always ON" at a predetermined fan speed, and "Cycle On/Off" at a predetermined speed. The chilled water valve and the heating water valve of these units receive commands from corresponding fan coil unit controller to modulate the water flow rates to maintain the room temperature within heating and cooling setpoints. A single outside air damper with a fixed position controls the ratio of outdoor air to recirculated air.

The room temperature setpoints for the zones were constant values of 68°F for heating and 72°F for cooling in most cases, and 70°F for heating and 74°F for cooling in other cases. The internal loads (people, equipment, computers, and lights) were scheduled so that they simulate normal office environments with higher loads during the day than at night, as shown in Figure 41-10.

Normally the test cases were set up in the mornings. Once a test case was set up, it continued to the next morning until the next test case was set up. There were no interruptions during a test case regardless whether room temperatures in the rooms under test were under control or not.

Test Cases

Case 1: Normal Operation

Chart patterns for a normal operation test day in November in the South-B room are illustrated in Figures 41-11 through 41-13. The "FCU-TEST" point in Figure 41-11 indicates whether the room was under "Setup: 0," "Normal: 1," or "Fault: 2" conditions as a function of time during the day. It can be seen that a system setup was implemented between 8:50am to 11:17am. Before and after the setup this fan coil unit ran at "Normal" operation mode indicated by value "1" for the "FCU-TEST" point. The difference between the two "Normal" modes were that outside air damper was used to control the fan coil unit's mixed air temperature "FCU-MAT" at 55°F during the cooling mode, and fixed 30% open control was implemented after the switchover. In both normal periods, the room temperature "RM-TEMP" was controlled properly between the room heating setpoint "RMHTGSPT" and the room cooling setpoint "RMCLGSPT" values. The unit discharge and mixing air temperatures "FCU-DAT" and "FCU-MAT" varied with cooling valve command "FCU-CVLV," heating valve command "FCU-HVLV," and outside air damper command "FCU-DMPR." All these variable patterns appeared normal, stable, with no or small oscillation.

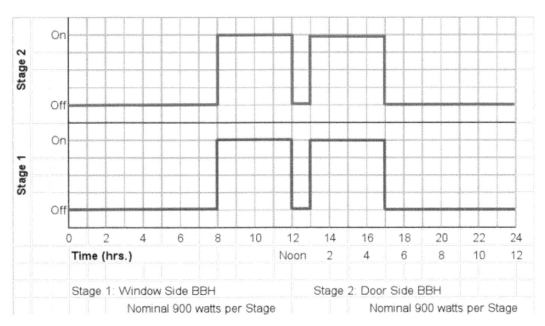

Figure 41-10. False Baseboard Heat Schedule in Test Rooms

The chart patterns include a combination of "Linear," "Exponential," and "Peaks."

Case 2: Fan Failure

A fan failure fault was instigated around 8:30am in an October day by turning off the fan coil units' fan power. Apparently, after the switchover, for the same process variables, chart patterns in Figures 41-14 and 41-15 are very different from that in Figure 41-11 and Figure 41-12. The room temperature and discharge air temperature on this day far exceeded the cooling setpoint most of the day, and did not seem to be in a "controlled" fashion. The chilled water valve command showed "Step Up" pattern and maxed out at 100% open for a long period of time. The unit's mixed air temperature showed a "Step Down" pattern due to fully opened chilled valve and

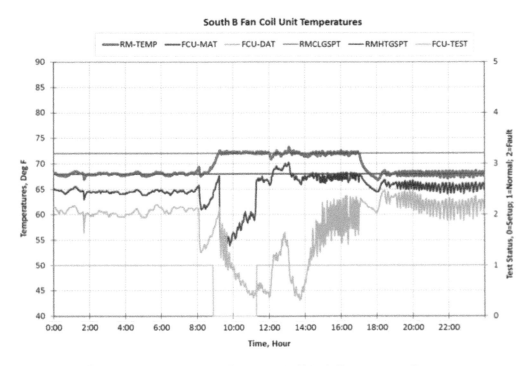

Figure 41-11. Case 1: "South-B" Fan Coil Unit Temperature Patterns

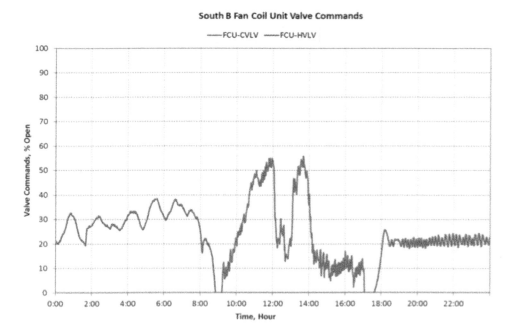

Figure 41-12. Case 1: "South-B" Fan Coil Unit Valve Command Patterns

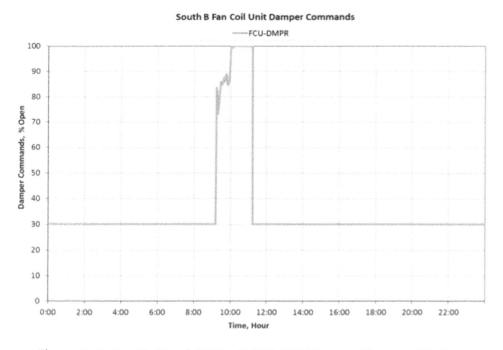

Figure 41-13. Case 1: "South-B" Fan Coil Unit OA Damper Command Patterns

maximum chilled water flow going through the cooling coil (not shown), even though there was no air flow because the fan did not run.

Case 3: Fan Outlet Blockage

Figure 41-16 and 41-17 show the key process variable's chart patterns for fan outlet blockage fault. A portion (50%) of the discharge air outlet area for the fan coil unit in East-B room was blocked in this case after around 9:00am on the following day of implementing Case 2. Since fan coil unit air flow rate is not normally measured, the performance analyses are based on the same set of monitoring/control points available in Case 1 and 2. Before 9:30am, the unit was still in Case 2 fan

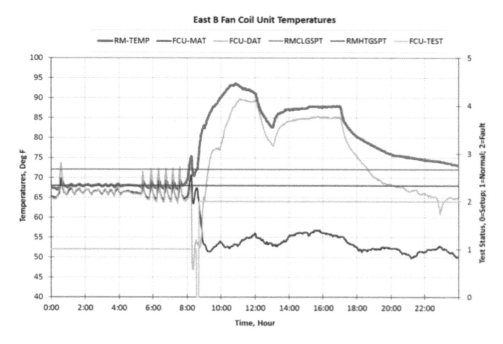

Figure 41-14. Case 2: "East-B" Fan Coil Unit Temperature Patterns

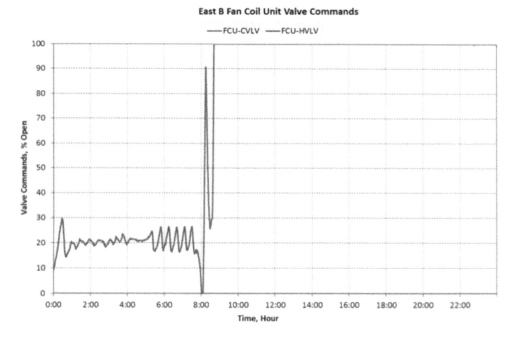

Figure 41-15. Case 2: "East-B" Fan Coil Unit Valve Command Patterns

failure fault mode where fan power was turned off as described in the previous case. The temperatures just drifted with ambient temperature but the maximum rate of chilled water went through unit cooling coil without fan running. At 8:00am, the scheduled internal loads (baseboard heat, lights, people, and a computer) were "ON" based on schedule and that caused the quick rise in all temperatures since no air was circulating in the room due to fan failure.

After the completion of setup of the new fan outlet fault around 10:50am, the room temperature was controlled normally, and the mixed air and discharge air temperature patterns also appeared "normal." However, the chilled water valve needed to open more in

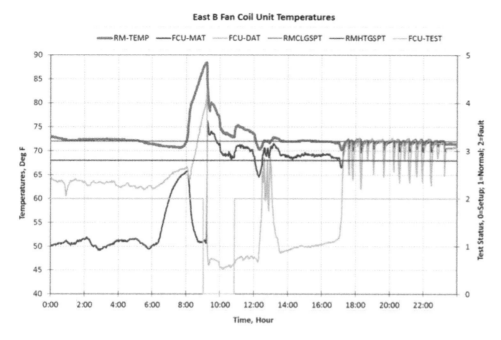

Figure 41-16. Case 3: "East-B" Fan Coil Unit Temperature Patterns

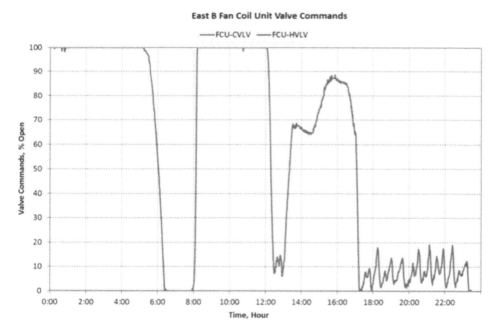

Figure 41-17. Case 3: "East-B" Fan Coil Unit Valve Command Patterns

Figure 41-17 compared to Figure 41-12 to compensate for the blocked discharge air outlet area. The small peaks in Figure 41-16 during the night period were due to frequent ON and OFF of the chilled water loop pump, which was automatically controlled based on the maximum chilled water valve position for these fan coil units and thresholds were set at 15% open or above (to turn

ON) and 5% open or below (to turn OFF).

Case 4: Restricted Heating Water Flow

In this case, the manual shutoff valve on the heating water return pipe from the fan coil unit in the East-B room was partially closed around 10:19am on a winter day. The maximum flow rate was restricted to about 30%

of the designed flow rate. It can be seen from Figures 41-18 through 41-20 that the room temperature could not be maintained at the heating setpoint of 68°F from 5:40pm to 8:40pm when heating was needed, even though the heating valve was 100% open. "Exponential," "Sinusoid" and "Step Up" patterns can be found in these temperatures and heating valve commands charts. The heating water flow rate was also shown a "Step Up" pattern on this night. It is worth to mention that before 10:19am on this day the "inadequate air flow" fault was instigated on this unit. The heating valve oscillation (and therefore heating water flow rate and various temperatures) seen in the early morning period in Figure 41-18 through 41-20 were due to oversized heating valve for the fan coil unit.

Case 5: Inadequate Air Flow

Figure 41-21 and 41-22 shows patterns of key variables when the West-B fan coil unit fan speed was manually reduced from "High" to "Medium" around 8:40am in a hot summer day. Everything seemed normal until around 3:00pm when the peak cooling load occurred and the room temperature could no longer be maintained below the cooling setpoint, even though the discharge air temperature was low enough. The chilled water valve quickly maxed out to 100% open—forming a near "Step Up" pattern shortly after 3:00pm.

Case 6: Restricted Chilled Water Flow

The manual shutoff valve on the chilled water return pipe from the fan coil unit in the West-B room was partially closed around 10:19am to instigate the restricted chilled water flow condition (maximum of 50% of designed flow rate) on a summer day. It can be seen from Figures 41-23 and Figure 41-24 that the room temperature could not be maintained at the cooling setpoint of 74°F from 2:00pm to 8:00pm, when the cooling load was the highest and the chilled water valve was already 100% open. Three temperatures had the similar chart pattern, and the chilled water valve showed a "Step Up" followed by a "Step Down" pattern.

Case 7: Heating Valve Stuck 100% Open

On a cold winter day, the heating valve for the fan coil unit in East-B room was instigated as stuck 100% open after 9:34am by disconnecting the heating valve control signal from the valve actuator and connecting the valve actuator to an independent 10 volt dc signal. The responses of key variables are shown in Figures 41-25 through 41-27. Since the heating water loop pump was not turned on from 8:00am to 7:43pm, due to no heating demand at all test rooms and all fan coil unit heating valves closed, there was no apparent abnormal pattern in these variables during this period. However, starting 7:43pm, the three temperatures showed abrupt changes or "Peak" pattern, while the heating valve com-

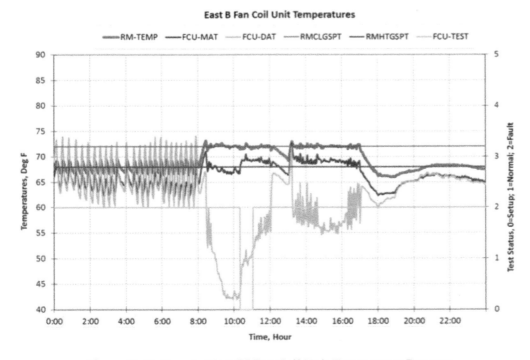

Figure 41-18. Case 4: "East-B" Fan Coil Unit Temperature Patterns

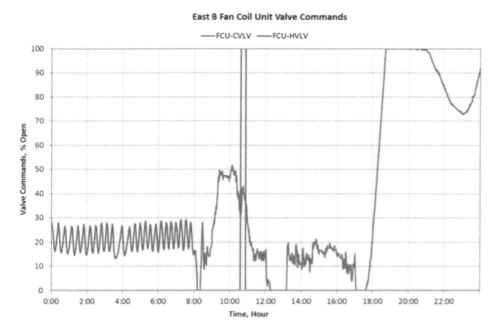

Figure 41-19. Case 4: "East-B" Fan Coil Unit Valve Command Patterns

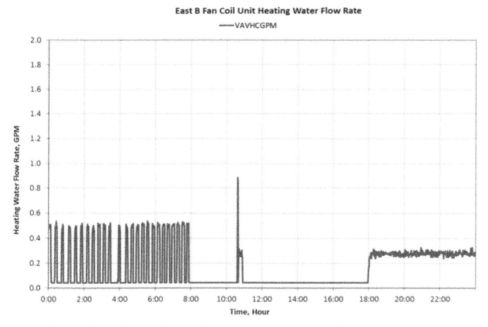

Figure 41-20. Case 4: "East-B" Fan Coil Unit Heating Water Flow Patterns

mand remained at 0% open. Looking at the actual fan coil unit heating water flow rate chart (Figure 41-27), the unusual intermittent heating water flow at maximum design flow rate of 1 gpm was observed. This was caused by the heating valve stuck 100% open fault at East-B room combined with automatic ON/OFF of the heating water loop pump due to heating valve position changes at other rooms (not shown). The oscillation pat-

terns before 9:00am in Figure 41-25 through 41-27 were the result of an oversized heating water valve.

Case 8: Heating Valve Stuck 100% Closed

This case shows that the heating valve in East-B room fan coil unit was instigated as stuck 100% closed in the following day when the Case 7 was implemented. It was accomplished by disconnecting the heating valve control signal from the valve actuator and connecting the valve actuator to an independent 0 volt dc signal around 9:00am. The responses of key variables are shown in Figures 41-28 through 41-30. Before 8:00am, the fan coil unit was still in heating valve stuck 100% open fault, and the heating water flow was constant 1 gpm but this room was in cooling mode because of the fault. The internal load was turned ON at 8:00am according to daily schedule, and other test rooms (not shown) changed from heating mode to cooling mode, and the heating water loop pump stopped automatically because of lack of heating demand. The fault switchover was completed at 9:30am and the heating valve was physically fully closed after that time. It can be seen from these figures that there was no apparent fault pattern after the test switchover until after 9:08pm the room temperature broke below the heating setpoint and could not maintain at that level. The heating valve command quickly rose from 0% open all day (due to cooling mode) to 100% open, forming a "Step Up" pattern, however, no heating water flow passed through heating coil due to valve stuck 100% closed fault and the room temperature continued to turn lower.

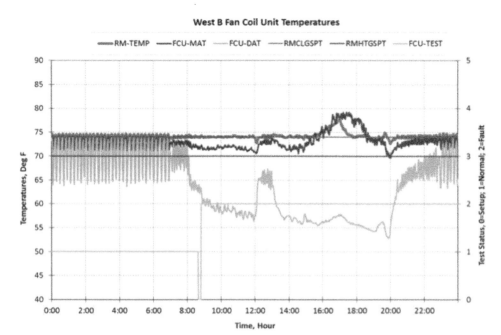

Figure 41-21. Case 5: "West-B" Fan Coil Unit Temperature Patterns

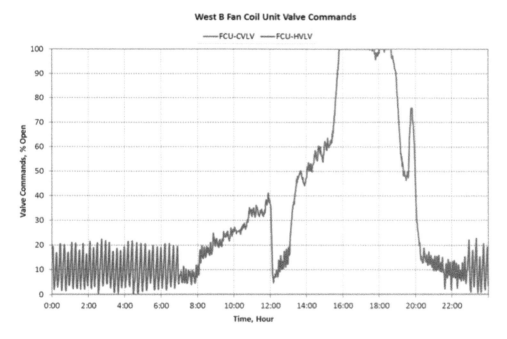

Figure 41-22. Case 5: "West-B" Fan Coil Unit Valve Command Patterns

Case 9: Cooling Valve
Stuck 100% Open

The East-B room fan coil unit cooling valve was instigated in a summer day as stuck 100% open by disconnecting the cooling valve control signal from the valve actuator and connecting the actuator to an independent 10 volt dc signal around 7:50am. The responses of key variables are shown in Figures 41-31 through 41-33. After the switchover, the room temperature dropped

continuously until the heating valve opened fully. The cooling valve command formed a "Step Down" pattern and heating valve command formed a "Step Up" pattern. There were simultaneous heating and cooling going on due to cooling valve stuck 100% open in this period. Because the chilled water loop pump was controlled to be turned ON and OFF based on chilled water valve positions in all test rooms, it was turned ON

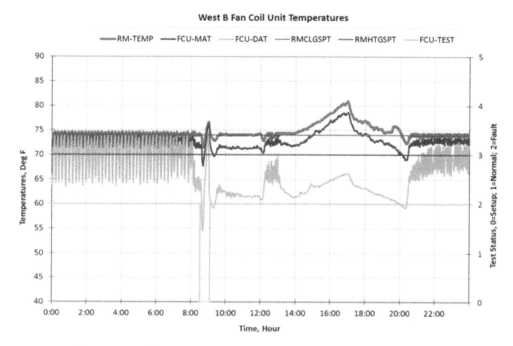

Figure 41-23. Case 6: "West-B" Fan Coil Unit Temperature Patterns

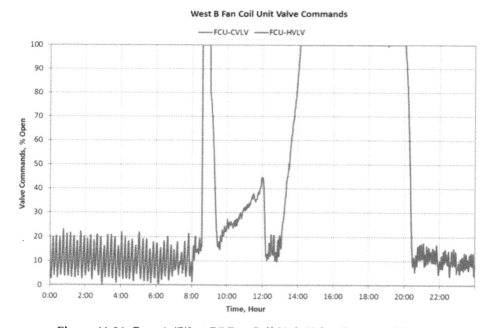

Figure 41-24. Case 6: "West-B" Fan Coil Unit Valve Command Patterns

rooms (Figure 41-31) showed "Peak" pattern. It can be seen that sometimes the fault patterns may be caused by the chilled water system or heating water system rather than the fan coil unit itself. It is worth to mention that before 7:50am, the fan coil unit in East-B rooms was instigated the cooling valve stuck 100% closed fault, so the temperatures were constantly exceeding the cooling setpoint. The rise in temperature around 6:40am was because of the solar radiation through east-facing exterior windows of that room.

Case 10: Cooling Valve
Stuck 100% Closed

The East-B room cooling valve was instigated stuck 100% closed in a summer by disconnecting the cooling valve control signal from the valve actuator and connecting the valve actuator to an independent 0 volt dc signal around 8:50am. The resulting responses of key variables are shown in Figures 41-33 and 41-34. The room temperature was not under control even though the cooling valve was wide open. The mixed air and discharge air temperature patterns also correlated very closely with the room temperature because no chilled water flowing through the coil. "Exponential" and "Step" patterns were seen from these chart patterns.

Case 11: Outdoor Air
Damper Stuck 100% Open

In this case the outside air damper was fixed at the 100% open position on a hot summer day. The responses of key variables are shown in Figures 41-37 through 41-39. It can be seen by comparing Figure 41-37 and Figure 41-38 that the mixed air temperature (FCU-MAT) trend was very similar to the outside air

and OFF frequently after 8:00pm. This was because the South-B room fan coil unit ran in normal control mode and the cooling valve command oscillated around ON/OFF thresholds of 5% and 15% open (Figure 41-34). The ON/OFF changes of the chilled water pump were the reason of the chilled water flow rate change for East-B room fan coil unit as the cooling valve was stuck fully open. Therefore the three temperatures in East-B

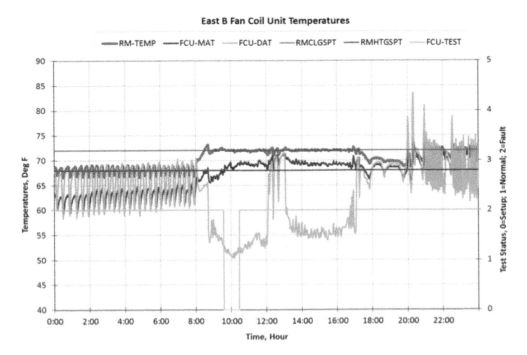

Figure 41-25. Case 7: "East-B" Fan Coil Unit Temperature Patterns

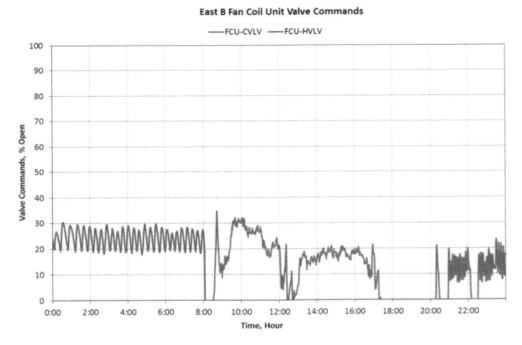

Figure 41-26. Case 7: "East-B" Fan Coil Unit Valve Command Patterns

temperature (OA-TEMP) pattern after the fault was being setup around 10:00am. Even though the room temperature was maintained pretty well below the cooling setpoint, but the energy was wasted by introducing unnecessarily large amount of hot outside air into the room. The oscillation on temperatures and cooling valve commands in Figure 41-38 and Figure 41-39 when running at normal condition before 10:00am was probably due to very small amount of cooling needed at night and the cooling PID control was tuned for better control of much larger amount of cooling loads during the day time.

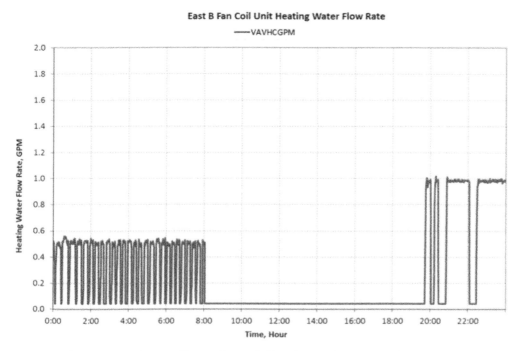

Figure 41-27. Case 7: "East-B" Fan Coil Unit Heating Water Flow Pattern

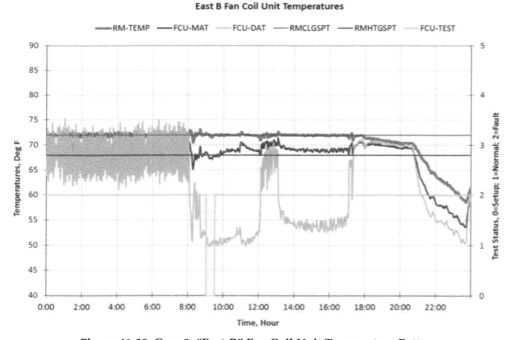

Figure 41-28. Case 8: "East-B" Fan Coil Unit Temperature Patterns

Case 12: Control Oscillation with
Outside Air Damper 100% Open

The heating and cooling control loops parameters were adjusted to instigate control oscillation when outside air damper is 100% open in this case. The cooling valve command positions (in a summer test; see Figures 41-40 and 41-41), heating valve positions (in a winter test; see Figures 41-42 and 41-43), and discharge air temperature (FCU-DAT) showed many sharp peaks indicating control oscillation. For the winter case (Figures 41-42 and 41-43), the control parameter changes was implemented around 10:00am. It can be seen higher frequency and larger amplitude for the heating valve command (FCU-HVLV) peaks comparing to that of be-

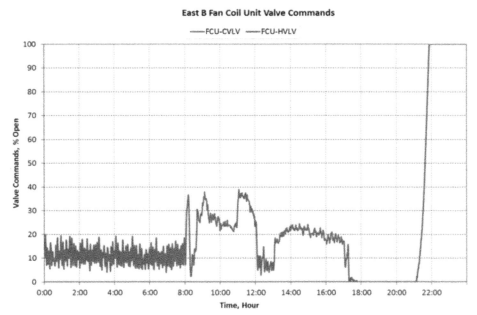

Figure 41-29. Case 8: "East-B" Fan Coil Unit Valve Command Patterns

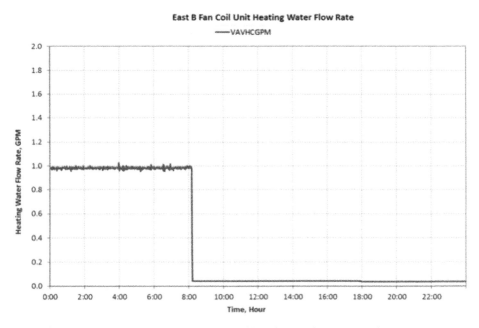

Figure 41-30. Case 8: "East-B" Fan Coil Unit Heating Water Flow Patterns

fore 10:00am, even though slight oscillation also existed before the change due to oversized heating valve.

Case 13: Control Oscillation with
Outside Air Damper Fully Closed

The control oscillation was instigated immediately following Case 12 summer case by fixing the outside air damper at 0% open position around 10:00am. Trends in time for key variables are shown in Figures 41-44 through 46. As with Case 12 summer case, the discharge air temperature and cooling valve command show many peaks and oscillations (comparing Figure 41-45 and 41-41). The chart patterns before 10:00am show the effect of Case 12 summer case (where the outside air damper was fixed at 100% open), which is not much different with those in this case. In Figure 41-46, FCU-DMPR is the fan coil unit outside air damper command and the pattern basically showed a straight line constant pattern.

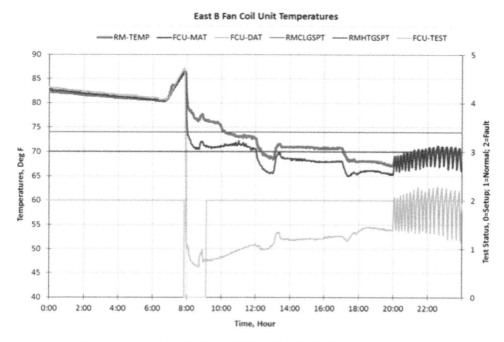

Figure 41-31. Case 9: "East-B" Fan Coil Unit Temperature Patterns

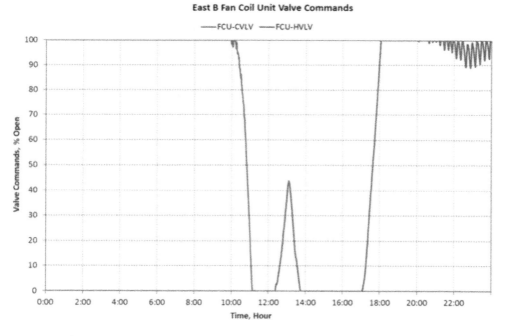

Figure 41-32. Case 9: "East-B" Fan Coil Unit Valve Command Patterns

FUTURE RESEARCH

Examples in the section on Test Cases show that for fan coil units, different structural patterns exist for different faults and many may be represented by a distinct combination of primitives proposed in the section Structural Pattern Recognition Techniques in HVAC Applications. To effectively applying the framework and

methodology proposed in that section, future research is suggested below:

• Develop effective feature extraction algorithm(s) that are suitable for extracting proposed primitives from HVAC system data. Many existing algorithms used in other fields described in that section can be tried and modified or new algorithms

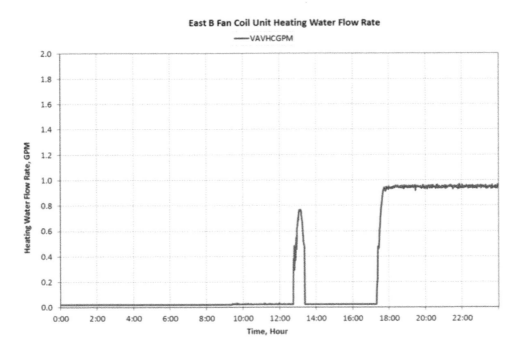

Figure 41-33. Case 9: "East-B" Fan Coil Unit Heating Water Flow Patterns

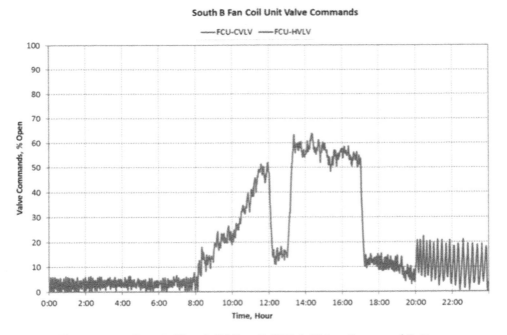

Figure 41-34. Case 9: "South-B" Fan Coil Unit Valve Command Patterns

developed. To accomplish this, experimental data covering a wide range of faulty cases for different HVAC system component is needed for testing the effectiveness of the algorithms.

• Based on proposed feature description methodology and coding schemes in that section, a library of templates for various faulty cases can be generated

using the feature extraction algorithms developed in the above step and experimental data for normal and faulty cases. Templates can be generated for different HVAC components as well: air handling unit, different types of variable-air-volume terminal units, etc. as long as experimental data can be obtained.

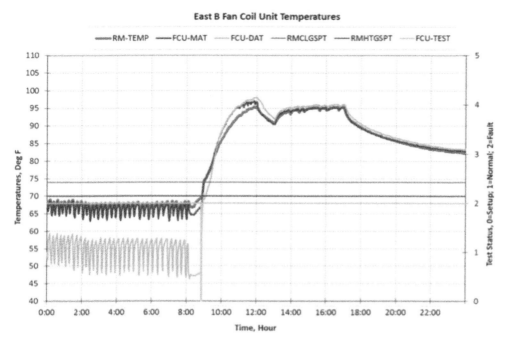

Figure 41-35. Case 10: "East-B" Fan Coil Unit Temperature Patterns

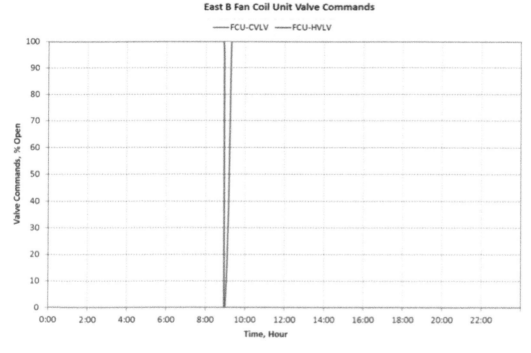

Figure 41-36. Case 10: "East-B" Fan Coil Unit Valve Command Patterns

• Test the effectiveness of template matching feature classification method described in that section by using field data. The effectiveness can be measure by the percentage of faults correctly detected and the percentage of specific faults correctly diagnosed. The test can be done on different HVAC systems (comparing to the system used in exper-

iments in generating the library of fault pattern templates), as long as the system design and component types are the same.

In conclusion, this chapter proposed a framework of applying a structural pattern recognition technique in automated fault detection and diagnostics of HVAC sys-

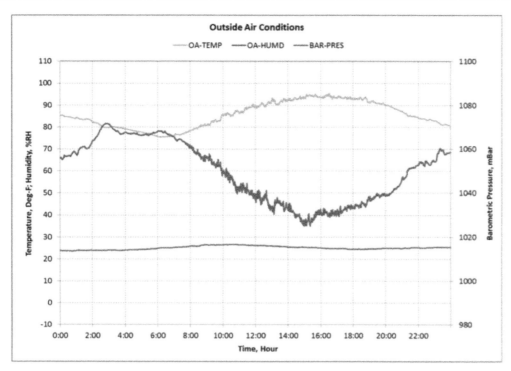

Figure 41-37. Case 11: Outdoor Air Temperature and Relative Humidity Patterns

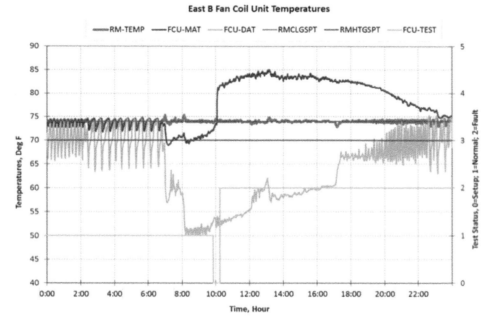

Figure 41-38. Case 11: "East-B" Fan Coil Unit Temperature Patterns

tem and component faults. This is an ongoing research and the suggested future research directions are given. Comparing to quantitative and qualitative model-based diagnostic methods or process history-based diagnostic methods, it is expected that the benefit of using structur-al pattern recognition approach include wider applicability because no historical field data will be need, more accurate fault diagnostic capability, more intuitive in understanding faults, and eventually more useful and effective in real-world environments.

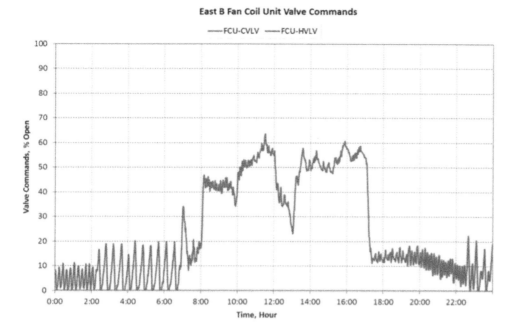

Figure 41-39. Case 11: "East-B" Fan Coil Unit Valve Command Patterns

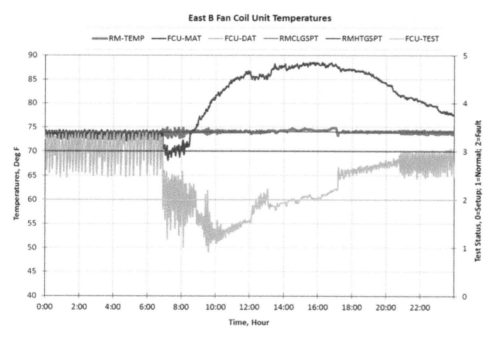

Figure 41-40. Case 12: "East-B" Fan Coil Unit Temperature Patterns for a Summer Test

References

1. Department of Energy, 2005. Advanced Sensors and Controls for Building Applications: Market Assessment and Potential R&D Pathways. DOE Report for DE-AC05-76RL01830.

2. Iowa Energy Center. 2010. 2010 ERS Technical Description. Iowa Energy Center. Ankeny, IA.

3. Katipamula, S. and M. R. Brambley, 2005. Methods for Fault Detection, Diagnostics, and Prognostics for Building Systems—A Review, Part I. HVAC&R Research 11(1):3–25.

4. Katipamula, S. and M. R. Brambley, 2005. Methods for Fault Detection, Diagnostics, and Prognostics for Building Systems—A Review, Part II. HVAC&R Research 11(2):169–187.

5. Koski, A., M. Juhola, and M. Meriste. 1995. Syntactic recognition of ECG signals by attributed finite automata. Pattern Recognition 28(12):1927–40.

6. Konstantinov, K.B., and T. Yoshida. 1992. Real time qualitative analysis of the temporal shapes of (bio) process variables. AIChE Journal 38(11):1703–15.

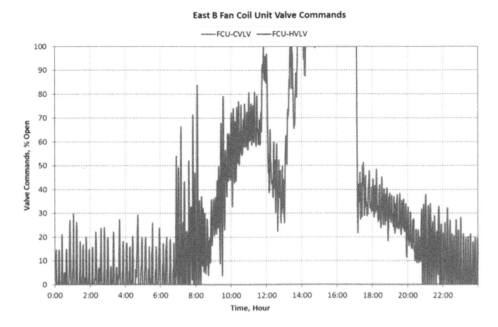

Figure 41-41. Case 12: "East-B" Fan Coil Unit Valve Command Patterns for a Summer Test

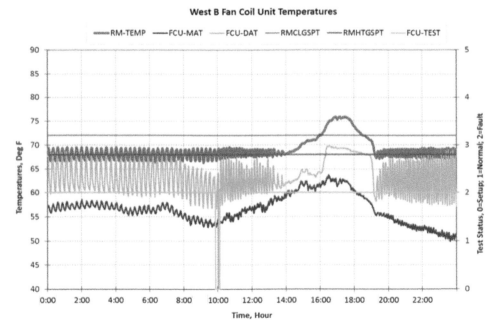

Figure 41-42. Case 12: "West-B" Fan Coil Unit Temperature Patterns for a Winter Test

7. Love, P.L., and M. Simaan, 1988. Automatic recognition of primitive changes in manufacturing process signals. Pattern Recognition 21(4):333–42.

8. Nadler, M., and E.P. Smith. 1993. Pattern Recognition Engineering. John Wiley & Sons INC., New York, NY

9. Olszewski, R.T., 2001. Generalized Feature Extraction for Structural Pattern Recognition in Time-Series Data, PH.D. Thesis, Carnegie Mellon University, Pittsburgh, PA

10. Rengaswamy, R. and V. Venkatasubramanian. 1995. A Syn-

tactic Pattern Recognition Approach for Process Monitoring and Fault Diagnosis, Engineering Applications of Artificial Intelligence, 8(1). pp. 35–51

11. Schein, J. and J. House, 2003. Application of Control Charts for Detecting Faults in Variable-Air-Volume Boxes. ASHRAE Transaction 109(2):1–12.

12. Stockman, G.C. and L.N. Kanal. 1983. Problem Reduction Representation for the Linguistic Analysis of Waveforms, IEEE Transactions on Pattern Analysis and Machine Intelli-

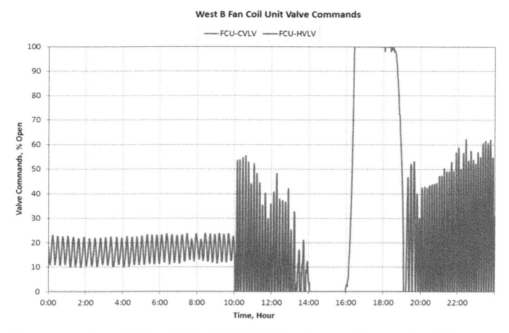

Figure 41-43. Case 12: "West-B" Fan Coil Unit Valve Command Patterns for a Winter Test

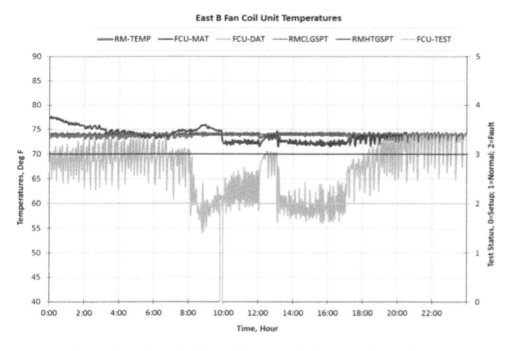

Figure 41-44. Case 13: "East-B" Fan Coil Unit Temperature Patterns

gence, PAMI-5(3). pp. 287-298

13. Trahanias, P. and E. Skordalakis. 1990. Syntactic Pattern Recognition of the ECG, IEEE Transactions on Pattern Analysis and Machine Intelligence, 12(7). pp. 648–657

14. Zhou, X. and R. Nelson, 2011. A plug-and-play Framework for an HVAC Air-Conditioning unit and temperature sensor auto recognition technique--part I: review of critical technol-

ogies. ASHRAE Transaction 117(2)

15. Zhou, X. and R. Nelson, 2011. A plug-and-play framework for an HVAC air-handling unit and temperature sensor auto recognition technique--part II: pattern recognition and tests. ASHRAE Transaction 117(2)

East B Fan Coil Unit Valve Commands

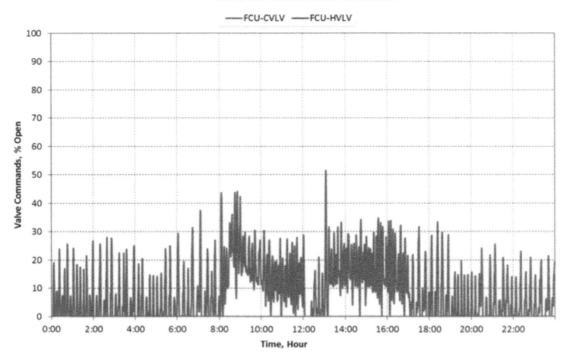

Figure 41-45. Case 13: "East-B" Fan Coil Unit Valve Command Patterns

East B Fan Coil Unit Damper Commands

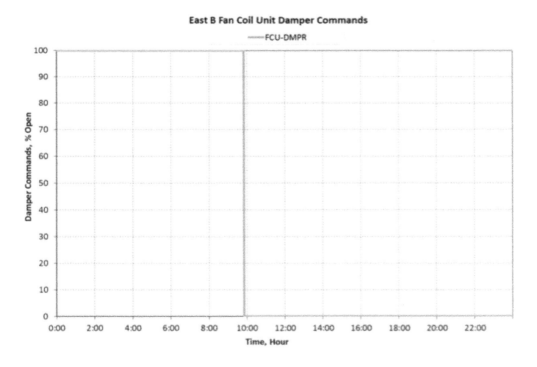

Figure 41-46. Case 13: "East-B" Fan Coil Unit Outside Air Damper Patterns

Fault Detection and Diagnostics for Rooftop Units: Standards are Key to Making it in the Market

Kristin Heinemeier and Jerine Ahmed

Moving the Technology into the Marketplace Requires More than Developing Products… it requires standardization.

FDD, which has the potential to identify sub-optimal performance and other operational problems, has been under development for quite a few years. Yet, its present impact on the market is still minimal. Several promising tools are beginning to emerge commercially, but moving the technology forward into the marketplace in a big way requires more than developing products; it requires things such as codes and standards, procurement guidelines, utility incentive programs, requirements in rating systems, modeling methods, data on fault incidences, design tools, standard nomenclature, market research, etc. For example:

- ASHRAE SSPC 90.1's Mechanical Subcommittee is beginning to consider including requirements for FDD. In order to do so, it too will have to develop definitions of key faults and test methods. Standard 90.1 would need to specify exactly what is meant by each of the key faults, although there is no specification in the industry to use as a model. Beyond this, if one is to evaluate whether an individual building has complied with the code or not, one must have more definition of what these faults mean and how they can be tested for. This can come either in the form of a field test of, or a certification of FDD functionality. Either of these requires a specification of the definition of the fault, as well as a method of testing.

- DOE has developed a specification for a "golden carrot" RTU (competition to encourage manufacturers to develop ultra-high efficiency models). The specification included requirements for FDD, and it included a list of faults to be detected and annunciated. However, lack of standardization in the industry led to dropping this requirement. In order for manufacturers to know what it is they are to develop and specification authors to be able to test

whether candidate units meet their specification, more specificity must be provided. EPA's Energy Star has also developed a specification for FDD.

- Utilities might provide incentives for FDD. But they need to know exactly what it is that they are incenting in order to assess available tools. They will need to model the savings available from these tools in order to get credit for the energy savings.

- Owners who are considering installing FDD need to know that a proposed tool actually works. But there are currently no metrics for performance of FDD.

- Modeling is crucial to utility program design, codes and standards development, compliance, design, marketing, and measurement and evaluation. Modeling cannot be done if there are no standard definitions of faults.

- Field data on fault incidence are difficult to obtain, and particularly difficult given that different research projects have different definitions of faults. If there were a standard definition, then researchers could collect comparable data, and meta-analyses could be conducted to appropriately estimate fault incidences and impacts in the field.

Once there is more standardization of this critical technology, it can be expected to achieve these milestones and overcome market barriers. This chapter describes two efforts at standardization in the industry that will hopefully lead to increased market penetration of FDD for rooftop units.

WHY IS FDD NEEDED?
AS AN EXAMPLE

Field studies have consistently shown that faults occur frequently in RTUs. Figure 42-1, for example, shows the results of a field study in California. There are quite a few types of faults that have serious energy and demand consequences, and occur in a substantial fraction of the installed units.

One of the faults leading to the largest energy waste is the airside economizer, so it is a good example of the potential for fault detection. Although economizers are an excellent energy-saving technology, in practice, out in the field, they are not performing well. Many different studies have identified the rates at which economizers, in new and existing RTUs, tend to fail. Table 42-1 shows the failure rates identified by different researchers, in different scenarios.

Failure modes in Jacobs' field study included dampers that were stuck or inoperable (38%), sensor or control failure (46%), or poor operation (16%). The average energy impact of inoperable economizers was about 37% of the annual cooling energy. Other than this study, there is not much documentation of what the specific failures are. To find out what some of the specific faults are, the authors conducted a survey of California commercial building contractors and found that economizer faults are perceived to be quite common. Table 42-2 shows the results when contractors were asked "What percent of existing commercial RTUs have the following faults?" The survey found that 30-40% of the time,

contractors find that the economizer is disabled and the outside air dampers are closed. This type of failure means that the economizer is not providing any savings, and that the building may not be bringing in any outside air.

Most of these economizer failures have energy penalties. For example, the magnitude of savings from an economizer depends directly on the high limit setpoint. Some economizers use enthalpy curves, selectable by choosing a dial setting of "A," "B," "C," or "D." The "A" setting results in the most hours of economizer cooling, although the Jacobs field study found, in addition to the 64% malfunction rate, that the high limit setpoints were set incorrectly in most cases. "A" was used only 28% of the time, and either the "C" or "D" setting was used 60% of the time. Figure 42-2 shows the potential for savings in a number of different California climate zones as a function of the high-limit setpoint. This was simulated by the Economizer Savings Estimator (Brandemuehl and Braun 1999). Clearly there are significant savings associated with ensuring that the setpoint is set correctly on day one, and that it is maintained at that setpoint. Yet contractors typically either are not aware of the optimal setpoint, they leave the economizer at the default setpoint, or they adjust it over time to alleviate comfort complaints.

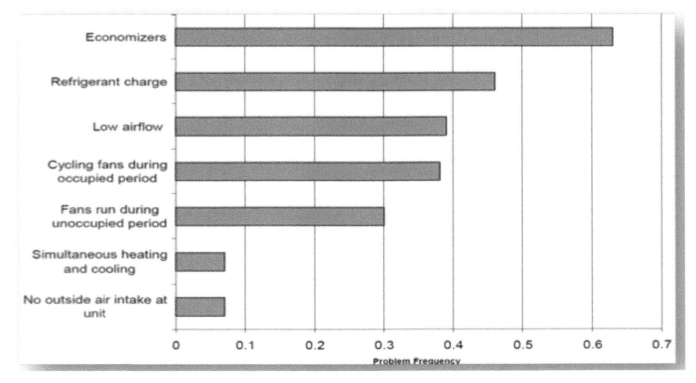

Figure 42-1: Results of a Field Study into the Prevalence of Faults in California Small Commercial Buildings. Source: Jacobs and Higgins, 2003

Table 42-1: Economizer Failure Rates Found in Various Studies

% Failure	Source	Notes
43%	AEC 2002.	Just damper faults.
50%	Mike Kaplan, Personal Communication with Dave Sellers, 1999.	New construction.
56%	HEC 1993.	Economizers up to two years old.
64%	Jacobs and Higgins 2003; and Jacobs et al. 2004.	124 RTUs 10 tons or less, with economizers.
64%	Jonathan Woolley, Personal Communication, 2013.	22 RTUs with economizers.
65%	Goody et al. 2003.	Small commercial RTUs.
70%	Davis et al. 2002.	Small number of RTUs.
70%	KEMA 2013	Economizers that had been fixed up to a year ago.
75%	Craig Hofferber, Personal Communication with Dave Sellers 2000.	Estimate from interviews with consultants, mechanical contractors, and commissioning agents.
80%	Felts and Bailey 2000.	Existing RTUs.
100%	Pratt, et al. 2000.	Four of four RTUs investigated.

Table 42-2: Results of Contractor Survey on the Prevalence of Various Economizer Faults

Economizer is disabled and dampers are closed	30-40%
Actuator/linkage broken, misaligned, or loose, due to normal wear and tear or lack of lubrication	20-30%
High/low limit setpoints incorrect, set by installing contractor	
Range/action setup incorrectly	
Min outside air is not set correctly: too low	
Actuator/linkage broken, misaligned, or loose, due to occupant/operation staff action	10-20%
Min outside air is not set correctly: too high	
High/low limit setpoints incorrect, set by factory	
Dampers mechanically forced open	
OA sensor (db, enthalpy) malfunction	
OA sensor (db, enthalpy) drift	
High/low limit setpoints incorrect, set by occupants/operating staff	5-10%
OA sensor (db, enthalpy) miscalibration	

When economizers fail, they often squander far more energy than they would ordinarily save if they worked properly. In some cases, wasted energy can be ten times as much as potential savings (Lunneberg 1999). Economizers that fail in the fully open position contribute to extremely high peak loads, because more heating and cooling energy is needed to condition the excess air the economizer lets in. For example, in a sim-

ulated building in Bakersfield, an economizer stuck in the fully open position would add 84% to the summer peak load (EDR 2011). In actual operation, it is unlikely (though possible) that a stuck economizer would waste quite as much energy as this simulation suggests, because cooling and heating in the building would be insufficient to keep occupants comfortable all year long with the damper fully open, and presumably the open

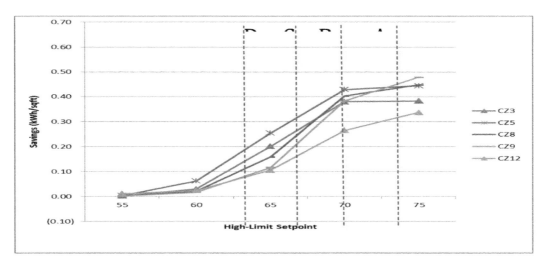

Figure 42-2: Economizer Savings in Different California Climate Zones, as a Function of High-limit Setpoint

damper would eventually be discovered and closed.

In one real example, when an economizer in a Texas building was stuck open, the economizer was taking in too much outside air during cold weather, costing the facility about $2,000 a year in additional energy (Liu, et al. 1994). It was disabled during the winter, and once the economizer was locked in place, steam consumption dropped dramatically. Costs involved with field repairing economizers have caused some researchers to recommend disabling rather than fixing broken economizers under some conditions. See Felts (2000) and Lunneberg (1999). A malfunctioning economizer can be worse than no economizer at all.

Why are economizers so prone to failure? The fact that they are "invisible," and often don't create comfort problems, contributes to an "out of sight, out of mind" attitude. They are also quite complex, although their basic principle of operation is simple. Most technicians do not understand all the complexities of how they work. Human factors also have a big influence in ensuring the best performance of economizers (Hunt et al. 2010). While there is little research into what the specific failures are in the field, there is no research into what causes these failures or what practices or attitudes allow them to stay broken. In some cases, the failures exist from the moment the economizer was installed (one contractor reported investigating the performance of an economizer, and discovering that the economizer was still in the packing position, meaning the installer never completely installed it or never ran power to it (Dale Rossi, Personal Communication 2014). In some cases, the economizer is intentionally "temporarily" disabled,

to address another pressing issue with the system. For example:

- A unit that is undersized and having difficulty meeting load during extreme heat and cold, leading to comfort complaints; hence a decision to keep the dampers closed.

- A customer that doesn't want to pay for something that doesn't affect comfort, so they won't pay to fix the economizer.

- Equipment that is in the vicinity of a lot of wood stoves/fireplaces during winter.

- During morning warm-up on gas packs you may get complaints of a gas smell from the tenant.

- RTUs are over grocery departments in the summer, when it's really humid.

- Bringing in undesirable outside odors, often truck fumes from units near loading docks.

- The fresh air intake is installed directly over a vent stack.

- Two units installed side by side and the gas exhaust of one is directed towards the economizer of the other.

The cost of repairing or replacing is often high, and when the financial proposition of fixing the unit is not well understood by the owner, it can be difficult to obtain authorization to do this work. If the quote isn't approved, "temporary" can become "permanent" quite easily. Units generally are serviced only when they stop delivering cooling (Jacobs et al. 2004), so any of a number of faults can appear and can be present for years before someone looking for energy savings opportunities looks carefully at the economizer. For this reason, economizers are a prime candidate for FDD.

FDD REQUIREMENTS ADOPTED BY BUILDING CODES

In 2012 the California Energy Commission adopted a requirement for economizer FDD in all newly installed air-cooled unitary direct-expansion units with mechanical cooling capacity of greater than or equal to 54,000 Btu/hr (CEC 2011). This includes packaged, split-systems, heat pumps, and variable refrigerant flow (VRF) systems. This requirement will go into effect on July, 2014.

Title 24 is California's energy code for new buildings and alterations. This set of requirements and processes is the most stringent in the country, and the California Energy Commission has a goal to reach zero net energy for residential buildings by 2020 and for commercial buildings by 2030. In order to reach this goal, HVAC will have to be much more efficient than it typically is today. However, HVAC will soon reach thermodynamic maximum efficiency, and the only way to get more energy savings is to remove operational mistakes, for example reducing energy waste through things like identifying faulty economizers.

For Title 24, the FDD system must be able to detect the following economizer faults:

- Air temperature sensor failure/fault. This failure mode is a malfunctioning air temperature sensor, such as the outside air, discharge air, or return air temperature sensor. This could include mis-calibration, complete failure either through damage to the sensor or its wiring, or failure due to disconnected wiring.

- Not economizing when it should. In this case, the economizer should be enabled, but for some reason it's not providing free cooling. This leads to an unnecessary increase in mechanical cooling energy. Two examples are the economizer high limit setpoint is too low, say 55°F, or the economizer is stuck closed.

- Economizing when it should not. This is opposite to the previous case of not economizing when it should. In this case, conditions are such that the economizer should be at minimum ventilation position but for some reason it is open beyond the correct position. This leads to an unnecessary increase in heating and cooling energy. Two examples are the economizer high limit setpoint is too high, say 82°F, or the economizer is stuck open.

- Damper not modulating. This issue represents a stuck, disconnected, or otherwise inoperable damper that does not modulate open and closed. It is a combination of the previous two faults: not economizing when it should, and economizing when it should not.

- Excess outdoor air. This failure mode is the economizer provides an excessive level of ventilation, usually much higher than is needed for design minimum ventilation. It causes an energy penalty during periods when the economizer should not be enabled, that is, during cooling mode when outdoor conditions are higher than the economizer high limit setpoint.

- During heating mode, excess outdoor air will increase heating energy.

It is acknowledged that these requirements are somewhat subjective, although this was deliberate, in order to allow manufacturers the latitude to develop capabilities in the way they see fit.

The FDD system can either be embedded in the RTU, or it can be a "strap on" after-market model, but faults must be reported to a fault management application accessible by day-to-day operating or service personnel, or annunciated locally on zone thermostats. In order to meet the requirements, an FDD tool must be certified by the manufacturer to be an effective, valid FDD tool. The requirements for factory certification were not established. Even an FDD tool that has been certified to be effective may not work correctly if it has not been installed and configured correctly, so simple field tests were defined to verify that the system is installed and configured correctly.

Cost effectiveness analysis was carried out, as is required for any measure included in Title 24. The approach that was taken was to model prototype buildings in Energy Plus, and to calculate a weighted average for the California building stock. The modeling assumed that all faults were present, which is certainly not always the case. Post-processing of the savings applies assumed probabilities for the fault occurring, for the fault being detected through other means, and for the fault being fixed. As is done for all Title 24 measures, "time dependent valuation" is used to establish the value of the benefits, and a benefit-cost ratio of 1.2 resulted (representing about 12% energy savings) (CEC 2011).

Manufacturers were involved in the committee that discussed potential approaches for this measure, and were generally supportive, although some of those that were not involved in the discussion were somewhat taken by surprise. The committee originally had a long list of faults that should be detected. Cost effectiveness and a requirement that the FDD capabilities must already be on the market caused the list to be whittled down to the five economizer-related measures that were eventually adopted. The group decided to propose the measure as a mandatory measure, which means that it cannot be traded off for other energy-saving measures as one could typically do with prescriptive measures. This was easier for manufacturers to implement.

Manufacturers would prefer not to make an RTU model with FDD just for California markets, so it is important to note that there has been action on the national front as well. The latest version of the ICC Energy Code includes requirements for FDD (IECC 2012), and it is being considered for inclusion in the Standard Mechanical Code (IAPMO 2013). Also, the ASHRAE Standard 90.1 Mechanical Subcommittee's Controls Working Group is watching what is happening in California, with an eye towards future FDD requirements in ASHRAE 90.1 and in ASHRAE 189.1 (ASHRAE 2010, 2009). They agree, however, that one cannot require FDD capabilities until one is able to fairly test proposed tools and verify that they will result in the performance that is expected. This lack of standardization is hurting the industry.

STANDARD METHOD OF TEST FOR FDD PERFORMANCE

Because of the need for a standard method of test of the performance of an FDD tool, the Smart Buildings Technical Committee at ASHRAE voted to establish a Standards Project Committee to develop a standard. The proposal, entitled "Laboratory Method of Test of Fault Detection and Diagnostics Applied to Commercial Air-Cooled Packaged Systems" was accepted by the ASHRAE Standards Committee and Board of Directors, and a Standard Project Committee 207P was established. This proposed standard would provide a method to define an FDD tool's function. This standard would also provide a method of laboratory test for the performance of FDD tools on commercial air-cooled packaged equipment.

This standard will apply to commercial air-cooled packaged air conditioning systems. The test is a physical laboratory test on a particular combination of diagnostic tool for each model of a unitary system. This standard applies to any on-board, after-market or hand-held hardware and/or software functionality that detects and/or diagnoses problems that lead to degraded performance such as, energy efficiency, capacity, increased maintenance costs or shortened equipment life.

The proposed Standard provides a Method of Test for testing the sensitivity of FDD tools to varying levels of faults. The tests defined are simple physical tests, such as identifying at what point faults are detected with different levels of charge, with a coil blocked off to varying degree, with physical obstruction of the economizer damper, or with a physical restriction in a refrigerant line of varying severity, or with some combination of these faults. A matrix of conditions and faults will allow for a balance between significance and feasibility.

It should be noted that it is left to building codes (such as Title 24) and other policies and programs to determine what fault intensity level should be detected. It is not the job of this standard to determine how accurate a FDD tool should be, but rather to provide a standard test method, a meter-stick if you will, to measure an FDD tool against.

The approach taken by this Standard is to verify the performance claims made by an FDD vendor and/or HVAC manufacturer. For example, if a vendor claims that their tool will be able to detect low refrigerant charge that causes a 5% decrease in EER, then the test will verify that indeed a 5% decrease in EER due to low refrigerant charge does cause an alarm. These tests will also gauge the reliability of detecting a fault. This is a combination of the probability of correctly detecting the fault when it is present, and the probability that a false alarm will be generated when some lower fault level is present.

The HVAC unit to be tested in the laboratory will be a unit with no known faults. The model will be representative of the line of HVAC units or controller families that are included in the manufacturer's claim. For example, if the vendor claims that FDD performs on a whole model series, from 2 tons thru 10 tons, then any unit in that model series within those sizes can be used for the test.

The driving conditions for the tests are determined by the manufacturer's claim. The manufacturer states within which range of conditions the FDD system is claimed to be able to perform (condenser inlet air temperature, return air wet bulb temperature). Tests are to be performed at these extreme points, as well as at two "wildcard" points within this range. These wildcard points are to be selected using a computer's random number generator to select a random number between the minimum and maximum claimed and must be selected by a neutral third-party.

For each fault claimed by the Vendor, the following tests are conducted:

1. No fault: confirm that no false alarms are generated

2. Introduce the Stated Fault (using techniques described in the standard) and increase its intensity until the system reaches the Stated False Alarm Impact Ratio: confirm that the Alarm Generation Rate remains below the Stated False Alarm Rate.

3. Continue to increase the fault intensity until you reach the Stated Fault Impact Ratio: confirm that the Alarm Generation rate is now above the Stated Alarm Rate.

4. Impose a Coincident Fault at the Stated Coincident Fault Level and repeat steps 1-3.

The tester can choose whether to conduct all of the fault level tests at a single set of environmental conditions before changing conditions, or to run through a range of environmental conditions with a single fault level. Steady state conditions must be achieved between each test. All testing should be done in an appropriate lab. The tests shall be completed by the manufacturer, an agent of the manufacturer, a neutral third party, or a standard-setting body. Testing may be witnessed by a neutral third party.

The results of the tests are to be provided on a test form that conforms to the requirements laid out in the Standard. Fault severity is measured by the Fault Impact Ratio: the impact that the fault has on a performance variable (such as EER, capacity, annual energy use, etc.), divided by the impact at 100% Fault Intensity.

The Standard-writing committee meets twice a year in person, with interim teleconferences between meetings. The timeline for the development of the Standard calls for releasing a version in 2015 for public review and publishing the Standard the following year. The California Energy Commission is following this activity carefully, and in a future version of Title 24, they will refer to the published ASHRAE Standard when it is available, for factory compliance with FDD requirements.

MOVING FORWARD

The California Title 24 and ASHRAE 207P efforts are works in progress. The California code goes into effect in July 2014. Before that time, there is a need to more clearly define requirements, flesh out acceptance tests (field verifications) and ensure that they are adequate and achievable by a technician. Manufacturers are now at work developing code-compliant tools, and only a handful are currently certified for compliance with Title 24. There is still work to be done to raise awareness of requirements among manufacturers. Of course, there is work to be done with end-users to raise awareness of the benefits of not just having FDD capabilities, but actually enabling them to diagnose problems, looking out for alerts, and then fixing the problems identified. Behavioral research is underway to determine how best to promote these capabilities.

Although FDD has been developing for decades, it has yet to break into the market in a significant way for RTUs. Standardization is key to "kick-start" the market. Once defining standards are available, utilities and the Consortium for Energy Efficiency plan to consider the requirements for utility incentive programs, ASHRAE Standard 90.1 and 189.1 will consider including FDD, consulting engineers will have more confidence in specifying systems with FDD, and OEMs will have more confidence in providing FDD.

The algorithms are already there, and the supporting technology and infrastructure are maturing. With standards like California's Title 24 and ASHRAE's SPC 207, the industry may be nearing a "tipping point," and we can expect that tools will start to fall into place in the marketplace. Industry-wide discussions are key to making sure this standardization happens in a deliberate way, consistent with the objectives and product cycles of FDD manufacturers and policies that promote the public good that is represented by improved energy performance of RTUs.

References

AEC. 2002. *Integrated Design of Small HVAC Systems.* Prepared for California Energy Commission by Architectural Energy Corporation. San Francisco, CA.

ASHRAE. 2010. *ANSI/ASHRAE/IES Standard 90.1-2010, Energy Standard for Buildings Except Low-Rise Residential Buildings.* American Society of Heating, Refrigerating and Air-Conditioning Engineers. Atlanta, GA.

ASHRAE. 2009. *ANSI/ASHRAE/USGBC/IES Standard 189.1-2009, Standard for the Design of High-Performance Green Buildings Except Low-Rise Residential Buildings.*: American Society of Heating, Refrigerating and Air-Conditioning, Atlanta, GA.

Brandemuehl, M. and J. Braun. 1999. "The Impact of Demand-Controlled and Economizer Ventilation Strategies on Energy Use in

Buildings." *ASHRAE Transactions*, V.105, Part 2.

CEC. 2011. *Working Draft Measure Information Template Light Commercial Unitary HVAC 2013 California Building Energy Efficiency Standards.* Prepared by the California Utilities' Statewide Codes and Standards Team. California Energy Commission, Sacramento, CA.

Davis, R., P Francisco, M. Kennedy, D. Baylon and B. Manclark. 2002. *Enhanced Operations & Maintenance Procedures for Small Packaged Rooftop HVAC Systems.* Prepared for Eugene Water and Electric Board. Seattle, WA.

EDR. 2011. *Energy Design Resources Design Brief: Economizers.* http://energydesignresources.com/media/2919091/edr_designbrief_economizers.pdf.

Felts, D. and P. Bailey. 2000. "The State of Affairs – Packaged Cooling Equipment in California," *Proceedings of the Summer Study on Energy Efficiency in Buildings*, 3:137-147. American Council for an Energy-Efficient Economy. Washington, D.C.

Goody, D., D. Banks, T. Haasl and J. Schwab. 2003. "New Service Protocol for Commercial Rooftop Units." *Proceedings of the National Conference on Building Commissioning*, Portland Energy Conservation, Inc., Portland, OR.

HEC 1993. *Commercial/Industrial Persistence Studies, Appendix M: Persistence of Savings from Mechanical System Measures Installed in the Energy Initiative and Design 2000 Programs.* Prepared for New England Electrical Systems, Foxboro MA.

Hunt, M., K. Heinemeier, M. Hoeschele, E. Weitzel. 2010. *HVAC Energy Efficiency Maintenance Study.* CALMAC Study ID SCE0293.01. Report to Southern California Edison, Rosemead, CA.

IAPMO, 2013. *2013 California Mechanical Code.* International Association of Plumbing and Mechanical Officials, Ontario, CA.

ICC. 2012. *2012 International Energy Conservation Code.* International Code Council, Washington, DC.

Jacobs, P., C. Higgins, and R. Schwon, 2004. "Upstream Solutions to Downstream Problems: Working with the HVAC and Efficiency Communities to Improve Field Performance of Small Commercial Rooftop Unit." *Proceedings, Summer Study on Energy Efficiency in Buildings.* American Council for an Energy Efficient Economy, Washington DC.

Jacobs, P., and C. Higgins. 2003. *Small HVAC Problems and Potential Savings Reports.* Report to the California Energy Commission, Sacramento, CA.

KEMA, 2013 (draft). *WO32 EM&V Interim Findings Memo for Commercial Quality Maintenance – Volume 1 – Field Observations.* Report to the California Public Utilities Commission. San Francisco, CA.

Liu, M., J. Houchek, A. Athar, A. Reddy, D. Claridge, and J. Haberl, 1994. "Identifying and Implementing Improved Operation and Maintenance Measures in Texas LoanSTAR Buildings," *Proceedings, Summer Study on Energy Efficiency in Buildings*, pp. 5.158-5.159. American Council for an Energy Efficient Economy, Washington DC.

Lunneberg, T. 1999. *When Good Economizers Go Bad.* Report ER-99-14, by E-Source, Boulder, CO.

Pratt R., S. Katipamula, M. Brambley, and S. Blanc. 2000."Field Results from Application of the Outdoor-Air/Economizer Diagnostician for Commissioning and O&M." *National Conference on Building Commissioning.* Portland Energy Conservation Inc., Portland, OR.

Chapter 43

Building Automation System Embedded HVAC System Energy Performance Degradation Detector*

Li Song, Ph.D., P.E., University of Oklahoma
Gang Wang, Ph.D., P.E., University of Miami
Michael Brambley, Ph.D., Pacific Northwest National Laboratory

ABSTRACT

By adopting virtual sensor measurements, the building automation system (BAS) embedded heating, ventilation and air-conditioning system (HVAC) performance degradation detector (PDD) works as an automation commissioning agent. The PDD adopts virtual energy meters for energy use measurements in critical HVAC energy use components such as coils, fans and pumps and uses simplified models deduced from the integration of physical models and data-driven models to predict reference energy uses. The deviations between measured and predicted energy consumption trigger alarms for system faults. The deviations can also be used to gauge the severity of the faults as energy/ economic losses that facility operators can use in making correction decisions. By providing real-time energy use monitoring and prediction of normal (without faults) system energy use, the PDD advances the typical open-loop commissioning process into a cyclic, closed-loop process that enables real-time continuous performance measurement and verification (M&V) after corrections are implemented.

INTRODUCTION

For buildings that consist of a central plant, which provides hot water and chilled water to air-handling units (AHUs), the number of AHUs can vary from one to over a hundred, depending on the size of the AHUs and the layout of buildings. Each AHU is equipped with a mandatory cooling coil, an optional heating coil, a supply fan, and sometimes a return-air fan. Coils in AHUs are the end users of the cooling and heating provided by the central plant, and the energy use by the AHU coils determines the electricity and gas consumed by central plants. Though energy measurements at a whole building level have been used for some time to evaluate building energy performance, and whole-building energy metering has increased in popularity recently (Torriti, 2014; Escriva-Escriva, 2011), determination of the root causes of deficiencies in energy efficiency and the ways that efficiency can be improved require an understanding of thermal energy distribution in AHUs. As a result, measurements in an AHU are critical for identifying energy savings opportunities and consequently for improving building energy efficiency. Because sufficient monitoring in HVAC systems is rarely in place, efforts to commission buildings and improve energy performance ordinarily use manual, short-term, on-site data measurements and collection by specially trained energy professionals. Today, energy measurements require professionals to travel to each facility with portable sensors and meters. With the large number of AHUs in many buildings, the detailed measurements in AHUs and the associated analysis are usually labor intensive, require specialized expertise and are costly. Therefore, by automating energy monitoring and diagnostics, the labor intensity of data collection would be decreased, measurements and monitoring made continuous, and thereby the effectiveness of commissioning and maintenance increased, while simultaneously lowering their costs. This technology would, as a result, lead to more energy efficient building systems operation.

Currently, rule-based and model-based methods are primarily employed in AHU FDD. One widely recog-

*This chapter will be presented at the ICEBO 2014, the International Conference for Enhanced Building Operations to be held on September 14-17, 2014, in Beijing, China, and published in the conference proceedings.

nized rule-based FDD method is the AHU Performance Assessment Rules (APAR) (House et al., 2001; Schein et al., 2003; Schein et al., 2006). APAR represents a first generation FDD methodology that can be embedded in building automation systems (BASs). The rules are based on knowledge of how AHUs physically operate. Faults are detected using qualitative (inequality) relationships that identify when physical conditions deviate from expectations based on that knowledge. Thresholds are used in these relationships (less than, equal to, or greater than) to help ensure that the occurrence of false alarms is minimized. The method, however, does not inherently provide indicators of the severity of faults or their impacts on energy use or costs, which are valuable indicators for building operators and managers to use in prioritizing maintenance and repairs.

Model-based FDD methods generally use component models to represent correct (unfaulted), normal behavior of each component in an AHU (Yang, 2013, Du et al., 2014). The reference estimates, which are obtained from these models, serve a critical role in model-based FDD because faults are detected based on differences (or residuals) between measurements and the values of variables from the models. Many different model types are used: physical models based on the first principles, neural network models, self-tuning models, auto-regressive models, fuzzy models, and system characteristic curves (Katipamula and Brambley, 2005a). Due to the complexity of these models, model-based methods often run on a separate computer platform in parallel with the BAS. Besides requiring additional hardware, setting up access to the BAS points required by the models is generally labor intensive, requiring mapping of points in the BAS to the corresponding points (or variables) in the software used for modeling. The process is simplified somewhat by standard communication protocols like BACnet (ASHRAE SSPC 135, 2013), which has an automated point discovery feature, but mapping of points must still be done manually. For BASs built on proprietary protocols, the process is even more labor intensive, and automated access to BAS points in real time might not even be possible. Both the communication configuration and computational demands of these models are key challenges faced in implementing model-based FDD (Wang et al., 2012).

Model-based methods may also require an installer or user knowledgeable of the system and methods for model calibration and tuning, which limits their practical viability. Because measurable operational variables in an AHU are not independent, FDD methods often cannot precisely specify the root causes of faults, especially when multiple faults occur (Wu and Sun, 2011; Bruton et. al, 2013).

Faced with the difficulty of embedding model-based approaches in today's BASs because of constraints on computational capacity and the need to map large numbers of points, another approach is needed to enable implementation of FDD in BASs. A near-term AHU FDD solution that avoids or minimizes the impact of these constraints while being BAS-embeddable should possess the following three features for successful application in practice: 1) The FDD itself should be easy to implement and operate; 2) Any models used should be simplified and impose sufficiently small computational burdens so that their execution does not slow down the routine control and data collection of BASs; 3) The system should report the energy, and ideally the cost impacts, of the faults detected, which can help facility managers and operators prioritize repairs and justify the costs of corrective action.

The BAS embedded energy performance degradation detector (PDD) presented in this chapter satisfies the three feature requirements through the following four key characteristics:

1) **Simple logic**—The approach detects HVAC faults using the direct comparison of the actual measured energy use and the estimated energy use that would have occurred in the absence of faults. The energy consumption of HVAC systems strongly depends on correct operations, i.e., the energy use of a faulty HVAC system deviates from the energy use during normal operation (Magoules et al., 2013). In addition, a cost-effective system for prompt detection and repair of faulty operations is more important for HVAC systems than operational reliability (Yoshida et al., 2001).

2) **Low cost**—The approach presented adopts virtual energy sensors to reduce or eliminate the costs of physical energy meter installations. The PDD includes three types of virtual flow meters—pump water flow meters in hot and chilled water systems for whole-building thermal metering, fan airflow meters in AHUs, and valve water flow meters in AHU coils for AHU thermal metering. The virtual measurements are made using existing mechanical equipment, such as pumps, fans and valves, to obtain water and air flow rates with no or minimum equipment installation.

3) **Simplified models**—Integration of physical and data driven models is used to reduce the number

of model input variables, decrease the required amount of training data, and attain adequate model accuracy.

4) **Alarm energy losses**—The energy impacts of faults, which are the differences between measured energy use and estimated reference energy use, are used to trigger alarms for system faults. The monetary values of those energy impacts are provided for facility operators to use to prioritize and justify their fault correction efforts.

Because the PDD is embedded in the BAS, it is capable of comparing actual operation measured by virtual meters in real-time with reference performance of the normal unfaulted system operation. Therefore, the PDD works as a BAS-embedded commissioning agent with improved features: 1) The PDD improves on the typical open-loop commissioning process by using a cyclical, closed-loop process as shown in Figure 43-1. The cyclic process allows real-time continuous performance measurement and verification (M&V) after corrections are implemented; 2) All phases in a typical commissioning process are conducted manually and additional manual M&V is usually required periodically after the commissioning is complete in order to verify that the improved system performance is sustained (or to detect degradation in performance when it occurs). The manual process is not only costly, but there is also a chance that performance degradation will not be identified quickly using it, which could lead to accumulation of energy waste for a long time before the fault is recognized and corrected. The automated, continuous process detects faults as they occur, providing information that enables fault correction quickly.

In this chapter, the mechanism and performance of three virtual energy sensors are first discussed along with experimental test results. Secondly, models used to estimate reference energy use are introduced individually for cooling coils, and fans and pumps. Finally, experimental results of diagnosing faults in outdoor air control are presented.

ENERGY MONITOR-BASED FAULT DETECTION

The energy consumption of HVAC systems is related to the HVAC system type and also strongly depends on correct operations. Through energy performance monitoring in AHUs and at the whole building level, the PDD is capable of detecting not only typical system faults caused by sensor drifting or malfunctioning equipment, but also sub-optimal operations. In built-up HVAC system operations, the major energy uses are cooling energy use in a cooling coil, heating energy use in heating coils/reheat coils and fan/pump power, whose consumption can be obtained from a virtual valve energy meter and a fan/pump variable frequency drive, respectively. Any faults including sensor drifting, malfunctioning equipment and sub-optimal operations can be detected by the PDD if they result in significant energy losses. For example, a broken outdoor air damper actuator can cause the outdoor air damper to be wide open. However, this fault will not be alarmed during economizer/mild seasons, when the cooling energy penalty is not significant. On the other hand, a leak in the preheat coil valve can be detected by excessive cooling energy use while space comfort is still maintained. Because the alarms are based on energy impacts, which

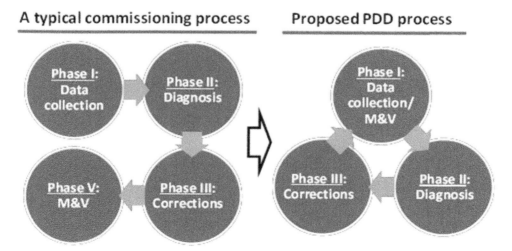

Figure 43-1: Process comparison between a typical commissioning and PDD.

are directly proportional to cost impacts, they provide a sound easy-to-understand basis for maintenance decisions by facility operators. Therefore, facility operators would be less frustrated by alarms, because minimum thresholds below which alarms will not be given can be determined not only statistically, but also by setting thresholds for energy loss impacts and costs that are used to filter out alarms that are too small for facility operators to care.

The mechanism of the detector is shown in Figure 43-2. The shaded box bordered by the dashed line represents the detector (a set of function modules) in a BAS. The two boxes on the left represent available sensor and command information in a BAS. The PDD uses the available BAS data plus virtual measurements at pumps, fans and valves to determine actual energy uses of key building systems and subsystems. The energy use of energy efficient operations for those systems, i.e. reference values of energy consumption for practices conforming to standards and guidelines by ASHRAE and other authorities (such as the static pressure reset requirements of ASHRAE 90.1), can be estimated through virtual flow measurements based on energy balances and/or the mechanisms of subsystems during unfaulted operation. As a result, the actual energy uses and reference energy uses for the outdoor air, the fan, the reheat, and the cooling coil in AHUs and pumps, chillers and boilers in chilled and hot water systems can be automatically obtained in Phase I, Data Collection. By comparing the actual performance (e.g., energy use) with the reference performance, the energy degradation can be determined in Phase II, Online Diagnosis. The severity of the performance degradation will trigger alarms and give information upon which facility operators can quickly

make decisions regarding correction of the faults. The correction process is represented by the unshaded box on the right of the detector module.

In the online diagnosis phase, alarms are not triggered until significant energy losses are detected. For example, in a single-duct variable-air-volume (SDVAV) AHU, the virtual valve energy meter is used to detect thermal energy related faults that are caused by temperature sensors and humidity sensors, economizer dampers, and cooling coil valves, while the fan power measurements along with the virtual fan flow meter are used to detect hydraulic faults in the air distribution system such as a fire damper stuck closed. Similarly, the pump power measurements along with the virtual pump flow meter are used to detect hydraulic faults in water distribution systems. The continuous energy monitoring enables the diagnosis algorithms to run only when necessary in order to conserve computational power of the BAS for its routine operations.

The core concept of this method is to use deviations between measured cooling, fan and pump power use and their reference energy uses as indicators of system faults. Therefore, low-cost virtual meters and determination of reference energy use for normal operation by a cooling coil, fan and pump are essential to the PDD.

VIRTUAL METERS

The virtual energy monitoring technology is a key component of the energy monitor-based PDD development. Three different types of virtual meters can significantly enhance the measurement capacity. Their roles in a building are shown in Figure 43-3 and are described as

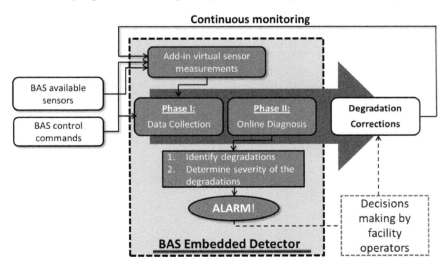

Figure 43-2: Flow chart of the embedded performance degradation detector.

follows: Type 1) the whole building level chilled water flow rate and hot water flow rate are virtually obtained using pump operational characteristics, namely a pump flow meter; Type 2) the airflow rate in each AHU is virtually obtained using fan operational characteristics, namely a fan flow meter; Type 3) the chilled water flow rate in each AHU is virtually obtained using control valve operational characteristics, namely a valve flow meter. The virtual sensors, built into the BAS, enable continuous and automatic performance monitoring, which is currently done manually by the use of portable meters during a typical commissioning or building energy assessment process.

Pump/Fan Flow Meter

Both the virtual pump and fan flow meters work according to the same principle. Power consumption is determined by useful mechanical work imparted to fluid (product of head and flow rate), fan/pump efficiency and motor efficiency. Theoretically, the fan/pump efficiency is a function of the ratio of head to flow rate squared, while motor efficiency is a function of power, frequency and voltage. Practically, head can be measured by a pressure differential sensor while voltage, power and frequency can be obtained from the existing variable frequency drive (VFD). The motor equivalent circuit, defined by six circuit parameters (IEEE 2004), can be applied to determine the motor efficiency under different frequencies and voltages. A break-through in this development is that Wang et al. (2013) successfully developed a method to estimate these six parameters

based on the published motor efficiency and power factor under rated frequency. Meanwhile, the fan/pump efficiency curve can be calibrated through experiments on each system. Using calibrated motor and fan/pump efficiencies, the flow rate can be obtained numerically, as shown in Equation (1). Wang et al. (2014) and Andiroglu et al. (2013) documented a comparison of the accuracy of the virtual meters, as shown in Figure 43-4, with conventional physical flow meters. The studies also show that the coefficient of determination or R-square for the entire validation period is 0.81 for the airflow meter and 0.97 for the water flow meter.

Virtual Valve Flow Meter

A cooling energy meter is typically not installed in an AHU because of high installation and maintenance costs and increased chilled water loop pressure resistance. It is impossible to install one correctly in existing AHUs due to space and system dimension limitations. The virtual valve meter uses the existing cooling coil control valve operational variables to indirectly obtain the water flow rates. Theoretically, the pressure drop through a valve is determined by valve position and flow rate for each specific valve, which is defined by a valve characteristic curve. The pressure drop can be measured by a water pressure differential sensor and the valve position can be obtained by either the valve command or a valve position sensor measurement fed back through a BAS. The valve characteristic curve needs to be obtained through a calibration process. Therefore, the flow rate can be obtained from the pressure drop, valve

$$\text{Flow} = \frac{\text{Power} \cdot \text{Efficiency}_{motor} (\text{Power, Voltage, Frequency}) \cdot \text{Efficiency}_{fan/pump} \left(\dfrac{\text{Head}}{\text{Flow-rate}^2} \right)}{\text{Head}} \qquad (43\text{-}1)$$

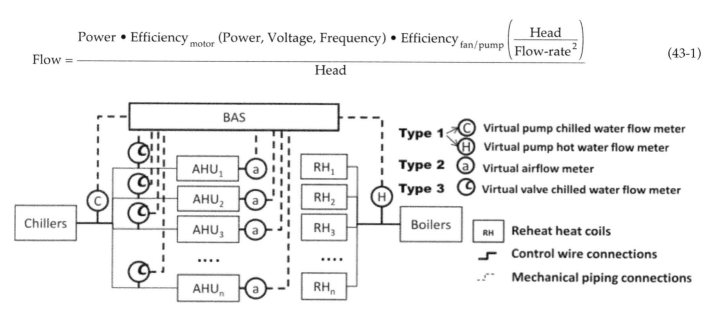

Figure 43-3: Added measurements by virtual meters in a typical HVAC operation

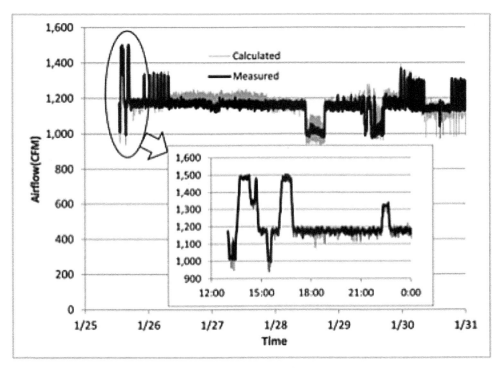

Figure 43-4(a) airflow

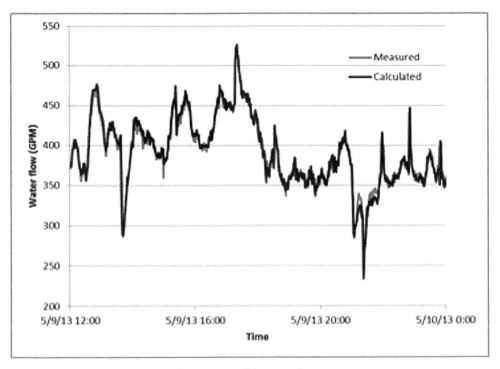

Figure 43-4(b) water flow

Figure 43-4: Measurement comparisons between virtual and conventional flow meters.

position, and the calibrated valve curve (Swamy et al. 2012; Song et al. 2011, 2012 and 2013), as shown in Equation (43-2).

Flow rate = Valve Curve function (Valve position)
$\times$ *pressure drop*$^{1/2}$ (43-2)

Song et al. (2013) found 0.46% uncertainty at 95% confidence for such a virtual meter compared with an ultrasonic meter, as shown as Figure 43-5, where the lines from the virtual and ultrasonic meters are essentially indistinguishable.

REFERENCE ENERGY BASELINES

Reference Cooling Energy Baseline

The IPMVP (2002) and ASHRAE Guideline 14 (ASHRAE 2002) provide guidance on measurement and verification of energy savings, including energy baseline determination on the whole building level. Due to the large number of variables potentially affecting whole building energy use, the whole building energy baseline is developed using either a regression model based on training data collected from the building for its simplicity or a sophisticated simulation model. The accuracy of the simulation results depends significantly on the accuracy of the input information. Several key inputs to simulations (e.g., actual schedules) are unavailable or values are unreliable, thus making simulation less

accurate and cost ineffective [Xu, 2005] [Wang and Xu, 2006]. For regression models, usually training data for one year are needed to obtain a reliable regression model that covers all seasons. The collection of an entire year of training data is considered by some as the largest obstacle in whole building energy baseline development [Effinger et al. 2011]. Since cooling energy performance in an AHU is also very much dependent on both weather and internal load, the same obstacle is also faced by the energy baseline development for a cooling coil, if a purely statistical model is adopted. The need to collect a year of data to determine an energy baseline would prohibit large scale application of energy monitor-based fault detection. Furthermore, a model based purely on physical principles could require more input variables than are typically measured in AHUs because sensors are only installed for basic rudimentary controls in HVAC systems [Katipamula and Brambley, 2005b].

To address the challenge of energy baseline development for AHUs, an integrated approach using both thermal balances and data driven models was investigated. The number of variables for AHU operations is much less than the number for the whole building and is, therefore, manageable (Song et al., 2014). The thermal balance model was first built based on the first law of thermodynamics to describe energy use of an AHU cooling coil. This model captures the equipment mechanism of operation and reveals the variables that impact cooling coil energy performance. To reduce the number of input variables needed, especially to avoid humidity measurements of the supply air and the return

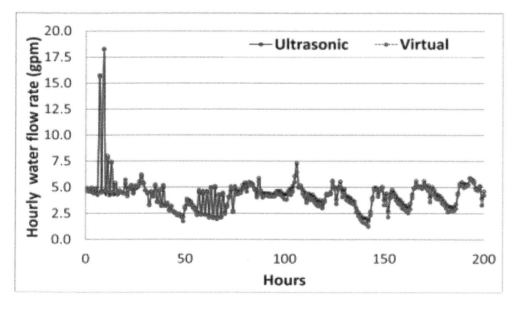

Figure 43-5: Measurement comparisons between virtual and ultrasonic flow meter

air, which are not readily available in most AHUs, the physical energy balance model is simplified to lumped temperature-based and enthalpy-based models. For dry coil operation the model is given by Equation (3) based on temperature, and for wet cooling coil operation, it is given by Equation (4) based on enthalpy, reducing the number of independent variables.

$$E_{cc} = \begin{cases} a_1^d \dot{m}_{sa} t_{oa} + a_2^d \dot{m}_{sa} t_{sa} + a_3^d \dot{m}_{sa} + a_4^d + E_{fan}, & Mode\ I \\ b_1^d \dot{m}_{sa} t_{oa} + b_2^d \dot{m}_{sa} t_{sa} + b_3^d + E_{fan}, & Mode\ II \\ 0, & Mode\ III \end{cases}$$

(43-3)

where Mode I=minimum outdoor air intake; Mode II=integrated economizer (β=1); Mode III= completely free cooling, and *a1c*, *a2d*, *a3d*, *b1d*, *b2d*, and *b3d* are coefficients of the lumped model under dry coil operation modes, which represent behavior for small fluctuations of outdoor air intake fraction (β) and room air temperature (Tra).

$$E_{cc} = \begin{cases} a_1^w \dot{m}_{sa} h_{oa} + a_2^w \dot{m}_{sa} + a_3^w + E_{fan}, & Mode\ I \\ b_1^w \dot{m}_{sa} h_{oa} + b_2^w \dot{m}_{sa} + b_3^w + E_{fan}, & Mode\ II \\ 0, & Mode\ III \end{cases}$$

(43-4)

where *a1w*, *a2w*, *a3w*, *b1w*, *b2w*, and *b3w* are the coefficients in the lumped model under wet coil operation modes, which represent average behavior for small fluctuations of outdoor air intake fraction (β), room air enthalpy (*hra*) and supply air enthalpy (*hsa*). The models given by Equations (43-3) and (43-4) apply well when the system is mechanically cooling (without economizing) and the room and supply air conditions are not changing much over time.

Regression coefficients in the lumped model are calculated using explicit equations, which are determined through searching for best fits using the least squares method with measured data for short time periods. Experiments using 1-day, 2-day, 4-day and 6-day training data on an operational AHU with a cooling capacity of 28kW (96,000Btu/hr) were conducted to validate the effectiveness of this approach with statistical analysis. The experimental results yielded an average estimated cooling consumption uncertainty of ±5.5% based on actual cooling consumption at 95% confidence, i.e. the method applied to the cooling energy baseline for this AHU had approximately ±2.5kBtu/hr (0.7kWh/hr) uncertainty with 95% confidence for an 96 kBtu/hr (28kW) AHU. Two days of training data at hourly intervals for obtaining the baseline is preferred because it was found to provide quantitative results similar to those obtained with 6 days of training data.

Reference Power Baseline for Fans and Pumps

Fans and pumps are additional energy users in HVAC systems. Thus, it is necessary to build power baselines for them for use in detecting operation faults and sub-optimal operations. An approach similar to the one used for the cooling coil load is adopted for development of the fan/pump power baseline. A model based on physical principles is first built; then a lumped model is established, and finally the lumped model is validated through experiments.

Using a fan as an example, a fan is an air pump that creates a pressure difference that causes air to flow. Figure 43-6 shows a system with a fan, a motor, and a VFD. The fan power includes the energy to deliver the pressure increase, described as fan head (H), for the required airflow rate (Q) as shown in Equation (43-5), and the power to overcome losses in each device captured by its efficiency: η_{fan} for the fan, η_{motor}) for the motor, and η_{VFD} for the VFD. Thus, the total energy consumed by a fan-motor-VFD system Winput is given by Equation (43-6).

$$W = H \cdot Q$$

(43-5)

$$W_{input} = H \cdot Q / (\eta_{fan} \cdot \eta_{motor} \cdot \eta_{VFD})$$

(43-6)

Total Input Power

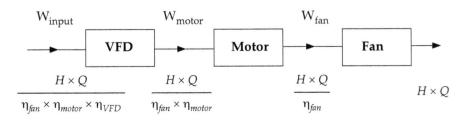

Figure 43-6: Electrical configuration of a fan system

As indicated in Equation (43-6), in order to determine fan power, the fan head, fan airflow rate, and fan, motor, and VFD efficiencies are needed. The steps by which these variables can be obtained follow.

1) The fan air flow rate, a measurable input, is determined by the cooling required to maintain the indoor temperature at its set point.

2) The fan head is a measureable variable, which is given by the virtual fan flow meter.

3) The fan efficiencies (η_{fan}) are determined by the fan head and efficiency curves from manufacturers. Usually, fan efficiency on the fan performance curve is given for the design fan speed. However, the VFD varies the fan speed under partial load conditions. According to fan performance characteristics (the fan law), fan efficiency is a function of the system resistance coefficient (S), which is defined in Equation (43-7), i.e., once the system resistance coefficient is fixed, the fan efficiency is fixed regardless of the fan speed. Fan efficiency is only related to fan head and fan airflow rate.

$$S = (H/Q^2) \tag{43-7}$$

4) Motor efficiency (η_{motor}) is related to motor load and motor speed, as expressed in Equation (43-8), where motor load is determined by the fan shaft work, fan efficiency, and the motor speed (ω) which corresponds to the fan speed. The motor load can be obtained from fan operation parameters, such as the fan airflow rate, fan head and fan efficiency. Recent motor efficiency curves published by motor manufacturers only give the motor efficiency as a function of motor load at the design speed. Thus, additional modeling and simulation are required to generate motor efficiency under variable speeds and partial loads. The result for one 3 hp (2.237 kW) motor with linear voltage-frequency relationship is presented in the study done by Wang et al. (2013).

$$\eta_{motor} = f(H \cdot Q / \eta_{fan}, \omega) \tag{43-8}$$

where ω is fan speed (e.g., in rpm), which can be obtained from the measured fan head and the fan airflow rate using fan laws.

5) The VFD losses should be considered while calculating power consumption since the VFD efficiency decreases as the motor load decreases. A VFD controls the rotational speed of a motor by adjusting the frequency and voltage of the power supplied to an AC motor. When the motor load is less than 20%, a dramatic VFD efficiency loss can be observed. The VFD efficiency loss is more pronounced for smaller horsepower motors. VFD efficiency is expressed in Equation (43-9).

$$\eta_{VFD} = f(H \cdot Q / \eta_{fan}, \eta_{motor}) \tag{43-9}$$

As a conclusion, fan head and fan airflow rate are two independent input variables that are used to determine the fan input power. Therefore, it is reasonable to simplify the relation for fan input power (Wfan) to Equation (43-10).

$$W_{fan} = Q \cdot H / (\eta_{fan} \cdot \eta_{motor} \cdot \eta_{VFD}) = f(H,Q) \tag{43-10}$$

According to Equation (43-10), it is possible to formulate a lumped parameter equation for the fan power baseline. Equation (43-11) was one of the lumped equations tested, and it showed the best results with the highest accuracy.

$$W_{fan} = a_1 H + a_2 Q_{sa} + a_3 \tag{43-11}$$

The experimental results (Shim, 2013) show the lumped fan power baseline generates ±0.05 kW uncertainty in fan power estimations with 95% confidence for a fan with a 3 HP (2.2 kW) motor.

Pump input power can be obtained using a lumped parameter model of the same form as Equation (43-11).

EXPERIMENTAL VALIDATIONS

In order to test the effectiveness of using the energy baseline to detect faults, an AHU operation fault was created in a test AHU. In normal AHU operations, the minimum outdoor air damper position is set at 20% open in order to maintain a 300 CFM (0.47 L/s) outdoor air flow rate to provide sufficient ventilation to ensure indoor air quality. In the experiment, a constant 50% damper command was used in order to detect the fault of the damper being at a position corresponding to a 50% damper position signal when it should be at the 20% position.

Experimental Apparatus

Figure 43-6 shows a schematic diagram of the experimental apparatus. The main components of the setup are: a test AHU, temperature sensors, humidity sensors, air flow meters, a portable ultrasonic flow meter, and a server computer, which is installed for control, real-time monitoring, data collection and data storage purposes. The AHU consists of a filter, preheat coil, cooling coil, and supply fan. The AHU serves a few faculty offices and a student lounge, and the capacity of the AHU is 28 kW (8 tons). The temperature and humidity sensors are used to measure the outdoor air temperature, supply and return chilled water temperature, and the outdoor air humidity. The accuracies of the temperature and humidity sensors are ±0.5°C (0.7°F) and ±1% of reading at 25°C (77°F). Two airflow meters (GP1 Gold Series Type B) are used to measure outdoor and supply air flow rates in the duct. The airflow meter consists of three sensor probes which are installed inside the duct and determine the air flow rate by averaging the air flow rates at each point. A portable ultrasonic flow meter (±0.5% of the rated flow rate), which is connected to a 25mm (1½") diameter pipe, measures the supply chilled water flow rate. The temperature and humidity sensors and airflow meter are all connected to the server computer, as shown in Figure 43-6. On the server computer, control software monitors the AHU under experimentation. The server computer records and stores data at one minute intervals.

Experimental Results

Based on the analysis introduced in the reference cooling baseline section, 2 days of hourly cool-

ing consumption data from June 15 to June 16, 2013, were used for the baseline generation. The outdoor air damper override was introduced at midnight on June 21, 2013. As shown in Figure 43-7, before the fault was introduced, the cooling energy baseline and measured cooling consumption match well despite some oscillations existing. Once the fault was introduced, instantaneous deviation can be observed between the measured cooling consumption and the cooling baseline. The daily peak deviation, i.e. maximum cooling energy loss each day, occurred between 3:00 pm and 4:00 pm when the outdoor air temperature reached its daily maximum (not shown in the figure because of space limitation). At nights, when the outdoor air temperature was lower, the cooling energy losses were smaller. However, the fault alarm was not enabled until the deviation persisted for 24 hours as shown by the solid red line in Figure 43-7, which was set to avoid false alarms possibly caused by sensor oscillations or unstable system operations. The cumulative energy losses, the integral between the cooling consumption and the cooling energy baseline curves, were also calculated and displayed, shown by the dotted dash line in Figure 43-7. Between issuance of the alarm and June 29, 2013, six days after the fault alarms were enabled, the cumulative energy waste was 1,507 kBtu (455 kWh), which can be used as incentive for building operators to take actions to correct the fault.

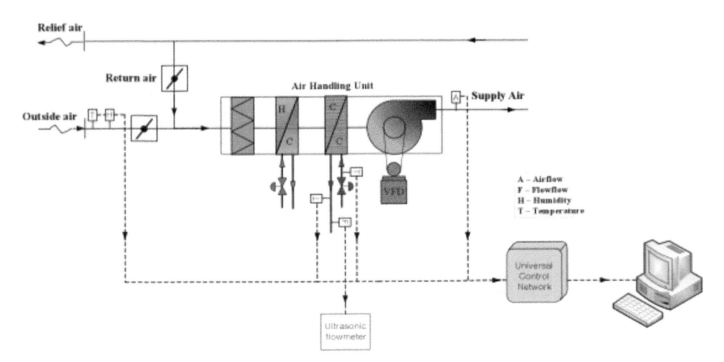

Figure 43-7: Schematic diagram of experimental setup.

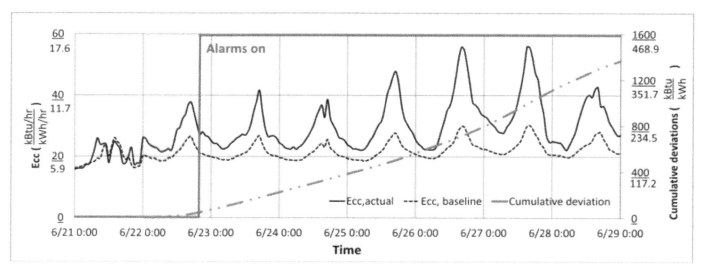

Figure 43-8: Measured cooling consumption, cooling energy baseline and cumulative deviation between the two.

CONCLUSIONS

The BAS embedded energy performance degradation detector (PDD) was introduced in this chapter using the theoretical and experimental results of three virtual meter studies as well as the investigations on simplified models to estimate reference energy use. The PDD is also tested through a preliminary experiment. A stuck outdoor air damper fault was introduced and instantaneously detected. The resulting cumulative energy losses were also successfully identified.

References

Andiroglu, E., Wang, G., Song, L. 2013. Development of a virtual water flow meter using pump head and motor power. Zero Energy Mass Customization Housing (ZEMCH2013) International Conference, Miami, FL

ASHRAE, 2002. ASHRAE Guideline 14-2002, Measurement of Energy and Demand Savings. American Society of Heating, Refrigerating and Air-Conditioning Engineers, Inc. Atlanta, GA.

ASHRAE SSPC 135. 2013. BACnet. Official Web Site of ASHRAE SSPC 135, http://www.bacnet.org/ Overview/index.html; accessed June 2013.

Bruton, K., Raftery P., Kennedy, B., Keane, M.M., O'Sullivan, D. T. J. 2013. Review of automated fault detection and diagnostic tools in air handling units, Energy Efficiency, DOI 10.1007/s12053-013-9238-2.

Du, Z., Bo, F., Chi, J., Jin X. 2014. Sensor fault detection and its efficiency analysis in air handling unit using the combined neural networks, Energy and Buildings 72:157-166.

Effinger, M., Anthony, J., Webster, L. 2011. Case studies in using whole building interval data to determine annualized electrical savings, Proceedings of 9th International Conference for Enhanced Building Operations, paper no: ESL-IC-09-11-23, Austin, TX.

Escriva-Escriva, G. 2011. Basic actions to improve energy efficiency in commercial buildings in operation, Energy and Buildings 43(11): 3106-3111.

House, J.M., and Vaezi-Nejad, H. Whitcomb, J.M. 2001. An expert rule set for fault detection in air-handling units, ASHRAE Transactions, 107(1):858-871.

IPMVP Committee. 2002. International Performance Measurement & Verification Protocol (IPMVP), Volume 1: Concepts and Options for Determining Energy and Water Savings, Revised March 2002. DOE/GO-102002-1554, U.S. Department of Energy, Washington, DC. Available electronically at http://www.nrel.gov/docs/fy02osti/ 31505.pdf; last accessed April 7, 2014.

Katipamula, S., Brambley, M.R., 2005a. Methods for fault detection, diagnosis, and prognostics for building systems—A review, part I, HVAC&R research 11(1):3-23.

Katipamula, S., Brambley, M.R., 2005b. Methods for fault detection, diagnosis, and prognostics for building systems—A review, part II, HVAC&R research 11(2):169-187.

Magoules, F., Zhao, H., Elizondo, D. 2013. Development of an RDP neural network for building energy consumption fault detection and diagnosis, Energy and Buildings 62:133-138.

Schein, J., Bushby, S.T., House, J.M. 2003. Results from laboratory Testing of Embedded Air Handling Unit and Variable Air Volume Box Fault Detection Tools. NISTIR 7036, National Institute of Standards and Technology, Gaithersburg, MD.

Schein, J., Bushby, S.T., Castro, N. S., House, J.M. 2006. A rule-based fault detection method for air handling units, Energy and Buildings 38:1485–1492.

Shim, G. 2013. Using integration of thermal-balanced and data-driven models to determine cooling load and fan power baseline for a single duct variable air volume systems, Master thesis, University of Oklahoma.

Song, L., Wang, G. Brambley, M. 2013. Uncertainty analysis for a virtual valve flow meter at an air handling unit, HVAC&R Research 19(3):335-345.

Song, L., Swamy, A., Shim, G., Wang, G. 2011. Feasibility study of developing a virtual chilled water flow meter at air handling unit level, Proceedings of International Conference of Enhanced Building Operation, ESL-IC-11-11-028, New York City.

Song, L., Wang, G., Swamy, A., Shim, G. 2012. In-situ resistance coefficient and experimental analysis of a virtual chilled water flow meter at air handling unit level, Proceedings of ASME 2012 International Mechanical Engineering Congress and Exposition, IMECE2012-87634, Houston TX.

Song, L., Wang, G., Shim, G. 2014. Integrated thermal-balance and data-driven methods to determine single duct variable air volume system cooling baseline in real-time for automatic energy audit, *Energy and Buildings*, **submitted in March 2014**, under review.

Swamy, A., Song, L., Wang, G. 2012. A virtual chilled water flow meter development at air handling unit level, ASHRAE Transactions, 118(1): 1013-1020.

Torriti, J. 2014. People or machines? Assessing the impacts of smart meters and load controllers in Italian office spaces, Energy for Sustainable Development, available online 31 January 2014, in press.

Wang, G., Song, L., Park, S. W. 2013. Estimation of induction motor circuit parameters and efficiency under variable frequencies, ASHRAE Transactions, 119(2):118-128.

Wang, G., Song, L. Andiroglu, E. Shim, G. 2014. Investigations on a virtual airflow meter using projected motor and fan efficiencies, HVAC&R Research 20(2):1-10.

Wang, H., Chen, Y., Chan, C.W.H., Qin, J., Wang, J. 2012. Online model-based fault detection and diagnosis strategy for VAV air handling units, Energy and Buildings 55:252-263.

Wang S.W., and Xu, X.H. 2006. Parameter estimation of internal thermal mass of building dynamic models using genetic algorithm, Energy Conversion and Management 47:13–14.

Wu, S., Sun, J. 2011, Cross-level fault detection and diagnosis of building HVAC systems, Building and Environment 46(8):1558-1566.

Xu, X. 2005. Model Based Building Evaluation and Diagnosis, PhD Dissertation, Department of Building Services Engineering, Hong Kong Polytechnic University.

Yang, X., Jin, X., Du, Z., Zhu, Y. , Guo, Y. 2013. A hybrid model-based fault detection strategy for air handling unit sensors, Energy and Buildings 57:132-143.

Yoshida, H., Kumar, S., Morita, Y. 2001. Online fault detection and diagnosis in VAV air handling unit by RARX modeling, Energy and Buildings 33(4): 391-401.

Chapter 44

Predictive Maintenance

Jim Lee

INTRODUCTION

Since Cimetrics introduced automated building analytics in the year 2000, there have been many advancements in the field. Much of the discussion has focused on energy savings, but many applications of big data analytics are not energy focused. This paper will discuss Predictive Maintenance in contrast to Preventive Maintenance, stressing the benefits of utilizing big data for predictive maintenance such as increased equipment life, improved reliability and lower labor cost.

The goal of Predictive Maintenance is to save money and increase equipment reliability. Money can be saved by only making repairs or servicing equipment when necessary. The risk of equipment failure can be reduced by continuous, automated analysis of equipment performance in order to identify faults before they become critical. Whereas Predictive Maintenance was once limited to high-value capital assets, modern automation systems allow us to collect and store vast amounts of data, and low-cost computing power makes it possible to analyze that data.

PREVENTIVE MAINTENANCE

Definition from Wikipedia:

Preventive maintenance (PM) has the following meanings:

1. *The care and servicing by personnel for the purpose of maintaining equipment and facilities in satisfactory operating condition by providing for systematic inspection, detection, and correction of incipient failures either before they occur or before they develop into major defects.*
2. *Maintenance, including tests, measurements, adjustments, and parts replacement, performed specifically to prevent faults from occurring.*

The primary goal of maintenance is to avoid or mitigate the consequences of failure of equipment. This may be by preventing the failure before it actually occurs which Planned Maintenance and Condition Based Maintenance help to achieve. It is designed to preserve and restore equipment reliability by replacing worn components before they actually fail. Preventive maintenance activities include partial or complete overhauls at specified periods, oil changes, lubrication and so on. In addition, workers can record equipment deterioration so they know to replace or repair worn parts before they cause system failure. The ideal preventive maintenance program would prevent all equipment failure before it occurs.

Here are some examples of routine scheduled maintenance of equipment:

- Oil Changes
- Belt Changes
- Filter Changes
- Linkage adjustments
- Valve seats replacement
- Steam traps replacement
- Boiler re-tubing
- Evaporator bundle cleaning
- Cooling tower water treatment chemical replenishment

Preventive Maintenance has been the back bone of mechanical and industrial equipment operation for decades. When systems are constructed, the designers take note of component life times, operating hours, wear parts, rated cycles and lubrication etc. Historically, elapsed time (run hours) has been used as the key driver for when maintenance activities should be performed. Although Preventive Maintenance is believed to be effective, in practice there are many shortcomings:

- The maintenance action is performed whether it is needed or not
- Labor is inefficiently deployed
- There is typically no verification of the repair action by observing system performance
- Deviation from normal operation is often limited to visual inspection

Preventive maintenance programs have frequently been automated by traditional Computer Maintenance Management Software (CMMS) packages, which require a user to understand the piece of equipment in order to create regular schedules for performing maintenance tasks. These are also coupled with spare parts inventory information, repair ticket tracking and enterprise accounting functionality. Most CMMS packages don't provide analytics capability and hence don't allow the user to gain institutional knowledge about the performance of assets over time. Furthermore, today's CMMS packages are unable to police the reliability of the repair, which is based on the skill level/training of the person performing the corrective action. In today's world of outsourced operations and repairs, how can one be certain that the repair technician knows what he is doing? By following up in the data and understanding the models and operations of the machine or system, we can physically measure whether or not the action has been performed and sometimes how well it has been performed.

PREDICTIVE MAINTENANCE

Definition from Wikipedia:

Predictive maintenance (PdM) techniques are designed to help determine the condition of in-service equipment in order to predict when maintenance should be performed. This approach promises cost savings over routine or time-based preventive maintenance, because tasks are performed only when warranted.

The main promise of Predictive Maintenance is to allow convenient scheduling of corrective maintenance, and to prevent unexpected equipment failures. The key is "the right information at the right time." By knowing which equipment needs maintenance, maintenance work can be better planned (spare parts, people, etc.) and what would have been "unplanned stops" are transformed to shorter and fewer "planned stops," thus increasing plant availability. Other potential advantages include increased equipment lifetime, increased plant safety, fewer accidents with negative impact on environment, and optimized spare parts handling.

A trivial example to compare Preventive versus Predictive maintenance would be in the area of air filters. Preventive Maintenance would attempt to calculate an average life of a filter and, perhaps enabled by a CMMS deploy maintenance staff to replace the filter at intervals that are shorter than this average life. This implies that some filters will be replaced prematurely and some

filters will be replaced too late. With Predictive Maintenance, we would be measuring the differential pressure across a filter. We can see it load up with dirt over time (differential pressure increasing) and hence trigger a maintenance action at the right time. When the filter is replaced, we can see the differential pressure drop and hence verify that the replacement was done correctly. By analyzing the pressure drop across the filters over time, we can better establish when and how often to change the filter and perhaps even glean information on which filter manufacturer sells a better product. Consider the benefits of this predictive approach, when the filter in question is serving a clean room manufacturing process area and any mistakes can translate to particulate contamination, production disruptions and potential product loss.

Although it is significantly more exotic than Preventive Maintenance, Predictive Maintenance is not a new topic. Historically, thermography and oil analysis have been done, temperature and pressures have been monitored and occasionally, vibration analysis has been applied to rotating equipment. What is new is the ability to gather and process much more physical data than in the past. By using modern big data approaches, which apply algorithms to system models, the effectiveness of Predictive Maintenance is greatly increased. Historically, Predictive Maintenance has been limited to individual pieces of equipment or "Islands of Automation,," but now with big data analytics, a systems level of Predictive Maintenance is possible. By having big data sets from sensors all around the process and equipment, we can build a composite view of systems operation and even correlate maintenance data to the comfort of the building or the integrity of a manufacturing process.

In the past, Predictive Maintenance was limited to high-value assets. What's different today is that we can automatically collect and analyze enough data so that Predictive Maintenance can even be applied to small end point devices such as variable air volume boxes and process utility connection points such as water for injection and compressed air connection points. The computer does the work, so automatic fault detection and diagnostics can be scaled down to the low-cost ubiquitous devices and sensors in a system.

Driven by automatic fault detection and diagnostics, these solutions can detect even minor anomalies and failure patterns to determine the assets and operational processes that are at the greatest risk of failure. This early identification of issues helps facility managers deploy limited maintenance resources more cost-effectively, maximize equipment uptime and enhance quality.

Predictive Maintenance can include:

- Observation of equipment performance drift
- Vibration analysis
- Critical process parameter monitoring
- Verification of repair (action)—by observing the data

WHAT IS REQUIRED TO DEPLOY A PREDICTIVE MAINTENANCE PROGRAM?

There are several elements needed to deploy a Predictive Maintenance program. First, a big data collection and analysis (Condition Monitoring) platform such as Cimetrics' Analytika solution, which can collect, model and perform automatic fault detection with root cause analysis. The analytics platform must comprise domain expertise so that the algorithms have a premeditated application to the system in question. The next critical element is Data Sufficiency—the availability of data from enough sensors, actuators and control parameters (e.g. set points) so that meaningful analysis can be performed. Then, equipment design information such as performance curves, rated cycles, design temperatures, design flow rates are essential to understanding how the machine works. A system is configured by mapping the sensor data to the model and by entering static data or metadata, which describe the physical characteristics of the system. Furthermore, we need sensor data and equipment specifications so that we can build a model of the system we would like to maintain. But unlike system simulation, this emulation of the process/equipment will take the model we created and drive the real time data we are gathering from the real sensors and actuators through that model. After analyzing the model

and real-time data with a series of algorithms, we can facilitate both equipment optimization and predictive maintenance notifications (alerts).

Figure 44-1 is a high level architecture for Analytika, Cimetrics' predictive analytics solution.

Many equipment makers have historically kept information about their design and operating characteristics proprietary. Now there is an opportunity for OEMs to differentiate their products by proving a complete operating model as well as sufficient sensors and actuators to provide the data for predictive maintenance analytics. We often hear that OEMs do not include sensors on systems because their customers don't perceive the value. Predictive Maintenance might provide the impetus to provide a new level of data sufficiency.

It is fascinating that the limited world view of original equipment manufacturers and consulting engineers dictates what they believe adds value—often eliminating sensors that could provide huge operational or reliability benefits. They often fail to appreciate that elements of a system rely on their individual piece of equipment. I recall a conversation with a manager at a chiller company in which he says: "My guys know everything there is to know about chillers. Ask them anything about the machine and they will tell you how to get 42 degree water. But they are completely ignorant of application and systems based issues of how their piece of equipment functions in a larger system."

There are several ways predictive maintenance platforms can work. In their most remedial form, these systems can be used to collect data when the machine or systems is working in steady state (regular operation) to create a baseline. Then statistical comparisons of current operation are compared to the baseline to determine whether equipment performance is drifting. This is useful for detecting a fault in the system, but generally not

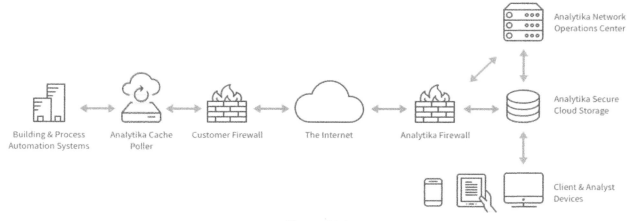

Figure 44-1

as good at determining the root cause of the fault.

With model-based automatic fault detection and diagnostics, the analytics system has a preconceived model for the machines' operating characteristics, and using real time data and sophisticated algorithms the system can predict where and sometimes when those failures might occur.

> Examples of simple statistical process control techniques for Predictive Maintenance:
> - Comparison of chiller performance curves to OEM predictions (design curves)
> - Monitoring of machine temperatures, oil viscosity, oil pressure, water pressure
> - Differential pressure measurements on filters and pumps.
> - Pump head pressure measurements
> - Vibration analysis
> - Infrared thermography

Figure 44-2 is an example of a Predictive Maintenance finding from Analytika.

Table 44-1 lists the top 10 pieces of equipment with the highest actuator rate of travel during the current monitoring period. A high rate of travel may lead to premature failure of the equipment and/or the control actuator. This application is also used to identify poorly tuned control loops.

Rate of Travel is defined as follows:

Rate of travel—The absolute change of a signal on an hourly basis

Relative average rate of travel—The average of the Rate of Travel (per hour) over the monitoring period

Cycles during monitoring period—The total number of actuator cycles during the monitoring period

Cycles to date—The total number of actuator cycles since the start of monitoring

Figure 44-3 shows an example of a piece of equipment with a high rate of travel.

BENEFITS

There are many benefits to a Predictive Maintenance program. Improved reliability and decreased risk of product loss and process disruptions are the most important benefits for mission critical applications such pharmaceutical production, healthcare or manufacturing. The data captured in the process of Predictive Maintenance analysis can be used for measurement and verification as well as providing data historian capability for compliance reporting. Increased equipment life and increase in Mean Time Between Failures (MTBF) can also be expected. In addition, labor savings are obtained by only servicing the equipment when necessary and dispatching repair crews with the necessary parts, therefore reducing truck rolls. Finally, in today's world, many maintenance functions are frequently outsourced to third parties. Predictive Maintenance analytics allow the verification of repairs using actual operating data, allowing verification of vendor and product performance.

A Predictive Maintenance program aims to identify the presence of a defect in such a way as to give sufficient time for the maintenance department to identify the root cause of the problem, efficiently order the parts, and schedule and complete the repair before a failure occurs.

> Predictive Maintenance program benefits:
> - Police outsourced services
> - Lower labor costs
> - Increased reliability
> - Risk mitigation
> - Increased equipment life

CONCLUSION

The big data analytics revolution is beginning to enable true predictive maintenance on a large scale. Users of predictive maintenance analytics can now enjoy the benefits of cost savings, increased reliability,

Date Opened	Priority	Status	Issue #	Category	Description
5/2013	High	Updated	ABC-010	Predictive Maintenance	Equipment Rate of Travel *Tune control loop(s) to prolong equipment life and reduce the possibility of premature equipment failure.*

Figure 44-2

Table 44-1

EQUIPMENT	Relative Average Rate of Travel	Cycles During Monitoring Period	Cycles to Date
VAV-157 Reheat Valve	259	1,864	21,434
VAV-150 Reheat Valve	146	1,050	11,344
VAV-223A Reheat Valve	132	952	9,424
VAV-197 Damper	99	713	6,207
VAV-North Office #1 Damper	81	581	6,212
AHU-15 Reheat Valve	78	562	6,750
VAV-North Office #2 Damper	67	481	4,236
VAV-023 Damper	59	424	3,436
AHU-02 Steam Humidifier Valve	58	415	4,025
Exhaust Fan EF-12 Damper	48	342	3,524

Zone 157 Reheat

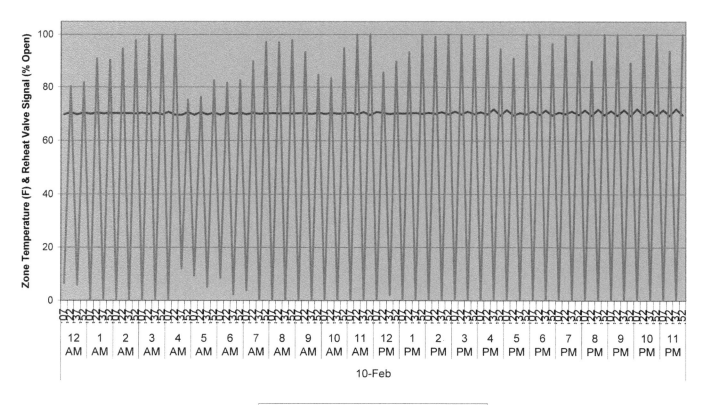

Figure 44-3

increased equipment life and reduced risk of process disruptions. A future challenge will be the integration of predictive maintenance systems with enterprise asset management systems. Although this would seem simple at first, there is a need for a human decision maker to add business insight as to what maintenance investments should be made.

Beyond Predictive Maintenance, there are many other benefits to big data analytics of building systems. For example, by combining automation data with shop floor data and quality data, we can begin to gain insights that enable enterprise level risk management and process optimization. These topics and others are beyond the scope of this chapter. They highlight that we are just scratching the surface of what is possible with physical world data analytics.

Chapter 45

What's Going on in the Trenches with Smart Buildings, Monitoring Based Commissioning, Automated Fault Detection and Diagnostics and Energy Regulatory Policy? What Does the Future Hold, When Does it Arrive?

Glenn T Remington

INTRODUCTION

Do the typical value engineering practices in the design process, and cutting installation corners by controls contractors or BAS programmers produce adequate system performance results and long term economic benefits to justify the expense or even meet the owners expectations? Can a (modern) planned, designed and installed moderate-to-large HVAC control system complete with the owners project requirements (OPR), well understood and considered during the design phase, effectively commissioned on installation, having a quality description of operation and O&M documentation package provided on closeout provide a fighting chance to operate the various (modern) mechanical (and lighting) systems to their optimum potential? Do we include access control too? Do ASHRAE 62 ventilation requirements count when staffs are in the building after hours and on weekends, while the fans are scheduled to run Monday through Friday 6:00 AM to 5:30 PM?

Figure 46-1: (Artwork credit: Johnson Control advertisement, Progressive Architecture magazine - July 1958).

Salesmen have always promised the champagne, while a lot of projects end up implemented on beer budgets.

Could that hypothetical modern HVAC control system be used for validation of the original design during commissioning and start up and provide measurement and verification for energy efficiency performance? It would be nice to provide forensic data when the envelope leaks or fan, pump, coil, valve, damper and VAV systems don't perform as designed. Better yet critical energy performance enhancing features like discharge air temperature resets or load shifting planned to meet various LEED or ASHRAE Standard 90 provisions or just plain avoid demand charges by design not just run blindly might be implemented. Yet those features are often cut in the latter stages of the project's design/build process or are not understood and implemented correctly during construction?

How often are those features or capabilities designed in? More often than not the owner may choose not to make the extra investment when doubtful about long term gains. But with a little planning the process could be managed better, the features thought out and communicated more clearly, and a few carefully planned modern capabilities might be deployed as pilots, then adopted in a more widespread fashion after gaining experience with them.

Building owners may or may not really understand the details of transformative technology. They may also underestimate the long term cost (penalty) of not renovating or properly operating a modern facility, let alone adopting next generation capabilities. And, rightfully so, owners may be apprehensive about making forward looking investments on the promise of energy savings, when all too often there are no verifiable track records of performance to demonstrate value.

But it doesn't have to be that way. What's missing is pragmatic, measured, and careful deployment of middle ground proven technologies and tools. What's needed is more and better delivery on promises made and less salesman bravado. We have smart grid planning, smart meters deployment, smart cars, and even smart water. We also have the federal and state governments and public utility regulators taking us down the path of not yet commercially deployed technology.

It may be that the time for smarter buildings is not just past due, but solutions for many of the difficulties owners experience are critically past due. The maintenance trades, various contractors, engineering firms, and equipment suppliers seem to be struggling specifying, installing and maintaining new systems as

we transition from a Happy Days era of "how we used to do things" to one more along the theme of the Jetsons that leaves us wondering if the investment is worth the risk.

WHAT NEEDS TO CHANGE?

It's always been the role of the building engineer, maintenance crafts, and other O&M staff to keep an eye on the building's operating controls. Over the years that's taken many forms. From simply making daily rounds and completing checklist entries, to noting and recording analog gauge positions, scanning a building automation system (BAS) system summary for "alarm flags" and reviewing the overnight printed alarm log, or anything in between, it's usually been the mechanics in the trenches with their hands on the pulse of the HVAC equipment and its controls who keep things running day in and day out. Taking it further along in complexity, the automated control system's "head end" is oftentimes being used at some distance remote from the building. Simple things like acknowledging alarms, reviewing trend logs, modifying schedules, tweaking set points and tuning control loops, and even reviewing energy consumption data t are more and more the normal course of business in a modern building. Yet, the mechanic on the ground is left out of the process. While that's all handled by the operations staff, and even sometimes with direct mechanic support, it's light years removed from the corporate board room and the people who make choices about funding projects.

However, at the same time operating buildings has become more and more complex and technologically intensive; the consequences of not running buildings well have become more severe. In this day and age, the escalating cost of energy is driving the need to operate buildings smarter and with resultant increased efficiency.

When control system communications left the realm of pressurized air running through arteries of copper tubing and entered the virtual land of packets traveling the information super highway, all bets were off and the game changed. Some would say, they changed forever, with pneumatic control systems even limited to actuating a slow device becoming obsolete. Electric actuation provides feedback, and also opens the door to embedded chip-based diagnostics at the device level.

We now have electronic card key access control, video surveillance for security and life safety, telecom and IT data, lighting controls, HVAC control, energy metering, and business processes, such as accounting functions, sharing the same LAN/VLAN/WAN, and over long distances the Internet, while utilizing server based data storage and computing resources. Now that building operations data and accounting data in some cases are sharing servers and physical and personal authentication security priorities, we also have external auditing requirements and government oversight from statutes like Sarbanes-Oxley. Once business finance and building controls shared the same IT resources and IT staff support (and security protocols), there was no turning back. Couple that complexity with the demographics of staff retirements, experienced staff chasing greener pastures during tough times, and headcount reductions due to budget cuts, and all too often we have fewer (less experienced) operators from a distance keeping any eye on increasingly complicated systems, with less skilled trades support on the ground in the building. Be it building operations or IT professionals, we have more complex systems requiring constant management and support and budget constraints not always aligned with the mission requirements.

As energy costs escalate, random equipment malfunctions increase from reduced skilled trades attention and smaller maintenance budgets, occupant complaints increase, and poorly running buildings have operational cost penalties both from energy cost and occupant discomfort perspectives - which are eventually noticed in the board room.

Without question and often justifiably so, the promise of new technologies often runs into strong headwinds from the previous systems perceived as working OK, the "way it was always done" even if it was not always done well. Oddly, this push back can at times come from both craft shops and leadership. Investment in continuing education and performance benchmarking is the only way to impact this drift toward staff obsolescence.

However, whether we like it or not, the age of relying on smart (or smarter) systems (and better trained staff) to help us operate buildings in as efficient a manner as possible is upon us. We can debate the definition of operating buildings "efficiently" later, but the mandate is there from the board room, and the higher level decision makers do still control the purse strings. At some point investing in technology will be a failure out of the gate without also investing in the people who use it. Recently, an industry discussion group was polled for feedback on what's happening in this important and quickly evolving market (air-conditioning controls, monitoring-based commissioning and FDD). Paraphrasing the question that was posed: Is anyone actually using any native BAS or 3rd party data mining tools that aggregate both BAS HVAC equipment points and energy metering data, then apply some sort of weather and occupancy normalization, crunch the data with some logic, and finally in some sort of routine eyes-on or automated process provide identification of "something amiss in the building or systems?" Can they then drill down (automatically) to identify a particular system, component, or control device as the root cause of the "fault" that needs attention?

Putting it in simpler terms, basic embedded "monitoring technologies at the field panel level, central data acquisition systems, data visualization, enhanced alarming based on many points instead of a single point, not to mention an automated analytical capability to monitor the performance of the systems while simultaneously applying logic to thousands of points in a large building and across dozens or hundreds of buildings, is quickly becoming an expected basic functionality of modern "control" systems.

These basic analytical tools are no longer rocket science, out on the fringe, or futuristic research lab only ideas, but rather fundamental best practices. Maybe the commercially available products aren't quite ready, but the sale reps sure say they are.

The basic question posed in the informal survey was how many folks are actually doing this in some form now? In response, only some folks said they were on the building owners or operator's side. Of course all the controls manufacturers and reps say that they can provide them. My impression was some folks are doing different things, but it appears the activity is fragmented.

Making it even more confusing, everyone seems to be calling their HMI a dashboard lately. With CPU horsepower and cost of memory less and less expensive every day and IT networks so prevalent in most buildings, the connected building is becoming more and more feasible, if not the norm. With information technology moving to virtual servers, data storage and computing horsepower is only limited by the creativity of those who would use building system data. Network communication bandwidth along with computing power and memory on aged HVAC control system field panels seem to be the remaining constraints in older building automation systems, but even that shouldn't be an overwhelming constraint to applying basic analytical performance monitoring methods. While off-the-shelf FDD products embedded in the native BAS still don't seem to be readily available (or at least having a sufficient track record in the field), a number of value-adding third party tools are entering the marketplace for enhanced trending, monitoring-based commissioning, and analytics. More importantly, analytical tools can be deployed in stages, beginning gradually. Analytic tools, however, generally apply to limited scopes. For example, tools are available for just energy consumption tracking, or just air conditioning sub-systems such as economizer cooling and ventilation air management. One example for data visualization in older buildings is a tool like ECAM [1} that lends itself primarily to energy meter data (at least currently). It is a perfect lead-in to energy consumption tracking and basic monitoring based commissioning tools for older non-DDC buildings that have the inherent limitation of no or aged computerized BASs. With a tool like ECAM for monitoring energy use, building owners can at least start out with the basics without wholesale control system retrofits.

Moving up to buildings with modern digital control systems, various work has been done by US NIST and various DOE research programs such as those at Pacific Northwest National Lab (PNNL), Lawrence Berkeley National Lab (LBNL) and other labs to develop fault detection and diagnostic software (FDD or automated FDD). At PNNL, a range of programs such as the "Whole Building Diagnostician" and the recent "Re-Tuning Program" have been proven to provide meaningful value. For entry level diagnostics, BAS trend data can be gathered, exported, visualized, and reviewed for basic operational faults. Simple things like validating occupied vs. non-occupied periods against scheduled periods or identifying systems in override run state is a good start. While requiring eyes for analysis, simple data trending and review can help spot these energy wasting occurrences. While some facility operations have the capability to custom program algorithms to "watch" for this mismatch in an automated fashion, many do not have programming access to proprietary control systems. Incrementally developing the operational staff's analytical skills from basic troubleshooting devices and control loops, to holistic building operational monitoring for improved fault detection will not only make life in the maintenance fields easier, but also result in energy savings and improved occupant satisfaction. Once this fundamental approach is mastered, additional FDD methods and tools will be that much easier to work with.

Researchers at LBNL PNNL and other institutions have done a lot of work in the monitoring-based commissioning (mCx) and fault detection and diagnostics (FDD) areas, and also automated utility to building demand response load shedding strategies are hitting the marketplace. Not long ago LBNL conducted a research project to study and benchmark the features of all the various energy tracking, mCx, and FDD products in the marketplace. An introduction to the findings can be reviewed at the link provided [2]. Perhaps just as valuable a resource, an important contribution to furthering these methods was to develop a roadmap to smarter buildings. LBNL also developed a "Specification" template for a "sample" building performance monitoring system. An overview is provided by [3]. Even the traditional proprietary control companies are increasingly offering add-on modules to improve data visualization.

NIST in conjunction with ASHRAE is also well along on developing a standard data model for Smart Grid/Smart Building communications [4]. This is just in time, as these tools may be code required in the not too distant future (like the first FDD requirements becoming part of energy code in California, which coincides with development of the proposed new ASHRAE Standard 207P "Laboratory Method of Test of Fault Detection and Diagnostics Applied Commercial Air-Cooled Packaged Systems") [5].

With not-so-large an investment in memory and computing capacity, the key is to have monitoring and/or trending for key points all the time even if it means rolling data off the bottom after a week, two weeks, or a month. By giving facility operators and maintenance staff access to pre-gathered trend data that can be instantly called up and reviewed, meaningful trouble shooting and accelerated corrective actions are enabled. This capability can provide proactive forward looking early warnings and accelerated service call response and resolution – before an alarm or customer complaint begins the process. This can be contrasted with the traditional model where there is little or no storage and computing horsepower on a small antiquated PC (usually having no updates and the monitor turned off for months), or a mechanic who has to ask for limited trending after a customer has already complained.

By saving the various data visualization templates for priority systems, they are pre-populated and ready for use in analysis when needed. Be it with the past 24 hours, past week, or past month's data ready for instant viewing from the historian. The trick is to either set up the native BAS database to be populated by the native trending tool or to at least provide the data for export to a data visualization tool for viewing. There would be no more need to ask for a trend to be set up, then coming back and checking it a week later.

Much of this has some dependency on retro-commissioning of existing sensors to shake out calibration and sensor placement issues or adding a few additional

It also requires a certain amount of IT/DDC programming savvy to connect the source of the data to the data-visualization tool (and any intermediate steps). While this analysis can be handled as a PM style periodic review conducted by various stakeholders (for example monitoring based or continuous commissioning activities), it's can also serve as a transformative daily operations and maintenance tool for staff, benefiting mechanics to central control operators to facility engineers to 3rd party service providers - the key is teaching folks to interpret the data and capture the value added.

Standardizing the analytical tool, output look and feel across systems and buildings, and training also helps to make the operations effort more efficient and communication of results easier. The idea is to easily pick out examples of faults such as fans running when you want them off and off when you want them on, economizer cooling "on" when its 90 degrees and humid outdoors, anomalies in static pressure control in VAV systems, and various other inconsistencies. All that's required is to apply operational staff's normal institutional knowledge of what's really supposed to be happening with the buildings and systems to enhanced analytical tools. As we all learn to work with more feature rich analytical tools, we also become more at home with incrementally more sophisticated automated systems and building operational strategies.
We've already established that the facility operations and IT teams need to work closer together anyway to meet regulatory requirements.

FDD tools are also becoming the norm for central plant chiller and boiler operation. Hand held FDD tools for non-BAS connected roof top units are becoming more and more available to deploy these "smart" technologies.

Large and small building owners can benefit from analytical software tools. The owner/operator of one building, a Community College campus or small complex of a dozen buildings, a university setting with 150 buildings, or a nation-wide chain of big box retail stores or restaurants all have the same vested interest in optimizing operations and maintenance programs, and minimizing energy consumption (operational cost savings and increased customer satisfaction is everyone's bread and butter).

Whether we are talking about the new age utility tariff and demand response driven automated load shedding, automated fault detection and diagnostics, monitoring-based commissioning, or simple daily operations and maintenance processes with enhanced data visualization, ALL these concepts require a few basic foundational starting points such as decent instrumentation, decent bandwidth and storage capacity, and basic software tools to monitor and operate buildings. If we start out slow with the basics, very soon the more complicated areas become enabled and easier to accomplish. The foundational technology and training for the basic capabilities are universal to the broader discussion.

Once the commitment is made to modernize the control system and analytical tools, we can debate which systems we wish to operate with what level of sophistication. It's prudent to bite off manageable steps and upgrade in incremental steps where the highest value added is easiest to obtain (for example ventilation air management and economizer cooling processes). At the end of the day we need reliable data, robust and targeted analytical tools, and continuing education and training programs for operational staff in order to realize the value from the analytical tools owners choose to deploy.

When participating in smart grid, smart metering, demand response meetings with regulatory policy makers in the Rust Belt region, I've asked how our policy makers expect older pneumatically controlled facilities (which may be in upwards of 80% or 90% of our commercial building stock) to participate in a modern smart meter/demand response/dynamic pricing model world without requiring significant/expensive control system upgrades. I got the deer in the headlights look from lawmakers and regulatory policy makers ("non-political" public service commissions).

I think it will be a rough sell during good times, let alone during tough economic "climates" to expect building owners to suddenly shoulder the cost of major control system upgrades, while the utility industry is being driven in that direction by the federal and state regulatory process.

Right now, demand response based tariffs with dynamic time-of-day pricing are implemented in various states. Right now, in California fault detection and diagnostics techniques will be required by code on new rooftop air conditioning systems. As a first step only the economizer cooling sub-system is targeted, but refrigeration cycle cooling systems are likely not far behind. It's not a stretch to suggest that as market penetration of various analytical tools increases, similar requirements may migrate into central station and air handler equipment too. Right now, various jurisdictions are requiring retro-commissioning results and coincident energy consumption improvements be reported. Right now the DOE and EPA have partnered and sent out a "regulatory policy framework" template to the energy regulators of all 50 states [6].

Right now, building operation and maintenance processes can be improved with off-the-shelf native BAS and 3rd party analytical tools, and putting some of these tools to use will start saving energy right now. Right now, basic trending tools can become part of a monitoring-based commissioning program, and right now building owners and operators need to be investing in training staff to use these tools.

Not only will a properly conceived and implemented building automation system provide important measurement and verification capability for more than just the air conditioning, lighting control systems and energy management metering instrumentation, but also will help identify, analyze and understand deficiencies in construction and building envelope in relation to weather sensitivities, recognize un-planned occupancy patterns not considered during the design and planning of a new facility, and help accommodate building use changes needing control program adjustments. In other words, a modern building information system helps understand and optimize how the facility operates as a whole, not just its individual parts and becomes an integral part of the commission process and follow up measurement and verification activities as a diagnostics tool.

At the end of the day, when considering how to optimize building and system maintenance operations and run the building as efficiently as possible so that it meets energy optimization and conservation goals, it's still the folks in the trenches and the tools they're provided that can make the biggest difference. The surest way for a facility, building system, or operations and maintenance staff to become obsolete is to fail to see and plan for the future that's coming.

Perhaps in a few generations (ten years from now) ASHRAE Standard 90 will require building automation systems to include these "futuristic" tools as standard equipment. Maybe this will be about the same time the Federal Energy Regulatory Commission, in conjunction with NIST/ASHRAE data model, will require the Smart Grid be able to interact with a Smart Building in an automated seamless fashion to shed load when needed, or automatically send out a cost penalty "traffic ticket" for running in an inefficient fashion.

CONCLUSION

One thing's for certain, we'll need facility management operations and maintenance staff backed by leadership/management trained to operate in this new world when it gets here, even as parts of it are knocking on our doorsteps now. March on, the world is changing fast and as the old saying goes, lead follow or get out of the way. It's a competitive world in nearly every aspect of enterprise management, and getting more so every day. Meanwhile, the utility meter is spinning...

References
[1]   http://www.cacx.org/PIER/ecam/
[2]   http://webcache.googleusercontent.com/search?q=-cache:http://eis.lbl.gov/eis-tech-case-studies.html
[3]   http://aceee.org/files/proceedings/2006/data/papers/SS06_Panel3_Paper10.pdf
[4]   http://www.bacnet.org/Bibliography/BACnet-Today-11/Bushby-2011.pdf
       https://osr.ashrae.org/Public%20Review%20Draft%20Standards%20Lib/ASHRAE%20201%20APR%20Draft.pdf
[5]   http://spc207.ashraepcs.org/
[6]   http://www1.eere.energy.gov/seeaction/

Chapter 46

Automated Diagnostics and Analytics for Wastewater Treatment Plants and Water Distribution Facilities

Michael Carroll

INTRODUCTION

Water is one of the world's most utilized natural resources, as clean and potable water is needed for everyday living. In the United States there were approximately 14,780 wastewater treatment facilities in 2008[*] and there were more than 155,000 water distribution facilities as of 2007[†]. These facilities play a crucial role in keeping our waterways clean and habitable for wildlife. Treatment plants generally consume the largest amount of energy for municipalities and local governments, accounting for 30-40% of the total energy consumed, and they also demand a large workforce to maintain operations. The general workforce includes technicians and operators whose day-to-day activities include monitoring statistics, documenting results, and maintaining equipment. The gathering of data and creation of reports is facilitated by automated diagnostics, while equipment can be effectively monitored to reduce any downtime or possible equipment failures.

WATER DISTRIBUTION FACILITIES

Water supply distribution facilities are the main source of potable water for most areas within the United States. These systems supply billions of gallons of water for not just consumption and general use in buildings and homes, but also for irrigation and emergency services via sprinkler systems and fire hydrants. Water storage tanks and distribution systems may vary depending on region, but effectively monitoring the water chemical levels is a constant concern.

Before water can be distributed for use, physical or chemical treatment and purification to remove natural contaminants is usually required. Legal regulations are

enforced by the Environmental Protection Agency (EPA) pertaining to contaminants, microorganisms, and the chemicals used for treatment. Based on where the water originates and other local factors, the specific methods and chemicals used may vary. One common way to treat water is to add chemicals to the water storage tank and test the water to assure all components are within the legal limits. This method requires that industrial pumps add small amounts of the required chemicals, such as hypochlorite, lye, or sodium hydroxide. Water authorities must employ operators who spend valuable time monitoring and adjusting chemical levels. Newer systems include multiple sensors and intelligent controllers that can take chemical readings within the water storage tank. These controllers can be interfaced with the chemical pumping system to assure that the correct amount of solution is added to optimize the quality of the water.

Many of these water facilities also have supervisory control and data acquisition (SCADA) systems. These systems are able to communicate over phone or ethernet lines with a central location. During the treatment process, sensor and transmitters detecting characteristics of the water can send real time information to the SCADA system. Within the SCADA system the operation can be programmed to automatically make adjustments. For instance, a sensor within the water storage tank can be set to detect a buildup in pressure and trigger a blow off valve to automatically open and relieve pressure in the tank. Such programming can save the water authority the time and effort that coincide with an operator manually determining that the pressure is dangerously high and then physically opening the appropriate valve. The possibility of damaging pumps and equipment is thereby reduced, which will lower material and labor costs as well as any valuable facility down time.

The use of SCADA systems and intelligent diagnostic equipment is integral to the operations of a water storage and distribution system. Through their use chemicals can be determined and reported to the

---

[*]ASCE 2013 Infrastructure Report Card
[†]EPA – Factoids: Drinking Water and Ground Water Statistics of 2007

appropriate authorities in a simpler fashion. Flow rates can be recorded and stored to determine the need for future growth, or to tell if there is a fault in the system. System pressure can also be monitored to assure the system is running safely and help extend the life of the equipment. A SCADA system can gather all of this information from a system that has multiple treatment and pumping locations throughout a town and bring them together at a single location. In effect, the time frame to obtain the information is shortened and the potential for human error is eliminated.

WASTEWATER TREATMENT FACILITIES

The treatment of wastewater is an indispensable process in society. Although it is complex, without this process the exposure to waterborne bacteria and disease could be deadly. A multitude of systems are available for treating wastewater. These systems can also vary in the type of water that they treat, from storm water to bio-waste-water treatment and in some cases both. The configuration and treatment methods can differ greatly depending on cost and location. One major consideration is the use of automation and diagnostics in the general practices and equipment, which may be used or adapted to multiple treatment scenarios.

The majority of all treatment processes requires pumps to move wastewater. These pumps are a critical component of the wastewater treatment process. They make it possible for used water to travel from a home or business through pumping stations to a treatment plant, pass through all the water cleaning stages of the plant, and then enter back into the environment. These pumping stations are located strategically throughout cities and towns that the wastewater treatment plant services. In order to be sure that these pumping stations are operating correctly operators have to spend time traveling from location to location to read meters and obtain this information. These stations can include SCADA systems which make it possible to monitor these remote locations at the central wastewater treatment plant. In the event that a pump fails, the SCADA system can automatically switch to the backup pump without losing any operational time. With the ever growing age of communications and information technology the newer SCADA systems can also be programmed to alert the central plant about the failed pump, call or email a technician to fix the pump, and simultaneously place an order for a replacement part. This greatly reduces any downtime the pumping station may face while helping

to streamline information about the problem directly to the people needed to fix it.

Some of the pumps in a wastewater treatment plant have large motors in the hundreds or thousands of horsepower range. These motors are very important to operations and have a long lead time as well as high cost for replacement. It is possible to fit these motors with multiple sensors that can relay the critical information to a computer system which can analyze the motor operation. Speed sensors, running time meters, vibration sensors, remote temperature sensors, voltage and amperage sensors are just a few of the sensors that can be added to a motor to obtain information while the motor is operating. This information can be analyzed by facilities' computer systems to determine if and when a motor will need maintenance. The pump control scheme can be programmed to determine which pumps run and for how long or to immediately stop the pump in the event of an unsafe condition. Consequently this can greatly extend the life of the motor and avert disaster. Analyzing motor operating statistics makes it possible to schedule replacement or general maintenance with greater ease. Both the proper workers and necessary parts are more readily available when work must be performed.

The pump operation information along with flow rates and wastewater chemical sensor data can enable operators to vary the required speed of the pump motor. By using a variable frequency drive with a large horsepower wastewater pump, the motor speed can be reduced when appropriate so it is not always operating at maximum speed. In a wastewater treatment plant motors generally account for the largest amount of electrical usage. By adding VFDs and installing sensors with appropriate programming, pump operation can be reduced by as much as 50% in some cases. These savings are directly related to the facilities' electrical consumption and can reduce a treatment plant's operating costs significantly while increasing the usable life of the motor.

There are a multitude of wastewater treatment plant operations that can utilize similar strategies. Sensors and transmitters can monitor process operations such as aeration, chemical analysis, UV filtration, aerobic digestion, and methane production to name a few. This information can be sent through the facility SCADA system to a central location for analysis. With the proper programming, this information can be used to create reports to determine if federal requirements are being met and improve plant operations. This analysis goes hand in hand with motorized valve and actuators that allow for plant operations from a central office. Prior to their use, plant operators had to read meters and analyze treatment situations manually. Upon discovering what changes were necessary, a team of operators would then open and close valves to divert flow or make the system corrections. With the integration of intelligent sensors and the use of motor operated valves and advanced system analysis software, a computer can determine when changes need to be made to a treatment process and automatically open and close valves to remedy the situation. These changes can be accomplished and reported to supervisory workers within just minutes.

Conclusion

The inclusion of SCADA systems, sensors, and transmitters is a valuable way to monitor water distribution and wastewater treatment systems. Through their usage, automated diagnostics of plant processes and automated operations of plant activities are possible. The ability to monitor system operations allows for a greater understanding of facility equipment; this helps individuals improve the allocation of resources to appropriate locations. As a result operators can use their time more efficiently, and spare parts can be stored and ordered more effectively. Equally significant, continually analyzing equipment and system operations reduces the man hours required in obtaining data. This data can be analyzed and used for maintenance scheduling and operational upgrades. All in all, automated diagnostics help reduce operating costs and maintenance time by enabling a safe, smooth running and well maintained water treatment facility.

References for Additional Information

Bertanza, Giorgio, et al. "How Can Sludge Dewatering Devices Be Assessed? Development Of A New DSS And Its Application To Real Case Studies." Journal Of Environmental Management 137.(2014): 86-92.

LIBHABER, MENAHEM, and ALVARO OROZCO-JARAMILLO. "Sustainable Treatment Of Municipal Wastewater." Water 21 Magazine Of The International Water Association (2013): 25-28.

Lingbo, Yang, et al. "Operational Energy Performance Assessment System Of Municipal Wastewater Treatment Plants." Water Science & Technology 62.6 (2010): 1361-1370.

Melo-Guimarães, Anemir, et al. "Removal And Fate Of Emerging Contaminants Combining Biological, Flocculation And Membrane Treatments." Water Science & Technology 67.4 (2013): 877-885.

Epa.org – This website is a resource for environmental legislation and best practices thorough the United States.

http://www.epw.senate.gov/water.pdf - Clean Water Act

SECTION VI

Conclusion and Author Bios

Conclusion

Barney L Capehart and Michael R Brambley

The exciting and rapidly advancing technologies of automated diagnostics and analytics for buildings should interest a broad range of stakeholders involved in the buildings enterprise—facility managers, energy managers, controls contractors, consultants, commissioning professionals, leaders in the building controls industry, utility program managers, and researchers in the area of building automation and control systems.

The editors of this book have done their best to deliver to you the latest descriptions and developments of the technologies of AFDD and analytics for buildings. They provide a comprehensive look at this field including:

- chapters that make the case for building analytics and AFDD in general

- current commercial products and service offerings from the whole-building and major systems perspectives as well as the benefits they provide

- case studies and data on the energy savings and operational benefits of utilizing these newer technologies

- chapters that focus specifically on existing tools and methodologies under development for heating, ventilating and air-conditioning systems, and

- the role of codes and standards in moving these technologies into common practice.

The field is rapidly advancing, and even better savings and better results are expected in the near future. It is time for existing building operators to embrace these technologies for their buildings and current building automation systems—time for those planning to construct new buildings to become more familiar with and consider these new technologies, to begin capturing the cost and performance benefits these technologies bring.

The editors wish to thank all the authors for graciously giving their time to write chapters which help provide so much valuable information to building owners and operators and to facility and energy managers, so they might operate their buildings much more energy efficiently at less cost. Their efforts yielded the information in this book, which we hope provides a convenient and up-to-date resource for building owners and operators, facility managers, and building energy managers, helping them to operate their buildings more successfully.

We also hope that building controls researchers, technology and product developers, and technical professionals involved in developing codes and standards will find value in this book as a resource which gives a broad perspective of the field of AFDD and building analytics, including current products and services as well as new methodologies and tools under development.

About The Authors

Paul J Allen, Walt Disney World

Paul J. Allen, P.E. is the Chief Energy Management Engineer at Reedy Creek Energy Services (a division of the Walt Disney World Co.) and is responsible for the development and implementation of energy conservation projects throughout the Walt Disney World Resort. Paul is a graduate of the University of Miami (BS degrees in Physics and Civil Engineering) and the University of Florida (MS degrees in Civil Engineering and Industrial Engineering). Paul is also a registered Professional Engineer in the State of Florida. The Association of Energy Engineers (AEE) inducted Paul into the Energy Managers Hall of Fame in 2003. (paul.allen@disney.com)

Roger Anderson

Dr. Roger Anderson is Senior Scientist at the Center for Computational Learning Systems (CCLS) in the Fu School of Engineering and Applied Science (SEAS) and Adjunct Professor at the Department of Earth and Environmental Sciences at Columbia University. In his 38+ years at Columbia, he is the inventor of 27 Patent applications and counting (10 granted so far), and has written 5 books, edited 4 others, and published 215+ peer-reviewed scientific papers, 52 editorially-reviewed papers, and 14 video productions. E-mail: anderson@ldeo.columbia.edu

Trevor E. Bailey

Dr. Trevor E. Bailey has been with UTC for 12 years and is currently the Program Manager within the UT Systems and Controls Engineering (UT SCE) organization. Dr. Bailey held a previous position at United Technologies Research Center (UTRC) from 2001 to 2013. While at UTRC, he contributed to several research and development activities spanning United Technologies aerospace and commercial divisions; Pratt & Whitney aircraft engines, Sikorsky helicopters, Hamilton Sundstrand power distribution systems, Otis elevators, UTC Power fuel cells, UTC Fire & Security systems, and Carrier heating, ventilation, air-conditioning, and refrigeration (HVAC/R) systems. In his current role within UT SCE, Dr. Bailey is leading a portfolio of programs within the UTC Comfort, Control & Security business to inject new technologies to improve product development quality and time to market. Dr. Bailey obtained a PhD degree in Mechanical Engineering in 2001 from McMaster University in Hamilton, Ontario, Canada.

Matthew Berbée—Director, Maintenance Management and Energy Services, California Institute of Technology

Matthew serves as the Director of Maintenance Management and Energy Services at the California Institute of Technology. He manages the business processes of the campus preventative and demand maintenance programs, the maintenance management system, the facilities service center, and the Caltech Energy Conservation Investment Program (CECIP). His responsibilities also include management of the campus energy retrofit and LEED (Leadership in Energy and Environmental Design) for Existing Building programs. Prior to joining Caltech, Matthew worked as the Energy Manager for the NASA Jet Propulsion Laboratory. Matthew is a Certified Energy Manager with an extensive background in building automation controls. He holds a Masters in Business Administration from Pepperdine University and a Bachelors of Science in Mechanical Engineering from California State Polytechnic University Pomona. Matthew is also a member of the California Commissioning Collaborative Advisory Council.

Klas Berglöf

Klas Berglöf is a Performance Analysis Engineer with ClimaCheck Sweden AB. He has been with ClimaCheck since 2004, and they developed the "Internal Method" for performance analyses based on a patent for the method from 1986. He is the inventor of several patented products that have been introduced commercially. The ClimaCheck Performance Analyser was awarded the "Refrigeration Product" of the Year at the RAC show in UK in February 2009 and ClimaCheck Sweden was appointed Climate Solver 2011 by the World Wildlife Found WWF. He has worked since 1984 in the Swedish refrigeration, air-conditioning and heat pump industry, and has had managing positions in distribution and manufacturing companies and worked between 1999 and 2004 as a consultant for manufacturers in development projects, fault analyses for insurance companies and in projects under the Montreal Protocol to phase out ODS in developing countries. Member of the steering committee of several national research projects for

"Alternative Refrigerants" and Energy efficient RAC systems. He has a Master of Science in applied thermodynamics and refrigeration at the Royal Institute of Technology in Stockholm.

Vaibhav Bhandari is a Research Associate at Columbia's Center for Computational Learning Systems

Sanjyot Bhusari

Sanjyot Bhusari is AEI's Market Leader for Intelligent Building Solutions and brings over 15 years experience in intelligent building design, systems integration, and commissioning. Sanjyot holds a Master of Science degree in mechanical engineering from the University of Florida; He is a Certified Energy Manager, a LEED Accredited Professional and holds a Professional Engineering license in the State of Ohio.

Jon A. Bickel—Fellow Engineer, Schneider Electric

Jon has over 25 years of experience in engineering, consulting, and product management in both the utility and manufacturing industries. He was originally employed by a large utility where he spent several years in the corporate Power Quality Services group. His other experiences at the utility include nuclear power generation, distribution engineering, account management, and marketing and sales. Jon joined Schneider Electric/Square D Company as a product manager where he was responsible for developing and sustaining sophisticated energy and reliability metering products. He developed new metering technologies, authored over a dozen articles, and trained and presented at many domestic and international venues. Jon also received company-wide recognition by Schneider Electric as a Senior Edison Expert. Jon is a graduate of Kansas State University with B.S. and M.S. degrees in Electrical Engineering, and is a registered professional engineer in the State of Texas. He has filed 30 patent applications with 22 granted to date. Jon is currently Vice President of Product Management at OptiSense Network in Plano, Texas.

Gene Boniberger

Gene joined the Rudin Organization in July of 1977 and has since served in various capacities including Chief Engineer and Building Manager in several different properties, eventually joining the Commercial Operations group as Operations Manager in 1987. Currently, as Director of Operations, he is responsible for the day to day operation of 15 million square feet of combined commercial and residential property. He is also involved in all aspects relating to MEPS in new development projects. E-mail: eboniber@rudin.com

Albert Boulanger

Albert Boulanger received a BS (1979) in physics at the University of Florida, Gainesville, Florida and a MS (1984) in CS at the University of Illinois, Urbana-Champaign, Illinois. He is a VP for the nonprofit World Team Now. He is a Senior Staff Associate at Columbia University's Center for Computational Learning Systems. E-mail: aboulanger@ccls.columbia.edu

Michael R. Brambley, Ph.D.—Co-Editor, Automated Diagnostics and Analytics for Buildings

Dr. Mike Brambley is a Staff Scientist at Pacific Northwest National Laboratory (PNNL), has over 35 years of research experience on energy technologies, and is a Fellow of the American Society of Heating, Refrigerating and Air-Conditioning Engineers (ASHRAE). For the last 26 years at PNNL, he has served in many roles including principal investigator, project and program manager, technical group leader, department chief scientist and leader of initiatives. For 6 years before joining PNNL, Dr. Brambley was a faculty member in the Engineering School at Washington University in St. Louis. Dr. Brambley has established himself as a leader in developing technologies for improving the actual operating efficiency of buildings through the use of automated fault detection, diagnostics, self-healing controls and other computer-based technologies. He has authored more than 75 peer-reviewed papers, many institution reports and book chapters, and numerous presentations. He served for 10 years as an Associate Editor of Energy, The International Journal, and presently serves as reviewer for several archival technical journals.

Jim Braun

Jim Braun is the Herrick Professor of Engineering at Purdue University and is a faculty of the School of Mechanical Engineering. He performs research at the Herrick Laboratories to study and improve the performance of building systems and related equipment. He has worked in both university and industrial settings, including working as a software developer for models of buildings and heating/cooling equipment and a developer of intelligent control algorithms for energy management and control systems at a large controls company."

Jim Butler—CTO—Cimetrics

Jim Butler provides technical oversight of Cimetrics' R&D activities and consulting services. He began his career at Cimetrics in 1990 as a contractor, later joining the company as Director of Software Engineering. Mr. Butler has been actively involved in the development of the BACnet® network protocol for more than 15 years. He organized the BACnet Testing Laboratories and served as its manager for 6 years. Jim earned his B.S. and M.S. degrees in Aeronautics and Astronautics from MIT. His thesis described the development of a software-based simulator for evaluating air traffic flow management strategies. After graduation, Butler worked as a contract software developer for several small companies and taught in the math and computer science department at Ithaca College. He is a member of ASHRAE and the IEEE Computer Society.

Barney L Capehart, PhD, CEM—Co-Editor, Automated Diagnostics and Analytics For Buildings

Dr. Barney L. Capehart is a Professor Emeritus of Industrial and Systems Engineering at the University of Florida, Gainesville, Florida, where he taught for 32 years. For the last thirty five years, energy systems analysis has been his main area of research and publication. He is the co-author of eleven books on energy topics, and over 50 energy research articles in scholarly journals. He worked with the Florida Legislature to write and pass the Florida Appliance Efficiency Act of 1987. He is given credit as the person most responsible for creating these appliance standards that have saved Florida electric and water utility customers over three billion dollars. He currently teaches energy management seminars around the country and around the world for the Association of Energy Engineers. He is a Fellow of the Association of Energy Engineers and a Member of the Hall of Fame of the Association of Energy Engineers; is listed in Who's Who in the World; and in 1988 he was awarded the Palladium Medal by the American Association of Engineering Societies for his work on energy systems analysis and appliance efficiency standards. He is also a Fellow of AAAS, IEEE, IIE and ASHRAE. He was the Editor of the Encyclopedia of Energy Engineering and Technology, 3 volumes, 190 articles, July 2007, and he is also the lead author of the Guide to Energy Management, 7th Edition, 2011. He has previously Edited four books in the area of Web Based Energy Information and Control Systems.

Michael Carroll, P.E., CEM

Michael Carroll is a professional engineer licensed in New York State. He received a B.S. in Electrical Engineering from Rensselaer Polytechnic Institute in 2005. During the beginning of his career he designed electrical distributions and building management systems for a multitude of construction projects in New York City. His career then progressed to electrical de-

sign and analysis for water and wastewater facilities, including the design of control and instrumentation systems. In his spare time Michael enjoys cooking and practicing virtual 3D construction.

Mattia Cavanna

Mattia is in charge of the New Business Initiatives Unit within Finmeccanica Group Strategy Department. His role is to identify and develop innovative solutions and business partnerships that, by leveraging the Group's wide technology portfolio and knowhow in the Aerospace, Helicopters, Space, Defense, Energy and Transport sectors, can successfully address sustainability related challenges in domains like smart energy, sustainable mobility, Earth observation, natural resources management and urban security. E-mail: Mattia.Cavanna@Finmeccanica.com

Edward G. Cazalet

Dr. Cazalet is a leader in the design and implementation of markets for electricity, the development of smart grid standards, and the analysis of transmission, generation, storage and demand management investments. Dr. Cazalet has decades of electric power and related experience as an executive, board member, consultant, and entrepreneur. He is a former Governor of the California Independent System Operator (http://www.caiso.com) and founder of TeMix Inc. (http://www.temix.com/), MegaWatt Storage Farms Inc. (http://www.megawattsf.com), The Cazalet Group (http://www.cazalet.com), Automated Power Exchange, Inc. (APX) (http://www.apx.com), and Decision Focus, Inc.

Dr. Cazalet has successfully promoted storage legislation and policy both in California and at the Federal level. He has advocated new electricity market designs to promote the integration of renewables and the use of price responsive demand as well as storage to support high penetration of variable renewables and efficient grid operation and investment by the grid participants including customers. He is co-chair of the OASIS Energy Market Information Exchange (EMIX) Technical Committee and a member of the OASIS Technical Committees on Energy Interoperation and Scheduling. Dr. Cazalet holds a PhD from Stanford University in economics, decision analysis and power system planning and engineering degrees from the University of Washington.

Gregory Cmar—Cofounder and CTO, Interval Data Systems, Inc.

Toby Considine

Toby Considine is a recognized thought leader in applying IT to energy, physical security, and emergency response. He is a frequent conference speaker and provides advice to companies and consortia on new business models and integration strategies. Toby has been integrating building systems and business processes for longer than he cares to confess. He has supported and managed interfaces to and between buildings, cogeneration plants, substations, chilled water plants, and steam and electrical distribution. This work led to Toby's focus on standards-based enterprise interaction with the engineered systems in buildings, and to his work in the Organization for the Advancement of Structured Information Standards (OASIS).

Toby is chair of the OASIS oBIX Technical Committee, a web services standard for interface between building systems and e-business, and of the OASIS WS-Calendar Technical Committee. He is editor of the OASIS Energy Interoperation and Energy Market Information Exchange (EMIX) Technical Committees and a former co-Chair of the OASIS Technical Advisory Board. Toby has been leading national smart grid activities since delivering the plenary report on business and policy at the DOE GridWise Constitutional Convention in 2005. He is a member of the SGIP Smart Grid Architecture Committee, and is active in several of the NIST Smart Grid Domain Expert Workgroups.

William Cox

William Cox is a leader in commercial and open source software definition, specification, design, and development.   He is active in the NIST Smart Grid Interoperability Panel and related activities. He contributed to the NIST conceptual model, architectural guidelines, and the NIST Framework 1.0. Bill is co-chair of the OASIS Energy Interoperation and Energy Market Information Exchange Technical Committees, past Chair of the OASIS Technical Advisory Board, a member of the Smart Grid Architecture Committee, and of the WS-Calendar Technical Committee. Bill has developed enterprise product architectures for Bell Labs, Unix System Labs, Novell, and BEA, and has done related standards work in OASIS, ebXML, the Java Community Process, Object Management Group, and the IEEE, typically working the boundaries between technology and business requirements. He earned a Ph.D. and M.S. in Computer Sciences from the University of Wisconsin-Madison.

Jerry R. Dennis—Energy Manager, DFW Airport

Jerry Dennis, CEM, CEP is the Energy Manager at the Dallas/Fort Worth International Airport. Mr. Dennis has 29 years of experience in the energy and utility industry. His expertise is in energy procurement and management.

Gregg Dixon—Senior Vice President of Marketing and Sales, EnerNOC

Gregg Dixon leads marketing and sales efforts worldwide to EnerNOC[1]s end-use customers. Prior to joining EnerNOC, Gregg was Vice President of Marketing and Sales for Hess Microgen and was also Partner at Mercer Management Consulting. Gregg graduated from Boston College with bachelor's degrees in Business Administration and Computer Science and he is a Certified Energy Manager, Certified Demand Side Management Professional, and Certified Sustainable Development Professional with the Association of Energy Engineers.

Dr. Bing Dong

Dr. Bing Dong is an Assistant Professor with the Department of Mechanical Engineering at the University of Texas at San Antonio. Before, he was a senior research scientist at United Technologies Research Center. His research interests are building energy modeling, building control systems design, and optimization, and HVAC system FDD. He received his Ph.D. in Building Performance and Diagnostics from Carnegie Mellon University in 2010. He is a member of ASHRAE, IBPSA, IEEE and ASME.

R. Neal Elliott, Ph.D., P.E., Associate Director for Research ACEEE

Neal Elliott coordinates ACEEE's overall research efforts and leads the Agricultural Program. He is an internationally recognized expert and author on energy efficiency, energy efficiency programs and policies, electric motor systems, combined heat and power (CHP) and clean distributed energy, and analysis of energy efficiency and energy markets, plus a frequent speaker at domestic and international conferences. He joined ACEEE in 1993.

Craig Engelbrecht—Director of Smart Services and Technology—Siemens

Craig has over 12 years of entrepreneurial experience within the building automation industries and pioneered energy information software products and solutions that spancommercial, government and education markets. For Siemens Infrastructure and Cities, Building Technologies Division, Craig is the Director of Remote Services and Technology working to identify market trends in remote services, energy efficiency and enterprise information management, while developing the vision and employing commercialization strategies to the future roadmap, and overseeing strategic business development that supports the business needs in the marketplace.

Jessica Forde, Ashwath Rajan, and Vivek Rathod

Jessica Forde, Ashwath Rajan, and Vivek Rathod are students at Columbia who participated in the Center for Computational Learning System's development of DiBOSS. Jessica Forde is a student at Columbia who participated in the Center for Computational Learning System's development of DiBOSS. Ashwath Rajan is a student at Columbia who participated in the Center for Computational Learning System's development of DiBOSS. Vivek Rathod is a student at Columbia who participated in the Center for Computational Learning System's development of DiBOSS.

Ashish Gagneja

Ashish Gagneja is a Research Associate at Columbia's Center for Computational Learning Systems

Dr. Jessica Granderson

Dr. Jessica Granderson is a Research Scientist and the Deputy of Research Programs for the Building Technology and Urban Systems Department at the Lawrence Berkeley National Laboratory. She is a member of the Commercial Buildings and the Lighting research groups. Dr. Granderson holds a PhD in Mechanical Engineering from UC Berkeley, and an AB in Mechanical Engineering from Harvard University. Her research focuses on intelligent lighting controls and building energy performance monitoring and diagnostics.

Nick Gayeski—Partner and Co-Founder at KGS Buildings

Nick Gayeski is a Partner and Co-Founder at KGS Buildings, LLC, responsible for product strategy, business operations, strategic partnerships, and key accounts. He works across systems integration, software and services teams to ensure that KGS technology and services constantly evolves with the changing marketplace. Nick holds a BA (2002) in physics from Cornell University, and an MS (2007) and PhD (2010) in Building Technology from the Massachusetts Institute of Technology.

John J. Gilbert III

John J. Gilbert III is COO/EVP/CTO of Rudin Management Company (RMC). RMC manages over 15 million square feet of commercial and residential space that is owned by the Rudin Family. RMC is the largest privately owned real estate company in NYC. He has been involved in every major project the company has built over the last 20 years including the redevelopment and creation of the world's first smart building at 55 Broad Street (1995- 1996), the development of the Reuters Building at 3 Times Square (1998-2001), the redevelopment of the former ATT Long Lines Building at 32 Avenue of the Americas (1999-2002), the redevelopment of 130 West 12th Street (2010-2011), as well as the redevelopment of the Greenwich Lane (2006-2014). He is acknowledged nationally as an industry thought leader in the integration of technology into the built environment and a co-inventor of DiBOSS. E-mail: jgilbert@rudin.com

Dr. Mikhail Gorbounov

Dr. Mikhail Gorbounov is a member of technical staff at United Technologies Research Center where he is responsible for development of thermal management solutions for components and systems of industrial products. He received his PhD in Mechanical Engineering from Bauman Moscow State Technical University. His current focus is in heat and mass transfer, thermodynamics and diagnostics in HVAC, refrigeration and thermal management of electronics. He has over 35 publications and holds 11 U.S. patents and is a member of ASME and ASHRAE.

Alex Grace—Director of Business Development, KGS Buildings

Alex Grace has extensive experience developing channel partner relationships based on integrated AFDD technology (SaaS) and engineering services. Prior to joining KGS Buildings, Alex held roles in operations and sales strategy with multiple fault detection and diagnostics companies. He has a Bachelor's from the University of Wisconsin-Madison and is based in Cambridge, MA.

David C. Green

David C. Green has combined experience in Intranet/Internet technology with database solutions and has developed programming for Energy Information Systems. David has been the president of his own software development consulting company, Green Management Services, Inc., since 1994. He has a Bachelor of Science degree in Chemistry and a Master of Arts degree in Computer Science. David retired from the United States Army February 1, 2014 as a Lieutenant Colonel after 30 years of military service. David has successfully completed major projects for The ABB Group, Cummins

Engine Company, ECI Telematics, The M.A.R.C. of the Professionals, Walt Disney World, Peregrine Energy Group, The Army Research Institute and Ohio State University. (dcgreen@dcgreen.com)

John Greenwell—President, CEPort

John Greenwell has vast experience in the design and implementation of complex building automation systems. He has been responsible for leading large project teams on multi-million dollar automation projects. Currently, Mr. Greenwell serves as President of CEPORT, LLC and is responsible for all aspects of daily operations; he oversees software development, sales and marketing, project management. He is passionate about energy management and protecting the limited natural resources we have. Mr. Greenwell founded CEPORT and built the BUILDINGWORX platform out of the need to do a better job of managing buildings by using data throughout organizations.

Kelsey Haas—Energy Manager, Ezenics, Inc.

Kelsey Haas is an Energy Manager for Ezenics, Inc. in Omaha, NE. K.haas@ezenics.com

Philip Haves

Philip Haves is the leader of the Simulation Research group in the Building Technologies and Urban Systems Department at Lawrence Berkeley National Laboratory. He has a BA in Physics from Oxford University and a PhD in Radio Astronomy from Manchester University. He is a Fellow of the American Society of Heating, Refrigerating and Air-conditioning Engineers (ASHRAE) and of the International Building Performance Simulation Association (IBPSA), a former chair of ASHRAE's Technical Committee 4.7 Energy Calculations and a former president of the US affiliate of IBPSA. He has particular interests in the use of simulation through the building life cycle and in the use of simulation to develop and test control strategies for low energy buildings and to detect and diagnose operational faults in real buildings. His current work includes extending DOE's EnergyPlus building energy simulation program and leading the project to develop Simergy, a comprehensive graphical user interface for EnergyPlus.

Kristin Heinemeier, P.E., Ph.D.

Kristin Heinemeier is a Principal Engineer with UC Davis' Western Cooling Efficiency Center. She is responsible for projects related to operations-phase energy efficiency in HVAC, as well as Human Behavior and its impacts on HVAC performance. Kristin has a

PhD in Building Science from UC Berkeley, and a Bachelors degree and PE in Mechanical Engineering. She has formerly worked for PECI, Texas A&M, Honeywell, and Lawrence Berkeley National Laboratory. She is chair of ASHRAE's Standards Project Committee on Method of Test for FDD in Packaged Rooftop HVAC Equipment, and a co-chair of the Western HVAC Performance Alliance's Automated FDD Committee.

Rusty T. Hodapp, PE, CEM, LEED AP

Rusty T. Hodapp, PE, CEM, LEED AP is Vice President for Energy, Transportation and Asset Management at the Dallas/Fort Worth International Airport. The Association of Energy Engineers admitted him to the "International Energy Managers Hall of Fame" in 2011 and named him "International Corporate Energy Manager of the Year" in 2003.

Srinivas Katipamula, PhD

Srinivas Katipamula, Ph.D., is a Staff Scientist with Battelle and has extensive technical experience in development of automated fault detection, diagnostics and commissioning techniques, building energy system simulations, analysis/evaluation of energy efficient technologies, analytical modeling techniques, tools and techniques to verify energy saving from building and equipment retrofits, and development and deployment of re-tuning.

Sila Kiliccote

Sila Kiliccote (M'10) is the acting group leader of the Grid Integration group and the Deputy of the PIER Demand Response Research Center and a program manager in the Energy Storage and Distributed Resources department at Lawrence Berkeley National Laboratory. Her research includes characterization of building loads and demand shaping, communication of demand response signals, demand responsive lighting systems, building systems integration and feedback for demand-side management. She has a master's degree in Building Science from Carnegie Mellon University and a Bachelor of Science in Electrical Engineering from University of New Hampshire. She has received the "Leadership in Smart Grid Acceleration Award" at GridWeek in October, 2010.

Joyce Jihyun Kim

Joyce Jihyun Kim received the B.Sc. degree in civil engineering from the University of Waterloo, Waterloo, ON, Canada in 2006 and the M.S. degree in Sustainable Design from Philadelphia University, Philadelphia,

PA in 2011. She is currently working toward the Ph.D. degree in building science and sustainability in the architecture department at the University of California, Berkeley, CA. Her research interests include design and implementation of automated price and demand response and analysis of customer economics under dynamic pricing.

Bill Koran, P.E.

Bill Koran, P.E.—Director of Energy Analytics, NorthWrite, Inc.

With expertise in energy analysis, diagnostics, data visualization, and training, Bill is the Director of Energy Analytics for NorthWrite. He spent twelve years designing aircraft HVAC systems, and has spent twenty-two working toward greater energy efficiency in buildings. He was primary author for four of Bonneville Power Administration's 2012 measurement and verification protocols. He is a voting member of ASHRAE Guideline Committee GPC 14, Measurement of Energy and Demand Savings.

Arthur Kressner

Arthur Kressner is a consultant and advisor for strategic planning and business operations of several early stage technology companies as well as working with major corporate clients. Mr. Kressner serves on several not-for-profit and for-profit boards dedicated to achieving business and environmental objectives that are responsive to community and national goals that are stewards of our resources. Retired from the Consolidated Edison Company in New York City with over 40 years of diverse and varied experience, most recently Director of Research and Development, he now focuses time consulting for the Columbia Center for Computational Learning Systems. His expertise on energy systems has been invaluable to the development of Di-BOSS. E-mail: artie.kressner@gmail.com

Sameer Kwatra

Sameer Kwatra conducts research and analysis to advance energy efficiency in commercial and residential buildings. His current research interests include miscellaneous energy loads in commercial and residential buildings; increasing utility participation in advancing building energy codes; and analyses to support DOE rulemaking on appliance and equipment standards. Sameer also leads the program development for the National Symposium on Market Transformation. He joined ACEEE in 2012.

Jim Lee—CEO, Founder—Cimetrics

Jim is the founder of Cimetrics and has acted as its CEO since its formation. Mr. Lee has been a leader in the embedded control networking and building automation community for over 20 years. As founder and former President of the BACnet Manufacturers Association (now BACnet International), the leading open systems networking consortium in the building automation industry, Mr. Lee's aggressive promotion of the BACnet open protocol standard has helped make Cimetrics a high-profile player in the arena. Mr. Lee has an earned B.A. in Physics from Cornell University.

Nate Maloney

Nate Maloney has been involved with Di-BOSS since March of 2013 developing the go-to-market strategy and commercialization of the operation system. Prior he headed the marketing team for Selex ES's law enforcement systems business. He holds a BA in English from Siena College in Loudonville, NY and a MA in Communication from the College of St. Rose in Albany, NY. E-mail: nate.maloney@elsag.com

Timothy Middelkoop—Director of Research Computing in IT, University of Missouri

Dr. Middelkoop is the Director of Research Computing Support Services in the Division of Information Technology at the University of Missouri. He is responsible for leading the evolution of the research computing infrastructure and facilitating the computational needs of research users campus-wide. Dr. Middelkoop also maintains a faculty position as an Assistant Teaching Professor in the department of Industrial and Manufacturing Systems Engineering at the University of Missouri. He received his Ph.D. from the University of Massachusetts Amherst in Industrial Engineering and Operations Research in 2006 and his M.S. and B.S. from the Florida State University in 1998 and 1996 respectively. His research interests include large-scale and multi-core scientific computing, clean energy optimization and control, distributed sensor networks, multi-agent systems, and integrated design systems. He has industry experience in designing web-based applications and embedded systems. He is a member of the Industrial Engineering honor society Alpha Pi Mu, the Institute for Operations Research and the Management Sciences (INFORMS), the Institute of Industrial Engineers (IIE), and is a Certified Energy Manager.

Danny Miller—President, Transformative Wave Technologies

Danny Miller is the President of Transformative Wave Technologies and the Performance Mechanical Group based in Kent, WA. He has over 37 years of experience in the commercial HVAC, controls, and energy efficiency industries. Danny is co-inventor of the patented CATALYST Efficiency Enhancing Controller, regarded as a landmark technology for reducing HVAC energy use in light and medium commercial facilities. He has been the visionary behind the use of digital web-based technologies to provide a new range of efficiency and performance assurance tools to commercial facility operators and companies with large portfolios. These include the "e-IQ: Energy Intelligence Platform," a fault detection, diagnostic, and energy monitoring system for CATALYST-equipped systems and the iHave suite of interactive HVAC asset management tools. Danny personally, Transformative Wave, and the CATALYST have won numerous awards including being named to the 2011 Washington Green 50 and selected "Technology Company of the Year" by the Seattle Business Magazine. The CATALYST has been named the "Game-Changing Technology of 2013" by E Source and Transformative Wave was named "Technology Deployment of the Year" by the Association of Energy Service Providers (AESP).

Nirosha Munasinghe

Nirosha is an industry recognized technology specialist and entrepreneur in product development and management in energy, building automation and smart grid industries. He has significant experience in innovative product development from proof of concept to mass market release having worked for high intensity start-ups around the world involved in successful go to market strategies. He has sound knowledge of the current building services and utility industry having been a contributing editor for North American online magazine AutomatedBuildings.com. For his work in the industry he has been recognized with Young Achiever Award and runner up in Future leader award by the Australian Institute of Refrigeration, Air Conditioning and Heating. He holds Masters in Business and Information Technology, Bachelors in Electrical Engineering with First Class Honors and Bachelor of Computer Science from The University of Melbourne.

Rob Murchison—Principal at Intelligent Buildings, LLC

Rob has over 20 years of experience in strategy consulting, design and implementation of information technology to real estate developers and commercial businesses. He is currently a principal at Intelligent Buildings, LLC, a smart real estate professional services

company that he co-founded in 2004. He has helped shape the industry by working with leading institutions including Lawrence Berkeley National Laboratories, Georgia Tech University and Harvard University where he is a continuing education instructor.

Vasanth Nadadur

Vasanth Nadadur worked for ACEEE as an intern for the Industry Program in the fall of 2012 while also working towards a Masters of Public Policy with a focus on energy and
environmental policy at American University in Washington, D.C. While at American, he was also a graduate assistant at the university's Academic Support Center and in the School of Professional and Extended Studies.

Willem "Bill" Nieuwkerk—COO of Selex ES

"Bill" is COO of Selex ES and responsible for the Smart Building Systems portfolio. He is the driving force behind the initiative between Selex ES, Rudin Management Company, and Columbia University to bring Di-BOSS to the commercial market. Bill has a solid background in management, operations, strategy and finance. Prior to his appointment at Selex ES, he turned-around an unprofitable manufacturing & technology company, re-established product lines for finance companies, founded a profitable VC-funded company supporting 5 Fortune 500 companies, and built a BPO business for a $100 million private company and sold it to an Indian based company for a profit. Bill is a graduate of Bentley College.

Patrick O'Neill, Ph.D.

Patrick O'Neill, Ph.D., is the co-founder and Chief Executive Officer of NorthWrite, Inc., an Internet-based facility, security, and energy management software company. Before NorthWrite, Patrick spent 10 years at Honeywell International, where he most recently served as Vice President of Technology and Development for e-Business. Prior to Honeywell, O'Neill worked at Battelle and the University of Illinois at Urbana-Champaign.

Zheng O'Neill

Zheng O'Neill is an Assistant Professor at the Department of Mechanical Engineering in The University of Alabama at Tuscaloosa where she has been since 2013. She received her Ph.D. in Mechanical Engineering from the Building and Environmental Thermal Systems Research Group at Oklahoma State University in 2004. From 2006 to 2013, she was a Principal Investigator in United Technologies Research Center (UTRC) where she was responsible for development and field-implement of advanced building optimal controls, building performance monitoring and energy diagnostics. She has more than 17 years of experience in building technology covering integrated building energy and control systems design, modeling and optimization, building commissioning, real time decision support system in buildings for fault detection and diagnostics, and low energy/net zero energy buildings. She has over 40 journal and conference papers published or in review. She is an active member of the American Society of Heating, Refrigerating and Air-conditioning Engineers (ASHRAE) and of the International Building Performance Simulation Association (IBPSA).

Paul Oswald—President, Environmental Systems, Inc

Paul Oswald is president of Environmental Systems, Inc (ESI). ESI is a professional services firm focused on building efficiency and systems integration. Paul has over 30 years of experience in building automation, system integration and energy management. His experience includes product strategy and development, business and channel development, and services.

Xiufeng Pang

Xiufeng Pang is a Sr. Scientific Engineering Associate in the Simulation Research Group, Lawrence Berkeley National Laboratory. He is a registered mechanical engineer in the State of California. His research includes integrating building performance simulation tools into research and practice, automating building commissioning and performance monitoring to improve the operation of buildings. He received his B.S. and M.S. degrees in HVAC from Harbin Institute of Technology, Harbin, China in 1999 and 2003 respectively; and earned his Ph.D. in Architectural Engineering from the University of Nebraska, Lincoln in 2008. He has published nearly 40 journal and peer-reviewed conference papers.

Mary Ann Piette—Department Director, Lawrence Berkeley National Laboratory

Mary Ann Piette is the head of the Building Technology and Urban Systems Department in the Environmental Energy Technologies Division at Lawrence Berkeley National Laboratory. The department has 7 groups covering building and industrial systems, windows, lighting and electronic technologies, simulation, and sustainable operations. She is also Director of the Demand Response Research Center which develops and evaluates technology to change loads on the electric grid. Ms. Piette is a former member of the National Institute

of Standards and Technology's Smart Grid Architecture Council and the lead on automated demand response communications, which is being used in 7 countries and is a national standard. She has won awards for her work in commissioning, energy information systems, and demand response. She has a Master's of Science Degree in Mechanical Engineering from UC Berkeley and a Licentiate in Building Services Engineering from the Chalmers University of Technology in Sweden.

John Petze, C.E.M.

John Petze, C.E.M., is a partner in SkyFoundry, the developers of SkySpark™, an analytics platform for building, energy and equipment data. John has over 28 years of experience in building automation, energy management and M2M, having served in senior level positions for manufacturers of hardware and software products including Tridium, Andover Controls, and Cisco Systems.

Ashwath Rajan

Ashwath Rajan is a student at Columbia who participated in the Center for Computational Learning System's development of DiBOSS.
Vivek Rathod is a student at Columbia who participated in the Center for Computational Learning System's development of DiBOSS.

Adam Regnier

Adam Regnier is a Ph.D. candidate with Drexel University's Building Science and Engineering Research Group. He has worked in various capacities relating to the construction and design of buildings in positions with construction management, litigation consulting, and renewable energy companies. Adam earned his M.S. in civil engineering from Drexel University and his B.S. in mechanical engineering from Lehigh University. He is actively engaged in researching the energy efficient operation of buildings via low-cost retrofits.

Glenn Remington—Consultant, Grand Rapids, MI

Glenn Remington has been involved with engineering project management, industrial hygiene, regulatory compliance, and facility O&M in predominantly mission critical, government, and academic campus facilities for nearly 25 years. Most recently he is consulting at a major Midwestern university for existing building commissioning and monitoring based commissioning. While working full time, Glenn earned a Bachelor of Science in HVAC Engineering from Ferris State University as a non-traditional evening program student. He has been active with ASHRAE at the local and Society levels since 1994, with ASHRAE TC 7.5 "Smart Building Systems" since 2001, and recently is volunteering time with ASHRAE SPC 207, "Laboratory Method of Test of Fault Detection and Diagnostics Applied to Commercial Air-Cooled Packaged Systems" while also enjoying family time, and for the first time in 20 years camping, canoeing, and playing golf again.

Rich Remke, Carrier Corporation

Rich Remke is a senior systems support engineer for Carrier/ALC Corporation based in Kennesaw, Ga. Rich is responsible for supporting the installation, design, and sale of Carrier control products and systems throughout North America. Rich is also involved with new product development. He holds a B.S. in information system management from the University of Phoenix. Rich has been in the HVAC and controls industry for over 30 years. Rich started his control work as a SCADA technician for Reedy Creek Energy Services at Walt Disney World, FL. He then moved into controls system engineering, project management, sales, and product management for United Technologies/ Carrier Corporation. Rich also spent several years supporting Carrier's Marine Systems group, providing controls technical support and system integration engineering. Rich has created several custom user applications, including a facility time schedule program, a DDE alarm interface, integration of Georgia Power real time pricing data to Carrier CCN, and a custom tenant billing application. (richard.remke@carrier.utc.com)

Doug Riecken

Doug Riecken is a Senior Research Scientist at Columbia's Center for Computational Learning Systems Ethan Rogers joined ACEEE in February 2012 and is responsible for directing the day-to-day activities of the industry program. The Industry Team conducts in-depth analysis of energy use and investments in efficiency by manufacturing, mining, agricultural, and construction facilities and systems. The research reports and white papers help policy makers and other stakeholders understand sector trends, motivations, and best practices.

Dale T. Rossi—Director of Special Operations, Field Diagnostic services, Inc.

Dale has been active in the HVAC industry for 35 years, first as a commercial service technician, then for nearly 20 years as a multi-state commercial service contractor serving national chain accounts, then as a founder of Field Diagnostic Services, Inc., a provider of fault

detection and diagnostic technology and services. As Director of Special Projects, he is responsible for the aspects of the product related to HVAC unit performance and the facility manager, contractor and technician user experience. Mr. Rossi is active in the Western HVAC Performance Alliance as the chairman of the ACCA 180 maintenance tasking work group and is a member of the ASHRAE SPC 207 committee working on a standard for on-board and in-field fault detection and diagnostic technology.

Dr. Soumik Sarkar

Dr. Soumik Sarkar is a senior scientist at United Technologies Research Center working on high-performance buildings and advanced decision-support technologies. He received his M.S. in Mechanical Engineering, M.A. in Mathematics in 2009 and Ph.D. in Mechanical Engineering in 2011 from Penn State. He has coauthored 35 peer-reviewed journal, conference papers and book chapters on complex system diagnostics and control. He is a member of ASME, IEEE and Sigma Xi.

Madhusudana Shashanka

Madhusudana Shashanka is a Staff Scientist with the Decision Support and Machine Intelligence Group at United Technologies Research Center (UTRC) in East Hartford, CT. Before joining UTRC, he worked as a Research Scientist at Mars, Inc. from 2007 to 2010. He is interested in all aspects of machine learning and has worked on applications in a variety of domains and datasets. His current research focus is towards machine learning applications for health monitoring of complex engineered systems such as helicopters, jet engines, HVAC systems and elevators. Dr. Shashanka has a PhD in computational neuroscience from Boston University and a bachelors degree in computer science from BITS Pilani, India. He is a senior member of the IEEE, holds 3 US patents and has published more than 30 peer-reviewed publications in leading conferences and journals.

Bruce Sher

Bruce Sher is co-founder of a start-up smart grid business, Viridity Energy that was involved as a partner with Rudin Management and Columbia University in the ConEd SGIG Demonstration Project. He has been leading the commercial development effort for the Di-BOSS smart building platform at Selex ES since June 2013. Bruce holds a MS in Energy Management from the University of Pennsylvania and a BS in Physics from University of Michigan. E-mail: bruce.sher@elsag.com

Ken Sinclair—Publisher/Owner AutomatedBuildings.com

Ken has been in the buildings automation industry for over 35 years as a service manager, building owner's representative, energy analyst, and consultant. He has been directly involved in more than 100 conversions to computerized controls. Ken is a founding member and a past president of both the local chapter of AEE and the Vancouver Island chapter of ASHRAE. The last eight years his focus has been on AutomatedBuildings.com, his online magazine. Ken also writes a monthly building automation column for Engineered Systems and has authored three industry automation supplements: Web-Based Facilities Operations Guide; Controlling Convergence; and Market Convergence.

Jim Sinopoli—Managing Principal, Founder—Smart Buildings, LLC

Jim is an innovator in the high performance building industry. For over 30 years, he has designed and engineered operationally efficient, intuitive and sustainable buildings through an integrated design matrix of building systems and technology. His design work can be found in many building types and uses throughout the US, Asia, Europe, the Middle East, South America and Africa. He has consulted and lectured government organizations, industry associations and Fortune 500 companies.

Tom Shircliff—Principal at Intelligent Buildings, LLC

Tom has over 20 years of experience in strategy consulting, design and implementation of information technology to real estate developers and commercial businesses. He is currently a principal at Intelligent Buildings, LLC, a smart real estate professional services company that he co-founded in 2004. He is founding Chairman of Envision Charlotte, a Clinton Global Initiative, a gubernatorial appointee for energy strategy and a collaborator and speaker and numerous universities and national laboratories.

Darrell Smith

Darrell Smith is the Director of Worldwide Energy and Building Technology for Microsoft's Real Estate and Facilities group. Darrell oversees the strategic direction for energy and sustainability across Microsoft's overall portfolio consisting of 34 Million square feet and over 650 sites. Darrell has 18 years of industry experience in Facilities, Data Center Operations, Energy, and Manufacturing.

David Solomon

David Solomon is a Product Development Manager for Selex ES in the Smart Buildings Division. He received his BA in Earth Science from Columbia University in 2012. Following graduation he worked at the Center for Computational Learning Systems developing predictive optimization algorithms for buildings until he joined Selex ES. E-mail: david.solomon@elsagna.com

Dr. Li Song (P.E.)

Dr. Li Song (P.E.) is an Assistant Professor in the School of Aerospace and Mechanical Engineering and Chair of the Building Energy Efficiency (BEE) Laboratory at the University of Oklahoma. Her research is focused on fault detection and diagnostics and optimal controls. Prior to this appointment, she worked as Vice President of Engineering Technology at Bes-Tech Inc. Song has served as a primary investigator for multiple federal and ASHRAE grants. She published 45 technical papers. She was selected for being one of five finalists in the 2011 ConocoPhillips Energy Prize.

Dr. Abhishek Srivastav

Dr. Abhishek Srivastav is a Machine Learning Researcher at GE Global Research Center, San Ramon, CA. He received his PhD in Mechanical Engineering from Penn State University, UP in 2009; dual masters degrees in Mathematics and Mechanical Engineering from Penn State in 2006; and his bachelor degree in Mechanical Engineering from Indian Institute of Technology Kanpur in 2003. His research interests are in the areas of machine-learning, PHM and decision-support.

Dr. Ashutosh Tewari

Dr. Ashutosh Tewari is a staff scientist at United Technologies Research Center. He develops machine learning algorithms and decision support tools for diagnostic and prognostics in complex engineering systems. He received his B.S. and M.S. in Material Science from Indian Institute of Technology, Bombay in 2003 and Ph.D. in Engineering Science & Mechanics from Penn State in 2008. He has coauthored 33 peer-reviewed journal, conference papers and book chapters is a member of IEEE and SIAM.

Steve Tom, PE, PhD

Steve Tom, PE, PhD, is the Director of Technical Information at Automated Logic Corporation, Kennesaw Georgia and has more than 30 years' experience working with HVAC systems. At ALC Steve has coordinated the training, documentation, and technical support programs, and frequently works with the R&D engineers on product requirements and usability. Steve also directed the development of www.CtrlSpecBuilder. com, a free web-based tool for preparing HVAC control system specifications, and EquipmentBuilder, an automated engineering tool for creating control logic. Steve is in charge of the Customer Applications Team, which develops HVAC control programming, Web applications, and third party integration drivers for Automated Logic products. Prior to joining Automated Logic, Steve was an officer in the U.S. Air Force where he worked on the design, construction, and operation of facilities (including HVAC systems) around the world. He also taught graduate level courses in HVAC Design and HVAC Controls at the Air Force Institute of Technology. Mr. Tom can be contacted at stom@automatedlogic.com.

Dan Trombley

Dan Trombley joined ACEEE in 2008. He works on assessing the energy efficiency potential of industries for state studies and characterizing emerging industrial technologies. He also helps develop and assess federal and state energy efficiency policies for the industrial sector.

Matthew Tyler, P.E.

Matthew Tyler, P.E. works as a mechanical engineer at PECI. He participated in California's 2013 Title 24 update by conducting research to support new and revised requirements for light commercial unitary HVAC equipment. This includes new mandatory requirements for economizer reliability and FDD systems installed on unitary equipment.

Daniel A Veronica—Engineer, National Institute of Standards and Technology

Dr. Daniel A. Veronica, P.E., is an Engineer with the Mechanical Systems & Controls Group; Building & Fire Research Laboratory; National Institute of Standards and Technology; in Gaithersburg, MD. He conducts federal government research transferring to the private sector new measurement technologies that automatically detect and diagnose faults in the mechanical systems of buildings. His graduate study was in the Karl Larson Building Energy Systems Laboratory at University of Colorado-Boulder. For the 15 preceding years, Dan had designed mechanical systems and controls for buildings, and earlier, was an engineering division officer aboard U.S. Navy nuclear submarines.

Gang Wang, Ph.D., P.E.

Gang Wang, Ph.D., P.E., is an Assistant Professor at the University of Miami (UM) in Coral Gables, FL. He earned PhD degrees in Heating, Ventilation and Air Conditioning in 1996 at Harbin Institute of Technology and in Architectural Engineering in 2005 at University of Nebraska-Lincoln. Prior to joining UM, he was a Research Associate Professor at the University of Nebraska-Lincoln from 2000 to 2008 and Assistant Professor at Texas A&M University- Kingsville from 2008 to 2011.

Jennifer Warnick—Microsoft Corporation

Jennifer Warnick is lead writer for Microsoft Stories, a platform for sharing in-depth, colorful stories about the company's products, people and ideas. She is a veteran storyteller with more than 13 years of professional reporting, writing and editing experience, including as an award-winning journalist for daily newspapers in Washington, New York and Idaho and as a technical writer and editor for two engineering firms.

Michael Watts—Market Leader for Commissioning Services AEI

Mike Watts has over 25 years of building automation and intelligent building design experience With more than 25 years in the HVAC industry, Mr. Watts currently serves our clients as a Project Manager and Intelligent Buildings Specialist. In this capacity, he works directly with contractors and Owners to coordinate specific aspects of complex facilities. Mr. Watts has managed, designed, serviced, programmed and/or commissioned numerous intelligent building systems including: vertical transportation, security, lighting, HVAC, and fire alarm. Mr. Watts has considerable experience of systems integration, energy management, critical monitoring, and direct digital control systems. His specialized technical abilities, project management skills, sensible solution, and ability to communicate effectively with the owner, facility users, and contractors have made him a key member of the AEI team. Mr. Watts is accredited and active member of the USGBC, ASHRAE, and BCA.

Guanghua Wei, PE—Engineering Director, Bes-Tech, Inc.

Guanghua Wei is Engineering Director at Bes-Tech, Inc. Mr. Wei has more than 11 years of hands-on experience in building commissioning. His expertise is in devising and implementing optimised operational and control schemes for building HVAC systems.

Dr. Jin Wen

Dr. Jin Wen is an Associate Professor in the Civil, Architectural, and Environmental Engineering Department at Drexel University and is actively involved with teaching and doing research in the building energy efficiency and indoor environmental quality areas. She has served as a principal investigator or co-principal investigator on many research and educational projects funded by ASHRAE, the U.S. Department of Energy(DOE), National Science Foundation (NSF), Department of Homeland Security, the Department of Education (DOEd) and the National Institute of Standards and Technology. She is an active ASHRAE member serving on several technical committees. Before she joined Drexel University in 2003, she conducted research in developing and validating operation and control methodologies for energy efficient buildings as a Ph.D. student in The University of Iowa and as a part-time research assistant at the Iowa Energy Center. Dr. Wen has also worked for Johnson Controls, Inc. (Beijing, China) as an application engineer designing control system and programming control strategies for building HVAC systems.

Leon Wu

Leon Wu has been researching smart building technologies since summer 2011 when he established the initial data feeds between BMS and the cloud-based databases. He prototyped the smart Human Machine Interface (HMI) components and designed various database schemas, system and workflow diagrams. Leon received his graduate degrees from Columbia University. E-mail: leon.wu@columbia.edu

Bahman Yazdani, PE

Bahman Yazdani, PE, is a research engineer and Associate Director at the Energy Systems Laboratory at Texas A&M University. Mr. Yazdani has over 30 years of experience in energy efficiency and building codes.

Rongxin Yin

Rongxin Yin received the B.S. degree in mechanical engineering from Xi'an University of Architecture and Technology, China in 2005, and the M.S. in mechanical engineering from Tongji University, China in 2009. He

is currently working toward the Ph.D. degree in building science and sustainability in the architecture department at the University of California, Berkeley, CA. His areas of interest include pre-cooling and demand response studies of commercial buildings, building energy modeling, and optimization of demand response control strategies in commercial buildings.

David Yuill

David Yuill is a PhD candidate in Mechanical Engineering at the Herrick Laboratories at Purdue University, with an anticipated graduation in 2014. He has previously worked in industry doing research, retro-commissioning and design of HVAC systems. He served as President of Building Solutions, Inc. from 2005-2009.

Dr. Xiaohui Zhou

Dr. Xiaohui Zhou is currently the Energy Efficiency Program Manager at the Iowa Energy Center. He has over 20 years of application and research experience in commercial building HVAC systems and building controls, and served as design engineer, application engineer, research scientist, and program manager. He holds a Bachelor of Science degree from Zhejiang University (China) and a Master of Science from the University of Connecticut, both in electrical engineering with concentration in controls. He also received his doctorate in mechanical engineering from Iowa State University with concentration in thermal energy systems. Dr. Zhou is actively involved in ASHRAE at local and national levels, participating Technical Committee TC 1.4 Control Theory and Application and TC 7.5 Smart Building Systems. He also participated in the writing of ASHRAE Standard 195—Method of Test for Rating Air Terminal Unit Controls, and involved with the development of ASHRAE Standard 211—Commercial Building Energy Audits. Beside multiple technical reports, he also published technical papers on ASHRAE Transactions, HVAC&R Research, and IEEE Transactions on Robotics and Automation."

Index

Printed and bound by CPI Group (UK) Ltd, Croydon, CR0 4YY

23/10/2024

01777682-0017